BIOMINERALIZATION SOURCEBOOK

BIOMINERALIZATION SOURCEBOOK

Characterization of Biominerals and Biomimetic Materials

Edited by
Elaine DiMasi and Laurie B. Gower

CRC Press
Taylor & Francis Group
Boca Raton London New York

CRC Press is an imprint of the
Taylor & Francis Group, an **informa** business

Cover Image: Original crystallographic orientation map of calcite fibers of the shell of the brachiopod, Terebratulina retusa, with overlay of Kikuchi patterns and data from EDS (energy dispersive spectroscopy) analysis. Courtesy of Professor Maggie Cusack and Peter Chung.

CRC Press
Taylor & Francis Group
6000 Broken Sound Parkway NW, Suite 300
Boca Raton, FL 33487-2742

Printed on acid-free paper
Version Date: 20140131

International Standard Book Number-13: 978-1-4665-1835-3 (Hardback)

Visit the Taylor & Francis Web site at
http://www.taylorandfrancis.com

and the CRC Press Web site at
http://www.crcpress.com

Contents

List of Figures

Foreword

The probing questions asked in biomineralization research are inevitably at the interface between two, three, or sometimes more disciplines. Observations are made at all length scales and involve bulk structures and interphases, macromolecules, cells and materials, and solid state transformations, and at the heart of this all, biomineralization is the product of complex cellular activity. It is therefore no surprise that the development of this field has closely tracked the development of analytical technology, literally over the last 300 years. Thus, a handbook devoted to methodology in the field of biomineralization is no doubt an invaluable resource.

ERA OF OPTICAL MICROSCOPY

Biomineralization products were among the very first objects to be examined in the second half of the seventeenth century, when the first optical microscopes were developed. One of the giants of early optical microscopy, Anton van Leeuwenhoek, studied, among many different objects, red coral, ivory, teeth, as well as cortical bone. Van Leeuwenhoek was in fact the first to describe bone osteons (van Leeuwenhoek, 1693) and at almost the same time, John Havers identified the lamellar structure of bone (Havers, 1691). The microscopes used by these pioneers had the capability of magnifying more than 200 times, and thus they could already characterize mineralized objects at a tissue level—cells and mineralized components and their relative distributions. The fantastic line drawings from this period and the accompanying texts are still some of the best documentations of the ultrastructure and the cellular activities that we have of mineralized tissue formation processes.

For biomineralization, the most significant technological advance in optical microscopy was probably the invention of the polarizing microscope in the late nineteenth century. This microscope made it possible to differentiate between minerals that had no atomic order (so-called isotropic or amorphous minerals) and crystalline minerals. The polarizing microscope could also be used to determine the orientations of the crystallographic axes of biogenic crystals. Some fundamental concepts of the field, such as the fact that many mineralized tissues are composed of aligned arrays of crystals, were established with this method. The polarizing microscope was also used to show that a significant number of biogenic minerals were, in fact, amorphous, including the metastable amorphous calcium carbonate identified in sponges in 1898 by Minchin (Minchin, 1898). The large body of knowledge that accumulated using optical microscopy was assembled in a landmark book by W.J. Schmidt entitled *Die Bausteine des Tierkorpers in Polarisiertem Lichte* published in 1924. This book is only available in German, and its translation into English would benefit the biomineralization community enormously.

Light microscopy use was extended enormously with the development of substrate-specific stains that enabled thin tissue sections to be examined in a variety of ways. This histological approach was also applied to mineralized tissues, with the production of wonderful studies of, for example, the growth plates of vertebrate bones. A key contribution to the field of vertebrate tissues, including mineralized tissues, is the book by Le Gros Clark (1945). Light microscopy has, in the last decade, undergone yet another reincarnation with the development of multiphoton microscopy and of the confocal microscope with its ability to produce 3D structures. With the help of specific fluorescent tags, it is now possible to dynamically track the development of processes in time and map the distributions of individual macromolecules. And all this can be carried out *in vivo*.

ERA OF X-RAY DIFFRACTION

X-ray diffraction was developed in the beginning of the twentieth century by von Laue, and subsequently by William Henry and William Lawrence Bragg (father and son), enabling crystal structures to be resolved. In fact, calcite was one of the first structures resolved in this way. In addition, determining the distribution of the structure of the component ions and molecules in the crystal, x-ray diffraction made mineral identification, as well as the determination of crystal axis orientation, a lot easier. In the 1930s, x-ray diffraction was used to confirm that the crystals of bone have a well-defined preferred orientation (Stuhler, 1937). In fact, this was already known based on much more challenging, and less compelling, observations using polarizing light microscopy (Schmidt, 1936). Looking back, the development of x-ray diffraction did not immediately impact the field of biomineralization. Over the years, its use has steadily increased, especially in recent years when new microspot capabilities were developed at synchrotron facilities, making the analysis of very small volumes possible. It can be predicted that when many proteins and other macromolecules that make up the scaffolds for mineral building blocks, or are directly involved in controlling crystal growth, can be crystallized and their structures determined, the impact will be enormous.

ERA OF ELECTRON MICROSCOPY

Transmission electron microscopy (TEM) became available for basic research in the late 1940s. Within just a few years, R.A. Robinson used TEM to show that the minute crystals of bone were plate shaped (Robinson, 1952), and then, together with Watson, showed the relation between the crystals and type I collagen fibrils (Robinson and Watson, 1952). The small size of bone crystals makes them ideal for direct observation using TEM where sample thickness is limited to around 50–70 nm. However, bone crystals are so thin (2–4 nm) that when observed lying face-on in a polymeric medium, as required for producing these sections, they are essentially invisible. They can only be seen edge-on as a projection, and this technical limitation led to more than 30 years of confusion about the shapes of bone crystals. Robinson, in his first paper where he analyzed dispersed crystals without embedding, was correct—they are indeed plate shaped (Robinson, 1952).

TEM studies in the 1950s and 1960s contributed enormously to the understanding of mineralized tissue structures at high resolution. TEM in many respects laid the foundation for the structural understanding of mineralized tissues, but this came along with some inevitable mistakes due to the electron beam causing radiation damage, microtoming that resulted in the loss of mineral components (often only recognized by the holes they left) or even shattering of crystals, and the resultant misinterpretation of structures. The visibility of the organic components was dependent on the addition of a contrasting agent in order to differentiate them from the embedding medium, and this too created a bias. Radiation damage could be reduced by cooling the specimen to liquid nitrogen temperatures, and one of the first studies using such radiation protection identified amorphous calcium phosphate in bone tissue (Gay and Schraer, 1975). In the 1980s, it was discovered how to freeze biological specimens so fast that ice crystal growth was prevented, and in essence, the specimens were observed in *frozen water* (so-called vitrified ice). This discovery enabled hydrated tissues to be studied at high resolution in the TEM (Dubochet et al., 1988), a major advantage for the field of biomineralization where imaging cells, unstable minerals, and the interface between them is often the focus. An exciting new development in TEM is the liquid cell, which enables observations to be made in solution, including the observation of growing crystals (Zheng et al., 2009).

The first commercial scanning electron microscope (SEM) was available for research in 1965. This, in a sense, magical tool produces wonderful 3D images. Some of the first applications of SEM were on mineralized tissues, providing a host of fascinating (and often aesthetic) images. In many respects, the SEM is the *binocular* of the field of biomineralization, as it provides the basic structural overview of the tissue under investigation. It is, however, very difficult to obtain a quantifiable analysis using the 3D capability of SEM. SEM images are often the source of many important questions in biomineralization but rarely provide conclusive answers. In the 1980s, the advent of atomic force microscopy (AFM) significantly extended the capabilities of electron microscopy, in a sense, with its capabilities of examining samples under water at very high resolution. SEM technology has undergone major advances in the last decade, with resolutions increasing enormously to just a few nanometers. And just like in TEM, the advent in the last few years of cryo-SEM capability offers for the first time an almost *in vivo* look at a mineralized tissue, including the cells, the mineralized structures, and the manner in which they are associated. Another very promising recently developed SEM is capable of examining the surfaces of biological samples in air (B-Nano Ltd., Israel, www.b-nano.com). The air SEM opens up many exciting possibilities for biomineralization, as imaging as well as elemental mapping at high resolution can be carried out on samples that have undergone no pretreatment.

Many mineralized tissues are not only hierarchically organized but have complex three-dimensional structures. Cryo-electron tomography is an effective tool for elucidating structures in a slice of less than a micrometer thick. Recently, a new and promising approach for 3D structure elucidation was developed using the dual beam electron microscope (Heymann et al., 2006). The serial surface view method involves using a heavy atom beam to slice off consecutive thin layers from an embedded sample, and then the electron beam for imaging the newly exposed surface. A stack of around a thousand images can be collected automatically over a 24 h period. This method has been used for elucidating bone lamellar structure (Reznikov et al., 2013).

SPECTROSCOPY

Infrared spectroscopy was developed in the middle of the nineteenth century but was rarely used for identifying minerals, x-ray diffraction being the method of choice. The early studies of Posner and Lowenstam (Lowenstam, 1972; Termine and Posner, 1966) demonstrated the major benefit of using infrared spectroscopy for studying biogenic minerals, namely, the possibility of characterizing both disordered and crystalline minerals. It was mainly the use of infrared spectroscopy that revealed the enormous diversity of differently ordered minerals in biology (Lowenstam and Weiner, 1989). New methodological developments make it possible to better separate the contributions to peak width of particle size and inherent atomic level disorder (Regev et al., 2010), thus revealing that marked differences in the extent of atomic order exist even in mature biogenic minerals.

Raman spectroscopy complements infrared spectroscopy well and has the major advantage of enabling spectral information to be obtained from the surfaces of wet specimens. The combination of Raman spectroscopy and confocal microscopy makes it possible to obtain spectra from within live tissues. X-ray absorption spectroscopy (XAS) has much potential in biomineralization studies, as it provides detailed information on the arrangement of atoms around an atom of choice, thus making it possible to obtain insights into biological control over atomic disorder in minerals. Spectroscopy combined with electron imaging and crystallographic information can be obtained from X-PEEM (x-ray photoelectron emission microscopy), providing the XAS spectrum of each pixel from the object at a resolution of down to 20 × 20 nm. Furthermore, elemental mapping of the surface can be achieved at a resolution comparable to the NanoSims. These x-ray-based spectroscopic tools are of much value in the field of biomineralization (Metzler et al., 2008).

This handbook includes chapters on the techniques mentioned earlier, as well as many other fantastic techniques that are used for better understanding biomineralization. The collection of all this know-how literally under one cover is an enormous asset for the field.

ONE OF THE CHALLENGES OF TODAY: THE CORRELATIVE APPROACH

Many questions in biomineralization relate to the type of mineral phase present at a specific location and at a specific time during development. Until the late 1990s, this was not thought to be a major issue. The consensus was that the mineral in the mature tissue formed directly from a saturated solution. However, the discovery that the first minerals deposited are often highly disordered and hence unstable, and may form inside vesicles within cells (Weiner and Addadi, 2011), has radically changed this simple view. Now the challenge is to characterize these minerals and to track their translocation from inside the cells to their final locations, as well as to document the transformation from disordered to ordered phases. All this occurs at the nanometer to micrometer scale.

The fact that we are dealing with highly unstable mineral phases highlights the need to obtain as much *in vivo* information as possible. So in a very real sense, it is back to optical microscopy. Confocal microscopes can be combined with spectroscopy such that spectra can be obtained from volumes of as little as tens of nanometers cubed. Probably the most versatile microspectroscopy method is Raman spectroscopy. Spectra can be obtained *in vivo* from a single pixel in a 3D stack of images. When combined with fluorescent imaging, the location of the mineral phase and its characterization can be determined in the context of the cell organization and activity. Optical microscopy observations can also be re-examined in exactly the same area, but at high magnification using the serial surface view method and the dual beam microscope.

It seems to us that we are now entering a new era, where an array of different types of information is obtained from exactly the same target area at more or less the same time. This correlative approach can be achieved by linking *in vivo* observations using optical microscopy with high-resolution serial surface view 3D reconstructions, for example, or with Raman spectra obtained directly through the confocal microscope. The air SEM is potentially another excellent tool for correlative observations. The correlative approach has the huge potential of being able to integrate structure, mineral characterization, and interphases—all in the context of *in vivo* observations—may be a dream come true for the field of biomineralization.

AUTHORS

Steve Weiner Department of Structural Biology, Weizmann Institute of Science, Rehovot, Israel
Lia Addadi Department of Structural Biology, Weizmann Institute of Science, Rehovot, Israel

REFERENCES

Dubochet, J., Adrian, M., Chang, J.-J., Homo, J.-C., Lepault, J., McDowall, A.W., and Schultz, P. 1988. Cryo-electron microscopy of vitrified specimens. *Q. Rev. Biophys.* 21, 129–228.

Gay, C.V. and Schraer, H. 1975. Frozen thin-sections of rapidly forming bone: Bone cell ultrastructure. *Calcif. Tissue Res.* 19, 39–49.

Havers, C. 1691. *Osteologia Nova.* Samuel Smith, London, U.K.

Heymann, J.A.W., Hayles, M., Gestmann, I., Giannuzzi, L., Lich, B., and Subramaniam, S. 2006. Site-specific 3D imaging of cells and tissues with a dual beam microscope. *J. Struct. Biol.* 165, 63–73.

Le Gros Clark, W.E. 1945. *The Tissues of the Body.* Clarendon Press, Oxford, U.K.

Lowenstam, H.A. 1972. Phosphatic hard tissues of marine invertebrates: Their nature and mechanical function, and some fossil implications. *Chem. Geol.* 9, 153–166.

Lowenstam, H.A. and Weiner, S. 1989. *On Biomineralization.* Oxford University Press, New York.

Metzler, R.A., Kim, W., Delak, K., Evans, J.S., Zhou, D., Beniash, E., Wilt, F. et al. 2008. Probing the organic–mineral interface at the molecular level in model biominerals. *Langmuir* 24, 2680–2687.

Minchin, E.A. 1898. Materials for a monograph of the ascons. I. On the origin and growth if the triradiate and quadriradiate spicules in the family Clathrinidae. *Q. J. Microsc. Sci.* 40, 469–587.

Regev, L., Poduska, K.M., Addadi, L., Weiner, S., and Boaretto, E. 2010. Distinguishing between calcites formed by different mechanisms using infrared spectrometry: Archaeological applications *J. Archaeol. Sci.* 37, 3022–3029.

Reznikov, N., Almany-Magal, R., Shahar, R., and Weiner, S. 2013. Three-dimensional imaging of collagen fibril organization in rat circumferential lamellar bone using a dual beam electron microscope reveals ordered and disordered sub-lamellar structures. *Bone* 52, 676–683.

Robinson, R. 1952. An electron microscope study of the crystalline inorganic component of bone and its relationship to the organic matrix. *J. Bone Joint Surg.* 34A, 389–434.

Robinson, R.A. and Watson, M.L. 1952. Collagen-crystal relationships in bone as seen in the electron microscope. *Anat. Rec.* 114, 383–410.

Schmidt, W.J. 1936. Uber die Kristallorientierung im Zahnschmelz. *Naturwissenschaften* 24, 361.

Stuhler, R. 1937. Uber den Feinbau des Knochens. *Fortscht. Geb. Rontgenstrahlen* 57, 231–264.

Termine, J.D. and Posner, A.S. 1966. Infra-red determination of the percentage of crystallinity in apatitic calcium phosphates. *Nature* 211, 268–270.

van Leeuwenhoek, A. 1693. An extract of a letter from Mr. Anth. Van. Leeuwenhoek, containing several observations on the texture of the bones of animals compared with that of wood: On the bark of trees: on the little scales found on the cuticula, etc. *J. R. Soc.* 838–843.

Weiner, S. and Addadi, L. 2011. Crystallization pathways in biomineralization. *Annu. Rev. Mater. Res.* 41, 21–40.

Zheng, H., Claridge, S.A., Minor, A.M., Alivisatos, A.P., and Dahmen, U. 2009. Nanocrystal diffusion in a liquid thin film observed by in situ transmission electron microscopy. *NanoLetters* 9, 2460–2465.

Preface

Biomineralization refers to the formation of minerals of biological origin, such as vertebrate bones and teeth, invertebrate shells and exoskeletons, and even mineral particles secreted by plants and bacteria. Biominerals can also arise from pathological conditions, such as the formation of kidney stones, gout, and atherosclerotic plaque. Regardless of whether the mineral is biologically controlled or pathological, biominerals are generally composites of mineral and organic phases, intertwined across the spectrum of length scales from nanometers to millimeters. They have been of interest as materials since humanity's earliest days of utilizing shell and bone to create objects and tools. Their high-performance properties compared to inorganic minerals and synthetic composites keep scientific interest high: new designs and applications are proposed for bioinspired materials every year! This sourcebook explores analytical materials science techniques that have contributed to our current understanding of biomineralization and that can further advance the field.

When scientists and engineers began to realize that superior properties could be found in biomineral composites, such as the combination of high strength and toughness, they approached the problem from the perspective of the first tenet of materials engineering: one must determine the relationships between structure, processing, and properties. Thus, much of the classic literature on biominerals was devoted to first characterizing the structures of these biological composites. Because of their complexity, which often consists of a hierarchical organization of structure across many length scales, biomineral researchers found they needed to be at the forefront in developing and using advanced characterization techniques. This is certainly exhibited in the older literature, where we find many beautiful papers that first deciphered the complex structure of bones, teeth, and mollusk shells. And it continues today, where newly discovered biomineral properties and structures continue to intrigue the materials engineer.

WHAT DOES IT MEAN TO BE AT THE FOREFRONT OF A CHARACTERIZATION TECHNIQUE?

It can mean *novel implementation of a suite of techniques*. Faced with a biomineralized tissue that exhibits incredible mechanical functions, researchers know that it is best addressed with collaborative efforts capable of combining specialized characterization methods—optical, structural, mechanical, and functional. Novel research can also require *pioneering the reach of a technique*, as when images resolve more detail, forces and chemistries are measured with increased sensitivity, spectra become accessible for new classes of sample systems, and computations access new length and time scales and more complexity. Finally, we believe that the frontiers of analytical methods have also created a scientific community with a *hierarchical attention span*: a consensus that the scientific story traverses the entire span from *molecular and atomic structure*, to finding new ways to *quantitatively visualize the composite*, to *probing entire organs and organisms* with new techniques.

With this sourcebook, we endeavor to emphasize the interplay between multiple techniques at their current frontiers and learn more about how such studies may be carried out. We know our audience shares the feeling of inspiration at the explosion of research in recent years, which in large part is due to the use of advanced techniques to characterize biominerals and biomimetic model systems. At the same time, we sometimes find it increasingly difficult to understand this new literature because of a general lack of expertise in these advanced techniques.

Such a large number of experimental approaches will not usually be mastered by any individual researcher. Thanks to collaboration and a lively schedule of conferences, participants often wish to know more about many of the techniques they encounter in their colleagues' work. Frequently, the researcher wants to know whether the technique can be applied to his or her particular system and what sort of information can be derived from the technique. The researcher also wants to know how difficult the technique is and whether he or she will be able to perform the experiments by himself or herself, or whether it would require a collaborator with that expertise. Finally, even though we may not have an immediate need for that technique in our own research, it is beneficial to have an adequate understanding in order to evaluate the ideas published and proposed by our colleagues.

Thus, the idea was born that researchers in the biomineral and biomimetics fields might benefit from some dedicated cross-training in the area of characterization. Biominerals pose their own unique challenges, which means that there is frequently a story to be told regarding how a technique must be adapted to their study. We have designed this sourcebook in such a manner that chapters will include elements of how-to, illustrate the power of combined techniques, showcase inspiring innovations, and motivate readers to pursue collaborations and new experiments at user facilities such as national labs and synchrotrons.

Parts I and II address *atomic and molecular structure*: how we describe it, detect it, and assess its importance. Included therein are diverse perspectives on the continuum of mineral crystallinity and the structured organic matrices. In Part III, we highlight additional measurements that are especially well suited for *imaging morphology and interfaces:* two- and three-dimensional systems with heterogeneous, if not hierarchical, structure. These systems enable particular aspects of biominerals and biomimetic models to be scrutinized. Parts IV and V present state-of-the-art methods to assess *properties of the composite* and discuss current approaches for measuring entire biological working structures while retaining as much fine-grained biophysical information as possible. In all these chapters, authors were asked to showcase discoveries from their own programs as well.

We believe this philosophy has created a book that will be at a level accessible to graduate students involved in comparable research and may be useful for relevant graduate courses. The book is also a snapshot of the state of the art in a spectrum of experimental techniques applied to a common interdisciplinary goal, where the ability to use the more advanced techniques often requires funding for collaboration and travel; thus, the book may be useful from a programmatic viewpoint as well.

It is our hope that this book will deepen the appreciation for the massive interdisciplinary effort underway, educate researchers across the field, and motivate new collaborations. Following is a sampling of the leading questions we asked our chapter contributors to address, as they saw fit:

1. *Benefits*. What are the specific advantages and/or historical uses of this technique?
2. *Applications*. What are its particular advantages when applied to biomineralization?
3. *Challenges*. What are the key challenges to consider in applying this technique?
4. *Innovation*. What was your innovation in overcoming those challenges? Examples: new chemical protocol, new analysis protocol, designed software, built new instrument, novel genetic or genomic discovery or design, and/or new computational technique or constraints.
5. *Motivation*. What was your main inspiration in designing/discovering this innovation? Examples: knowledge of proteins or other details of biological system, unresolved question raised elsewhere in the field, deep knowledge of a (nonbiomineral) materials science issue that was applicable, or accidental discovery in the lab!

We know that readers will find that our authors went well beyond the call of duty to address these questions, and we found their responses to be highly innovative; we thank them for this.

When we first embarked upon this project, our goal was to develop a general sourcebook on biomineralization. But when we realized the list of authors doing exciting work in this area was threatening to become too exhaustive, we decided that this project might best be approached as a series. From this volume addressing advanced characterization techniques, we envision two more: a volume on advances in mineral–organic synthesis and a volume on biology and biotechnology methods. Only time will tell whether such a series will be as welcome as we think it should be.

With this, we acknowledge our chapter contributors once again and thank our publisher, Taylor & Francis Group, for initiating and supporting this work.

Looking forward,
The editors

Editors

Elaine DiMasi is a physicist and synchrotron x-ray scattering expert and has made her career at Brookhaven National Laboratory (BNL) since 1996. Research for her PhD (University of Michigan, Ann Arbor) and postdoctoral appointment (BNL) focused on structure and electronic properties in metallic condensed matter systems. Since 1999, she has investigated numerous aspects of biomineralization, including mineralization at Langmuir films, assembly and mineralization of extracellular matrix proteins, structures of organics assembled on mineral surfaces, and microbeam diffraction mapping of mineral–organic composites and biological minerals. More recent areas of interest include lipid–mineral interactions and soft x-ray microspectroscopy. At the National Synchrotron Light Source II, DiMasi is currently engaged in building a state-of-the-art small- and wide-angle x-ray scattering facility, which is dedicated to soft materials and biomaterials, specializing in aqueous interfaces that allow the measurement of the hierarchical structures of biominerals over a wide range of length scales in realistic material and biomimetic environments.

Laurie B. Gower is an associate professor in the Department of Materials Science and Engineering and supervisor of the Biomimetics Laboratory at the University of Florida. She received her master's degree in bioengineering from the University of Utah in 1990 and her doctoral degree in polymer science and engineering from UMass at Amherst in 1997. In the latter case, her dissertation was focused on biomineralization, making use of model systems to examine the interactions between polypeptides and crystal growth and correlating features observed in the *in vitro* systems to those observed in biominerals. Most of her research has focused on examining potential mechanisms involved in biomineralization. She has also discovered a novel crystallization process that relies on a polymer-induced liquid-precursor (PILP) phase and was one of the first to suggest that biominerals might be formed from a hydrated amorphous precursor. She has built a line of evidence to suggest that this polymer-directed crystallization process may play a fundamental role in both calcium carbonate (marine exoskeletons) and calcium phosphate (bones and teeth) biomineralization, as well as calcium oxalate precipitation in kidney stones.

Contributors

Elia Beniash
Department of Oral Biology
University of Pittsburgh
Pittsburgh, Pennsylvania

Karim Benzerara
Institut de Minéralogie et de Physique des Milieux Condensés
CNRS and University Pierre et Marie Curie
Paris, France

Dominique Blamart
Laboratoire des Sciences du Climat et de l'environnement
CEA-CNRS-UVSQ/IPSL
Gif-sur-Yvette, France

Adele L. Boskey
Hospital for Special Surgery
Weill Medical College
Cornell University
New York, New York

Markus J. Buehler
Department of Civil and Environmental Engineering
Center for Computational Engineering
and
Center for Materials Science and Engineering
Massachusetts Institute of Technology
Cambridge, Massachusetts

Peter Chung
School of Geographical & Earth Sciences
University of Glasgow
Glasgow, United Kingdom

Sungwook Chung
The Molecular Foundry and Physical Biosciences Division
Lawrence Berkeley National Laboratory
Berkeley, California

Julie Cosmidis
Institut de Minéralogie et de Physique des Milieux Condensés
CNRS and University Pierre et Marie Curie
Paris, France

Qiang Cui
Department of Chemistry and Theoretical Chemistry Institute
University of Wisconsin-Madison
Madison, Wisconsin

Maggie Cusack
School of Geographical and Earth Sciences
University of Glasgow
Glasgow, United Kingdom

Yannicke Dauphin
UMR IDES
Université de Paris-Sud
Orsay, France

James J. De Yoreo
Physical Sciences Division
Pacific Northwest National Laboratory
Richland, Washington

Archan Dey
Laboratory of Materials and Interface Chemistry
and
Soft Matter CryoTEM Unit
Eindhoven University of Technology
Eindhoven, the Netherlands

Leon S. Dimas
Department of Civil and Environmental Engineering
Massachusetts Institute of Technology
Cambridge, Massachusetts

Elaine DiMasi
Photon Sciences Division
Brookhaven National Laboratory
Upton, New York

Melinda J. Duer
Department of Chemistry
University of Cambridge
Cambridge, United Kingdom

Pulak Dutta
Department of Physics & Astronomy
Northwestern University
Evanston, Illinois

Francis Esmonde-White
Department of Chemistry
University of Michigan
Ann Arbor, Michigan

Karen Esmonde-White
Department of Internal Medicine
University of Michigan
Ann Arbor, Michigan

John Spencer Evans
Division of Basic Sciences and Craniofacial Biology
Laboratory for Chemical Physics
New York University College of Dentistry
New York, New York

Raymond W. Friddle
Sandia National Laboratories
Livermore, California

Pupa U.P.A. Gilbert
Department of Physics
and
Department of Chemistry
University of Wisconsin-Madison
Madison, Wisconsin

Orestis L. Katsamenis
Faculty of Engineering and the Environment
University of Southampton
Southampton, United Kingdom

Flavia Libonati
Department of Civil and Environmental Engineering
Massachusetts Institute of Technology
Cambridge, Massachusetts

Arun K. Nair
Department of Civil and Environmental Engineering
Massachusetts Institute of Technology
Cambridge, Massachusetts

Yael Politi
Department of Biomaterials
Max-Planck Institute of Colloids and Interfaces
Potsdam, Germany

Zhao Qin
Department of Civil and Environmental Engineering
Massachusetts Institute of Technology
Cambridge, Massachusetts

S. Roger Qiu
Physical and Life Sciences Directorate
Lawrence Livermore National Laboratory
Livermore, California

Claire Rollion-Bard
Department of Geochemistry
Centre de Recherches Pétrographiques et Géochimiques
Vandoeuvre-lès-Nancy, France

Nita Sahai
Department of Polymer Science
University of Akron
Akron, Ohio

Murielle Salomé
European Synchrotron Radiation Facility
Grenoble, France

Ron Shahar
Koret School of Veterinary Medicine
The Hebrew University of Jerusalem
Rehovot, Israel

Nico A.J.M. Sommerdijk
Laboratory of Materials and Interface Chemistry
and
Soft Matter CryoTEM Unit
Eindhoven University of Technology
Eindhoven, the Netherlands

Stuart R. Stock
Department of Molecular Pharmacology and Biological Chemistry
Feinberg School of Medicine
Northwestern University
Evanston, Illinois

Benjamin D. Stripe
Department of Physics and Astronomy
Northwestern University
Evanston, Illinois

Philipp J. Thurner
Faculty of Engineering and the Environment
University of Southampton
Southampton, United Kingdom

Adam F. Wallace
Department of Geological Sciences
University of Delaware
Newark, Delaware

James C. Weaver
Wyss Institute
Harvard University
Cambridge, Massachusetts

Xianghui Xiao
Advanced Photon Source
Argonne National Laboratory
Argonne, Illinois

Zhijun Xu
Department of Polymer Science
University of Akron
Akron, Ohio

Yang Yang
Department of Chemistry and Biochemistry
Rowan University
Glassboro, New Jersey

Paul Zaslansky
Julius Wolff Institut
and
Berlin-Brandenburg Center for Regenerative Therapies
Charité-Universitätsmedizin Berlin
Berlin, Germany

Ivo Zizak
Institute Nanometre Optics and Technology
Helmholtz-Zentrum Berlin für Materialien und Energie GmbH
Berlin, Germany

Part I

Characterization of atomic and molecular structure: Diffraction and scattering

1 Synchrotron x-ray scattering: Probing structure for the structure–function relationship

Elaine DiMasi

Contents

1.1 INTRODUCTION

Which, in science, appeals to us more: complexity or simplicity? We often feel that biological materials are dauntingly complex yet somehow simpler than the artificial world humanity has created. Both complexity and simplicity are revered in science. Scholarly works about biomineralization often note that complex materials from nature, like wood and animal bones, were the first materials used by humankind (leaving out rocks, which we compare when we want to highlight biomineral complexity). As simple ores were worked into metal alloys with deliberately manipulated grain structures and natural fibers were woven into larger-scale strands and textiles, the ability for hierarchical organization to produce new function was demonstrated perfectly. Now we pursue both structure and function—and the ability to measure them—back down through micro- and nanoscales.

This is the theme of the present handbook as well as the theme of this chapter, which concerns itself with what structure is, how we describe it and detect it with x-rays, and how modern instrumentation has made x-ray diffraction (XRD) into an imaging method. Seeing structure has a central place in science. We imagine that if we could look inside a material or a mechanism, there the understanding lies. However, atomic details could be an embarrassment of riches. Is the structure simply the list of all positions of the $\sim 10^{24}$ atoms we have in this gram of interesting stuff? We know that the atoms push and pull on one another. Is the function some kind of list of total internal and external forces? No, we don't want these lists of ten-to-the-twenty-four things. Surely there's a way to make it simple to see that the seashell should be made of layers and bent into a spiral and the tooth should be made of nanoscale crystals and fibers woven into textures supporting a distinctive shape. It's part of the magic of scientific understanding that this does take place when an explanation is complete—simplicity and complexity as two sides of the same design.

In this chapter, we relate simplicity and complexity in the context of x-ray scattering, beginning with the fundamentals of how we describe and quantify structure. The interaction of x-rays with a material makes several classes of structural information measurable, and interpretations can be calculated very thoroughly or simplified for insight. Analysis of the entire scattering pattern for as much detail as possible leads to quantitative crystallography, discovery of protein structures, and so on. These techniques have been nothing short of revolutionary. On the other hand, it is often possible to extract something simple from a diffraction pattern—one molecular spacing, for example—and probe many positions on a specimen, or a large series of specimens, or a series of externally applied conditions, to see how that quantity may change. This approach begets XRD imaging, high-throughput structure studies, and *in situ* structure studies. *In situ* studies are a particular strength of x-rays, which are very forgiving of the constraints of sample enclosures and simultaneous complimentary probes.

Emphasizing the latter types of experiments, the second section of this chapter will show how synchrotron x-ray scattering has been applied to address important structure–function questions in biomineralization. Accessible questions are wide ranging: mineral or organic? Crystal or amorphous phase? Micro- or nanoscale? Mechanically active or passive? X-ray analysis has been applied to biominerals for over 100 years, but the bright beams and analytical savvy inherent in recent synchrotron-based studies have enormous impact now and point the way to developing even better experiments in future.

The enduring interest of these specific structure–function questions, and of biominerals generally, will be evident in the chapters that follow. It's exciting to see the diverse skills, insights, and enthusiasms from many fields come together in the collaborative research we collectively appreciate. Thanks to all involved in making this handbook an informative reference and a fun read.

1.2 HOW WE MEASURE STRUCTURE WITH X-RAY SCATTERING

1.2.1 FUNDAMENTALS: HOW WE DESCRIBE AND QUANTIFY A STRUCTURE

Ideally, we describe structure—where the atoms are—not by writing down the coordinates of each and every atom but by making simplifying generalizations that tell us the most important things about how the material is connected. Some solid materials commonly form as crystals, which are regular arrangements of atoms; a small and easily sketched unit cell depicts the symmetry, the molecular connections, and other aspects (magnetic moment could be another example) that would characterize the entire, macroscale material. Figure 1.1 shows (a) a piece of the common mineral calcite and (b) its rhombohedral unit cell. The calcium, carbon, and oxygen atoms have known locations. The positions and bonds between atoms are understood to prevail throughout a crystal of arbitrary size. The structure has been described.

Similarly, we might describe an organic material with a sketch of the neighbors in a single molecule and an indication that the molecules chain together to form a polymer, which specifies how molecules relate structurally to their neighbors. As with crystals, information about bonds and therefore the

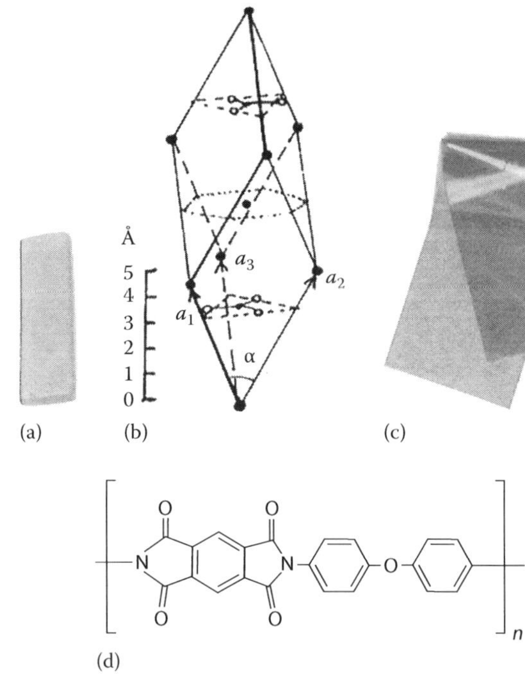

Figure 1.1 (a) A geological calcite crystal about 20 mm long. (b) A depiction of the rhombohedral unit cell of the calcite crystal structure. (c) A folded piece of Kapton (polyimide) film. (d) A scheme of the structure of poly-oxydiphenylene-pyromellitimide (Kapton K).

charge and chemistry of the bulk material is implied. Our model polymer in Figure 1.1c and d is Kapton, a polyimide material that is fairly transparent to x-rays and a familiar sight at synchrotrons, in use as windows to separate the sample environment from beamline vacuum. A good simple material, the Kapton film is isotropic, so it can be mounted in the beam in any direction equivalently, and the x-ray transparency of a thin piece versus a thick piece can be related by a simple equation inputting the film thickness, with no structural surprises at the macroscale.

Materials science becomes interesting when structures deviate from these ideals—ceramics with grains that interrupt the fixed alignments and polymers with molecules aligned along a particular macroscopic direction. Some minerals and organics are far from being crystalline; they are glassy or liquid-like. Some minerals and organics are far from being simple, as they may form new structures at larger scales when the smaller-scale building blocks are arranged in special ways. Many materials combine crystalline and amorphous components.

We can and do describe materials, including biominerals, as deviations from simple ideals, with cartoons depicting the building-blocks-upon-building-blocks models. But to look at a material with the more impartial x-ray vision, and turn conceptual models into established working theories and even facts, we quantify these structures mathematically.

1.2.1.1 Pair density function: How we can describe structure in real space

Let's look at a simple 2D crystal model. Its atoms form a rectangular lattice. There are atoms at the corner and in the

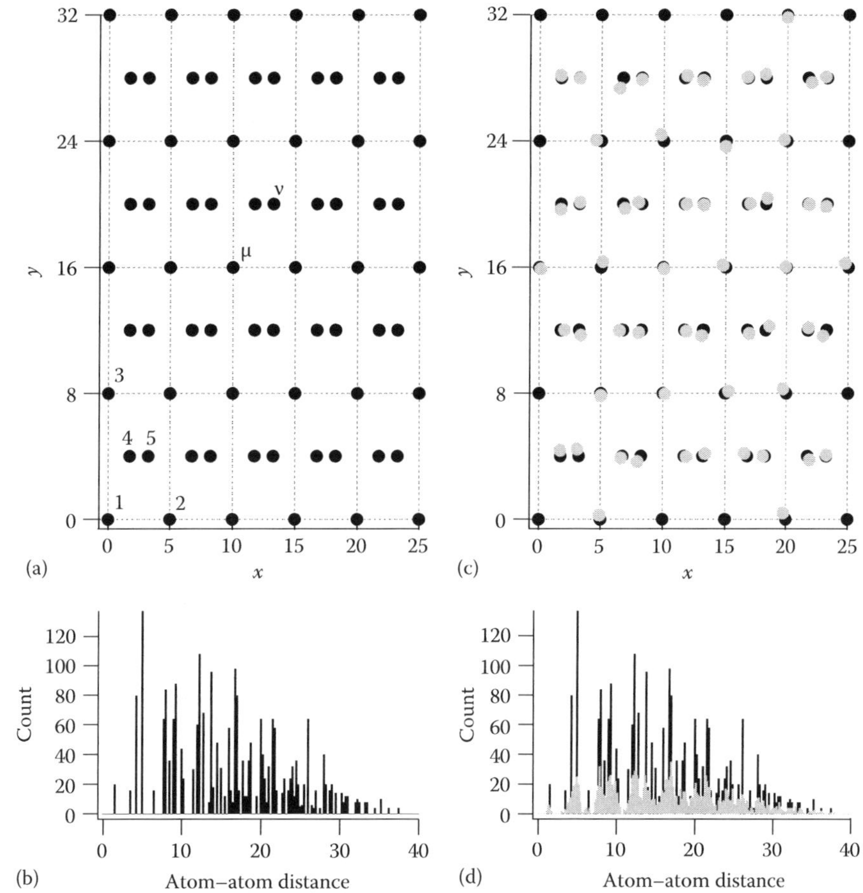

Figure 1.2 (a) Atoms in a crystal showing 20 unit cells, with a few atoms labeled for reference. (b) Histogram of atom–atom spacings in the structure of (a). For example, the peak at 5 results from the great many atoms with a neighbor 5 units away in these 5-unit-wide cells. (c) Gray circles that have been given a random (Gaussian-distributed) position away from their nominal positions (black circles, same as (a)). (d) Gray bars show the histogram of the disordered cell in (c) (black bars, same as (b)).

middle of the unit cell. They are the same kinds of atoms, indistinguishable except for position in the cell. The model is finite, 20 unit cells plus a few extra surface atoms. Figure 1.2a shows the atoms as markers plotted at their x and y positions. A few of the atoms at the lower left are numbered for reference.

We can glance at this picture and see an ideal regular crystal, but how would we say this mathematically? One way is in a function that examines how each atom relates to its neighbor. This is one of the most important things we care about, since bonds relate to chemical and physical properties. Start with atom 1 at (0, 0) and note the distance to atom 2 at (5, 0). Their atom–atom distance is 5. The distance from atom 1 to 3 at (8, 0) is 8. By counting every such atom–atom distance in the crystal, we create the histogram in Figure 1.2b.

The histogram is not a diffraction pattern, but it has features reminiscent of a diffraction pattern. It has well-separated peaks at low values of atom–atom spacing and more closely spaced peaks with decreasing amplitude at higher values. This graph reflects the distribution of atom–atom distances. There is a spike at 1.5: this is the distance between the pair of atoms 4 and 5, occurring 20 times in our 20 unit cells. The next spike is at 3.5, the distance from atom 5 to the atom in the next cell to its right, occurring 16 times in our sample. The third spike, at 4.25, captures the diagonal distance from atom 1 to 4. The spike at 5.0 represents the unit cell length of 5 (such as atom 1–2). This results in many counts and is analogous to a strong peak of a diffraction pattern.*

What happens when the crystal structure becomes nonideal? Figure 1.2c shows an overlay of atoms in gray that have been given small random deviations from their original positions. Their histogram, Figure 1.2d, shows peak counts broadened in distribution and diminished in amplitude. You can imagine that if we made the model glassy, with atoms only approximating a regular distance from each other, the distribution of atom–atom distances would smear out even more. The atom–atom distance distribution contains an enormous amount of information about the structure, *without questioning whether the sample is amorphous or crystalline.*

There is a fundamental reason that scattering experiments can be related to these functions that describe structure in general terms—scattering is sensitive to the density of scattering particles within a sample (Egami and Billinge 2003). The structure can be

* That spike also includes the distance 5.15 diagonally from atom 1 to 5. The two spikes are within our histogram resolution of 0.25 bin width. This is a good example of the concept of resolution, since we could have histogrammed with finer spacing and resolved them separately.

described by formal density of atomic pair distances, $\rho(r)$, which has a value at any possible atom–atom distance r we consider:

$$\rho(r) = \rho_0 g(r) = \frac{1}{4\pi N r^2} \times \sum_{\mu,\nu} \delta(r - r_{\mu\nu})$$

In the equation, ρ_0 is the number density of atoms in the system of N atoms. This provides a normalization (into units of length related to r) of atoms in the volume that contains them. The quantity $g(r)$ created here is a unitless density function. Here, r is a length (not a vector), and the expression sums up atom–atom pairs in all directions of a 3D system.

The sums are over every pair of atoms in the volume; $r_{\mu\nu}$ is the distance between atom ν and atom μ. The Dirac delta function $\delta(r - r_{\mu\nu})$ is the histogram counter. This function is zero everywhere except at all $r = r_{\mu\nu}$, where it is infinite such that the integral over it is equal to unity. This correctly counts each instance of $r = r_{\mu\nu}$ that describes a pair of atoms.

The atomic pair density function can be written in different ways depending on the problem. If we want to keep track of directions in a 3D crystal, we could make r the vector $\vec{\mathbf{r}}$. If we want to keep track of different atom species, we can write more terms that add up distances between like and unlike species separately. If we want to bundle bigger chunks of density than atoms, we can count molecules, micron-size colloidal particles, or anything in between. The key is that the x-ray scattering experiment is a probe of density correlations. For small-angle x-ray scattering (SAXS) and wide-angle x-ray scattering (WAXS) from biominerals, we choose the form of equation that is appropriate for the experiment in hand.

1.2.1.2 Structure–function: How we can describe structure in reciprocal space

Scattering probes such as x-rays, neutrons, and electrons have wavelike properties, and when they interact with a sample, the result is an interference pattern that can be analyzed. The key to connecting the interference pattern to the structure is the Fourier transform. A general Fourier transform $F(q)$ of a function $f(x)$ looks like this:

$$F(q) = \sum_j f(x_j) e^{-iqx_j}$$

(as a discrete sum) or

$$F(q) = \int f(x) e^{-iqx} dx$$

(as a continuous integral), with the sum over all the instances of x_j or range of x in the system. (Here x is now a vector, along a particular direction; the rigorous treatment of the 3D system requires either including all three x, y, z coordinates or making a spherical average, both of which we avoid here for simplicity.) The exponential can be written like a sum of waves, $\cos(qx_j) + i\sin(qx_j)$, which shows how the x_j become multipliers of frequencies, picking out the most intense of them. Going back to our model crystal and its histogram, $f(x_j)$ is the collection of spikes in the histogram at meaningful atom–atom distances; $f(x_j)$ equals zero at x_j values that do not correspond to any atom–atom distances. Think how many multiples of the unit cell sizes 5 and 8 there would be in a much larger crystal. This real-space structure produces a structure in the q space, or wave-vector space, that describes the same system through its Fourier transform. When we transform variables from x to q in this manner, we are working in reciprocal space.

Did you ever wonder why XRD peaks show inverse d-spacings of atom planes? Suppose $x = d$ is the only spike in the histogram $f(x_j)$. Then, the only wave in the $F(q)$ is $\cos(qd)$. This special value $q = 1/d$ (or as we more often scale it, $q = 2\pi/d$) is the spatial frequency for which we will observe constructive interference in scattered waves.

The $g(r)$ function defined earlier, when written in its Fourier transform, turns into a form directly related to the scattering equation. Rewrite $\rho_0 g(r)$ from earlier (returning to a spherically averaged expression) into another form called $G(r)$ (no new variables); it turns out to be a Fourier transform of what we can introduce as a structure-function, $S(q)$:

$$G(r) = 4\pi r\rho_0[g(r) - 1] = \left(\frac{2}{\pi}\right)\int_0^\infty q[S(q) - 1]\sin(qr)dq$$

This is fabulous, because, math-a-magically, the $S(q)$ function is exactly what you get by measuring the intensity of x-rays scattered from a material sample. This is the reason scattering techniques are powerful. Mathematically, it doesn't matter whether we talk about real space or q space when describing a structure. If total knowledge of the Fourier transform over all space could be known, we could reconstruct everything about the sample including its macroscopic shape.

In real scattering experiments, the intensity we measure is only a partial knowledge of the Fourier transform, so our job is to choose equations of a form that is most useful. Every experiment is a combination of directly measurable quantities and assumptions. In the example earlier, $S(q)$ is measured to a finite q value, and other things, such as an isotropic collection of crystal grains, are assumed. This example has been chosen to demonstrate the relationship between a scattering intensity and a histogram of atom–atom distances, which is easy to draw. However, there is more to the story, such as when alignments of structural units are related to the material function. We make different assumptions and handle the equations in different ways, as further handbook chapters will show. Many different types of scattering experiments are possible, if we know our way around q space. So now we need to answer: What is scattering? And is q something real? What the heck is it?

1.2.1.3 Scattering and the wave-vector transfer q

Scattering in physics refers to a particle (a quantum mechanical particle, having wavelike properties) traveling toward, and interacting with, another particle (for us, an atom in the sample). Something changes in the aftermath: the particles may trade some energy or change their direction of travel. Chapters in Part I of this handbook will describe scattering by x-ray photons, infrared photons, electrons, and ions. Some chapters will describe spectroscopy techniques, analyzing changes of the energy of the scattered particles. Sometimes the scattered particle is the same type: photon in → photon detected or electron in → electron detected. Sometimes the particles are different: for example, photon in → electron detected.

This chapter describes x-ray scattering. (An excellent fundamental reference is Warren 1969.) The incoming and scattered particles are photons in the x-ray wavelength regime. The scattering is elastic (no energy is transferred). Instead, the direction of travel of the photon changes, described by its momentum. The everyday momentum we are familiar with is the product of mass and velocity. The photon's momentum is a function of its direction of travel and its energy, which is inversely proportional to its wavelength. We indicate both the wavelength and the travel direction of x-ray beams by the vector $\vec{\mathbf{k}}$. The incoming wave vector $\vec{\mathbf{k}}_{in}$ has a direction in real space—whichever way the beam is really coming through the equipment onto the sample. Some scattered photons will come out of the sample: $\vec{\mathbf{k}}_{out}$. The magnitude of the vector is given by $|\vec{\mathbf{k}}| = k = 2\pi/\lambda = |k_{in}| = |k_{out}|$. For the wavelength λ, we'll use units of Å (= 10^{-10} m). Many readers are familiar with the 1.54 Å wavelength of x-rays from a laboratory diffractometer.

Now we can demystify this q, which is the momentum transfer or wave-vector transfer of the scattering event: $q = |\vec{\mathbf{q}}|$, where

$$\vec{\mathbf{q}} = \vec{\mathbf{k}}_{out} - \vec{\mathbf{k}}_{in}$$

This is a vector subtraction. The change in direction of the scattered ray is captured in the $\vec{\mathbf{q}}$ vector. Textbook treatments of Bragg's law in crystals relate q to lattice spacing in crystals with diagrams such as the one in Figure 1.3a, with x-rays $\vec{\mathbf{k}}_{in}$ incident at an angle θ to some nicely aligned atom planes. The scattered rays $\vec{\mathbf{k}}_{out}$ are reflecting at the same angle θ. Work out the vector geometry for $\vec{\mathbf{q}} = \vec{\mathbf{k}}_{out} - \vec{\mathbf{k}}_{in}$ and the result is $q = (4\pi/\lambda)\sin(2\theta/2)$.* This captures the relationship between scattering angle and wavelength for a given q. But please note that the q equation is quite general. Figure 1.3b is another perfectly

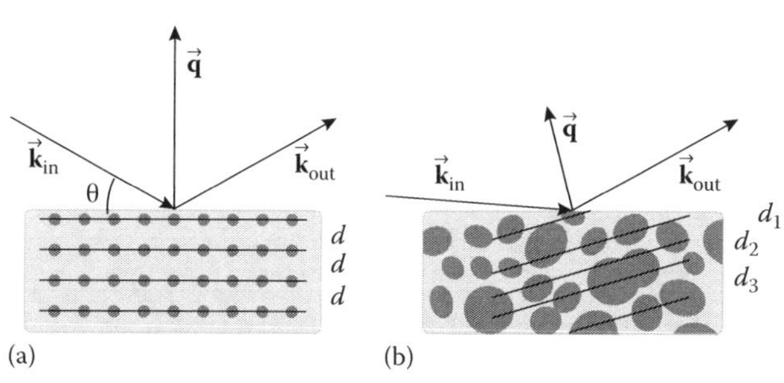

Figure 1.3 (a) X-rays with incident momentum $\vec{\mathbf{k}}_{in}$ and outgoing momentum $\vec{\mathbf{k}}_{out}$ are scattered from a sample at an angle θ so that momentum transfer $\vec{\mathbf{q}}$ will probe the nicely aligned lattice planes, which have separation d. (b) A different $\vec{\mathbf{q}}$ has been chosen, so as to probe this less structured sample and determine the distribution of distances such as d_1, d_2, and d_3 that may describe the density variations.

* Your student can prove this. They have to write the vector components, for example, $\vec{\mathbf{k}}_{in} = |\vec{\mathbf{k}}|\cos(\theta)\hat{\mathbf{x}} - |\vec{\mathbf{k}}|\sin(\theta)\hat{\mathbf{y}}$, and similar for $\vec{\mathbf{k}}_{out}$, then subtract them and use $|\vec{\mathbf{k}}| = k = 2\pi/\lambda$.

valid scattering geometry. We know that the experimentalist chose funny angles of $\vec{\mathbf{k}}_{in}$ and $\vec{\mathbf{k}}_{out}$ on purpose, suspecting that something about the spacing of the density blobs, in the direction indicated, is of interest.

A typical biomineral sample will have many real-space features of interest, on different length scales: atomic plane spacings in crystals, boundaries between grains, organization of larger macromolecule structures, and further hierarchical structures that lead to special functions. All these features, analogous to atom–atom spacings on their different levels, relate to $\vec{\mathbf{q}}$ vectors of interest: frankly, important $\vec{\mathbf{q}}$ vectors are sticking out from the sample in all directions like sea urchin spines! The relationship $q \sim 2\pi/r$ is always meaningful, where r can be a magnitude in any direction, because there is always a 3D density function $\rho(\vec{\mathbf{r}})$ describing the sample. The inverse $q \sim 2\pi/r$ relationship means that $\vec{\mathbf{q}}$ vectors are short for long-length correlations and long for atomic-scale correlations. See Figure 1.4a for the spiny sea urchin in real space, Figure 1.4b for some of the intense $\vec{\mathbf{q}}$ **spines** of a calcite crystal, and Figure 1.4c for the scattering $\vec{\mathbf{q}}$ vectors of an aligned collagen fibril.

It is up to the experimentalist to capture these $\vec{\mathbf{q}}$ vectors, knowing the incoming beam vector $\vec{\mathbf{k}}_{in}$ and setting the detector in the right position for $\vec{\mathbf{k}}_{out}$. In further sections of this chapter, we will take specific examples of real-space structure in biominerals and relate them to important $\vec{\mathbf{q}}$ features. To complete the present section, we'll look at what the experimenter, or their host the beamline scientist, does to set the experiment up.

1.2.2 SMALL- AND WIDE-ANGLE X-RAY SCATTERING AT THE SYNCHROTRON

1.2.2.1 Basic methodology

A synchrotron is a machine, usually constructed at a national laboratory or a university, which accelerates electrons in a ring to produce beams of infrared, ultraviolet, and x-ray light. The photon beams are used to study samples, with resident staff managing diverse experimental stations and helping the visiting researchers to conduct their experiments. Visit http://www.lightsources.org for information and news provided collaboratively by 25 synchrotron facilities worldwide.

If you are conducting a SAXS/WAXS experiment at a synchrotron, you will be assigned to a particular station (a beamline), where the staff arrange for an x-ray wavelength and beam size to suit your experiment. From your point of view, the beam is emerging horizontally from a pipe in the wall. You hand over your sample, and it goes onto a sample positioner between the pipes delivering the $\vec{\mathbf{k}}_{in}$ and the detectors capturing the $\vec{\mathbf{k}}_{out}$. X-rays scattered at small angles will go into a detector 1–2 m from the sample. The **wide**-angle detector may be only 20 cm away.

Figure 1.4d shows WAXS and SAXS angles on a realistic scale for wavelength λ = 1 Å. The equation $q = (4\pi/\lambda)\sin(2\theta/2)$ relates λ and q to the angle 2θ of scattered rays. If there would be a collagen d-banding structure in the sea urchin spine with $d \sim 70$ nm, then peaks would appear at $q \sim 2\pi/d \sim 0.01$ Å^{-1}, and 2θ = 0.04°. Rays traveling to a detector 1 m away will strike it 1–2 mm above the line of the direct beam. The measurement of several orders (multiples) of this d-spacing may go out to 2θ = 2° or further. (The staff will calibrate the detector positions against a known standard, to relate pixel positions to λ, 2θ, and q.) If we

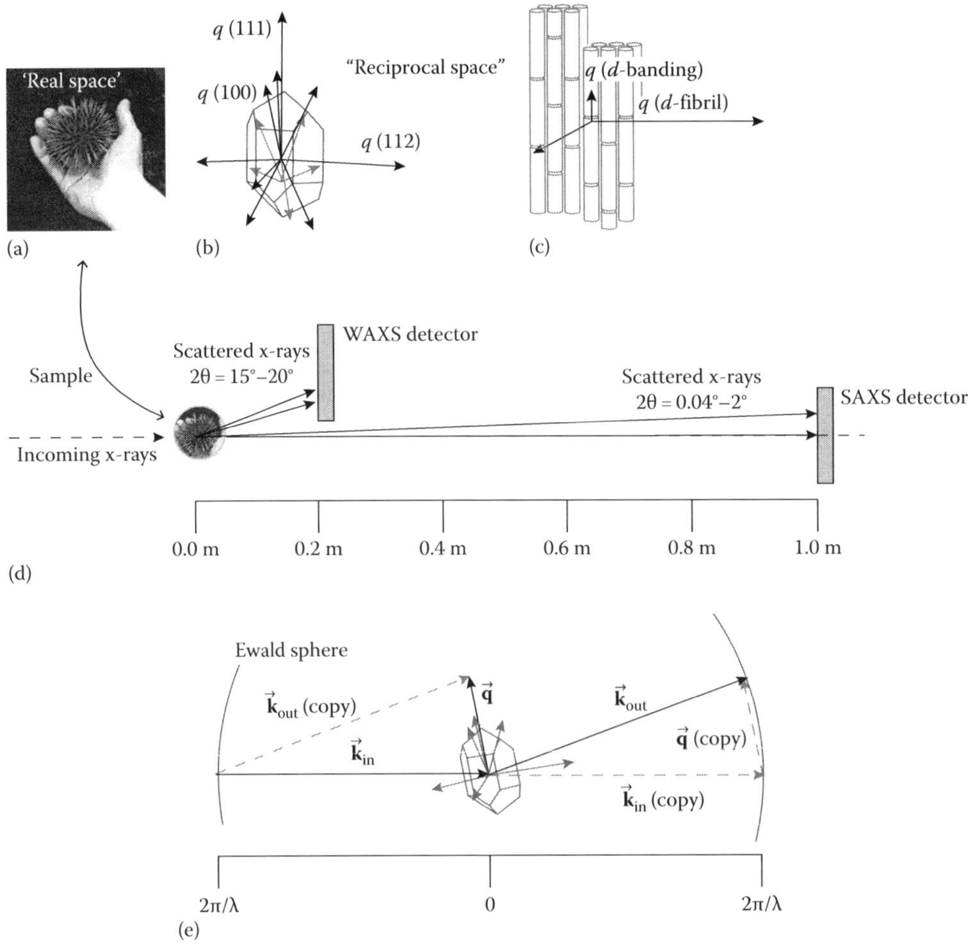

Figure 1.4 (a) This is a sea urchin, with spines sticking out in all directions in real space. (b) Biomineral components such as calcite crystals and (c) collagen fibrils create $\vec{\mathbf{q}}$ vectors sticking out in all directions in reciprocal space. (d) A typical SAXS/WAXS experiment requires placing the sample in an x-ray beam. Photons scattering at *wide* angles are captured in an area-resolving detector placed close to the sample. *Small* angles require a camera of order meters away. (e) The Ewald sphere, with a radius 2π/λ, is the meterstick of reciprocal space. Many $\vec{\mathbf{q}}$ vectors can be examined, by rotating the sample to align with the right x-ray angles and making sure the outgoing ray hits a detector.

also want to measure, say, Mg calcite lattice parameters in this sample (Long et al. 2011), we would seek out the strong Bragg peaks with d-spacings ~2 to 4 Å. This is a q range 1.6–3.1 Å^{-1}, scattering angles ~15° to 20°. This real-space picture illustrates the large range of length scales accessible to x-ray scattering. Observing intensity as a function of the *q magnitude* is one important consideration in scattering.

At the same time, the experimenter is thinking in reciprocal space. The orientations of important $\vec{\mathbf{q}}$ vectors need to be aligned so that the rays enter the detectors. Figure 1.4e shows how one q of interest was chosen to line up with the beam, given the $\vec{\mathbf{k}}_{in}$ and $\vec{\mathbf{k}}_{out}$ constraints. Light gray **copies** of the vectors are shown, to emphasize the $\vec{\mathbf{k}}_{in} + \vec{\mathbf{q}} = \vec{\mathbf{k}}_{out}$ relationship and to show a useful geometric method called the Ewald sphere. The radius of this sphere is 2π/λ, which is the meterstick of reciprocal space. Any q that has its endpoints on the sphere's surface can be measured, which requires pivoting the sample around the incoming beam. If the sample is made up of many grains in different orientations, all the possible $\vec{\mathbf{k}}_{out}$ vectors will form a cone, and when we look with a detector, we will see a ring of scattering. If the collection of grains are at a few orientations, we will see arcs or partial rings. Hence, observing intensity as a function of the *orientation* of the $\vec{\mathbf{q}}$ vector is a second important consideration.

1.2.2.2 Technical considerations

Generally two separate detectors, or a movable detector position, are used to capture all the length scales of interest in a biomineral (Paris et al. 2007, Paris 2008). Modern detectors are usually rectangular arrays of pixels that count x-rays as a function of position. The pixels are 0.050–0.175 mm in size, in an array 1000–2000 pixels across or larger. The pixels are quite sensitive. The intense direct beam will be blocked by a stop, since it is 1,000–1,000,000 times more intense than the scattered rays. Limitations at the low-q end may come from the beam size itself (anywhere from <10 μm to >1 mm), angular divergence (can be 0.05–5 mrad), or path length (some beamlines extend 10–15 m!). Limitations on maximum q usually come from detector size, available position, or inability to go to very short wavelengths (q must be less than the Ewald sphere size). Individual beamlines have typical q ranges where they work well and extremes suitable for exceptional experiments. Some detectors can take 30,000 images per second. Others require 5 s between measured points. Some experiments might require 60–120 s counting times and others will count for subsecond times.

In short, detector capabilities are equally important as the beamline capabilities in determining how to make a given experiment feasible. If the variations in $\rho(\vec{\mathbf{r}})$ are extremely small,

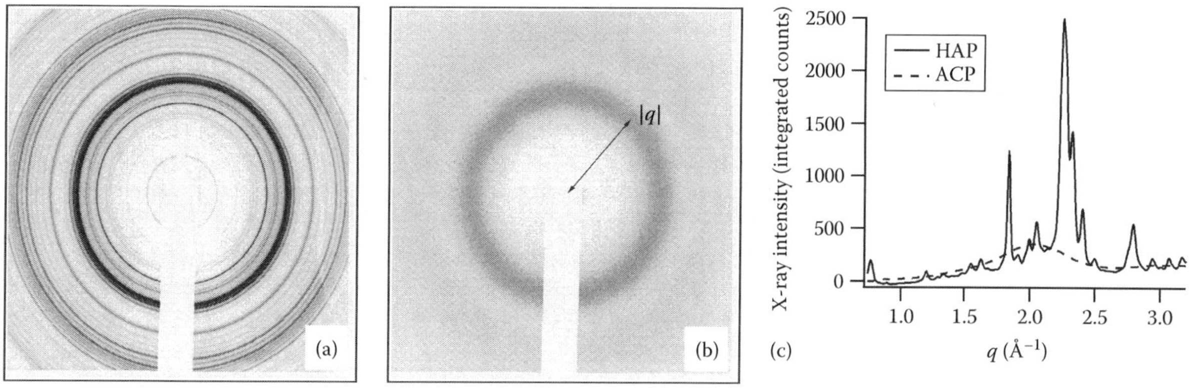

Figure 1.5 (a) X-ray WAXS pattern of a HAP (calcium phosphate, from Sigma-Aldrich) mineral standard (λ = 0.65 Å). (b) WAXS pattern of an ACP taken on the same scale. (c) Intensities $I(q)$ (scaled by thickness of each sample for direct comparison) integrated around the rings in (a) and (b) show the Bragg peaks of HAP (solid line) and the amorphous structure factor of ACP (dashed line).

we call this a sample with low contrast, and it will likely need long count times, a vacuum environment with minimal windows to reduce background scattering, and perhaps extra protection from beam damage. On the other hand, a well-ordered, robust sample may be of interest because of fast time-dependent reactions. Here, a fast detector is required.

What about a heterogeneous sample that varies from region to region? It can be mapped, if the beam is small compared to the regions of interest, creating an x-ray structural image. The beamline designers may have chosen beamline and detector parameters to optimize certain types of experiments, whether it is large or small beam, large or small detector pixels, and so on. Whether it is timing, mapping, or something else, the third important scattering capability is the one where the scientist fills in the blank: intensity as a function of *the most important variable that the experimenter can think of.*

1.2.3 MAGNITUDE AND ORIENTATION: TWO SOURCES OF STRUCTURAL INFORMATION

The biomineral literature now boasts many examples of synchrotron diffraction. Even though diffraction has been applied to these minerals almost since the discovery of x-rays, the brighter beams and creative cross-disciplinary perspectives at the synchrotrons frequently enable new discoveries.

Two examples separating q magnitude and q orientation applications will offer glimpses into this literature and set the stage for the multidimensional approach and the research to be highlighted in more detail.

1.2.3.1 Information from the q magnitude

Peaks at discrete q magnitudes are related to spacings between atom positions or between larger electron-dense regions. In many cases, orientational information is either not available or not important to the question at hand. Examples lacking orientation information include powders of crystal grains with random orientations, amorphous materials (organics or mineral glasses), and nanoscale molecules or nuclei suspended in liquid. An example of orientation that is present but not necessarily pertinent could be a study of Mg content in calcite or of a biological apatite mineral, where the crystal structure is known but changes in lattice parameters can indicate the amounts of impurities.

Isotropic crystalline powders and amorphous phases both show as rings on the detector, but the difference between the two is dramatic. Figure 1.5a shows the WAXS (or XRD) pattern of a crystalline hydroxyapatite (HAP) powder. The intense peaks (dark rings) are sharp in q and their q positions reflect the very many sharply defined d-spacings in the crystal, a complicated version of the one shown in Figure 1.3a. Integrated around the rings, the diffraction pattern $I(q)$ is shown as a solid line in Figure 1.5c. This information can be used for phase identification: "Yes, this is really calcium phosphate, in the HAP crystal structure." When measured carefully, with references to the literature, the pattern can be used to infer carbonate substitution for phosphate, in HAP (Zapanta-LeGeros 1965), Mg for Ca in calcite (Tsipursky and Buseck 1993, Long et al. 2011), and other substitutions that affect the crystal lattice planes. The $I(q)$ data can be used to identify presence of multiple phases in a biomineral (Wang et al. 2013) and to identify perturbations of the lattice such as finite grain size and strain (see Almer and Stock 2005, 2007, Chapter 2 of this handbook, and references therein).

In principle, we could also distinguish different mixtures of amorphous phases. This is more difficult—a structure like the one shown in Figure 1.3b has density correlations over a blend of d-spacings distributed around an average value. The diffraction pattern of Figure 1.5b illustrates the case of a synthetic amorphous calcium phosphate (ACP) (Sigma-Aldrich, St. Louis, MO). There is a clear maximum in the $I(q)$ (the pattern looks like a donut and not a sphere), but this is only evident because of very careful subtraction of sample holder background and the purity of the sample. If this phase were mixed with disordered organics or were shown as raw data together with low-q air or water scattering instead, identification would be very difficult. The structural organization of an amorphous material is reduced below that of a crystal, but it is present, even for liquids like water. Amorphous phases are almost always distinguished by a number of near-neighbor atomic distances, though this is hard to measure. Mixtures of amorphous and crystalline mineral are very important in biomineralization and very often distinguished by other experiments such as Raman and infrared light spectroscopies, rather than scattering (see Chapters 4 and 5 of this handbook).

Another special case of amorphous systems is the scattering from mineral nuclei or protein droplets that may be suspended in

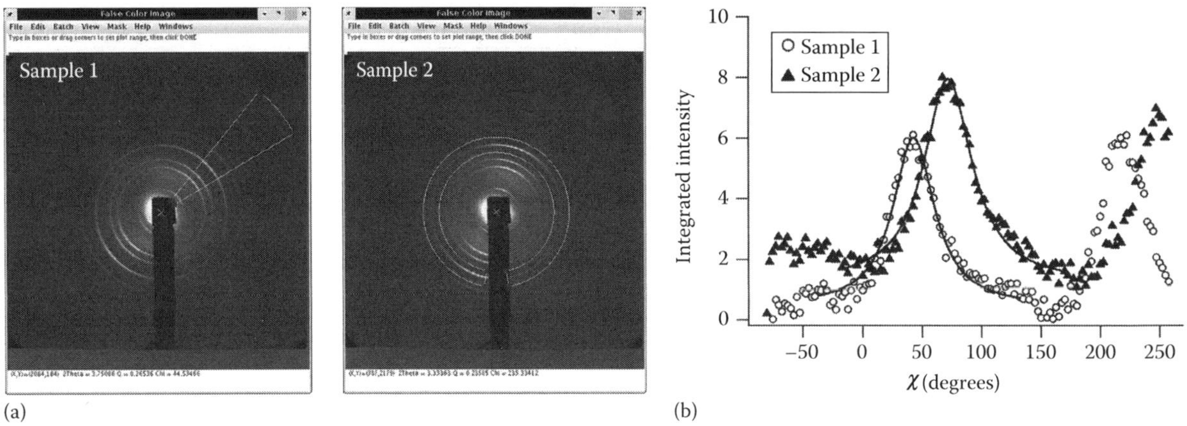

Figure 1.6 (a) Screenshots from the Datasqueeze software showing x-ray patterns of two collagen films. The data frames show arcs of differing orientations and widths along the azimuthal angle. (b) Intensity integrated in a small q region around the intense peaks, plotted along the azimuthal angle χ to reveal differences in fibril alignment.

a liquid phase. At extremely small values of q, the $\rho(r)$ function of nanosized particles adds up to a correlation that can be measured, giving us a way to measure nanodroplet sizes, size distributions, and the shapes of folded proteins. Protein solution scattering is increasingly important to biology since many proteins do not readily form crystals for diffraction. For biomineralization, one application has been to study amelogenin proteins, which form nanoscale aggregates that are believed to affect formation of tooth enamel (Aichmayer et al. 2010). These studies require comparing theoretical $I(q)$ curve shapes to the data very close to $q = 0$ (Schneidman-Duhovny et al. 2012).

1.2.3.2 Information from the $\vec{\mathbf{q}}$ vector orientation

Crystal lattice planes, organic fibers, and other organized structures have a direction in real space in the physical sample, and from this arises an orientation dependence of the $\vec{\mathbf{q}}$ vectors as seen on the detector when the sample is placed in the beam. This can be exploited to learn how subparts of the biomineral relate to each other and to boundary surfaces, and can lead to insight into material formation. Examples are the crystal twinning and angular deviations seen in the aragonite crystals of nacre and the arrangements of crystals at organic interfaces, among others, as will be seen in further chapters of this handbook.

One of the most famous biomineral–organic alignments is that between the apatite platelets in bone and the fibril axis of the supporting collagen. X-ray structural studies of collagen in bone and cartilage have become extremely sophisticated. Chen et al. (2005) studied methods of demineralizing shad bone and used SAXS/WAXS to show that disruption of the collagen supermolecular packing inhibited remineralization with calcium phosphate. In this fish bone, the mineral/fibril packing has been elucidated, and the collagen fibril superstructure has been determined using exquisite analysis of the shapes of meridional SAXS reflections (Zhou et al. 2007, Burger et al. 2008a,b). Rat tail tendon is so well ordered that since decades ago, meridional scattering patterns going out to 41 orders or more could be analyzed in terms of amino acid constituencies, interpreting the electron densities of the gap/overlap bands in the structure (Hulmes et al. 1977, Bradshaw et al. 1989). More recent work analyzes axial patterns for 3D depictions of the fibrillar packing (Orgel et al. 2000, 2001). Literature on stress-induced molecular rearrangements (Mosler et al. 1985) and the organization of collagen in the cornea (Hayes et al. 2007) is also available.

To illustrate how the collagen orientation can be measured, we show in Figure 1.6 some data from two synthetic collagen constructs (molecularly crowded collagen samples kindly supplied by Nima Saeidi and J. Ruberti, Northeastern University: Saeidi et al. 2012). The screenshots are from the Datasqueeze software (http://datasqueezesoftware.com), which is one of several available software packages to calibrate and analyze such patterns. The data frame for sample 1 shows diffraction arcs (primarily the third-, sixth-, seventh-, and ninth-order peaks) that have their maximum intensity at an azimuthal angle (χ) about 45° from the horizontal axis. A pie-wedge region in χ is outlined, where integration of an $I(q)$ curve would be the highest quality. In sample 2, it is clear that the χ angle has shifted, and the arcs also seem to be broader. By summing the intensity in a thin band of q and plotting over a long range of χ, as shown by the outlines drawn in the screenshot, we quantify the χ dependence, as the plots at right show. Quick peak fits to the $I(\chi)$ curves allow us to say that in sample 1, the collagen fibril axis has position $\chi = 42.9°$ with a distribution of ±21.3°. The sample 2 fit gives $\chi = 72.0° \pm 24.6°$. The measurement reveals a 15% broader distribution of collagen fibril axis alignments in the second case.

1.2.3.3 Multidimensional scattering data

Hence, the SAXS/WAXS measurement is very sensitive to variations in $\vec{\mathbf{q}}$ vector orientation as well as magnitude and intensity. It is up to the experimenter to survey the sample in the right ways to determine the significance of any results. The samples 1 and 2 just discussed are actually two different points located about 6 mm apart, on a single thin collagen sheet. The illuminated beam spot, 0.3 mm × 0.3 mm, determines how much sample volume we average over for the pattern. By measuring at different points, we can see how the orientations and orientational distributions vary. One could equally well imagine measurements as a function of temperature, of applied stress, or of time during some chemical reaction. The patterns can be observed for relative or absolute changes of intensity as a function of just q, just χ, or both.

Thus, SAXS/WAXS can be a powerful imaging tool, providing a visualization of many types of parameter spaces. One of the most-demanded capabilities, therefore, is to make SAXS/WAXS a literal imaging tool for biomineralization by making the beam spot as small as possible, so that the instrument becomes a kind of microscope. The state of the art at present for hard x-rays (wavelengths 0.5–2 Å) is to create a beam spot of dimension <0.2 × 0.2 μm (Stock et al. 2011) (although microbeams of order 1 μm are available at a greater number of facilities worldwide). Scattering can beautifully capture the hierarchical structures of biominerals: The micron scale and larger is captured by scanning the sample in a microbeam, the 40 nm to 1 μm features are identified by SAXS, and the 40 nm to subatomic length scales are analyzed by WAXS (Paris 2008).

In the following, we illustrate uses of synchrotron SAXS/WAXS to obtain chemical content, physical structure, and parameter space images to answer our scientific questions about biominerals.

1.3 QUESTIONS AND ANSWERS IN BIOMINERALIZATION FROM SAXS/WAXS

1.3.1 BONE: HOW CAN CHANGES IN MINERAL COMPOSITION BE CORRELATED WITH MORPHOLOGY?

To illustrate this question, we review a study of strontium incorporation into newly formed bone mineral crystals during strontium ranelate treatment (Li et al. 2010). Strontium ranelate is a treatment for postmenopausal osteoporosis patients to increase bone mass and reduce fracture risk. Methods are required to determine the effects of such treatment on bone tissue characteristics and quality. Scanning SAXS/WAXS has the ability to probe the size and aspect ratio of the nm-scale Ca phosphate mineral platelets, as well as obtain a high-resolution measurement of the apatite lattice constant, which has a linear dependence on Sr content incorporated into the structure. Moreover, by scanning thin slices of biopsied bone with a 15 μm beam spot, the team could distinguish regions of newly formed bone from areas of old bone formed before the ranelate treatment began.

In each diffraction pattern, Figure 1.7a, the HAP (002) peaks can be seen to have an angular orientation and distribution, corresponding to the alignment of the mineral *c*-axis within the bone. The width Δq of the peak is used to extract a length $L \sim 2\pi/\Delta q$ of the apatite crystal, knowing from the literature that the (002) is a $\vec{q}$ vector lying along the length of the platelike particles. Because precise measurements of the *c* lattice parameter are desired, the samples were sputtered with Au, with the Au (111) powder ring present in the diffraction pattern as a reference. Note that the SAXS signal at the center has an elliptical contour, with the shorter dimension aligned toward the maximum in HAP (002) peak. This shape is due to the particle anisotropy and allows a typical mineral plate thickness T and a degree of alignment ρ to be extracted at every sample point (Fratzl et al. 1994, Rinnerthaler et al. 1999).

Mapping the T, ρ, and χ orientation parameters across a sample, such as the placebo-treated specimen shown in Figure 1.7b through d, comprises an assessment of bone tissue structure that can be compared between treated and untreated subjects. With these measurements, the study was able to conclude that the crystal thickness T in biopsies from patients treated with Sr ranelate has no significant difference from the placebo group (Figure 1.7e). Note that because of the efficiency of the scanning microbeam method, the numbers of measurements obtained from the biopsy range in the hundreds, providing good statistics and a clear picture of the deviations within each specimen, compared to that across treatment groups. This is important in biological samples where individual differences can obscure trends. The state of the art currently achieves tens of thousands of diffraction patterns in comparable studies.

Most important, studies like these can distinguish how treatment affects old versus new bone. SAXS/WAXS mapping enables the position-resolved Sr content to be measured, through subtle changes in apatite lattice constant. This measurement has an advantage over fluorescence methods (also employed, for complementary information): only Sr incorporated into the bone crystals will be detected, as distinguished from Sr complexed or adsorbed into soft tissues. Figure 1.7f through i shows data obtained from a subject treated with Sr ranelate for 36 months. The T and L parameters are uniform, as implied by Figure 1.7e, meaning that the mineral crystals have similar size and aspect ratio. By contrast, the *c* lattice parameter (Figure 1.7h) and, hence, the Sr content in the crystal (Figure 1.7i) are distinctly greater at the left of the measured region. This region is known from its morphology in the accompanying visible light microscopy (not shown in this review) to be a new osteon, adjacent to an older one formed before Sr ranelate treatment. Using the SAXS/WAXS data, this study was able to conclude that Sr incorporated only into newly formed bone crystals and showed no signs of affecting mineral ultrastructure.

In the foregoing discussion, mention has not been made of the equally important effects of strain within materials, which also broaden diffraction peaks, which must be analyzed with care to properly interpret changes as arising from strain or crystallite size. See Almer and Stock (2005, 2007) for x-ray studies of the micromechanical responses of mineral and collagen phases in bone.

1.3.2 LOBSTER CUTICLE: WHAT CAN WE LEARN ABOUT MINERAL/ORGANIC COORGANIZATION?

Part of the puzzle of any instance of biomineralization is to figure out whether the mineralization of a particular phase has evolved for a clear purpose or is a happenstance of metabolism. Crustaceans, for example, have shells, mandibles, and other organs that are mineralized to various extents. The usual majority phase, amorphous calcium carbonate (ACC), strengthens the shell and can also serve as a calcium reservoir, one that can be recycled by resorption into the organism's body during molt cycles that shed the exoskeleton. The organic matrix of the exoskeleton (cuticle) is frequently mineralized with smaller amounts of additional phases: calcite, ACP, or apatite. Might it be possible to find specific relationships between the minority minerals and specific structural elements of the exoskeleton and propose additional biomineral functions?

This interesting question has been asked for the case of the American lobster *Homarus americanus* (Al-Sawalmih et al. 2008), with a SAXS/WAXS examination of structure of organic and mineral in the cuticle. The prominent structural subunits of the cuticle matrix are understood to be nanofibrils consisting of α-chitin and protein, aligned in layers a few hundred nm thick that are stacked upon one another in helical fashion to form a

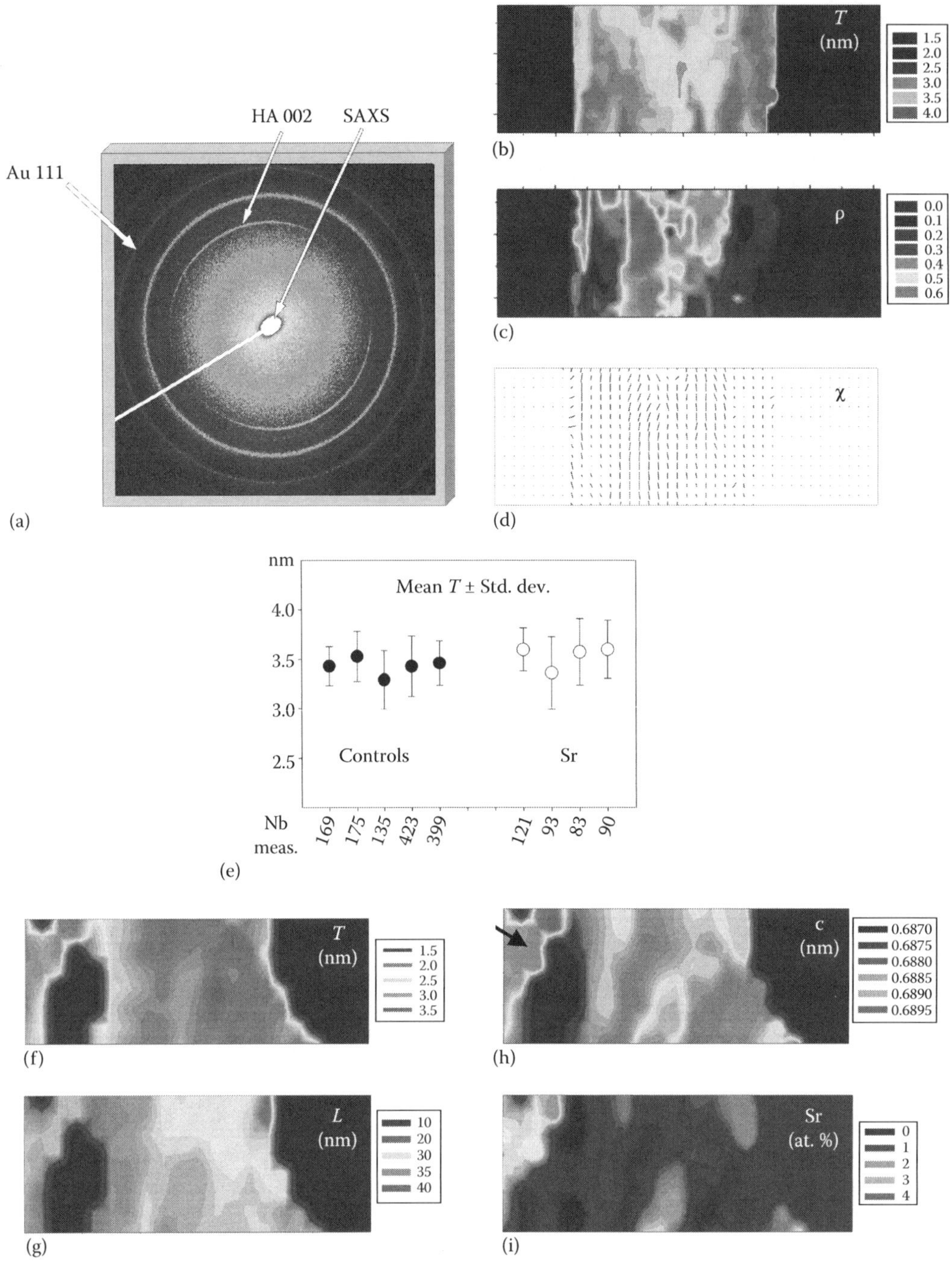

Figure 1.7 (See color insert.) (a) Exemplary SAXS/WAXS pattern from a bone section acquired at the μSpot beamline at BESSY II, Berlin. The HAP peak (HA 002) appears in arcs that provide mineral orientation information. The Au 111 ring arises from an Au film sputtered onto the bone for calibration. (b–d) Parameters determined from SAXS data, mapped over a rectangular region of bone sampled from a placebo-treated patient: (b) typical mineral crystal thickness *T*, which is fairly constant across the sample; (c) degree of alignment ρ; and (d) orientations of the elongated crystals show how crystals are grouped into *packets*. (e) Dependence of the *T* parameter on sample and on Sr content. (f–i) Parameter maps of a sample from a Sr-treated patient: (f) the similar mineral thickness *T* from SAXS; (g) mineral crystal length *L* from WAXS; (h) the HA *c* lattice parameter obtained from WAXS, where higher values indicate Sr incorporation; (i) Sr concentration derived from analysis of the *c* lattice parameter. (Adapted from Li, C. et al., *J. Bone Miner. Res.*, 25, 968, 2010. With permission.)

twisted plywood pattern. Lobster cuticle also has a micron-scale system of pores running perpendicular to the cuticle layers. Associated with these are additional α-chitin/protein fibers that run perpendicular to the cuticle surface.

The chitin fibrils are expected to have fiber symmetry, meaning that a particular axis of the α-chitin crystal structure is aligned preferentially, with arbitrary rotation around that axis. To determine axis alignments and symmetries, sections of cuticle were measured by x-ray scattering over sample orientations that aligned the $\vec{\mathbf{q}}$ vector to the sample normal $\vec{\mathbf{N}}$ and to lateral axes $\vec{\mathbf{X}}$ and $\vec{\mathbf{Y}}$. Plotting specific diffraction spot intensities on a circle designating $\vec{\mathbf{N}}$ at the pole and $\vec{\mathbf{X}}$ and $\vec{\mathbf{Y}}$ on the equator, as shown in Figure 1.8, produced pole figures both for chitin peaks and for the crystalline calcite peaks that were also observed.

Figure 1.8a shows that the calcite crystals have *c*-axis alignment along the cuticle normal direction: the peak with

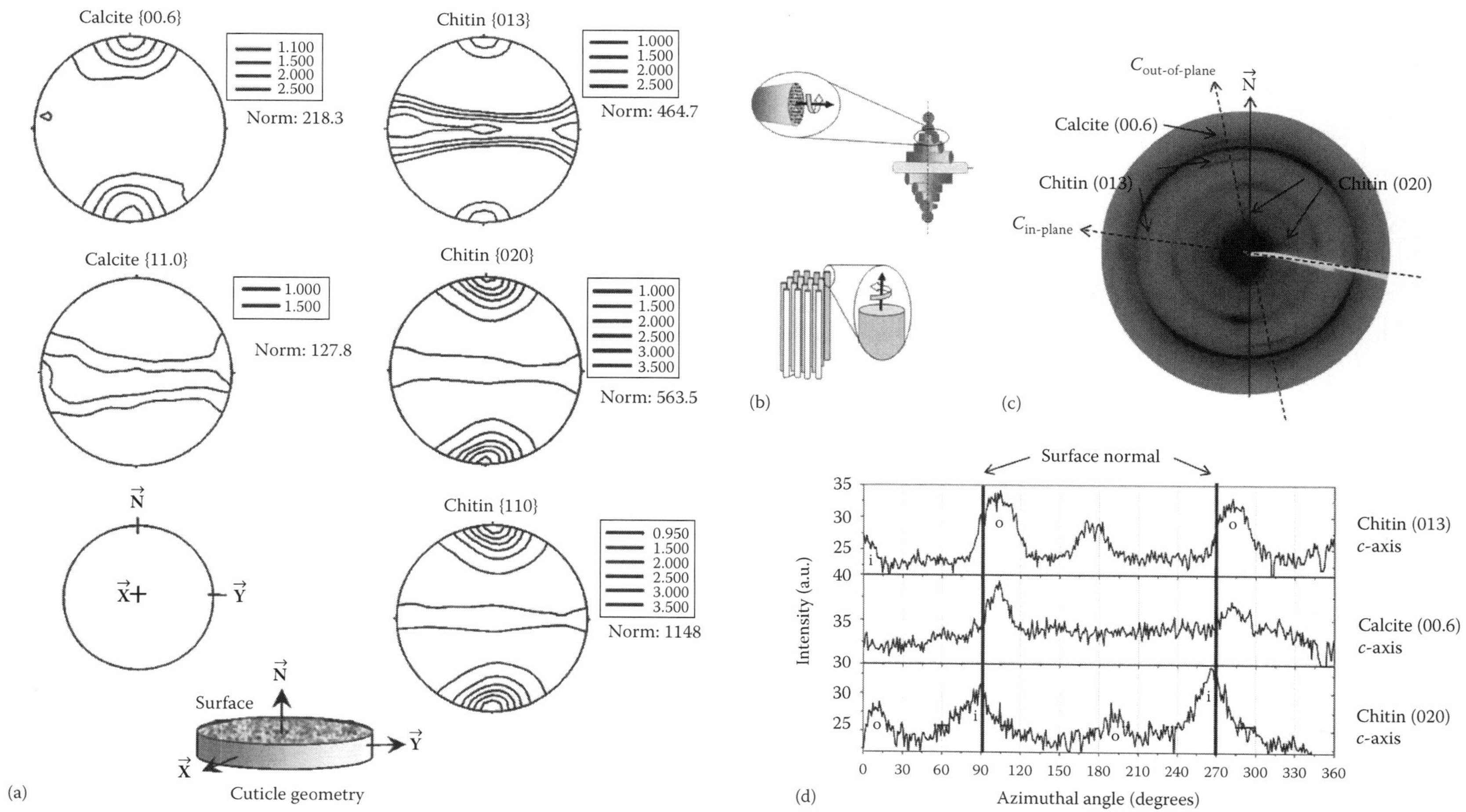

Figure 1.8 (a) Pole figures (oriented WAXS patterns) from lobster cuticle, arranged to underline the orientation relationship between calcite and α-chitin, $\vec{N}$ being the direction normal to the cuticle plane. (b) In-plane fibers (top) and out-of-plane fibers (bottom) in the cuticle both have fiber symmetry about the cuticle normal due to the helicoidally stacked fiber planes. (c) Microbeam WAXS with the beam parallel to the cuticle reveals arcs that show preferential orientation. (d) Intensity plotted against azimuthal angle for out-of-plane chitin (upper), calcite (middle), and in-plane chitin (lower) indicates that the calcite crystals co-align with out-of-plane chitin. (Adapted from Al-Sawalmih, A. et al., *Adv. Funct. Mater.*, 18, 3307, 2008. With permission.)

intensity at the poles is the {006} (the *c*-axis component is nonzero). Bands of scattering with a contour around the equator characterize the {110} peak (*a*- and *b*-axis components nonzero); arbitrary rotations of the calcite crystals around the *c*-axis scatter uniformly into this band. When the chitin peaks are examined, it is found that both polar and equatorial intensities are observed, consistent with the large proportion of chitin fibrils lying in the cuticle plane in the plywood layers, with a lesser fraction of chitin fibers running normal to the cuticle (out of plane) along the pores (Figure 1.8b) (see Al-Sawalmih et al. 2008 for details).

In light of this complexity, would it be possible to distinguish whether the calcite alignment might be associated with the in-plane versus the out-of-plane chitin fibrils? It is, thanks to the capability for the microbeam to sample small sections of the cuticle with the beam lying in the cuticle plane and the high resolution in orientation distinguishing normal from equatorial directions. Figure 1.8c shows arcs of diffraction from aligned calcite and chitin that demonstrate that the in-plane and out-of-plane chitin fibers are not strictly perpendicular, but appear about 15° apart in azimuthal angle. Plotting the intensities of in-plane chitin, out-of-plane chitin, and calcite peaks, Figure 1.8d demonstrates that the calcite *c*-axis is co-aligned with the out-of-plane chitin, suggesting its association with the fibers extending along the pores.

Further experiments probed mineral localization, confirming that the calcite exists only in a thin region of the exocuticle, while the bulk of the cuticle is mineralized with ACC (Al-Sawalmih et al. 2009). The *c*-axis alignment normal to the shell is shared by other biominerals, such as the prismatic layer in mollusk shells and the palisade of avian eggshells. Yet, the calcite of the lobster shell differs from these, whose entire hard coatings are calcite phase. It is conceivable that the calcite-stiffened vertical fibers play a specific role to reinforce the impact direction for the shell, while the ACC phase beneath it plays a dual role in being more accessible for dissolution on molting and for dissipating impact energy. These hypotheses, made specific thanks to SAXS/WAXS observations, set the stage for more pointed questions about force and function in mineralized tissues.

1.3.3 STOMATOPOD DACTYL CLUB: HOW DO CRYSTALLINE AND AMORPHOUS REGIONS WORK TOGETHER?

The stomatopod, or mantis shrimp, is an aggressive predator, and among its prey are mollusks, housed in shells that are very well known and studied as models of tough biominerals. The impressive performance of a smashing club that breaks these damage-tolerant materials surely has lessons of its own to impart.

The functional club of this crustacean is composed of the two terminal segments of its second thoracic appendage, the knobby dactyl folded back into a groove in the propodus, leading to a somewhat horseshoe-shaped cross section perpendicular to the direction of impact. Examinations with both x-rays and electrons confirm that the dactyl club is the most electron-dense region of the stomatopod exoskeleton and has three prominent chemically and mechanically distinct regions within it (Weaver et al. 2012).

The morphology of a cross-sectioned dactyl can be imaged with x-ray transmission. In the WAXS geometry shown in Figure 1.9a, the diffracted beam scatters to the area detector, while the transmitted direct beam is captured by a diode. The intensity measured in the diode indicates the absorption, and hence electron density, of the sample. Here, the sectioned dactyl was probed by scanning the sample position in the beam, with a spatial resolution (beam size and scan step size) of 5 μm × 5 μm. Figure 1.9b shows that the density is highest in the club's impact region (purple tones, see color insert), which encloses less electron-dense regions where transmitted x-ray intensity is higher (blue and green areas).

Since the impact of the smashing club is in the direction perpendicular to its front surface, of particular interest is to investigate changes in structure along the symmetry axis. Panels (c–f) in Figure 1.9 show WAXS patterns along the symmetry line. In the impact region, we find well-crystallized HAP, with sharper peaks (larger well-crystallized regions) than found, for example, in mammalian bone. Peaks from α-chitin, as expected in arthropod exoskeletal material, are also seen. The α-chitin peaks appear as single or double arcs due to the alignment of chitin fibrils into layers parallel and perpendicular to the organ's surface. The helicoidal chitin layers, very similar to those of the lobster cuticle discussed earlier, have a periodicity of ~75 μm, large enough to be resolved with the microbeam, in contrast to other arthropods. These chitin layers continue further into the club, and as Figure 1.9e and f show, the crystalline HAP gives way in this periodic region to an amorphous mineral phase characterized by a single broad diffraction ring.

Mechanical deformation and fracturing experiments, combined with dynamic finite element analysis modeling of the dactyl club, suggest that one of the most important structural features of the dactyl club is the existence of the softer amorphous mineral region, which, in combination with the hard crystalline impact layer, presents a significant mismatch of elastic modulus across the interface, which acts to deflect cracks away from the organ's surface and prevents catastrophic failure (Weaver et al. 2012). To completely survey the organ's structure, we would prefer to be able to map this amorphous component spatially, using diffraction rather than electron density methods to distinguish subtle changes in structure.

As has been shown earlier, this is challenging and particularly so when a specimen of millimeter dimensions is to be mapped with micron spatial resolution. Fluorescence methods show that the amorphous mineral in the stomatopod dactyl club contains both carbon and phosphorous, and the observed WAXS peak appears at a q value distinct from that of pure amorphous Ca phosphate or Ca carbonate standards. This lack of structural distinction, plus the presence of variable amounts of crystalline HAP and chitin diffracting within the same volume, makes it a daunting task to analyze the data. Figure 1.10 shows a series of integrated WAXS patterns $I(q)$ taken at points along the symmetry axis.

The HAP peaks broaden, shift, and diminish progressively as the amorphous content increases. It is an enormous task to fit multiple peaks to such data for the thousands of frames we acquire. Is there a simpler figure of merit to analyze? One could track, for example, the intensity at $q = 2.15$ Å^{-1}, between HAP peaks, as a measure of amorphous content. However, since the raw intensity changes with density of the sample, this

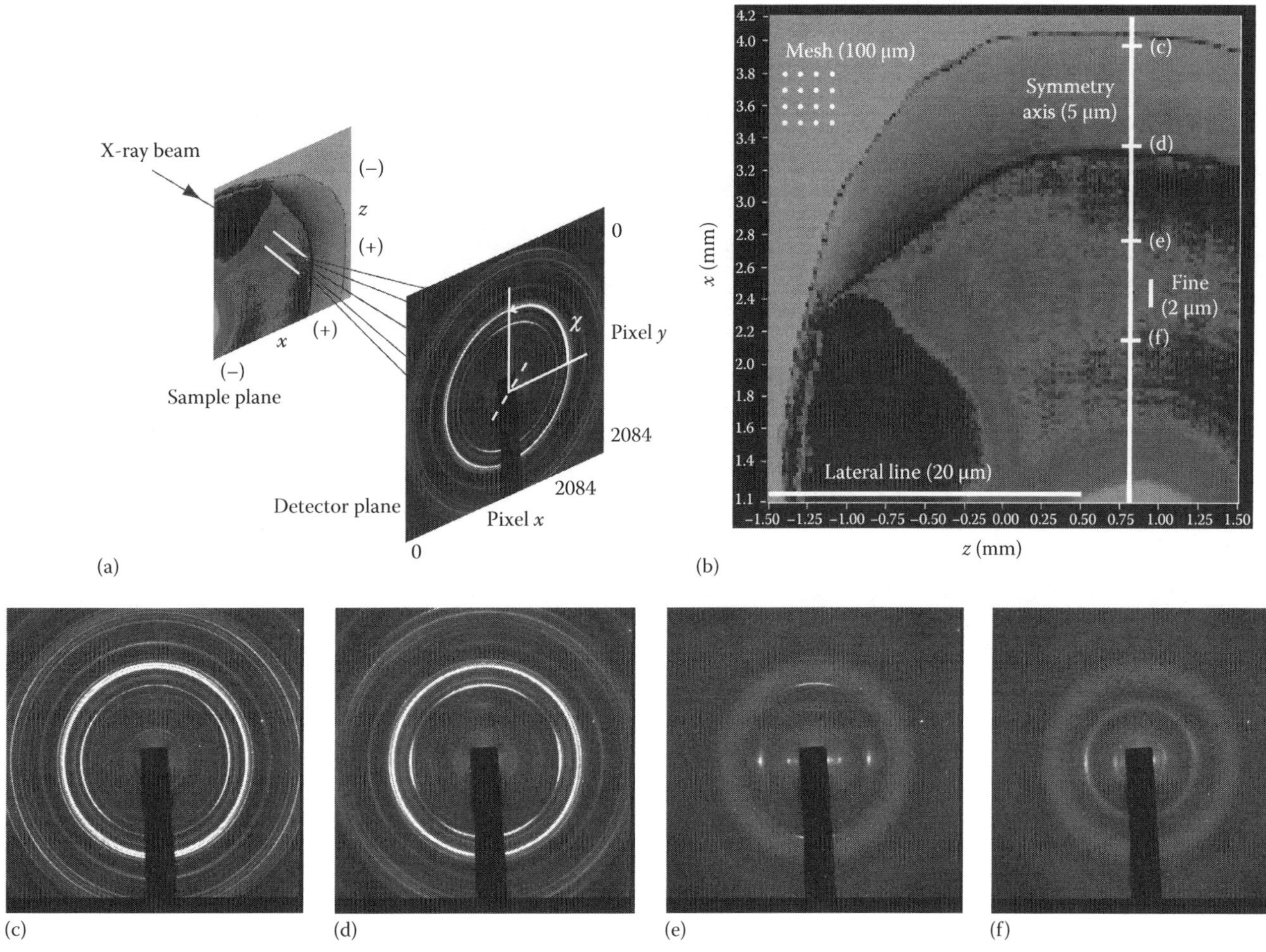

Figure 1.9 (See color insert.) (a) Representation of the transmission mapping experiment. Aligned crystal planes (white lines in the sample plane) create diffraction peaks along a particular line in detector azimuth χ (dashed line in detector plane). (b) X-ray transmission map of sectioned stomatopod dactyl club. Regions of decreasing density lead from purple to blue to green (red color is the surrounding epoxy region). White lines and points indicate areas and spacings of diffraction maps taken. (c) WAXS pattern in mineralized impact region. (d) Oriented mineral at the boundary region. (e and f) WAXS patterns dominated by amorphous mineral and chitin.

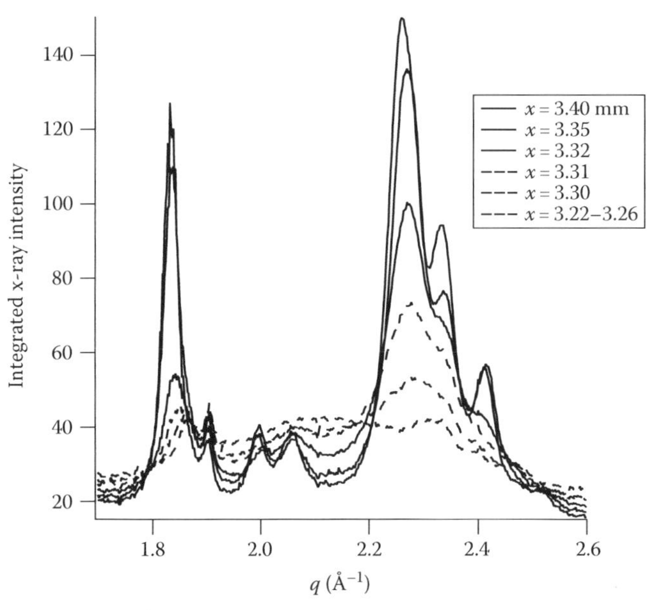

Figure 1.10 Radially integrated intensity $I(q)$ from WAXS patterns along the symmetry line of the dactyl club, at x positions according to the scale in Figure 1.9b.

measure is not accurate. It would be better to fit some aspect of the amorphous peak, since this is the only reliable signature of the amorphous phase. Success in such tasks is what makes synchrotron SAXS/WAXS an imaging method worthy of the name, hence our emphasis in this final section on demonstrating how this can be done.

In Figure 1.11a, we illustrate features of a single-peak fit to the complicated data (which are not characterized by a single peak). The panels show how the narrow chitin peak can pull a fit to the broad peak toward the lower q values, while HAP peaks pull the fit toward higher q (top and middle panels). As we run off the edge of the sample (bottom plot), the fit runs away in both position and width. Although none of these fits are correct, we can make use of these trends to ask a simple yes or no question of the data: In which of the (several hundred) diffraction patterns is there appreciable amorphous phase?

After some experimentation, we constrained the single-peak fits to be fixed in position, and we examined the effect on the widths of the fit Gaussian peaks. The results for all frames along the symmetry axis line scan (2 mm long, 5 μm steps) are shown in Figure 1.11b. The scallops are not noise, but rather the effect of the periodic chitin layers, as their intensity oscillates and causes

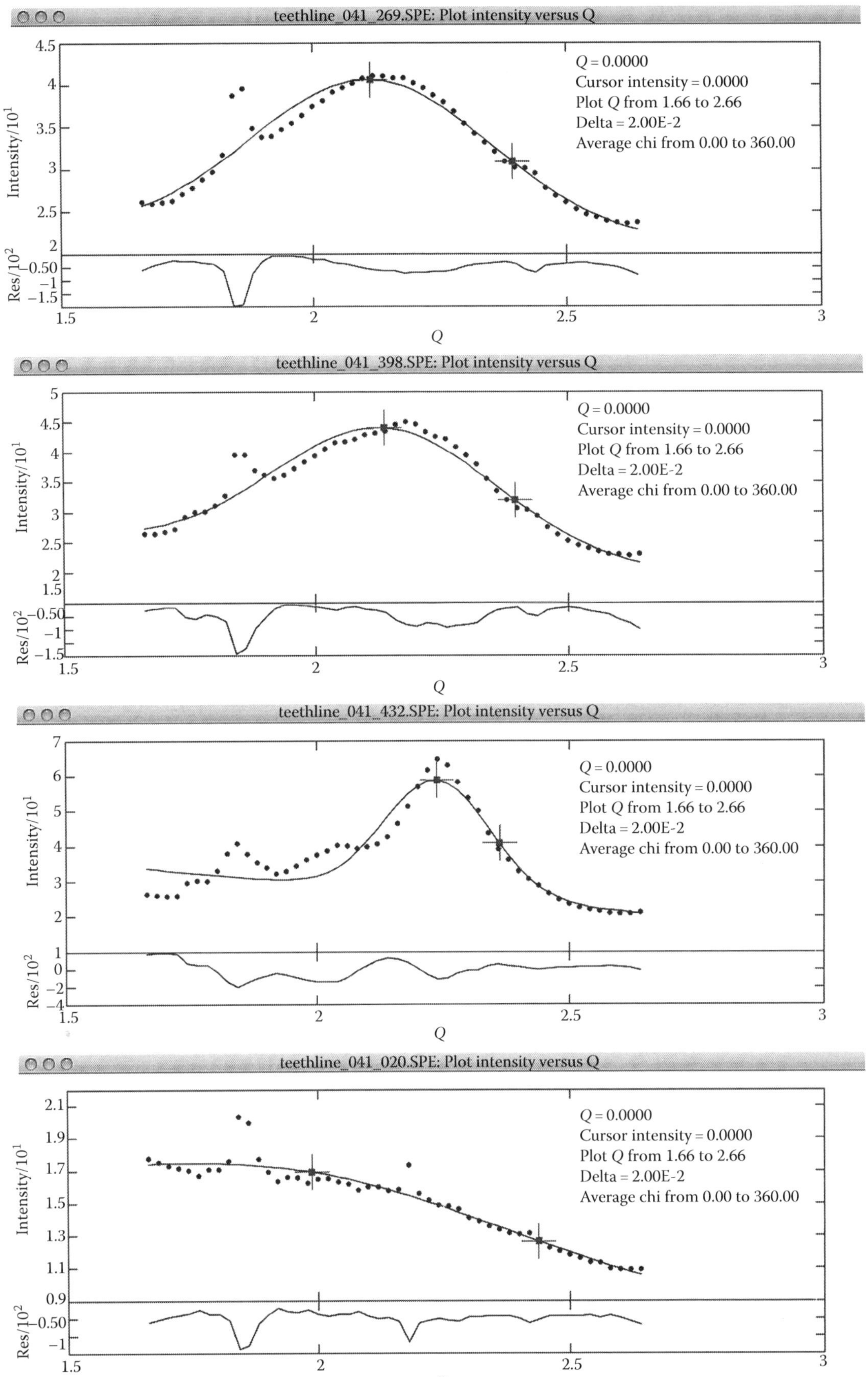

(a)

Figure 1.11 (a) Four examples of how the $I(q)$ data (points) can be fit to a single broad peak, with the existence of sharper chitin and HAP peaks pulling the fit (solid line, with center and width indicated by crosses) in different directions.

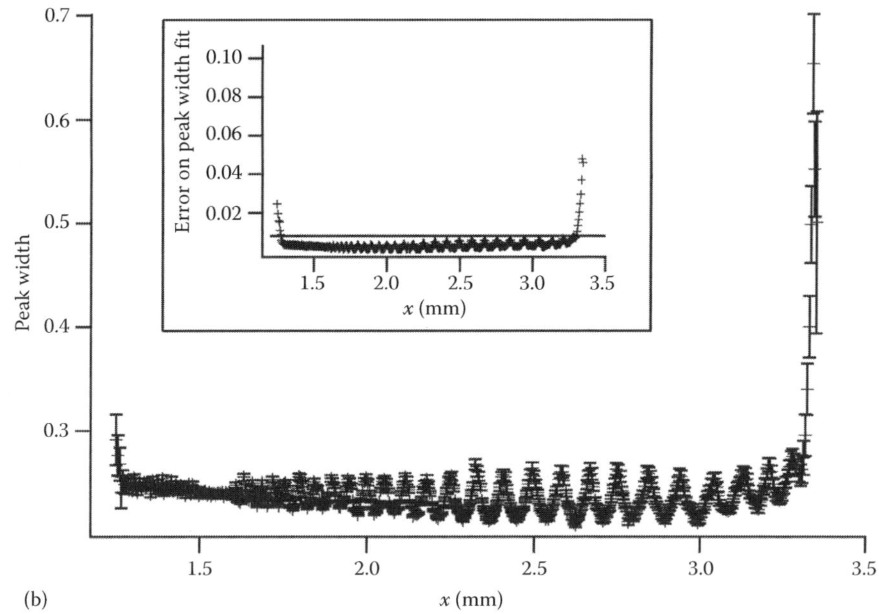

Figure 1.11 (continued) (b) A plot along the entire dactyl club symmetry line of the resultant peak widths and (inset) peak width error bars, when this fit is performed. Positions containing amorphous mineral by inspection have errors <0.08 and can easily be distinguished in a batch programming mode from the outliers at the extremes of the amorphous region.

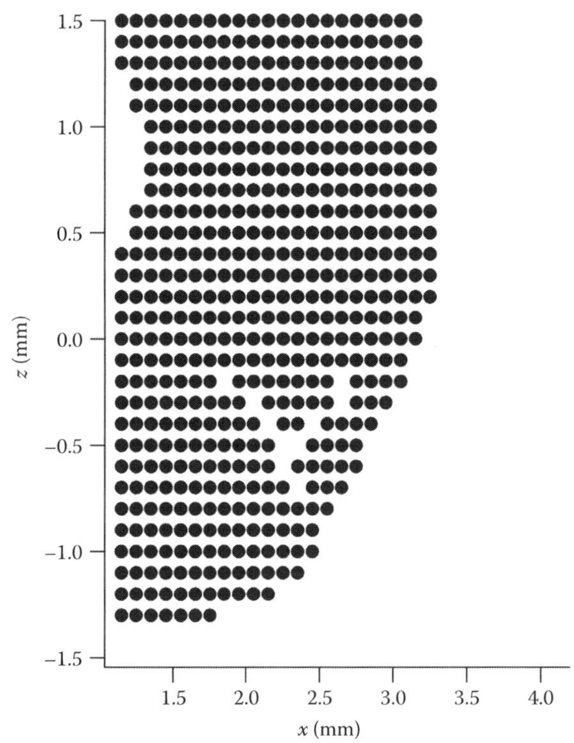

Figure 1.12 Black points indicate points in x and z on the dactyl club sample where amorphous mineral phase is identified using the criterion illustrated in Figure 1.11, where the error bar on a single broad peak fit exceeds a threshold value of 0.08. This creates a map of amorphous mineral in the sample, even accurately identifying the V-shaped region at the side (compare to the color figure of Figure 1.9b). Nine hundred WAXS patterns were processed to create this map.

the fit peak to broaden very slightly. (Our process, as a bonus, gives an extremely simple and reliable measure of the period of the chitin layers, 85 μm in this specimen, less ambiguous than an electron micrograph.) At the edges of the scanned region, the peak widths jump above 0.25, but by less than a factor of 3. However, as the inset shows, the *standard deviation* of the fit parameter jumps by orders of magnitude. Hence, our solution is to accept only low values of the standard deviation of the peak width fit. Now we can apply this criterion to the entire 100 μm mesh of points over the entire dactyl club section (the spacing indicated by white dots labeled mesh in Figure 1.9b). The result is Figure 1.12: a yes/no map identifying amorphous mineral regions, to be further analyzed with more quantitative peak fitting as the science case demands. By combining rigorous analysis of diffraction patterns with these kinds of figure of merit images, synchrotron SAXS/WAXS achieves its potential to get to the heart of structure–function questions in these hierarchical, multicomponent biomaterials.

1.4 OUTLOOK: WHAT ARE THE FRONTIERS?

1.4.1 SMALLER BEAMS AND EVEN BETTER DETECTORS

Biominerals have structure on the 1 and 10 nm scales, and present-day synchrotrons are pursuing nanometer-scale beams, by building very well collimated electron storage rings, long beamlines extending into outbuildings with as much vibration-free technology as possible, and new optics to focus x-rays down to such small spots. With beams focused so intensely onto soft and damage-prone organic materials, it becomes more important than ever to measure every photon, to take the data quickly, and to make detector pixel sizes smaller so the narrow beams aren't blurred on detection.

But, bringing the experimental probe to the nanoscale is only half the battle! We, the scientists in the field, must bring increased sophistication to the analysis. We should not expect that a nanoparticle consisting of 4 or 5 unit cells in each dimension will produce a diffraction pattern that matches those of standards in the literature. Peaks will be broadened by the finite crystal size and microstrain and altered by the surface reconstruction of atoms that move under the forces at the interfaces. Proper theory and language will need to be applied to tackle these problems and even more collaboration between experts in structure measurement and experts in the biological systems.

1.4.2 REVOLUTION IN SOFTWARE FOR DATA PROCESSING

Software for data processing must perform many tasks. Calibration is required to relate the beam, sample, and detector positions and orientations to the scattering angles and finally $\vec{q}$ vectors being probed. Special orientations must be identified, such as a flat surface of reference or a fiber axis. The data are acquired in frames of millions of pixels, as rapidly as 200 times per second. These data frames, 20–50 MB, will be acquired for each point of interest in parameter space. To scan a 500 nm area in 10 nm steps, not an outrageous near-future request, requires capturing 2500 frames, producing >100 GB of data just for one image. Synchrotron SAXS/WAXS beamlines could already be acquiring a terabyte of data daily, placing our science squarely beside that of the enormous detectors constructed for high energy and nuclear physics, with 400-person collaborations who process the data.

Why do we not yet typically acquire a TB of data daily at more than a select few of the dozens of beamlines worldwide? The answer is that there is no diffraction software that magically gets the answer from these datasets. Software products on the market are modular and have small scope, each performing only some of the required calibration and analysis tasks. Other packages have been lovingly tended by small groups but have difficulty finding traction to wider audience and wider application. Many packages are not designed to be used with automated scripts, but by hand in a mouse-driven user interface. These can be well suited for introducing a student to the analysis steps but almost unusable when confronted with several hundred frames. In the case of simultaneous SAXS/WAXS with two detectors, the available software can usually only handle one detector's frame in its process, providing an artificial separation between SAXS and WAXS datasets, even if the spectrometer was designed so that their q ranges overlap.

Hence, analysis becomes a bottleneck, proceeding at a crawl, especially if the synchrotron scientist—whose job it may be to support a new user group every 3 days—is required also to perform the several months' worth of programming required to customize data analysis for the experiment. As our example of the stomatopod dactyl club's amorphous mineral phase map showed, each new experiment may demand a new, creative way to tease out the features of scientific interest. To make progress on this frontier, a joint, collaborative clamor, from biomaterials scientists and synchrotron scientists together, will be required. With a unified voice, this group must approach funding agencies to direct funds specifically toward the analysis bottleneck problem and motivate computer engineering and applied mathematics talent to join the effort. When this occurs, our prospects to see into biomineral structures over a dazzling range of length scales, and answer scientific questions with formidable theoretical sophistication, will become reality.

ACKNOWLEDGMENT

Brookhaven National Laboratory is supported under US DOE Contract No. DE-AC02-98CH10886.

REFERENCES

Aichmayer, B., F.B. Wiedemann-Bidlack, C. Gilow et al. 2010. Amelogenin nanoparticles in suspension: Deviations from spherical shape and pH-dependent aggregation. *Biomacromolecules* 11:369–376.

Almer, J.D. and S.R. Stock. 2005. Internal strains and stresses measured in cortical bone via high-energy diffraction. *J. Struct. Biol.* 152:14–27.

Almer, J.D. and S.R. Stock. 2007. Micromechanical response of mineral and collagen phases in bone. *J. Struct. Biol.* 157:365–370.

Al-Sawalmih, A., C. Li, S. Siegel et al. 2008. Microtexture and chitin/calcite orientation relationship in the mineralized exoskeleton of the American lobster. *Adv. Funct. Mater.* 18:3307–3314.

Al-Sawalmih, A., C. Li, S. Siegel et al. 2009. On the stability of amorphous minerals in lobster cuticle. *Adv. Mater.* 21:4011–4015.

Bradshaw, J.P., A. Miller, and T.J. Wess. 1989. Phasing the meridional diffraction pattern of type I collagen using isomorphous derivatives. *J. Mol. Biol.* 205:685–694.

Burger, C., H.-W. Zhou, I. Sicŝ et al. 2008a. Small-angle X-ray scattering study of intramuscular fish bone: Collagen fibril superstructure determined from equidistant meridional reflections. *J. Appl. Crystallogr.* 41:252–261.

Burger, C., H.-W. Zhou, H. Wang et al. 2008b. Lateral packing of mineral crystals in bone collagen fibrils. *Biophys. J.* 95:1985–1992.

Chen, J., C. Burger, C.V. Krishnan, B. Chu, B.S. Hsiao, and M. Glimcher. 2005. In vitro mineralization of collagen in demineralized fish bone. *Macromol. Chem. Phys.* 206:43–51.

Egami, T. and S.J.L. Billinge. 2003. *Underneath the Bragg Peaks: Structural Analysis of Complex Materials*. Oxford, U.K.: Elsevier.

Fratzl, P., P. Roschger, J. Eschberger, B. Abendroth, and K. Klaushofer. 1994. Abnormal bone mineralization after fluoride treatment in osteoporosis: A small-angle x-ray scattering study. *J. Bone Miner. Res.* 11:248–253.

Hayes, S., C. Boote, S.J. Tuft, A.J. Quantock, and K.M. Meek. 2007. A study of corneal thickness, shape, and collagen organization in keratoconus using videokeratography and X-ray scattering techniques. *Exp. Eye Res.* 84:423–434.

Hulmes, D.J.S., A. Miller, S.W. White, and B.B. Doyle. 1977. Interpretation of the meridional X-ray diffraction pattern form collagen fibres in terms of the known amino acid sequence. *J. Mol. Biol.* 110:643–666.

Li, C., O. Paris, S. Siegel et al. 2010. Strontium is incorporated into mineral crystals only in newly formed bone during strontium ranelate treatment. *J. Bone Miner. Res.* 25:968–975.

Long, X., Y. Ma, and L. Qi. 2011. In vitro synthesis of high Mg calcite under ambient conditions and its implication for biomineralization process. *Cryst. Growth Des.* 11:2866–2873.

Mosler, E., W. Folkhard, E. Knörzer, H. Nemetschek-Gansler, and Th. Nemetschek. 1985. Stress-induced molecular rearrangement in tendon collagen. *J. Mol. Biol.* 182:589–596.

Orgel, J.P., T.J. Wess, and A. Miller. 2000. The in situ conformation and axial location of the intermolecular cross-linked non-helical telopeptides of type I collagen. *Structure* 8:137–142.

Orgel, J.P.R.O., A. Miller, T.C. Irving, R.F. Fischetti, A.P. Hammersley, and T.J. Wess. 2001. The in situ supermolecular structure of type I collagen. *Structure* 9:1061–1069.

Paris, O. 2008. From diffraction to imaging: New avenues in studying hierarchical biological tissues with x-ray microbeams (review). *Biointerphases* 3:FB16–FB26.

Paris, O., C. Li, S. Siegel et al. 2007. A new experimental station for simultaneous x-ray microbeam scanning for small- and wide-angle scattering and fluorescence at BESSY II. *J. Appl. Crystallogr.* 40:s466–s470.

Rinnerthaler, S., P. Roschger, H.F. Jacob, A. Nader, K. Klaushofer, and P. Fratzl. 1999. Scanning small angle x-ray scattering analysis of human bone sections. *Calcif. Tissue Int.* 64:422–429.

Saeidi, N., K.P. Karmelek, J.A. Paten, R. Zareian, E. DiMasi, and J.W. Ruberti. 2012. Molecular crowding of collagen: A pathway to produce highly-organized collagenous structures. *Biomaterials* 33:7366–7374.

Schneidman-Duhovny, D., S.J. Kim, and A. Sali. 2012. Integrative structural modeling with small angle X-ray scattering profiles. *BMC Struct. Biol.* 12:17.

Stock, S.R., A. Veis, A. Telser, and Z. Cai. 2011. Near tubule and intertubular bovine dentin mapped at the 250 nm level. *J. Struct. Biol.* 176:203–211.

Tsipursky, S.J. and P.R. Buseck. 1993. Structure of magnesian calcite from sea urchins. *Am. Mineral.* 78:775–781.

Wang, Q., M. Nemoto, D. Li et al. 2013. Phase transformations and structural developments in the radular teeth of *Cryptochiton stelleri*. *Adv. Funct. Mater.* 23:2908–2917. doi: 10.1002/adfm.201202894.

Warren, B.E. 1969. *X-Ray Diffraction*. Reading, MA: Addison-Wesley; 1990 edition, Mineola, NY: Dover Publications.

Weaver, J.C., G.W. Milliron, A. Miserez et al. 2012. The stomatopod dactyl club: A formidable damage-tolerant biological hammer. *Science* 336:1275–1280.

Zapanta-LeGeros, R. 1965. Effect of carbonate on the lattice parameters of apatite. *Nature* 206:403–404.

Zhou, H., C. Burger, I. Sicŝ et al. 2007. Small-angle X-ray study of the three-dimensional collagen/mineral superstructure in intramuscular fish bone. *J. Appl. Crystallogr.* 40:s666–s668.

2 *In situ* x-ray scattering from molecular templates and nucleating minerals at organic–water interfaces

Benjamin D. Stripe and Pulak Dutta

Contents

2.1 INTRODUCTION

Biomineralization presents mysteries at every length scale, but a central mystery is at the nanoscale: how and why do organisms grow precisely oriented inorganic crystals, often expressing crystal faces different from those found in abiotic minerals? *Why* is a question for biology, but *how* is a surface science problem, and the tools of surface science (highly developed over recent decades through studies of a variety of other systems) are likely to reveal much more about the crystal growth process than is possible from bulk studies. In this chapter, we discuss the use of surface x-ray scattering techniques to determine the structures of ordered mineral crystals and organic surfaces, to detect amorphous phases, and to investigate the roles of surface charge and stereochemistry.

The crystals in biominerals are grown in specific environments: in particular, the crystal faces are not exposed to air but are in contact with biomolecules. It is reasonably assumed that these molecules do not merely stabilize faces that would not be stable otherwise but also guide the nucleation and growth of oriented crystals in some undetermined fashion. An elegant way to simulate this process in the laboratory is to make supersaturated solutions, put them in contact with a suitable organic surface, and see what nucleates.

Langmuir (floating) monolayers are ideal templates for such investigations: They are already in contact with water, they are ordered, and their structures are in many cases well known and can be varied by changing pressure, temperature, etc. (Kaganer et al. 1999). It must be acknowledged from the outset that this is an imperfect simulation of the biological process. It is impractical to reproduce perfectly, within a floating or substrate-supported monolayer, the multicomponent organic layers within real biominerals. Most, although not all, studies have used single-component monolayers. Moreover, these experiments involve single flat interfaces, whereas the inorganic crystals in biominerals are in contact with biomolecules on all sides. Nonetheless, such model studies provide a simple way to make organic–inorganic interfaces accessible to surface science probes, and they offer opportunities to explore in controlled ways the effects of various individual factors on the growth of oriented crystals relevant to biomineralization.

There have been many such studies of nucleation under Langmuir monolayers (for reviews, see Fricke and Volkmer 2007; Sommerdijk and de With 2008). For example, Mann and coworkers have reported the oriented growth of calcium carbonate (Mann et al. 1991; Walker et al. 1991; Heywood and Mann 1994; Litvin et al. 1997) and other materials (Zhao et al. 1992; Yang and Fendler 1995; Backov et al. 2000; Talham et al. 2006; Wang et al. 2006) under a variety of monolayers. These studies were performed by picking up the crystals floating just under the monolayer and looking at them with x-rays, electrons, etc. Only unoriented nucleation in the bulk of the solution is observed without a monolayer. When there is oriented crystal growth, it is reasonable to expect that there is epitaxy, and thus, one looks for lattice matches. In Walker et al. (1991), for example, epitaxial growth is suggested by the close match between the surface structure of the nucleating crystal and the (10) plane spacing of the arachidic acid monolayer lattice, but the latter spacing was obtained by assuming a hexagonal lattice with an area/molecule of 20 Å^2. Similarly inexact lattice-matching arguments are given in many other papers.

In surface science, a determination of epitaxy requires knowing the structures of both the monolayer template and the structures and orientations of the crystals nucleated. However, in the system under discussion, the actual structure of the organic template cannot credibly be determined from the isotherm area or from studies under different conditions. Because the monolayer is soft and its structure is easily changed (in other words, it is a *compliant substrate*), the only way to know the actual structure of the organic template is to observe it *in situ*. Moreover, determining the crystal species and orientation after *harvesting* (e.g., lifting up with electron microscope grids) is problematic for many reasons. The supporting substrate may align the crystals (e.g., platelike crystals will probably lie flat on the support even if they were randomly oriented before transfer). If amorphous phases are transferred, they may crystallize on the support as the water dries, but such crystals are not biomimetically grown. Further, examination using various imaging methods (microscopies) may result in neglecting irregular clumps, invisibly small crystals, or otherwise unidentifiable objects and drawing conclusions instead from attractive but nonrepresentative images (*selection bias*).

Therefore, it is not surprising that the role of epitaxy in oriented crystal growth is not universally accepted. For example, Fricke and Volkmer (Volkmer et al. 2006; Fricke and Volkmer 2007) have found that monolayers of macrocyclic amphiphiles, which are *fluid*, nucleate well-oriented crystals of calcium carbonate; they argue that monolayer charge, rather than monolayer structure, drives the process. However, the relationship between mechanical properties and structure in soft materials is complex and poorly understood, and easy deformability does not prove that there is no lateral order. Only scattering probes can determine whether there is lateral order. Further, it is known that in floating monolayers, there are many partially ordered mesophases (Kaganer et al. 1999); it should not be assumed that monolayers must be either crystalline solid or disordered fluid.

It is also possible to use self-assembled (substrate-supported) monolayers, rather than floating monolayers, as templates. Aizenberg and coworkers (Aizenberg et al. 1999) reported [012]-oriented calcite growth on carboxylic acid terminated alkanethiol self-assembled monolayers on silver substrates and [105]-oriented growth on carboxylic acid-terminated alkanethiol self-assembled monolayers on gold substrates. These results were interpreted in terms of stereochemical matching of the carbonate groups in the crystal with the terminal acid groups in the monolayer. However, the orientations of the monolayer terminal groups were obtained from prior publications rather than from the system under study.

The examples earlier do not constitute a comprehensive review; they were chosen to make the point that if structures are not determined in the system at hand under actual experimental conditions (in other words, *in situ*), then arguments for and against epitaxy are all speculative. These important studies reflect the state of the art at the time of their publication, but it is now possible to do better. Scattering of synchrotron x-rays permits the observation of both template structure (2D lattice and molecular orientation) and mineral structure directly during the nucleation process. This allows both the template structure and the crystal structure and orientation to be determined *in situ* under nucleation conditions. Moreover, in the low-incidence angle geometries to be described later, the x-ray beam illuminates a broad swath (*footprint*) of the interface, and thus, the data are necessarily statistically significant averages over many crystals. This means that there is no possibility of selection bias.

As we shall show later, x-rays can be used not only to elucidate the role of lattice matching but to characterize amorphous phases and to explore the roles played by surface charge and stereochemistry. Therefore, this probe should be an essential part of the model biomineralization toolkit.

The two most widely used techniques for the study of surfaces are grazing incidence x-ray diffraction (GID) and x-ray reflectivity (XRR) measurements. GID gives details of ordering in the plane of a surface or interface, while XRR tells us about the electron charge density perpendicular to the surface/interface. We describe each of these techniques in turn and summarize the results of some recent x-ray scattering experiments that probe not just the template or just the nucleated crystals but both.

2.2 SURFACE-SPECIFIC X-RAY TECHNIQUES

2.2.1 GRAZING INCIDENCE X-RAY DIFFRACTION

The fact that the refractive indices of materials are always slightly *less than* 1 for x-rays is fortuitous: it means that the x-rays can undergo total *external* reflection at surfaces. When the angle of incidence onto the surface is smaller than the critical angle for total external reflection, the x-rays do not penetrate into or scatter from the bulk material, except for an evanescent wave that penetrates a few nanometers. Thus, scattering is primarily surface-sensitive, with a relatively weak background contribution from the substrate.

The basic principles of elastic GID are as follows. The incident wave vector $\mathbf{K}$ is defined as a vector of magnitude $2\pi/\lambda$, where λ is the x-ray wavelength; the fixed direction of this vector is that of the incident x-ray beam. The outgoing wave vector $\mathbf{K}'$ has the same magnitude, but its variable direction is that in which the scattered x-rays are detected. Scattered intensity (I) data are collected as functions of $\mathbf{q} \equiv \mathbf{K} - \mathbf{K}'$. In an ordered crystal, when $\mathbf{q}$ matches a reciprocal lattice vector of the crystal, there is a maximum in the scattered intensity. Another way to say it is that the diffraction maxima occur when $\mathbf{q}$ is normal to any lattice plane and has a magnitude of $2\pi/d$ where d is that spacing between planes. More generally, one can say that $I(\mathbf{q})$ depends on the Fourier transform of the correlation function describing the statistical average of the product of the electron density at one point and at a distance $\mathbf{r}$ from that point. All nontrivial correlations give rise to scattering with maxima (peaks) and minima. Ideal periodic crystals are simply an extreme case of such density correlations. The scattering peak positions and widths can be used to calculate the periodicity and the correlation length, respectively.

In the case of an ordered monolayer, where there is periodicity in the monolayer (x–y) plane but not normal to the plane (z-direction), the same conditions may be restated as follows: If $\mathbf{q}$ has a component $\mathbf{q}_{xy}$ in the monolayer plane and q_z normal to it, then $\mathbf{q}_{xy}$ must satisfy the Bragg conditions described earlier, with *lattice planes* replaced by *lattice lines* since this is a 2D lattice. The Bragg peak intensity has only a broad maximum as q_z is varied; the q_z values at the maxima determine the molecular tilt angle

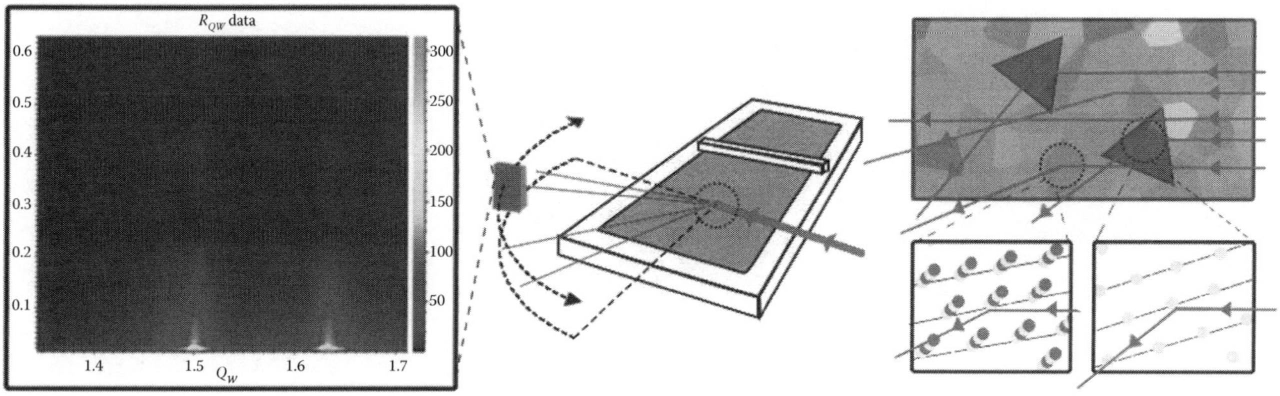

Figure 2.1 Langmuir monolayer GID setup. The central figure shows the Langmuir and grazing incidence scattering geometry. The left side is an example contour plot of the scattered intensities as collected by an area detector. The right side illustrates the scattering of the x-rays from the monolayer domains and the nucleating crystals. Keep in mind that both the monolayer and crystals are both powders in the plane of the water's surface (all lattice planes are horizontal, but different regions have different orientations about the plane normal).

and direction (Kaganer et al. 1999), while the widths of the peaks as function of q_z depend on the thickness of the monolayer.

Figure 2.1 shows a schematic setup for GID from a Langmuir monolayer and crystals growing under it. At the center is a Langmuir trough, which holds the subphase, usually ultrapure water (18.2 Mohm-cm) supersaturated with the biomineral whose nucleation is to be studied. For example, many studies use supersaturated $CaCO_3$ subphases made by mixing equimolar solutions of $CaCl_2$ and $NaHCO_3$, by bubbling a solution of excess $CaCO_3$ with CO_2 and then filtering, or by gaseous diffusion of crushed ammonium bicarbonate into a solution of $CaCl_2$. Once prepared, the subphase is placed in the trough, which is usually made of polytetrafluoroethylene. The trough is filled such that the water level is higher than the edge of the trough, so that the x-rays can be incident on the surface at grazing angles. In addition, the trough must have proportions large enough that the meniscus, while curved at the edges, is flat in the center to give the x-rays a flat interface from which to reflect.

Once the subphase is placed in the trough, the organic monolayer is spread on the surface. The monolayer material is first dissolved in a volatile solvent (e.g., hexane or chloroform) that *wets* water, which means that when the solution is dropped on the water using a micropipette, the drops will quickly spread across the surface of the water and then the solvent will evaporate. The monolayer material remains, and the monolayer is then compressed by a mechanical barrier to the desired surface pressure. The temperature and pressure dependence of Langmuir monolayers of simple fatty acid molecules have been extensively studied (Kaganer et al. 1999) and are known to have quite complex phase diagrams on pure water.

The monochromatic incident x-ray beam (thick red line) comes in at a small downward angle (~0.2°) so that it undergoes total reflection. Slits and focusing optics are used to shape the beam such that the beam footprint on the surface of the water covers the entire length of the trough in the direction of the beam path. Several factors go into determining the width of the footprint perpendicular to the beam path. Footprints can be kept small to focus as many of the available photons into the angular acceptance of the detector. Sometimes, wider footprints are used in order that there may be more surface area across which to average the signal, or in order to reduce radiation damage in cases where it is a problem.

Scattered intensities are recorded over the range of directions as appropriate; they may be plotted as shown at the left, where the axes of the intensity contour plot are the in-plane component of $\mathbf{q}(\mathbf{q}_{xy})$ and the normal component q_z. The x and y directions are equivalent because Langmuir monolayers tend to be 2D powders (multiple domains with all possible orientations about the z-axis). The positions of the intensity maxima along the $\mathbf{q}_{xy}$ axis tell us the lattice spacings; the q_z positions contain information on the tilts of the molecules; the monolayer peak widths in the z-direction are large (see Figure 2.1) and tell us the thickness of the monolayer. In addition, there will, in general, be intensity maxima from the nucleating crystals. If there are only *rings*, this indicates that the crystals grow with all possible orientations, whereas *spots* (maxima that are sharp in all directions) indicate that the crystals have a specific orientation that can be determined from the positions of the spots seen. Recall that intensities are additive, from the many crystals having different orientations, within the footprint illuminated by the x-ray beam.

Originally, simple point detectors were used to collect GID data. These detectors can only collect data in a specific direction at one time, but collimating (*soller*) slits can be used to pick up scattering in this direction from a large section of the footprint. Contour plots like the one shown in Figure 2.1 can then be created by moving the detector both vertically and horizontally and collecting an array of data points. Linear and area detectors greatly reduce the time needed to capture data since they allow for the collection of multiple points at one time. The trade-off is a loss of resolution when the x-ray footprint is long or wide. Recently, the use of area detectors with a *pinhole* geometry has been developed (Lin et al. 2003; Meron et al. 2009) to provide real-time imaging with high resolution. This has reduced the time needed to collect contour plots by several orders of magnitude and allowed recent studies to focus on kinetic effects that could not be seen previously.

2.2.2 X-RAY REFLECTIVITY

XRR is a one-dimensional (1D) scattering technique: it tells us the density profile normal to an interface, averaged in the x- and y-directions. It is applicable quite generally to interface studies whether or not there is lateral order, and it has proved to be a powerful probe in a wide variety of systems (Braslau et al. 1988;

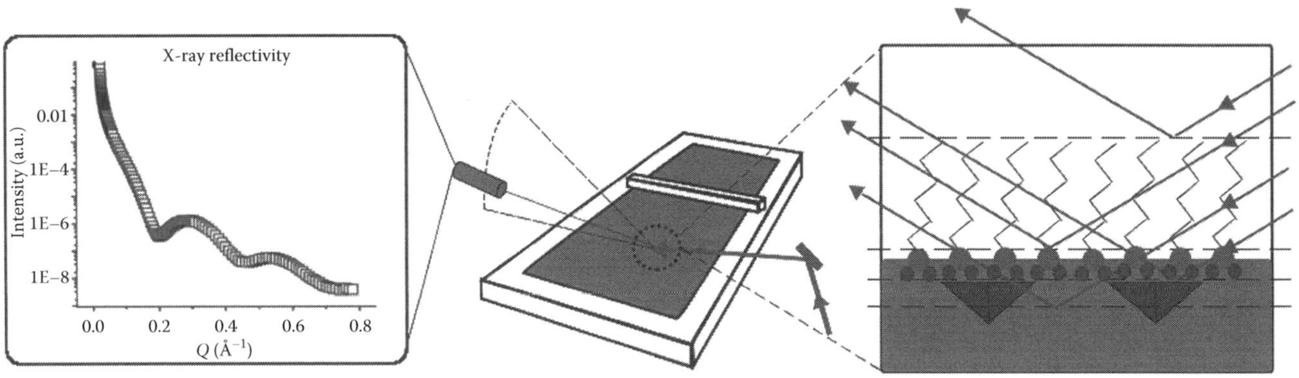

Figure 2.2 The center schematic diagram shows the geometry for reflectivity studies from the surface of water. The incident beam is brought down on the water by the use of a Bragg reflection from a steering crystal. The left side of the figure gives an example of the specular reflectivity as recorded by the detector. The right side of the figure illustrates x-ray scattering from changes in electron density at each interface.

Alsnielsen et al. 1994). Its application to water surfaces is made difficult by the fact that the water surface is always horizontal and cannot be oriented in the x-ray beam. Since the water surface cannot be rotated, the beam must be brought down to the surface using the Bragg reflection from a steering crystal that can be tilted. To scan a range of incident angles requires both varying the tilt of the steering crystal and moving the Langmuir trough both vertically and horizontally to catch the x-ray beam. This requires a specialized liquid diffractometer setup. Sample preparation is the same as described earlier in the section on GID.

The experimental geometry is illustrated in Figure 2.2. Since for specular reflection the incident and scattered beams make equal angles with the surface, we have $\mathbf{q}_{xy} = 0$; there is only the scalar variable q_z. The scattered intensity as a function of q_z depends on

$$R(q_z) = R_F(q_z)\left|\frac{1}{\Delta\rho}\int \frac{d\rho(z)}{dz} e^{iq_z z} dz\right|^2$$

The known function $R_F(q_z)$ is the *Fresnel* reflectivity from an ideally sharp interface, and $\Delta\rho$ is the total electron density change across the interface (in the systems under discussion, it is the electron density of water). The previous equation shows that the reflected intensity depends on the Fourier transform of the spatial derivative of the electron density, making reflectivity data sensitive to interfaces since that is where the electron density changes. The determination of $\rho(z)$ across a given interface is an *inverse problem* (Parratt 1954), since it cannot be calculated directly and uniquely from the finite-range, finite-accuracy experimental data. However, starting with a reasonable model for $\rho(z)$, one can vary the model parameters until a good fit to the observed data is obtained. This method is particularly useful in cases where there is no lattice order, making diffraction useless.

2.3 EXAMPLES OF THE USE OF *IN SITU* X-RAY SCATTERING

The following examples of the use of x-ray scattering from biological templates and biominerals are not a comprehensive review. It is impractical to include all studies of biological templates and nucleation of biologically relevant crystals. While the previous sections described *how* to make use of x-ray scattering at these interfaces, this section is designed to illustrate *when* and *why* to use the techniques. Most of the examples in the following sections focus on the growth of calcium carbonate because of the current depth of the literature.

2.3.1 GID STUDIES OF TEMPLATED NONBIOMINERAL GROWTH

Before we discuss the growth of biominerals using liquid surface scattering techniques, it is useful to look at one example of the kind of information that has been gathered by these techniques in a nonbiomineral system. There exist a number of reports of the templated growth of oriented crystals such as barium fluoride (Kmetko et al. 2003) and lead carbonate (Kewalramani et al. 2005). We show in Figure 2.3, as a vivid example of what can be seen, data from lead carbonate crystals growing under heneicosanoic acid monolayers. The absence of powder rings indicates oriented crystal growth. The large number of spots include some from the monolayer, some from the oriented bulk $PbCO_3$ peaks, and some fractional-index peaks that indicate the presence of a reconstructed surface superlattice (such surface reconstruction has previously been seen only in ultraclean surfaces observed under ultrahigh vacuum). The surface-reconstructed layer helps *mediate* between the monolayer template and the bulk crystal, because its structure is commensurate with both the monolayer and the crystal plane.

Figure 2.3b shows a grazing incidence contour plot centered on the organic monolayer peaks. The peaks are known to be from the organic monolayer because they correspond closely to the peak locations for the monolayer on pure water, the peaks are broader than crystalline peaks, and the width in the q_z direction gives the correct length for the molecule. From the positions of the peaks, it is possible to calculate the monolayer lattice. Figure 2.3a shows a simple in-plane scan ($\mathbf{q}_{xy}$ varied at $q_z = 0$). It is important to note that the leftmost peak is strongest out of plane (Figure 2.3b); however, the peak is still visible at $q_z = 0$.

The intensity maxima seen in Figure 2.3c can be interpreted as follows. It is easy to separate the bulk crystalline (integer index) peaks from the reconstructed surface (fractional-index) peaks both because the lattice spacings of the bulk crystals are previously known and because the surface peaks are weaker. The reconstructed peaks are very broad in q_z, indicating that the surface reconstruction layer is very thin, similar to

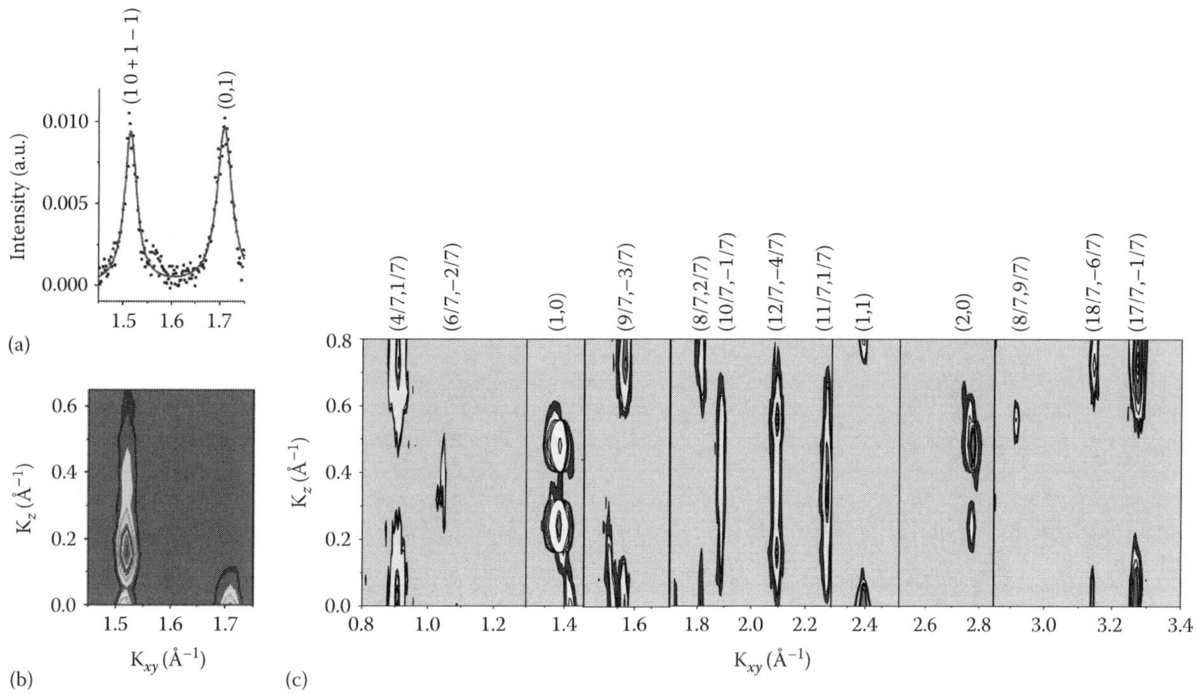

Figure 2.3 Diffraction peaks from bulk and surface inorganic lattices during the templated growth of lead carbonate. (a) In-plane x-ray diffraction data from the monolayer as nucleation begins. (b) GID contour plot: it can be seen that the (10) + (1–1) peak from the monolayer is just out of the horizontal plane. (c) Intensity contours showing diffraction peaks from the bulk and surface of the nucleating crystals. The six peaks labeled with integer indices correspond to bulk hydrocerussite. The fractional order peaks (all of which are very broad in the *z*-direction) are from a $\sqrt{7} \times \sqrt{7}$ interfacial superlattice ~40 Å thick. (Reprinted from *Surf. Sci. Lett.*, 591(1–3), Kewalramani, S., Evmenenko, G., Yu, C.J., Kim, K., Kmetko, J., and Dutta, P., Evidence of surface reconstruction during 'bioinspired' inorganic nucleation at an organic template, L286–L291, Copyright 2005, with permission from Elsevier.)

the monolayer. Using the peak locations in $\mathbf{q}_{xy}$, it is possible determine that the reconstructed layer is a $\sqrt{7} \times \sqrt{7}$ supercell of the surface unit cell of the bulk crystal. Further analysis of the intensities yields the detailed structure of the reconstructed layer (Kewalramani et al. 2005).

The same scattering technique can be used to obtain information about the templated growth of biominerals, as discussed later.

2.3.2 GID STUDIES OF CALCITE NUCLEATION

The first study to present GID-oriented calcite growth under Langmuir monolayers was by Kewalramani et al. (2008). Figure 2.4a shows grazing incidence scattering data establishing that [001]-oriented calcite crystals nucleate under a monolayer of arachidic sulfate. If the crystals were not oriented, the intensity from the scattered x-rays would form powder rings, which would extend uniformly across the contour plot such that $q_{xy}^2 + q_z^2 = \text{constant}$. In fact, however, the peaks are localized with a small angular spread. By calculating the angles between the [001] direction and the visible {110}, {012}, {1–13}, and {202} peaks, it is trivial to verify that this is indeed dominantly [001]-oriented growth. The scattering peaks from the arachidic sulfate monolayer (not shown here) indicate a 5 Å hexagonal unit cell, nearly identical to the 4.99 Å hexagonal surface unit cell of the (001) calcite plane. In other words, there is epitaxy.

Similar GID data from Stripe et al. showed [104]-oriented calcite growth under a monolayer of heneicosanol (a neutral alcohol terminal monolayer) in the presence of Mg (Stripe et al. 2011). However, in this case, no direct epitaxial match was seen. It had long been assumed that calcite grew [014] oriented under alcohol monolayers (Mann et al. 1991; Walker et al. 1991; Heywood and Mann 1994; Aizenberg et al. 1999), but no *in situ* evidence of [104]-oriented growth has been obtained in the absence of Mg. Simulations and calculations by Duffy et al. suggest that these results could be explained by simple charge balancing (Duffy and Harding 2004a,b). The fact that there is well-oriented growth in the presence of Mg, but not without it, demonstrates that even in the absence of evidence of epitaxy, GID can be extremely informative.

X-ray scattering can also be done from crystals grown at solid substrate-supported self-assembled films (Aizenberg et al. 1999; Travaille et al. 2002, 2003; Han and Aizenberg 2003; Kwak et al. 2005). These studies were mainly done to verify the orientation of the crystals and, due to the roughness of the substrates, contain no *in situ* data on the structure of the self-assembled templates.

2.3.3 X-RAY REFLECTIVITY STUDIES

Going a step beyond the example in the previous section where there was no visible epitaxy, we illustrate that it is still possible to obtain important information using x-rays in the absence of crystals or in the absence of ordered monolayers. In its simplest form, the XRR technique can be used to probe the vertical structure of the Langmuir monolayer and its interaction with ions on the solution. DiMasi et al. published the first study containing XRR data on the interaction of stearic acid monolayers with Ca^{2+} ions in solution. Figure 2.5a shows the normalized XRR intensity as a function of q_z. Normalizing the data consists of dividing $R(q_z)$ by $R_F(q_z)$ (for definitions, see the reflectivity methods section); this removes the strongest part of the q_z

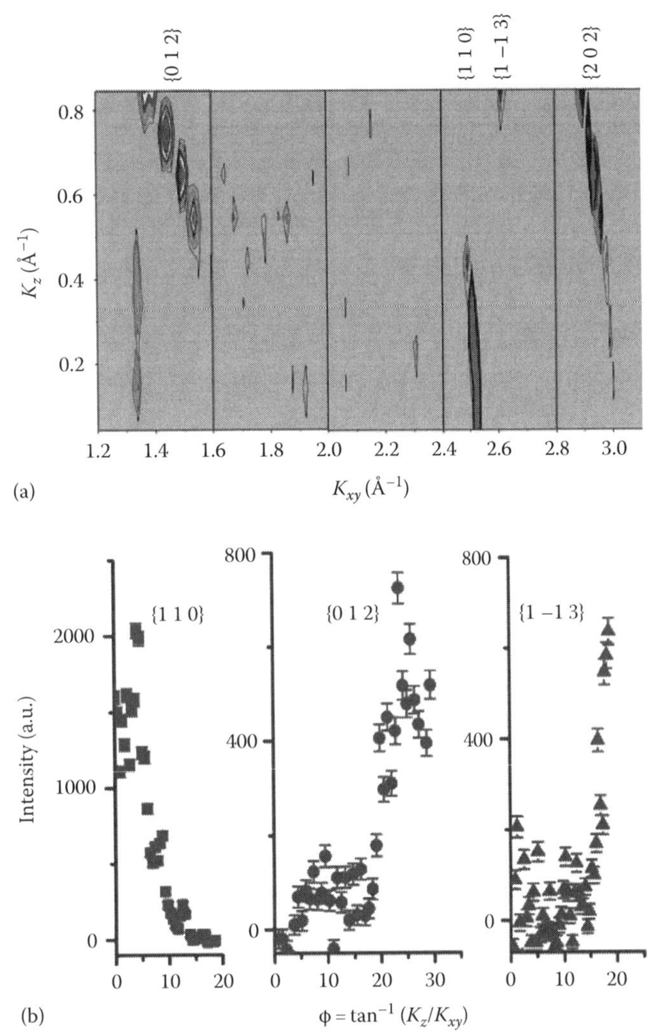

Figure 2.4 Calcite growth under a sulfate monolayer. (a) Contour plots derived from x-ray scattering data collected during late crystal growth stages show four strong diffraction peaks from (001)-oriented calcite crystals. (b) Debye ring scans (equivalent to rocking curves) for the three strongest calcite peaks. The position of the maxima in the ring scans unambiguously indicates that calcite is (001) oriented on average, with a misorientation of ±5° FWHM. (Reprinted with permission from Kewalramani, S., Kim, K., Stripe, B., Evmenenko, G., Dommett, G.H.B., and Dutta, P., Observation of an organic-inorganic lattice match during biomimetic growth of (001)-oriented calcite crystals under floating sulfate monolayers, *Langmuir*, 24(19), 10579–10582. Copyright 2008 American Chemical Society.)

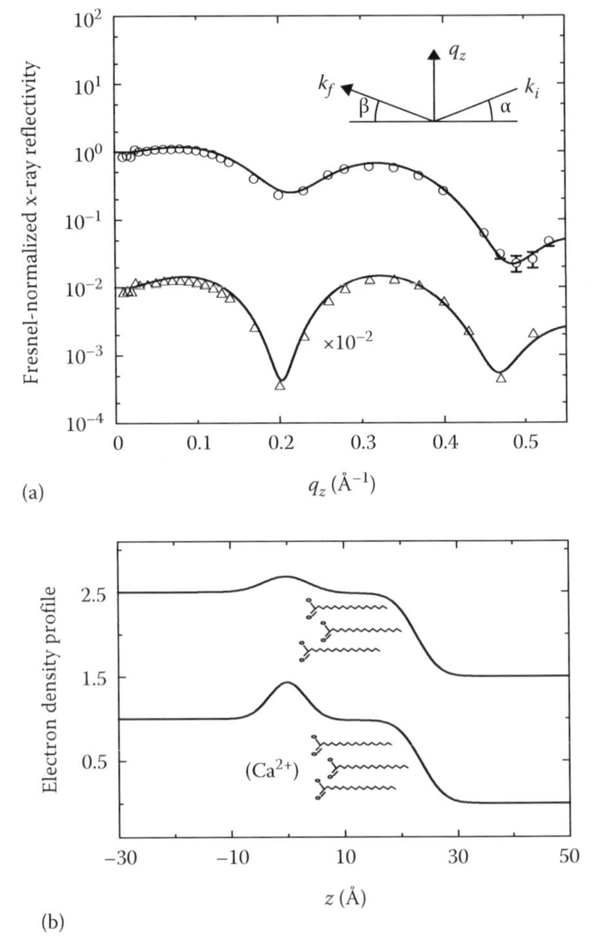

Figure 2.5 (a) Fresnel-normalized XRR $R(q_z)/R_F(q_z)$ from C18 monolayers on water (O) and calcium bicarbonate solution (Δ) at surface pressures of 20–25 mN m^{-1}. Lines are calculated from models in (b). Inset: reflection geometry. (b) Surface-normal electron density profiles $\rho(z)$, normalized to the density of the water subphase. Upper curve: best fit to data on water. Lower curve: calcium bicarbonate subphase. Schematics show the positions of molecules at the surface. Increased density in the head group region, modeled by a Gaussian, shows that calcium collects at the interface of the bicarbonate subphase. The Ca^{2+} binding was computed by subtracting the Gaussian contribution from the water measurement and assuming that excess electrons come from a combination of Ca^{2+} and H_2O. Taking the volume ratios of these two species in the range 0.4–1.0, the authors estimate 4–8 molecules per cation. (DiMasi, E., Olszta, M.J., Patelbc, V.M., and Gower, L.B., When is template directed mineralization really template directed?, *CrystEngComm*, 5, 346–350. Reproduced by permission of The Royal Society of Chemistry.)

dependence of the reflectivity and highlights the features in the data that are of interest. The electron density profiles in Figure 2.5b were generated by fitting as described previously.

For example, starting with a simple two-box model for the tail group and head group, in addition to an infinite layer of air above and water below, it is possible to fit these data fairly well. Allowing for the addition of surface roughness in each layer leads to the smooth electron density models seen in Figure 2.5b. The difference between a stearic acid monolayer on water and on Ca^{2+} subphase can clearly be seen in the higher electron density of the modeled head group in the presence of Ca^{2+}. The density change in the head group was then related to the electron density of the Ca^{2+}, and it was calculated that there was roughly 1 Ca^{2+} ion per 4–8 stearic acid molecules.

In another example, DiMasi et al. used XRR where the subphase is supersaturated with $CaCO_3$ and also contains dissolved polyacrylic acid (PAA) (DiMasi et al. 2006). Figure 2.6a shows the normalized reflectivity data after 7.3 h of mineralization. Figure 2.6b shows the difference between the electron density profiles of the film at 1.5 and 7.3 h, demonstrating the growth of the amorphous layer. Figure 2.6c and d shows the raw and normalized data in a selected region close to the critical angle, at multiple time steps. The focus near the low angles close to the critical angle is important because it gives the most information on the relatively thick amorphous film. Fitting of these oscillations requires some care, as discussed in DiMasi et al. (2006). This study clearly established the time-dependent growth of an amorphous calcium carbonate (ACC) layer beneath the stearic acid monolayer.

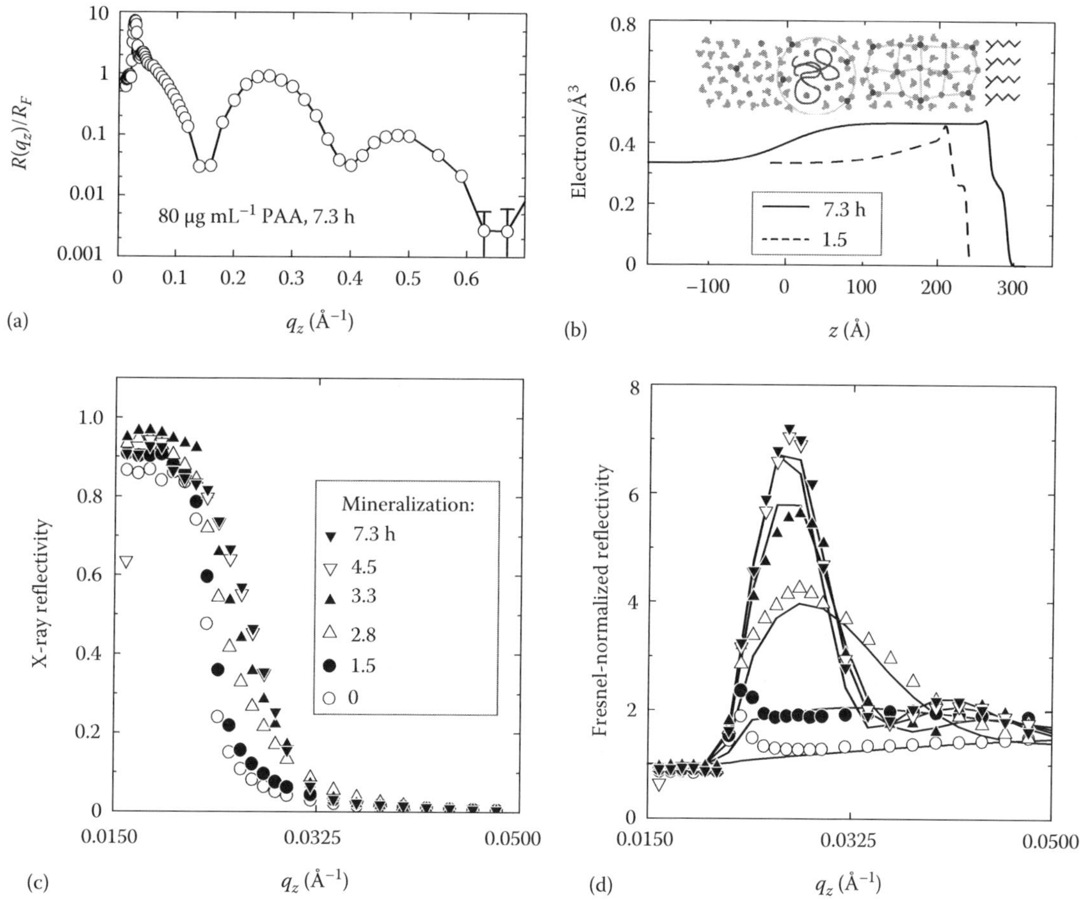

Figure 2.6 Representative data from monolayer prepared on 80 μg mL⁻¹ PAA in subphase. (a) Full R_F-normalized reflectivity data set from sample mineralized for 7.3 h. (b) Comparison of real-space model profiles for 7.3- and 1.5-h mineralization. Schematic shows proposed ACC formation process as discussed in the text. (c) Raw reflectivity data time series (symbols), with R_F for comparison (solid line). (d) Selected data from (c) normalized by R_F. (Reprinted with permission from DiMasi, E., Kwak, S.Y., Amos, F.F., Olszta, M.J., Lush, D., and Gower, L.B., Complementary control by additives of the kinetics of amorphous $CaCO_3$ mineralization at an organic interface: In-situ synchrotron x-ray observations, *Phys. Rev. Lett.*, 97(4), 045503. Copyright 2006 by the American Physical Society.)

XRR is limited in some ways with respect to biomineralization because it requires some degree of uniformity normal to the plane. This means once the crystals start growing and the sample is no longer as uniform, the XRR signal begins to degrade. While this is not ideal, it is still an extremely useful tool for examining these systems. As illustrated earlier, it can still be used to probe the density profile of the monolayer, the interactions with ions, and to detect the presence of amorphous phases.

2.3.4 USE OF GID TO PROBE LANGMUIR TEMPLATES

So far we have discussed the use of GID to study the orientation of crystals and the use of XRR reflectivity to study the electron density normal to the monolayer surface. However, perhaps the most important use of GID is the study of the structure of the biological templates. Various types of oriented calcite growth have been observed on self-assembled films (Aizenberg et al. 1999; Travaille et al. 2002, 2003; Han and Aizenberg 2003; Kwak et al. 2005). However, far fewer stable orientations have been seen under Langmuir monolayers and most of them date to the original work done by Mann et al. (Mann et al. 1991; Walker et al. 1991; Heywood and Mann 1994; Litvin et al. 1997). This inability to replicate the same level of controlled growth using Langmuir monolayers has led to several new theories in regard to the growth of calcite at organic templates. Some theories address changes in pH or HCO_3 inclusion in the monolayer structure (Duffy et al. 2005), while others suggest that the most fluid or adaptable monolayers are the most likely to grow oriented calcite (DiMasi et al. 2007; Popescu et al. 2007; Lendrum and McGrath 2010). This idea has been supported by the work of Popescu et al. looking at complex monolayers capable of making directional specific hydrogen bonds (DiMasi et al. 2007; Popescu et al. 2007). This study makes use of both GID and XRR.

Figure 2.7a shows a schematic diagram of the ordering of the molecules on the surface. Because the tail groups are offset from each other (Figure 2.7b), the scattering from tail groups will not be in the horizontal plane. Given this information, it is not surprising that the GID peaks in Figure 2.7c through f are all out of plane. What is surprising though is in Figure 2.7d, which corresponds to the larger head group valine surfactant molecule (VAL) molecule in the presence of $CaCl_2$, shows only one out of plane peak. The disappearance of the second peak is interpreted as showing that the VAL molecule becomes ordered in only 1D, and is unconstrained in the second direction. This was correlated with scanning electron microscope (SEM) evidence demonstrating that this molecule was capable of modifying the calcite faces while the more rigid glycine surfactant molecule (GLY) molecule

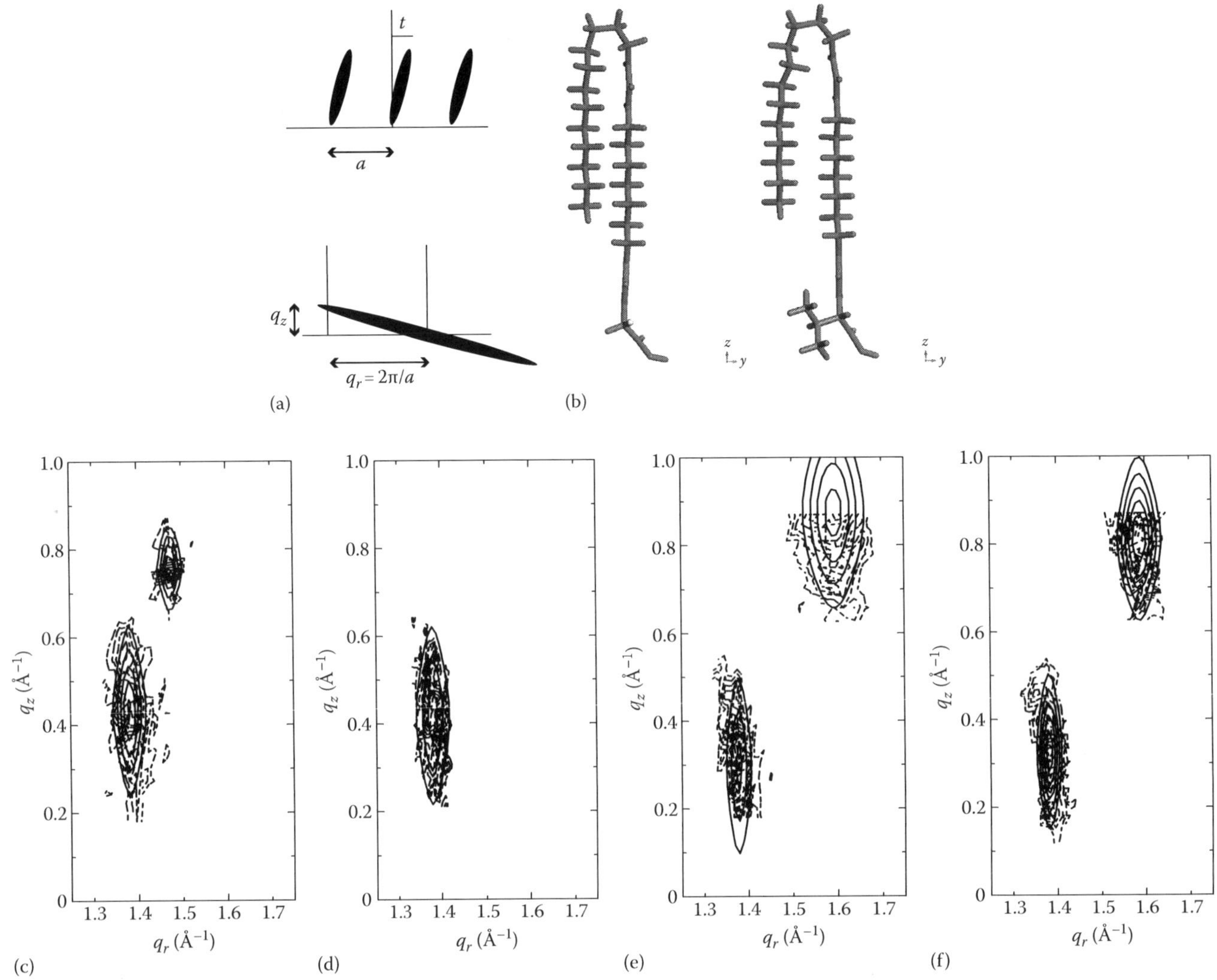

Figure 2.7 (a) Top: lattice of rod-shaped molecules with spacing *a* and tilt angle *t* relative to the surface normal. Bottom: map of the corresponding reciprocal space intensity for this structure, with vertical Bragg rods spaced at $2\pi/a$ and molecular form factor intersecting the Bragg rod at $(2\pi/a)$ tan *t*. (b) Folded molecules for GLY and VAL surfactant molecules, drawn in Accelrys Molecular Modeling software. (c) VAL, H_2O. (d) VAL, 0.01 M $CaCl_2$. (e) GLY, H_2O. (f) GLY, 0.01 M $CaCl_2$. (—) Experimental intensity contours. (—) Contours from 2D Lorentzian fit peaks. (DiMasi, E., Kwak, S.Y., Pichon, B.P., and Sommerdijk, N.A.J.M., Structural adaptability in an organic template for $CaCO_3$ mineralization, *CrystEngComm*, 9(12), 1192–1204. Reproduced by permission of The Royal Society of Chemistry.)

was not. This work is an excellent example of the combination of structural information available from GID and evidence from SEM images. However, other recent papers contain no additional *in situ* structural information (Popescu et al. 2007; Lendrum and McGrath 2010). The idea of *fluidity* or relaxation of the template structure is theorized to be the driving mechanism even though the actual structure of the template is not known. Fluidity or lack thereof deduced from isotherms or other secondary methods is not evidence of the lack of template structure.

In contrast, recent work by Stripe et al. has demonstrated well-oriented growth of calcite under a system of miscible Langmuir monolayers (Stripe et al. 2012). Not only are the monolayers miscible and well ordered but they are extremely stable, in fact so stable that mixtures of dioctadecyldimethylammonium bromide (DODAB) and dodecylsulfate were used to demonstrate giant folds in Langmuir monolayers (Coppock et al. 2009). Figure 2.8 shows SEM images of crystals grown under varying mixtures of heneicosanoic acid and DODAB. It can be seen in images a through e, representing mixtures of 100%–70% acid, that the crystals grow to be very uniform in both morphology and orientation. In addition to the SEM images, *in situ* GID data were collected from the monolayer templates.

Using monolayer mixtures of heneicosanoic acid (negatively charged) and dioctadecyldimethylammonium bromide, DODAB (positively charged), the monolayer charge could be varied over a wide range. The GID data demonstrated that this changed the monolayer structure and the molecular tilt as well. This gave an excellent window into the interplay between charge, epitaxy, and stereochemistry as possible controlling factors in the oriented nucleation of calcite.

The fact that the charge, tilt, and lattice spacings of the monolayer were changing together made decoupling the results somewhat difficult. Nonetheless, the results were unambiguous. Oriented calcite crystals were nucleated, with different orientations under different monolayer compositions, but the most highly charged calcite face seen, the (012), forms under a less

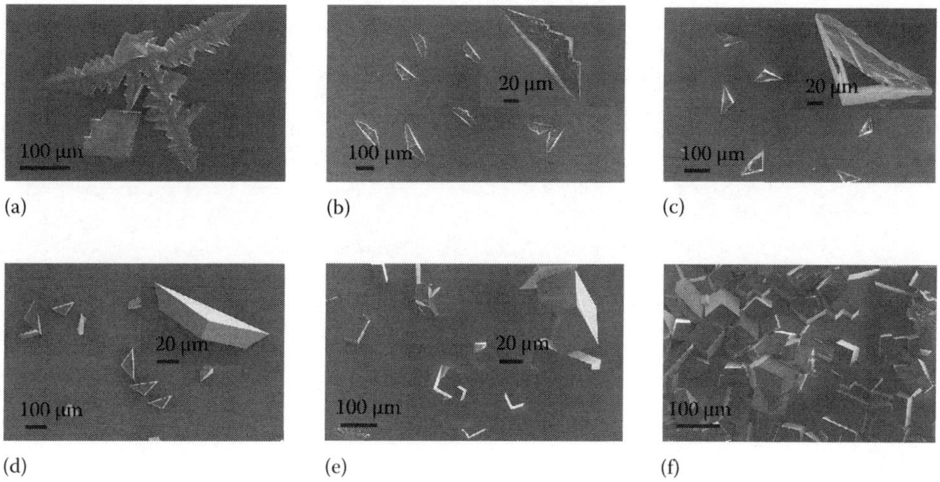

Figure 2.8 SEM images of calcite crystals nucleated under mixed monolayers of DODAB and heneicosanoic acid. Image (a) is under pure heneicosanoic acid, while image (b) is 95% heneicosanoic acid, (c) is 90%, (d) is 80%, (e) is 70%, and (f) is seen for 60% and less. (Reprinted with permission from Stripe, B., Uysal, A., Lin, B., Meron, M., and Dutta, P., Charge, stereochemistry, or epitaxy? Toward controlled biomimetic nucleation at mixed monolayer templates, *Langmuir*, 28(1), 572–578. Copyright 2012, American Chemical Society.)

charged monolayer compared to where the (1–10) face nucleates. There was also no observable correlation between the monolayer tilt angle and the calcite face that is nucleated. However, the nucleated crystal orientation changed when the structure of the monolayer changed, and this is the only consistent correlation.

This study demonstrates the use of GID in biomineralization studies not merely to see whether or not there is a lattice match but also to explore the roles of other commonly proposed (but rarely verified) factors.

2.4 FACILITIES AND NEW METHODS

Although synchrotron radiation facilities are designed to serve a broad range of users, readers should note that the experiments described here are difficult and require long-range planning. Users will need to consult, and in most cases seek to collaborate with, experts associated with beamlines that have the requisite capabilities.

In the United States, one facility that is particularly well set up for studies such as those described here is the ChemMatCARS Sector 15 at the Advanced Photon Source (APS). The National Synchrotron Light Source (NSLS) at Brookhaven National Laboratory has also hosted many liquid surface studies over the past decade. In 2014, NSLS will cease operations and its successor NSLS-II is expected to come online. It is designed to deliver extremely bright and stable x-ray beams for studying smaller samples. The Soft Matter Interfaces beamline will have a liquid reflectometer and other relevant x-ray scattering capabilities.

Current developments in techniques and methods are quickly changing the time scales and accuracy of both liquid surface GID and liquid surface XRR. ChemMatCARS Sector 15 at the APS has pioneered two particularly relevant techniques. The first is the use of a pinhole geometry with an area detector (Meron et al. 2009). This technique allows for the acquisition of GID contour plots orders of magnitude quicker than previously possible. For example, Uysal et al., looking at the growth of

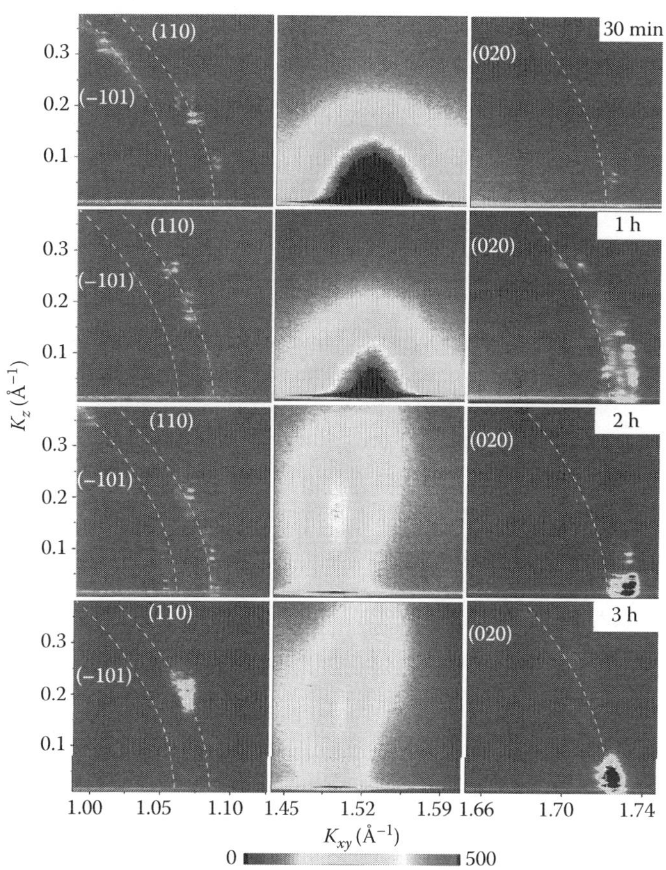

Figure 2.9 *In situ* GID intensity contours for the first 3 h of calcium oxalate nucleation and growth under heneicosanoic acid monolayer. Dashed lines represent Debye rings for the bulk crystal. The peaks not along dashed lines are from the monolayer template. (Uysal, A., Stripe, B., Kim, K., and Dutta, P., Epitaxy driven interactions at the organic-inorganic interface during biomimetic growth of calcium oxalate, *CrystEngComm*, 12(7), 2025–2028. Reproduced by permission of The Royal Society of Chemistry.)

calcium oxalate, were able to study the changes of the monolayer and crystal structures during the nucleation process based on the time-dependent positions of multiple peaks (Uysal et al. 2010). Figure 2.9 is an example of the GID data they were able to collect. While the time scale for this process may seem slow, it is actually the ability to watch the changes in these peaks in real time that allowed the authors to easily identify the time-dependent changes.

The second new method of interest is grazing incidence off-specular x-ray scattering (Dai et al. 2011). This is a technique in which the electron density normal to the surface is determined from an analysis of the off-specular scattering when x-rays are incident at a fixed small angle (below the critical angle for total reflection). This method is faster than standard XRR because the incident angle never needs to be changed. Some aspects of the data analysis are not yet fully understood, and the development of fitting procedures is still ongoing. However, the technique promises the ability to monitor changes in the electron density normal to the surface in real time, using a linear detector. This greatly extends the range of kinetics and dynamics that can be monitored *in situ*, much as the pinhole geometry does for GID.

2.5 SUMMARY

The first stage of biomineral nucleation takes place at interfaces, at and below the nanoscale. There is simply no probe other than x-rays that has the necessary resolution to characterize this stage and that nondestructively reveals so much about what goes on during chemical or physical processes occurring at the relevant interfaces. Scattering of x-rays at surfaces will probably never become a *mail-in* characterization technique; nonetheless, it is no longer reasonable today to speculate on the basis of indirect evidence or from the examination of macroscopic final products. The information that can be obtained using synchrotron x-rays has the potential to teach us what living organisms already know about crystal growth and design, and to use it not only to control bone and shell growth processes but to find new ways to grow every kind of material biomimetically under ambient (*mild*) conditions.

REFERENCES

Aizenberg, J., A.J. Black, and G.H. Whitesides. 1999. Oriented growth of calcite controlled by self-assembled monolayers of functionalized alkanethiols supported on gold and silver. *Journal of the American Chemical Society* 121(18):4500–4509. doi: 10.1021/ja984254k.

Alsnielsen, J., D. Jacquemain, K. Kjaer, F. Leveiller, M. Lahav, and L. Leiserowitz. 1994. Principles and applications of grazing-incidence X-ray and neutron-scattering from ordered molecular monolayers at the air-water-interface. *Physics Reports-Review Section of Physics Letters* 246(5):252–313.

Backov, R., C.M. Lee, S.R. Khan, C. Mingotaud, G.E. Fanucci, and D.R. Talham. 2000. Calcium oxalate monohydrate precipitation at phosphatidylglycerol Langmuir monolayers. *Langmuir* 16(14):6013–6019. doi: 10.1021/la991684v.

Braslau, A., P.S. Pershan, G. Swislow, B.M. Ocko, and J. Alsnielsen. 1988. Capillary waves on the surface of simple liquids measured by X-ray reflectivity. *Physical Review A* 38(5):2457–2470. doi: 10.1103/PhysRevA.38.2457.

Coppock, J.D., K. Krishan, M. Dennin, and B.G. Moore. 2009. Fluorescence microscopy imaging of giant folding in a catanionic monolayer. *Langmuir* 25(9):5006–5011. doi: 10.1021/la803773y.

Dai, Y., B. Lin, M. Meron, K. Kim, B. Leahy, and O.G. Shpyrko. 2011. A comparative study of Langmuir surfactant films: Grazing incidence x-ray off-specular scattering vs. x-ray specular reflectivity. *Journal of Applied Physics* 110(10):102213.

DiMasi, E., S.Y. Kwak, F.F. Amos, M.J. Olszta, D. Lush, and L.B. Gower. 2006. Complementary control by additives of the kinetics of amorphous $CaCO_3$ mineralization at an organic interface: In-situ synchrotron x-ray observations. *Physical Review Letters* 97(4):045503. doi: 10.1103/Physrevlett.97.045503.

DiMasi, E., S.Y. Kwak, B.P. Pichon, and N.A.J.M. Sommerdijk. 2007. Structural adaptability in an organic template for $CaCO_3$ mineralization. *CrystEngComm* 9(12):1192–1204. doi: 10.1039/B711153c.

DiMasi, E., M.J. Olszta, V.M. Patelbc, and L.B. Gower. 2003. When is template directed mineralization really template directed? *CrystEngComm* 5:346–350.

Duffy, D.M. and J.H. Harding. 2004a. Growth of polar crystal surfaces on ionized organic substrates. *Langmuir* 20(18):7637–7642. doi: 10.1021/La049551j.

Duffy, D.M. and J.H. Harding. 2004b. Simulation of organic monolayers as templates for the nucleation of calcite crystals. *Langmuir* 20(18):7630–7636. doi: 10.1021/la049552b.

Duffy, D.M., A.M. Travaille, H. van Kempen, and J.H. Harding. 2005. Effect of bicarbonate ions on the crystallization of calcite on self-assembled monolayers. *Journal of Physical Chemistry B* 109(12):5713–5718. doi: 10.1021/Jp044594u.

Fricke, M. and D. Volkmer. 2007. Crystallization of calcium carbonate beneath insoluble monolayers: Suitable models of mineral-matrix interactions in biomineralization? In *Biomineralization I: Crystallization and Self-Organization Process*, ed. K. Naka, pp. 1–41. Berlin, Germany: Springer.

Han, Y.J. and J. Aizenberg. 2003. Face-selective nucleation of calcite on self-assembled monolayers of alkanethiols: Effect of the parity of the alkyl chain. *Angewandte Chemie International Edition* 42(31):3668–3670. doi: 10.1002/anie.200351655.

Heywood, B.R. and S. Mann. 1994. Molecular construction of oriented inorganic materials—Controlled nucleation of calcite and aragonite under compressed Langmuir monolayers. *Chemistry of Materials* 6(3):311–318.

Kaganer, V.M., H. Mohwald, and P. Dutta. 1999. Structure and phase transitions in Langmuir monolayers. *Reviews of Modern Physics* 71(3):779–819.

Kewalramani, S., G. Evmenenko, C.J. Yu, K. Kim, J. Kmetko, and P. Dutta. 2005. Evidence of surface reconstruction during 'bioinspired' inorganic nucleation at an organic template. *Surface Science Letters* 591(1–3):L286–L291. doi: 10.1016/j.susc.2005.06.086.

Kewalramani, S., K. Kim, B. Stripe, G. Evmenenko, G.H.B. Dommett, and P. Dutta. 2008. Observation of an organic-inorganic lattice match during biomimetic growth of (001)-oriented calcite crystals under floating sulfate monolayers. *Langmuir* 24(19):10579–10582. doi: 10.1021/La802124v.

Kmetko, J., C. Yu, G. Evmenenko, S. Kewalramani, and P. Dutta. 2003. Organic-template-directed nucleation of strontium fluoride and barium fluoride: Epitaxy and strain. *Physical Review B* 68(8):085415. doi: 10.1103/Physrevb.68.085415.

Kwak, S.Y., E. DiMasi, Y.J. Han, J. Aizenberg, and I. Kuzmenko. 2005. Orientation and Mg incorporation of calcite grown on functionalized self-assembled monolayers: A synchrotron X-ray study. *Crystal Growth & Design* 5(6):2139–2145. doi: 10.1021/Cg050164x.

Lendrum, C. and K.M. McGrath. 2010. Toward controlled nucleation: Balancing monolayer chemistry with monolayer fluidity. *Crystal Growth & Design* 10(10):4463–4470. doi: 10.1021/cg1006764.

Lin, B.H., M. Meron, J. Gebhardt, T. Graber, M.L. Schlossman, and P.J. Viccaro. 2003. The liquid surface/interface spectrometer at ChemMatCARS synchrotron facility at the Advanced Photon Source. *Physica B—Condensed Matter* 336(1–2):75–80. doi: 10.1016/s0921-4526(03)00272-2.

Litvin, A.L., S. Valiyaveettil, D.L. Kaplan, and S. Mann. 1997. Template-directed synthesis of aragonite under supramolecular hydrogen-bonded Langmuir monolayers. *Advanced Materials* 9(2):124–127. doi: 10.1002/adma.19970090205.

Mann, S., B.R. Heywood, S. Rajam, and J.B.A. Walker. 1991. Structural and stereochemical relationships between Langmuir monolayers and calcium-carbonate nucleation. *Journal of Physics D—Applied Physics* 24(2):154–164. doi: 10.1088/0022-3727/24/2/011.

Meron, M., J. Gebhardt, H. Brewer, J.P. Viccaro, and B. Lin. 2009. Following transient phases at the air/water interface. *European Physical Journal—Special Topics* 167:137–142. doi: 10.1140/epjst/e2009-00949-0.

Parratt, L.G. 1954. Surface studies of solids by total reflection of x-rays. *Physical Review* 95(2):359–369. doi: 10.1103/PhysRev.95.359.

Popescu, D.C., M.M.J. Smulders, B.P. Pichon, N. Chebotareva, S.Y. Kwak, O.L.J. van Asselen, R.P. Sijbesma, E. DiMasi, and N.A.J.M. Sommerdijk. 2007. Template adaptability is key in the oriented crystallization of CaCO(3). *Journal of the American Chemical Society* 129(45):14058–14067. doi: 10.1021/Ja075875t.

Sommerdijk, N.A.J.M. and G. de With. 2008. Biomimetic $CaCO_3$ mineralization using designer molecules and interfaces. *Chemical Reviews* 108(11):4499–4550. doi: 10.1021/cr078259o.

Stripe, B., A. Uysal, and P. Dutta. 2011. Orientation and morphology of calcite nucleated under floating monolayers: A magnesium-ion-enhanced nucleation study. *Journal of Crystal Growth* 319(1):64–69. doi: 10.1016/j.jcrysgro.2011.01.102.

Stripe, B., A. Uysal, B. Lin, M. Meron, and P. Dutta. 2012. Charge, stereochemistry, or epitaxy? Toward controlled biomimetic nucleation at mixed monolayer templates. *Langmuir* 28(1):572–578. doi: 10.1021/la2037422.

Talham, D.R., R. Backov, I.O. Benitez, D.M. Sharbaugh, S. Whipps, and S.R. Khan. 2006. Role of lipids in urinary stones: Studies of calcium oxalate precipitation at phospholipid Langmuir monolayers. *Langmuir* 22(6):2450–2456. doi: 10.1021/la052503u.

Travaille, A.M., J. Donners, J.W. Gerritsen, N. Sommerdijk, R.J.M. Nolte, and H. van Kempen. 2002. Aligned growth of calcite crystals on a self-assembled monolayer. *Advanced Materials* 14(7):492–495. doi: 10.1002/1521-4095(20020404)14:7<492::aid-adma492>3.0.co;2-l.

Travaille, A.M., L. Kaptijn, P. Verwer, B. Hulsken, J.A.A.W. Elemans, R.J.M. Nolte, and H. van Kempen. 2003. Highly oriented self-assembled monolayers as templates for epitaxial calcite growth. *Journal of the American Chemical Society* 125(38):11571–11577. doi: 10.1021/Ja034624r.

Uysal, A., B. Stripe, K. Kim, and P. Dutta. 2010. Epitaxy driven interactions at the organic-inorganic interface during biomimetic growth of calcium oxalate. *CrystEngComm* 12(7):2025–2028. doi: 10.1039/b926751d.

Volkmer, D., N. Mayr, and M. Fricke. 2006. Crystal structure analysis of $Ca(O_3SC18H37)(2)(DMSO)(2)$, a lamellar coordination polymer and its relevance for model studies in biomineralization. *Dalton Transactions* (41):4889–4895. doi: 10.1039/b608760d.

Walker, J.B.A., B.R. Heywood, and S. Mann. 1991. Oriented nucleation of $CaCO_3$ from metastable solutions under Langmuir monolayers. *Journal of Materials Chemistry* 1(5):889–890. doi: 10.1039/jm9910100889.

Wang, B.J., X.T. Zhang, Z.H. Xue, S.X. Dai, Y.C. Li, Y.B. Huang, and Z.L. Du. 2006. Synthesis of CaF_2 single crystal induced by Langmuir monolayers. *Journal of Inorganic Materials* 21(1):12–16.

Yang, J.P. and J.H. Fendler. 1995. Morphology control of PbS nanocrystallites, epitaxially under mixed monolayers. *Journal of Physical Chemistry* 99(15):5505–5511. doi: 10.1021/j100015a038.

Zhao, X.K., J. Yang, L.D. McCormick, and J.H. Fendler. 1992. Epitaxial formation of PbS crystals under arachidic acid monolayers. *Journal of Physical Chemistry* 96(24):9933–9939. doi: 10.1021/j100203a065.

3 Electron backscatter diffraction for biomineralization

Maggie Cusack and Peter Chung

Contents

3.1 INTRODUCTION

3.1.1 ELECTRON BACKSCATTER DIFFRACTION AS A TOOL FOR INVESTIGATING BIOMINERALS

In biomineral studies, electron backscatter diffraction (EBSD) is an electron-beam technique that can be used to simultaneously identify mineralogy, crystallographic orientation, and crystal size at high spatial resolution. EBSD thus provides a means of determining crystallographic orientation in context, which greatly enhances our understanding of the biological control exerted on biomineral formation. The technique can be used, for example, to compare mineralogy and crystallographic orientation through ontogeny or in different parts of biomineral structures.

3.1.2 HISTORY OF EBSD

The fundamental observations necessary for the conception of EBSD were made in 1928. Thomson (Thomson 1928) observed diffraction of an electron beam by celluloid film and by metals. Thomson's observation was confirmed by Nishikawa and Kikuchi (Nishikawa and Kikuchi 1928b) who detected a diffraction pattern when they passed an electron beam through a sheet of mica and later repeated the experiment with calcite (Nishikawa and Kikuchi 1928a). Nishikawa and Kikuchi observed patterns on a photographic plate that they had placed behind a calcite crystal, normal to the incident electron beam. The patterns comprised intersecting lines of different thicknesses. Nishikawa and Kikuchi provided the following explanation for the formation of the pattern by stating "…if electrons could penetrate into the crystal, undergoing a multiple scattering without an appreciable loss of energy, then the electrons scattered by the crystal atoms will form divergent rays emerging from a point source in the crystal itself. These are regularly reflected by the net planes in the crystal according to Bragg conditions, and the cones of reflected rays thus formed intersecting with the photographic plate…" This depiction by Nishikawa and Kikuchi (Nishikawa and Kikuchi 1928a), of the pattern of what we now term *Kikuchi bands* and the explanation presented in 1928, laid the foundation for EBSD.

For several decades, the development of EBSD was a research area in its own right (Alam et al. 1954, Venables and Harland 1973, Venables and Binjaya 1977). During its development, the technique has been known by several names including Kikuchi electron diffraction (Thomas 1965), backscatter Kikuchi diffraction (Babakishi and Dingley 1989), and high-angle Kikuchi diffraction (Alam et al. 1954). This range of terminologies may cause some confusion in the older literature, but in general, the terminology has become more, although not entirely, consistent and the term EBSD is now well established. The development of EBSD as a research technique initially saw the majority of applications in materials science and metallurgy. This is entirely logical given that a technique based on diffraction of electrons is more straightforward in conductive materials and the importance of the application in these areas relates to the influence that crystallographic orientation has on material properties. EBSD continues to be an important tool in materials science and metallurgy, for example, Schwartz (2009), Glage et al. (2012), Moy et al. (2012). More recent development and expansion of EBSD has seen the technique applied to research questions in other areas including structural geology (Ebner et al. 2010, Okudaira and Shigematsu 2012), mineralogy (Inoue and Kogure 2012), paleontology (Poole and Lloyd 2000, Perez-Huerta et al. 2012), planetary science (Lee and Nicholson 2009, Lindgren et al. 2011), and physics (Zou and Zhang 2010, Kumar and Pollock 2011). EBSD can be carried out in transmission electron microscopy (TEM), for example, Farooq et al. (2008), as well as in scanning electron microscopy (SEM). The expansion of EBSD into a range of research areas was, to a large extent, due to the development of automated EBSD systems that became integrated into SEMs.

In the field of biomineral research, EBSD became available when the rigor of careful sample preparation enabled EBSD analysis to be applied to these insulators. It soon became evident that EBSD could be used to answer questions posed by a wide range of biominerals including brachiopods (Schmahl et al. 2004, England et al. 2007a), coccoliths (Saruwatari et al. 2006), bivalve mollusks (Dalbeck et al. 2006), and avian eggshells (Dalbeck and Cusack 2006). As EBSD has become more established in biomineral research, the range of biomineral systems has expanded rapidly to include a plethora of modern biominerals such as gastropods (Perez-Huerta et al. 2011), red coral sclerites (Vielzeuf et al. 2010), echinoderms (Moureaux et al. 2010), oyster cement (MacDonald et al. 2010), and earthworm granules (Lee et al. 2008) as well as fossil biominerals including coccoliths (Saruwatari et al. 2011), brachiopods (Balthasar et al. 2011), conodonts (Perez-Huerta et al. 2012), and trilobite eyes (Lee et al. 2012). In all cases, patience at the stage of sample preparation is crucial in obtaining an excellent EBSD analysis.

3.2 ELECTRON BACKSCATTER DIFFRACTION

3.2.1 WHAT EBSD DOES

The technique of EBSD involves an electron beam being fired at a polished sample, the beam is diffracted by the first few lattice planes, and the diffracted beam is backscattered to the detector (phosphor camera) where a series of Kikuchi bands are formed. These Kikuchi bands enable identification of the mineral as well as its orientation at that analysis point. EBSD therefore provides data on crystallographic orientation and mineralogy *in situ* at high spatial resolution. This information is a valuable asset in developing our understanding of biomineral formation. A key factor in successful EBSD analysis of biominerals is in establishing a rigorous approach to sample preparation. In brief, EBSD requires that biomineral samples are cut carefully and slowly to avoid deformation; that they are well polished since EBSD is, in effect, a surface technique; and since biominerals are insulators, that they are adequately coated to avoid charging while minimizing loss of the backscattered signal. This section on the practical aspects of EBSD will be presented in chronological order, beginning with the importance of sample preparation.

3.2.2 HOW TO DO EBSD

3.2.2.1 Sample preparation

There are no shortcuts in sample preparation for EBSD. Biominerals, for example, shell samples, are cut to manageable size under running water using a diamond saw cutter such as an IsoMet 5000 Precision Cutter from Buehler. It is important to cut slowly since rapid cutting introduces heating and vibration that deforms the sample and subsequently would require that the deformed layer be removed by grinding and polishing. It is far better to cut the sample slowly and minimize such deformation. Typically, a slice would be cut from a large shell such as an adult common blue mussel, *Mytilus edulis*, of around 6 cm long in about 20 min. The steps involved are summarized in Figure 3.1, which shows that the cut sections are then embedded in resin in molds. Comparison of a range of resins including conductive resins that had carbon incorporated within the resin found that resin improved EBSD analysis and that epoxy resin was the most suitable of those tested (Perez-Huerta and Cusack 2009). The embedded shell section then goes through a series of grinding and polishing steps as follows.

Specimens are ground using the following grit papers for 3 min each, P180 (82 μm), P320 (46 μm), P800 (21 μm), P1200 (15 μm), and P2500 (8 μm), and then P4000 (<5 μm) for 5 min. EBSD analyses is suitable for small samples or small portions of samples. Samples in resin formed in standard holders lend themselves to automated grinding since the holder will accommodate such standard sizes of molds. However, excess resin should be removed to avoid restricting geometry within the sample chamber when tilting the sample and ensuring that interaction with the beam and sample accommodates maximum signal for the detector. It is therefore recommended that excess

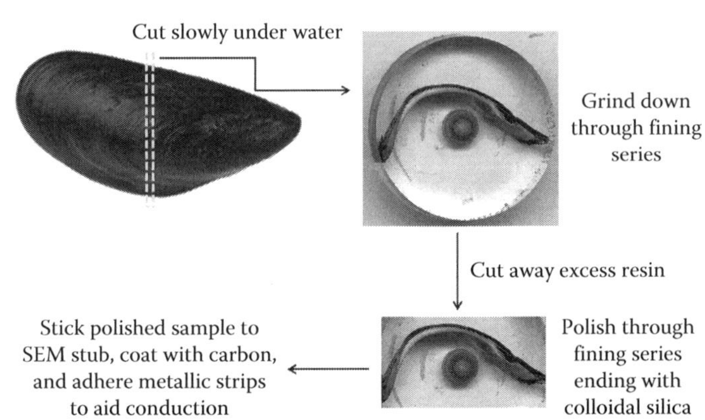

Figure 3.1 Sample preparation for EBSD. Slice cut from shell very slowly. Shell slice then embedded in resin block. Embedded cut surface taken through increasingly finer grinding stages. Excess resin cut away to maximize maneuverability in SEM chamber. The sample is then polished by hand. After cleaning in a sonicating water bath and air-drying, the sample is stuck to a SEM stub, coated with carbon and metal strips attached to aid conduction.

resin be cut away after the grinding stages and then the polishing stages be completed manually. Polishing stages then continue the pattern of increasing fineness using alpha aluminum oxide at 1 μm and then 0.3 μm and finally a 5 min polish using 0.06 μm colloidal silica. Since an entirely flat surface is required for EBSD analysis, care must be taken to ensure that the polished sample is flush with the resin surface and that no relief is introduced into the sample. Frequent checks of the sample and resin surface, using light microscopy, are required throughout the grinding and polishing stages. The polished sample is then cleaned in a sonicating water bath to remove colloidal silica, air-dried, and then coated with carbon. A comparison of the Kikuchi patterns detected with increasing thickness of carbon revealed that an even coating of 2.5 nm is ideal (Perez-Huerta and Cusack 2009). An ion beam coater such as the Precision Etching–Coating System (Model 682) from Gatan Inc. ensures that the coating is even (Perez-Huerta and Cusack 2009). The sample block is stuck to the SEM stub using silver paint (and not carbon stickers), and strips of metal tape are then applied to form a conductive pathway from the shell sample to the metallic SEM stub.

3.2.2.2 EBSD analysis

The sample is mounted in the SEM and the stage tilted to 70° before the EBSD camera is inserted into the chamber (Figure 3.2). Adhesion by carbon stickers is not sufficiently strong for samples tilted at 70°, and since any sample slippage would inevitably result in damage to the phosphor screen, adhesion of samples to stubs by silver paint is recommended. EBSD can be carried out in low and high vacuum (Perez-Huerta and Cusack 2009) although the data shown here are all from analyses at high vacuum. By way of a starting point, we suggest an accelerating voltage of 20 kV, working distance of 10 mm, aperture of 50 nm, and step size of 0.25 μm. All of these parameters need to be optimized for individual sample requirements and will depend on the sensitivity of the specific sample to the electron beam, the spatial resolution required, and the analysis time available. If a sample is very beam-sensitive, then small step size, high kV analyses will introduce too much energy and cause damage. Larger step size may suffice, but if fine spatial detail is required, then small step sizes (many spot analyses in a given area) may be necessary and, in such cases, lower kV values may be the solution (Griesshaber et al. 2012).

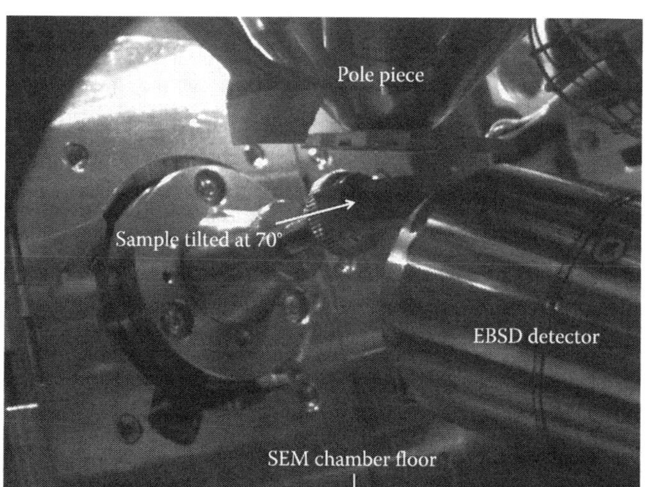

Figure 3.2 Arrangement of sample and detector for EBSD analysis. Sample within SEM chamber showing stage tilted to 70° and the position of the EBSD detector (phosphor screen).

SEM images are usually collected with the sample in a horizontal position and then again once the sample has been tilted by 70°. In horizontal samples, the working distance is constant, but tilted samples, such as the situation with EBSD, require that dynamic focus mode be employed.

With the sample tilted, the electron beam is controlled from the EBSD system. EBSD software systems include a tilt correction feature that should be applied before collecting images of tilted samples (Figure 3.2). Once the sample is tilted, the EBSD camera is then inserted in the SEM chamber.

Since EBSD requires a polished surface, the secondary electron image of the sample is generally featureless. Backscatter electron imaging may reveal compositional differences that may help with identification of specific regions of interest within a structure. A forescatter detector on the EBSD camera can also be used to identify features. The EBSD camera with forescatter detector can be withdrawn by 10 mm to receive the crystallographic signal or inserted closer to collect more compositional information from backscattered electrons. The settings on the forescatter detector are then adjusted to optimize brightness and contrast.

Spot analyses are used to establish that the sample is diffracting, to identify the mineral(s) present, and to optimize analysis conditions. Each spot analysis generates a Kikuchi pattern that can be analyzed against commercially available database files to determine the quality of the fit (Figure 3.3). The mineral phases that are likely to occur, for example, calcite and aragonite, are entered in the setup and the step size and analysis area selected. Data on diffraction intensity, mineral phase, crystallographic orientation, and grain size are collected at each step or analysis point and combine to form the EBSD map. These data can be presented as separate maps or in combination (Figures 3.4 and 3.5). EBSD software will match the data to the phases available using a voting system. It is therefore necessary to process the data to remove data with a low confidence index (CI) and to state the minimum CI value.

3.2.2.3 Combining EBSD and EDS

Most SEM systems with EBSD also have elemental analyses in the form of energy dispersive spectroscopy (EDS) if not wavelength dispersive spectroscopy (WDS). It is often helpful to carry out rapid EDS analyses to identify which elements are present, and this information is becoming more routinely integrated into EBSD systems so that, in cases where differences in chemistry can help in the selection of possible mineral phases, this approach can increase efficiency at this stage. Having both chemistry and crystallography in combined EDS/EBSD maps can be very informative. In order to combine EDS and EBSD, it is important to remember that EBSD is for small samples or small portions of samples. Excess resin should be removed to allow analyses at a working distance suitable for both EDS and EBSD, ensuring that maximum signals are detected by the x-ray detector and phosphor screen, respectively. Large samples or those with excess resin surrounding them restrict maneuverability of the samples within the chamber.

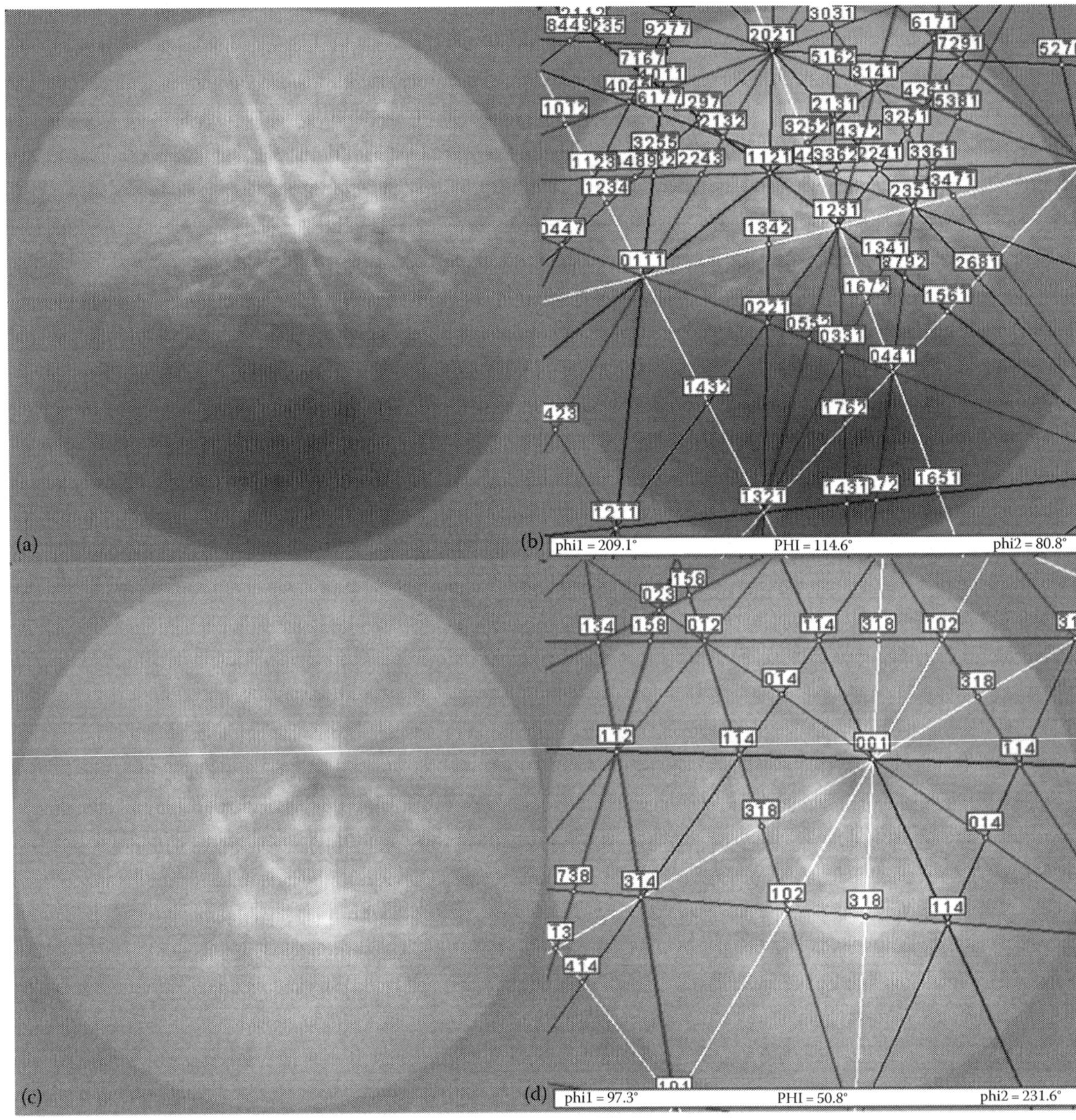

Figure 3.3 Kikuchi patterns. Kikuchi patterns generated by spot analyses of (a) calcite and (c) aragonite with diffraction planes identified for calcite (b) and aragonite (d).

3.2.2.4 EBSD data output

In many instances, spot analyses (Figure 3.3) provide the required information although, in most instances, a series of spot analyses is obtained over a specified area with a predetermined spacing or step size, and these data combine to form an EBSD map (Figure 3.4) that depicts diffraction intensity, mineralogy, and crystallographic orientation all together or as separate maps (Figure 3.5). EBSD mapping is a powerful technique in biomineral research since it provides so much information together in spatial context.

3.3 APPLICATIONS OF EBSD IN BIOMINERAL RESEARCH

3.3.1 CASE STUDY 1: EBSD INVESTIGATION OF BRACHIOPOD SHELL CRYSTALLOGRAPHY

Brachiopods were one of the first biominerals to be examined using EBSD (Schmahl et al. 2004, Griesshaber et al. 2007, Perez-Huerta et al. 2007). The phylum Brachiopoda first appeared around 550 million years ago and there are modern brachiopods living in all of the world's oceans. This evolutionary longevity and the fact that rhynchonelliform brachiopods have simple shells composed of stable low-Mg calcite (Williams et al. 1996) make them attractive models for the study of biomineralization. *Terebratulina retusa* is a modern rhynchonelliform brachiopod that has punctae (natural perforations) of complex treelike structures (Perez-Huerta et al. 2009). Both valves of the shell have an outer (primary) and inner (secondary) layer. Both layers consist of calcite and the secondary layer calcite is in the form of fibers of about 10 μm wide. EBSD analysis demonstrates that, while the fibers are roughly parallel with the shell exterior, the *c*-axis of calcite is perpendicular to the fiber axis and each fiber is effectively a single crystal (Cusack et al. 2008a), even as they bend and flex around the punctae (Perez-Huerta et al. 2009) (Figure 3.6).

Novocrania anomala is another example of a modern brachiopod, this time from a different subphylum: the Craniiformea. *N. anomala* shell comprises high-Mg calcite, for

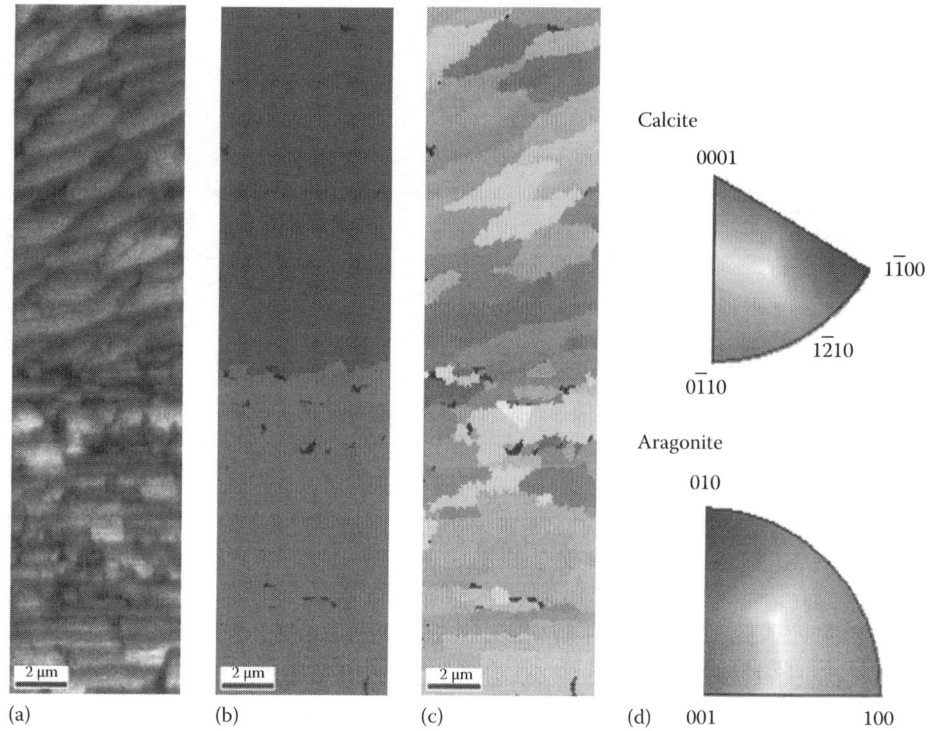

Figure 3.4 (See color insert.) Output from EBSD analyses. EBSD analysis of the interface between calcite (top) and aragonite (bottom) in the shell of the common blue mussel, *M. edulis*. Data obtained using accelerating voltage of 20 kV, working distance of 10 mm, aperture of 50 nm, and step size of 0.25 μm The main aspects of data output from EBSD analyses are (a) diffraction intensity with brighter regions indicating more diffraction, (b) phase, with each assigned a separate color, which in this case is calcite (red) and aragonite (green) and (c) crystallographic orientation with a color key (d) for calcite (top) and aragonite (bottom) indicating which crystallographic planes are normal to the view. Scale bar = 2 μm.

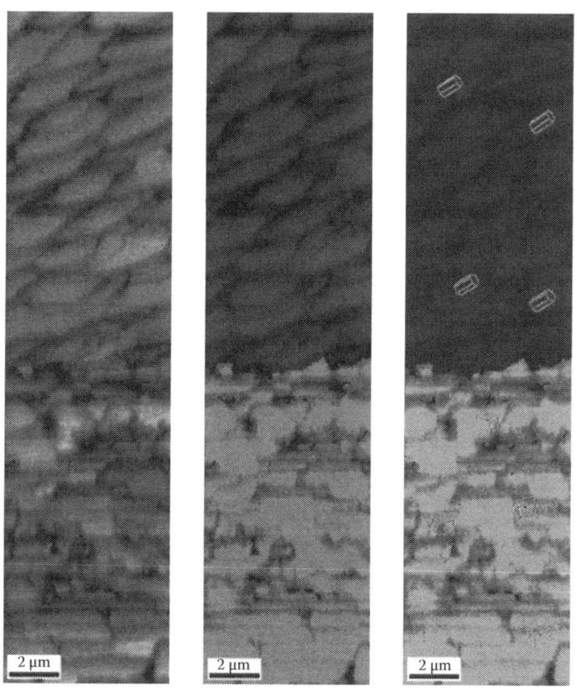

Figure 3.5 (See color insert.) Some options for EBSD data presentation. These examples depict some of the options available for presenting EBSD analyses using the data from *M. edulis* from Figure 3.4. (a) Diffraction intensity map sits behind the crystallographic orientation map. (b) Phase map from 4B with diffraction intensity behind and then in (c) with wire frames depicting crystallographic orientation. Scale bar = 2 μm.

example (Williams et al. 1996, England et al. 2007b), with a secondary layer of semi-nacre. EBSD analyses indicate that the *c*-axis of the semi-nacre is roughly parallel with the shell exterior, undulating with the semi-nacreous layers (England et al. 2007a, Checa et al. 2009b) (Figure 3.7).

While the biological control on secondary layer (fiber or semi-nacre) formation is evident from the ultrastructure of the shell alone, biological control on crystallographic orientation is exerted from the primary layer as well as the secondary layer. The overall crystallographic orientation of the primary and secondary layers are the same as demonstrated for *T. retusa* and *N. anomala* (Cusack et al. 2010) (Figure 3.8) and *Gryphus vitreus* in which the primary layer is described as an interdigitating 3D jigsaw (Goetz et al. 2011).

3.3.2 CASE STUDY 2: EBSD AS A TOOL FOR INVESTIGATING CORAL DIAGENESIS

Scleractinian corals store detailed climate information that has provided key insights into tropical climate variability and implications for global climate change, for example, Wilson et al. (2010), Duprey et al. (2012), Toth et al. (2012), Welker (2012). The climate information such as sea surface temperature (SST) is stored in proxy form, that is, the $\delta^{18}O$ or Sr/Ca ratio of the coral aragonite. Accurate interpretation of such proxy data requires that original aragonite is analyzed, sampling the influence of environmental conditions at the time of coral growth. Scleractinian corals comprise aragonite fibers radiating out from centers of calcification (COCs). Dalbeck et al. (2011) used EBSD to identify nonbiogenic calcite that mimics aspects of the original aragonite coral structure, the dissepiments (Figure 3.9). X-ray diffraction (XRD)

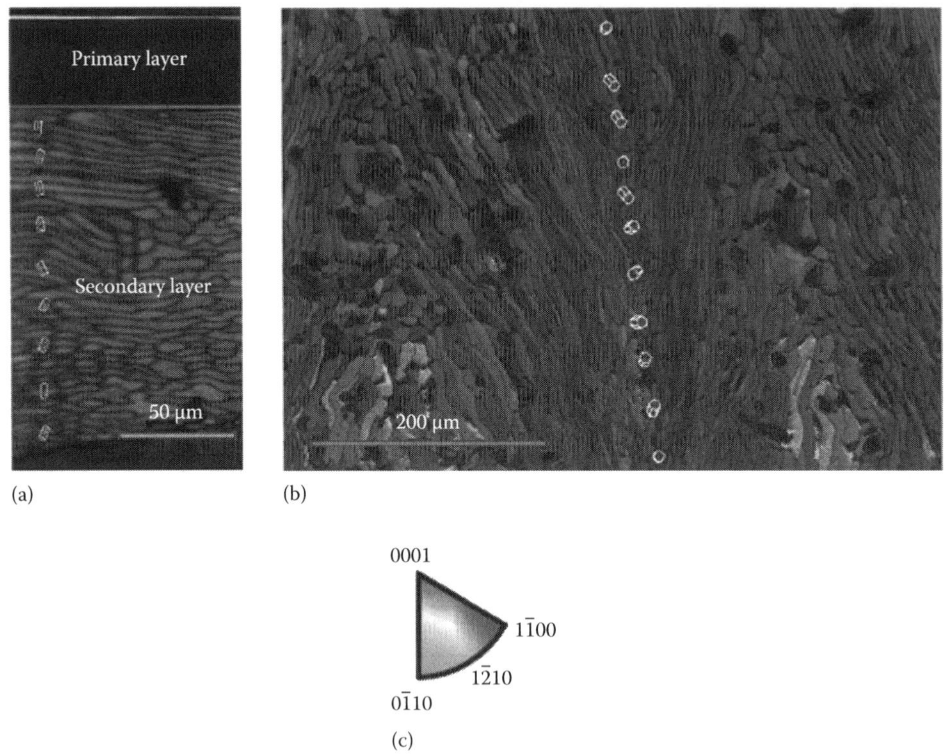

Figure 3.6 Crystallographic orientation map of calcite fibers of the shell of the brachiopod, *T. retusa*. (a) Crystallographic orientation map of fibers viewed *side on*, that is, shell exterior to top. Scale bar = 50 μm. (b) Crystallographic orientation map at 90° to (a), that is, looking down on shell with the primary layer removed. Scale bar = 200 μm. (c) Color key used in (a) and (b). (Reprinted from *J. Struct. Biol.*, 164, Cusack, M., Dauphin, Y., Chung, P., Perez-Huerta, A., and Cuif, J.P., Multiscale structure of calcite fibres of the shell of the brachiopod *Terebratulina retusa*, 96–100, Copyright 2008, with permission from Elsevier.)

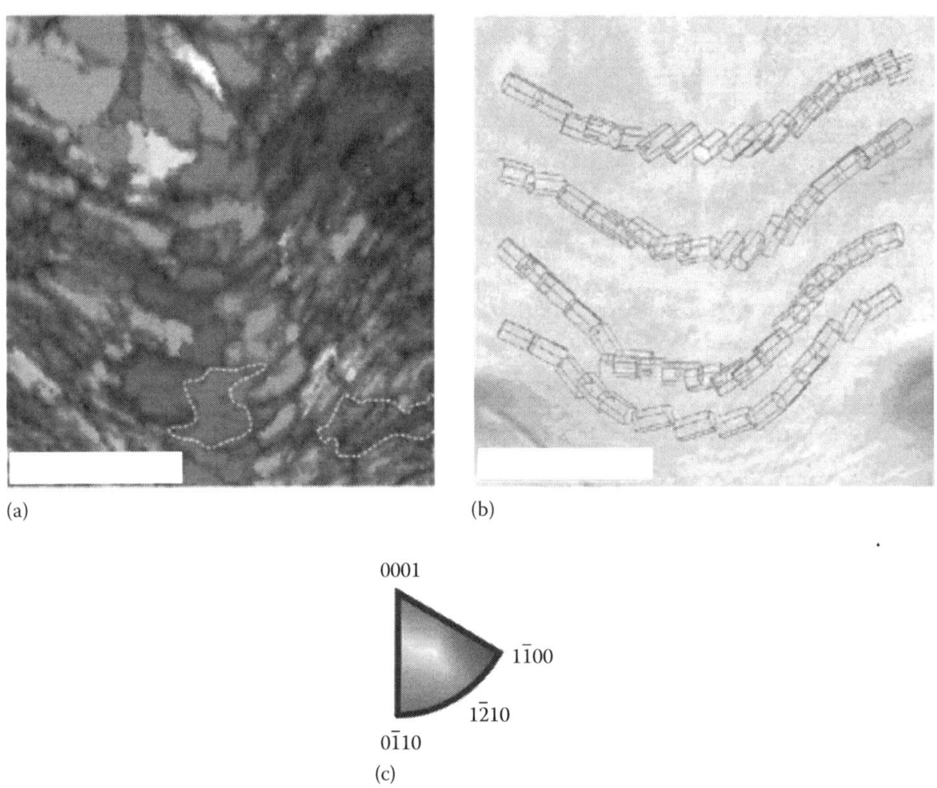

Figure 3.7 (See color insert.) EBSD mapping of calcite semi-nacre of the shell of the brachiopod, *N. anomala*. (a) Crystallographic orientation map of the secondary layer of the shell of *Novocrania huttoni*. Scale bar = 15 μm. (b) Wire frame unit cells superimposed on SEM image of *N. huttoni* laminae indicating that the *c*-axis of calcite is coincident with the undulations of the laminae and more or less parallel with the shell exterior. Scale bar = 10 μm. (c) Color key used in (a). (Reprinted with permission from England, J., Cusack, M., Dalbeck, P., and Perez-Huerta, A., Comparison of the crystallographic structure of semi nacre and nacre by electron backscatter diffraction, *Cryst. Growth Des.*, 7, 307–310. Copyright 2007 American Chemical Society.)

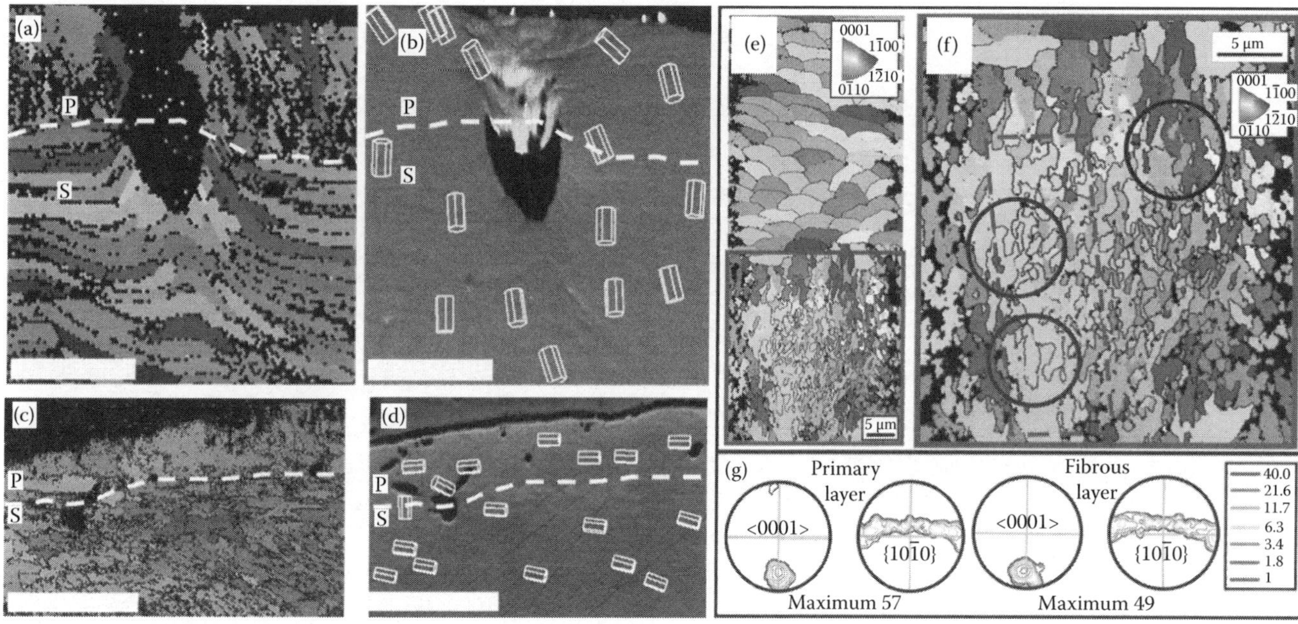

Figure 3.8 (See color insert.) EBSD analyses of primary (outer) layer of calcite-shelled brachiopods. (a) Crystallographic orientation of primary (P) and secondary (S) layers of (a) *Terebratalia transversa* and (c) *N. anomala* with corresponding secondary electron images in (b) and (d), respectively, with wire frames indicating crystallographic orientation. Scale bars = 50 μm, 50 μm, 100 μm, and 100 μm for (a–d), respectively. (e) Crystallographic orientation map of *Gryphus vitreus*, with area of primary layer having interdigitating grains highlighted in the red box and enlarged in (f). (g) Pole figures of the primary and secondary layers of *G. vitreus*. (a–d: From Cusack, M., Chung, P., Dauphin, Y., and Perez-Huerta, A.: Brachiopod primary layer crystallography and nanostructure. *Evolution and Development of the Brachiopod Shell*. Special Papers in Palaeontology, 2010, 84, 99–105. Copyright Wiley-VCH Verlag GmbH & Co. KGaA. Reproduced with permission from the Palaeontological Association; e–g: From Goetz, A.J., Steinmetz, D.R., Griesshaber, E., Zaefferer, S., Raabe, D., Kelm, K., Irsen, S., Sehrbrock, A., and Schmahl, W.W.: Interdigitating biocalcite dendrites form a 3-D jigsaw structure in brachiopod shells. *Acta Biomater.* 2011. 7. 2237–2243. Copyright Wiley-VCH Verlag GmbH & Co. KGaA. Reproduced with permission.)

can detect such secondary mineralization that is different from the original, that is, significant quantities of calcite would be detected in an originally aragonite skeleton. However, the quantity of calcite present here is below the detection limit of XRD (Cusack et al. 2008b). Inadvertent inclusion of as little as 1% of the nonbiogenic calcite within the original aragonite could result in calculated SST elevated by 1.2°C. EBSD was highly effective in identifying the mineralogy and crystallographic orientation of the original aragonite and replacement calcite. Atomic force microscopy (AFM) indicated that while the original coral aragonite was composed of nanogranules as in the mesocrystal model (Cölfen and Mann 2003, Cölfen and Antonietti 2005), the nonbiogenic nature of the replacement calcite is confirmed by the lack of nanocrystals since the pseudomorphic calcite is formed by nonbiogenic processes (Dalbeck et al. 2011).

The model proposed to explain the formation of the pseudomorphic calcite dissepiments is depicted in Figure 3.10.

The secondary calcite nucleates on the original aragonite dissepiments. Through capillary action, the thin, horizontal dissepiments hold pore fluids in a meniscus on either side of what is, in effect, a horizontal shelf. The original aragonite dissepiment can then dissolve with recrystallization of a pseudomorphic calcite structure. EBSD is therefore not simply a tool to identify diagenesis but can be used with other techniques to help us understand the processes of diagenesis.

While XRD can detect secondary mineralization that is different from the original and present in sufficient quantities, XRD cannot distinguish between original and secondary mineralization if the mineralogy is the same. Cusack et al. (2008b) demonstrated that EBSD can be used to distinguish between secondary aragonite and original primary aragonite in Porites coral (Figure 3.11). Figure 3.11a presents the featureless, polished section of the coral with the diffraction intensity depicted in Figure 3.11b in which a circular darker region is apparent, that is, an area of lower diffraction intensity than the rest of the coral. The phase map in Figure 3.11c indicates that only aragonite is present, while the map of crystallographic orientation (Figure 3.11d with color key in Figure 3.11e) clearly shows that the crystallographic orientation of aragonite in this circular region is completely different from the original aragonite fibers and is therefore identified as secondary aragonite.

In this instance, a microboring, most likely of fungal hyphae, possibly occurring while the coral was alive, has resulted in reprecipitation of aragonite that is in a different crystallographic orientation from that of the original coral aragonite. Secondary aragonite is very likely to have a different $\delta^{18}O$ and Sr/Ca ratio than the original, primary aragonite and, if a sizeable portion of the aragonite present, would therefore distort proxies for SST.

3.4 OUTLOOK

3.4.1 WHAT ARE THE FRONTIERS?

With careful sample preparation, EBSD can continue to expand as a research tool in biomineralization. While biogenic calcite tends to diffract more readily than aragonite, there is a growing number of EBSD applications in calcium carbonate biominerals

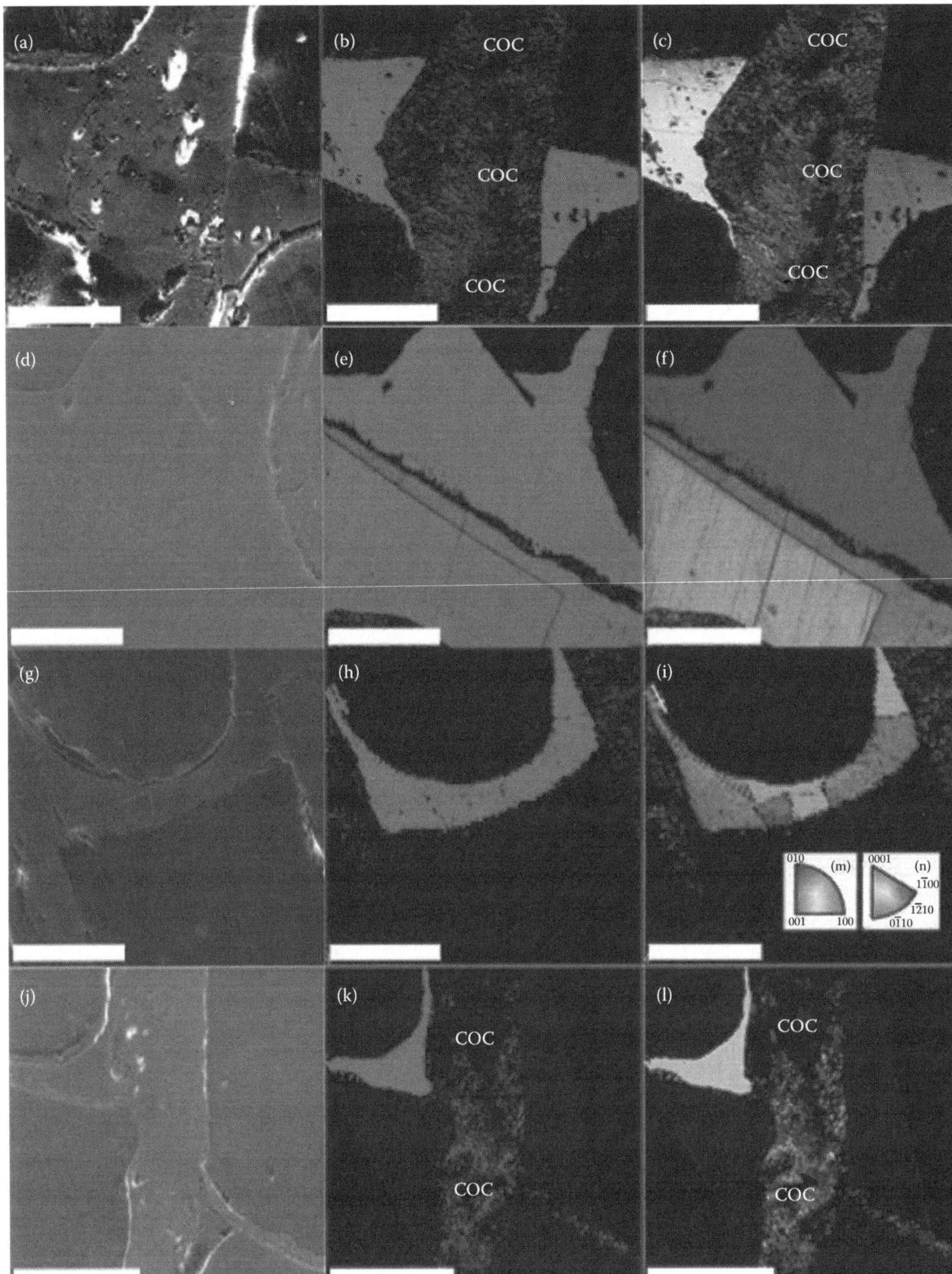

Figure 3.9 (See color insert.) EBSD analysis of fossil coral revealing relatively large crystals of calcite resembling horizontal structures in pristine corals. Secondary electron images (a, d, g, j) of the areas analyzed by EBSD. Phase maps of these areas (b, e, h, k) where aragonite is indicated in red and calcite in green. Crystallographic orientation maps (c, f, i, l) of these same regions with crystallographic orientation indicated by color key (m and n) insets in (i) for aragonite and calcite, respectively. Scale bars for a–c = 50 μm, d–f = 25 μm, g–i = 50 μm, and j–l = 80 μm. (Reprinted from *Chem. Geol.*, 280, Dalbeck, P., Cusack, M., Dobson, P.S., Allison, N., Fallick, A.E., Tudhope, A.W., and EIMF, Identification and composition of secondary meniscus calcite in fossil coral and the effect on predicted sea surface temperature, 314–322, Copyright 2011, with permission from Elsevier.)

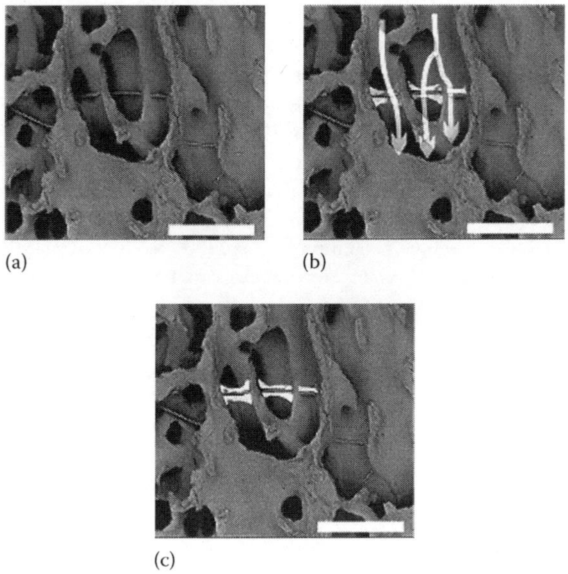

Figure 3.10 Diagram representing the formation of *meniscus* secondary calcite within coral. (a) Primary, unaltered sample showing slender, horizontal aragonite dissepiments. (b) Arrows indicate pore water running through the sample, with some water being held on imperfections by capillary action and surface tension. (c) Calcite (white) is deposited on the aragonite dissepiments with the new crystals assuming the form of the water meniscus. Scale bar = 400 μm. (Reprinted from *Chem. Geol.*, 280, Dalbeck, P., Cusack, M., Dobson, P.S., Allison, N., Fallick, A.E., Tudhope, A.W., and EIMF, Identification and composition of secondary meniscus calcite in fossil coral and the effect on predicted sea surface temperature, 314–322, Copyright 2011, with permission from Elsevier.)

with examples in the different polymorphs including calcite (Checa et al. 2009a), aragonite (Checa and Harper 2010), and vaterite (Frenzel et al. 2012). In general, calcium carbonate biominerals tend to lend themselves to EBSD analyses. Biominerals with a high mineral content and larger crystals tend to diffract well, while samples with higher concentrations of organic components and more finely grained mineral tend to be more challenging. This explains why phosphatic biominerals such as bone (Wang et al. 2006) or phosphatic-shelled brachiopods (Williams et al. 1994, Merkel et al. 2007) are a challenging set of materials for EBSD.

3.4.2 FUTURE POSSIBILITIES

3.4.2.1 3D EBSD

The possibility of obtaining crystallographic data in 3D would be very informative especially when considering key interfaces such as those between different shell layers or polymorphs. Current approaches include carrying out EBSD on one surface and then using a focused ion beam (FIB) to slice through the sample perpendicular to that section and then carry out EBSD analyses on the perpendicular surface. Repeated cycles of FIB and EBSD can be used to construct crystallographic orientation data in 3D. This approach is being developed for metals and alloys (Ferry et al. 2012a,b, Gholinia et al. 2012) where the approach is possible but not without challenges. A system that streamlines this approach and enables it to become more routine and suitable for insulators would make EBSD a very powerful tool in biomineral research.

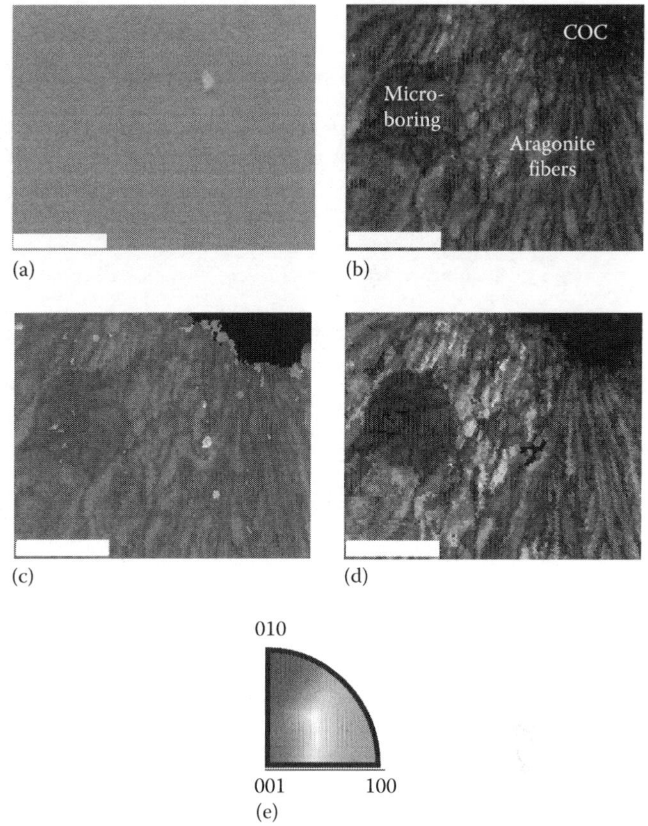

Figure 3.11 (See color insert.) EBSD analysis of Porites coral revealing secondary aragonite within original primary aragonite. Images (a–d) depict the same area within the coral. Scale bar = 10 μm. (a) Secondary electron image of the coral region that is analyzed by EBSD in (c–d). (b) Diffraction intensity map. Dark area at top right indicates the poor diffraction of the COC. (c) Combined diffraction intensity and phase map with green indicating aragonite. (d) Map of combined diffraction intensity and crystallographic orientation of the same area (obtained simultaneously), with mainly blue and green color coding with reference to the key (e), indicating that the {010} and {100} planes of aragonite are normal to the plane of view indicating that the {001} is concurrent with the fiber axis. Secondary aragonite in what is likely to be a microboring has a different crystallographic orientation to the original aragonite fibers with the 001 plane normal to the plane of view, which is why the secondary aragonite appears red in the image. (Reprinted from *Coral Reefs*, 27, Cusack, M., England, J., Dalbeck, P., Tudhope, A.W., Fallick, A.E., and Allison, N., Electron backscatter diffraction (EBSD) as a tool for detection of coral diagenesis, 905–911, Copyright 2008, with permission from Elsevier.)

REFERENCES

Alam, M.N., M. Blackman, and D.W. Pashley. 1954. High-angle Kikuchi patterns. *Proceedings of the Royal Society of London Series A—Mathematical and Physical Sciences* 221:224–242.

Babakishi, K.Z. and D.J. Dingley. 1989. Application of backscatter Kikuchi diffraction in the scanning electron microscope to the study of NiS_2. *Journal of Applied Crystallography* 22:189–200.

Balthasar, U., M. Cusack, L. Faryma, P. Chung, L.E. Holmer, J. Jin, I.G. Percival, and L.E. Popov. 2011. Relic aragonite from Ordovician–Silurian brachiopods: Implications for the evolution of calcification. *Geology* 39:967–970.

Checa, A.G., F.J. Esteban-Delgado, J. Ramirez-Rico, and A.B. Rodriguez-Navarro. 2009a. Crystallographic reorganization of the calcitic prismatic layer of oysters. *Journal of Structural Biology* 167:261–270.

Checa, A.G., D. Gaspard, A. Gonzalez-Segura, and J. Ramirez-Rico. 2009b. Crystallography of the calcitic foliated-like and seminacre microstructures of the brachiopod novocrania. *Crystal Growth & Design* 9:2464–2469.

Checa, A.G. and E.M. Harper. 2010. Spikey bivalves: Intra-periostracal crystal growth in anomalodesmatans. *Biological Bulletin* 219:231–248.

Cölfen, H. and M. Antonietti. 2005. Mesocrystals: Inorganic superstructures made by highly parallel crystallization and controlled alignment. *Angewandte Chemie International Edition* 44:5576–5591.

Cölfen, H. and S. Mann. 2003. Higher-order organization by mesoscale self-assembly and transformation of hybrid nanostructures. *Angewandte Chemie International Edition* 42:2350–2365.

Cusack, M., P. Chung, Y. Dauphin, and A. Perez-Huerta. 2010. Brachiopod primary layer crystallography and nanostructure. *Evolution and Development of the Brachiopod Shell. Special Papers in Palaeontology* 84:99–105.

Cusack, M., Y. Dauphin, P. Chung, A. Perez-Huerta, and J.P. Cuif. 2008a. Multiscale structure of calcite fibres of the shell of the brachiopod *Terebratulina retusa*. *Journal of Structural Biology* 164:96–100.

Cusack, M., J. England, P. Dalbeck, A.W. Tudhope, A.E. Fallick, and N. Allison. 2008b. Electron backscatter diffraction (EBSD) as a tool for detection of coral diagenesis. *Coral Reefs* 27:905–911.

Dalbeck, P. and M. Cusack. 2006. Crystallography (electron backscatter diffraction) and chemistry (electron probe microanalysis) of the avian eggshell. *Crystal Growth & Design* 6:2558–2562.

Dalbeck, P., M. Cusack, P.S. Dobson, N. Allison, A.E. Fallick, A.W. Tudhope, and EIMF. 2011. Identification and composition of secondary meniscus calcite in fossil coral and the effect on predicted sea surface temperature. *Chemical Geology* 280:314–322.

Dalbeck, P., J. England, M. Cusack, M.R. Lee, and A.E. Fallick. 2006. Crystallography and chemistry of the calcium carbonate polymorph switch in *M. edulis* shells. *European Journal of Mineralogy* 18:601–609.

Duprey, N., C.E. Lazareth, T. Correge, F. Le Cornec, C. Maes, N. Pujol, M. Madeng-Yogo, S. Caquineau, C.S. Derome, and G. Cabioch. 2012. Early mid-Holocene SST variability and surface-ocean water balance in the southwest Pacific. *Paleoceanography* 27:PA4207.

Ebner, M., S. Piazolo, F. Renard, and D. Koehn. 2010. Stylolite interfaces and surrounding matrix material: Nature and role of heterogeneities in roughness and microstructural development. *Journal of Structural Geology* 32:1070–1084.

England, J., M. Cusack, P. Dalbeck, and A. Perez-Huerta. 2007a. Comparison of the crystallographic structure of semi nacre and nacre by electron backscatter diffraction. *Crystal Growth & Design* 7:307–310.

England, J., M. Cusack, and M.R. Lee. 2007b. Magnesium and sulphur in the calcite shells of two brachiopods, *Terebratulina retusa* and *Novocrania anomala*. *Lethaia* 40:2–10.

Farooq, M.U., R. Villaurrutia, I. MacLaren, H. Kungl, M.J. Hoffmann, J.J. Fundenberger, and E. Bouzy. 2008. Using EBSD and TEM-Kikuchi patterns to study local crystallography at the domain boundaries of lead zirconate titanate. *Journal of Microscopy* 230:445–454.

Ferry, M., M.Z. Quadir, N.A. Zinnia, L. Bassman, C. George, C. McMahon, W.Q. Xu, and K. Laws. 2012a. The application of 3D-EBSD for investigating texture development in metals and alloys. In *Textures of Materials, Pts 1 and 2*, eds. A. Tewari, S. Suwas, D. Srivastava, I. Samajdar, and A. Haldar, pp. 469–474. Stafa-Zurich, Switzerland: Trans Tech Publications Ltd.

Ferry, M., W.Q. Xu, M.Z. Quadir, N.A. Zinnia, K. Laws, N. Mateescu, L. Robin et al. 2012b. 3D-EBSD studies of deformation, recrystallization and phase transformations. In *Recrystallization and Grain Growth IV*, eds. E. J. Palmiere and B.P. Wynne, pp. 41–50. Durnten-Zurich, Switzerland: Trans Tech Publications Ltd.

Frenzel, M., R.J. Harrison, and E.M. Harper. 2012. Nanostructure and crystallography of aberrant columnar vaterite in *Corbicula fluminea* (Mollusca). *Journal of Structural Biology* 178:8–18.

Gholinia, A., I. Brough, J. Humphreys, and P. Bate. 2012. A 3D FIB investigation of dynamic recrystallization in a Cu-Sn bronze. In *Recrystallization and Grain Growth IV*, eds. E.J. Palmiere and B.P. Wynne, pp. 498–501. Durnten-Zurich, Switzerland: Trans Tech Publications Ltd.

Glage, A., S. Martin, S. Decker, C. Weigelt, M. Junghanns, C.G. Aneziris, U. Martin, L. Kruger, and H. Biermann. 2012. Cyclic deformation of powder metallurgy stainless steel/Mg-PSZ composite materials. *Steel Research International* 83:554–564.

Goetz, A.J., D.R. Steinmetz, E. Griesshaber, S. Zaefferer, D. Raabe, K. Kelm, S. Irsen, A. Sehrbrock, and W.W. Schmahl. 2011. Interdigitating biocalcite dendrites form a 3-D jigsaw structure in brachiopod shells. *Acta Biomaterialia* 7:2237–2243.

Griesshaber, E., W.W. Schmahl, R. Neuser, T. Pettke, M. Blum, J. Mutterlose, and U. Brand. 2007. Crystallographic texture and microstructure of terebratulide brachiopod shell calcite: An optimized materials design with hierarchical architecture. *American Mineralogist* 92:722–734.

Griesshaber, E., H.S. Ubhi, and W.W. Schmahl. 2012. Nanometer scale microstructure and microtexture of biological materials revealed by high spatial resolution (15 to 5 kV) EBSD. In *Textures of Materials, Pts 1 and 2*, eds. A. Tewari, S. Suwas, D. Srivastava, I. Samajdar, and A. Haldar, pp. 924–927. Stafa-Zurich, Switzerland: Trans Tech Publications Ltd.

Inoue, S. and T. Kogure. 2012. Electron backscatter diffraction (EBSD) analyses of phyllosilicates in petrographic thin sections. *American Mineralogist* 97:755–758.

Kumar, A. and T.M. Pollock. 2011. Mapping of femtosecond laser-induced collateral damage by electron backscatter diffraction. *Journal of Applied Physics* 110:083114.

Lee, M.R., M.E. Hodson, and G.N. Langworthy. 2008. Crystallization of calcite from amorphous calcium carbonate: Earthworms show the way. *Mineralogical Magazine* 72:257–261.

Lee, M.R. and K. Nicholson. 2009. Ca-carbonate in the Orgueil (CI) carbonaceous chondrite: Mineralogy, microstructure and implications for parent body history. *Earth and Planetary Science Letters* 280:268–275.

Lee, M.R., C. Torney, and A.W. Owen. 2012. Biomineralisation in the Palaeozoic oceans: Evidence for simultaneous crystallisation of high and low magnesium calcite by phacopine trilobites. *Chemical Geology* 314:33–44.

Lindgren, P., M.R. Lee, M. Sofe, and M.J. Burchell. 2011. Microstructure of calcite in the CM2 carbonaceous chondrite LON 94101: Implications for deformation history during and/or after aqueous alteration. *Earth and Planetary Science Letters* 306:289–298.

MacDonald, J., A. Freer, and M. Cusack. 2010. Attachment of oysters to natural substrata by biologically induced marine carbonate cement. *Marine Biology* 157:2087–2095.

Merkel, C., E. Griesshaber, K. Kelm, R. Neuser, G. Jordan, A. Logan, W. Mader, and W.W. Schmahl. 2007. Micromechanical properties and structural characterization of modern inarticulated brachiopod shells. *Journal of Geophysical Research—Biogeosciences* 112: G02008.

Moureaux, C., A. Perez-Huerta, P. Compere, W. Zhu, T. Leloup, M. Cusack, and P. Dubois. 2010. Structure, composition and mechanical relations to function in sea urchin spine. *Journal of Structural Biology* 170:41–49.

Moy, C.K.S., M. Weiss, J.H. Xia, G. Sha, S.P. Ringer, and G. Ranzi. 2012. Influence of heat treatment on the microstructure, texture and formability of 2024 aluminium alloy. *Materials Science and Engineering A—Structural Materials Properties Microstructure and Processing* 552:48–60.

Nishikawa, S. and S. Kikuchi. 1928a. Diffraction of cathode rays by calcite. *Nature* 122:726–726.

Nishikawa, S. and S. Kikuchi. 1928b. Diffraction of cathode rays by mica. *Nature* 121:1019–1020.

Okudaira, T. and N. Shigematsu. 2012. Estimates of stress and strain rate in mylonites based on the boundary between the fields of grain-size sensitive and insensitive creep. *Journal of Geophysical Research—Solid Earth* 117:B03210.

Perez-Huerta, A. and M. Cusack. 2009. Optimizing electron backscatter diffraction of carbonate biominerals-resin type and carbon coating. *Microscopy and Microanalysis* 15:197–203.

Perez-Huerta, A., M. Cusack, and J. England. 2007. Crystallography and diagenesis in fossil craniid brachiopods. *Palaeontology* 50:757–763.

Perez-Huerta, A., M. Cusack, S. McDonald, F. Marone, M. Stampanoni, and S. MacKay. 2009. Brachiopod punctae: A complexity in shell biomineralisation. *Journal of Structural Biology* 167:62–67.

Perez-Huerta, A., M. Cusack, and C.A. Mendez. 2012. Preliminary assessment of the use of electron backscatter diffraction (EBSD) in conodonts. *Lethaia* 45:253–258.

Perez-Huerta, A., Y. Dauphin, J.P. Cuif, and M. Cusack. 2011. High resolution electron backscatter diffraction (EBSD) data from calcite biominerals in recent gastropod shells. *Micron* 42:246–251.

Poole, I. and G.E. Lloyd. 2000. Alternative SEM techniques for observing pyritised fossil material. *Review of Palaeobotany and Palynology* 112:287–295.

Saruwatari, K., N. Ozaki, H. Nagasawa, and T. Kogure. 2006. Crystallographic alignments in a coccolith (*Pleurochrysis carterae*) revealed by electron back-scattered diffraction (EBSD). *American Mineralogist* 91:1937–1940.

Saruwatari, K., Y. Tanaka, H. Nagasawa, and T. Kogure. 2011. Crystallographic variability and uniformity in Cretaceous heterococcoliths. *European Journal of Mineralogy* 23:519–528.

Schmahl, W.W., E. Griesshaber, R. Neuser, A. Lenze, R. Job, and U. Brand. 2004. The microstructure of the fibrous layer of terebratulide brachiopod shell calcite. *European Journal of Mineralogy* 16:693–697.

Schwartz, A.J. 2009. *Electron Backscatter Diffraction in Materials Science.* Boston, MA: Springer Science & Business Media.

Thomas, G. 1965. Kikuch electron diffraction and dark-field techniques. *Journal of Applied Physics* 36:2610.

Thomson, G.P. 1928. Experiments on the diffraction of cathode rays. *Proceedings of the Royal Society of London Series A—Containing Papers of a Mathematical and Physical Character* 117:600–609.

Toth, L.T., R.B. Aronson, S.V. Vollmer, J.W. Hobbs, D.H. Urrego, H. Cheng, I.C. Enochs, D.J. Combosch, R. van Woesik, and I.G. Macintyre. 2012. ENSO drove 2500-year collapse of eastern Pacific coral reefs. *Science* 337:81–84.

Venables, J.A. and R. Binjaya. 1977. Accurate micro-crystallography using electron backscattering patterns. *Philosophical Magazine* 35:1317–1332.

Venables, J.A. and C.J. Harland. 1973. Electron backscattering patterns—New technique for obtaining crystallographic information in scanning electron-microscope. *Philosophical Magazine* 27:1193–1200.

Vielzeuf, D., N. Floquet, D. Chatain, F. Bonnete, D. Ferry, J. Garrabou, and E.M. Stolper. 2010. Multilevel modular mesocrystalline organization in red coral. *American Mineralogist* 95:242–248.

Wang, L.J., G.H. Nancollas, Z.J. Henneman, E. Klein, and S. Weiner. 2006. Nanosized particles in bone and dissolution insensitivity of bone mineral. *Biointerphases* 1:106–111.

Welker, J.M. 2012. ENSO effects on delta O-18, delta H-2 and d-excess values in precipitation across the U.S. using a high-density, long-term network (USNIP). *Rapid Communications in Mass Spectrometry* 26:1893–1898.

Williams, A., S.J. Carlson, C.H.C. Brunton, L.E. Holmer, and L. Popov. 1996. A supra-ordinal classification of the Brachiopoda. *Philosophical Transactions of the Royal Society of London, B.* 351:1171–1193.

Williams, A., M. Cusack, and S. Mackay. 1994. Collagenous chitinophosphatic shell of the brachiopod *Lingula. Philosophical Transactions of the Royal Society of London, B* 346:223–266.

Wilson, R., E. Cook, R. D'Arrigo, N. Riedwyl, M.N. Evans, A. Tudhope, and R. Allan. 2010. Reconstructing ENSO: The influence of method, proxy data, climate forcing and teleconnections. *Journal of Quaternary Science* 25:62–78.

Zou, H.F. and Z.F. Zhang. 2010. Application of electron backscatter diffraction to the study on orientation distribution of intermetallic compounds at heterogeneous interfaces (Sn/Ag and Sn/Cu). *Journal of Applied Physics* 108:103518.

Part II

Characterization of atomic and molecular structure: Spectroscopy and spectromicroscopy

4 Infrared spectroscopy and imaging

Adele L. Boskey

Contents

4.1 INTRODUCTION

Infrared (IR) spectroscopy is a vibrational spectroscopy technique [1] for chemical composition or structure determination that was first applied to the identification of calcium phosphate mineral phases in bones [2] and teeth [3] more than 50 years ago. It still remains an advantageous technique when small amounts of sample are available or when samples are not homogeneous. In the IR experiment, light interacts with matter; the photons that interact with a sample are either absorbed or scattered; photons of specific energy are absorbed and the absorption pattern provides information, or a fingerprint or characteristic signature, on the molecules that are present in the sample. The IR radiation wavelength 0.78–1000 μm impinges on the molecules in the sample being analyzed; exciting *allowed* molecular vibrations that absorb light when the frequency of the vibration is the same as the IR *light* frequency. The frequency at which any given vibration occurs is determined by the strengths of the chemical bonds involved and the mass of the component atoms. The spectra are obtained over a range of IR frequencies, revealing the vibrations in the component molecules. Vibrations that result in changes in dipole moments are the strongest, and the wavelength (given in wavenumbers or reciprocal wavelength) is dependent on the environment in which the vibrating molecule is situated. IR analyses can be performed in the gas, liquid, and solid phases, providing fingerprint spectra. IR spectra, reported in three regions, the near IR (1–2.5 μm or 10,000–4,000 cm^{-1}), the mid-IR (2.5–50 μm or 4,000–200 cm^{-1}), and the far IR (50–1,000 μm or 200–10 cm^{-1}), can be used to provide information about the nature of the chemical bonds in pure compounds or mixtures. Each isolated compound subjected to IR analysis provides *fingerprints* that can be used for component identification (Figure 4.1). In mixtures, peak intensities or peak areas can be used to approximate concentrations. Shifts in band position can be interpreted in terms of changes in the environment around a vibrating ion or in terms of changes in conformation of the molecule. Most vibrations of biominerals are studied in the mid-IR, which will be the focus of the following discussion.

In terms of biominerals, IR spectroscopy has been applied to characterize calcium phosphates and carbonate substitution in enamel [4], the bacterial layer coating teeth [5], and pathologic calcium phosphate deposits [6,7]. It has also been used to identify calcium carbonates [8–11] and other mineral phases (e.g., oxalates in plants [12], calcium oxalates in the otoconia of the inner ear [13]), which have distinct fingerprint patterns enabling identification of individual phases. In the early studies, the contribution of the organic phase to the biomineral was mainly ignored, but more recently, as spectra–structure correlations have evolved, the organic matrix of biominerals has also been characterized by IR. More recently, FTIR analyses of eggshell membranes made permeable to allow entry of substrates for silicification or calcium phosphate formation have provided important insight into mineralization mechanisms, but the FTIR components of these studies were not spatially resolved [14].

The paradigm shifts in IR spectroscopy came with the coupling of a light microscope with an IR spectrometer early in the 1980s (FTIR microspectroscopy [FTIRM]) and the addition of a focal plane array (FPA) detector later in the 1980s (FTIR imaging [FTIRI] microscopy), but commercial units were not readily available until 10 years later. These instruments made it possible to collect spectra at specific locations in thin specimens, allowing *specimen photographs* that portrayed spectral features or calculated parameters at specific *x*, *y* locations. The imaging instrumentation (Figure 4.2) produces multidimensional (hyperspectral) chemically specific images while simultaneously

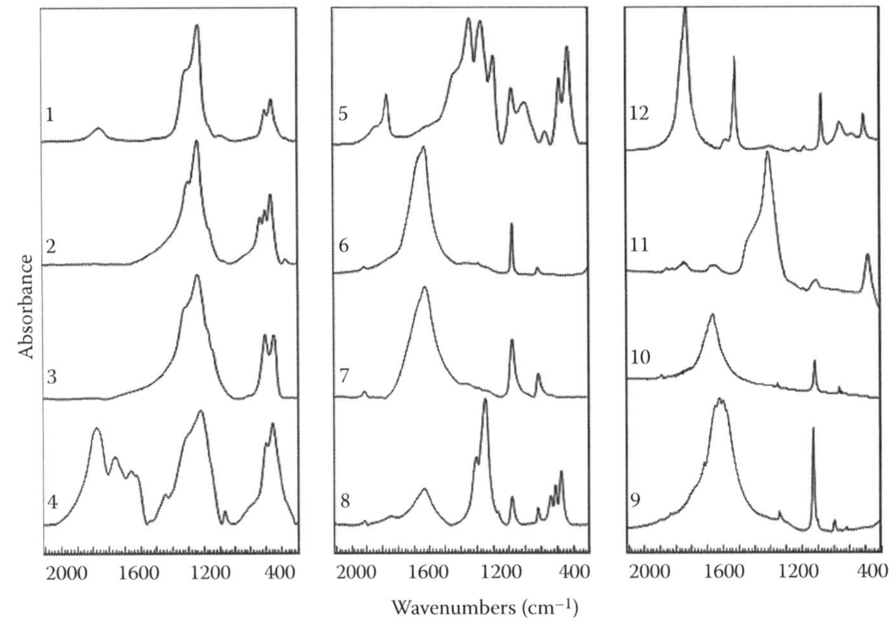

Figure 4.1 IR fingerprint analysis in the mid-IR range (400–2000 cm^{-1}) can be used to identify the composition of various minerals or mixtures thereof. Spectrum of (1) poorly crystalline HA, (2) highly crystalline HA, (3) β-tricalcium phosphate (TCP), (4) mouse bone, (5) brushite ($CaHPO_4$)·H_2O, (6) calcite ($CaCO_3$), (7) seashell (calcite), (8) calcite + HA mixture, (9) vaterite ($CaCO_3$), (10) aragonite ($CaCO_3$), (11) diatomite (silica), and (12) whewellite (calcium oxalate monohydrate). Spectra 1–5 show different forms of calcium phosphate; spectra 6, 7, 9, and 10 show different forms of calcium carbonates; spectra 11–12 shows other biominerals; and spectra 8 shows how a mixture of two compounds can be identified. (Images 9–12 were acquired from Dr. Steve Weiner's IR spectra library, available online from http://www.weizmann.ac.il/kimmel-arch/InfraredStandards.zip.)

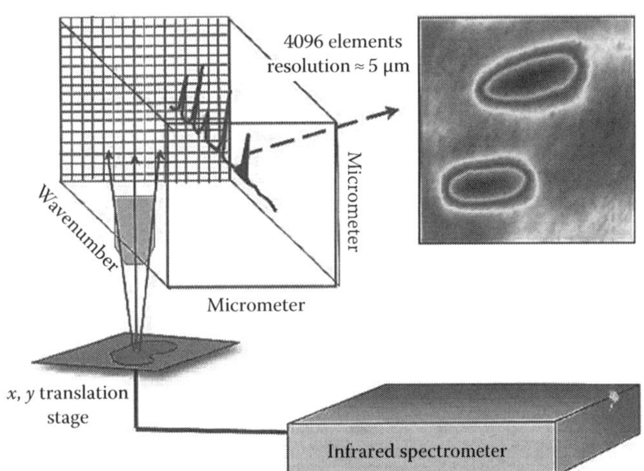

Figure 4.2 How IR imaging is performed. Schematic diagram of IR imaging instrumentation showing a typical image for osteonal bone, where the centers of the oval images represent a blood vessel and the surrounding bone mineral and matrix. (Courtesy of Dr. A. Gericke, Worcester Polytech, Worcester, MA.) The IR beam is directed through a microscope onto the focal plane array (FPA) detector, which in this illustration consists of 4096 detector elements. A spectral hypercube is obtained with each detector element representing an entire IR spectrum. Integration or ratioing of specific bands yields a 250 × 250 μm colored image (adjacent images can be stitched together to obtain larger images). Alternatively, statistical methods such as principal component or cluster analysis can be employed to obtain the image. The tissue sample is cut to about 2 μm thickness and is placed on a BaF_2 window that sits on the stage of the instrument.

obtaining high-resolution spectra for each detector pixel. The spatial resolution of the images approaches the diffraction limit for mid-IR wavelengths, while the actual spectral resolution is determined by the interferometer associated with the instrument, and generally is 2–8 cm^{-1} or greater. With synchrotron radiation, which is highly polarized and has a stronger source, the resolution improves to 1–2 μm [15].

The synchrotron provides higher overall illumination and a higher signal-to-noise ratio than does the thermal IR light source *globar* used in standard FTIRI instruments. As reviewed elsewhere, synchrotron FTIRI had been used for analysis of cancer cells and other pathologic cell changes [16,17] when our research group started to look at FTIRI to learn about the site-specific variations in bone composition. Because the detector on the synchrotron enabled us to analyze regions of the spectra not accessible to standard instruments, we were able to use synchrotron radiation to characterize bone acid phosphate content based on the 630–650 cm^{-1} region not accessible in the FTIRI instrument [18]. The limitation to the use of synchrotron FTIRI is that while most synchrotron sources have an FTIRI instrument, beam time is limited, so the advantage of the gain in resolution is offset by disadvantage of the delay in analysis time.

4.2 OVERVIEW OF THE EXPERIMENT

Two fundamental concepts are important for understanding what is needed to perform the FTIRM or FTIRI experiment that allows investigation of the spatial localization of chemical composition and compositional changes. This localization makes the study distinct from just grinding up a shell or a bone and doing an IR analysis of a homogenized powder. While this type of experiment will provide information on the components within the powder, it will not provide any spatially resolved information.

For the spatially resolved FTIRM or FTIRI experiment, one must remember that the experiment works in transmittance—thus, the sample has to be thin enough to transmit light. This means that a thin culture section or a histological tissue sample can be examined. IR spectroscopy can be applied to solutions and gases as well as solids, but IR microspectroscopy and IR microspectroscopic imaging are generally applied to tissue sections (1–3 μm thick). These sections can be acquired by lifting an adherent cell culture onto a window (Figure 4.3a), using tissues that are naturally transparent, such as zebra fish bone (Figure 4.3b), or

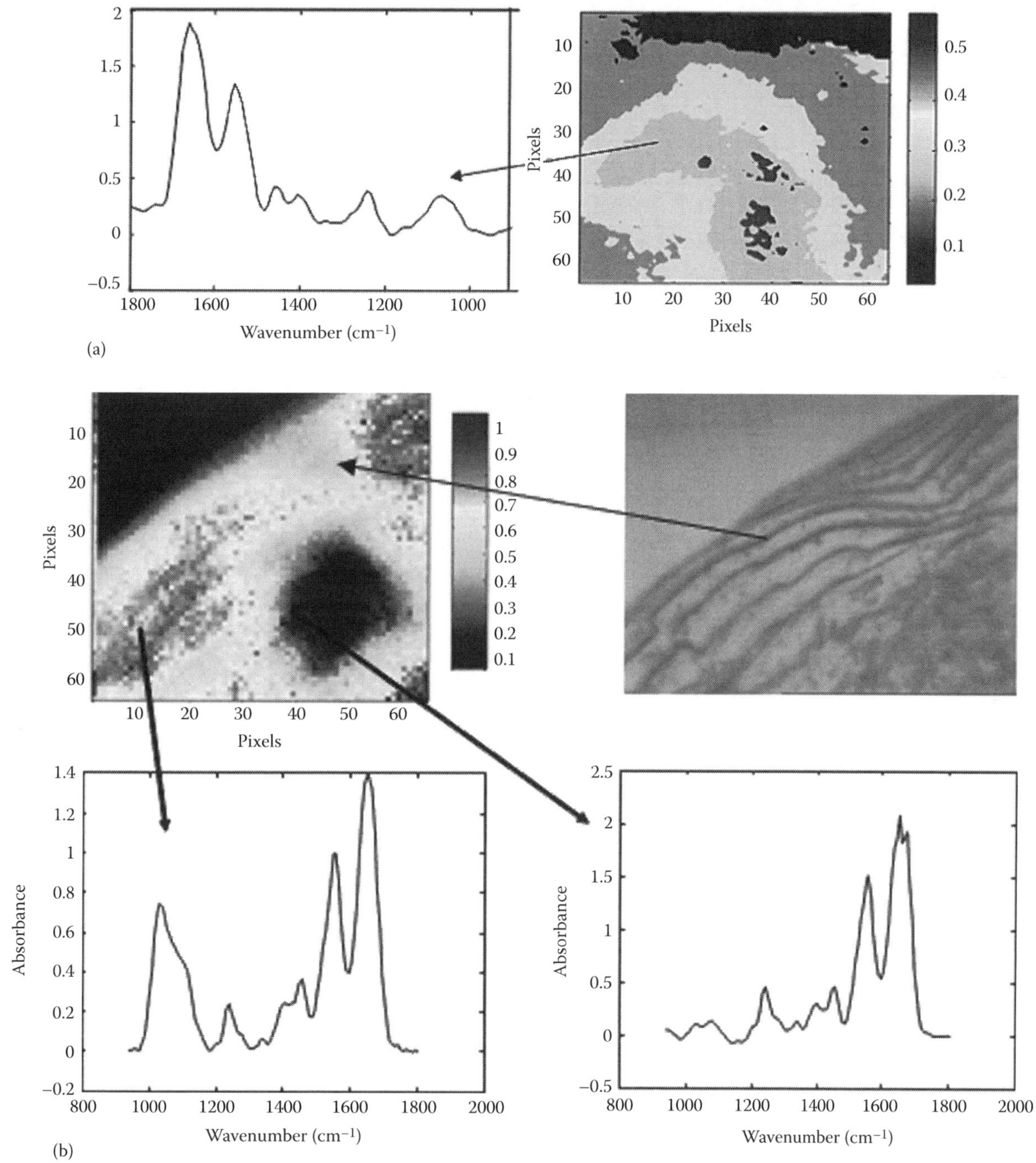

Figure 4.3 IR images can be obtained from thin tissues without fixation. (a) Early mineralization in a differentiating chick limb bud micromass culture system. The image on the left shows the mineral-to-matrix ratio; the spectrum corresponding to the head of the arrow is shown on the right. (b) Unprocessed image of the scales of a zebra fish, imaged on the mineral peak. The figure on the top right is a photograph of the tissue; the one on the left is the unprocessed image. The spectra shown come from the area indicated by the tail of the arrow; a mineralized region and a nonmineralized region. One pixel on each of the images corresponds to ~6.25 μm.

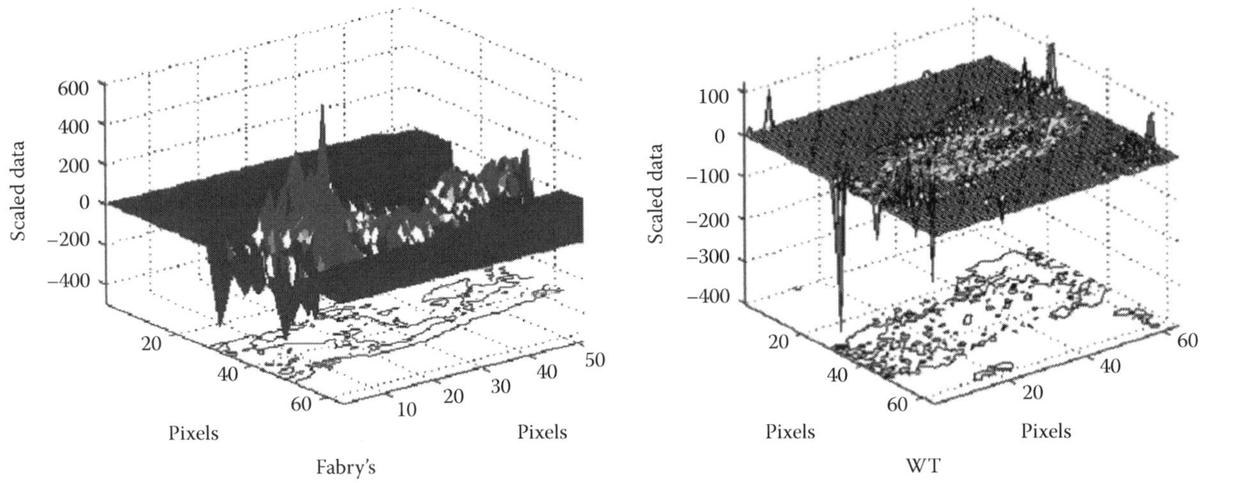

Figure 4.4 (See color insert.) IR images can be used to illustrate the lipid distribution in healthy and diseased bone. A 3D projection of the lipid/protein ratio in the long bone of a mouse with Fabry's disease (a lipid storage disease that results in the accumulation of globotriaosylceramide in tissues) compared to a WT mouse. Notice the increased accumulation of lipid in Fabry's mouse (note that the scale for Fabry's mouse is 6× that of the WT).

embedding the tissue in a medium that is as hard as the tissue and then cutting it with a microtome [19,20]. Second, the technique, depending on the IR source and the wavelength in question (globar or synchrotron), can give a spatial resolution of 2–10 μm.

The two key advantages of the FTIRM or FTIRI techniques are that they are nondestructive (no staining required) and compositional information can be obtained from a single section about a multitude of vibrating molecules including secondary structure of proteins [21], membrane lipid composition [22] and distribution (Figure 4.4), changes in cell cycle and apoptosis [23–26], as well as mineral structure in naturally occurring biominerals ranging from bone [27] to shells [28] and pathologic mineral deposits [29]. There are a few limitations of FTIRM and FTIRI that must be considered when planning to use this technique. Because water interferes in the spectrum, the specimens must either be dried or substituted with D_2O [29]. The specimens must also not be too absorbing, requiring thin sectioning (1–8 μm) of most biologic materials. Finally, the spatial resolution is not as good as Raman (see Chapter 5) although signal intensities for the most part are stronger in FTIRI than in Raman imaging.

The data obtained by FTIRM and FTIRI techniques can be provided as spatial (x, y) maps of the intensities of a specific contribution (spatial map) or hyperspectral image, where the z axis is the intensity or a value acquired by manipulating the spectra. Multivariate analyses enable the distribution of factors, clusters of factors, or derivatives to be displayed on the x, y plane. This can be used to compare the general spectral appearance in different samples (Figures 4.5 through 4.7), the variation in the distribution of particular components (Figures 4.5c through 4.7), or the contribution of different underlying spectra components (factor) to a given structure (Figure 4.5d). Processing the data usually involves subtracting the embedding media's contribution to the overall image, correcting for the presence of water vapor, and correcting for any other contributions to the spectral background (Figure 4.5b).

4.3 APPLICATIONS TO BIOMINERALIZATION

4.3.1 FTIR MICROSCOPY APPLICATIONS

Data derived from biominerals, as discussed in the succeeding text, show the advantages of FTIR microscopy and FTIRI for illustrating chemical complexity and for providing statistical data on sample heterogeneity. While IR had long been used to identify the phases present in biominerals not plentiful enough for x-ray diffraction analyses [30], these samples had to be homogenized (in KBr or Nujol mulls) and thus lost any information related to spatial localization of these phases. IR microspectroscopy and later on IR spectroscopic imaging provided the opportunity to define site-specific changes in mineral compositional parameters.

The first study using FTIR microscopy to evaluate changes in tissue mineral properties was performed in 1989 using normal and vitamin D-deficient (rachitic) rodent growth plates [31]. The growth plates at the ends of developing bones provide a good example of a gradient of changes that occur as mineralization commences. During the process of endochondral ossification that occurs both within the growth plate and during fracture healing, the cells differentiate, deposit a matrix, and regulate the mineralization of that matrix. Thus, this analysis enabled the first visualization of changes during normal and impaired growth and development at ~20 μm resolution. The second study using FTIR microscopy at 50 μm spatial resolution extended these observations on mineral maturation by including polarized light to measure collagen matrix orientation and characterize the progression of mineralization in the calcifying turkey tendon [32]. Similarly, another group used FTIR microscopy to study calcified and noncalcified atherosclerotic plaque identifying the presence of cholesterol, lipids, and carbonated hydroxyapatite (HA) [33]. Bone formed *in situ* on macroporous calcium phosphates was also identified as being distinct in properties from the calcium phosphate phase in tissue regeneration and repair study using

FTIRM [34] as were osteoporotic rat, monkey, and human bones, when contrasted with bones from normal and drug-treated individuals [35–37].

An FTIRM study of horse dentin was also reported [38]. There were also several FTIR microscopic studies comparing the mineral properties of bones in genetically modified animals to their wild-type (WT) controls, each at 20 μm spatial resolution (Table 4.1). These studies as summarized in the table provided new and confirmatory insights into the functions of these proteins, functions that were not detectable from light microscopy or radiographic studies.

Other than physiologically deposited calcium phosphates, important FTIR microscopy studies have included the use of the carbonate content of breast calcifications to identify cancer

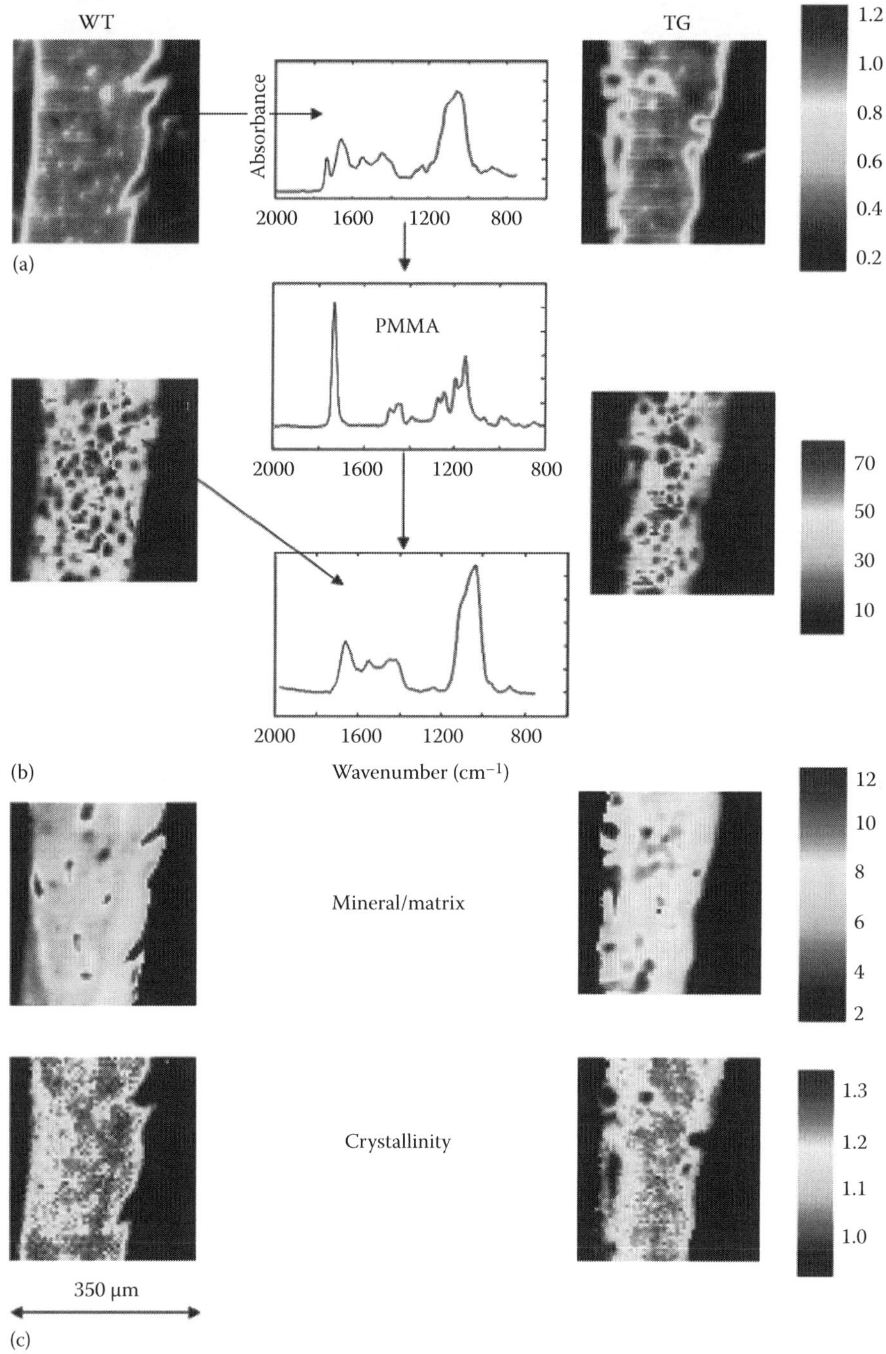

Figure 4.5 (See color insert.) Imaging processing comparing a 2-month-old transgenic (TG) mouse overexpressing a mutant collagen (TG) and its background-matched WT control. (a) Unprocessed image of a TG and age- and background-matched WT mouse bone as seen on the spectrometer. The spectrum in the center corresponds to the point indicated by the arrowhead. (b) Corrected image of the data shown in (a). These data were corrected for background, PMMA, and water vapor, indicated in the adjacent spectrum by spectral subtraction, resulting in the processed spectrum and image. (c) Mineral-to-matrix ratio and crystallinity distribution in the same bone. The color scales are valid for each set of images from the TG and WT mice. The scale bar indicates the *x* and *y* dimensions of the figure.

(continued)

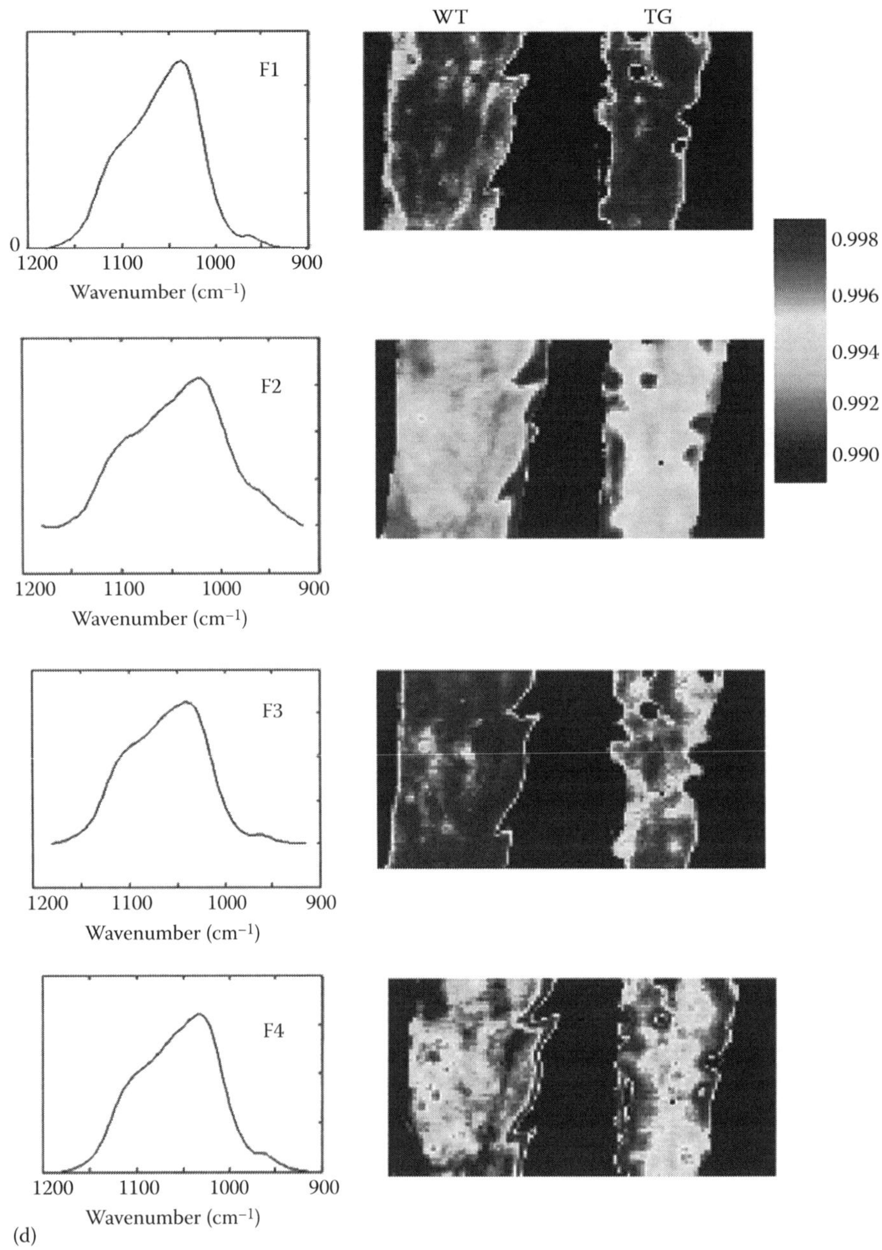

Figure 4.5 (continued) Imaging processing comparing a 2-month-old transgenic (TG) mouse overexpressing a mutant collagen (TG) and its background-matched WT control. (d) Factor analysis and score images of the phosphate bands in the same bones. The color scales are valid for each set of images from the TG and WT mice. The scale bar indicates the *x* and *y* dimensions of the figure.

risk [42], the demonstration that the distribution of calcium carbonates in gall stones of children is distinct from that in adults [43], the first indication of the nature of the composition of a corneal calcification showing that it was a poorly crystalline HA [44], and the first spectroscopic identification of the nature of the different crystalline phases in urine of patients with *crystalluria* [29]. In addition, FTIR microscopy has been used to characterize the properties of calcite in marine organisms [45] and the effects of magnesium on the distribution of amorphous and crystalline calcium carbonates in calcite [28].

In my laboratory, studying micromass cultures of chick limb buds in the presence of inorganic phosphate and ascorbate, we could identify mineralization occurring, as had our colleague Itzhak Binderman in his initial publication, by von Kossa staining or ^{45}Ca uptake [46]. Characterizing the nature of the biomineralization in tissue culture was challenging because a large number of identical cultures had to be prepared to obtain 10 mg of initial mineralized phase for x-ray diffraction examination. Collaboration with Dr. Richard Mendelsohn (Department of Chemistry, Rutgers, Newark, NJ) enabled us to examine the entire high-density micromass culture spot lifted from the culture dish and placed on a nonreactive transparent material (IR window, generally barium fluoride). Using FTIR microscopic mapping, we could then determine the spatial distribution of phosphate vibrations throughout the culture [47]. We confirmed the significance of these changes by studies

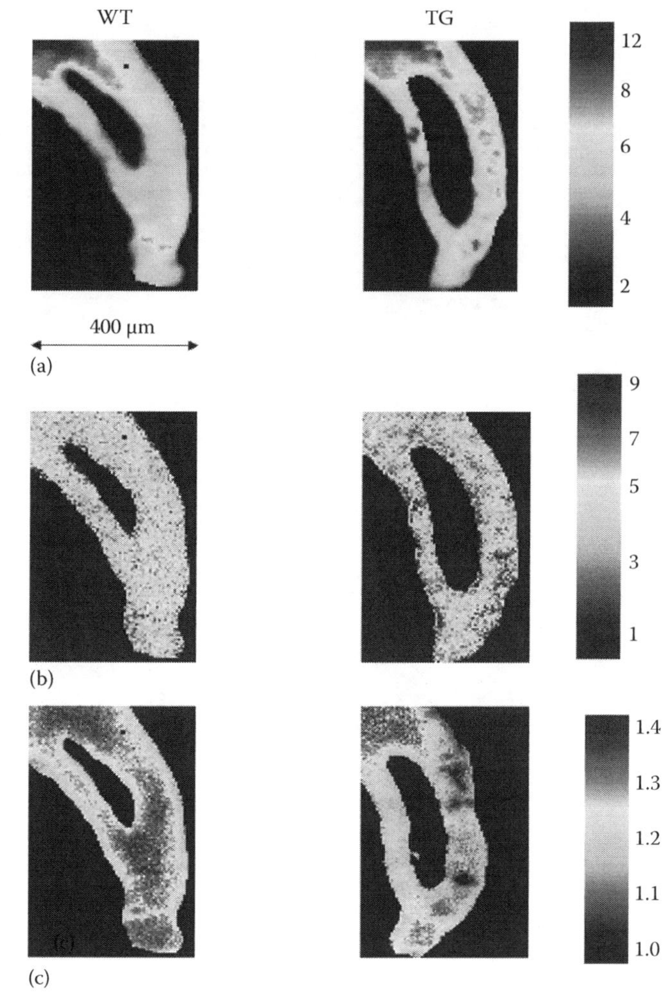

Figure 4.6 (See color insert.) Images of the dental root in the same animals whose bones are shown in Figure 4.5 representing (a) mineral/matrix ratio, (b) carbonate/phosphate ratio, and (c) crystallinity.

of model calcium phosphate compounds and comparison of the earlier reported analysis of the changes throughout the epiphyseal growth plate [31].

The limitation of FTIRM is that while thin sections (or in some cases, ground dispersed powders) were used to provide spatial resolution, the stage of the microscope (or the sample) had to be moved by some distance and then the next scan repeated; thus, the procedure of collecting data was slow, and the spatial resolution of the data was limited by the size of the aperture and how well a position on a bright field photograph could be matched to the position scanned. Nonetheless, these studies provided extremely important information, demonstrated the heterogeneity of the tissues, and enabled analysis of the changes in the properties of the mineral phases analyzed to be determined.

4.3.2 FTIRI APPLICATIONS

The introduction of an array detector enabled the more rapid collection of data with more pixels per unit area than the point-by-point measurements made in FTIRM. We first used FTIRI in 1998 to describe the same changes in mineral properties about the osteons that we had reported using FTIRM, only the imaging studies were done on dog alveolar bone [48]. Prior to that time, FTIRI had only been used in the characterization of polymers, liver, and skin samples and had not been applied to any biominerals.

FTIRI was next applied to map the spatial changes in carbonate substitution for phosphate in the bone matrix [49] and to map the changes in collagen maturity during trabecular bone remodeling [50]. The collagen study was important because it demonstrated that using thin sections, matrix changes could be analyzed without the need for decalcification of the specimens, a finding that should be applicable to any biomineral. These studies provided the framework for other studies of human and animal bones that revealed changes in bone mineral and matrix compositional properties associated with drug treatment in healthy and diseased subjects [51–55] and the association of these changes with fracture risk [56].

Additional studies described the spatial changes in dentin properties in a variety of animal models [62,63] (Figure 4.6) and the dentin–adhesive interface [64]. FTIRI was also used to characterize the properties of a variety of pathologic calcifications [6,65,66]. FTIRI was also used to monitor the progress of calcification in the chick limb bud cell culture model previously described [67] and to study the mineral composition in diatoms [68], mollusk shells [69], and the distribution of HA mineral in the shell of a living fossil from the Early Paleozoic Era, the brachiopod *Glottidia pyramidata*, demonstrated by x-ray diffraction to contain HA rather than calcium carbonate.

4.3.3 KEY CHALLENGES TO CONSIDER IN APPLYING FTIRI

There are several challenges that must be recognized when selecting FTIRI over FTIRM or even either system over other techniques. First, because these techniques are transmittance-based techniques, thin samples that can transmit light are required. While new mineral formation in cultures and thin structures such as newly formed calvaria or the fins of zebra fish (as shown in Figure 4.2) can be scanned without further processing, most materials require embedding in a material that is harder than they are (e.g., polymethyl methacrylate [PMMA], Spurr's medium, or araldite) [19]. To embed samples in these materials requires fixation, and any fixation method, required for the sectioning of thin sections of mineralized tissue, always changes the morphology and the molecular contents of tissues and may distort the structure of cells. For example, immersion of tissues in lipid solvents, such as xylene, dissolves the tissue lipids. This treatment causes alterations for any further analysis. The partial solution to this is to characterize the changes caused by each fixation and embedding step and be consistent in the way these steps are applied for all specimens in a given study. The background of the embedding material can be subtracted from the spectra being analyzed, but the user must be aware that subtle changes due to the embedding process could have occurred.

Once the mineralized material is thin sectioned, there is also the challenge of knowing that the thin section sampled is representative of the tissue and not just of the surface of the

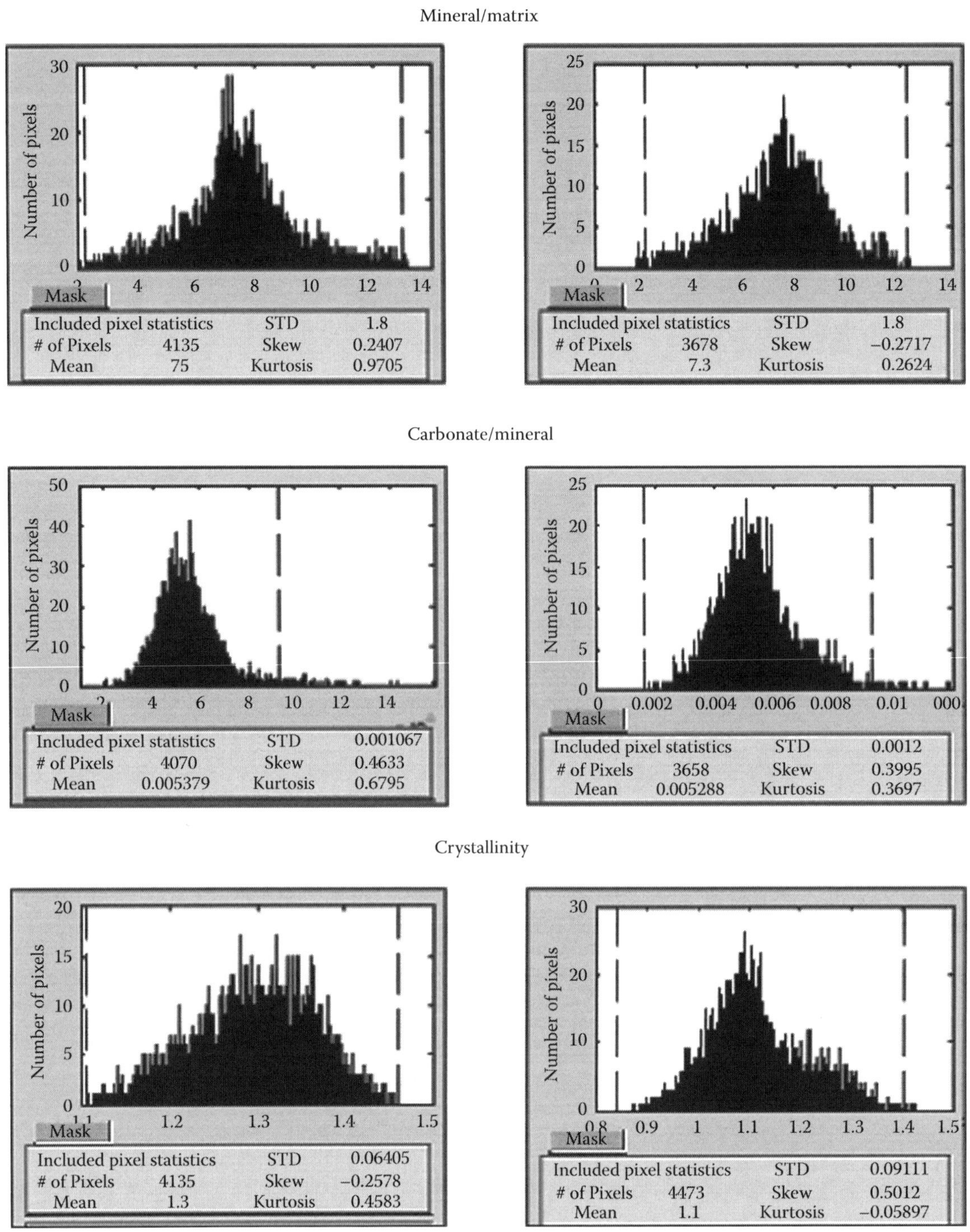

Figure 4.7 Image histograms can be used to show the distribution of the data as well as its mean data, as illustrated here for two femoral neck samples, one from a patient with a fracture (fracture) and one an age-matched necropsy control (control). Note the sharpening of the distribution of crystallinity pixels in the fracture case. The images corresponding to these histograms are on the right.

tissue. To circumvent this challenge, multiple sections, separated by a representative distance, and sections from different sites in the tissue under investigation must be analyzed. The image data can then be portrayed as 3D images, averaged, or presented as individual depth-dependent data. Chemometrics can be used to process data and learn about the variation in spectral content.

Finally, the structures studied in biomineralizing tissues have heterogeneity in the nanometer range. This means that when the spatial resolution is ~10 μm as it is in FTIRI as determined by the diffraction limit [70], the results are averaged over a region that might be greater than this heterogeneity. We have characterized this heterogeneities at this resolution to provide some insight into the variations (Figure 4.7) and report the heterogeneity of each image based on the line width at half maximum of each pixel image. Heterogeneities among samples or among images can then be described. However, if a unique phase or component occurs in a region that is less than 10 μm in breadth, this component might be missed.

Table 4.1 Selected examples of FTIR analyses of knockout and transgenic mouse bones

MODEL	FTIRM OR FTIRI	FINDINGS	REFERENCE
Type X collagen knockout Col X	FTIRM	Altered mineral distribution without change in mineral crystal sizes implied role for type X in matrix organization but not regulation of crystal growth	[40]
Osteopontin knockout OPN	FTIRM	Increased bone mineral content and crystal size implies role of OPN in bone remodeling	[41]
Osteocalcin knockout OCN	FTIRM	Increased size of mineral crystals as a function of age suggested that OCN played a role in formation rather than the originally postulated role in bone formation	[39]
Phosphaturic hormone on the X chromosome (PHEX) transgenic	FTIRI	Mineral properties rescued in the hypophosphatemic male mouse's bones and teeth	[57]
Dentin sialophosphoprotein (DSPP) knockout	FTIRI	Knockout of this dentin-specific protein affects bone mineral	[58]
Tissue-specific alkaline phosphatase (TNAP)	FTIRI	Decreased mineral content in young mice supported histochemical and other morphologic data	[61]
Mutant SH3BP2 adaptor protein associated with cherubism	FTIRI	Decreased mineral content and altered composition associated with osteoporotic phenotype in these mice	[59]
Dentin matrix protein-1 (DMP1) knockout	FTIRI	Decreased mineral content and increased crystal size implicated DMP1 in mineral formation	[60]

4.4 OBTAINING AND INTERPRETING FTIRM AND FTIRI DATA

Many laboratories do not have FTIR microscopes or FTIRI microscopes, but they can be found in most polymer laboratories, in some pathology and chemistry departments, and by searching the web for FTIRI cores or centers. The first topic the user should consider when approaching these cores is what question is being asked. Is it just the composition of the entire biomineralized structure? If so, imaging or microscopy is not required. If the question concerns the spatial distribution of different phases or of compositional variation, then length scales and access to equipment will be the first determining factor. Time and desired resolution will also have an influence as FTIRM is about 1000× slower for scanning a 400 × 400 image, but low-resolution 50 × 50 μm scans can be collected in minutes by FTIRM. The second question is how to prepare thin sections for these transmission experiments. Can the tissue be used as is (e.g., zebra fish) or can it be embedded in some media whose parameters do not overlap the desired peaks. Dr. Steve Weiner has a useful public domain database of IR spectra of a library of different biominerals (http://www.weizmann.ac.il/kimmel-arch/InfraredStandards.zip), and the embedding media spectra have been published [19] or can be obtained by taking a small block of the media itself. Once an appropriate way for obtaining thin specimens is determined, the specimens must be mounted on transparent IR windows (often made of barium fluoride, calcium fluoride, or silver chloride). The thin specimens are easily damaged, so covering them with a second window or placing them in small containers is useful. A typical specimen should then be tested on the instrument planned for use to make sure the absorbance bands of interest are not supersaturated (meaning a thinner section is needed) or are too weak (indicating a need for a thicker section). After specimen thickness, embedding media, and other qualifications have been determined, data should then be collected. For FTIRM or FTIRI, the data should be presented as representative spectra (for purpose of identification) and as images, as this is the best way to illustrate the variation of any property imaged. Global averages can also be reported, but such global averages mask the significance of the imagining/microscopic experiment, unless they are reported in specific areas of the tissue.

Processing of FTIRM or FTIRI data involves baselining the spectra, correcting for the background contributed by the window material, water vapor, and embedding media. A variety of public domain, commercial, and instrument-provided software can be used to produce the processed data, or spectral subtraction can be done manually. Histograms generated by the software or by standard image processing routines indicate the mean, standard deviation, and kurtosis or skewness of the data in each image. These can be averaged to provide similar information about the specimen heterogeneity. An informative way to present the data is as images, as have been presented through this text. But additional information about complex images and overlapping spectra can be obtained by applying chemometrics or multivariate analyses. Chemometric describes the mathematical and statistical treatment of analytical data to detect patterns and uses principal component analysis (PCA) followed by a neural network to identify spectral features and map their distributions, a variety of regression analyses including classical least squares (CLS), inverse least squares, principal component regression (PCR), and partial least squares to measure the amounts of components in unknown spectra based on standard data sets [71,72]. With the exception of a few unpublished studies that used chemometrics to study bone mineral, there are not yet reports of chemometric analysis of mineral composition, but at the rate the field is advancing, they may appear in the near future.

4.5 FUTURE

In 2012, a group of investigators reported "nano-FTIR chemical mapping of minerals in biological materials" [73]. By combining near-field optical microscopy (s-SNOM) with a mineralized atomic force microscope (AFM) tip that was used as the light-concentrating antenna, samples were probed with a nanofocused light field. The reported spatial resolution was 20 nm, as the IR diffraction limit was no longer applicable. In this initial study, calcium carbonate species in different layers of a shell could readily be identified, as could the tubules in dentin and the enamel/dentin interface. However, IR spectra that were produced had peaks that are nonuniformly and substantially shifted from those in the typical IR experiment, and substantial work will have to be done to establish new correlations. But even in this early stage of development, the technique can be used to identify the heterogeneity of composition in biominerals.

The other new technique that may be relevant for biominerals is attenuated total reflectance (ATR) imaging [74], which allows images to be obtained in the presence of water or other fluids. This technique has already been applied to characterize mineral changes during HA dissolution [75], to characterize the presence of different minerals in paints in Mayan ruins [76], and to determine the nature of mineral deposits in kidney biopsies [77].

ACKNOWLEDGMENTS

Dr. Boskey's work described in this chapter was supported by NIH grants AR041325, DE04141, AR037661, and AR046121. Dr. Boskey is grateful to Ms. Mila Spevak who performed the majority of the FTIRM and FTIRI studies and to Dr. Richard Mendelsohn, Rutgers University, for his prior collaboration and review of this manuscript.

REFERENCES

1. McDonald, R.S. 1986. Review: Infrared spectrometry. *Anal Chem* 58:1906–1925.
2. Stutman, J.M., Termine, J.D., and A.S. Posner. 1965. Vibrational spectra and structure of the phosphate ion in some calcium phosphates. *Trans NY Acad Sci* 27:669–675.
3. Posner, A.S. and G. Duyckaerts. 1954. Infrared study of the carbonate in bone, teeth and francolite. *Experientia* 10:424–425.
4. Rey, C., Renugopalakrishnan, V., Shimizu, M., Collins, B., and M.J. Glimcher. 1991. A resolution-enhanced Fourier transform infrared spectroscopic study of the environment of the $CO_3(2-)$ ion in the mineral phase of enamel during its formation and maturation. *Calcif Tissue Int* 49:259–268.
5. Rohanizadeh, R. and R.Z. Legeros. 2005. Ultrastructural study of calculus-enamel and calculus-root interfaces. *Arch Oral Biol* 50:89–96.
6. Pachman, L.M., Veis, A., Stock, S. et al. 2006. Composition of calcifications in children with juvenile dermatomyositis: Association with chronic cutaneous inflammation. *Arthritis Rheum* 54:3345–3350.
7. Niu, D.M., Lin, S.Y., Li, M.J., Cheng, W.T., Pan, C.C., and C.C. Lin. 2011. Idiopathic calcinosis cutis in a child: Chemical composition of the calcified deposits. *Dermatology* 222:201–205.
8. de Paula, S.M. and M. Silveira. 2009. Studies on molluscan shells: Contributions from microscopic and analytical methods. *Micron* 40:669–690.
9. Lowenstam, H.A. and D.P. Abbott. 1975. Vaterite: A mineralization product of the hard tissues of a marine organism (Ascidiacea). *Science* 188:363–365.
10. Weiss, I.M., Tuross, N., Addadi, L., and S. Weiner. 2002. Mollusc larval shell formation: Amorphous calcium carbonate is a precursor phase for aragonite. *J Exp Zool* 293:478–491.
11. Evans, L.A., Macey, D.J., and J. Webb. 1992. Calcium biomineralization in the radular teeth of the chiton, *Acanthopleura hirtosa*. *Calcif Tissue Int* 51:78–82.
12. Monje, P.V. and E.J. Baran. 2002. Characterization of calcium oxalates generated as biominerals in cacti. *Plant Physiol* 128:707–713.
13. Dror, A.A., Politi, Y., Shahin, H. et al. 2010. Calcium oxalate stone formation in the inner ear as a result of an Slc26a4 mutation. *J Biol Chem* 285:21724–21735.
14. Li, N., Niu, L.N., Qi, Y.P. et al. 2011. Subtleties of biomineralisation revealed by manipulation of the eggshell membrane. *Biomaterials* 32:8743–8752.
15. Chalmers, J.M., Everall, N.J., Hewitson, K. et al. 1998. Fourier transform infrared microscopy: Some advances in techniques for characterisation and structure–property elucidations of industrial material. *Analyst* 123:579–586.
16. Carr, G.L., Chubar, O., and P. Dumas. 2006. Multichannel detection with a synchrotron light source: Design and potential. In: *Spectrochemical Analysis Using Infrared Multichannel Detectors*, eds. Bhargava, R. and Levin, I.W., pp. 56–84. Oxford, U.K.: Blackwell Publishing.
17. Miller, L.M. and P. Dumas. 2010. From structure to cellular mechanism with infrared microspectroscopy. *Curr Opin Struct Biol* 20:649–656.
18. Miller, L.M., Vairavamurthy, V., Chance, M.R. et al. 2001. In situ analysis of mineral content and crystallinity in bone using infrared micro-spectroscopy of the nu(4) PO(4)(3–) vibration. *Biochim Biophys Acta* 1527:11–19.
19. Aparicio, S., Doty, S.B., Camacho, N.P. et al. 2002. Optimal methods for processing mineralized tissues for Fourier transform infrared microspectroscopy. *Calcif Tissue Int* 70:422–429.
20. Yeni, Y.N., Yerramshetty, J., Akkus, O., Pechey, C., and C.M. Les. 2006. Effect of fixation and embedding on Raman spectroscopic analysis of bone tissue. *Calcif Tissue Int* 78:363–371.
21. Manning, M.C. 2005. Use of infrared spectroscopy to monitor protein structure and stability. *Expert Rev. Proteomics* 2:731–743.
22. Zhang, G., Moore, D.J., Flach, C.R., and R. Mendelsohn. 2007. Vibrational microscopy and imaging of skin: From single cells to intact tissue. *Anal Bioanal Chem* 387:1591–1599.
23. Kahn, T.R., Fong, K.K., Jordan, B. et al. 2009. An FTIR investigation of flanking sequence effects on the structure and flexibility of DNA binding sites. *Biochemistry* 48:1315–1321.
24. Sahu, R.K., Mordechai, S., and E. Manor. 2008. Nucleic acids absorbance in mid IR and its effect on diagnostic variates during cell division: A case study with lymphoblastic cells. *Biopolymers* 89:993–1001.
25. Krummel, A.T. and M.T. Zanni. 2006. DNA vibrational coupling revealed with two-dimensional infrared spectroscopy: Insight into why vibrational spectroscopy is sensitive to DNA structure. *J Phys Chem B* 110:13991–14000.
26. Matthäus, C., Boydston-White, S., Miljković, M., Romeo, M., and M. Diem. 2006. Raman and infrared microspectral imaging of mitotic cells. *Appl Spectrosc* 60:1–8.
27. Paschalis, E.P. 2012. Fourier transform infrared imaging of bone. *Methods Mol Biol* 816:517–525.

28. Long, X., Nasse, M.J., Ma, Y., and L. Qi. 2012. From synthetic to biogenic Mg-containing calcites: A comparative study using FTIR microspectroscopy. *Phys Chem Chem Phys* 14:2255–2263.
29. Verdesca, S., Fogazzi, G.B., Garigali, G., Messa, P., and M. Daudon. 2011. Crystalluria: Prevalence, different types of crystals and the role of infrared spectroscopy. *Clin Chem Lab Med* 49:515–520.
30. Bertinetti, L., Drouet, C., Combes, C. et al. 2009. Surface characteristics of nanocrystalline apatites: Effect of Mg surface enrichment on morphology, surface hydration species, and cationic environments. *Langmuir* 25:5647–5654.
31. Mendelsohn, R., Hassankhani, A., DiCarlo, E., and A. Boskey. 1989. FT-IR microscopy of endochondral ossification at 20 mu spatial resolution. *Calcif Tissue Int* 44:20–24.
32. Gadaleta, S.J., Camacho, N.P., Mendelsohn, R., and A.L. Boskey. 1996. Fourier transform infrared microscopy of calcified turkey leg tendon. *Calcif Tissue Int* 58:17–23.
33. Manoharan, R., Baraga, J.J., Rava, R.P., Dasari, R.R., Fitzmaurice, M., and M.S. Feld. 1993. Biochemical analysis and mapping of atherosclerotic human artery using FT-IR microspectroscopy. *Atherosclerosis* 103:181–193.
34. Bohic, S., Heymann, D., Pouëzat, J.A., Gauthier, O., and G. Daculsi. 1998. Transmission FT-IR microspectroscopy of mineral phases in calcified tissues. *C R Acad Sci III* 321:865–876.
35. Bohic, S., Rey, C., Legrand, A. et al. 2000. Characterization of the trabecular rat bone mineral: Effect of ovariectomy and bisphosphonate treatment. *Bone* 26:341–348.
36. Gadeleta, S.J., Boskey, A.L., Paschalis, E. et al. 2000. A physical, chemical, and mechanical study of lumbar vertebrae from normal, ovariectomized, and nandrolone decanoate-treated cynomolgus monkeys (*Macaca fascicularis*). *Bone* 27:541–550.
37. Boskey, A.L., DiCarlo, E., Paschalis, E., West, P., and R. Mendelsohn. 2005. Comparison of mineral quality and quantity in iliac crest biopsies from high- and low-turnover osteoporosis: An FT-IR microspectroscopic investigation. *Osteoporos Int* 16:2031–2038.
38. Magne, D., Pilet, P., Weiss, P., and G. Daculsi. 2001. Fourier transform infrared microspectroscopic investigation of the maturation of nonstoichiometric apatites in mineralized tissues: A horse dentin study. *Bone* 29:547–552.
39. Boskey, A.L., Gadaleta, S., Gundberg, C., Doty, S.B., Ducy, P., and G. Karsenty. 1998. Fourier transform infrared microspectroscopic analysis of bones of osteocalcin-deficient mice provides insight into the function of osteocalcin. *Bone* 23:187–196.
40. Paschalis, E.P., Jacenko, O., Olsen, B., Mendelsohn, R., and A.L. Boskey. 1996. Fourier transform infrared microspectroscopic analysis identifies alterations in mineral properties in bones from mice transgenic for type X collagen. *Bone* 19:151–156.
41. Boskey, A.L., Spevak, L., Paschalis, E., Doty, S.B., and M.D. McKee. 2002. Osteopontin deficiency increases mineral content and mineral crystallinity in mouse bone. *Calcif Tissue Int* 71:145–154.
42. Baker, R., Rogers, K.D., Shepherd, N., and N. Stone. 2010. New relationships between breast microcalcifications and cancer. *Br J Cancer* 103:1034–1039.
43. Stringer, M.D., Soloway, R.D., Taylor, D.R., Riyad, K., and G. Toogood. 2007. Calcium carbonate gallstones in children. *J Pediatr Surg* 42:1677–1682
44. Chen, K.H., Cheng, W.T., Li, M.J., and S.Y. Lin. 2006. Corneal calcification: Chemical composition of calcified deposit. *Graefes Arch Clin Exp Ophthalmol* 244:407–410.
45. Rahman, M.A. and T. Oomori. 2009. Analysis of protein-induced calcium carbonate crystals in soft coral by near-field IR microspectroscopy. *Anal Sci* 25:153–155.
46. Binderman, I., Greene, R.M., and J.P. Pennypacker. 1979. Calcification of differentiating skeletal mesenchyme in vitro. *Science* 206:222–225.
47. Boskey, A.L., Camacho, N.P., Mendelsohn, R., Doty, S.B., and I. Binderman. 1992. FT-IR microscopic mappings of early mineralization in chick limb bud mesenchymal cell cultures. *Calcif Tissue Int* 51:443–448
48. Marcott, C., Reeder, R.C., Paschalis, E.P., Tatakis, D.N., Boskey, A.L., and R. Mendelsohn. 1998. Infrared microspectroscopic imaging of biomineralized tissues using a mercury-cadmium-telluride focal-plane array detector. *Cell Mol Biol (Noisy-le-grand)* 44:109–115.
49. Ou-Yang, H., Paschalis, E.P., Mayo, W.E., Boskey, A.L., and R. Mendelsohn. 2001. Infrared microscopic imaging of bone: Spatial distribution of $CO_3(2-)$. *J Bone Miner Res* 16:893–900.
50. Paschalis, E.P., Verdelis, K., Doty, S.B., Boskey, A.L., Mendelsohn, R., and M. Yamauchi. 2001. Spectroscopic characterization of collagen cross-links in bone. *J Bone Mine Res* 16:1821–1828.
51. Paschalis, E.P., Burr, D.B., Mendelsohn, R., Hock, J.M., and A.L. Boskey. 2003. Bone mineral and collagen quality in humeri of ovariectomized cynomolgus monkeys given rhPTH(1-34) for 18 months. *J Bone Miner Res* 18:769–775.
52. Paschalis, E.P., Boskey, A.L., Kassem, M., and E.F. Eriksen. 2003. Effect of hormone replacement therapy on bone quality in early postmenopausal women. *J Bone Miner Res* 18:955–959.
53. Paschalis, E.P., Glass, E.V., Donley, D.W., and E.F. Eriksen. 2005. Bone mineral and collagen quality in iliac crest biopsies of patients given teriparatide: New results from the fracture prevention trial. *J Clin Endocrinol Metab* 90:4644–4649.
54. Durchschlag, E., Paschalis, E.P., Zoehrer, R. et al. 2006. Bone material properties in trabecular bone from human iliac crest biopsies after 3- and 5-year treatment with risedronate. *J Bone Miner Res* 21:1581–1590.
55. Faibish, D., Gomes, A., Boivin, G., Binderman, I., and A. Boskey. 2005. Infrared imaging of calcified tissue in bone biopsies from adults with osteomalacia. *Bone* 36:6–12.
56. Gourion-Arsiquaud, S., Faibish, D., Myers, E. et al. 2009. Use of FTIR spectroscopic imaging to identify parameters associated with fragility fracture. *J Bone Miner Res* 24:1565–1567.
57. Boskey, A., Frank, A., Fujimoto, Y. et al. 2009. The PHEX transgene corrects mineralization defects in 9-month-old hypophosphatemic mice. *Calcif Tissue Int* 84:126–137.
58. Verdelis, K., Ling, Y., Sreenath, T., Haruyama, N., MacDougall, M., van der Meulen, M.C., Lukashova, L., Spevak, L., Kulkarni, A.B., and A.L. Boskey. 2008. DSPP effects on in vivo bone mineralization. *Bone* 43(6):983–990.
59. Wang, C.J., Chen, I.P., Koczon-Jaremko, B. et al. 2010. Pro416Arg cherubism mutation in Sh3bp2 knock-in mice affects osteoblasts and alters bone mineral and matrix properties. *Bone* 46:1306–1315.
60. Ling, Y., Rios, H.F., Myers, E.R., Lu, Y., Feng, J.Q., and A.L. Boskey. 2005. DMP1 depletion decreases bone mineralization in vivo: An FTIR imaging analysis. *J Bone Miner Res* 20:2169–2177.
61. Anderson, H.C., Sipe, J.B., Hessle, L. et al. 2004. Impaired calcification around matrix vesicles of growth plate and bone in alkaline phosphatase-deficient mice. *Am J Pathol* 164:841–847.
62. Verdelis, K., Crenshaw, M.A., Paschalis, E.P., Doty, S., Atti, E., and A.L. Boskey. 2003. Spectroscopic imaging of mineral maturation in bovine dentin. *J Dent Res* 82:697–702.
63. Sloofman, L.G., Verdelis, K., Spevak, L. et al. 2010. Effect of HIP/ribosomal protein L29 deficiency on mineral properties of murine bones and teeth. *Bone* 47:93–101.
64. Spencer, P., Wang, Y., Katz, J.L., and A. Misra. 2005. Physicochemical interactions at the dentin/adhesive interface using FTIR chemical imaging. *J Biomed Opt* 10:031104.

65. Zhao, Y., Urganus, A.L., Spevak, L. et al. 2009. Characterization of dystrophic calcification induced in mice by cardiotoxin. *Calcif Tissue Int* 85:267–275.
66. Kavukcuoglu, N.B., Li, Q., Pleshko, N., and J. Uitto. 2012. Connective tissue mineralization in Abcc6(-/-) mice, a model for *Pseudoxanthoma elasticum*. *Matrix Biol* 31:246–252.
67. Boskey, A.L., Paschalis, E.P., Binderman, I., and S.B. Doty. 2002. BMP-6 accelerates both chondrogenesis and mineral maturation in differentiating chick limb-bud mesenchymal cell cultures. *J Cell Biochem* 84:509–519.
68. Kammer, M., Hedrich, R., Ehrlich, H., Popp, J., Brunner, E., and C. Krafft. 2010. Spatially resolved determination of the structure and composition of diatom cell walls by Raman and FTIR imaging. *Anal Bioanal Chem* 398:509–517.
69. Falini, G., Sartor, G., Fabbri, D. et al. 2011. The interstitial crystal-nucleating sheet in molluscan *Haliotis rufescens* shell: A bio-polymeric composite. *J Struct Biol* 173:128–137.
70. Lasch, P. and D. Naumann. 2006. Spatial resolution in infrared microspectroscopic imaging of tissues. *Biochim Biophys Acta* 1758:814–829.
71. Lavine, B. and J. Workman. 2008. Chemometrics. *Anal Chem* 80:4519–4531.
72. Wang, L. and B. Mizaikoff. 2008. Application of multivariate data-analysis techniques to biomedical diagnostics based on mid-infrared spectroscopy. *Anal Bioanal Chem* 391:1641–1654.
73. Amarie, S., Zaslansky, P., Kajihara, Y. et al. Nano-FTIR chemical mapping of minerals in biological materials. *Beilstein J Nanotechnol* 3:312–323.
74. Kazarian, S.G. and K.L. Chan. 2006. Applications of ATR-FTIR spectroscopic imaging to biomedical samples. *Biochim Biophys Acta* 758:858–867.
75. Kazarian, S.G., Chan, K.L., Maquet, V., and A.R. Boccaccini. 2004. Characterisation of bioactive and resorbable polylactide/bioglass composites by FTIR spectroscopic imaging. *Biomaterials* 25:3931–3938.
76. Goodall, R.A., Hall, J., Sharer, R.J., Traxler, L., Rintoul, L., and P.M. Fredericks. 2010. Micro-attenuated total reflection spectral imaging in archaeology: Application to Maya paint and plaster wall decorations. *Appl Spectrosc* 62:10–16.
77. Gulley-Stahl, H.J., Bledsoe, S.B., Evan, A.P., and A.J. Sommer. 2010. The advantages of an attenuated total internal reflection infrared microspectroscopic imaging approach for kidney biopsy analysis. *Appl Spectrosc* 64:15–22.

5 Raman spectroscopy in biomineralization

Karen Esmonde-White and Francis Esmonde-White

Contents

5.1 INTRODUCTION

Raman spectroscopy is an optical method that enables nondestructive analysis of chemical composition and molecular structure. First discovered by C.V. Raman in 1928, Raman scattering is an inelastic form of optical scattering due to interactions between light and the symmetric vibrational modes of molecules (Raman and Krishnan 1928). Few photons are Raman scattered as compared to other optical interactions such as elastic scattering, fluorescence, absorption, or reflection. As a result, early Raman experiments were time-consuming and required intense excitation sources to overcome the inefficiency of Raman scattering. Advances in instrumentation and data processing enabled rapid nondestructive analysis and now Raman spectroscopy is feasible for many applications. Furthermore, Raman spectroscopy allows intrinsic compositional information to be measured in light scattering and hydrated samples unlike other optical methods. As discussed later in the chapter, modern Raman instrumentation includes a bright laser source, efficient optical filters and gratings, and imaging detectors with high quantum efficiency and low noise. Applications of Raman spectroscopy in polymer, pharmaceutical, and biomedical analysis have surged over the past three decades as improvements in the instrumentation and data analysis have rendered Raman a practical laboratory methodology. In large part because of instrumentation advances, Raman spectroscopy is used for routine analysis in academic, government, and industrial laboratories (Lyon et al. 1998; Mulvaney and Keating 2000).

Biological applications followed less than 10 years after the original discovery of Raman spectroscopy and remain a vibrant area of study (Koenig 1972). Early studies of biological macromolecules established the utility of Raman spectroscopy for measuring structure and conformation of peptides, proteins, nucleic acids, and carbohydrates. It was recognized very quickly that low water signal, conformational sensitive bands, and minimal sample preparation were attractive features of Raman spectroscopy. Raman spectroscopy does not require sample dehydration, fixing, or embedding, enabling measurement of molecular composition under in situ conditions. In mineralized tissues, Raman spectroscopy can be used to measure the carbonate and phosphate components of biominerals, as well as the collagen protein conformation and secondary structure.

Because of its ubiquity in the human body, collagen was one of the first proteins to be studied by Raman spectroscopy. In the early 1970s, the Koenig laboratory pioneered Raman measurements in many biological molecules, including collagen (Vasko et al. 1971; Frushour and Koenig 1975; Cael et al. 1976; Lin and Koenig 1976). These earliest studies established Raman as an excellent technique for analyzing protein secondary and tertiary structure in various physiochemical states. Local chemical

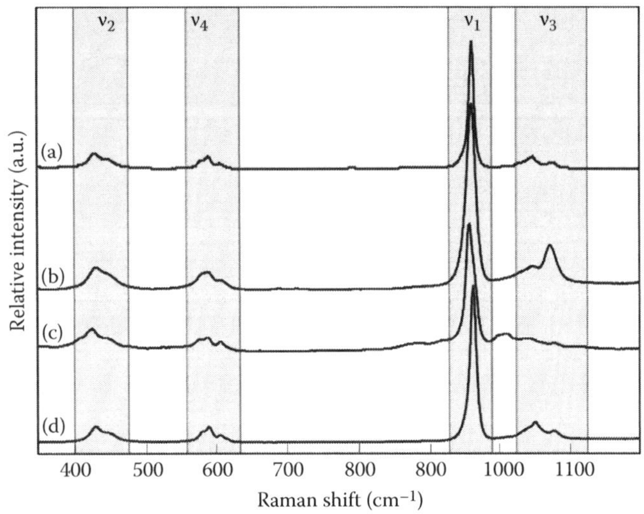

Figure 5.1 Raman spectrum of (a) hydroxyapatite (<0.5% carbonate), (b) carbonated apatite (7.3% carbonate), (c) octacalcium phosphate, and (d) fluorapatite in the 350–1500 cm^{-1} spectral region, measured using a near-IR Raman system (λ = 785 nm). (Spectra are courtesy of Professor Mary Tecklenburg from Central Michigan University, Mount Pleasant, MI.)

environments, such as hydration state, pH, and ionic strength, affect both the hydrogen bonding and cross-linking of collagen that maintains both the alpha-helix structure and the spatial arrangement of collagen in tissue. In biological specimens, collagen secondary and tertiary structures are also affected by disease and age. In one of the first papers, Frushour concluded that "collagen remains one of the most difficult systems to study with Raman spectroscopy" (Frushour and Koenig 1975). Ongoing research in interpreting collagen Raman spectra underscores the complex structure of collagen in mineralized and nonmineralized tissues.

5.1.1 BIOMINERALS

Raman spectroscopy of geological minerals was another early application because minerals have very strong Raman scattering properties (Griffith 1987). Extension of Raman to biological minerals continues to focus on hard tissues because they contain a strongly scattering mineral moiety, such as hydroxyapatite. Raman studies of biominerals fall under three broad classes: (1) identification of pathological mineralization, (2) identification of transient phases during early stages of biomineralization, and (3) quantification of mineral composition in synthetic or natural apatites. Calcium phosphate minerals have a very strong Raman signal in the 940–980 cm^{-1} spectral region corresponding to the symmetric phosphate vibrational mode. The symmetric P–O stretch, called ν_1 in the Raman literature, is the most widely used phosphate band to identify mineral phase and measure mineral crystallinity and stoichiometry. Band position, width, and intensity of ν_1 reflect the chemical environment of the phosphate ion. A comparison of synthetic apatite minerals is shown in Figure 5.1 to demonstrate how spectral features are changed with apatite lattice composition. All three metrics are commonly used to identify not only the type of calcium phosphate species present but also information on mineral crystallinity or mineral stoichiometry. Table 5.1 shows band assignments for several calcium phosphate minerals (Blakeslee and Condrate 1971; Cornilsen 1984; Sauer et al. 1994; de Aza et al. 1997; Carden and Morris 2000; Gunasekaran et al. 2006). This sensitivity enables Raman identification of different calcium phosphate species in tissue. Although bone and teeth are the most commonly examined mineralized tissues, calcium phosphate minerals can be found in other anatomic locations.

Raman in biomineral analysis is a vibrant area of research, making valuable contributions toward understanding disease pathophysiology with potential translation for clinical

Table 5.1 Raman bands, assignments, and relative intensities for various calcium phosphate minerals

MINERAL SPECIES	RAMAN BANDS (cm^{-1})	REFERENCE
Calcium hydroxyapatite $Ca_{10}(PO_4)_6(OH)_2$	431, 447: PO_4 ν_2 (m); 581–608: PO_4 ν_4 (m); 945 (sh): 962: PO_4 ν_1 (s); 1031, 1048, 1076: PO_4 ν_3 (w–m)	Blakeslee and Condrate (1971)
Fluorapatite $Ca_5(PO_4)_3F$	431, 447: PO_4 ν_2 (m); 581–608: PO_4 ν_4 (m); 945 (sh): 962: PO_4 ν_1 (s); 1031, 1048, 1076: PO_4 ν_3 (w–m)	Blakeslee and Condrate (1971)
Carbonated apatite B type	420–450: PO_4 ν_2 (m); 560–620: PO_4 ν_4 (m); 945 (sh): 958: PO_4 ν_1 (s); 1070: CO_3 ν_1 (m); 1076: PO_4 ν_3 (w–m)	Carden and Morris (2000)
Octacalcium phosphate $Ca_8H_2(PO_4)_6 \cdot 5H_2O$	400–440: PO_4 ν_2 (m); 570–590: PO_4 ν_4 (m); 957: PO_4 ν_1 (s); 1060–1080: PO_4 ν_3 (w–m)	Sauer et al. (1994)
Tricalcium phosphate $Ca_3(PO_4)_2$	370–505 PO_4 ν_2 (m–s); 530–645 PO_4 ν_4 (m); 949: PO_4 ν_1 (s); 970: PO_4 ν_1 (sh); 995–1120: PO_4 ν_3 (m)	de Aza et al. (1997)
Dicalcium phosphate dihydrate $CaHPO_4 \cdot 2H_2O$	400–440: PO_4 ν_2 (m); 570–590: PO_4 ν_4 (m); 985: PO_4 ν_1 (s); 1060–1080: PO_4 ν_3 (w–m)	Sauer et al. (1994)
Amorphous dicalcium phosphate $CaHPO_4$	400–440: PO_4 ν_2 (m); 570–590: PO_4 ν_4 (m); 952: PO_4 ν_1 (s); 1060–1080: PO_4 ν_3 (w–m)	Sauer et al. (1994)
Calcium pyrophosphate $Ca_2P_2O_7 \cdot 2H_2O$	486–491: PO_4 δ (m); 755: POP ν_1 (m–s); 1052: PO_4 ν_1 (s)	Cornilsen (1984), McGill et al. (1991)
Calcium carbonate $CaCO_3$	710: PO_4 ν_4; 1088: PO_4 ν_1; 1430: PO_4 ν_3	Gunasekaran et al. (2006)

Note: s, strong; sh, shoulder; m, moderate; w, weak.

applications. The first Raman spectrum of bone was reported in 1970 by Walton et al., establishing Raman as a complement to infrared (IR) bone studies (Walton et al. 1970). Previous to Raman and IR studies, bone mineral content was measured using destructive techniques such as ash weight or histopathology. IR and Raman analysis of bone provides composition, similar to ash weight. But, only vibrational spectroscopy can provide molecular structure and orientation information for the collagen matrix and carbonated apatite mineral. Moreover, the high spatial resolution of spectroscopic techniques, as compared to ash weight or histopathology, enabled investigation of compositional heterogeneity in bone tissue structures such as trabeculae, microcracks, or osteons.

5.2 INSTRUMENTATION OVERVIEW

5.2.1 RAMAN SPECTROGRAPH

The instrument used to collect Raman spectra (interchangeably called a spectrometer or spectrograph) includes a light source, optics, and a detector. Modern Raman instrumentation almost exclusively uses lasers as excitation light sources due to the high-intensity monochromatic light that can be produced. The excitation light is delivered to the sample and Raman-scattered light is collected back into the spectrograph by some intermediate optics. These intermediate optics can include lenses, mirrors, and filters. Only a small fraction of the light used to excite the sample is Raman scattered. The majority of the excitation light is scattered inelastically. One or more optical filters are used to reject elastically scattered excitation light to minimize stray light that otherwise overwhelms the detector.

The spectrograph is the detection module. The spectrograph includes an optical detector to record intensity with respect to wavelength (in nanometers or micrometers). Raman scattering causes photons to gain or lose specific amounts of energy based on molecular vibrations—regardless of the excitation energy. Raman spectra are reported with wavenumber units (photon energy in inverse centimeters, or cm^{-1}), which are independent of the excitation frequency. This allows spectra collected at different excitation frequencies to be compared. After correcting for variable detector efficiency, spectra collected using different excitation frequencies will usually be similar in shape, though at some excitation frequencies, resonance enhancement can lead to amplification of particular bands (resonance Raman spectroscopy). Spectrographs use either dispersive (with reflective or transmissive gratings) or interferometric (Fourier transform, or FT) methods for resolving light of different frequencies. Reflective gratings have the benefit of not having chromatic aberration, while transmissive systems have a higher throughput (greater efficiency). Early Raman systems used scanning systems to turn reflective gratings, measuring each point in the spectrum sequentially using a single detector element. Single-point detectors are less expensive but require that the spectral data be collected sequentially. Scanning dispersive Raman systems are no longer widely used because there is an advantage in collecting light for longer periods of time. Line and array imaging detectors allow light of different wavelengths to be measured simultaneously. Instead of spending a small amount of time at many separate points, these systems allow light to be collected across the entire spectrum for the whole measurement time. Scanning interferometric spectrometers vary a path length difference rather than a grating angle and are more efficient than scanning dispersive systems because they measure all frequencies at the same time. Some FT Raman systems also used imaging detectors to collect data from different spatial positions simultaneously. FT systems are most widely used in IR spectroscopy. Despite the many positive attributes of FT Raman instrumentation, most Raman instrumentation is now based on dispersive systems using imaging detectors. The main reason for this is that Raman spectra are typically limited by shot noise, due to few Raman-scattered photons being collected. In FT Raman, all wavelengths are collected at once, and the noise arising from the background can obscure the Raman bands of interest.

5.2.2 MICROSCOPY

Raman spectroscopy can be performed on a standard microscope that has been coupled to a Raman spectrometer or spectrograph. Both dispersive and FT Raman instruments can be coupled to microscopes and are available as commercial microscopy instruments. FT Raman spectroscopy is typically performed at higher wavelengths (λ = 1064 nm) to reduce fluorescence background. FT Raman will not be discussed here but the reader is pointed to several excellent technical, instrumentation, and application reviews (Chase 1986; Hendra et al. 1991; Chase and Rabolt 1994; Naumann 2001). Dispersive Raman spectroscopy is performed in the ultraviolet, visible, and near-IR regions of the wavelength range (λ = 200–1064 nm). Although details vary by instrument, a laser, a spectrograph equipped with a notch filter, and a high-performance low-noise camera are standard components in dispersive Raman systems. For detailed discussion on instrumentation for dispersive Raman spectroscopy, including considerations for sampling, overall performance, lasers, instrumentation accuracy, and signal to noise, we refer the reader to the following books (McCreery 2000; Lewis and Edwards 2001).

Depending on the application, the excitation laser can be shaped into either a spot or line shape on the sample. A spot-shaped laser is most often used because it can be used for a variety of measurements, ranging from point spectroscopy to high-resolution confocal mapping or imaging. Confocal configurations using a tightly focused laser excitation spot is used to collect Raman maps with high spatial resolution (~1–5 μm). Line-shaped laser excitation has the advantage of scanning over a larger area than a spot-shaped laser at once allowing the user to decrease the laser power for an equivalent mapping time. However, the spatial resolution of line imaging systems is typically lower than if a spot-shaped laser is used.

In addition to providing spatially resolved measurements, Raman microspectroscopy has been adapted for polarized Raman measurements. Raman scattering is sensitive to the molecular orientation relative to the incident excitation light. In the instrument, this requires maintaining a well-polarized excitation beam and including some polarization selective optics prior to the spectrograph. Polarization optics are added to the excitation and collection beam paths. Each of these optics can be oriented independently to set the polarization state either parallel or perpendicular features of the sample. Measuring Raman depolarization requires collection of four measurements.

The depolarization ratio is calculating by taking the Raman spectral intensity recorded with opposite polarization states set on the two paths and dividing by the Raman spectral intensity recorded with the same polarization state set. Raghavan et al. have demonstrated that orientation of particular molecular species can be quantitatively assessed using this method (Raghavan et al. 2010).

A common question by those familiar with microscopy, but unfamiliar with Raman spectroscopy, is "What is the spatial resolution of this Raman microscope?" Nominally, it should be possible to measure the Raman spectrum from a region approximately half the size of excitation light wavelength. Unfortunately, in practice, there is no precise answer in most cases where Raman microscopy is applied to biomedical tissues. Light scattering causes the excitation light to be diffused, and multiple scattering can greatly broaden the region from which Raman-scattered light is collected. Together these effects typically increase the size of the sampled region. This is also important because light scattering is not uniform through tissue. As a result, as a sample is imaged, the interrogated volume will change. Finally, microscopes typically have much finer resolution in the plane of the stage than along the optical imaging axis. As a result of these many confounding issues, the sampling volume is not necessarily symmetric (Everall 2000; Everall et al. 2007).

Optimal spatial resolution is obtained using a confocal Raman microscope with a well-focused single-mode laser beam, a very high-magnification objective, and a pinhole or other small aperture to limit collection of out-of-focus light. The speed of acquiring a map (or image) must be balanced with the spatial resolution of the system. In practice, fast imaging is not possible with low excitation power in confocal imaging configurations. In many applications, high spatial resolution is not required. For experiments that require either rapid signal collection or scanning over a large region, semiconfocal systems are a compromise. These systems are equipped with a line-shaped laser and a slit, rather than a pinhole. We note that the spatial resolution along the slit is degraded substantially, especially when compared to a pinhole. While the results of most Raman microscopy is not compromised by spatial resolution degradation, results of polarized Raman spectroscopy can be strongly affected by multiple optical scattering. To avoid these effects in polarized Raman spectroscopy, a high-magnification objective and small pinhole or slit are typically used.

5.2.3 FIBER OPTICS

A variety of fiber optic–coupled Raman instruments are commercially available. These include large benchtop spectrographs, miniature spectrographs, and handheld portable instruments. While low-cost ultraportable instruments are also available, these systems typically use sets of optical filters to solve specific problems and are not ideal for primary research. For research purposes, multipurpose benchtop spectrographs are typically employed due to their high performance and reliability. Fiber optics are often used to deliver the excitation light to the sample and to deliver collected Raman-scattered light back to the instrument.

There are several challenges inherent in using fiber optics. Nearly all Raman fiber-optic probes use multimode silica (glass) fiber optics, though hollow metallic waveguides have also been investigated. While silica fibers are very efficient across the visible and near-IR portions of the spectrum, glass also generates a spectral background due to the Raman scattering of the silica and a fluorescence background. This background can be reduced through the use of low hydroxyl content optical fibers, though broad silica bands will still be generated in the fiber. To further suppress this spontaneous background, optical filters are often used at the end of the probe. The excitation laser light is filtered using a short-pass or notch filter, while the collected Raman-scattered signal is filtered with a notch filter or a long-pass filter. The optical paths for excitation and collection can be overlapped (combined onto a single axis) using a dichroic or notch filter, either of which can be used to reflect the laser beam but transmit the Raman-scattered light.

5.3 SPECTRAL INTERPRETATION

Bone is the most widely studied mineralized tissue by Raman spectroscopy, and the principles used in interpreting bone Raman spectra can be applied toward other mineralized tissues. Figure 5.2 shows a typical Raman spectrum of cortical and cancellous bone tissue in the fingerprint region of 600–1800 cm^{-1}, with the intensity normalized to the phosphate ν_1 at 958 cm^{-1}. Band assignments of the bone spectrum in Figure 5.2 are presented in Table 5.2, using reference material analysis of collagen, nonmineralized tissues, and synthetic calcium phosphate minerals in the biophysics, mineralogy, and geology literature (Frushour and Koenig 1975; Goheen et al. 1978; Griffith 1987; Sauer et al. 1994; Edwards and Carter 2001; Wopenka and Pasteris 2005; Awonusi et al. 2007). Tooth enamel and dentin mineral is similar to bone mineral and shares major mineral bands at 958 and 1070 cm^{-1} with bone; it is more crystalline. The main spectral contributions in bone arise from the carbonated apatite mineral and type I collagen matrix that comprises over 90% of the organic content of bone tissue. Although there are most certainly spectral contributions from noncollagenous proteins, lipids, and cellular material, these contributions are of much lower intensity and cannot easily be discriminated. In contrast to the 400–1800 cm^{-1}

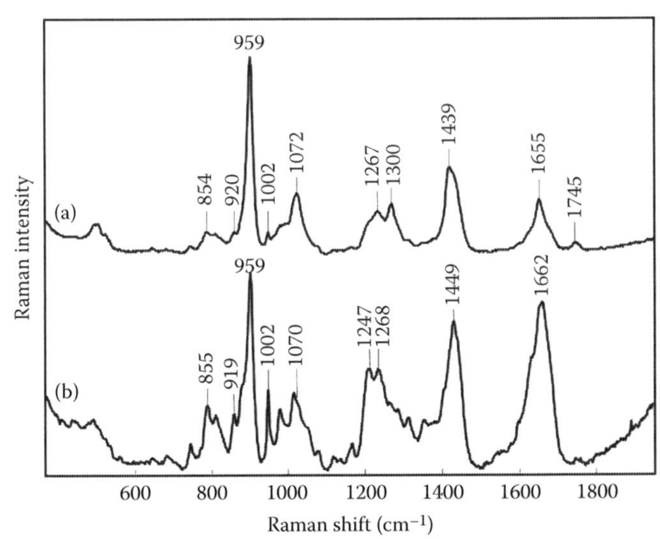

Figure 5.2 Raman spectrum of human (a) cancellous and (b) cortical bone.

Table 5.2 Raman band assignments for spectra of human cortical bone and cancellous bone shown in Figure 5.3

RAMAN SHIFT (cm^{-1})	BAND ASSIGNMENT	MOLECULAR SPECIES
590	ν_4 PO_4^{3-}	Carbonated apatite mineral
854	ν_{CC} hydroxyproline	Collagen matrix
878	ν_{CC} hydroxyproline	Collagen matrix
920	Proline	Collagen matrix
945	Disordered apatite	Carbonated apatite mineral
959	ν_1 PO_4^{3-}	Carbonated apatite mineral
1002	Phenylalanine	Collagen matrix
1067	**$\nu_{C\text{–}C}$ skeleton, trans**	**Lipid**
1070	ν_1 CO_3^{2-}	Carbonated apatite mineral
1079	**$\nu_{C\text{–}C}$ skeleton, random**	**Lipid**
1247	Amide III (random coil)	Collagen matrix
1268	Amide III (α-helix)	Collagen matrix
1300	**CH_2 deformation**	**Lipid**
1439	**CH_2 deformation**	**Lipid**
1448	CH_2 wag	Collagen matrix
1655–1666	Amide I	Collagen matrix
1745	**ν (C=O)**	**Lipid**

Note: Bands unique to untreated cancellous bone, with marrow lipids intact, are shown in bold.

region, the 2800–3600 cm^{-1} spectral region is sparse and has two main spectral features: a C–H stretching band from 2880 to 2980 cm^{-1} and a broad shapeless band centered ~3500 cm^{-1} assigned to water. While it is expected that an O–H stretch from apatite hydroxyl ions would be observed at ~3500 cm^{-1}, earlier reports have shown that the broad water feature in bone and dentin obscures this signal (Penel et al. 1998).

Bands assigned to phosphate and substitution ions in the apatite mineral are found in the 420–620 cm^{-1} and 900–1100 cm^{-1} spectral region. Weaker asymmetric ν_2 and ν_4 phosphate modes are found at 420–450 cm^{-1} and 560–620 cm^{-1} and are rarely used in analyzing bone composition. The symmetric ν_1 phosphate stretch is the most intense phosphate band in bone Raman spectra. Features of ν_1 such as position, width, and intensity of ν_1 are commonly used to identify not only the type of calcium phosphate species present but also information on mineral crystallinity or mineral stoichiometry. A shoulder at 945 cm^{-1} was assigned to a disordered apatite by Tarnowski et al. (2002). Carbonate is substituted in the apatite lattice and bone mineral is typically called a carbonate apatite for this reason. In bone Raman spectra, ν_1 of carbonate is located at ~1070 cm^{-1}. A weak ν_3 phosphate band is found at ~1076 cm^{-1} and overlaps with the carbonate ν_1. This overlap appears as a broadband centered at ~1072 cm^{-1} and may potentially confound spectral interpretation. However, a detailed study in synthetic carbonated apatites showed that the carbonate band is the dominant band at carbonate concentrations greater than 3% (Awonusi et al. 2007). Lipid bands are found at ~1060 and 1080 cm^{-1}, which may also affect spectral interpretation for bone specimens that have not been defatted. This issue is especially pertinent in clinical measurements, where lipids are ubiquitous throughout the body in the form of fat tissue and bone marrow and it is not possible to completely remove the lipids. In our experience with human cadaveric cortical and cancellous bone tissues, a band at ~1300 cm^{-1} can be used as a marker of lipid content because it does not overlap with bone collagen bands. Additional spectral evidence of lipid content is a narrowing amide I envelope and a shift in the CH_2 wag envelope to ~1439 cm^{-1} (Esmonde-White et al. 2011).

Bands and envelopes assigned to collagen matrix are found in the 1000–1800 cm^{-1} region of the spectrum. Historically, the amide III and amide I envelopes have been the most widely used bone collagen bands. However, there are recent studies reporting the use of the hydroxyproline bands at 852, 878, and 920 cm^{-1}. The amide III (1200–1300 cm^{-1}) and amide I (1600–1700 cm^{-1}) envelopes in Raman spectra are used to examine collagen secondary structure and cross-linking. The amide III envelope is assigned to peptide amide bonds and is observed as a broad feature in the Raman spectra of proteins. Collagen spectra, from mineralized or unmineralized tissue, exhibit two maxima within the amide III envelope corresponding to α-helix and random coil content in the protein's secondary structure (Chi et al. 1998). The amide I envelope is assigned to amide carbonyl stretches and is also observed as a broad envelope with a maximum ~1666 cm^{-1} with visible shoulders at ~1640 and 1685 cm^{-1}, depending on the chemical environment of the collagen (Frushour and Koenig 1975; Sane et al. 1999).

Spectral features of bandwidth, relative area, and relative intensity can serve as spectroscopic markers of tissue composition and molecular structure. Table 5.3 shows typical Raman markers of bone quality. The mineral-to-matrix ratio (MTMR) provides a spectroscopic marker of tissue mineralization. The MTMR is similar to bone mineral density in that the relative amount of calcium mineral is ratioed against a marker of tissue volume, where either band area or band intensity is used in the ratio. Typically, phosphate ν_1 at ~958 cm^{-1} is used as the mineral band. Historically, the amide I band was used as the matrix band, but other studies have also reported the use of the 1448 cm^{-1} CH_2 envelope or hydroxyproline bands at 852 and 878 cm^{-1} as matrix markers because they are insensitive to protein conformation (Crane et al. 2005; Dehring et al. 2006; Kohn et al. 2009).

Mineral composition is measured using several metrics. Carbonate substitution into the apatite lattice is measured spectroscopically using the carbonate-to-phosphate ratio (CTPR), where the area or intensity of the carbonate ν_1 at 1072 cm^{-1} is ratioed against the phosphate ν_1 at 958 cm^{-1}. Crystallinity is a spectroscopic marker of mineral size, shape, and stoichiometry. Mineral crystallinity is measured by the full width at half max of the phosphate 958 cm^{-1} band. A more crystalline material is reflected by a narrower band, and less crystalline band is reflected by band broadening (Morris and Mandair 2011). Typically, carbonate substitution into the apatite lattice reduces mineral crystallinity (Awonusi et al. 2007).

Because of the sensitivity of Raman to small spectral changes, the amide III (1220–1280 cm^{-1}) and amide I envelopes

Table 5.3 Raman spectroscopic markers commonly used to examine bone tissue composition and mineral molecular structure

SPECTROSCOPIC MARKER	CALCULATION OF MARKER	TISSUE PROPERTIES
MTMR	Intensity or area ratio of 958 cm^{-1}/matrix band	Tissue mineralization
CTPR	Intensity or area ratio of 1070 cm^{-1}/958 cm^{-1}	Apatite crystal size Mineral stoichiometry
Width of phosphate ν_1	Full width at half maximum of phosphate 958 cm^{-1} band	Mineral crystallinity Apatite crystal size
Amide III	Intensity or area ratio of 1247 cm^{-1}/1270 cm^{-1}	Collagen secondary structure
Amide I	Intensity or area ratio of 1685 cm^{-1}/1667 cm^{-1}	Collagen secondary structure Collagen cross-linking

Note: Most recent studies have found that Raman markers correlate with material properties of bone including resistance to fracture, modulus, and stiffness.

(1600–1720 cm^{-1}) are commonly used to gauge protein secondary structure (Pelton and McLean 2000). The band positions and relative intensities of these band systems depend on the protein conformation, sample orientation, and the local chemical environment. In a Raman study of molecular changes in proteins, changes in the secondary structure, due to environmental, mechanical, or chemical stresses, will give rise to changes in band position, band area, or both. The amide III envelope reflects peptide amide bonds and is observed as a broad feature in the Raman spectra of proteins. Two maxima are typically observed within the amide III envelope. The wavenumber maximum at a lower Raman shift (range 1230–1240 cm^{-1}) is indicative of a lower α-helix content in the protein's secondary structure, and the maximum at a higher Raman shift (range 1260–1270 cm^{-1}) is indicative of a higher α-helix content in the protein secondary structure (Chi et al. 1998). The amide III ratio is measured as an intensity or area ratio of the 1247 cm^{-1}:1270 cm^{-1} bands. An increase in the amide III ratio indicates increased disorder of the collagen fiber.

The amide I envelope (1600–1720 cm^{-1}) arises from amide carbonyl stretches and is widely used as a Raman and IR marker of collagen structure in bone. Interpretation of the amide I envelope in bone Raman spectra draws on experimental and theoretical Raman studies of proteins and soft tissues. The number of discrete bands in the amide I envelope is an issue of contention and depends on the protein examined (Krimm and Bandekar 1986; Sane et al. 1999). The amide I envelope is also used to study bone collagen cross-linking. Interpretation of amide I in Raman bone spectra as spectroscopic markers of collagen cross-linking draws from the IR literature (Paschalis et al. 2001, 2004, 2011; Gamsjaeger et al. 2011). IR studies showed that the 1660 cm^{-1} band corresponds to mature pyridinoline (Pyr) cross-links and the 1690 cm^{-1} band corresponds to immature dehydrodihydroxylysinonorleucine (de-DHLNL) cross-links. Application of these assignments has been successful in Raman bone spectra (Carden et al. 2003; Gamsjaeger et al. 2012).

5.4 FIGURES OF MERIT

When designing methods for collecting Raman spectra, several figures of merit should be taken into account. These include signal-to-noise, signal-to-background, and Raman band intensity. Signal-to-noise is defined as the intensity of a Raman band of interest with respect to the noise level in the spectrum. A quick estimate of the noise level can be calculated by calculating the standard deviation over many subregions within a spectrum and selecting the region with the lowest standard deviation (and ideally no discernable Raman bands). Signal-to-background is calculated as the intensity of a Raman band of interest with respect to the background offset due to the luminescence background at that band. When collecting spectra, these three metrics should be maximized to the extent possible within experimental constraints.

Most commercially available Raman spectrometers include software that performs automated calibration and spectral preprocessing. It is preferable that commercial software packages include the ability to regularly calibrate the instrument against luminescence standards traceable to primary standards. Luminescence standards for Raman instrumentation include broadband white light sources for intensity calibration, atomic emission lamps for wavelength axis calibration, and fluorescent glasses (NIST Standard Reference Materials, SRM 2241-2243) for calibrating the absolute luminescence intensity. The laser wavelength is measured by collecting the Raman spectrum of a known material (a Raman shift frequency standard), such as cyclohexane. Data preprocessing steps include correcting for the dark current, removal of cosmic ray signals, correction of image distortion, intensity calibration, wavelength calibration, and laser/wavenumber calibration.

Once data have been calibrated, it can be processed in a number of ways. First, the background signal is removed, then band fitting, univariate, or multivariate data processing methods are applied to recover the specimen compositional properties. Advanced data processing methods can also be employed to improve noisy data, but the Raman bands of interest should be clearly distinguishable above the background noise prior to data treatment. Background signals in Raman spectroscopy of biological specimens typically include both a fluorescent background, and some extraneous Raman spectral features from interfering chemical species. Other chemical species can contaminate the recorded spectrum with unwanted Raman spectral features, such as microscope slides, fiber optics, embedding media, or tissue stains. It is often possible to subtract measured reference signals from the collected data, but this requires sufficient signal intensity.

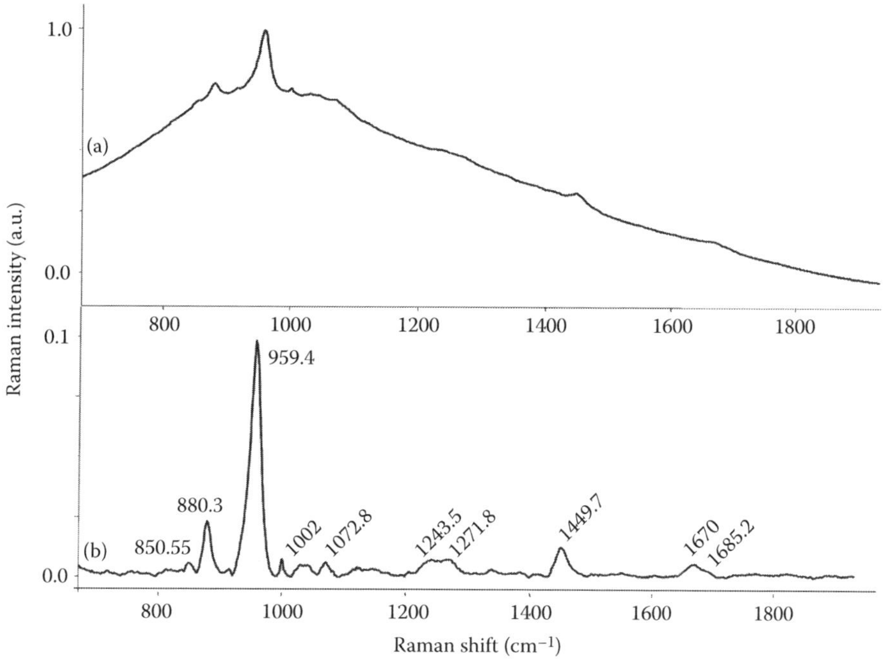

Figure 5.3 (a) Comparison of raw spectrum and (b) spectrum corrected for fluorescence background. (Reprinted from Dehring, K.A. et al., *Appl. Spectrosc.*, 60(10), 1134. Copyright 2006 with permission from the Society of Applied Spectroscopy.)

Raman bands of biological materials are composed of a continuum of molecular species with slightly different molecular environments. Closely spaced Raman shifts arising from similar molecular components cannot be resolved, leading to band envelopes or band *shoulders* (distortions that make bands asymmetric). Envelopes and asymmetry are taken into account by fitting several overlapped functions, such that the sum minimizes the unfitted spectral residual. Raman metrics of band intensity, area, and width are most commonly determined by band fitting, where idealized mathematical functions are used to empirically fit the recorded data. Commercial software, such as GRAMS (a software package by Thermo Scientific), is typically used to fit the data. In theory, Raman bands should follow a Voigt profile, a convolution of a Lorentzian function with a Gaussian function. In practice, most spectroscopists use a pure Lorentzian, a pure Gaussian, or a weighted sum of Gaussian and Lorentzian profiles to fit Raman bands. There is ambiguity in selecting the number of underlying bands in broad envelopes. By calculating the first and second derivative spectra, inflection points can be identified that indicate the *true* number of bands present. In most cases, the number of underlying bands is subjective. Not all underlying bands can be resolved because either spectral resolution is insufficient or the second derivative is too noisy. Band intensity is affected by many factors, and so intensities are instead reported as ratios between two bands measured in a single spectrum.

5.5 STRATEGIES TO OPTIMIZE DATA QUALITY

A major obstacle in many biological Raman measurements is the fluorescence background. Early Raman spectra of biological molecules were often obscured by a large fluorescence background. The fluorescence background is often more than 10 times the magnitude of the Raman signal, and so methods for eliminating or correcting for the fluorescence background are important in obtaining high-quality measurements. As shown in Figure 5.3, fluorescence backgrounds often obscure Raman spectral features. While correction methods do not reduce the additional noise associated with the fluorescence background, they greatly simplify interpretation of the Raman spectrum.

Methods for eliminating the fluorescence background include photobleaching, performing measurements at higher temperatures, using spatial offsets, and recording time-domain spectra. By eliminating a portion of the fluorescence background, the noise can be decreased while maintaining the Raman signal intensity. In photobleaching, the sample is illuminated at a high intensity. Fluorescence decreases with time during exposure to light, though this can be a slow process and clearly involves an ambiguous compositional change. Spatial offsets between the excitation laser and the optical collection region permit rejection of surface fluorescence, particularly with transcutaneous measurements. This is often called *spatially offset Raman spectroscopy* (SORS) and can have several benefits for measuring subsurface features or suppressing interfering signals originating at the surface of the sample (Matousek 2006; Schulmerich et al. 2006). Time-domain measurements are used to discriminate between Raman and fluorescent light because Raman scattering is virtually instantaneous, while fluorescence takes nanoseconds to microseconds.

Software correction methods involve some data processing to separate the Raman signal from the fluorescence background. The most common approaches for fluorescence background correction include piecewise subtraction using manually selected points on the baseline between Raman bands and polynomial background fitting (Lieber and Mahadevan-Jansen 2003; Mazet et al. 2005; Brandt et al. 2006; Cao et al. 2007).

5.5.1 DATA REDUCTION TECHNIQUES

Raman spectra are used for both qualitative identification of materials and can also be used for quantitative measurements of compositional properties. The position, width, and height of bands are indicative of bond frequency, heterogeneity, and composition. Many spectral bands overlap, especially in the 1000–1200 cm^{-1} region, which may be resolved using data processing techniques such as band fitting or multivariate analysis. The excitation and collection sampling regions are both affected by sample orientation and optical scattering, rendering absolute quantification by Raman spectroscopy very difficult and likely impossible in most biological applications. Instead, quantitative Raman spectra are reported by normalizing the spectra to a band present in the spectrum. A common approach is to normalize the spectrum to the maximum intensity in the spectrum. Alternately, if the sample has a component with constant composition, a portion of the spectrum from the constant component can be used as an internal reference.

5.6 APPLICATIONS IN MICROSCOPY AND FIBER OPTICS

Raman microspectroscopy is a powerful tool for examining molecular composition on the micrometer and millimeter scales. We direct the reader to the review by Morris et al. for a detailed treatment of Raman and IR imaging of bone and cartilage, including imaging applications in adaptations to genotype modifications, osteoporosis, external mechanical loading, and aging (Morris et al. 2008). Carden and Morris reviewed IR and Raman spectroscopy of mineralized tissues, including dental specimens and archeological samples (Carden and Morris 2000). Tsuda and Arends reviewed normal and polarized Raman microscopy for dental sample analysis (Tsuda and Arends 1997). In this chapter, we focus on these exciting applications of Raman microspectroscopy: correlating Raman spectral outcomes with biomechanical or material properties in bone, polarized Raman microspectroscopy to examine molecular orientation, bone composition in erosive bone diseases, and pathological crystal formation in soft tissues and biological fluids.

5.6.1 CORRELATING RAMAN OUTCOMES WITH BONE MATERIAL PROPERTIES

Quantitative risk models or bone density measurements inadequately estimate fracture risk because they do not account for the underlying metabolic status of bone (*Consensus Development Conference Statement: Osteoporosis Prevention, Diagnosis, and Therapy* 2000). Together with bone mineral density, biomechanical, architectural, and physiochemical properties comprise *bone quality*—a comprehensive descriptor of bone strength and its resistance to fracture. The fingerprint of data obtained from high-resolution imaging, biomechanical testing, and physiochemical analysis provides a more robust metric of bone strength than bone mineral density measurements alone. Biomechanical properties describe bone material properties such as elastic modulus, ultimate strength, and toughness. Physiochemical properties of bone describe molecular composition or structure including collagen cross-linking, mineral crystallinity, and mineral stoichiometry. Biomechanical and physiochemical properties are widely viewed as important contributions to bone quality. Understanding the relationship between biomechanical and physiochemical properties may lead to improved diagnosis of fracture risk assessment. While most studies examine both mineral and matrix properties, some have focused on either the mineral or matrix. We note that compositional properties do not vary monotonically, even within one study. We have found that examination of Raman spectra for alterations in both the mineral and matrix provides a comprehensive description of compositional properties. In some cases though, such as transcutaneous measurements, overlapping signal from soft tissue may interfere with bone collagen signal and affect calculation of amide III or amide I ratios. In those special cases, we believe that alterations to bone mineral can be a suitable surrogate for whole-tissue composition because earlier studies suggest that mineral properties are highly associated with biomechanical function.

Many studies, using either IR or Raman, have demonstrated that bone compositional properties are associated with altered biomechanical performance. In a 2004 study, the mid-diaphysis of femurs from young, skeletally mature, and old rats were subject to three-point bending and Raman microspectroscopy (Akkus et al. 2004). Tissue mineralization, carbonate substitution, and mineral crystallinity were compared against biomechanical properties including stiffness, yield stress, yield strain, bending modulus, and resilience. They found that compositional properties were highly correlated, and possibly causative, with decreased elastic deformation capacity. A later study of human femora showed that mineral properties accounted for a significant amount of variability in fatigue properties, dependent on anatomic site (Yerramshetty and Akkus 2008).

There is variability in observed compositional changes, depending on the specimen examined. A study of the SAMP6 mouse model of skeletal fragility found normal mineral composition but changes to collagen biochemistry (Silva et al. 2006). And a combined nanoindentation/Raman study of fibrillin 2 (*Fbn2*)-deficient mice found little compositional differences between the normal and *Fbn2* mice even though nanoindentation tests revealed significant reductions in the elastic modulus and hardness of *Fbn2* mice. Exercise was found to affect both the mineral and matrix in addition to improving bone strength (Kohn et al. 2009). Raman-derived metrics correlated with biomechanical properties measured at the whole-bone and tissue level in wild-type and transgenic MMP2 (–/–) mice (Bi et al. 2011). Altered collagen properties and bone mineral stoichiometry are strongly correlated with increased fragility, even in patients with no clinical evidence of osteoporosis (McCreadie et al. 2006; Misof et al. 2012). Correlating localized compositional properties with reduced mechanical integrity and understanding bisphosphonate treatment efficacy are active research fields with ramifications in the clinical treatment of osteoporosis (Gamsjaeger et al. 2011; Paschalis et al. 2011). Finally, we are excited about the extension of a combined Raman and mechanical approach to other musculoskeletal tissues. A recent Raman study of mineral distribution and composition at the tendon–bone interface has exciting prospects for understanding tendon mechanical properties (Schwartz et al. 2012).

5.6.2 ORIENTATION–COMPOSITION EFFECTS

Polarized Raman spectroscopy is used to determine molecular-level orientation in tissue and can be used to quantify orientation between different molecular components in mineralized tissues. All photons carry a polarization state, reflecting the orientation of the electric field (linear polarization) and the phase difference between the oscillating electric and magnetic fields (circular or elliptical polarization). In polarized Raman spectroscopy, a polarization optical filter is used to select Raman-scattered light with the same (parallel) or opposite (perpendicular) polarization state of a polarized laser source. Commonly reported polarized Raman metrics include the depolarization ratio and anisotropy for selected Raman bands. Depolarization ratios are calculated as the intensity of the perpendicular Raman-scattered light relative to the parallel Raman-scattered light. A high depolarization ratio indicates a small difference in the intensities between parallel and perpendicular polarization. A small depolarization ratio indicates a large difference. Small depolarization ratios indicate a high degree of molecular alignment. Polarization anisotropy is calculated as the difference between the parallel and perpendicular polarized Raman intensities relative to the summed (or depolarized) Raman intensities. A high anisotropy value indicates a high degree of molecular alignment. The interaction strength of a photon with a molecular vibration depends on the orientation of the photon and the molecular vibration. In structured tissue with well-defined molecular orientation (such as long parallel collagen fibers and oriented mineral crystallites), the Raman scattering efficiency will depend strongly on relative orientation between the light and the tissue under study. If a sample is kept fixed, individual Raman bands will vary in height relative to each other as the polarization of the incident light is changed, due to the different orientations of the molecular vibrations.

Polarized Raman spectra for hydroxyapatite single crystals and human enamel crystallites as well as for fluorapatites have been reported (Tsuda and Arends 1994; Leroy et al. 2000). Polarized Raman imaging has been used to elucidate the orientation of bone in osteonal tissue (Kazanci et al. 2006, 2007) and to quantify mineral and collagen orientation in a murine model of osteogenesis imperfecta (Raghavan et al. 2010). Quantification of orientation by polarized Raman microspectroscopy has also been used in the analysis of stress in collagen of tendon (Masic et al. 2011). Polarization anisotropy of the 959 cm^{-1} phosphate Raman band has been demonstrated to differentiate between carious and intact human dental enamel (Ko et al. 2006).

5.6.3 BONE COMPOSITION IN EROSIVE BONE DISEASES

Erosive bone diseases have a significant impact on human health. Osteoarthritis, osteomyelitis, rheumatoid arthritis, and cancer skeletal metastasis are leading causes of disability, deformity, nontraumatic amputations, and fractures (Myasoedova et al. 2010; Melton et al. 2011). Despite extensive studies of the biological signaling governing bone remodeling, the compositional properties in these diseases remain poorly understood. Current radiographic imaging, including bone mineral density, techniques remain inadequate for providing a predictive measure of joint deformity, identifying early-stage bone loss, and monitoring the efficacy of therapeutic interventions (Heide et al. 1995; Lipsky 1997; Wakefield et al. 2000; Szkudlarek et al. 2004; Lasker et al. 2007). Compositional properties are widely viewed as important contributors to bone quality in metabolic bone diseases such as osteoporosis, and these properties are now studied in other diseases that affect bone.

Increased fracture risk has been reported in patients with cancer bone metastasis or rheumatoid arthritis. However, it is unclear if the increased susceptibility results from reduced bone tissue volume, altered material properties, or a combination of both. Raman spectroscopy of bone in these diseases can help to address how bone remodels in response to the presence of foreign inflammatory or cancerous cells. In cancer bone metastasis, cell culture and animal model studies suggest that the bone microenvironment and tissue composition are affected in prostate cancer bone metastasis. Lin et al. examined molecular pathways that induce formation of osteoblastic lesions in a C4-2B prostate cancer cell line (Lin et al. 2001). A prominent band at ~958 cm^{-1} in spectra from C4-2B cultures is consistent with apatite mineral, suggesting that prostate cancer cells can produce bone-like mineral *in vitro*. Raman microspectroscopy of metastatic bone tissue specimens has been reported by Bi et al. and Esmonde-White et al. (Bi et al. 2010, 2012; Esmonde-White et al. 2012). Bone mineral composition was altered in both osteolytic and osteoblastic tumors, as measured by mineral crystallinity and mineral stoichiometry. Tissue mineralization decreased in both osteolytic and osteoblastic tumors. Decreased tissue mineralization, in addition to a poorly crystalline mineral, suggests the presence of immature or woven bone in osteoblastic tumors (Esmonde-White et al. 2012). Early reports suggest that bisphosphonate use has a protective effect against osteolytic lesions, preserving tissue mineralization and mineral crystallinity (Bi et al. 2012). Alterations to bone mineral stoichiometry, tissue mineralization, and collagen cross-linking have been reported by Maher et al. and Takahata et al. in a mouse model of rheumatoid arthritis (Maher et al. 2011; Takahata et al. 2012). In their studies, biomechanical and compositional properties of bone in mice overexpressing tissue necrosis factor (TNFα) were compared against wild-type mice. The effect of glucocorticoid treatment, a commonly used steroid to treat rheumatoid arthritis, in the TNFα and wild-type mice was also reported. The genotype and treatment factors independently reduced tissue mineralization and increased mineral stoichiometry, consistent with stiff bone formation (Akkus et al. 2004; Bailey et al. 2004). These studies demonstrate that compositional properties are consistent with reduced bone strength, suggesting that bone remodeling is affected by cancer tumor cells or inflammatory cells.

5.6.4 PATHOLOGICAL CRYSTAL OR MINERAL DEPOSITION

Myriad diseases induce pathological deposition of calcium phosphate minerals. Often these diseases do not affect hard tissues but rather are found in unexpected locations such as the breast, urinary tract, articular joints, muscle, or coronary arteries. In particular, Raman may be useful in noninvasive identification of pathological mineralization in rheumatic diseases. Crystal deposition diseases such as gout or pseudogout can cause deposition of monosodium urate, calcium pyrophosphate, or hydroxyapatite crystals into cartilage, synovium, or surrounding tendon in articular joints. Unambiguous Raman identification of pathological minerals in articular tissues or synovial fluid may improve discrimination of these diseases in a clinical context. One of the first applications of Raman spectroscopy toward this analysis was in synovial fluid and urinary stones (Kodati et al. 1991; McGill et al. 1991). Light microscopy, or polarized light microscopy, of synovial fluid smears is a widely used method to identify calcium phosphate or sodium urate crystals associated with gout, pseudogout, or osteoarthritis. However, this technique is limited in its sensitivity because it uses crystal morphology as its criterion. Raman spectroscopy may improve the sensitivity of synovial fluid tests for crystal deposition diseases because it can differentiate calcium phosphate from sodium urate crystals based on chemical composition. The first IR study of synovial fluid crystals in 1983 identified apatite, octacalcium phosphate, and tricalcium phosphate (McCarty et al. 1983). Since 1991, several groups have reported Raman identification of calcium pyrophosphate, basic calcium phosphate, and sodium urate crystals in synovial fluid (Yavorskyy et al. 2008; Cheng et al. 2009). We recently reported the development of a fiber-optic arthroscopic Raman instrument for translation into the clinical field (Esmonde-White et al. 2011).

In addition to exciting prospects in a clinical context, Raman and IR spectroscopies have made important contributions to understanding mineralization in cartilage and other joint tissues (Morris et al. 2008). Mechanisms of pathologic mineral deposits in osteoarthritic matrix vesicles of articular cartilage were examined *in vitro* using enzyme assays, radiometric assay of ^{45}Ca, Fourier-transform infrared (FTIR) spectroscopy, and polarized light microscopy by Derfus et al. (1998). FTIR spectra identified calcium pyrophosphate dihydrate (CPPD) and apatite mineral crystals that were being produced in osteoarthritis matrix vesicles, which are relevant in osteoarthritis progression. Heger et al. proposed a mechanism that laser reshaping of cartilage induces tissue mineralization that is responsible for long-term structural stability (Heger et al. 2006; Ignatieva et al. 2007). Raman spectra of laser-reshaped rabbit auricular (ear) cartilage showed calcium sulfate and calcium carbonate crystals intracellularly and calcium phosphate crystals were found in extracellular spaces. The authors proposed that, in addition to a stress relaxation of cartilage tissue, laser treatment caused apoptotic cells to release matrix vesicles and the released matrix vesicles mediated mineralization. Raman mapping of fixed and dehydrated cartilage sections showed the presence of calcium carbonate in the middle cartilage zone (Bonifacio et al. 2010).

Pathologic mineralization can occur in unexpected anatomic locations including muscle, arteries, and breast tissues. Mineralization of soft tissues, called heterotopic ossification, is a serious complication of traumatic injury particularly for blast wounds obtained in the course of combat. Heterotopic ossification is only treated through surgical removal, sometimes requiring multiple revision surgeries. Identification of mineralizing muscle tissue in heterotopic ossification by Raman spectroscopy has recently been reported by Crane and Elster (2012). Interestingly, Raman identified the presence of an apatite-like mineral in sites deemed normal by palpation. Mineralized atherosclerosis lesions may carry a higher risk of clot formation, and identification of mineralized lesions before or during surgery can potentially improve angiography outcomes. Brennan et al. identified hydroxyapatite in several lesions from coronary artery specimens (Brennan et al. 1997). Microcalcifications in breast tissues have been proposed as markers of benign or metastatic disease (Cox and Morgan 2013). Discrimination between calcium oxalate found in benign lesions from calcium hydroxyapatite found in malignant lesions is not possible with current radiographic tests. Stone et al. reported the use of time-gated and transmission Raman spectroscopy for discriminating calcifications in simulated breast tissue (Baker et al. 2007; Stone and Matousek 2008). The unique and strong Raman signal of calcium phosphate minerals is easily detected in soft tissues, and we anticipate many Raman reports of pathologic calcification in other soft tissues.

5.7 FUTURE TRENDS IN RAMAN SPECTROSCOPY

The future is bright for new applications of Raman spectroscopy in mineralized tissues. We see two exciting trends emerging from the past 5 years of research. The first trend is Raman microscopy approaching nanoscale spatial resolution. Tip-enhanced Raman spectroscopy (TERS) is a technique that adapts surface-enhanced Raman spectroscopy to scanning near-field optical microscopy (Stockle et al. 2000). The fusion of nanometer spatial resolution with compositional information is a powerful tool for detailed analysis of macromolecular structure. An attractive feature of TERS is the capability to probe secondary structure in precise locations. For collagen, the normally broad amide I would present as discrete bands, enabling detailed characterization of protein secondary structure on the collagen fibril surface (Gullekson et al. 2011). Application to mineralized collagen would be expected to yield similarly exciting results.

The second trend is translation of Raman spectroscopy toward noninvasive mineral detection in a clinical setting using SORS. Since the first reports in 2006 by Matousek et al. and Schulmerich et al., technological advances in SORS have enabled Raman spectroscopy of buried tissue or layers in highly turbid systems. Measurements of mineral through centimeters of overlying soft tissue are now possible, enabling Raman analysis of bone or mineral without collecting a biopsy. Translation of the SORS technology to clinical use will need to address safety and efficacy requirements that are standard for regulatory approval of medical devices. Pilot clinical studies by us and other groups are establishing laser safety parameters, identifying suitable spectroscopic disease markers, and adapting technology for use in a clinician's office or surgical suite. It is exciting to envision a Raman intraoperative or bedside tool, providing real-time compositional analysis of tissue.

ACKNOWLEDGMENTS

Raman microscopy and fiber-optic spectroscopy research by the Esmonde-White's is funded by grant R21EB101026 (NIH/NIBIB), R01AR055222, and R01AR047969 (NIH/NIAMS). KEW acknowledges training grant T32AR007080 (NIH/NIAMS) and a career development grant from the University of Michigan CTSA UL1RR024986 (NIH/NCRR). We thank Professor Blake Roessler, Professor Michael Morris, and Miss Daphne Esmonde-White for their support during the manuscript preparation.

REFERENCES

Akkus, O., F. Adar, and M.B. Schaffler. 2004. Age-related changes in physicochemical properties of mineral crystals are related to impaired mechanical function of cortical bone. *Bone* 34 (3):443–453.

Awonusi, A., M. Morris, and M. Tecklenburg. 2007. Carbonate assignment and calibration in the Raman spectrum of apatite. *Calcified Tissue International* 81 (1):46–52.

Bailey, A.J., J.P. Mansell, T.J. Sims, and X. Banse. 2004. Biochemical and mechanical properties of subchondral bone in osteoarthritis. *Biorheology* 41 (3–4):349–358.

Baker, R., P. Matousek, K.L. Ronayne, A.W. Parker, K. Rogers, and N. Stone. 2007. Depth profiling of calcifications in breast tissue using picosecond Kerr-gated Raman spectroscopy. *Analyst* 132 (1):48–53.

Bi, X., J. Nyman, C. Morrissey, M. Roudier, A. Dowell, and A. Mahadevan-Jansen. 2012. Using Raman spectroscopy to characterize bone metastasis and to evaluate treatment response in prostate cancer patients. Paper read at *Biomedical Optics (BIOMED)*, Miami, FL.

Bi, X., C.A. Patil, C.C. Lynch, G.M. Pharr, A. Mahadevan-Jansen, and J.S. Nyman. 2011. Raman and mechanical properties correlate at whole bone- and tissue-levels in a genetic mouse model. *Journal of Biomechanics* 44 (2):297–303.

Bi, X., C. Patil, C. Morrissey, M.P. Roudier, A. Mahadevan-Jansen, and J. Nyman. 2010. Characterization of bone quality in prostate cancer bone metastases using Raman spectroscopy. Paper read at *SPIE Photonics West*, San Francisco, CA.

Blakeslee, K.C. and R.A. Condrate. 1971. Vibrational spectra of hydrothermally prepared hydroxyapatites. *Journal of the American Ceramic Society* 54 (11):559–563.

Bonifacio, A., C. Beleites, F. Vittur, E. Marsich, S. Semeraro, S. Paoletti, and V. Sergo. 2010. Chemical imaging of articular cartilage sections with Raman mapping, employing uni- and multi-variate methods for data analysis. *Analyst* 135 (12):3193–3204.

Brandt, N.N., O.O. Brovko, A.Y. Chikishev, and O.D. Paraschuk. 2006. Optimization of the rolling-circle filter for Raman background subtractions. *Applied Spectroscopy* 60 (3):288–293.

Brennan, J.F., T.J. Romer, R.S. Lees, A.M. Tercyak, J.R. Kramer, and M.S. Feld. 1997. Determination of human coronary artery composition by Raman spectroscopy. *Circulation* 96 (1):99–105.

Cael, J.J., D.H. Isaac, J. Blackwell, J.L. Koenig, E.D.T. Atkins, and J.K. Sheehan. 1976. Polarized infrared spectra of crystalline glycosaminoglycans. *Carbohydrate Research* 50:169–179.

Cao, A., A.K. Pandya, G.K. Serhatkulu, R.E. Weber, H. Dai, J.S. Thakur, V.M. Naik et al. 2007. A robust method for automated background subtraction of tissue fluorescence. *Journal of Raman Spectroscopy* 38 (9):1199–1205.

Carden, A. and M.D. Morris. 2000. Application of vibrational spectroscopy to the study of mineralized tissues (review). *Journal of Biomedical Optics* 5 (3):259–268.

Carden, A., R.M. Rajachar, M.D. Morris, and D. Kohn. 2003. Ultrastructural changes accompanying the mechanical deformation of bone tissue: A Raman imaging study. *Calcified Tissue International* 72:166–172.

Chase, D.B. 1986. Fourier transform Raman spectroscopy. *Journal of the American Chemical Society* 108 (24):7485–7488.

Chase, D.B. and J.F. Rabolt, eds. 1994. *Fourier-Transform Raman Spectroscopy: From Concept to Experiment*. San Diego, CA: Academic Press.

Cheng, X., D.G. Haggins, R.H. York, Y.N. Yeni, and O. Akkus. 2009. Analysis of crystals leading to joint arthropathies by Raman spectroscopy: Comparison with compensated polarized imaging. *Applied Spectroscopy* 63 (4):381–386.

Chi, Z., X.G. Chen, J.S.W. Holtz, and S.A. Asher. 1998. UV resonance Raman-selective amide vibrational enhancement: Quantitative methodology for determining protein secondary structure. *Biochemistry* 37:2854–2864.

Cornilsen, B.C. 1984. Solid state vibrational spectra of calcium pyrophosphate dihydrate. *Journal of Molecular Structure* 117:1–9.

Cox, R.F. and M.P. Morgan. 2013. Microcalcifications in breast cancer: Lessons from physiological mineralization. *Bone* 53 (2):437–450.

Crane, N.J. and E.A. Elster. 2012. Vibrational spectroscopy: A tool being developed for the noninvasive monitoring of wound healing. *Journal of Biomedical Optics* 17 (1):010902.

Crane, N.J., M.D. Morris, M.A. Ignelzi, Jr., and G. Yu. 2005. Raman imaging demonstrates FGF2-induced craniosynostosis in mouse calvaria. *Journal of Biomedical Optics* 10 (3):031119–031119.

de Aza, P.N., C. Santos, A. Pazo, S. de Aza, R. Cusco, and L. Artus. 1997. Vibrational properties of calcium phosphate compounds. 1. Raman spectrum of B-tricalcium phosphate. *Chemistry of Materials* 9 (4):912–915.

Dehring, K.A., N.J. Crane, A.R. Smukler, J.B. McHugh, B.J. Roessler, and M.D. Morris. 2006. Identifying chemical changes in subchondral bone taken from murine knee joints using Raman spectroscopy. *Applied Spectroscopy* 60 (10):1134–1141.

Derfus, B., S. Kranendonk, N. Camacho, N. Mandel, V. Kushnaryov, K. Lynch, and L. Ryan. 1998. Human osteoarthritic cartilage matrix vesicles generate both calcium pyrophosphate dihydrate and apatite *in vitro*. *Calcified Tissue International* 63 (3):258–262.

Edwards, H.G.M. and E.A. Carter. 2001. Biological applications of Raman spectroscopy. In *Infrared and Raman Spectroscopy of Biological Materials*, eds. H.-U. Gremlich and B. Yan. Vol. 24, Practical Spectroscopy Series. New York: Marcel Dekker Inc.

Esmonde-White, K.A., F.W.L. Esmonde-White, M.D. Morris, and B.J. Roessler. 2011. Fiber-optic Raman spectroscopy of joint tissues. *Analyst* 136 (8):1675–1685.

Esmonde-White, K.A., J. Sottnik, M. Morris, and E. Keller. 2012. Raman spectroscopy of bone metastasis. Paper read at *Photonic Therapeutics and Diagnostics VIII*, San Francisco, CA.

Everall, N., J. Lapham, F. Adar, A. Whitley, E. Lee, and S. Mamedov. 2007. Optimizing depth resolution in confocal Raman microscopy: A comparison of metallurgical, dry corrected, and oil immersion objectives. *Applied Spectroscopy* 61:251–259.

Everall, N.J. 2000. Confocal Raman microscopy: Why the depth resolution and spatial accuracy can be much worse than you think. *Applied Spectroscopy* 54 (10):1515–1520.

Frushour, B.G. and J.L. Koenig. 1975. Raman scattering of collagen, gelatin, and elastin. *Biopolymers* 14 (2):379–391.

Gamsjaeger, S., B. Buchinger, R. Zoehrer, R. Phipps, K. Klaushofer, and E.P. Paschalis. 2011. Effects of one year daily teriparatide treatment on trabecular bone material properties in postmenopausal osteoporotic women previously treated with alendronate or risedronate. *Bone* 49 (6):1160–1165.

Gamsjaeger, S., S. Robins, D.N. Tatakis, K. Klaushofer, and E.P. Paschalis. 2012. Identification of trivalent collagen cross links by Raman microspectroscopy. *Bone* 51 (6):S18.

Goheen, S.C., L.J. Lis, and J.W. Kauffman. 1978. Raman spectroscopy of intact feline corneal collagen. *Biochimica et Biophysica Acta* 536:197–204.

Griffith, W.P. 1987. Advances in the Raman and infrared spectroscopy of minerals. In *Spectroscopy of Inorganic-Based Materials*, eds. R.J.H. Clark and R.E. Hester. New York: John Wiley & Sons, Inc.

Gullekson, C., L. Lucas, K. Hewitt, and L. Kreplak. 2011. Surface-sensitive Raman spectroscopy of collagen I fibrils. *Biophysical Journal* 100 (7):1837–1845.

Gunasekaran, S., G. Anbalagan, and S. Pandi. 2006. Raman and infrared spectra of carbonates of calcite structure. *Journal of Raman Spectroscopy* 37 (9):892–899.

Heger, M., S. Mordon, G. Leroy, L. Fleurisse, and C. Creusy. 2006. Raman microspectrometry of laser-reshaped rabbit auricular cartilage: Preliminary study on laser-induced cartilage mineralization. *Journal of Biomedical Optics* 11 (2):024003.

Heide, A.V.D., C.A. Remme, D.M. Hofman, J.W.G. Jacobs, and J.W.J. Bijlsma. 1995. Prediction of progression of radiologic damage in newly diagnosed rheumatoid arthritis. *Arthritis & Rheumatism* 38 (10):1466–1474.

Hendra, P., C. Jones, and G. Warnes. 1991. *Fourier Transform Raman Spectroscopy: Instrumentation and Chemical Applications*. New York: Ellis Horwood.

Ignatieva, N., O. Zakharkina, G. Leroy, E. Sobol, N. Vorobieva, and S. Mordon. 2007. Molecular processes and structural alterations in laser reshaping of cartilage. *Laser Physics Letters* 4 (10):749–753.

Kazanci, M., P. Roschger, E.P. Paschalis, K. Klaushofer, and P. Fratzl. 2006. Bone osteonal tissues by Raman spectral mapping: Orientation-composition. *Journal of Structural Biology* 156 (3):489–496.

Kazanci, M., H.D. Wagner, N.I. Manjubala, H.S. Gupta, E. Paschalis, P. Roschger, and P. Fratzl. 2007. Raman imaging of two orthogonal planes within cortical bone. *Bone* 41 (3):456–461.

Ko, A.C.T., L.-P. Choo-Smith, M. Hewko, M.G. Sowa, C.C.S. Dong, and B. Cleghorn. 2006. Detection of early dental caries using polarized Raman spectroscopy. *Optics Express* 14 (1):203–215.

Kodati, V.R., G.E. Tomasi, J.L. Turumin, and A.T. Tu. 1991. Raman spectroscopic identification of phosphate-type kidney stones. *Applied Spectroscopy* 45 (4):581–583.

Koenig, J.L. 1972. Raman spectroscopy of biological molecules: A review. *Journal of Polymer Science: Macromolecular Reviews* 6 (1):59–177.

Kohn, D.H., N.D. Sahar, J.M. Wallace, K. Golcuk, and M.D. Morris. 2009. Exercise alters mineral and matrix composition in the absence of adding new bone. *Cells Tissues Organs* 189 (1–4):33–37.

Krimm, S. and J. Bandekar. 1986. Vibrational spectroscopy and conformation of peptides, polypeptides, and proteins. *Advances in Protein Chemistry* 38:181.

Lasker, J.M., C.J. Fong, D.T. Ginat, E. Dwyer, and A.H. Hielscher. 2007. Dynamic optical imaging of vascular and metabolic reactivity in rheumatoid joints. *Journal of Biomedical Optics* 12 (5):052001.

Leroy, G., N. Leroy, G. Penel, C. Rey, P. Lafforgue, and E. Bres. 2000. Polarized micro-Raman study of fluorapatite single crystals. *Applied Spectroscopy* 54 (10):1521–1527.

Lewis, I.R. and H.G.M. Edwards, eds. 2001. *Handbook of Raman Spectroscopy*. New York: Marcel Dekker, Inc.

Lieber, C.A. and A. Mahadevan-Jansen. 2003. Automated method for subtraction of fluorescence from biological Raman spectra. *Applied Spectroscopy* 57 (11):1363–1367.

Lin, D.-L., C.P. Tarnowski, J. Zhang, J. Dai, E. Rohn, A.H. Patel, M.D. Morris, and E.T. Keller. 2001. Bone metastatic LNCaP-derivative C4-2B prostate cancer cell line mineralizes *in vitro*. *The Prostate* 27:212–221.

Lin, V.J.C. and J.L. Koenig. 1976. Raman studies of bovine serum albumin. *Biopolymers* 15 (1):203–218.

Lipsky, B.A. 1997. Osteomyelitis of the foot in diabetic patients. *Clinical Infectious Diseases* 25:1318–1326.

Lyon, L.A., C.D. Keating, A.P. Fox, B.E. Baker, L. He, S.R. Nicewarner, S.P. Mulvaney, and M.J. Natan. 1998. Raman spectroscopy. *Analytical Chemistry* 70:341R–361R.

Maher, J.R., M. Takahata, H.A. Awad, and A.J. Berger. 2011. Raman spectroscopy detects deterioration in biomechanical properties of bone in a glucocorticoid-treated mouse model of rheumatoid arthritis. *Journal of Biomedical Optics* 16 (8):087012.

Masic, A., L. Bertinetti, R. Schuetz, L. Galvis, N. Timofeeva, J.W.C. Dunlop, J. Seto, M.A. Hartmann, and P. Fratzl. 2011. Observations of multiscale, stress-induced changes of collagen orientation in tendon by polarized Raman spectroscopy. *Biomacromolecules* 12 (11):3989–3996.

Matousek, P. 2006. Inverse spatially offset Raman spectroscopy for deep noninvasive probing of turbid media. *Applied Spectroscopy* 60 (11):1341–1347.

Mazet, V., C. Carteret, D. Brie, J. Idier, and B. Humbert. 2005. Background removal from spectra by designing and minimising a non-quadratic cost function. *Chemometrics and Intelligent Laboratory Systems* 76 (2):121–133.

McCarty, D.J., J.R. Lehr, and P.B. Halverson. 1983. Crystal populations in human synovial fluid. Identification of apatite, octacalcium phosphate, and tricalcium phosphate. *Arthritis & Rheumatism* 26 (10):1220–1224.

McCreadie, B.R., M.D. Morris, T.-C. Chen, D.S. Rao, W.F. Finney, E. Widjaja, and S.A. Goldstein. 2006. Bone tissue compositional differences in women with and without osteoporotic fracture. *Bone* 39 (6):1190–1195.

McCreery, R.L., ed. 2000. *Raman Spectroscopy for Chemical Analysis*. Vol. 157, Chemical Analysis. New York: John Wiley & Sons, Inc.

McGill, N., P.A. Dieppe, M. Bowden, D.J. Gardiner, and M. Hall. 1991. Identification of pathological mineral deposits by Raman microscopy. *The Lancet* 337:77–78.

Melton, L.J., M.M. Lieber, E.J. Atkinson, S.J. Achenbach, H. Zincke, T.M. Therneau, and S. Khosla. 2011. Fracture risk in men with prostate cancer: A population-based study. *Journal of Bone and Mineral Research* 26 (8):1808–1815.

Misof, B.M., S. Gamsjaeger, A. Cohen, B. Hofstetter, P. Roschger, E. Stein, T.L. Nickolas et al. 2012. Bone material properties in premenopausal women with idiopathic osteoporosis. *Journal of Bone and Mineral Research* 27 (12):2551–2561.

Morris, M.D. and G.S. Mandair. 2011. Raman assessment of bone quality. *Clinical Orthopaedics and Related Research* 469 (8):2160–2169.

Morris, M.D., M.V. Schulmerich, K.A. Dooley, and K.A. Esmonde-White. 2008. Vibrational spectroscopic imaging of hard tissue. In *Infrared and Raman Spectroscopic Imaging*, eds. R. Salzer and H.W. Siesler. Weinheim, Germany: Wiley-VCH.

Mulvaney, S.P. and C.D. Keating. 2000. Raman spectroscopy. *Analytical Chemistry* 72 (12):145–158.

Myasoedova, E., C.S. Crowson, H.M. Kremers, T.M. Therneau, and S.E. Gabriel. 2010. Is the incidence of rheumatoid arthritis rising?: Results from Olmsted County, Minnesota, 1955–2007. *Arthritis & Rheumatism* 62 (6):1576–1582.

National Institutes of Health. 2013. *Consensus Development Conference Statement: Osteoporosis Prevention, Diagnosis, and Therapy*, March 27–29, 2000. Bethesda, MD: National Institutes of Health [cited April 29, 2013]. Available from http://consensus.nih.gov/2000/2000osteoporosis111html.htm.

Naumann, D. 2001. FT-infrared and FT-Raman spectroscopy in biomedical research. *Applied Spectroscopy Reviews* 36 (2):239–298.

Paschalis, E.P., E. Shane, G. Lyritis, G. Skarantavos, R. Mendelsohn, and A.L. Boskey. 2004. Bone fragility and collagen cross-links. *Journal of Bone and Mineral Research* 19 (12):2000–2004.

Paschalis, E.P., D.N. Tatakis, S. Robins, P. Fratzl, I. Manjubala, R. Zoehrer, S. Gamsjaeger et al. 2011. Lathyrism-induced alterations in collagen cross-links influence the mechanical properties of bone material without affecting the mineral. *Bone* 49 (6):1232–1241.

Paschalis, E.P., K. Verdelis, S.B. Doty, A.L. Boskey, R. Mendelsohn, and M. Yamauchi. 2001. Spectroscopic characterization of collagen cross-links in bone. *Journal of Bone and Mineral Research* 16 (10):1821–1828.

Pelton, J.T. and L.R. McLean. 2000. Spectroscopic methods for analysis of protein secondary structure. *Analytical Biochemistry* 277:167–176.

Penel, G., G. Leroy, C. Rey, and E. Bres. 1998. Micro Raman spectral study of the PO_4 and CO_3 vibrational modes in synthetic and biological apatites. *Calcified Tissue International* V63 (6):475–481.

Raghavan, M., N.D. Sahar, R.H. Wilson, M.-A. Mycek, N. Pleshko, D.H. Kohn, and M.D. Morris. 2010. Quantitative polarized Raman spectroscopy in highly turbid bone tissue. *Journal of Biomedical Optics* 15 (3):037001.

Raman, C.V. and K.S. Krishnan. 1928. A new type of secondary radiation. *Nature* 121 (3028):501–502.

Sane, S.U., S.M. Cramer, and T.M. Przybycien. 1999. A holistic approach to protein secondary structure characterization using amide I band Raman spectroscopy. *Analytical Biochemistry* 269 (2):255–272.

Sauer, G.R., W.B. Zunic, J.R. Durig, and R.E. Wuthier. 1994. Fourier transform Raman spectroscopy of synthetic and biological calcium phosphates. *Calcified Tissue International* 54:414–420.

Schulmerich, M.V., K.A. Dooley, M.D. Morris, T.M. Vanasse, and S.A. Goldstein. 2006. Transcutaneous fiber optic Raman spectroscopy of bone using annular illumination and a circular array of collection fibers. *Journal of Biomedical Optics* 11 (6):060502.

Schwartz, A.G., J.D. Pasteris, G.M. Genin, T.L. Daulton, and S. Thomopoulos. 2012. Mineral distributions at the developing tendon enthesis. *PLOS One* 7 (11):e48630.

Silva, M.J., M.D. Brodt, B. Wopenka, S. Thomopoulos, D. Williams, M.H.M. Wassen, M. Ko, N. Kusano, and R.A. Bank. 2006. Decreased collagen organization and content are associated with reduced strength of demineralized and intact bone in the SAMP6 mouse. *Journal of Bone and Mineral Research* 21 (1):78–88.

Stockle, R.M., Y.D. Suh, V. Deckert, and R. Zenobi. 2000. Nanoscale chemical analysis by tip-enhanced Raman spectroscopy. *Chemical Physics Letters* 318 (1–3):131–136.

Stone, N. and P. Matousek. 2008. Advanced transmission Raman spectroscopy: A promising tool for breast disease diagnosis. *Cancer Research* 68 (11):4424–4430.

Szkudlarek, M., E. Narvestad, M. Klarlund, M. Court-Payen, H.S. Thomsen, and M. Østergaard. 2004. Ultrasonography of the metatarsophalangeal joints in rheumatoid arthritis: Comparison with magnetic resonance imaging, conventional radiography, and clinical examination. *Arthritis & Rheumatism* 50 (7):2103–2112.

Takahata, M., J.R. Maher, S.C. Juneja, J. Inzana, L. Xing, E.M. Schwarz, A.J. Berger, and H.A. Awad. 2012. Mechanisms of bone fragility in a mouse model of glucocorticoid-treated rheumatoid arthritis: Implications for insufficiency fracture risk. *Arthritis & Rheumatism* 64 (11):3649–3659.

Tarnowski, C.P., M.A. Ignelzi, and M.D. Morris. 2002. Mineralization of developing mouse calvaria as revealed by Raman microspectroscopy. *Journal of Bone and Mineral Research* 17 (6):1118–1126.

Tsuda, H. and J. Arends. 1994. Orientational micro-Raman spectroscopy on hydroxyapatite single crystals and human enamel crystallites. *Journal of Dental Research* 73 (11):1703–1710.

Tsuda, H. and J. Arends. 1997. Raman spectroscopy in dental research: A short review of recent studies. *Advances in Dental Research* 11 (4):539–547.

Vasko, P.D., J. Blackwell, and J.L. Koenig. 1971. Infrared and Raman spectroscopy of carbohydrates: Part I: Identification of O–H and C–H-related vibrational modes for D-glucose, maltose, cellobiose, and dextran by deuterium-substitution methods. *Carbohydrate Research* 19 (3):297–310.

Wakefield, R.J., W.W. Gibbon, P.G. Conaghan, P. O'Connor, D. McGonagle, C. Pease, M.J. Green, D.J. Veale, J.D. Isaacs, and P. Emery. 2000. The value of sonography in the detection of bone erosions in patients with rheumatoid arthritis: A comparison with conventional radiography. *Arthritis & Rheumatism* 43 (12):2762–2770.

Walton, A.G., M.J. Deveney, and J.L. Koenig. 1970. Raman spectroscopy of calcified tissue. *Calcified Tissue Research* 6 (1):162–167.

Wopenka, B. and J.D. Pasteris. 2005. A mineralogical perspective on the apatite in bone. *Materials Science and Engineering C* 25 (2):131–143.

Yavorskyy, A., A. Hernandez-Santana, G. McCarthy, and G. McMahon. 2008. Detection of calcium phosphate crystals in the joint fluid of patients with osteoarthritis—Analytical approaches and challenges. *Analyst* 133 (3):302–318.

Yerramshetty, J.S. and O. Akkus. 2008. The associations between mineral crystallinity and the mechanical properties of human cortical bone. *Bone* 42 (3):476–482.

6 Chemical mapping with x-ray absorption spectroscopy

Yannicke Dauphin and Murielle Salomé

Contents

6.1 INTRODUCTION AND TYPE OF INFORMATION

Crystalline or amorphous, biominerals always comprise mineral and organic fractions, whatever the mineralogy. It is now well established that the shape and composition of biogenic minerals differ from those of their nonbiogenic equivalents. Chemical compositions are known using atomic absorption, inductively coupled plasma, etc., with good precision. Nevertheless, samples must be dissolved, and the spatial distribution of the chemical elements is unknown. In such conditions, it is not possible to associate structures and chemical compositions. As most shells and skeletons are composed of multiple regions, the mineralogy, structure, and composition of which differ, to decipher the relationships between growth process and physicochemical character is a major question. Spatially resolved analyses have been made possible at micro- and nanoscales thanks to the development of electron and ion microprobes. Such systems allow quantitative as well as detailed maps when the organic components of biominerals are well-structured layers, such as the interprismatic envelopes of some bivalve molluscs or the interlamellar membranes of the nacreous layers.

Another major question is: Where is a chemical element? Is it in the organic or mineral fraction, or both? Shells and skeletons are not simple successions of organic and mineral layers. *Growth lines* reflect changes in mineral–organic ratios and/or chemical changes throughout the life of the organisms. SEM and AFM observations have shown that the mineral–organic interplay exists at a nanoscale, within a *crystal*. Electron and ion microprobes are not able to solve this problem, despite their high spatial resolution and sensitivity. Moreover, the speciation of a chemical element can be related to the presence of a mineral. It can be also indicative of a biogenic or sedimentary origin for enigmatic structures. The presence of organic components in fossil organisms can help to understand their taphonomic history, with broader implications such as phylogenetic or paleoenvironmental reconstructions. Finally, detection of some metallic elements (Zn, Hg, etc.) is used to understand the behavior of dental amalgams and bone replacement for health purposes.

To investigate the composition and chemical specificity of biominerals, x-ray absorption spectroscopy and microspectroscopy are very attractive tools, which can provide information both on their mineral and organic components. While x-ray fluorescence (XRF) gives access to quantitative information about the elemental composition of a sample, x-ray absorption near-edge structure (XANES) spectroscopy is sensitive to the oxidation state and the chemical environment of an element of interest. When combined with x-ray microscopy capabilities, the so-called micro-XRF allows the mapping of elements in a sample in 2D or even 3D (in fluo-microtomography) and the study of their colocalization, which may contribute to the understanding of their interactions. Access to spatially resolved information with high resolution is essential for the analysis of finely structured and heterogeneous materials as biominerals usually are. XRF is an extremely sensitive technique, suitable for the detection of trace elements, since detection limits down to the ppm can be reached depending on the atomic number of the element of interest.

To go further than the elemental composition of the sample, XANES provides in addition chemical information about the binding of a given element, which is of great interest to unravel the organic/mineral interface at the molecular level in biominerals. 2D maps of chemical molecules can also be obtained. When measured in fluorescence mode, XANES benefits from the high sensitivity of XRF and can be used to assess the chemical state of even diluted species. The high penetration depth of x-rays confers to XANES the possibility to probe hidden atoms in the sample matrix and not only the sample surface. Importantly, no long-range ordering of the investigated material is required; hence, the technique is well suited also for noncrystalline and highly disordered materials, like organic matter, and even solutions. Typically, in biominerals, XANES will allow us to identify molecular groups constituting the organic matrix and the chemical bonds they form with mineral material.

XANES necessitates an accurate scanning of the energy of the impinging x-ray beam and is therefore usually performed at synchrotron facilities, where high photon flux and beam energy tunability are available. Most synchrotron facilities have several beamlines optimized for bulk x-ray absorption spectroscopy and additionally microprobe beamlines allowing microspectroscopy.

6.1.1 HISTORY

The history of the development of XANES is closely related to the development of extended x-ray absorption fine structure spectroscopy (EXAFS). The first observations and attempts to explain EXAFS structures were initiated in the 1920s (Kossel 1920, Kronig 1931), but possibility to obtain good data was limited with conventional x-ray tubes. In the 1970s, rapid advances were made in the theory of XAFS by Lytle, Sayers, and Stern who introduced a new Fourier transform analysis of EXAFS (Sayers et al. 1971, Lytle et al. 1982). With the advent of synchrotron sources, EXAFS quickly became a quantitative tool for structure determination in a wide range of applications. For a detailed retrospective of EXAFS history, please refer to Stern (2001). Later on, it appeared that the basic phenomena in EXAFS and XANES were the same, but with more complexity in XANES, because of multiple scattering effects. First papers reporting XANES measurements appear in the 1980s (Bianconi et al. 1988).

The technique in bulk or focused mode is now widespread, with applications in many scientific fields ranging from geochemistry, materials science, and chemistry to biology and environmental sciences.

6.1.2 PHYSICAL PRINCIPLE

6.1.2.1 X-ray fluorescence

In the multi-keV energy range, the x-ray absorption cross section is dominated by the photoelectric effect. This is the process by which a photon is absorbed by an atom and its energy transferred to an inner shell electron. XRF occurs when x-rays strike an atom with an energy sufficient to eject one of its inner shell electrons. The resulting vacancy needs to be filled to allow the atom to return to its stable ground state. The relaxation process can be achieved through a subsequent reorganization of the electronic shells of the atom with the transition of an electron from a higher energy level to the core-hole level. This transition is accompanied with the emission of a fluorescence photon, having an energy corresponding to the gap between the initial energy level E_i of the transition electron and its final energy E_f. The energy of the fluorescence photon $\Delta E = E_i - E_f = h\nu$ is characteristic of the electronic configuration of the emitting atom. Tables of the energies of the allowed transitions in the atom are available and allow the identification of their emitting elements (Thompson et al. 2009). Fluorescence emission lines are labeled by the destination shell (K, L, M, N) of the transition electron and a Greek letter (α, β, γ) denoting transitions from higher shells (Figure 6.1). A further numbering (α1, α2, β1, …) distinguishes transitions from different orbits of the same shell. For example, Kα lines result from the transition of an L-shell electron to fill a vacancy in the K-shell. This is the most frequent transition and, thus, the most intense fluorescence emission line. Kβ lines result from M-shell electron transitions to the K-shell. XRF is the primary relaxation process; however, as the atom returns to its ground state, instead of emitting a fluorescence photon, the excitation energy can alternatively be transferred to an outer electron that is ejected. These so-called Auger electrons are more probable for low Z elements.

Fitting of the fluorescence spectra can be performed using dedicated software such as, among others, PyMca (Solé et al. 2007) or AXIL (Vekemans et al. 1994) to deconvolve the different fluorescence emission lines and evaluate the area under the peak, which is proportional to the elemental concentration (Figure 6.2). Quantitative measurements can be obtained, provided suitable calibration samples have been measured. Those standards should

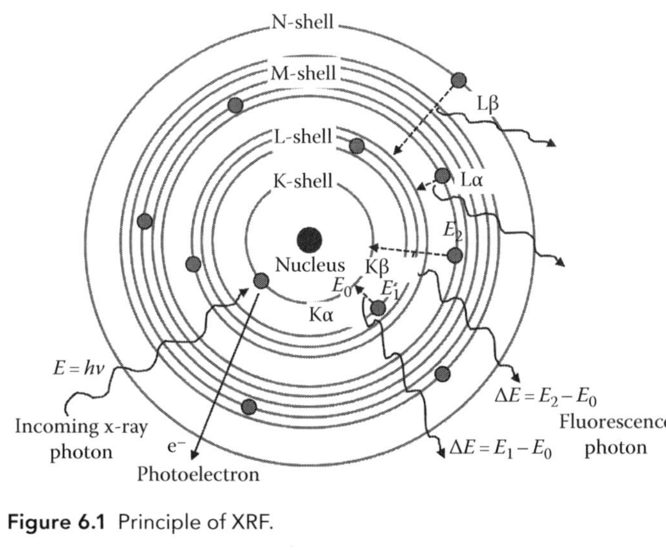

Figure 6.1 Principle of XRF.

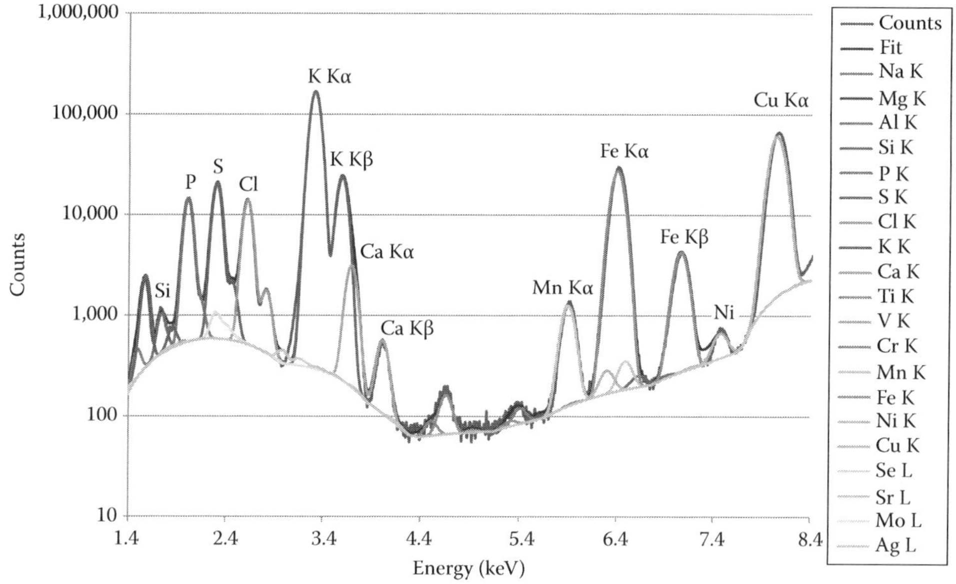

Figure 6.2 (See color insert.) Example of fitting of a fluorescence spectrum using PyMca (Solé et al. 2007).

exhibit the same matrix composition as the sample (so that self-absorption phenomena are comparable) and contain known concentrations of the element of interest. Standard concentrations should be precisely known. They can be synthesized in the laboratory or if a high level of accuracy is required, purchased as certified reference materials. Fluorescence spectra of the standards are acquired in the same experimental conditions as the real samples, fitted and used to generate a calibration plot of the fluorescence peak intensity versus the concentration.

6.1.2.2 X-ray absorption near-edge structure

Absorption edge spectroscopy techniques are based on the measurement of the absorption coefficient variation by tuning the energy of the probing photons through an absorption edge of an element. This variation is physically related to the excitation cross section of the core electrons into unoccupied electronic states or into vacuum continuum, following dipole selection rules ($\Delta l = \pm 1$). The spectral features observed close to the absorption edge—known as XANES or near-edge x-ray absorption fine structure (NEXAFS)—reflect the molecular environment of a given absorbing atom and provide the basic mechanism for imaging with chemical sensitivity (Bianconi 1988, Stöhr 1992). Information on different chemical states within systems having the same elemental composition is therefore possible.

Figure 6.3 shows a typical x-ray absorption spectrum acquired in transmission through a thin Fe foil. As the energy of the incident x-ray beam is scanned near and above the binding energy of the core level electrons of an element, a sharp jump in absorption, known as the absorption edge, is observed and accompanied with complex structures that provide information about the molecular environment of the atom. XANES allows probing transitions to unoccupied or partially filled bound electronics states or to the continuum and is sensitive to oxidation state and coordination environment.

The x-ray absorption spectrum can be divided into three different regions. (1) The so-called pre-edge region, preceding the absorption edge, corresponds to incident beam energies below the ionization threshold of the core electrons of the absorbing atom. When the x-ray energy reaches a sufficient level to provoke the transition of a core electron to an unoccupied bound state, these transitions appear as pre-edge peaks in the spectrum. (2) Past the sharp absorption edge, the energy of the incident beam becomes higher than the core binding energy; the photoelectron is thus ejected with a low kinetic energy and subject to multiple scattering by the electron shells of the neighboring atoms, giving rise to resonance peaks. The XANES domain conventionally includes the pre-edge region and the absorption edge and extends over a few tens of eV post-edge. The energy position of the absorption edge and the shape and position of the pre-edge peaks depend on the oxidation state and the geometry of the coordination sphere of the absorbing atom.

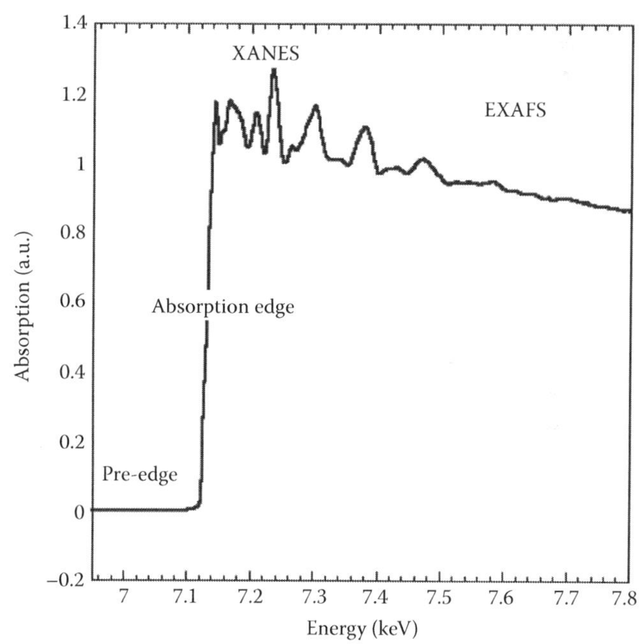

Figure 6.3 X-ray absorption spectrum of a thin Fe foil in transmission.

An increase of the oxidation state raises the binding energy of the core levels (and thus the position of the absorption edge) by a few eV for each electron removed from the valence shell. These properties confer to XANES its chemical sensitivity. (3) The higher energy region up to 1000 eV above the absorption edge corresponds to another regime, referred to as the EXAFS domain, and exhibiting oscillatory structures that result from dominating single backscattering of the ejected photoelectron (described as a wave emanating from the absorbing atom) by neighboring atoms. EXAFS provides information about the distances, coordination number, and species of the neighbors of the absorbing atom (Argarwal 1979). (See Chapter 7 of this handbook.)

Collection of the x-ray absorption spectrum can be achieved in transmission, in fluorescence mode or in total electron yield (TEY) mode. This latter mode requires a conductive sample surface (which can be achieved by a suitable coating).

In transmission mode, the absorption is obtained directly by measuring the transmitted beam intensity (I) through the sample and normalizing it to the incident beam intensity (I_0):

$$I = I_0 e^{-\mu(E)t}$$

$$\mu(E)t = -\ln\left(\frac{I}{I_0}\right)$$

where

$\mu(E)$ is the linear absorption coefficient at the x-ray energy E

t is the sample thickness (see also Chapter XX of this handbook)

In fluorescence mode, it is actually the relaxation of the absorbing atom that is monitored. The intensity of the fluorescence emission line (I_f) normalized to the incident beam intensity (I_0) provides a measurement of the absorbance:

$$\mu(E) \propto \left(\frac{I_f}{I_0}\right)$$

When possible in terms of sample thickness and concentration, measurements in transmission geometry are preferred since they provide a higher signal-to-noise ratio.

It is complicated to draw a full physical interpretation of XANES features because they cannot be described by a simple analytical model as the EXAFS structures do. However, *ab initio* calculations of XANES spectra are possible to provide a more detailed interpretation of atomic bounds and electronic structure, using the *ab initio* self-consistent multiple scattering FEFF code, for example (Rehr et al. 2010). On the other hand, XANES allows an easier crude interpretation than EXAFS using a fingerprint approach, even if peaks cannot always be directly assigned to specific molecular structures.

Classically, XANES experiments start with the collection of XANES spectra from known reference compounds exhibiting selected molecular environments of the element of interest. These spectra will be used as fingerprints for the identification of the phases detected in the sample. Indeed, the XANES spectrum resulting from a mixture of compounds containing the element of interest in different chemical states can be decomposed as a linear combination of the spectra of the single compounds. A semiquantitative evaluation of the sample composition is thus possible, provided the library of reference spectra collected for calibration is relevant. Curve fitting consists of summing normalized reference spectra weighted by the proportion of the compound in the mixture sample. Fitting can also be performed by modeling the resonances with Gaussian or Lorentzian curves and the edge jump with arctangent baselines (Prietzel et al. 2003). Principal component analysis can also be applied to the XANES spectra to identify the number of chemical species present in the sample (Wassermann 1997).

Before interpretation and use as fingerprints, spectra need to be normalized. The normalization follows two main steps; first, the fitting and subtraction of a linear background in the pre-edge region, which reflects the absorption from higher shells and also from other elements composing the sample matrix; second, the normalization of the so-called edge jump to unity to allow quantitative comparison of spectra. The energy position of the absorption edge is usually taken as the inflection point of the main absorption jump, identified from the derivative of the spectrum.

Energy calibration is also fundamental, since it depends on the monochromator settings and may slightly differ from one beamline to another. In that purpose, the spectrum of a pure compound, generally a thin metal foil, is collected and allows the correction of the energy shift between different datasets.

Several software packages allow the normalization and fitting of XANES spectra like the IFEFFIT-based programs Athena and Artemis (Ravel and Newville 2005), SixPack (Webb 2005), or the EXAFSPAK suite of programs (George et al. 2001).

6.1.2.3 First illustration

Sulfur K-edge micro-XANES is an example of choice for the explanation and illustration of XANES chemical mapping. Figure 6.4 presents a series of standard spectra from sulfur-bearing reference compounds. Sulfur K-edge can be considered as a friendly absorption edge for XANES chemical mapping, because of the richness and great variability of the spectra. This can be credited to the sharpness of the resonance peaks and the significant energy shift of the most intense resonance, usually referred to as the *white line*, toward the high energies as a function of the oxidation state of sulfur, which covers a large range from −2 to +6 (Pickering et al. 1998, Vairavamurthy 1998, Prange et al. 1999, Prietzel et al. 2003). For example, the energy absorption threshold is shifted by +10 eV for sulfate groups (SO_4) compared to elemental sulfur. Empirical relationships relating the oxidation state to the white line position can be drawn to assess the oxidation state of sulfur in unknown samples (Vairavamurthy 1998). This particularity strongly eases the identification of molecular bonds and their spatially resolved mapping in a sample. A detailed overview of spectroscopic analysis of sulfur in natural samples can be found in Jalilehvand (2006).

At the sulfur K-edge, the XANES structures are attributed to dipole-allowed transition of a 1s core electron to unoccupied molecular orbitals involving sulfur p-orbitals. This s → p electronic transition gives rise to an intense and well-defined white line.

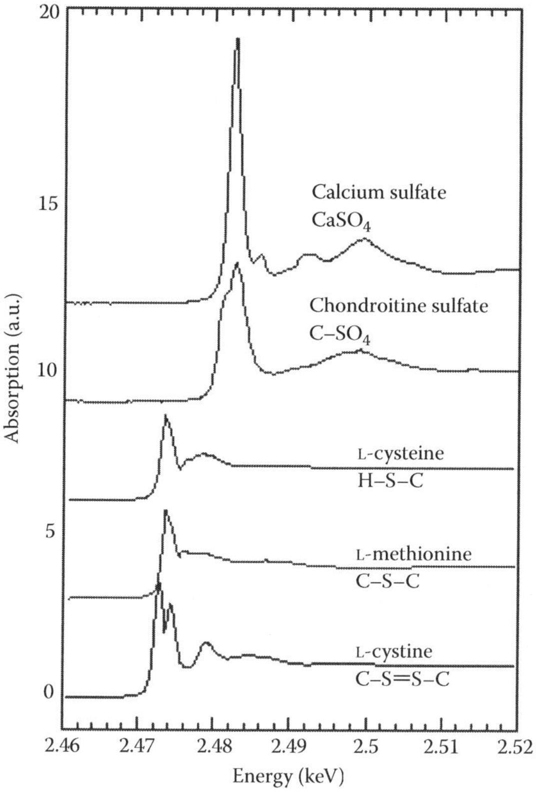

Figure 6.4 XANES spectra of sulfur-containing mineral and organic reference compounds.

The set of sulfur reference spectra shown in Figure 6.4 was collected during a micro-XANES experiment carried out at the x-ray microscopy beamline (ID21) of the European Synchrotron Radiation Facility (ESRF) and aiming at mapping the different organic mineralizing matrices in mollusc shells (Dauphin et al. 2003a,b). In such matrices, chemical binding of sulfur is possible in four forms: the C–S and C–SH forms in the sulfur-containing amino acids (methionine and cysteine), the SO_4 form in sulfated sugars, and the S=S disulfur form. Using pure reference compounds, fingerprint spectra of the following atomic environments of sulfur were characterized: C–S–C molecular bonds in methionine, C–SH bonds in cysteine, C–S=S–C disulfide bonds in cystine, and C–SO_4 bonds in chondroitin sulfate. Calcium sulfate was used as a secondary reference for energy calibration. The spectra exhibit very distinct features and in particular, the presence of disulfide bridges S=S in cystine yields a typical double-peak white line. The first resonance peak is assigned to a transition to a molecular orbital $1s \rightarrow \sigma^*$ (S–S) (Hitchcock et al. 1989, Pickering et al. 1998) and the second peak at higher energy to a $1s \rightarrow \sigma^*$ (S–C) transition. Cysteine and methionine, although they are very closely related compounds, nevertheless show distinguishable spectra. Their white line results from a main contribution from a $1s \rightarrow \sigma^*$ (S–C) transition (Dezarnaud et al. 1990) and possibly a $1s \rightarrow \pi^*$ (CH_2) transition. For thiols, the energy of the $1s \rightarrow \sigma^*$ (S–H) transition is very close to that of $1s \rightarrow \sigma^*$ (S–C). The strong shift of the sulfate white line toward higher energy is very clear in the spectra of chondroitin sulfate and calcium sulfate.

The XANES structures that are related to the chemical environment and the oxidation state of the element of interest, when coupled with microbeam focusing, provide a very smart tool to obtain spatially resolved maps of different chemical moieties.

A series of maps can be acquired, at energies carefully selected based on the energy position of the white lines of the reference spectra. Figure 6.5 shows an example of such a chemical mapping. The sample is a polished transverse cut through the calcitic layer of the shell of a *Pinna nobilis*. On the x-ray images, the borders of the

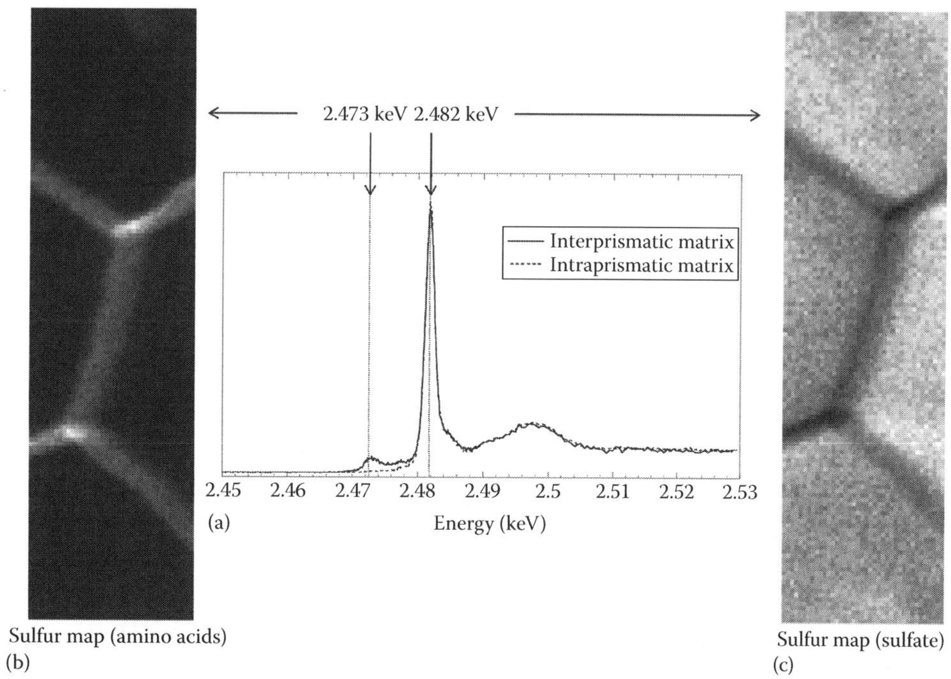

Figure 6.5 Micro-XANES chemical mapping of a polished transverse section through the calcitic layer in *P. nobilis*. (a) Micro-XANES spectra acquired in intraprismatic matrix and in the surrounding matrix walls. (b) Distribution map of the sulfur in the amino acid form acquired at 2473 eV. (c) Distribution map of the sulfur in the sulfate form acquired at 2482 eV. Maps are 20 × 70 μm wide and were acquired with 0.5 μm step size.

polygonal calcitic prisms forming the shell are clearly recognizable. An organic control is exerted on the growth of these perfectly organized mineral units, which surprisingly exhibit a single crystal structure despite their evident organic content. The purpose of the experiment was to analyze the organic matrices driving this crystalline growth based on their sulfur content. Here, the idea was to separately map the distribution of sulfated sugars identified by their sulfate content and proteins identified by their thiol group (–SH). For the sulfate map, the beam energy was set to 2482 eV, the energy of the sulfate white line, as observed on the reference spectrum. In a similar way, a map of proteins was collected at the –SH white line energy ~2473 eV. This measurement made evident the existence of two matrices with different chemical compositions: an intraprismatic matrix rich in sulfated sugars, which is suspected to form a template for the crystalline nucleation and growth of the prism, and a more proteinic outer matrix membrane circumventing the boundaries of the prisms.

As illustrated in the example in Figure 6.5, qualitative chemical mapping can be obtained. The map acquired at 2473 eV, which corresponds to the characteristic energy of the thiols white line, depicts an outer organic membrane rich in amino acids surrounding the prisms. The second map collected at the sulfate peak energy 2482 eV shows an intraprismatic matrix rich in sulfated sugars and delineated by the darker outer matrix in which sulfate content is lower. A depletion in sulfate is observed in the mineralization center in the middle of the prism.

For more quantitative results, normalization can be performed by combining micro-XANES maps acquired at different energies (Cotte et al. 2006). As illustrated in Figure 6.6, in that purpose, micro-XANES maps are acquired at four energies: below the absorption edge I (2.460 keV), at the amino acid sulfur energy I (2.473 keV), at the sulfate energy I (2.482 keV), and at an energy further above the edge I (2.520 keV). Normalized maps can then be calculated as

$$\left(\frac{(I(2.473\text{ keV})-I(2.460\text{ keV}))}{(I(2.520\text{ keV})-I(2.460\text{ keV}))}\right)$$

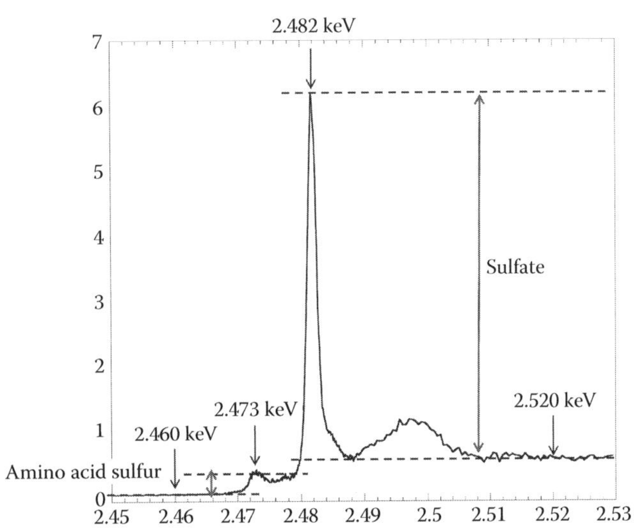

Figure 6.6 Example of chemical map normalization.

for the amino acids and

$$\left(\frac{(I(2.482\text{ keV})-I(2.520\text{ keV}))}{(I(2.520\text{ keV})-I(2.460\text{ keV}))}\right)$$

for the sulfate.

6.2 TYPICAL MICROSPECTROSCOPY BEAMLINE

The typical layout and components of a microspectroscopy beamline are shown in Figure 6.7. The x-ray beam originating from the synchrotron radiation source usually first passes through a primary mirror to tailor its spectrum. A monochromator ensures the energy selection and scanning. A focusing optic concentrates the beam down to a small microprobe, in which the sample is raster scanned. A fluorescence detector collects the fluorescence photons emitted by the sample. Owing to the polarization of the synchrotron beam, this detector is usually positioned perpendicular to the x-ray beam axis, where the scattering contribution is minimized. Photodiodes or gas detectors are used to monitor the incoming and the transmitted beam intensities through the sample, provided this latter is thin enough. Microspectroscopy at low energy requires operation of the end station under vacuum to avoid the strong absorption of air. In this case, the microscope is enclosed in a vacuum chamber.

6.2.1 SOURCE AND X-RAY ENERGY SELECTION

Synchrotron radiation is produced in a ring by the deflection of relativistic electrons by a magnetic field. During this deflection, part of the electrons' energy is lost and emitted tangentially to the ring as synchrotron radiation with a large spectrum covering from infrared to hard x-rays. The beam trajectory is steered using bending magnets and, at third-generation facilities, using specific magnetic devices, called insertion devices, including multiple magnets, which enhance synchrotron radiation emission by forcing the electron beam to a wavy trajectory. Synchrotron sources offer a high brightness, a coherent beam, a small source size, and energy tunability perfectly suited to the needs of spectroscopic microprobes.

The first optical element on spectroscopy beamlines is usually a single or a double mirror in which coating and grazing angles have been optimized to cut the high-energy harmonics of the beam. XANES requires a monochromatic beam and an accurate scanning of the energy of the incident beam; hence, a key element of a spectroscopy beamline is the monochromator. The principle of the monochromator is based on Bragg's law and allows the selection of energy with a narrow bandwidth using the diffraction of the beam on a crystal lattice:

$$n\lambda = 2d \sin\theta$$

where

- n is an integer representing the order of the reflection ($n = 1$ is the fundamental, $n > 1$ corresponds to the harmonics)
- λ is the wavelength of the selected x-ray beam
- d is the atomic lattice spacing of the crystal
- θ the glancing incidence angle of the beam on the crystal (see Chapter 7)

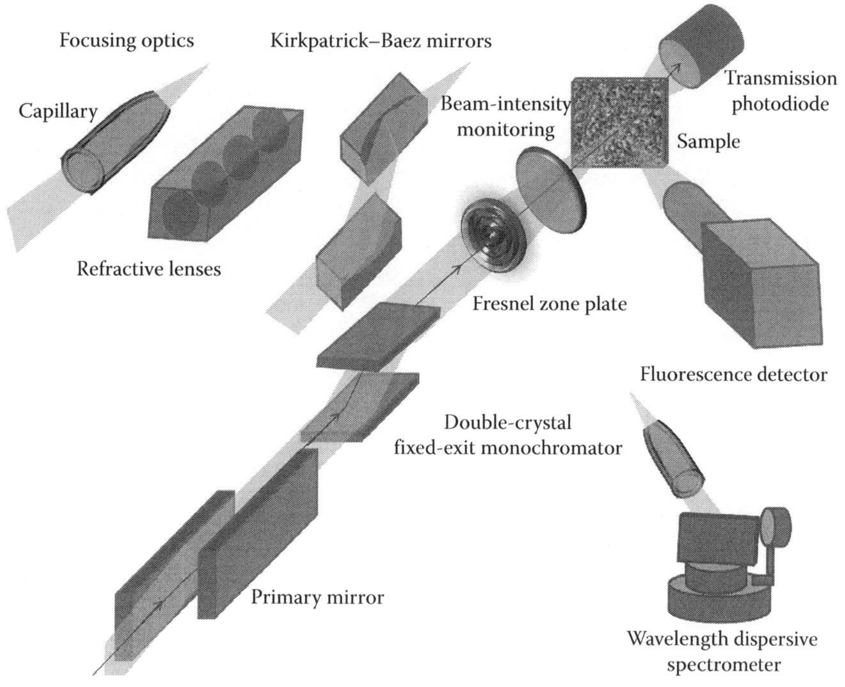

Figure 6.7 Typical layout and components of a synchrotron microspectroscopy beamline. Beam travels from the primary mirror, through monochromator and focusing optics, to the sample. Focusing optics (upper left) include capillary, refractive lens, and KB mirrors. Detector systems (right) include transmission diodes, fluorescence detectors, and wavelength dispersive systems.

As can be observed from this equation, several wavelengths can satisfy this Bragg condition; this is the reason why efficient rejection of high energies by the primary mirror is essential to avoid contamination of the fundamental beam by higher harmonics. The energy scan is ensured by an accurate rotation of the crystal. The monochromator can be single or double crystal, the advantage of the latter one being its possible fixed exit, that is, the direction of the emerging beam does not change with the energy. Several sets of crystals can equip the monochromator to optimize their accessible energy range and resolution as a function of the energy of the analyzed absorption edge. Typically, a resolution of $\Delta E/E \sim 10^{-4}$ is achieved with Si(111) crystal monochromator, which represents a resolution of 0.25 eV at the sulfur K-edge, for example.

6.2.2 FOCUSING COMPONENTS

X-ray focusing optics can be classified into three groups according to their optical behavior: diffractive optics like Fresnel zone plates (FZPs), reflective optics like Kirkpatrick–Baez (KB) mirrors and capillaries, or refractive optics like compound refractive lenses (CRLs). Review of the x-ray lenses can be found in Snigirev and Snigireva (2008).

FZPs are circular diffraction gratings. Detailed description of their properties and use can be found in Howells et al. (2006). Zone plates are extensively used in soft x-ray microscopes, but also extended to the multi-keV energy range, thanks to advances in their fabrication, usually by means of electron lithography, which enables the writing of high-aspect-ratio structures (meaning thin and deep) necessary to achieve simultaneously high spatial resolution and high diffractive efficiency (David et al. 2000, Kang et al. 2008, Charalambous 2011, Gorelick et al. 2011). ZPs nowadays reach spatial resolution in the 15 nm range (Chao et al. 2005, 2009, Vila-Comamala et al. 2011). The drawback of ZPs for microspectroscopy is their chromaticity, for example, the ZP focal length varies with x-ray beam energy. During energy scans, the zone plates needs to be scanned along the beam axis synchronously with the beam energy so as to maintain its focus in the sample plane. This additional translation of the optics requires a high-quality stage and a perfect alignment, because it can be source for a drift of the microbeam introducing distortions in the micro-XANES spectra. (See also Chapter 7.)

KB mirrors (Kirkpatrick and Baez 1948) consist of an assembly of two crossed bent mirrors reflecting at grazing incidence and focusing the beam one in vertical and the other in the horizontal direction. These mirrors are coated with an appropriate high-Z material or a multilayer to offer an optimized reflectivity in their working energy range. They can be either dynamically bent allowing optimization of their curvature for a given energy or fixed curvature when the energy range to be covered is small enough. This latter geometry results in a more compact device, providing shorter focal length and increased mechanical stability. KBs have seen a strong development over the last few years owing to their excellent performances in lateral resolution enabled by advances in mirror figuring techniques and mechanics and their high reflective efficiency, hence high photon flux. Beam spots as small as <100 nm have been demonstrated recently and routinely used for microspectroscopy (Hignette et al. 2005, Mimura et al. 2010). The achromaticity of KBs is obviously a clear asset for microbeam stability during energy scans.

Capillary and polycapillary optics are composed of single or multiple hollow glass capillaries in which the x-rays are guided by multiple total reflections on the inner surface of the capillary. They provide a lateral resolution in the micron range and can be used either for microbeam focusing or for fluorescence photon collections in confocal geometry (Vincze et al. 2004) and wavelength dispersive spectrometers (WDSs) as will be described in the detectors' section (Szlachetko et al. 2010).

The refractive index of materials for x-rays being slightly smaller than unity, compound refractive lenses (CRL) look like inverted lenses, that is, they consist of a series of concave lenses drilled in a low-absorbing material like beryllium or aluminum (Snigirev et al. 1996, Lengeler et al. 2001). To obtain a suitable focal length, many lenses (up to 100) are stacked in a linear array. Due to the absorption of the beam through the lens material, these optics are more efficient at high energy. Recently, a spot size below 100 nm has been obtained from lithographically fabricated silicon CRLs (Schroer et al. 2005).

6.2.3 DETECTION METHODS

Energy dispersive fluorescence solid-state detectors (EDS) consist of a nonconducting or semiconducting material polarized between two electrodes. XRF photons striking the detector material ionize it causing it to become momentarily conductive. The charges generated by the fluorescence photons are accelerated by the detector polarization and collected by the electrodes. The height of the resulting current pulse is proportional to the energy level of the incoming x-ray photon. These current pulses are processed by an amplifier and a so-called multichannel analyzer (MCA) electronics, which will cast them as a function of their height. The output is a histogram of the fluorescence photons as a function of their energy, the so-called fluorescence spectrum. This spectrum exhibits well-defined peaks at specific energies corresponding to the fluorescence emission lines of the atoms contained in the sample. A *region of interest* (ROI) can be defined around a fluorescence line and be used as a counter providing direct integration of the number of counts issued from a given element. In the case of 2D mapping, such ROIs are defined for several elements and stored in a 2D array directly reflecting the distribution of the element in the sample. Advances in MCA electronics and data acquisition rates now allow the collection of a full fluorescence spectrum for each 2D data point. 3D elemental distributions in samples can also be generated in tomographic (Golosio et al. 2004) or confocal geometry (Vincze et al. 2004, Erko and Zizak 2009) using specific reconstruction software.

Fluorescence spectra fitting can then be performed as postprocessing to generate deconvolved distribution maps of any element detected in the spectrum. For micro-XANES, the ROI is used to monitor the x-ray absorption signal of the element of interest.

Classical EDS detectors are Si(Li) or Ge crystal detectors that offer excellent energy resolution (around 130 eV) and peak-to-valley (P/V) ratio (several thousands) but require cryogenic cooling. Over the last few years, silicon drift diodes (SDDs) have known a considerable development and are now more and more used at microspectroscopic beamlines for the high count rate they can handle and the competitive energy resolution and P/V they now offer. Their integrated cooling and compacity also greatly facilitate their integration in the microspectroscopy setups and their daily operation. A clear trend is also observed toward the development of large area and/or multiple element detectors (Letard et al. 2006, Ryan et al. 2010) that provide increased solid angle and thus a better covering of the isotropic 4π fluorescence emission sphere to collect as much signal as possible. This is fundamental for the optimization of the signal-to-noise ratio versus dose, especially for radiation-sensitive samples.

For higher energy resolution and/or high sensitivity to trace elements, another approach is required. The principle of WDSs is based on a diffractive element such as a crystal or multilayer to accurately select a wavelength and thus a fluorescence emission line with an energy resolution in the range of a few eV. A capillary optic can be used to collect the XRF emerging from the sample with a large solid angle and direct it as a quasi-parallel beam onto the analyzer crystal. A large area detector records the photons selected by the crystal. Both crystal and detector are mounted on rotation stages, synchronized in $\theta/2\theta$ angular position, to adjust the incidence angle on the crystal to the required wavelength. Since the diffracted wavelength is much more intense than other wavelengths that scatter the crystal, an excellent rejection of scattering background and contribution from other elements' fluorescence lines is achieved, hence allowing the detection of trace elements even in a highly fluorescence-emitting matrix. Given its high energy resolution, WDS is a very powerful detection scheme to separate neighboring emission lines of different elements, which EDS detectors are unable to resolve. Such instruments are also perfectly suited for XANES measurement with high selectivity and low background, since the element of interest can be isolated from other contributions, with an even higher efficiency than an EDS ROI would provide. Compact WDS devices can now be integrated in microprobes (Szlachetko et al. 2010).

Apart from scanning microprobes, 2D chemical mapping can also be performed in full-field geometry. Full field allows collecting direct 2D hyperspectral datasets by acquiring radiographic images of a sample at different energies across the absorption edge of the element of interest with a 2D x-ray camera. Spatial resolution down to the submicron range can be achieved. The resulting stack of images as a function of energy can be seen as a 3D matrix, in which a full XANES spectrum is available for each pixel of the image. This technique is only applicable to thin, relatively concentrated samples, since it is a transmission measurement (De Andrade et al. 2011a).

Most synchrotron facilities worldwide comprise beamlines optimized for microspectroscopy. In the multi-keV and hard x-ray energy range, one can cite the following, among others:

In Europe: the micro-XAS (Borca et al. 2009) and PHOENIX beamlines at the Swiss Light Source, the LUCIA beamline at SOLEIL (France) (Flank et al. 2006), the ID21 (Salomé et al. 2009) and ID22 (Martinez-Criado et al. 2012) beamlines at the European Synchrotron Radiation Facility (France), the P06 (Schroer et al. 2010) at PETRA III (Germany), the I18 beamline at the Diamond Light Source (United Kingdom), and the MySpot beamline (Erko et al. 2009) at BESSY II (Germany). In project or under construction are the Nanoscopium beamline (Somogyi et al. 2011) at SOLEIL (France), the NINA beamline at ESRF (France), the P11 beamline at PETRA III, and the I08 beamline at Diamond.

In North America: the 2-ID-B (McNulty et al. 1998), 2-ID-D, 2-ID-E, and 26-ID-C (Maser et al. 2006) beamlines at the Advanced Photon Source (United States), the 10.3.2 beamline at the Advanced Light Source (United States), the BL 14.3 and BL 2.3 beamlines at Stanford Synchrotron Radiation Lightsource (United States), and the SXRMB beamline at the Canadian Light Source. Under construction is the SRX-KB beamline at NSLS-II (United States) (De Andrade et al. 2011b).

In Asia: the BL15U-HXMF beamline at Shanghai Synchrotron Radiation Facility (Xu et al. 2011) in China and the BL37XU (Hayakawa et al. 2001) at Spring 8 (Japan).

In Australia: the XFM beamline at the Australian Synchrotron (Paterson et al. 2011).

6.3 SAMPLE PREPARATION

XANES can be measured in a wide range of sample environment and conditions allowing in situ experiments such as hydrated environments, low or high temperatures, under high pressure, and in an electrochemical cell.

In the case of solid samples, flat surfaces are needed for quantitative analysis. Polishing surfaces is also mandatory to minimize scattering effects and avoid topological effects due to sample roughness. Surfaces oriented toward the fluorescence detector will exhibit a higher fluorescence count rate than those oriented backward, which will be shadowed. This effect induces artificial signal fluctuation in the map, even for homogeneously concentrated samples. Fragile samples can be first embedded in resin before polishing. If microtome cutting of the sample is possible, it is often interesting to work on thin sections. Due to the penetration of x-rays, in-depth structures of the sample may overlap in the acquired map. Depending on the x-ray energy and the sample absorption, this penetration depth is very variable and may be highly detrimental to the final spatial resolution. In such cases, thin sections allow a better match between the lateral microbeam resolution and the in-depth resolution.

Sealed wet cells can be used to analyze liquid or hydrated samples in near-native state, even under vacuum. It is important to note that spectra of the same compound in solid or solution forms may differ, as described at the sulfur K-edge by Pickering et al. (1998). They demonstrated also that the pH conditions could introduce variations in the S K-edge XANES spectra. Solid spectra may exhibit richer secondary structures in the post-edge region due to long-range ordering that generates multiple scattering effects. Spectra of reference compounds are thus preferably acquired in diluted liquid form in the case of liquid samples to be identified. These considerations show the importance of collecting the standard reference XANES spectra in the same conditions in terms of preparation, environment, and acquisition setup as the real sample.

Biological samples (tissues, cell cultures) are usually much too radiation sensitive to be analyzed at room temperature in their hydrated state in the multi-keV energy region. Hence, they are often deposited or grown on a suitable, contamination-free (from the XRF point of view) substrate like silicon nitride windows, TEM grids, or polymer films and freeze dried. However, freeze drying may modify the ionic distribution and the chemical speciation of the element of interest in the sample. In such cases, cryopreservation is preferred. The cryo samples are transferred to a cooled stage inserted in the microscope and kept at low temperature during the whole data acquisition. Cryo cooling allows the analysis of the sample in a stable hydrated state; limits its oxidation by air during its manipulation, since the samples are kept in a dry nitrogen atmosphere; and preserves its morphological integrity under the beam. Analysis in the frozen state also limits the diffusion of free radicals (as can be observed in liquid samples) and limits secondary radiation damage. More details about cryogenic cooling techniques can be found in Chapter 7.

The preparation of the reference samples for XANES is crucial, in particular if they are to be used as a fitting library. Standards are usually purchased as powder samples. These latter must be finely ground and can be either prepared as thin layers on adhesive tape or as pressed pellets. Grinding can minimize scatter and self-absorption effects due to particle size. Mixing with neutral cellulose or boron nitride powder allows a dilution of the compound to minimize self-absorption effects. Pressing as pellets compacts more of the sample into the analysis area and ensures a uniform density. Whenever possible, measurements in transmission are preferred since they provide a higher signal-to-noise ratio than in fluorescence mode. Software allowing calculation of the optimal pellet concentration/thickness for an optimized edge jump is available, for example, Hephaestus (Ravel and Newville 2005). To get an optimized signal, the sample thickness t is usually taken so that $\Delta\mu(E)t \approx 1$. Sample preparation requirements for XANES and EXAFS being very similar, techniques described in Chapter 7 are also highly relevant for XANES analysis.

Depending on the nature of the sample and its preparation, several effects can distort the XANES spectrum. Standard XANES spectra are often collected in *unfocused* beam condition, with a larger beam simply collimated by a pinhole. This provides a higher signal-to-noise ratio, since a larger sample area and thus a larger number of atoms can be probed. In such unfocused transmission mode, the sample thickness and homogeneity are crucial to avoid pinhole effect, that is, sample-free area through which the beam travels without interaction and that induces a nonlinear relation between the measured intensity and the atom absorbance.

For XANES in fluorescence mode (with focused or nonfocused beam), the assumption that the x-ray absorption cross section is proportional to the fluorescence signal normalized by the incident beam intensity is only valid for thin or diluted samples. For thick or too concentrated samples, self-absorption phenomena may induce a damping of the resonances of the spectrum (Pickering et al. 1998). Self-absorption problems are more pronounced for soft x-ray XANES. In the case of samples for which this effect cannot be avoided, self-absorption correction algorithms can be applied, provided a complete knowledge of the sample composition.

Another important point to keep in mind is that, depending on the nature of the sample, beam radiation damage may arise and modify the XANES spectrum at the time it is being acquired. A thorough description of the different types of radiation damages that might occur under the beam is given in Chapter 7. Obviously, morphological degradation can be observed on fresh nonfixed biological samples. But although biominerals usually are hard materials, their chemical speciation may also be altered by photoreduction effects. A few examples can be described at the sulfur K-edge, which is a sensitive edge for radiation damage. The breakage of disulfide bonds into thiols under irradiation is a clear illustration of photoreduction. In some particular environments, one can also sometimes observe the reduction of sulfates to sulfites (Métrich et al. 2009, Chalar et al. 2012). A decrease of the sulfate peak with a concomitant increase of the sulfite white line can be noticed. The strategies to overcome this limitation are (1) the acquisition and summing of a series of fast single spectra that allow monitoring of the sample evolution under the beam; (2) the comparison to a global XANES spectrum acquired in nonfocused mode, hence with a much lower photon density that by comparison decreases photoreduction; and (3) the building of a calibration curve of the photoreduction effect with time, for example, the decrease of the sulfate peak as a function of time (Métrich et al. 2009). Such effects are not only confined to soft organic materials but can also be unexpectedly observed on hard materials like minerals. This issue will become even more critical with the development of high-brightness nanoprobes, in which the photon density will be orders of magnitude higher.

Due to the synchrotron beam polarization, XANES is also sensitive to the crystallinity and crystalline orientation of the sample that may modify the spectra. This property can be exploited to assess crystalline orientations in different neighboring phases of a sample (Pérez-Huerta et al. 2008, De Andrade et al. 2011).

6.4 CASE STUDIES IN BIOMINERALIZATION

As previously said, problems are diverse, depending upon the samples and their context. The selected examples are an attempt to reflect this diversity and complexity. Despite the variety (modern and fossil examples, unicellular and pluricellular organisms, vegetal and animal, etc.), they are not a comprehensive set of what is done with XANES. Some other cases will be studied in the next chapter.

6.4.1 LOCALIZATION OF MINERALS AND ORGANIC MATRICES AT A MICROSCALE

Ca is abundant in invertebrate shells and skeletons, as well as in bones and teeth. $CaCO_3$ is found as amorphous calcium carbonate (ACC) or crystalline phases. One of the main problems is the initial nature of the biomineral: is it amorphous or crystalline? Thanks to diverse analyses including XANES spectra at the Ca L-absorption edge, Politi et al. (2008) have shown that sea urchin skeleton goes through three phases: hydrated ACC, followed by an intermediate form of ACC, and finally, the crystalline calcite. ACC and crystalline phases coexist at a submicrometric scale, suggesting a complex process (Figure 6.8). ACC and crystals are spatially close. A similar pattern was observed in the calcitic prisms of the pearl oyster (Baronnet et al. 2008). Ca K- and L-edge spectra are also used for Ca phosphates (bioapatites as dentine, bone and enamel, and synthetic minerals) to directly determine the maturity of poorly crystalline apatite (Eichert et al. 2005) and to distinguish crystalline from amorphous phases (Beniash et al. 2009). Sr^{2+} cations in physiological and pathological calcifications (bones and kidney stones) were shown to be linked to Ca phosphate apatites (Bazin et al. 2011) (see Section 6.4.4).

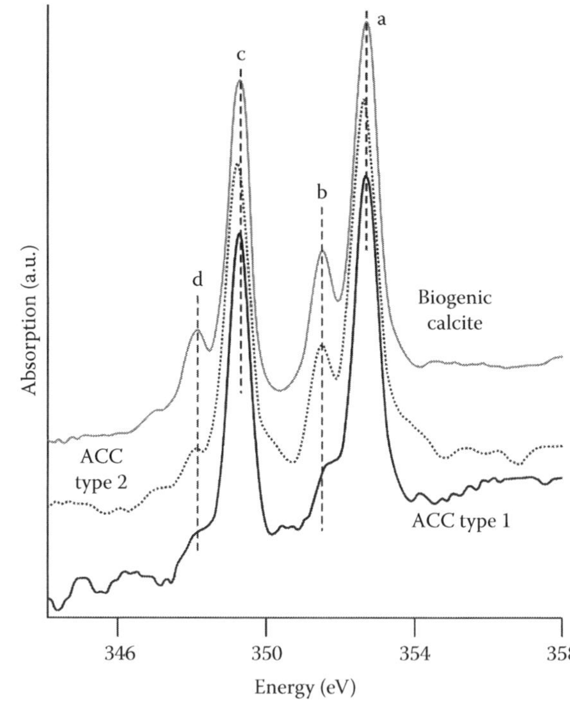

Figure 6.8 Ca L-edge spectra of ACC and calcite of a sea urchin spicule. (Redrawn from Politi, Y. et al., *Proc. Nat. Acad. Sci. USA*, 105, 17362, 2008.)

Another problem with biominerals is to determine the topographic relationships between mineral and organic components at a micrometer scale. Organic matrices are composed of carbon, oxygen, hydrogen, and nitrogen. Hydrogen is the lightest chemical element so that spatially resolved analyses are usually not available. Carbon and oxygen are present in some minerals constituting shells and skeletons, so that maps are not very useful. Nitrogen is abundant and *characteristic* of organic matrices but only in proteins. Nitrogen is absent in sugars and lipids. The classical opinion is that proteins are the most abundant organic components in biominerals and that they exert the main controls on the biomineralization processes. Thus, most papers deal with proteins. Nevertheless, it has been shown that sugars play a major role in the mechanisms (Wada 1964, 1980; Addadi et al. 1987) and that they are abundant, if not dominant, in some taxa. Thus, mapping nitrogen is not *the* solution and another chemical element must be found.

Quantitative electron microprobe analyses have shown that sulfur is often abundant in biogenic Ca carbonates (Cuif and Dauphin 1998, Dauphin and Cuif 1999). On the other hand, electrophoreses of the soluble organic matrices extracted from these samples have evidenced that acidic sulfated polysaccharides are present (Dauphin 2001, 2003). Sulfur also exists in some amino acids (cysteine, cystine, and methionine), but they are destroyed during hydrolyses used in the common analytical methods. At last, mineral sulfur is rare as a primary biogenic component. Thus, sulfur is a potential marker for organic matrices, because XANES spectra allow us to know whether S is mineral or organic, in protein or in sugar.

Corals are marine animal, solitary or living in colonies. Numerous species are able to secrete a calcareous skeleton, usually in the aragonite form of $CaCO_3$. They have a radial

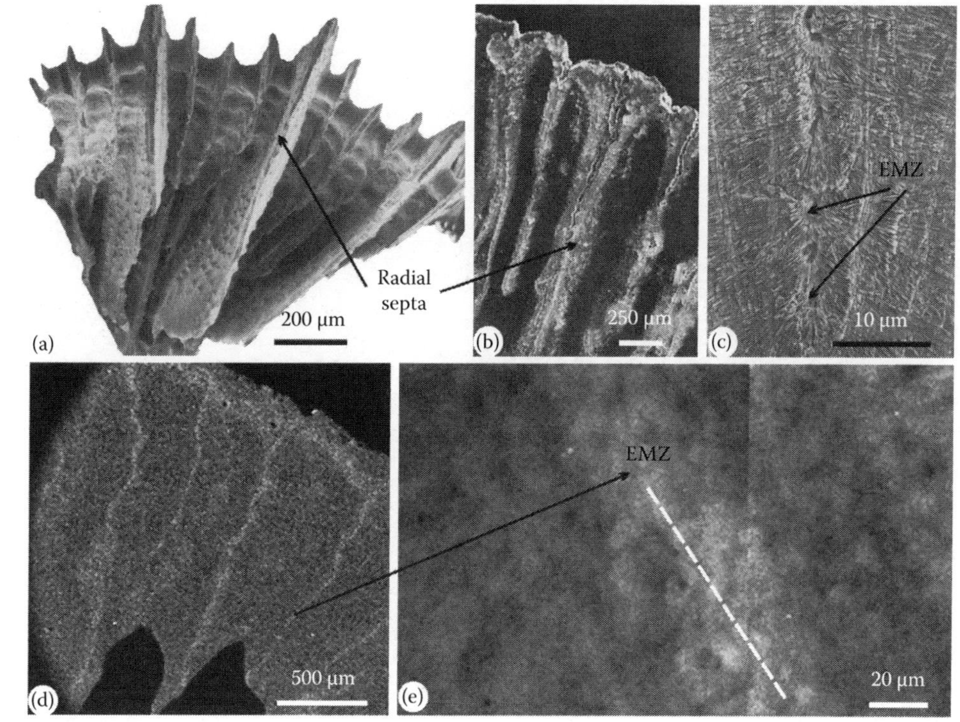

Figure 6.9 *Caryophyllia* coral skeleton. (a) Fragment showing the radial septa. (b) SEM image of a polished section showing the median line in the radial septa. (c) Polished section of a septa, showing EMZ and the surrounding fibers. (d) XANES sulfate map showing the high content of the median zone of septa, composed of EMZ. (e) Detailed sulfate map of the central part of a septa.

symmetry, with an external wall and inner septa dividing a calyx. From a microstructural point of view, the skeleton comprises two main zones: centers of calcification or early mineralization zones (EMZs) in the median part of septa, surrounded by fibers (Figure 6.9a through c). A microstructural and chemical banding growth pattern is visible in fibers (Cuif and Dauphin 1998, Cuif and Sorauf 2001, Meibom et al. 2004, Nothdurft and Webb 2007, Sandeman 2008). Sulfate has been detected in modern powdered coral skeletons using K-edge XANES spectra (Pingitore et al. 1995). From a comparison with anion chromatography and standards, they conclude that sulfur is present only as sulfate. They note that coral XANES spectra differ from dolomite spectra and conclude that sulfate substitutes for carbonate. More recently, spectra and maps of various coral species have confirmed that sulfate is present in both EMZ and fibers (Cuif et al. 2003) (Figure 6.9d and e). Nevertheless, a comparison with organic and mineral standards and with electrophoresis data of the extracted soluble organic matrices shows that sulfate is linked to acidic polysaccharides (Dauphin 2001). Sulfur amino acids are rare or very weak in modern aragonitic coral skeletons.

As for corals, chemical growth lines are visible in mollusc and brachiopod shells (Dauphin et al. 2003a,b; Cusack et al. 2008; Nouet et al. 2012). In both structural and chemical images, zonations are synchronous across adjacent structural units, showing the strict physiological control of the biomineralization processes. Nevertheless, the comparison of chemical maps shows that the rhythm and intensity differ according to the element. Thus, dealing with quantification and/or ratios of chemical elements to estimate paleosalinity or paleotemperature without taking into account compositional changes at a microscale gives us only *average* estimation and reconstructions. Moreover, results obtained from different chemical elements differ because of the nonsuperposition of the chemical bandings.

For geologists, microorganisms are important because they accumulate in sediments of considerable thickness. Among them, unicellular silica algae such as diatoms are known since the Jurassic period. In diatoms, C is a marker for organic components. XANES spectroscopy near the carbon edge (290 eV) on diatom samples corresponds to aromatic/unsaturated and carbonyl structures. The carbonyl peak probably results from carboxylic acid in proteins. Maps show that organic components are not spatially homogeneous within a diatom skeleton and the physical structure is identifiable (Abramson et al. 2009).

Whatever the mineralogy, heterogeneous distributions of chemical elements throughout a skeleton have been observed. These distributions are both related to the microstructure and growth layers. How the heterogeneity affects the preservation of biogeochemical signals in ancient and fossil organisms used for paleoenvironmental reconstruction is still unknown.

6.4.2 SPECIFICITY OF THE COMPOSITION OF DIFFERENT STRUCTURES WITHIN AN ORGANISM

The elemental chemical differences between EMZ and fibers (Figure 6.9d) and the chemical banding pattern have been repeatedly shown in coral skeletons (Cuif and Sorauf 2001, Meibom et al. 2004). Mollusc shells are also composed of several layers, the mineralogy of which can differ. Electron microprobe analyses of mollusc shells have shown that S is more abundant in the calcitic layer (Dauphin and Cuif 1999). *P. nobilis* and pearl oyster (*Pinctada margaritifera*) have an outer calcitic prismatic layer and an inner aragonitic nacreous layer. Prisms are

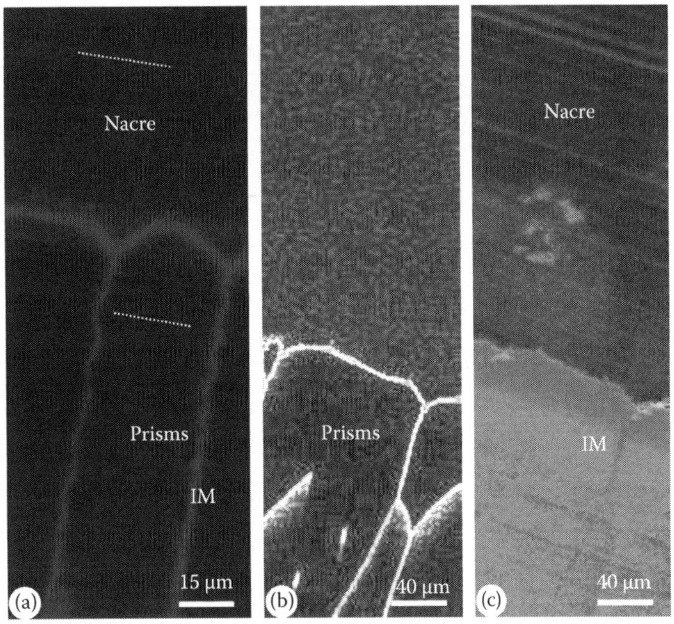

Figure 6.10 Longitudinal section across the shell of *P. margaritifera*. (a) Confocal image showing the growth lines (dotted white lines) in the calcitic prisms and the nacreous aragonitic layer. (Image by A.D. Ball, Natural History Museum, London, U.K.) (b) S amino acid XANES map showing the high contents of the organic membranes, $E = 2.473$ keV. (c) SO_4 XANES map showing the high contents at the end of the calcite prisms, $E = 2.482$ keV. (From Dauphin, Y. et al., *Anal. Bioanal. Chem.*, 390, 1659, 2008. With permission.)

separated by organic membranes. Besides the fact that organic sulfate content is higher in the calcitic layers, in both taxa, sulfur speciation showed that the protein S content of the interprismatic membranes is higher than the SO_4 content, whereas the reverse is true for the intraprismatic structures. A detailed examination of the transition between nacre and prisms shows that the composition and mineralogical and structural changes are not synchronous. Sulfate content is modified at the end of the prism and is not stable at the beginning of the aragonitic layers (Dauphin et al. 2003b) (Figure 6.10).

Similar observations have been done on other mollusc shells with an outer *prismatic* calcitic layer and an inner aragonitic one, whatever the structure of the layers. Organic sulfate content is higher in both layers than that of S amino acids and is higher in calcite than in aragonite (Dauphin et al. 2005, 2012; Nouet et al. 2012). Again, in these shells, the irregular boundary between the calcite and aragonite observed with SEM and Fourier transform infrared spectroscopy (FTIR) is visible in XANES map.

Some shells are built of several calcitic layers or aragonitic layers. In both cases, the structure and the chemical composition of the layers differ within an individual. In contrast to most bivalve molluscs, the shells of which exhibit similar microstructures on both valves, the two valves of oyster shells differ (Orton and Amirthalingam 1927). Both left and right valves are calcitic and comprise a foliated (internal) layer, but only the right valve (the upper one) has an outer thin prismatic calcitic layer. In addition, oysters exhibit an additional specificity in shell mineralization: they are the only known mollusc to produce a chalky layer, an irregular calcareous structure whose formation creates more or less hollow chambers within the foliated layers. In such samples, only in situ analyses allow us to investigate the chemical composition of these imbricated layers. XANES maps show that despite a common mineralogy, the chemical contents of the layers within a valve differ (Figure 6.11a through c). Similarly, the sulfur contents of the aragonitic prismatic and nacreous layers of another mollusc shell differ (Figure 6.11d through g). Thus, it can be said that the differences in sulfur amino acids and sulfate in mollusc shells are not correlated only with the mineralogy (Dauphin et al. 2013).

Brachiopod shells are also composed of two calcitic layers, the structures of which differ. A higher sulfate concentration has been observed in the thin primary (outer) layer. Sheaths surrounding the calcite fibers of the secondary (inner) layer contain sulfur as thiol, confirming the presence of protein. Within the fibers themselves, S occurs as sulfate (Cusack et al. 2008).

Pearls are a by-product of mollusks. They are the result of a grafting process: a piece of the mantle tissue of a pearl oyster is inserted simultaneously with a mineral nucleus in another oyster. The usual theory is that the obtained pearl is a reverse shell. However, the examination of the structure and composition of pearls using electron microprobes and XANES shows that the grafted tissue secretes structure unknown in the shell: aragonitic *prisms* (Cuif et al. 2008, 2011). The organic sulfate contents differ in the three structures (nacre, calcitic prisms, and aragonitic *prisms*), the calcitic layers showing the highest contents.

Sulfur is a good marker for organic matrices in XANES analyses. In calcareous biominerals, organic sulfate is usually higher than sulfur in amino acids, and cystine double bonds are rare or very weak. From other analyses, it can be said that the organic sulfate is associated to acidic polysaccharides. Growth lines, organic layers, or envelopes, as well as transition layers, are displayed at a microscale using XANES maps. Such data help to improve our understanding of the biomineralization processes. The organomineral relationships play also a role in the behavior of molluscs and corals submitted to environmental changes such as ocean acidification and during the fossilization mechanisms.

6.4.3 IS A CHEMICAL ELEMENT PRESENT IN THE ORGANIC OR MINERAL PHASE?

According to Pingitore et al. (1995), sulfate substitutes to carbonate in coral skeletons. However, previous analyses of the soluble organic matrices extracted from the coral skeleton have shown that organic sulfates are present. FTIR spectra display bands for sugars and sulfates, whereas specific staining electrophoresis is indicative of acidic sulfated polysaccharides (Cuif and Dauphin 2003). A comparison of an organic sulfated sugar and mineral sulfate standards has demonstrated that sulfate in corals is organic. No mineral sulfate has been found using Raman spectrometry. On the other hand, amino acid and monosaccharide analyses have shown that the composition of these sugars–proteins assemblages is taxonomy dependent (Cuif et al. 1999). Thus, it is suggested that the various characteristics of coral skeletons are linked to the biochemically driven crystallization process. A two-step growth model places coral skeleton among the typical *matrix-mediated* structures (Cuif and Dauphin 2005): a biomineralization cycle starts by the secretion of a mineralizing organic matrix rich in acidic sulfated

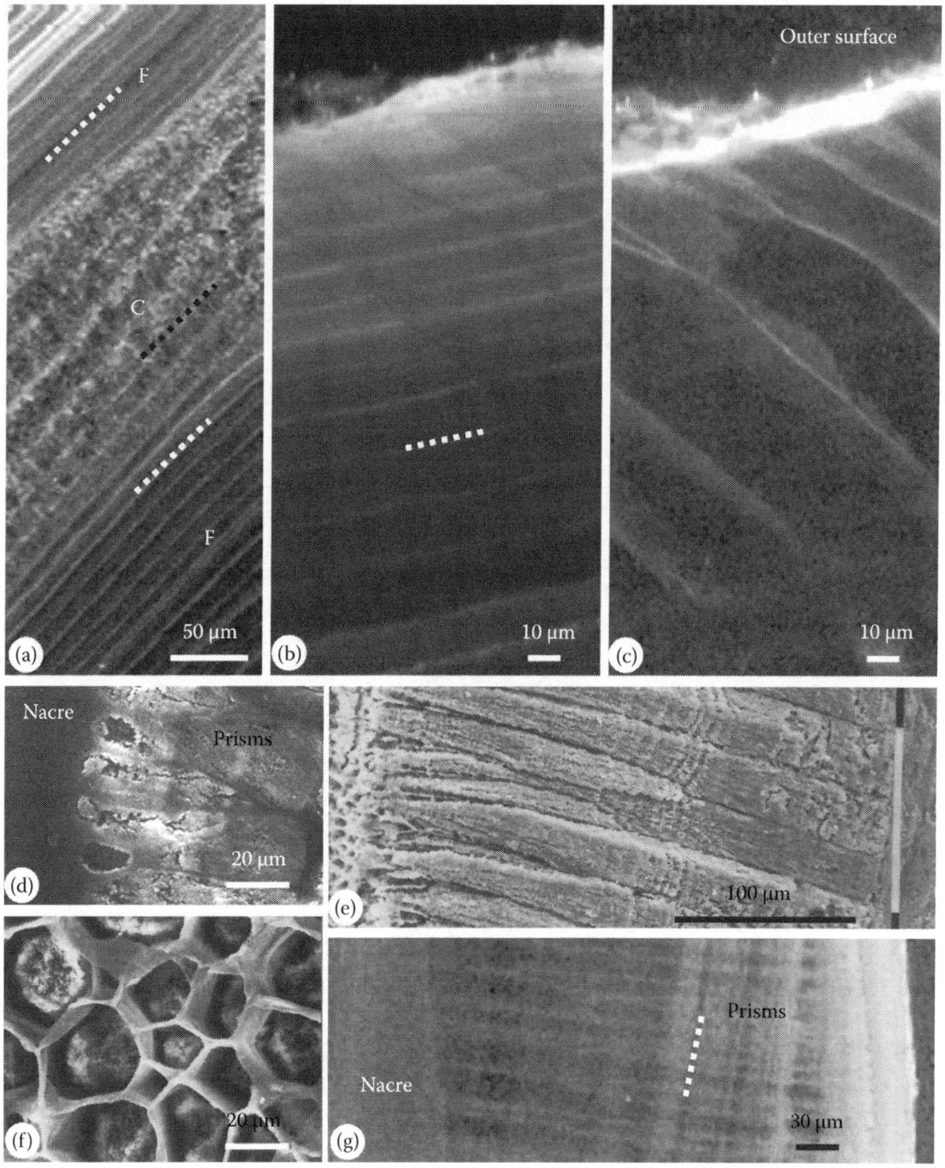

Figure 6.11 (a–c) Calcitic shell of the common oyster (*Crassostrea*). (a) Growth lines (dotted lines) are visible in the foliated (F) and chalky layers (C) in a sulfate map. In the thin prismatic layer, growth lines are visible in the sulfate map (b), while interprismatic envelopes are visible in sulfur amino acid maps (c). (d–g) Aragonitic shell of *Margaritifera margaritifera*. (d) Polished surface showing the aragonitic nacreous and prismatic layers; SEM image. (e) Detailed SEM images of a longitudinal section of the prismatic layer, showing some growth lines. (f) Decalcified transversal section of the prisms showing the organic interprismatic envelopes. (g) XANES map showing the high contents in sulfur amino acids of the interprismatic organic envelopes and growth lines.

polysaccharides. The final step is the crystallization phase, during which mineral material grows onto this organic framework.

Once it has been shown that a chemical element is present in the organic component in a biomineral, it is interesting to know whether this element is present in the soluble or in the insoluble matrix. The comparison of the spectra obtained on the lyophilized insoluble and soluble organic matrices extracted from a sea star shows that sulfate is the main component in both extracts (Figure 6.12a). Sulfur amino acids are more abundant in the insoluble matrix. Similar results were obtained on the organic matrices extracted from the nacreous layer of the aragonitic shell wall of *Nautilus* (a mollusc) (Figure 6.12b). It must be noticed that these spectra differ from those of *Asteria*, so that we can suggest that they depend on the taxonomy or structure, or both factors, as shown by a comparison of the soluble matrices extracted from the nacre and prisms of another mollusc shell: *Pinctada* (Figure 6.12c).

As for other skeletons, the respective role of the soluble and insoluble organic matrices in the biomineralization process is not yet deciphered. Similarly, our knowledge of the role of the quantities and composition of the organic matrices in relation to the differential preservation of the shell layers during the fossilization is still rather limited.

Sulfur is not the only chemical element the position of which is still questionable. Sr is abundant and used as a marker for aragonite, while Mg is abundant and a marker for calcite. The classical hypothesis is that Sr and Mg substitute for Ca. Because corals and molluscs are widely used as paleoenvironmental recorders, to know the state of preservation of ancient and fossil skeletons is important. The initial structure, mineralogy,

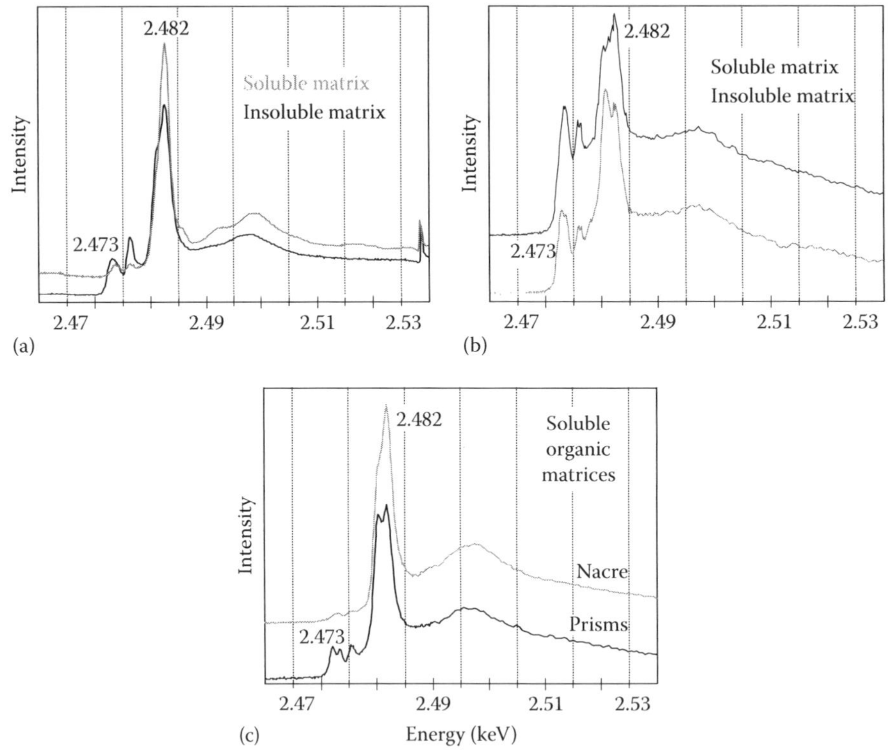

Figure 6.12 Comparison of XANES spectra at the S K-edge of the lyophilized organic matrices extracted from calcareous biominerals. (a) Calcitic skeleton of a sea star. (b) Nacreous layer of a mollusc cephalopod. (c) Aragonite nacre and calcitic prisms of the pearl oyster (*Pinctada*).

and biogeochemical composition play a role in the diagenetic processes. Thus, XANES has been used to improve our knowledge of these skeletons. XANES spectra of aragonitic coral skeletons at the Sr K-edge are similar to those of Sr in $SrCO_3$. Sr substitutes for Ca in the aragonite, but 40% of the total Sr are in strontianite (Greegor et al. 1997). More recently, from a series of spectra related to the coral skeletal structure on thin sections, Finch and Allison (2003) and Finch et al. (2003) conclude that "Sr K-edge EXAFS of all the coral samples refine, within error, to an ideally substituted Sr in aragonite, in contrast to previous studies, in which significant strontianite was reported" (2003a). "We have found no evidence for any coordination in Sr other than single-phase aragonite in any of the samples analyzed" (2005). X-ray diffraction diffractograms obtained on powdered coral skeletons have repeatedly shown the presence of aragonite, but strontianite has never been detected. However, because coral skeletons are not compact, secondary minerals can be found (usually calcite or clays). Thus, this discrepancy between Greegor et al. and Finch and colleagues may be due to (1) a nonbiogenic mineral crystallized in the pores of the skeleton and (2) the improvement of beamlines and databases.

Few species are studied using XANES, and results are sometimes contrasted. Using Mg K-edge spectra in modern corals (Figure 6.13a), Finch and Allison (2007) conclude that "Mg in coral skeletons is not substituted into aragonite… skeletal Mg

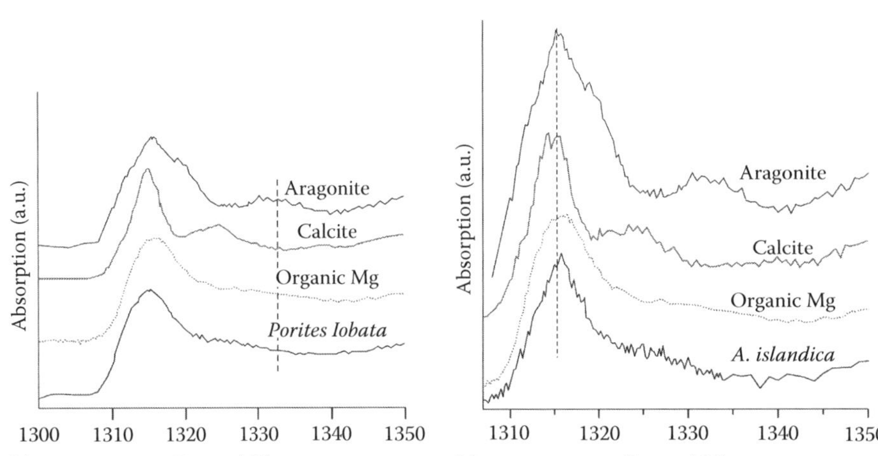

Figure 6.13 Mg K-edge XANES spectra showing the environment of Mg in a coral skeleton (a) and mollusc shell (b), compared to abiogenic aragonite and calcite, and an organic component. (Redrawn from Finch, A.A. et al., *Geochim. Cosmochim. Acta*, 67, 1189, 2003; Allison, N. et al., *Geochim. Cosmochim. Acta*, 69, 3801, 2005.)

may be present in an organic form" or "may occur in a highly disordered inorganic phase," may be as ACC. In molluscs, Mg K-edge spectra and maps show that Mg is mainly associated in the mineral part of the calcitic shell of *Pecten*. However, Mg is mainly located in the organic matrix of another bivalve calcitic shell (*Mytilus*) (Clarke et al. 2009). The position of Mg in mollusc aragonitic shell is more difficult to assess. Despite a distinct Mg zonation in the shell, XANES spectra indicate that Mg is not substituted into aragonite. It seems to be hosted by a disordered phase such as organic matrices or inorganic nanoparticles (Foster et al. 2008) (Figure 6.13b). In biogenic silica, Al is often abundant. In diatom frustules, Al has a complex behavior. In cultured algae, Al is tetrahedrally coordinated by oxygen as Si is in the silica network. It is suggested that Al is incorporated into the silica during the synthesis of the silica network. Al and Si are associated at the atomic level in the diatom silica structure (Beck et al. 2002).

Mn K-edge XANES spectra on powdered samples of the nacreous layer of seven freshwater molluscs have shown that Mn is incorporated as Mn^{2+} in the carbonate fraction and absent in the organic matrix. However, although the nacre is aragonite, "Mn in the shell material is in a calcite microenvironment" (Soldati et al. 2009). In *Diplodon*, provenance, age, and seasonal deposition seem not to influence the speciation (Soldati et al. 2010).

Cu and Fe have been examined in coral skeletons. These elements associated with coral skeletal aragonite are present in their fully oxidized states: Fe as Fe (III) and Cu as Cu (II). Cu substitutes for Ca in the aragonite structure, but Fe location is more questionable (Pingitore et al. 2002a). Uranium is used as a marker for paleotemperature (U/Ca) and radiometric determination of geological age, mainly in coral skeletons. U concentrations are very low (about 3 ppm), but Pingitore et al. (2002b) were able to obtain XANES spectra at the U L-II edge in powdered coral skeletons. In a modern coral (*Pavona*), U is in the form of U6+. In contrast, a fossil coral (Pleistocene) shows the spectral features of U4+. It was suggested that the U4+ results from a diagenetic alteration of the skeleton.

Shells and skeletons are not the only hard tissues in invertebrates. Jaws and radula are partly or fully mineralized. Some worms (polychaetes) have jaws, the tip of which is enriched in metal. According to Zn K-edge XANES and EXAFS spectra, electron microprobes analyses, and nanoindentation measurements, Zn and Cl are correlated. However, Zn seems to be linked with the proteins (Lichtenegger et al. 2003).

Despite some examples seem anecdotic or of minor importance, improvements in our knowledge of topographic organomineral relationships, as well as in the composition of biominerals, lead to improvements in our understanding of biomineralization and fossilization processes. The broader purposes are to understand phylogeny and to avoid biases in reconstructing paleoenvironments.

6.4.4 DETECTION OF DIAGENESIS BY ANALYZING COMPONENTS IN FOSSILS

A mineralized skeleton is a factor that favors fossilization. Fossil organisms are used to unravel taxonomy, phylogeny, and paleoenvironmental conditions.

Corals are widely used to reconstruct paleotemperature, salinity, and pH of ancient seawater. However, they are subject to diagenesis that alters mineralogy, structures, and composition. Triassic samples from Austria and Turkey are still preserved with their primary mineralogy (high-Sr aragonite) and structure (Cuif 1980). However, they are not devoid of diagenesis (Dauphin et al. 1996, Cuif et al. 2011). XANES S K-edge maps confirm the general good preservation, the persistence of the organic sulfate, and a high Sr content. The main alteration is seen in the centers of calcification or EMZ of the septa (Figure 6.14). These EMZs are composed of small granules embedded in an organic matrix and are more soluble that the surrounding fibers so that diagenetic changes are stronger in EMZ.

Arthropod shields are a mixture of organic and phosphate carbonate minerals. Depending on the taxa, the organomineral ratio varies. Moreover, the organic component is a complex mixture of chitin and proteins. In fossil samples, diagenesis usually increases the mineral contents and modifies the organic components. XANES spectra have shown the cuticles of a fossil scorpion (310 My old) and a eurypterid (417 My old) are still composed of organic materials as *nitrogen-rich, chitin–protein complex* (Cody et al. 2011). However, the composition of the fossil cuticles clearly differs from that of a modern sample, due to diagenetic alteration (Figure 6.15).

Modern bones are pink pale or white, whereas archaeological samples are often colored. The origin of the color is the result of diagenetic alteration or burning. Such samples are often used in manufactured objects, because of the colors. Odontolite is a semiprecious stone, actually a blue-colored ivory (= dentine, the inner part of a tooth). Various blue samples of bones and ivory were studied through FTIR, XRD, electron microprobes, and PIXE analyses, as well as Mn K-edge XANES spectra. All samples were *apatites*, and the blue color was due to Mn5+. Moreover, the presence of Mn5+ is due to a heating process of the samples diagenetically enriched in Mn (Reiche and Chalmin 2008). Ancient cultural habits can be known using such analyses.

6.4.5 MEDICAL AND HEALTH PURPOSES

Eggs are a main food resource: about 76.2 billion eggs were consumed in 2009. Eggs are protected by a mineral calcitic eggshell. Antimicrobial compounds have been found in chicken eggshells (Jonchère et al. 2010). Nevertheless, salmonella bacteria can be found in eggs with cracked shells, so that the eggshell mechanical properties are important. Mechanical characteristics of the eggshell depend on the thickness, mineralogy, structure, and composition. A typical avian eggshell comprises a thin outer organic cuticle (with antibacterial properties), a thick palisade layer, a mammillary layer with a spherulitic arrangement, and a thick inner organic membrane (Figure 6.16). Electron microprobe (WDS) maps have shown that the spatial distribution of the chemical elements is heterogeneous (Cusack et al. 2003, Dauphin et al. 2006). WDS and XANES maps are concordant and confirm the presence of sulfur across the thickness of the eggshell and the high S content of the organic layers. In *Gallus* and *Numida* eggshells, the inner membranes are rich in both sulfur amino acids and organic sulfate. These results are concordant with immunofluorescence using keratan and dermatan sulfates (Arias and Fernandez 2003). A better understanding of the formation

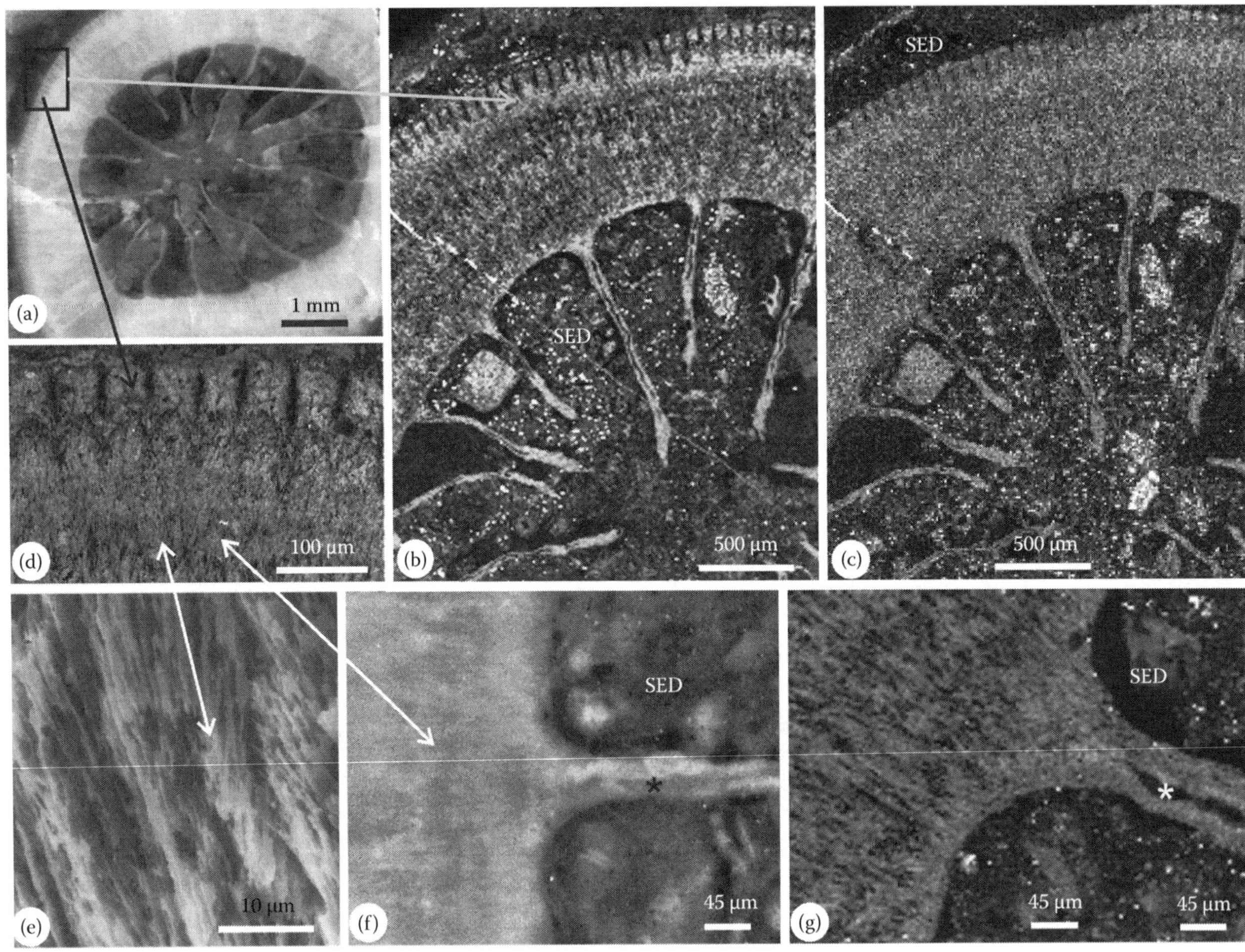

Figure 6.14 (a) Transversal section of the calyx of *Pachysolenia*, a Triassic coral, showing the thick external wall and radial septa; SEM image. (b) XANES sulfate maps of *Pachysolenia*; in this sample, diagenetic processes have modified the composition of the middle zone of septa, composed of EMZ; the structure of the thick wall and the heterogeneous sediment (SED) are well visible. (c) Sr maps of the same zone. (d) Detail of the outer part of the external wall. (e) SEM image of the aragonitic fibers of the external wall. (f) Polished surface showing the structure of the external wall and septa; black star: diagenetically modified central line; (g) XANES sulfate detailed map of the outer wall and the beginning of a radial septa, showing the diagenetic modified central line (white star).

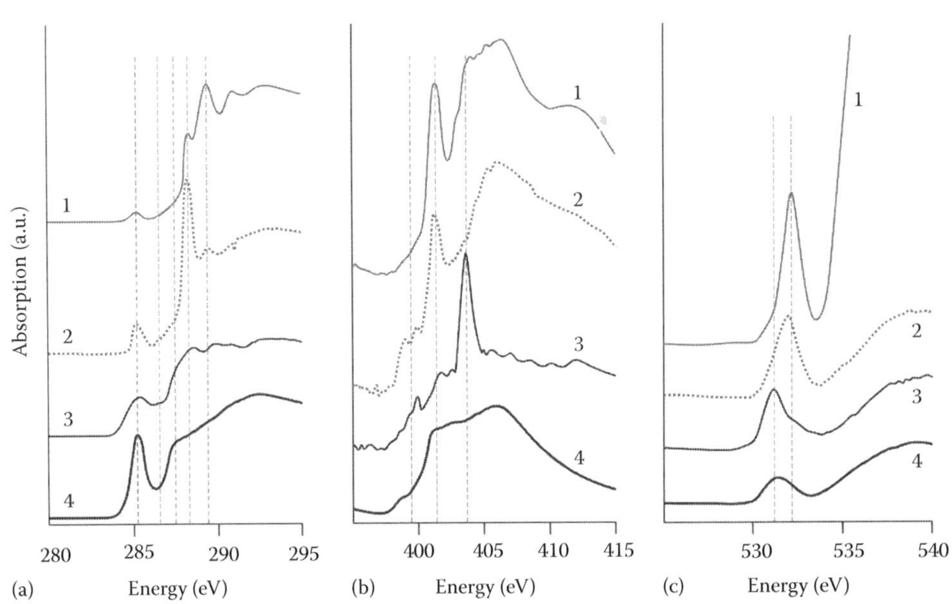

Figure 6.15 Carbon (a), nitrogen (b), and expanded view of the carbonyl 1s-π^* transition (c) spectra. (1) Chitin, (2) ancient scorpion, (3) modern scorpion, and (4) fossil eurypterid. (Redrawn from Cody, G. D. et al., *Geology*, 39, 255, 2011.)

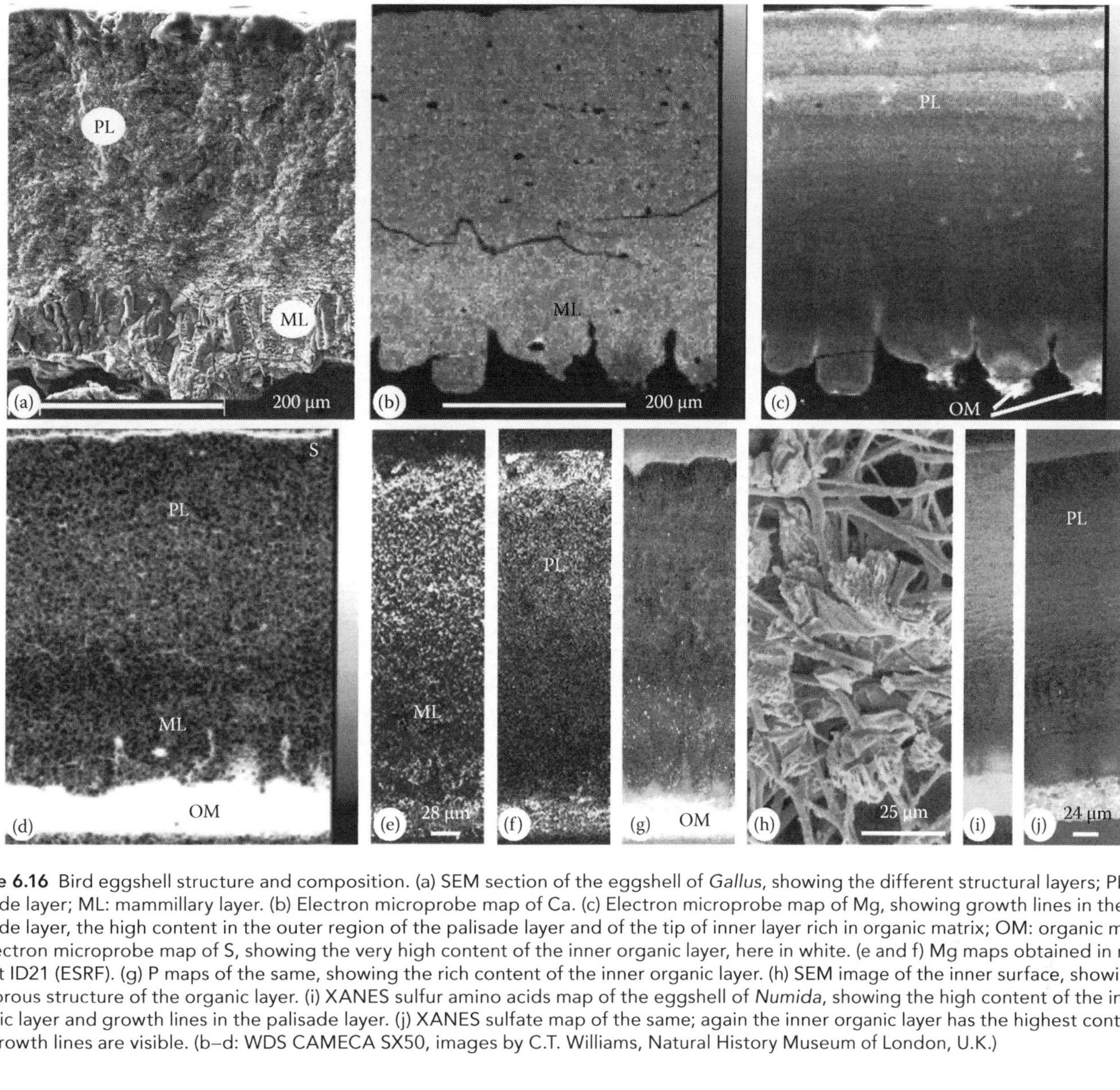

Figure 6.16 Bird eggshell structure and composition. (a) SEM section of the eggshell of *Gallus*, showing the different structural layers; PL: palisade layer; ML: mammillary layer. (b) Electron microprobe map of Ca. (c) Electron microprobe map of Mg, showing growth lines in the palisade layer, the high content in the outer region of the palisade layer and of the tip of inner layer rich in organic matrix; OM: organic matrix. (d) Electron microprobe map of S, showing the very high content of the inner organic layer, here in white. (e and f) Mg maps obtained in micro-XRF at ID21 (ESRF). (g) P maps of the same, showing the rich content of the inner organic layer. (h) SEM image of the inner surface, showing the fibrous structure of the organic layer. (i) XANES sulfur amino acids map of the eggshell of *Numida*, showing the high content of the inner organic layer and growth lines in the palisade layer. (j) XANES sulfate map of the same; again the inner organic layer has the highest content and growth lines are visible. (b–d: WDS CAMECA SX50, images by C.T. Williams, Natural History Museum of London, U.K.)

and resistance of the eggshells results from these analyses. These data can be used to improve the quality of eggshells and to reduce broken eggs for health safety and commercial interests.

XANES maps and spectra are also used to verify the results of medical treatments (Nguyen et al. 2011). Takatsuka et al. (2005) have shown that ZnO inhibited dentine demineralization. Using Zn K-edge spectra, they show that most of the zinc atoms detected were attached to hydroxyapatite and not to collagen. The long-term safety of dental amalgams containing mercury, associated with Zn and Cu, has been investigated (Harris et al. 2008). Maps show that the spatial distribution of Zn and Hg differs, but the possible presence of Hg in bloodstream was not evidenced. Toxic effects of Pb on cartilage metabolism and bone have been also studied (Meirer et al. 2011).

6.5 CONCLUSION AND PERSPECTIVES

Qualitative and quantitative chemical analyses, as well as distribution maps, can be obtained using electron and ionic microprobes. The spatial resolution and sensitivity of such analytical systems are regularly improved, and recent instruments such as NanoSIMS or atom probe are now able to provide chemical or isotopic maps at a nanoscale (3D maps for the atom probe tomography, CAMECA). However, they are not able to provide data on the coordination environment of a chemical element.

Both spectra and distribution maps are useful for a better understanding of the composition and biomineralization processes in Ca carbonates, Ca phosphate, and Si samples, as well as the fossilization mechanisms. In situ analyses provide valuable data for a better knowledge of the interplay of the mineral and organic components at various scales of observation. Moreover, they allow to differentiate the composition of irregularly imbricated skeleton layers as in the abalone shell or coral skeletons. When shell or skeletal layers are easy to separate (enamel and dentine in large teeth, most mollusc shells), it is possible to extract and separate the insoluble and soluble organic components. XANES spectra on these samples provide data on the possible presence of sulfur amino acids, destroyed in the usual preparative processes for amino acid analyses using chromatography. As shown by the aforementioned examples, XANES analyses range from fundamental research to commercial interest.

The cases of Sr and S in coral skeletons clearly demonstrate the improvement of both databases and beamlines dealing with XANES.

The continuous development of existing and new microspectroscopy endstations worldwide, optimized in different energy ranging from soft to hard x-rays and thus giving access to many different elements of interest in biomineralization, opens up a large variety of applications in this field. In addition, the development of nanoprobe beamlines, going further in terms of spatial resolution thanks to the great advances in focusing optics, will also provide new tools for the spectroscopic investigation of mineralized tissues at the nanoscale. A clear trend is also observed toward the multimodal analysis of sample, that is, beamlines combining multiple analysis tools on the same sample like microspectroscopy, microdiffraction, Raman spectroscopy, and FTIR. Such combination of techniques can provide complementary information about both the mineral and organic fractions of biominerals. Last, an important aspect to consider is also the improvement made in terms of sample environment (hydrated, frozen, temperature regulated, etc.) to analyze samples in conditions as close as possible to their native state.

ACKNOWLEDGMENTS

The authors are grateful to all people who through the years have contributed to this research (Dr. J. Susini ESRF, Dr. J.P. Cuif, Dr. A. Denis and Dr. J. Doucet Université Paris XI, Dr. A.D. Ball and Dr. C.T. Williams Natural History Museum London) and A. Lethiers (Université Paris VI) for his help with illustrations.

REFERENCES

Abramson L., S. Wirick, C. Lee et al. 2009. The use of soft X-ray spectromicroscopy to investigate the distribution and composition of organic matter in a diatom frustule and a biomimetic analog. *Deep-Sea Res. II* 56: 1369–1380.

Addadi L., J. Moradian, E. Shay et al. 1987. A chemical model for the cooperation of sulfates and carboxylates in calcite crystal nucleation: Relevance to biomineralization. *Proc. Natl. Acad. Sci. USA* 84: 2732–2736.

Allison N., A.A. Finch, M. Newville et al. 2005. Strontium in coral aragonite: 3. Sr coordination and geochemistry in relation to skeletal architecture. *Geochim. Cosmochim. Acta* 69: 3801–3811.

Argarwal B.K. 1979. *X-Ray Spectroscopy*. Berlin, Germany: Springer.

Arias J.L. and M.S. Fernandez. 2003. Biomimetic processes through the study of mineralized shells. *Mater. Charact.* 50: 189–195.

Baronnet A., J.P. Cuif, Y. Dauphin et al. 2008. Crystallization of biogenic Ca-carbonate within organo-mineral micro-domains. Structure of the calcite prisms of the pelecypod *Pinctada margaritifera* (Mollusca) at the submicron to nanometer ranges. *Mineral. Mag.* 72: 617–626.

Bazin D., M. Daudon, Ch. Chappard et al. 2011. The status of strontium in biological apatites: A XANES investigation *J. Synchrotron. Radiat.* 18: 912–918.

Beck L., M. Gehlen, A.M. Flank et al. 2002. The relationship between Al and Si in biogenic silica as determined by PIXE and XAS. *Nucl. Instrum. Methods Phys. Res.* B189: 180–184.

Beniash E., R.A. Metzler, R.S.K. Lam et al. 2009. Transient amorphous calcium phosphate in forming enamel. *J. Struct. Biol.* 166: 133–143.

Bianconi A. 1988. XANES spectroscopy. In *X-Ray Absorption: Principles, Applications, Techniques of EXAFS, SEXAFS and XANES*, eds. D.C. Koningsberger and R. Prins, pp. 573–662. New York: John Wiley.

Borca C.N., D. Grolimund, M. Willimann et al. 2009. The microXAS beamline at the Swiss Light Source: Towards nano-scale imaging. *J. Phys.: Conf. Ser.* 186: 012003.

Chalar C., M. Salomé, M. Senorale-Pose et al. 2012. A high resolution analysis of the structure and chemical composition of the calcareous corpuscles from *Mesocestoides corti*. *Micron* 44: 185–192. http://dx.doi.org/10.1016/j.micron.2012.06.008.

Chao W., B.D. Harteneck, J.A. Liddle et al. 2005. Soft X-ray microscopy at a spatial resolution better than 15 nm. *Nature* 435: 1210–1213.

Chao W., J. Kim, S. Rekawa et al. 2009. Demonstration of 12 nm resolution Fresnel zone plate lens based soft X-ray microscopy. *Opt. Express* 17: 17669–17677.

Charalambous P. 2011. Developments in the fabrication of zone plates and other nanostructures. *AIP Conf. Proc.* 1365: 65–68.

Clarke L.J., A.A. Finch, and T. Huthwelker. 2009. XANES determination of magnesium speciation in shells of two marine bivalve molluscs. *Geophys. Res. Abstr.* 11: EGU2009-9011-1.

Cody G.D., N.S. Gupta, D.E.G. Briggs et al. 2011. Molecular signature of chitin-protein complex in Paleozoic arthropods. *Geology* 39: 255–258.

Cotte M., J. Susini, N. Métrich et al. 2006. Blackening of Pompeian cinnabar paintings: X-ray microspectroscopy analysis. *Anal. Chem.* 78: 7484–7492.

Cuif J.P. 1980. Microstructure *versus* morphology in the skeleton of Triassic scleractinian corals. *Acta Palaeontol. Polon.* 25: 361–374.

Cuif J.P., A.D. Ball, Y. Dauphin et al. 2008. Structural, mineralogical, and biochemical diversity in the lower part of the pearl layer of cultivated seawater pearls from Polynesia. *Microsc. Microanal.* 14: 405–417.

Cuif J.P. and Y. Dauphin. 1998. Microstructural and physico-chemical characterizations of the "centers of calcification" in the septa of some recent Scleractinian corals. *Paläont. Z.* 72: 257–270.

Cuif J.P. and Y. Dauphin. 2005. The environmental recording unit in coral skeletons—A synthesis of structural and chemical evidences for a biochemically driven, stepping-growth process in fibres. *Biogeosciences* 2: 61–73.

Cuif J.P., Y. Dauphin, J. Doucet et al. 2003. XANES mapping of organic sulfate in three scleractinian coral skeletons. *Geochim. Cosmochim. Acta* 67: 75–83.

Cuif J.P., Y. Dauphin, B. Farre et al. 2008. Distribution of sulphated polysaccharides within calcareous biominerals indicates a widely shared layered growth-mode for the invertebrate skeletons and suggest a two-step crystallization process for the mineral growth units. *Mineral. Mag.* 72: 233–237.

Cuif J.P., Y. Dauphin, A. Freiwald et al. 1999. Biochemical markers of zooxanthellae symbiosis in soluble matrices of skeleton of 24 *Scleractinia* species. *Comp. Biochem. Physiol.* A123: 268–278.

Cuif J.P., Y. Dauphin, L. Howard et al. 2011. Is the pearl layer a reversed shell? A re-examination of the theory of pearl formation through physical characterizations of pearl and shell developmental stages in *Pinctada margaritifera*. *Aquat. Living Resour.* 24: 411–424.

Cuif J.P., Y. Dauphin, A. Meibom et al. 2008. Fine-scale growth patterns in coral skeletons: Biochemical control over crystallization of aragonite fibres and assessment of early diagenesis. In *Biogeochemical Controls on Palaeoceanographic Environmental Proxies*, eds. W.E.N. Austin and R.H. James, pp. 87–96. London, U.K.: Geological Society, London, Special Publication.

Cuif J.P., Y. Dauphin, and J.E. Sorauf. 2011. *Biominerals and Fossils through Time*. Cambridge, U.K.: Cambridge University Press.

Cuif J.P. and J.E. Sorauf. 2001. Biomineralization and diagenesis in the *Scleractinia*: Part I, biomineralization. *Bull. Tohoku Univ. Museum* 1: 144–151.

Cusack M., Y. Dauphin, J.P. Cuif et al. 2008. Micro-XANES mapping of sulphur and its association with magnesium and phosphorus in the shell of the brachiopod, *Terebratulina retusa*. *Chem. Geol.* 253: 172–179.

Cusack M., A.C. Fraser, and T. Stachel. 2003. Magnesium and phosphorus distribution in the avian eggshell. *Comp. Biochem. Physiol.* B134: 63–69.

Dauphin Y. 2001. Comparative studies of skeletal soluble matrices from some Scleractinian corals and Molluscs. *Int. J. Biol. Macromol.* 28: 293–304.

Dauphin Y. 2003. Soluble organic matrices of the calcitic prismatic shell layers of two Pteriomorphid bivalves: *Pinna nobilis* and *Pinctada margaritifera*. *J. Biol. Chem.* 278: 15168–15177.

Dauphin Y., A.D. Ball, H. Castillo-Michel et al. 2013. In situ distribution and characterization of the organic content of the oyster shell *Crassostrea gigas* (Mollusca, Bivalvia). *Micron* 44: 373–383.

Dauphin Y., A.D. Ball, M. Cotte et al. 2008. Structure and composition of the nacre—Prism transition in the shell of *Pinctada margaritifera* (Mollusca, Bivalvia). *Anal. Bioanal. Chem.* 390: 1659–1669.

Dauphin Y. and J.P. Cuif. 1999. Relation entre les teneurs en soufre des biominéraux calcaires et leurs caractéristiques minéralogiques. *Ann. Sci. Nat.* 2: 73–85.

Dauphin Y., J.P. Cuif, M. Cotte et al. 2012. Structure and composition of the boundary zone between aragonitic crossed lamellar and calcitic prism layers in the shell of *Concholepas concholepas* (Mollusca, Gastropoda). *Invert. Biol.* 131: 165–176.

Dauphin Y., J.P. Cuif, J. Doucet et al. 2003a. In situ chemical speciation of sulfur in calcitic biominerals and the simple prism concept. *J. Struct. Biol.* 142: 272–280.

Dauphin Y., J.P. Cuif, J. Doucet et al. 2003b. In situ mapping of growth lines in the calcitic prismatic layers of mollusc shells using X-ray absorption near-edge structure (XANES) spectroscopy at the sulphur edge. *Mar. Biol.* 142: 299–304.

Dauphin Y., J.P. Cuif, M. Salomé et al. 2005. Speciation and distribution of sulfur in a mollusk shell as revealed by in situ maps using X-ray absorption near-edge structure (XANES) spectroscopy at the S K-edge. *Am. Mineral.* 90: 1748–1758.

Dauphin Y., J.P. Cuif, M. Salomé et al. 2006. Microstructures and chemical composition of giant avian eggshells. *Anal. Bioanal. Chem.* 386: 1761–1771.

Dauphin Y., P. Gautret, and J.P. Cuif. 1996. Evolution diagénétique de la composition chimique des aragonites biogéniques chez les spongiaires, coraux et céphalopodes triasiques du Taurus lycien (Turquie). *Bull. Soc. Géol. Fr.* 167: 247–256.

David C., B. Kaulich, R. Barrett et al. 2000. High-resolution lenses for sub-100 nm x-ray fluorescence microscopy. *Appl. Phys. Lett.* 77: 3851–3853.

De Andrade V., J. Susini, M. Salomé et al. 2011a. Submicrometer hyperspectral X-ray imaging of heterogeneous rocks and geomaterials: Applications at the Fe K-edge. *Anal. Chem.* 83: 4220–4227.

De Andrade V., J. Thieme, P. Northrup et al. 2011b. The sub-micron resolution X-ray spectroscopy beamline at NSLS-II. *Nucl. Instrum. Meth.* A649: 46–48.

Eichert D., M. Salomé, M. Banu et al. 2005. Preliminary characterization of calcium chemical environment in apatitic and non apatitic calcium phosphates of biological interest by X-ray absorption spectroscopy. *Spectrochim. Acta* 60: 850–858.

Erko A. and I. Zizak. 2009. Hard X-ray micro-spectroscopy at the Berliner Elektronenspeicherring für Synchrotronstrahlung II. *Spectrochim. Acta Part B* 64: 833–848.

Finch A.A. and N. Allison. 2003. Strontium in coral aragonite: 2. Sr coordination and the long-term stability of coral environmental records. *Geochim. Cosmochim. Acta* 67: 4519–4527.

Finch A.A. and N. Allison. 2007. Coordination of Sr and Mg in calcite and aragonite. *Mineral. Mag.* 71: 539–552.

Finch A.A., N. Allison, S.R. Sutton et al. 2003. Strontium in coral aragonite: 1. Characterization of Sr coordination by extended absorption X-ray fine structure. *Geochim. Cosmochim. Acta* 67: 1189–1194.

Flank A.M., G. Cauchon, P. Lagarde et al. 2006. LUCIA, a microfocus soft XAS beamline. *Nucl. Instrum. Methods Phys. Res.* B246: 269–274.

Foster L.C., A.A. Finch, N. Allison et al. 2008. Mg in aragonitic bivalve shells: Seasonal variations and mode of incorporation in *Arctica islandica*. *Chem. Geol.* 254: 113–119.

George G.N., S.J. George, and I.J. Pickering. 2001. *EXAFSPAK: A Suite of Computer Programs for Analysis of X-Ray Absorption Spectra*. Menlo Park, CA: Stanford Synchrotron Radiation Laboratory (SSRL). http://www-ssrl.slac.stanford.edu/exafspak.html.

Golosio B., A. Somogyi, A. Simionovici et al. 2004. Nondestructive three-dimensional elemental microanalysis by combined helical X-ray microtomographies. *Appl. Phys. Lett.* 84: 2199–2201.

Gorelick S., J. Vila-Comamala, V.A. Guzenko et al. 2011. High-efficiency Fresnel zone plates for hard X-rays by 100 keV e-beam lithography and electroplating. *J. Synchrotron. Radiat.* 18: 442–446.

Greegor R.B., N.E. Pingitore Jr., and F.W. Lytle. 1997. Strontianite in coral skeletal aragonite. *Science* 275: 1452–1454.

Harris H.H., S. Vogt, H. Eastgate et al. 2008. Migration of mercury from dental amalgam through human teeth. *J. Synchrotron. Radiat.* 15: 123–128.

Hayakawa S., N. Ikuta, M. Suzuki et al. 2001. Generation of an X-ray microbeam for spectromicroscopy at SPring-8 BL39XU. *J. Synchrotron. Radiat.* 8: 328–330.

Hignette O., P. Cloetens, G. Rostaing et al. 2005. Efficient sub 100 nm focusing of hard x rays. *Rev. Sci. Instrum.* 76: 063709.

Hitchcock A.P., S. Bodeur, and M. Tronc. 1989. Sulfur and chlorine K-shell spectra of gases. *Physica B* 158: 257–258.

Howells M., C. Jacobsen, and A. Warwick. 2006. *Principles and Applications of Zone Plate X-Ray Microscopes*. Berlin, Germany: Springer.

Jalilehvand F. 2006. Sulfur: Not a "silent" element any more. *Chem. Soc. Rev.* 35: 1256–1268.

Jonchère V., S. Réhault-Godbert, C. Hennequet-Antier et al. 2010. Gene expression profiling to identify eggshell proteins involved in physical defence of the chicken egg. *BMC Genomics* 11: 57 (19pp.).

Kang H.C., H. Yan, R.P. Winarski et al. 2008. Focusing of hard x-rays to 16 nanometers with a multilayer Laue lens. *Appl. Phys. Lett.* 92: 221114-1–221114-3.

Kirkpatrick P. and A.V. Baez. 1948. Formation of optical images by X-rays. *J. Opt. Soc. Am.* 38: 766–774.

Kossel W. 1920. Zum Bau der Rontgenspektren. *Zeit. Phys.* 1: 119–134.

Kronig R. de L. 1931. Zur Theorie der Feinstruktur in den Röntgenabsorptionsspektren. *Zeit. Phys.* 70: 317–323.

Lengeler B., C.G. Schroer, B. Benner et al. 2001. Parabolic refractive X-ray lenses: A breakthrough in X-ray optics. *Nucl. Instrum. Methods Phys. Res. A* 467–468(Part 2): 944–950.

Letard I., R. Tucoulou, P. Bleuet et al. 2006. A multi-element Si(Li) detector for the hard X-ray micro-probe at ID22 (ESRF). *Rev. Sci. Instrum.* 77: 063705-1–063705-8

Lichtenegger H.C., T. Schöberl, J.T. Ruokolainen et al. 2003. Zinc and mechanical prowess in the jaws of *Nereis*, a marine worm. *Proc. Natl. Acad. Sci. USA* 100: 9144–9149.

Lytle F., D. Sayers, and E. Stern. 1982. The history and modern practice of EXAFS spectroscopy. In *Advances in X-Ray Spectroscopy*, eds. C. Bonnelle and C. Mande, pp. 267–286. New York: Pergamon Press.

Martínez-Criado G., R. Tucoulou, P. Cloetens et al. 2012. Status of the hard X-ray microprobe beamline ID22 of the European synchrotron radiation facility. *J. Synchrotron. Radiat.* 19: 10–18.

Maser J., R. Winarski, M. Holt et al. 2006. The hard X-ray nanoprobe beamline at the advanced photon source. *Proceedings of 8th International Conference on X-Ray Microscopy*, Himeji, Japan, IPAP Conference Series, Vol. 7, pp. 26–29. Tokyo, Japan: Institute of Pure and Applied Physics.

McNulty I., S.P. Frigo, C.C. Retsch et al. 1998. Design and performance of the 2-ID-B scanning X-ray microscope. *SPIE Proc.* 3449: 67.

Meibom A., J.P. Cuif, F. Hillion et al. 2004. Distribution of magnesium in coral skeleton. *Geophys. Res. Lett.* 31: L23306. doi: 10.1029/2044GL021313.

Meirer F., B. Pemmer, G. Pepponi et al. 2011. Assessment of chemical species of lead accumulated in tidemarks of human articular cartilage by X-ray absorption near-edge structure analysis. *J. Synchrotron. Radiat.* 18: 238–244.

Métrich N., A.J. Berry, H.S.C. O'Neill et al. 2009. The oxidation state of sulfur in synthetic and natural glasses determined by X-ray absorption spectroscopy. *Geochim. Cosmochim. Acta* 73: 2382–2399.

Mimura H., S. Handa, T. Kimura et al. 2010. Breaking the 10 nm barrier in hard-X-ray focusing. *Nat. Phys.* 6: 122–125.

Nguyen C., H.K. Ea, D. Thiaudière et al. 2011. Calcifications in human osteoarthritic articular cartilage: Ex vivo assessment of Ca compounds using XANES spectroscopy. *J. Synchrotron. Radiat.* 18: 475–480.

Nothdurft L.D. and G.E. Webb. 2007. Microstructure of common reef-building coral genera *Acropora*, *Pocillopora*, *Goniastrea* and *Porites*: Constraints on spatial resolution in geochemical sampling. *Facies* 53: 1–26.

Nouet J., M. Cotte, J.P. Cuif et al. 2012. Biochemical change at the setting-up of the crossed-lamellar layer in *Nerita undata* (Mollusca, Gastropoda). *Minerals* 2: 85–99.

Orton J.H. and Amirthalingam G. 1927. Notes on shell-depositions in oysters. *J. Mar. Biol. Assoc. UK* 14: 935–953.

Paterson D., M.D. de Jonge, D.L. Howard et al. 2011. The X-ray fluorescence microscopy beamline at the Australian synchrotron. *AIP Conf. Proc.* 1365: 219–222.

Perez-Huerta A., M. Cusack, M. Janousch et al. 2008. Influence of crystallographic orientation of biogenic calcite on in situ Mg XANES analyses. *J. Synchrotron. Radiat.* 15: 572–575.

Pickering I.J., R.C. Prince, T. Divers et al. 1998. Sulfur K-edge X-ray absorption spectroscopy for determining the chemical speciation of sulfur in biological systems. *FEBS Lett.* 441: 11–14.

Pingitore Jr. N.E., A. Iglesias, A. Bruce et al. 2002a. Valences of iron and copper in coral skeleton: X-ray absorption spectroscopy analysis. *Microchem. J.* 71: 205–210.

Pingitore Jr. N.E., A. Iglesias, F. Lytle et al. 2002b. X-Ray absorption spectroscopy of uranium at low ppm levels in coral skeletal aragonite. *Microchem. J.* 71: 261–266.

Pingitore Jr. N.E., G. Meitzner, and K.M. Love. 1995. Identification of sulfate in natural carbonates by x-ray absorption spectroscopy. *Geochim. Cosmochim. Acta* 59: 2477–2483.

Politi Y., R.A. Metzler, M. Abrecht et al. 2008. Transformation mechanism of amorphous calcium carbonate into calcite in the sea urchin larval spicule. *Proc. Natl. Acad. Sci. USA* 105: 17362–17365.

Prange A., I. Arzberger, C. Engemann et al. 1999. In situ analysis of sulfur in the sulfur globules of phototrophic sulfur bacteria by X-ray absorption near edge spectroscopy. *Biochim. Biophys. Acta* 1428: 446–454.

Prietzel J., J. Thieme, U. Neuhausler et al. 2003. Speciation of sulphur in soils and soil particles by X-ray spectromicroscopy. *Eur. J. Soil Sci.* 54: 423–433.

Ravel B. and M. Newville. 2005. Athena, Artemis, Hephaestus: Data analysis for X-ray absorption spectroscopy using IFEFFIT. *J. Synchrotron. Radiat.* 12: 537–541.

Rehr J.J., J.J. Kas, F.D. Vila et al. 2010. Parameter-free calculations of x-ray spectra with FEFF9. *Phys. Chem. Chem. Phys.* 12: 5503–5513.

Reiche I. and E. Chalmin. 2008. Synchrotron radiation and cultural heritage: Combined XANES/XRF study at Mn K-edge of blue, grey or black coloured palaeontological and archaeological bone material. *J. Anal. Atom. Spectrom.* 23: 799–806.

Ryan C., R. Kirkham, R. Hough et al. 2010. Elemental X-ray imaging using the Maia detector array: The benefits and challenges of large solid-angle. *Nucl. Instrum. Methods A* 619: 37–43.

Salomé M., P. Bleuet, S. Bohic et al. 2009. Fluorescence X-ray micro-spectroscopy activities at ESRF. *J. Phys.: Conf. Ser.* 186:012014-1–012014-3.

Sandeman I.M. 2008. Fine banding in the septa of corals. *Proceedings of 11th International Coral Reef Symposium*, Ft Lauderdale, FL, Session 3, pp. 67–71.

Sayers D., E. Stern, and F. Lytle. 1971. New technique for investigating noncrystalline structures: Fourier analysis of the extended X-ray—Absorption fine structure. *Phys. Rev. Lett.* 27: 1204–1207.

Schroer C.G., P. Boye, J.M. Feldkamp et al. 2010. Hard X-ray nanoprobe at beamline P06 at PETRA III. *Nucl. Instrum. Methods Phys. Res. A* 616: 93–97.

Schroer C.G., O. Kurapova, J. Patommel et al. 2005. Hard X-ray nanoprobe based on refractive X-ray lenses. *Appl. Phys. Lett.* 87: 124103–124105.

Snigirev A., V. Kohn, I. Snigireva et al. 1996. A compound refractive lens for focusing high-energy X-rays. *Nature* 384:49–51.

Snigirev A. and I. Snigireva. 2008. Hard X-ray microoptics. In *Modern Developments in X-Ray and Neutron Optics*, Springer Series in Optical Sciences, eds. A. Erko, M. Idir, T. Krist, and A.G. Michette, pp. 255–285. New York: Springer.

Soldati A.L., P. Glatzel, J. Geck et al. 2009. X-ray emission spectroscopy for paleoclimatology: Studies of the trace element distribution in fossil mussel shells. ESRF Report Experiment (EC-571) 42388_A.

Soldati A.L., J. Goettlicher, D.E. Jacob et al. 2010. Manganese speciation in *Diplodon chilensis patagonicus* shells: A XANES study. *J. Synchrotron. Radiat.* 17: 193–201.

Solé A., E. Papillon, M. Cotte et al. 2007. PyMCA: A multiplatform code for the analysis of energy-dispersive X-ray spectra. *Spectrochim. Acta* B62: 63–68. http://pymca.sourceforge.net/.

Somogyi A., C.M. Kewish, F. Polack et al. 2011. The scanning nanoprobe beamline Nanoscopium at Synchrotron Soleil. *AIP Conf. Proc.* 1365: 57–60.

Stern E.A. 2001. Musings about the development of the XAFS. *J. Synchrotron. Radiat.* 8: 49–54.

Stöhr J. 1992. *NEXAFS Spectroscopy*. Berlin, Germany: Springer.

Szlachetko J., M. Cotte, J. Morse et al. 2010. Wavelength-dispersive spectrometer for X-ray microfluorescence analysis at the X-ray microscopy beamline ID21 (ESRF). *J. Synchrotron. Radiat.* 17: 400–408.

Takatsuka T., J. Hirano, H. Matsumoto et al. 2005. X-ray absorption fine structure analysis of the local environment of zinc in dentine treated with zinc compounds. *Eur. J. Oral Sci.* 113: 180–183.

Thompson A., I. Lindau, D. Attwood et al. 2009. *X-Ray Data Booklet.* Berkeley, CA: Center for X-Ray Optics and Advanced Light Source, Lawrence Berkeley National Laboratory, University of California. http://xdb.lbl.gov.

Vairavamurthy A. 1998. Using X-ray absorption to probe sulfur oxidation states in complex molecules. *Spectrochim. Acta* A54: 2009–2017.

Vekemans B., K. Janssens, L. Vincze et al. 1994. Analysis of X-ray spectra by iterative least squares (AXIL): New developments. *X-Ray Spectrom.* 23: 278–285.

Vila-Comamala J., S. Gorelick, E. Färm et al. 2011. Ultra-high resolution zone-doubled diffractive X-ray optics for the multi-keV regime. *Opt. Express* 19: 175–184.

Vincze L., B. Vekemans, F.E. Brenker et al. 2004. Three-dimensional trace element analysis by confocal X-ray micro-fluorescence imaging. *Anal. Chem.* 76: 6786–6791.

Wada K. 1964. Studies on the mineralization of the calcified tissue in molluscs. IV. Selective fixation of ^{45}Ca into or onto the metachromatic matter in the processes of shell mineralization. *Bull. Jpn. Soc. Sci. Fish* 30: 393–399.

Wada K. 1980. Initiation of mineralization in bivalve molluscs. In *The Mechanisms of Biomineralization in Animals and Plants*, eds. M. Omori and N. Watabe, pp. 79–92. Tokyo, Japan: Tokai University Press.

Wasserman S.R. 1997. The analysis of mixtures: Application of principal component analysis to XAS spectra. *J. Phys. IV* 7: 203–205.

Webb S.M. 2005. SIXpack: A graphical user interface for XAS analysis using IFEFFIT. *Phys. Scripta.* T115: 1011–1014.

Xu H., X. Yu, and R. Tai. 2011. X-ray microscopy beamlines at SSRF—Present status and future plan. *AIP Conf. Proc.* 1365: 52–56.

7 Local structure development: Characterization of biominerals using x-ray absorption spectroscopy

Yael Politi and Ivo Zizak

Contents

7.1 INTRODUCTION

Extended x-ray absorption fine structure (EXAFS) probes the energy-dependent absorption cross section of a selected atom of interest. EXAFS analysis provides details on the short-range (up to a few Å) atomic arrangement around a central absorbing atom (the *absorber*), including the type and number of nearest neighbors, their distances to the absorber, and the degree of local disorder. A number of features make EXAFS a key analytical tool in many fields of science and especially in biomineralization: First, it can be applied to a wide range of samples and sample environments, including amorphous minerals, liquids, and solutions. Second, it is element specific and thus suitable for the study of complex multicomponent systems, even when the element of interest is present only in low concentration.

In the field of biomineralization, EXAFS has filled a gap in the ability to structurally characterize a wide range of amorphous biominerals (Taylor et al. 1993; Levi-Kalisman et al. 2000; Becker et al. 2003), which could not have been studied by conventional x-ray or electron diffraction, the routinely used methods to obtain structural information at the atomic scale of crystalline minerals. Moreover, the high degree of spatial resolution in determining bond lengths around the absorber atom makes EXAFS analysis specifically sensitive to small structural changes occurring, for example, during the amorphous to crystalline mineral transitions, that are a widespread biomineralization mechanism (Addadi et al. 2003). EXAFS was thus often used for time-dependent studies of mineralization mechanisms in both biological and synthetic systems (Marxen et al. 2003; Politi et al. 2006; Lam et al. 2007; Gebauer et al. 2010; Baumgartner et al. 2013). Last but not least, x-ray absorption spectroscopy (XAS), which includes both the EXAFS spectra described here and the x-ray absorption near edge structure (XANES) spectra described in Chapters 6 and 8 has been used in structural characterization of dopant atoms present in a wide range of biominerals. Examples for this approach are the use of Mg and Sr K-edge XAS to study their incorporation in $CaCO_3$ (Allison et al. 2005; Finch and Allison 2007; Cusack et al. 2008b; Politi et al. 2010) and incorporation of different metal ions in bone (Korbas et al. 2004; Laurencin et al. 2010), teeth (Matsunaga et al. 2009; Laurencin et al. 2010), and implant ceramics (Rokita 2000; Korbas et al. 2004; Terra et al. 2009). Particularly interesting is also the employment of S K-edge XAS as a marker for inter- and intracrystalline organic macromolecules as described in Chapter 6 (Cuif et al. 2003; Cusack et al. 2008a).

This chapter will begin with a short introduction for EXAFS including a simple theoretical description. We then highlight important aspects of the technical requirements of an EXAFS experiment followed by a basic description of data analysis approaches, and we finish with a few examples from EXAFS studies in the field of biomineralization. An in-depth introduction to EXAFS can be found in classical compilation edited by Koningsberger and Prins (1988). For a thorough technical and practical overview, we recommend the more recent book by Bunker (2010).

7.1.1 SHORT HISTORY

The history of EXAFS and XANES is closely related, as described in Chapter 6, with the first experiments performed already in the 1920s (Fricke 1920; Koningsberger and Prins 1988). The introduction of photoelectron scattering to explain the EXAFS oscillations and the recognition of the phenomena as a short-range order effect occurred already in the 1930s (Kronig 1931, 1932a,b). The next major step in EXAFS understanding and interpretation came about only at the beginning of the 1970s when Stern, Lytle, and Sayers (Sayers et al. 1971; Stern et al. 1975) showed that the Fourier transform (FT) of the EXAFS equation with respect to the photoelectron wave number is related to the atomic distance of the neighboring atoms. With further development of the experimental and analysis methods in parallel with increasing availability of synchrotron radiation sources, the technique is nowadays widely used in many fields of science including, for example, chemistry, material science, geology, and biology (Sayers et al. 1971; Rehr et al. 1978). The first experiments to use EXAFS analysis for the study of biomineralization were performed already in the early 1980s and focused on the mineral structure of mature and developing mice bone as compared to synthetic calcium phosphate phases (Eanes et al. 1981; Miller et al. 1981; Binsted et al. 1982). Interestingly, two similar studies were published in the same year (1981) by groups working in the United States (Eanes et al. 1981) and the United Kingdom (Miller et al. 1981), mirroring the fast developments in EXAFS application and analysis and its introduction into biological research shortly before (Shulman et al. 1978).

7.1.2 THEORY

A thorough description of the photoelectric effect, relaxation process, and the origin of the XAS spectra is given in Chapter 6. Briefly, XAS measures the attenuation of the absorption cross section of an element with increasing excitation energy (Figure 7.1). While EXAFS can be described in terms of scattering theory, XANES is additionally influenced by electronic transitions to partially or unfilled orbitals. In terms of scattering theory, the ejected photoelectrons are backscattered by neighboring atoms around the central absorbing atom. Using the de Broglie relation, the electrons can be described as waves, in which the momentum of the electron is related to its wavelength λ. The interference of the outgoing and backscattered electrons modulates the probability of absorption at the central atom.

7.1.2.1 XANES vs. EXAFS

With increasing incident photon energy (E) above the continuum threshold, that is, the edge energy (E_0), the photoelectron is ejected with increasing kinetic energy ($E_K = E - E_0$), leaving a *core hole* behind. In wave description, the photoelectron travels as a wave with wave number (k) depending on the kinetic energy (Equation 7.1):

$$k = \sqrt{\frac{2m_e(E - E_0)}{\hbar^2}} \tag{7.1}$$

where

m_e is the electron mass
$\hbar$ is the Planck constant

The relation between the electron wave number to its de Broglie wavelength is given by $k = 2\pi/\lambda$.

At low kinetic energy, the photoelectron is likely to be scattered multiple times by the surrounding atoms around

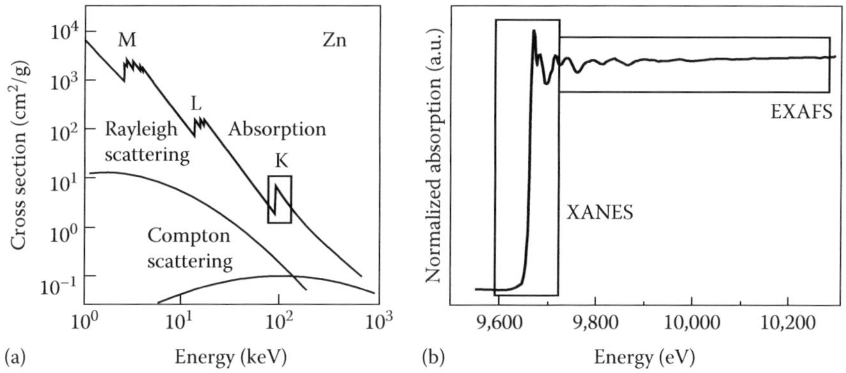

Figure 7.1 (a) Absorption and scattering cross section for Zn. Up to about 15 keV, the absorption has a larger cross section than elastic (Rayleigh) and inelastic (Compton) scattering. At higher energies, the Compton scattering becomes comparable with the absorption and has to be taken into account. Notice the sharp absorption edges marked M, L, and K. (b) Zn K-edge absorption spectrum. The XANES and EXAFS regions are designated in boxes. Note the oscillatory pattern at the EXAFS region and its decay with increasing energy.

the absorber (XANES region, up to about 50 eV above E_0). As the photoelectron E_K is increased, single-scattering events are more common (EXAFS region, ~100–1000 eV above E_0). The border between the XANES and the EXAFS regions is frequently defined as the energy where the wavelength of the photoelectron equals the distance between the absorber and the nearest neighbors (Koningsberger and Prins 1988). Since the XANES region is dominated by multiple-scattering events, it is specifically sensitive to the geometry around the absorber, while single-scattering events at the EXAFS region contain quantitative information on the atomic distance of the scattering atoms from the absorber (Bianconi and Benfatto 1988). Importantly, a number of physical phenomena come into play in the XANES regions, such as excitations to unoccupied bound states and resonances that make the interpretation of this region more complicated (Koningsberger and Prins 1988). The multiple-scattering description of the XANES region is used to unify the treatment of XANES and EXAFS and can be sufficient in many, but not all, cases (Ankudinov et al. 1998; Kelly et al. 2008). Figure 7.2 shows a comparison between the Ca K-edge XANES and EXAFS spectra of calcite, aragonite, and amorphous calcium carbonate (ACC). Calcite and aragonite are two crystalline polymorphs of $CaCO_3$, with the same chemical formula but different coordination geometry, and ACC is an amorphous phase with the formula $CaCO_3{\cdot}H_2O$.

7.1.2.2 EXAFS spectra: Interference of photoelectron and backscattered waves

The modulation of the absorption coefficient is described by the EXAFS function, $\chi(E)$, defined as

$$\chi(E) = \frac{\mu(E) - \mu_0(E)}{\Delta\mu_0(E)} \tag{7.2}$$

where

$\mu(E)$ is the measured absorption coefficient

$\mu_0(E)$ represents the absorption contribution from an isolated atom described as a smooth background function

$\Delta\mu_0(E)$ is the measured jump in absorption at the edge energy E_0

The modulation of the absorption coefficient is caused by the interference between the outgoing photoelectron waves and the backscattered waves, as determined by the phase shifts between them. The relative phases in turn are determined mainly by

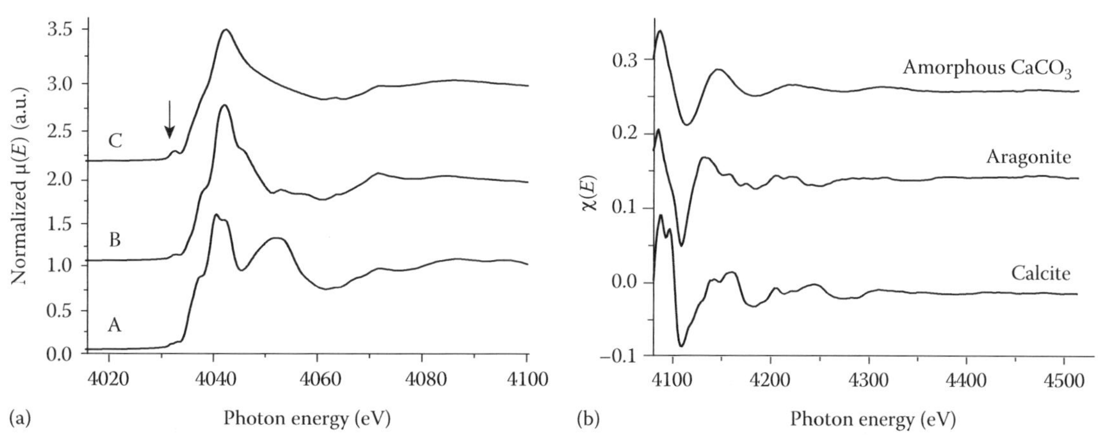

Figure 7.2 Comparison between the XAS spectra of three $CaCO_3$ polymorphs (A) calcite, (B) aragonite, and (C) ACC $CaCO_3{\cdot}H_2O$. (a) The XANES region of the spectra. The pre-edge peak (arrow) attributed to 1s–3d transition is forbidden in centrosymmetric systems (e.g., calcite) (Levi-Kalisman et al. 2000); the peak is enhanced in ACC and in aragonite; both structures deviate from this coordination geometry. (b) The EXAFS part of the spectra, showing the oscillatory nature of the signal, which diminishes with increasing energy. The ACC spectrum in both the XANES and EXAFS regions contains less-pronounced spectral features, as compared with the spectral richness of the two crystalline polymorphs, calcite and aragonite. The spectra were acquired in fluorescence mode from powder samples.

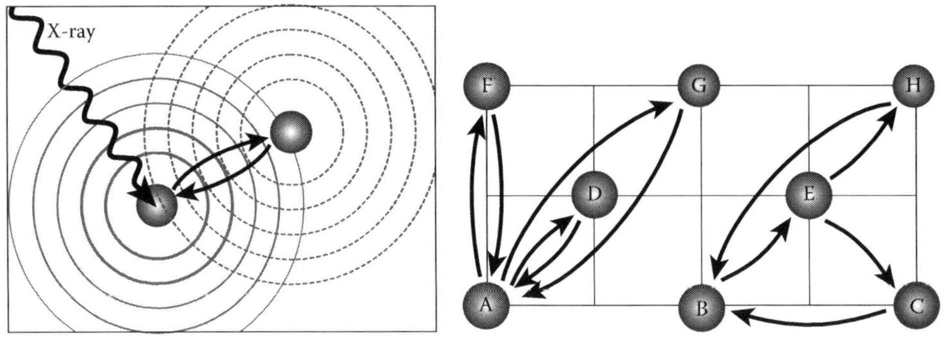

Figure 7.3 (a) Illustration of the photoelectron-scattering event. The wave emanating from the absorber atom, represented by solid line, is backscattered by a neighboring atom, dashed line. Wiggled arrow represents the incoming x-ray beam, and the curved arrows represent the photoelectron *path*. (b) Depiction of single- and multiple-scattering paths. All the paths on the left side (A–F, A–D, and A–G) represent single-scattering events (as in part *a* of this figure), and the paths on the right (B–E–H and B–E–C) represent *two-leg* multiple-scattering paths.

(1) the photoelectron wavelength, (2) the interatomic distances between the absorber and the scattering atoms, and (3) the scattering potential of the neighboring atoms. At each excitation energy, the outgoing and backscattered waves may interfere constructively or destructively, thus increasing or reducing the absorption probability at the central atom, resulting in the EXAFS oscillations. These oscillations therefore hold quantitative information on the type of scattering atoms and their distance from the absorber; each scattering event is represented by a *path* of the photoelectron emanating from the absorber and backscattered by a coordinating atom (Figure 7.3), and the EXAFS equation is a summation of all these scattering paths (j):

$$\chi(k) = \sum_j \frac{N_j f_j(k) e^{-2k^2\sigma_j^2} e^{-2R_j/\lambda(k)}}{kR_j^2} \sin[2kR_j + \delta_j(k)] \quad (7.3)$$

where

N is the number of neighboring atoms at distance R from the absorber

σ^2 is the mean square displacement in the absorber–neighbor bond distance

$f(k)$ and $\delta(k)$ are the scattering properties (amplitude and phase shift) of the neighboring atoms

$\lambda(k)$ is the mean free path of the photoelectron, that is, the mean distance the photoelectron travels before the core hole is filled, and depends on the lifetime of the core hole

The scattering properties, that is, the amplitude $f(k)$ and the phase shift $\delta(k)$, depend on the element atomic number (Z) and can be calculated by computer software such as FEFF (Rehr and Albers 2000; Rehr et al. 2009). The photoelectron mean free path varies between 5 and 30 Å, which makes EXAFS sensitive to only a short range around the absorbing atom (Figure 7.4). In this range, the EXAFS equation allows us to determine N, R, and σ^2 of the surrounding neighbors around the absorber with high accuracy. All atoms within similar distance (R) from the absorber atom constitute a *coordination shell*, with the *coordination number* being the number of atoms at this distance (N). With small changes in the interpretation, the same EXAFS formula can be applied to multiple scattering, where R represents the half length of the path travelled by the photoelectron from the absorber through the different scatterers and back to the absorber atom.

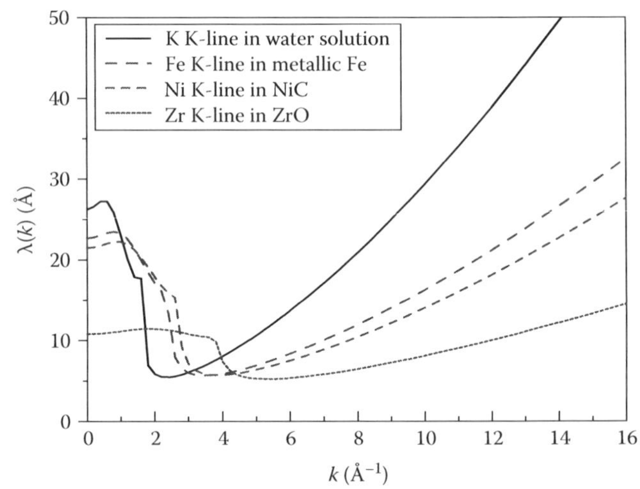

Figure 7.4 Calculation (FEFF) of the k dependence of the photoelectron mean free path for different atoms in various structures.

7.2 PERFORMING AN EXAFS EXPERIMENT

7.2.1 MEASUREMENT DESIGN

7.2.1.1 Transmission vs. fluorescence

The most direct method to measure the absorption cross section is by measuring the intensity of transmission of an x-ray beam through the sample (I_t) and normalizing it to the incident beam (I_0). Other more indirect methods are used to calculate absorption through measurements of relaxation processes like the total electron yield (TEY) or fluorescence yield (FLY).

The choice of the detection method depends on many factors, but mostly on the geometry of the sample and the concentration of the absorbing element in the sample. Transmission measurements are usually fast and have the highest signal to noise ratio when the sample thickness (x) fulfills $\exp(-x\Delta\mu_0) \approx 1$ with respect to the absorber element of interest. When the sample contains other strongly absorbing elements, the total absorption should be taken into account, as it reduces the intensity of the transmitted beam. Similarly, if the concentration of the element of interest is very low, the sample thickness becomes too large to allow transmission. In these cases, it is better to measure the TEY or fluorescence. The ratio of electron to FLY depends on the atomic number; therefore, TEY method is usually used for light elements (up to Mg) and fluorescence for heavier elements.

The major difficulty in fluorescence mode experiments is the so-called *self-absorption* effect. Self-absorption is the energy-dependent change of the mean penetration depth of the incident beam that changes the probability of the fluorescent photons to be reabsorbed by the sample, reducing the intensity of the EXAFS oscillations (Stern and Kim 1981). Self-absorption occurs in samples that are too thick or too concentrated.

7.2.1.2 Detectors

The standard measurement setup involves three ionization chambers, called simply I_0, I_1, and I_2. The sample is placed between I_0 and I_1, and a standard for energy calibration is placed between I_1 and I_2. The chambers are filled with gas to optimize the absorption according to the energy used. For transmission measurements, I_0, I_1, and I_2 are aligned along the beam axis, while for fluorescence measurements, I_f is placed in the horizontal plane 90° to the incident beam and the sample is tilted 45° with respect to both the beam and I_f. Since fluorescence intensity is much lower than the primary beam and the transmitted beam, other detector types are used. Photosensitive diodes are simple to use, fast, and efficient as they integrate current induced by all photons. This however makes them energy insensitive. When the sample contains absorbing elements, other than the element of interest, an energy-dispersive detector (EDS) has to be used to select the interesting fluorescence only. Chapter 6 covers the principle of EDSs such as Ge, Si(Li) crystal detectors, multichannel analyzer (MCA), and wavelength-dispersive spectrometers (WDSs). For obtaining high-quality EXAFS data, special attention has to be given to the signal to noise ratio and detector dead time.

Signal to noise ratio: In counting detectors (as opposed to integrating detectors), the statistics obey the Poisson distribution, so that the standard deviation of the signal equals the square root of the number of counted photons. To obtain the signal to noise ratio of about 1000 (recommended minimum for EXAFS at $k > 12$ Å^{-1}), at least 10^6 photons in the fluorescence of the studied element need to be collected. Thus, the detector and sample properties will determine the time needed for the measurement. The determinations of measurement parameters such as counting time and number of scans are discussed in the succeeding text. If necessary, the experiment has to be repeated until the statistics is satisfactory.

Detector dead time: Maximal count rate in modern fast counting detectors is in the order of 10^5–10^6 counts/s. This count rate includes all the photons arriving at the detector. Large input on the detector due to high concentration of absorbing elements in the sample saturates the detectors, thus distorting the XAFS signal. Sample concentration should be adjusted to ensure low dead time.

7.2.1.3 Beamline properties

EXAFS requires a fine variation of the excitation energy over a large energy range (~1 keV). Although the first XAFS experiments were performed using laboratory sources, it is nowadays only performed at synchrotron radiation facilities with the excitation energy selected by a monochromator as described in Chapter 6. The standard x-ray monochromator consists of two crystals that select the x-ray energy by diffraction. For maximum flux, the two crystals are parallel, transmitting the same energy. The bandwidth depends on the type of the crystals and the Bragg reflection used and is usually on the order of 1 eV. Normally, at every synchrotron source, there are several beamlines dedicated to performing different XAFS techniques. A list of synchrotron light sources and XAFS beamlines can be found on the web page of the International X-ray Absorption Society (www.ixasportal.net) or the XAFS community website (http://www.xafs.org).

While the most important beamline property for XANES is the energy bandwidth of the monochromatic beam, the most important properties for EXAFS are the purity of the incident beam with respect to higher harmonics and its intensity stability over the required large energy range. If a higher harmonic runs over an absorption edge, it can spoil the measurement. Suppression of the higher harmonics in the beamline is done either by slightly detuning the second monochromator crystal or by using x-ray mirrors with coatings that do not reflect the higher harmonics. A somewhat larger problem is the intensity stability. When the energy is scanned by rotating the monochromator crystals, the beam position changes. Since the intensity over the beam cross section is not uniform, this results in variations in beam intensity at the sample. Another source of instability is absorption or Bragg scattering by the optical elements in the beamline that can create at certain energies positive or negative glitches in intensity. These intensity fluctuations are orders of magnitude weaker than the primary beam intensity, but may be on the order of the EXAFS fluctuations, especially at large k values. To overcome this, monochromator stabilization devices are frequently used in many beamlines. They use the input from the I_0 signal as feedback for tuning the second monochromator crystal during the measurement so that the intensity is kept constant with fluctuations smaller than 0.1%.

7.2.2 SPECIAL TECHNIQUES

7.2.2.1 Time-resolved EXAFS

A standard EXAFS measurement takes typically 30–60 min. For most chemical and biological reactions, this time scale is much longer than the reaction time. In order to perform real time–resolved *in situ* experiments, measurement time must be much shorter than the reaction time. To answer a growing demand of fast XAFS measurements, two approaches were developed: quick EXAFS (QEXAFS) and energy-dispersive EXAFS (ED-EXAFS) (Newton et al. 2002).

Quick EXAFS: Ordinary EXAFS experiments are performed by stepping the monochromator to set the energy for each measurement point. The mechanical stepping of the motors, resetting the detectors, and aligning the beam are the rate-limiting steps of the scan. This time can be greatly reduced if the monochromator is moved continuously through the energy window. In such so-called QEXAFS (Dent 2002; Stötzel et al. 2008)

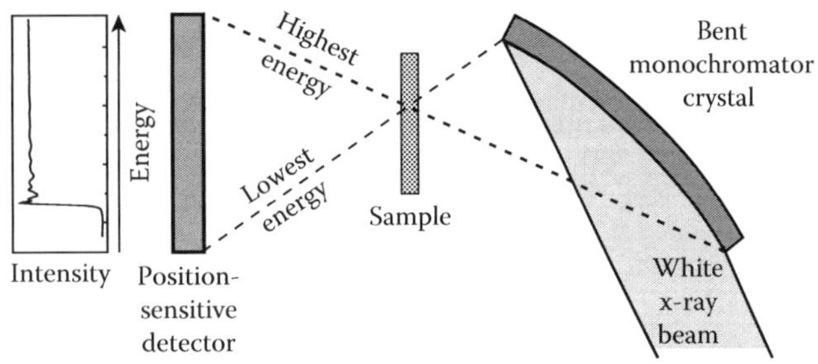

Figure 7.5 Bent monochromator used for ED-EXAFS measurements. The incoming white x-ray beam is reflected by the bent monochromator such that the energy span is dispersed in space within a focused beam of a few microns. The beam passes through the sample and gets detected by a position-sensitive detector.

experiment, one spectrum can be acquired in time scale of ms. The limiting factors in QEXAFS are (1) the relatively low signal to noise levels that often require averaging a few spectra thus reducing the time resolution and (2) mechanical instabilities that affect monochromator reproducibility with respect to energy calibration. Nevertheless, in the last years, a great improvement has been achieved in the mechanics, software, and hardware of monochromator elements, and QEXAFS is by now widely used in many EXAFS-dedicated beamlines (see list in http://www.xafs.org).

Energy-dispersive EXAFS (ED-EXAFS): Even faster measurement is achieved if the complete spectrum is acquired at once. ED-EXAFS (EDE) uses a polychromatic instead of a monochromatic beam to illuminate the sample with the entire energy spread at once (Dent 2002; Teramura et al. 2008; Newton 2009). Using a bent monochromator (Figure 7.5), photons with different wavelengths are spatially dispersed over the sample, and the transmission is measured by a position-sensitive detector. The complete spectrum is thus acquired at once. The dispersion of the beam through the sample imposes a few limitations on the experiments. First, as the fluorescence signal is emitted in all directions, only transmission measurements are possible, limiting sample concentration and thickness. Second, since each energy point of the spectrum probes a slightly different position on the sample, the sample must be homogenous, with domain size larger than the size of the beam.

For these reasons, although ED-EXAFS might provide better time resolution, QEXAFS, being more flexible, is more widely used today and is particularly suited for the time scale of most biomineralization processes (seconds to hours).

7.2.2.2 Polarized EXAFS

The EXAFS amplitude is dependent on the absorber–scatterer bonding orientation with respect to the polarization of the incident beam, with higher amplitude for scattering paths that are aligned along the polarization direction (Manceau et al. 1997; Manceau and Schlegel 2001; Pérez-Huerta et al. 2008). Thus, P-EXAFS can pick up specific bonding in a system depending on their orientation. In systems where the absorber occupies different atomic sites in a complex lattice, or is present in more than one form, P-EXAFS can help in resolving the structure around each of these sites, provided they have different polarization dependence.

7.2.2.3 Diffraction anomalous fine structure

The EXAFS function can be measured indirectly by any method that depends on the absorption cross section, scattering, as well as diffraction (Stragier et al. 1992; Sorensen et al. 1994). The atomic scattering factor is a complex function comprised of a real term describing the Thompson (elastic) scattering and an imaginary term that is the energy-dependent anomalous contribution. These terms are interrelated by Kramers–Kronig relations (Sorensen et al. 1994). In diffraction anomalous fine structure (DAFS), the diffraction intensity profile of a selected Bragg reflection is studied as a function of the incident photon energy. Since different crystallographic sites and different subsets of atoms contribute to different Bragg reflections, the EXAFS function for these particular positions can be extracted. Although promising, and fast developing, the requirements on the experimental setup and the evaluation of DAFS signals are still rather complicated (Favre-Nicolin et al. 2000).

7.2.3 EXAFS EXPERIMENT FROM THE SAMPLE POINT OF VIEW

7.2.3.1 Radiation damage

Due to the long measurement time and the accumulated exposure, radiation damage is an important issue in XAFS. Different types of damage might occur during x-ray exposure:

Photoreduction/photooxidation: Photoexcited species might lose electrons to (photooxidation) or receive them from (photoreduction) the surrounding medium. An example for spectral changes upon photoreduction of Fe^{3+} to Fe^{2+} during x-ray illumination is shown in Figure 7.6.

Free-radical formation and water radiolysis: Free radicals are produced when x-rays strike organic material and hydrated samples. The migrating free radicals degrade the organic material. Water may disintegrate and form oxidative species such as hydrogen peroxide, oxygen, and radicals such as superoxide.

Crystallization or amorphization of minerals: Changes in mineral structure, such as altering crystal structure or amorphization, are known to occur at high energies (tens to hundreds of keV) (e.g., Sickafus et al. 1999). At the K-edge energy of most biomineralization-relevant elements, this process is not very likely. The organic phase on the other hand is much more vulnerable to radiation damage. If the organic phase within or around the mineral has a role in stabilizing an otherwise unstable mineral phase, then its degradation might promote changes in the mineral structure.

Radiation damage can rarely be omitted, but precautions can be taken to minimize its effect on the measurement. For example, measuring at low temperature slows down the migration of formed free radicals. Frequent sample changes or scanning along the sample ensures that a fresh nondamaged sample is probed. Other precautions include, when suitable, radical traps and measuring in the presence of an oxidant or a reductant. In any case, it is a good practice to assess the sensitivity of the sample to radiation damage by comparing the spectra of repeated quick XANES scans at the same position on the sample.

7.2.3.2 Sample environment

The flexibility of EXAFS in terms of the type of samples to which it can be applied is also reflected by a wide range of sample environments from vacuum chambers, gas-filled chambers,

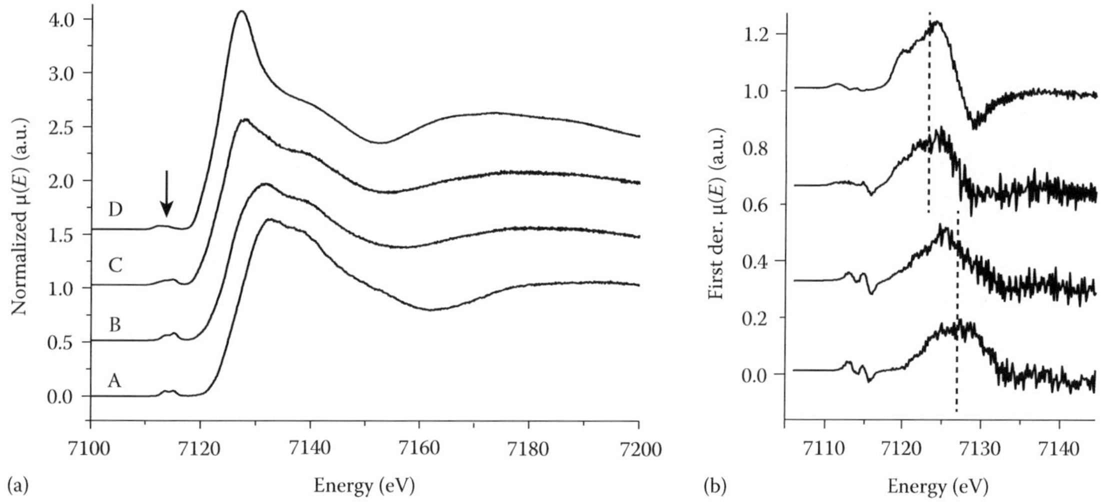

Figure 7.6 Quick XANES (30 s) spectra of iron chlorides dissolved in water (0.1 M). (A) $FeCl_3$ in aqueous solution measured at room temperature with constant steering; (B–D) iron chlorides dissolved and frozen in water containing 30% glycerol as cryoprotectant, measured at 20 K; (B) first scan of $FeCl_3$; and (C) second scan (+30 s) of the same sample and (D) $FeCl_2$. Arrow points at the pre-edge doublet peak. This pre-peak shifts about 1.7 eV between Fe^{2+} (sample d) and Fe^{3+} (sample a). (b) First derivative of the edge region of spectra in (a) showing the edge position as the global maximum. The edge position is shifted by 3.7 eV between Fe^{2+} and Fe^{3+}. Radiation damage in the form of photoreduction is clearly seen in samples (B) and (C) as indicated by shifts of the edge position towards lower energies and changes in the line shape, even though the samples are measured in cryo-conditions. Note that sample (A), which is the least susceptible to radiation damage, was measured under continuous steering and without organic cryoprotectant. Continuous steering ensures that fresh material is probed at all times. (Image courtesy of J. Baumgartner and G. Morin.)

cryo-stages, and freestanding samples. The choice will depend on the sample requirements and on the x-ray energy used.

Vacuum chambers: When measuring at low energies, for example, at the Ca K-edge (4038.5 eV), absorption by argon is a major concern. Although the concentration of argon in air is low, the x-ray beam travels a long way through the air, irradiating a volume much larger than the studied sample volume. Keeping the entire beam path and the sample under vacuum chamber is a straightforward solution for solid-state samples. Measurements of light elements (up to Mg, P) require low-energy x-rays that cannot be transmitted with useful intensity through any windows in the beam path. Therefore, soft x-ray beamlines have the so-called windowless beamline construction, with continuous ultrahigh vacuum from the synchrotron ring through beam delivery to the sample chamber and detectors.

Helium chamber: Another way to reduce Ar absorption is to fill the chamber by an inert low-Z element gas like He. To prevent sample dehydration, the He gas can be bubbled through a water container. When measuring fluorescence at higher energies, the chambers are rather disruptive. Even very small amounts of stray radiation induce fluorescence from the chamber walls normally made of alloys containing a multitude of different elements. In these cases, a collimator in front of the detector or polycapillary lenses are used to select only the fluorescence of the sample.

Cryo-XAS: Cryo-XAS is often used to protect unstable samples from crystallization or degradation, or for quenching a chemical reaction as discussed further in the *sample preparation* section. Additionally, the disorder in the sample, reflected by the mean square displacement of bond length (σ^2), is composed of static disorder (a distribution of bond lengths) and of a temperature-dependent factor (Debye–Waller factor). Measuring at cryogenic temperatures minimizes the Debye–Waller factor and helps identifying the degree of structural disorder in the system. In terms of sample damage, cryo-measurements can help in minimizing radiation damage by slowing down migration of radicals. On the other hand, as seen in Figure 7.6, processes like photoreduction/oxidation cannot be prevented.

Several different methods can be used to cool the sample. The simplest method, borrowed from protein crystallography, uses a cold gas stream of evaporated liquid nitrogen. The temperature at the sample is measured to regulate the stream intensity, keeping the temperature constantly around −130°C. The cryo-stream is shielded by another stream of warm nitrogen or dry air to prevent water condensation. This method is applicable to samples up to about 1 mm in size as larger samples create turbulence in the lamellar cooling streams resulting in water condensation on the sample. For larger samples, cooled chambers are used. The sample holder is internally cooled using controlled flow of liquid helium or nitrogen or by multiple Peltier-cooling stages. Stacked Peltier coolers can decrease the temperature by 60°C below room temperature, while liquid nitrogen cooling can in principle reach −196°C and liquid helium cryostats are held at around −270°C. In all cases, it is important to have a thermoelement positioned close to the sample for controlling the temperature. To thermally isolate the sample from the environment, chambers are often evacuated or filled with helium.

7.2.3.3 Sample preparation

Being a probe of a short range around the absorber element, no long-range translational order in samples is required for an EXAFS measurement. This opens the door to the investigation of many kinds of samples such as amorphous solids, liquids, biological tissues, and reaction solution with minimal if any sample preparation. Many biological samples are characterized by relatively low concentration of the element of interest and large background absorption from the tissue or the medium. Additionally, biomineralization specimens have often a complex microstructure that results in a nonhomogeneous absorber concentration. If such samples were to be measured by transmission, regions of low absorber concentration will directly transmit the beam through the sample distorting the absorption spectrum (pinhole effect) (Stern and Kim 1981; Goulon et al. 1982). For these reasons, most biological EXAFS experiments in general and specifically in biomineralization are performed in fluorescence mode.

Powder samples: Many biominerals can safely be extracted from the organism and ground into powder without major alteration of their structure. Powder samples are useful for eliminating effects of polarization dependence in the XAS signal. The powder is then smeared on some sticky tape like a carbon, polyimide (PI), or polyethylene terephthalate tapes (PETs). An important parameter for a powder sample is the grain size. For transmission measurement, the grain size should be smaller than the absorption length (x), to allow transmittance through the mineral. Notably, smaller grains can be more densely packed thus avoiding large spacing between the grains that behave as pinholes.

In fluorescent mode, the grain size should also be smaller than the absorption length, and the concentration of the element of interest should also be kept low to avoid self-absorption. When necessary, powder samples can be diluted by mixing with a low-absorbing powder.

Samples in solution: Solutions and liquids can be mounted between two window frames made of metal or polycarbonate, Teflon, etc., covered by a low-absorbing film. A spacer between the windows is used to obtain the right thickness for transmission measurements. In fluorescence measurements, capillaries or cuvettes containing windows covered with low-absorbing film can be used. One problem of measuring an aqueous solution is the formation of gas bubbles that behave as pinholes. This can occur, for example, due to radiolysis of water or gas diffusion into the sample. Measuring a large reservoir, continuous mixing, or within flow chambers are ways to overcome this problem.

Whole cells/tissue samples: When studying biominerals, the soft tissue is often critical for the stability of the mineral phase. This is especially true for transient mineral phases that might convert to the mature mineral phase upon extraction. In these cases, it is often beneficial to measure the mineral *in situ*, that is, embedded within the tissue. In such samples, it is important to also measure the signal of the tissue without the mineral, for example, before the onset of mineralization. The large amount of organic material in the sample however will make it prone to radiation damage.

Frozen samples: Cryo-measurements are advantageous for minimizing damage from sample preparation or radiation and for time-dependent series of measurements (Grossman et al. 2011; Grossman and Sagi 2012). When performing cryo-measurements, it is important to note that freezing the sample might introduce structural changes in the probed material. The sample should be frozen as fast as possible or with a cryoprotectant such as glycerol to prevent ice formation that can damage the sample. When mounting a frozen sample, it is important to avoid ice formation from air humidity on the sample surface, as it absorbs and scatters the fluorescent signal.

7.2.3.4 Standards and references

In every EXAFS experiment, a series of standards and references should be included for energy calibration, as fingerprints for identifying structural features, and for constraining fitting parameters.

Energy calibration: Measuring a reference in parallel or in between measurements is required for calibration of energy shifts caused by slight monochromator misalignments. Only after energy calibration, it is possible to determine energy shifts in the sample that result from changes in the oxidation state or the effective charge of the absorber. These shifts can range between a few tenths of eV to a few eV.

Understanding the XAS line shape: Standards may help in identifying specific electronic transitions or structural motifs (e.g., pre-edge peaks, position of white lines) in the sample by either visual inspection or peak-fitting analysis. Linear combination fitting (LCF) often used in XANES analysis requires a series of relevant standards to be combined to construct the spectrum of the unknown sample.

EXAFS modeling: Modeling the EXAFS spectrum of a standard measured at the same condition as the sample is a critical step in an EXAFS experiment. It allows the evaluation of certain parameters used for the fitting and can help building the structural model of the unknown sample, as will be discussed further in the *EXAFS analysis* part of the chapter.

A standard has usually much higher absorber concentration than the sample, and care should be taken to avoid self-absorption. Finally, it is noteworthy that biominerals are often considerably different from their inorganically formed counterparts (Lowenstam and Weiner 1989). They may contain different inorganic additives and concentration gradients even within a single object (Wang et al. 1997). The incorporation of organic molecules inside the mineral at the nanometer scale creates interfaces that affect the local atomic structure around the absorber and impose strains on the mineral (Berman et al. 1993; Pokroy et al. 2004). Finally, the formation of many biominerals via amorphous precursor phases often results in a disordered mature mineral phase, and different species have been shown to form minerals that are crystalline to different degree (Politi et al. 2006, 2008; Zolotoyabko et al. 2010; Gong et al. 2012). Accordingly, the spectra of a geogenic or synthetic mineral may be slightly different from that of a similar biogenic mineral.

7.2.4 RECORDING AN EXAFS SPECTRUM: PARAMETER FOR DATA ACQUISITION

Choosing the parameters for data acquisition depends on the sample properties (i.e., absorber concentration, tendency for radiation damage) and the need to operate an FT to the EXAFS function. The relation $\Delta R = \pi/2\Delta k$ derived from the properties of Fourier transformation implies that in order to probe a large r distance (higher shells), one needs to measure with small step size (Δk), and vice versa, in order to obtain high resolution in r-space; the k-range has to be sufficiently large. In addition, the energy range and energy resolution will determine the number of degrees of freedom (DOF) available for the fitting analysis (see in the succeeding text).

Data range: The data range should be as wide as possible as determined by the presence of other edges, photon flux, and signal to noise ratio. The ability to probe longer distances around the absorber ion depends also on the mean free path of the electron (Figure 7.4); thus, to probe larger distances, the measurement should extend at least up to $k = 10$ Å^{-1} for the first shell and higher up to $k = 16$ Å^{-1} for higher shells.

As the spatial resolution of the results is determined by the length of the data range used in the FT ($\Delta R = \pi/2\Delta k$), a wide energy range is desirable. In practice, the data range depends very much on the beamline, on the experimental conditions, and on the samples. Note that some of the most interesting edges in the study of biomineralization are relatively soft x-ray energies (e.g., Ca K-edge = 4038.5 eV, Mg K-edge = 1303 eV, S K-edge = 2472 eV). At these energy ranges, the available data range is often much shorter than for heavier atoms due to neighboring edges and technical difficulties in the soft x-ray regime.

Energy stepping: The XAS spectra are usually divided into several subregions of different step sizes to reduce measurement time without compromising energy resolution where it is most important. The most common way to divide the subregions is to determine (1) a pre-edge region for background removal; (2) a XANES region including positions of pre-edge peaks, the edge jump, and the large post-edge features; and (3) the EXAFS region. In region (1), large steps (5–10 eV) are sufficient to define the baseline. Region (2) requires the highest resolution to resolve pre- and post-edge peaks and determine edge position (0.2–0.5 eV depending on monochromator energy resolution). The EXAFS oscillations in region (3) have lower frequency as the energy is increased; thus, energy step can increase during the measurement and is therefore defined as a function of k: $\Delta E = k^2\hbar/2m_e$ (Equation 7.1), with normally about 20 measurement points per 1 Å^{-1}.

Exposure (integration) time depends on the method and detector type used, as well as on the signal intensity, and usually ranges between 1 and 5 s. Counting time is often also increased with k to improve the statistics of the weaker oscillations at higher energies. Due to monochromator instabilities over time and accumulated radiation damage, it is in most cases better to average several measurements of short integration time rather than to increase integration time.

7.3 DATA ANALYSIS

Even the simplest EXAFS evaluation involves numerical methods like Fourier transformation and least square fitting. More advanced methods include ab initio calculations of atomic potentials and electron-scattering functions. An especially useful and rather detailed practical description of data analysis including examples and tips for data analysis is found in Kelly et al. (2008).

7.3.1 SOFTWARE

Most EXAFS analysis software packages are built around a core program that calculates electron scattering in solid from first principles or semiempirical tabulated data. One of the most widely used software for scattering evaluation is the FEFF code (Ankudinov et al. 1998). Different versions of FEFF are used to account for different required accuracy. Version 6 is mostly used for evaluating EXAFS data, while versions 8 and 9 include more advanced multiple-scattering codes (Newville et al. 2009; Rehr et al. 2010) and better theoretical potentials and relativistic effects necessary for XANES simulation.

One approach for EXAFS modeling is used by the IFEFFIT software package for data processing and analysis. In this approach, the software generates individual electron paths with multiplicity for degeneration, assigning a set of parameters to each path. It uses FEFF6 to calculate the scattering function, from which a model is built and refined to fit the measured data (De Leon et al. 1991). The combination of the IFEFFIT and user interface *horae* (Ravel and Newville 2005) is a simple to use and freely available package used by many EXAFS users for data evaluation.

Another approach is demonstrated by the software package GNXAS (http://gnxas.unicam.it/ and the online version http://gnxas2.unicam.it/), which tries to group the electron paths and assign similar parameters to whole groups (Filipponi and Di Cicco 1995; Filipponi et al. 1995).

More analysis packages are mostly freely available on the Internet and provide a nice evaluation environment and use FEFF or other codes for data evaluation in the background. A lengthy list of available software is maintained at the www.xafs.org web pages.

7.3.2 EXAFS DATA PROCESSING AND ANALYSIS

A good practice for EXAFS analysis and interpretation is to start by critical evaluation of the XANES part of the spectrum, as it may hold valuable information that can help in constructing a valid model to be fitted during EXAFS analysis. Specific features like pre-peaks, shoulders, positions of white lines, and multiple-scattering peaks can indicate the presence of specific chemical species or structural motives of the sample. This inspection can be accompanied by peak-fitting analysis or LCF. Both these methods were described in detail in Chapter 6. For XAS interpretation of complex, multicomponent samples, LCF can be complemented by principal component analysis (PCA), to estimate the number of components in the sample, and residual phase analysis (RPA), which is used to extract the residual spectra of an unknown species from a mixture.

LCF, PCA, and RPA rely on the fact that the absorption spectrum with respect to E or k is a weighted average of all the components within the sample. It can therefore be described as a linear combination of the spectra from each species in the sample. PCA is used when a series of spectra are related to each other, for example, in time-dependent measurements or composition gradients within a sample, to extract the minimum *number* of components within the series (Kelly et al. 2008; Grossman and Sagi 2012). When a sample contains known as well as unknown components, it is possible to subtract the spectrum of the known component from the measured spectrum and obtain a residual spectrum that represents the unknown component(s). The residual EXAFS spectra can then be analyzed independently. This approach of combining PCA with RPA was developed by Sagi et al. (Frenkel et al. 2002) and used extensively for the analysis of time-dependent series of measurements of metalloenzyme-catalyzed reaction (reviewed in Grossman and Sagi 2012).

7.3.2.1 Data preprocessing

The set of acquired scans of the same sample needs to be averaged to increase the signal to noise ratio. Before averaging, it is important to regard each of the individual spectra, when needed to *clean* it from instrumental noise (deglitching) and oversampling (rebinning), and determine the useful range for the analysis. Self-absorption effects can be corrected to some extent provided that the exact composition of the sample is known. It is important to check for differences between spectra that may be indicative of radiation damage, sample dehydration, etc. The last step, energy alignment, is of specific importance to avoid loss of energy resolution. Often, a reference standard is used in parallel to the sample and used to define the absolute energy scale for the measured sample spectra. Energy calibration is specifically important for XANES data analysis, whereas for EXAFS analysis, it is sufficient to align spectra to a common, relative energy scale (Kelly et al. 2008).

7.3.2.2 Background removal, normalization, and *k*-weighting

Extracting the EXAFS, $\chi(E)$, from the measured XAS, $\mu(E)$, spectrum begins with two steps: (1) removal of a smooth pre-edge function that contains instrumental background and absorption from lower edges, and (2) normalization to a smooth post-edge background function that approximates the absorption of an isolated atom, $\mu_0(E)$ (Equation 7.2). In practice, the pre-edge line is a linear function extrapolated from the pre-edge region to the entire measurement range and subtracted from the data $\mu(E)$, so that the pre-edge portion of the data is set on the $y = 0$ axis. The pre-edge subtracted data are then divided by a constant, the *edge step* $\mu_0(E_0)$ (Figure 7.7a and b). The edge step is determined as the absorption difference between the pre- and post-edge lines at the absorption edge energy, E_0. Thus, in the normalized spectrum, the edge step is 1. In Athena, the preprocessing software inside IFEFFIT, a method called AUTOBK (Newville et al. 1933), is used in addition, which uses *r*-space information from the Fourier-transformed spectrum to construct the background function. An additional *flattening* function is applied in order to ensure that the post-edge region oscillates about one (Newville 2001).

Determination of the edge energy (E_0): The absorption edge is taken as the energy at which the spectrum shows a sharp jump in the absorption; it is usually approximated using the global maximum point of the first derivative of the spectrum. The value of E_0 determines the *k*-scaling of the EXAFS and is used to align the experimental spectrum to the theoretically calculated one (Koningsberger and Prins 1988).

Displaying the EXAFS spectrum in k space: The normalized and background-removed spectrum is written as a function of *k* using Equation 7.3. As the EXAFS oscillations decay with increasing *k*, it is common to multiply the signal by *k* at the power of 1–3, to emphasize the high *k* oscillations (*k*-waiting) (Figure 7.7c and d).

7.3.2.3 Converting the EXAFS spectrum to real space: Fourier transform

Fourier transformation extracts the independent frequencies that compose the sum of waves of the EXAFS equation and separates them in *r*-space. It allows us to identify the phase and amplitude of each of the sine waves that contribute to the EXAFS signal and thereby extract the structural information as these are directly related to the type of scattering atom and its distance from the absorber.

The magnitude of the FT of the EXAFS data (Figure 7.7e), sometimes termed the radial structure function (RSF), is related to the atomic positions of the coordinating atoms; it is *not*, however, a radial *distribution* function (RDF) (the RDF is closely related to the concept of pair distribution functions described in Chapter 1). The peak positions in the FT data do not correspond to the bond length, mainly due to (1) the phase shift, $\delta_j(k)$, in the EXAFS equation in the sine term ($\sin[2kR_j + \delta_j(k)]$), (2) the signal containing different contributions that might increase or decrease the amplitude such as thermal disorder and multiple-scattering paths, and (3) the decay of the EXAFS signal at larger *r* due to the photoelectron mean free path. While the real and imaginary parts of the FT are less intuitively understood like the FT magnitude, they are very useful in the fitting analysis as the contribution of different paths is easily resolved (Figure 7.8d).

FT windows: To avoid artificial signal (FT ripples) in the Fourier-transformed data, caused by the abrupt drop of the data to zero at both ends of the chosen *k*-range, the data are multiplied by a window function that creates a gradual transition to zero. The window is defined between the *k*-range used for the analysis, and the width of the transition zone is defined by the parameter d*k* (Figure 7.7d).

k-weighting: *k*-weighting is applied to enhance the EXAFS oscillations at higher *k* values, as described before. The *k* dependence of the EXAFS signal is related to the type of scattering atoms, such that heavier atoms scatter stronger at higher *k* values relative to lighter atoms. Thus, plotting the data with different *k*-weighting can help in identifying contributions from different atom types.

Back FT q(k): The Fourier-transformed *r*-space data can be back-transformed to produce a *k*-dependent spectrum *q*(*k*) (Figure 7.7f). The back-transformation can help in understanding the relation of different contributions in *r* and *k* space.

Performing FT and back-FT of the data during the experiment at the beamline can be used to monitor the degree of noise and help refine the experimental conditions.

7.3.3 EXAFS MODELING AND FITTING ANALYSIS

The most common way to evaluate the EXAFS function is to create a theoretical reference model, calculate its EXAFS function, and vary the model parameters until the least square difference between the model and measured data is minimal. Quite a few specialized EXAFS analysis software packages are available as mentioned earlier; in our description, we follow the routine used by Artemis that is part of the IFEFFIT software package.

7.3.3.1 Generating absorber–scatterer paths

The first step is creating an atomic model from which the paths will be calculated. In Artemis, this can be done by specifying the crystalline lattice and symmetries. The *Atoms* program creates a model cluster, with the absorbing atom in the center, from the crystalline structure. The cluster is forwarded to FEFF, which assigns the scattering potentials to atoms and creates a list of possible electron paths. Alternatively, when working with noncrystalline material, the cluster can be created using a macromolecular visualization application

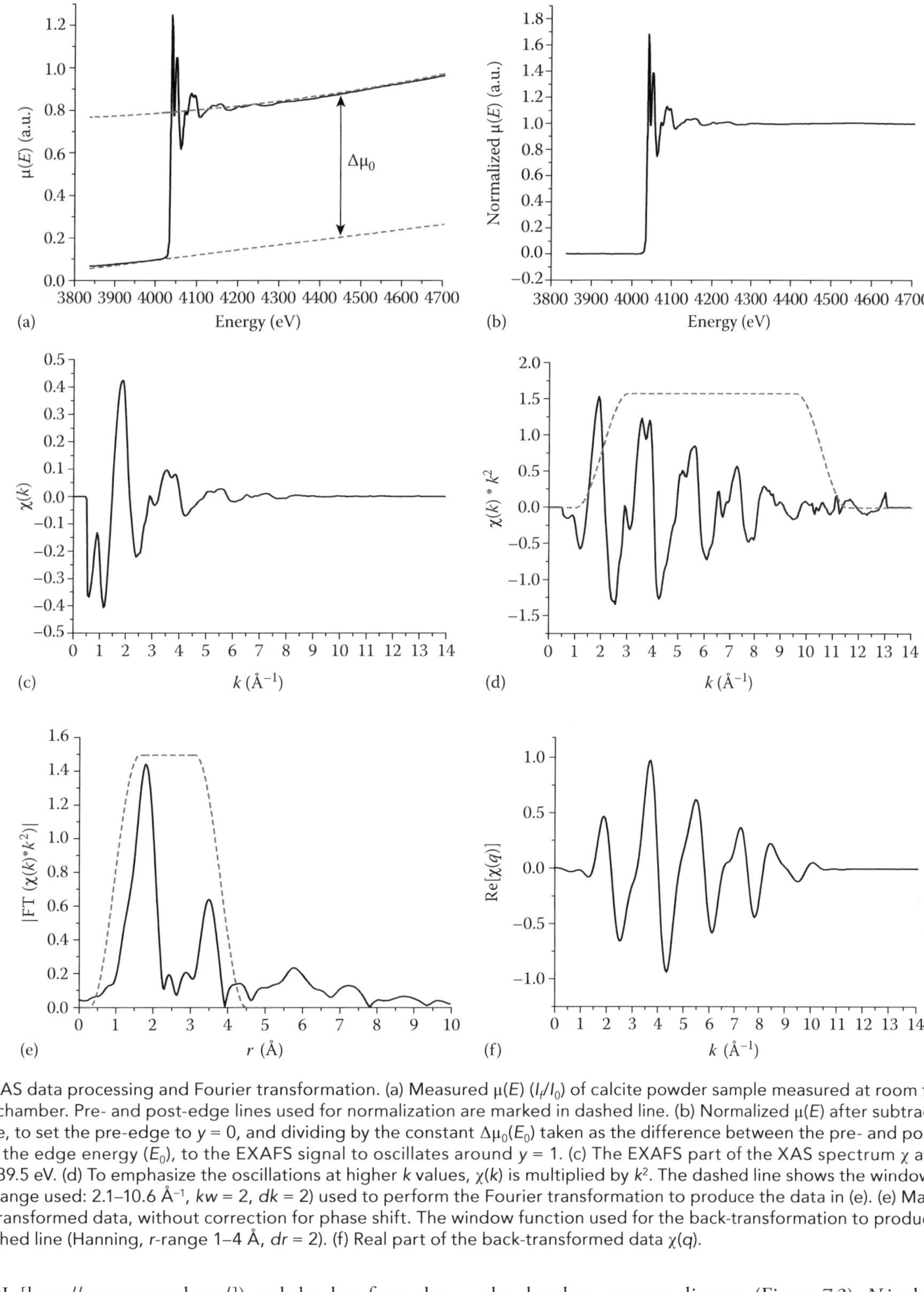

Figure 7.7 XAS data processing and Fourier transformation. (a) Measured $\mu(E)$ (I_f/I_0) of calcite powder sample measured at room temperature in a vacuum chamber. Pre- and post-edge lines used for normalization are marked in dashed line. (b) Normalized $\mu(E)$ after subtraction of the pre-edge line, to set the pre-edge to $y = 0$, and dividing by the constant $\Delta\mu_0(E_0)$ taken as the difference between the pre- and post-edge lines ($\Delta\mu_0$) at the edge energy (E_0), to the EXAFS signal to oscillates around $y = 1$. (c) The EXAFS part of the XAS spectrum χ as a function of k. $E_0 = 4039.5$ eV. (d) To emphasize the oscillations at higher k values, $\chi(k)$ is multiplied by k^2. The dashed line shows the window function (Hanning, k-range used: 2.1–10.6 Å^{-1}, $kw = 2$, $dk = 2$) used to perform the Fourier transformation to produce the data in (e). (e) Magnitude of the Fourier-transformed data, without correction for phase shift. The window function used for the back-transformation to produce $\chi(q)$ is shown in dashed line (Hanning, r-range 1–4 Å, $dr = 2$). (f) Real part of the back-transformed data $\chi(q)$.

(like PyMOL [http://www.pymol.org/]) and the data from the protein data bank (Kleifeld et al. 2000, 2001; Rosenblum et al. 2003). It is important that the cluster size is larger than the maximum radius used for modeling the data.

The next step is to calculate the effective scattering functions, $f(k)$, the phase shift $\delta(k)$ of each scattering event, and the relevant mean free path, $\lambda(k)$. The output of the FEFF calculation is a series of scattering paths, where each path is associated with one $R_{effective}$, $f(k)$, $\delta(k)$, and N. R_{eff} is the half length of the electron path. For single-scattering paths, R_{eff} is the absorber–scatterer distance (Figure 7.3). N is the degeneracy of the path, that is, the number of equivalent paths in the given cluster. Some of these paths will be chosen and summed to build the model during the fitting procedure described in the succeeding text.

7.3.3.2 Building and fitting a model

Equation 7.3, as published by Sayers et al. (1971), already contained most of the parameters necessary for EXAFS evaluation; however, the meaning of some parameters has

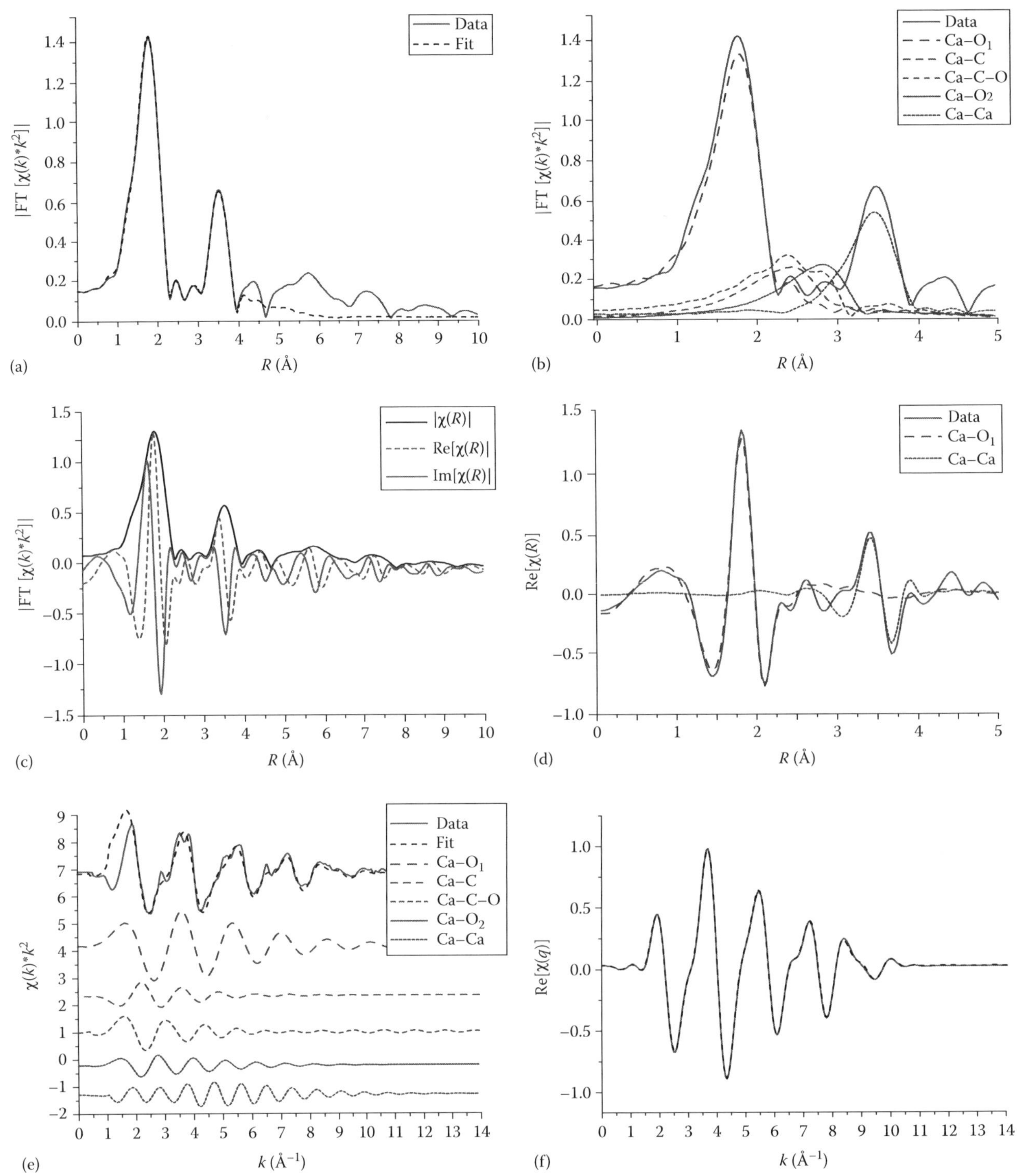

Figure 7.8 Fitting calcite data. (a) r-space data (solid line) and fit (dashed line). Fitting range between 1 and 4 Å. (b) Calcite data (solid line) and scattering paths used to build the model. Four single-scattering paths and one multiple-scattering path were used. First shell was constructed using Ca–O_1 path ($N = 6$); second (Ca–C) and third (Ca–O_2) shells are not clearly visible in the FT magnitude, as this region has also a contribution of the multiple-scattering path (Ca–C–O). This is an example why FT magnitude cannot be regarded as RDF. The fourth shell was constructed by the Ca–Ca path ($N = 6$). (c) Fourier-transformed components: real part (dashed line), imaginary part (grey line), and the FT magnitude (black line). The FT magnitude is an envelope of the real and imaginary parts. (d) Real part of the FT data (solid line) and two paths, Ca–O and Ca–Ca paths (dashed lines), showing the additive property of the real part of the spectra. Each path peaks at different *r* position with minimal contribution at other *r*-ranges. This makes the real part of the FT useful in identifying and selecting the relevant paths for the model. The imaginary part of FT has the same behavior. (e) Data (solid line) and fitted curve (dashed line) in *k* space. The individual contribution of each of the paths used for fitting is shown as a stack in the succeeding text. Note that the paths involving light elements (C, O) peak at lower *k* values ($2 < k < 4$ Å^{-1}), while in the Ca–Ca paths, the oscillations peak at $k > 4$ Å^{-1}. This difference is mainly due to the atomic potential of these atoms. (f) Back-transformation of data (solid line) and fit (dashed line).

changed with increasing understanding of electron-scattering processes. The fully parameterized EXAFS equation, based on Equation 7.4, becomes

$$\chi(k) = \sum_j \frac{S_0^2 N_j f_j(k) e^{-2\sigma_j^2 k^2} e^{-2R_j/\lambda(k)}}{k(\Delta r + R_j)^2} \sin\left[2k(\Delta r + R_j) + \delta_j(k)\right] \quad (7.4)$$

An important parameter S_0^2, the amplitude reduction factor, is added. This factor depends only on the absorber atom and accounts for amplitude reductions due to intra-atomic processes including electronic transitions of outer shell electrons into unoccupied states (shake-up) or emissions of low-energy electrons from outer shells (shake-off processes). Although Equation 7.4 was defined as a sum running over all neighboring shells, in practice, it sums the contribution of selected electron-scattering *paths*. The model is built by summing selected paths (j) from the FEFF calculation and using the calculated $f_j(k)$, $\delta_j(k)$, and $\lambda_j(k)$ in Equation 7.4 (Figure 7.8b, d, and e). The varied parameters in the fit are those associated with the path length, the path degeneracy, and the degree of disorder, R_j, N_j, and σ_j^2, respectively.

R is defined as $R = R_{eff} + \Delta r$. R_{eff} is taken from the FEFF calculation and Δr is varied during the fitting analysis. N_j, the number of equivalent paths, that is, the degeneracy of the path, is fitted to determine the number of alike neighboring atoms at a certain distance. The mean square variation in the bond distance, σ^2, describes the thermal and structural disorder in the system. The term ΔE_0 is used to align the k scale between the experiment and the theoretical model, by allowing small shifts in the energy origin E_0: $k^2 = 2m_e(E - E_0 + \Delta E_0)/\hbar$.

The maximal number of the fitting parameters available (DOF) is determined by $N_{max} = \pi \Delta R \Delta k$. If all fitting parameters (N, Δr, σ^2, S_0^2, and ΔE_0) would be varied for each path, the calculation would exceed the number of DOF. It is useful to understand where reducing the number of parameters is possible. As the S_0^2 factor deals with processes within the absorber atom, it is taken as a single parameter for all fitted paths. It is often determined by first fitting a standard material with the same (or similar) composition, measured at the same conditions. In principle, one ΔE_0 could be used for all paths, but in practice, this is rarely the case. It is nevertheless important to keep ΔE_0 small and to group similar paths with one ΔE_0. Δr and σ^2 are usually varied independently for each path. In this case, the choice of paths is important. For example, if two scatterers of the same atom type have similar distance to the absorber, it is better to combine them into one path and increase σ^2 accounting for the structural disorder.

It is common to start the analysis by fitting only the first shell, by constraining the fit range to take into account only the first peak in the FT. When a reasonable fit is achieved, the first shell parameters are fixed and more paths are added to describe higher coordination shells. It is important to note that the first peak in the FT includes contributions from other shells and that the first coordination shell might also affect the FT amplitude at higher r values. Therefore, the fitting process would include iterative refinements of the first and higher shells until the best results are obtained.

Correlation between parameters and accuracy of determining their values: The parameters can be grouped as those that contribute to the amplitude of the EXAFS spectrum (S_0^2, N, and σ^2) and those that determine the phase (E_0, which determined the k-grid and R). It comes out from Equation 7.4 that there exists an intrinsic correlation between all the parameters within the same groups. These correlations are especially high within a given shell, but can be also significant between shells. For this reason, it is usually estimated that at best case, accuracy within about 10% can be obtained for coordination numbers and 1% for bond lengths (Koningsberger and Prins 1988). The accuracy of determining the atomic number of the scattering atom is about ±10. This makes it impossible to distinguish, for example, between N and O or between Cl and S without additional knowledge on the sample based on different techniques or comparison to standards with known structure and composition.

The practice of fitting analysis includes these five steps:

1. Choosing scattering paths calculated by FEFF.
2. Parameterizing each path and grouping parameters.
3. Assigning initial guesses or setting (fixing) the parameter values.
4. Running the fit—the software will vary the guessed parameters to match the measured $\chi(k)$ function.
5. Inspecting the results in terms of (a) statistical parameters (goodness of fit: usually determined as χ^2, reduced χ^2, and R-factors), (b) the fitted parameters that make physical sense and the uncertainty defined during the fit that is reasonable, and (c) the graphical result in k-, r-, and q-space (Figure 7.8).

This process is repeated while modifying the initial guesses, examining contributions from different paths and rerunning the fits until the best results are obtained.

When reporting the best fit results, it is important to describe the original crystal structure from which the paths have been calculated; the k- and r-ranges and the windows used for the analysis, if a parameter was set or fixed; and the uncertainty in determining its value and the goodness of the fits using statistical parameters.

7.4 APPLICATIONS OF EXAFS IN BIOMINERALIZATION

In the following, we discuss a few examples of EXAFS studies in biomineralization that make use of the main advantage that EXAFS has to offer, namely, (1) probing the local structure when the absorber is part of the host mineral in (a) amorphous biomineral phases, (b) using time-dependent structural analyses of biomineralization process, and (2) probing the local structure when the absorber is a dopant in the mineral.

7.4.1 PROBING THE LOCAL STRUCTURE OF THE HOST MINERAL

7.4.1.1 Ca K-edge EXAFS in amorphous calcium carbonate

Ca K-edge EXAFS has been used to determine the local structure around the calcium ion in various biogenic and synthetic ACC and amorphous calcium phosphate (ACP) samples. In the context of biomineralization, synthetic ACP along with hydroxyapatite (HAP) and bone were the first minerals to be characterized using EXAFS analysis (Eanes et al. 1981; Miller et al. 1981).

EXAFS has played a pivotal role in the ability to characterize the atomic structure of amorphous minerals and therefore in the understanding of their properties in general and their stability in particular. ACC has very high solubility that makes it prone to crystallization or dissolution during sample preparation (Addadi et al. 2003). The use of EXAFS in this regard is highly advantageous as the sample can be measured with minimal sample preparation. To date, the short-range structure of biogenic ACC from eight different species was determined using Ca K-edge EXAFS analysis (Table 7.1). The XANES spectra of all these minerals are characterized by one main broad peak after the edge and a pre-edge peak (Figure 7.2), with only minor differences between them. The pre-edge peak attributed to the 1s–3d electronic transition is indicative of a deviation from centrosymmetric geometry, while a single broad feature above the edge reflects the decreased ordering around the Ca ion, meaning a wide range of absorber–scatterer bond lengths and poorly defined coordination geometry. Small shifts in the edge energy are observed between different samples indicating changes in the total effective charge of the Ca^{2+} ion for each of these phases. In $CaCO_3$, such changes were correlated with changes in the averaged coordination number with +0.7 ± 0.1 eV shift for each additional coordinating oxygen in the first shell (Sowrey et al. 2004). Analyses of the EXAFS region of the spectra revealed two main interesting observations: First, each of these *amorphous* phases contains different, species-specific short-range orders, and second, the extent of short-range ordering, determined as the number of resolved coordination shells in the FT magnitude, is reversely correlated to the stability of the mineral (Levi-Kalisman et al. 2002; Addadi et al. 2003). Similar to biogenic ACC, synthetic ACC was also shown by Ca K-edge XAS studies to comprise short-range order that depends on the synthesis method. ACC formed in the presence of polyaspartic acid had short-range structure resembling vaterite, while Mg-stabilized ACC had short-range structure resembling aragonite (Lam et al. 2007). ACC precipitated without stabilizing additives had different short-range structures depending on the conditions of the precipitating solutions (Günther et al. 2005; Politi et al. 2006; Gebauer et al. 2010).

A simplified summary of the structural differences obtained in these studies is presented in Table 7.1. In many of the reported structures, the first coordination shell is described by two scattering paths with slightly different R and Ns. In Table 7.1, all paths contributing to one coordination shell are averaged and represented by one Ca–O distance, with averaged σ^2 and N is the total coordination number. Therefore, the table does not represent the *best fit results* of each mineral but is rather an overview of the different structures obtained.

7.4.1.2 Time-dependent studies of mineral formation and crystallization

As yet, no real time–resolved studies have been performed in the context of biomineralization. Nevertheless, several time-dependent EXAFS studies on biomineral formation and *in vitro* crystallization are reported in which mineralization processes were monitored by quenching the reaction at different time points.

Ca K-edge EXAFS in developing bone mineral: In an attempt to answer the question whether bone forms via a precursor ACP phase, Binsted et al. (1982) performed an XAS study on femur bones (extracted and powdered) of developing mice (aged 3 days, 1 week, 1, 2, and 7 months). The authors showed that at early developmental stages, the Ca environment is significantly disordered, while during maturation, the order around the Ca ions increases (Eanes et al. 1981; Miller et al. 1981; Binsted et al. 1982). As XAS averages over all calcium ions present in the sample, it could not have been ruled out that the disorder seen at early mineralization stages is not a contribution of cellular Ca ions. In this case, the signal might in fact reflect an increase in the ratio between mineralized to non-mineralized calcium.

Ca K-edge EXAFS in developing mollusk larva: Ca K-edge analysis in combination of XRD was used to determine whether a precursor amorphous or crystalline phase exists during the initial stage of mineralization in developing embryos of *Biomphalaria glabrata* that then transforms to aragonite (Marxen et al. 2003). The first aragonite crystallites were detected by XRD in embryos about 72 h old, although most of the mineral was still x-ray amorphous. EXAFS analysis of the material at this stage indicated that the transient amorphous mineral phase initially formed has a short-range order resembling aragonite, and no other crystalline form could be detected neither by XRD nor by XAS (Hasse et al. 2000; Marxen et al. 2003). This study was the first to indicate that the precursor phase in biomineralization of $CaCO_3$ might have an incipient short-range structure related to the final crystalline form, which was later confirmed to be a more general phenomenon.

Ca K-edge EXAFS in developing sea urchin spicules: The single-crystalline calcite spicules of sea urchin larvae are formed via a precursor ACC phase, which gradually transforms into calcite during growth and maturation of the spicules (Beniash et al. 1997). The mineral phase in early stages of spicule development is intrinsically unstable and tends to crystallize in an uncontrolled manner upon extraction, hindering proper structural characterization. The transformation of ACC into calcite in sea urchin larval spicules was thus studied in a time-dependent manner without extracting the spicules from their environment (Politi et al. 2006). Sea urchin larvae from synchronized cultures were harvested at different developmental stages, suspended in glycerol solution and quickly frozen with liquid nitrogen. Thus, the spicules were in fact measured embedded in their own cellular environment. The XANES spectra series obtained showed that the Ca environment in the sample gradually changed from considerably disordered to an environment similar to calcite. Since Ca is an abundant ion in all animal cells, also in those that do not mineralize, it is likely that some part of the signal, especially at young stages of mineralization, arises from cellular Ca ions. A frozen $CaCl_2$ solution was thus used as control in addition to ACC and calcite. If the relative concentrations of Ca in solution or within the organic matrix relative to the Ca in the mineral were known, linear combination and RPA could have been used to determine the structure of the mineral phase. In this study, however, this ratio could not be obtained, and in addition, the signal to noise ratio of the EXAFS spectra was too low to carry out a detailed analysis. A way to overcome both these problems was to measure concentrated samples of extracted spicules. While during extraction, some of the spicule samples crystallized, some remained mostly amorphous, and the ratio between calcite and ACC in the sample was determined by IR analysis. By subtracting increasing fractions of calcite spectra from the spicule

Table 7.1 Summary of reported structures of biogenic and synthetic ACC

MINERAL TYPE		SPECIES	FIRST SHELL O			SECOND SHELL C			(THIRD) SHELL O			(FOURTH) SHELL CA			REFERENCE
			N	d (Å)	σ^2 (Å^2)	N	d (Å)	σ^2 (Å^2)	N	d (Å)	σ^2 (Å^2)	N	d (Å)	σ^2 (Å^2)	
Biological *stable* ACC	Plant cystoliths	*Ficus retusa*	8	2.48	0.022[b]	1.5	3.2	0.0015							Taylor et al. (1993)
		Ficus microcarpa	6	2.32	0.009[b]	2	3.46	0.02				4	3.79	0.015	Levi-Kalisman et al. (2002)
	Lobster carapace	*Homarus americanus*	6	2.35	0.035[b]	4	3.47	0.014				2	3.8	0.014	
	Ascidian spicules	*Pyura pachydermatina*	7.4	2.37	0.009	4.5	3.25	0.002							Levi-Kalisman et al. (2000)
	Wood lice sternal deposits	*Porcellio scaber*	3.8	2.38	0.014	1.4	3.00	0.007	0.4	3.72	0.001	6	3.89	0.109	Becker et al. (2003)
	Wood lice cuticle	*Armadillidium vulgare*	6.6	2.37	0.012	2.4	3.00	−0.023[c]	0.8	3.64	0	6	4.02	0.035	Becker et al. (2005)
		P. scaber	5.1	2.37	0.010	2.2	2.99	0.048	1	3.64	0	6	4.05	0.038	
Biological *transient* ACC	Snail larval shell	*B. glabrata* 96 h	9	2.38	0.015							6	3.92	0.06	Marxen et al. (2003)
		B. glabrata 120 h	9	2.45	0.013							6	3.91	0.014	
	Sea urchin larva	*Paracentrotus lividus* 40 h	6[a]	2.38	0.002[b]	6	3.26	0.003	2.4	3.59	0.013				Politi et al. (2006)
		P. lividus 58 h	6	2.4	0.011	6	3.29	0.012	6	3.6	0.022	6	4.07	0.015	
Synthetic ACC		s. ACC-polyAsp	6	2.36	0.023										Lam et al. (2007)
		9:1 Mg/Ca 240 min	9	2.33	0.022[c]										
		s. ACC no additives	5.3	2.41	0.01										Günther et al. (2005)
		s. ACC no additives	7	2.43	0.014	3	3.09	0.013	4	4.07	0.019				Politi et al.(2006)
Geologic mineral standards		Calcite	6	2.36	0.006	6	3.21	0.015	6	3.52	0.02	6	4.11	0.01	Levi-Kalisman et al. (2000)
		Aragonite	9	2.47	0.033	3	3.27	0.035	3	3.28	0.007	6	3.94	0.028	Marxen et al. (2003)
		Monohydrocalcite	7.4	2.37	0.009	4.5	3.25	0.002							Becker et al. (2003)

Note: N is a coordination number denoting the sum of all paths of similar bond length; D is the distance of the scatterers to the absorbing calcium ion; σ^2 is the Debye–Waller factor, represents the degree of disorder, and the largest σ^2 reported for the coordination shell is presented.

[a] Two alternative fits are reported.

[b] Multiple paths are summed; thus, σ^2 presented in the table is an underestimate.

[c] Negative value for σ^2 does not have a physical meaning.

spectra, as in RPA approach, it could be determined that the spicule spectrum is not a simple average of a distinct amorphous phase and calcite, but rather that the spicules are composed of an amorphous phase that has short-range order resembling calcite. The combination of studying both *in situ* spicules and extracted spicules allowed overcoming the disadvantage of each of the individual experiments. This example also emphasizes the importance of using additional techniques for complementing the EXAFS analysis whenever possible. In this work, IR analysis was pivotal in determining the degree of crystallization of the spicule mineral, strengthening the conclusion that the amorphous phase has calcitic short-range order.

Fe K-edge XANES and EXAFS in magnetotactic bacteria: Quasi-time-resolved XAS studies were performed on the magnetotactic bacterial strains *Magnetospirillum magneticum* AMB-1 and *M. gryphiswaldense* MSR-1 to elucidate the formation of biogenic magnetite in these organisms (Baumgartner et al. 2013; Fdez-Gubieda et al. 2013). The intracellularly formed iron mineral serves as a geomagnetic field sensor and the presence or absence of precursor structures had been a longstanding unresolved question. Baumgartner et al. and Fdez-Gubieda et al. could independently demonstrate the existence of an amorphous ferric phosphate precursor phase by XANES and EXAFS. However, minor ferric iron(oxyhydr)oxide intermediates, which were only transiently present at very low concentrations, could not be measured by XAS. These additional phases were only observed by high-resolution transmission electron microscopic methods (Baumgartner et al. 2013).

Crystallization of synthetic ACC in vitro: In dry state, many synthetic ACC phases are relatively stable. They can be thus studied by XAS without additional precaution once extracted from the precipitating solutions. Lam et al. (2007) precipitated synthetic ACC with different additives and studied their crystallization during incubation in aqueous solution, extracting the precipitate at different times (minutes–days) of incubation. They showed that the short-range structure around Ca was synthesis dependent and that the coordination environment around the Ca ions in the first precipitated ACC phase is retained to a large extent during recrystallization dictating the structure of the final crystalline polymorph. Thus, Mg-stabilized ACC was initially more similar to either aragonite or monohydrocalcite both having high coordination numbers ($N = 8$ and 9, respectively), while ACC stabilized by polyaspartic acid had short-range order resembling calcite or vaterite ($N = 6$). By correlating EXAFS analysis, XRD, and IR, these experiments also nicely showed that the first stage of recrystallization in all samples was similar, involving the expulsion of structural water without any significant changes in the immediate short-range structure around the calcium ion.

7.4.2 PROBING SHORT-RANGE ORDER AROUND DOPANT IONS IN THE MINERAL

Being element specific, EXAFS analysis is an ideal tool to study the local effect of dopant ions and additives on the host mineral. Often in these cases, however, the concentration of the element of interest is too low to allow measurements in transmission mode. However, fluorescence measurements provide the required high sensitivity and are therefore preferably used.

7.4.2.1 Probing incorporation of metal ions in bone

The use of metal ions in various medical applications is often accompanied, intentionally or not, by these ions being incorporated in the mineral phase of bone and teeth. It is therefore of high importance to study the effect of these incorporations on the mineral structure. For example, strontium-based drugs are often used for the treatment of osteoporosis (Aaseth et al. 2012). It has been shown in postmenopausal osteoporosis patients that Sr ions accumulate in the bone, increase bone mass, and reduce the risk of fractures. Gallium ions on the other hand are used in certain cancer treatments; this results in Ga ion accumulation in bone, especially in regions of high remodeling activity. XAS was used to address the question on how Ga^{3+} and Sr^{2+} are incorporated within the mineral phase in several studies on synthetic HAP precipitates or in cell culture–formed bone tissue (Korbas et al. 2004). These measurements showed that while Sr^{2+} is incorporated within the mineral phase without significant distortion, Ga ions cause significant distortion in the local structure, preventing the precursor mineral to transform into the poorly crystalline HAP crystal characteristic of mature bone tissue. In addition, Ga ions introduced to the cell culture at an advanced stage of biomineralization were found to adsorb on the surface of preexisting HAP crystallites preventing their further growth.

7.4.2.2 Magnesium ions in amorphous calcium carbonate

Magnesium ions are frequently found in many biominerals with tissue- and species-specific concentrations. Magnesium has thus been suspected to play an important role both in the formation of the mineral and in determining the properties (e.g., mechanical) of the final functional mineral. In paleoclimatic research, the Ca/Mg ratio in various fossilized biominerals is used as a past temperature proxy. Studying the Mg K-edge (E_0 = 1303 eV) poses a significant challenge as its energy is in the higher end of the soft x-ray range (Batchelor et al. 2001). It is thus hard to reach at many synchrotron sources, especially within the full energy range needed for quantitative high-resolution EXAFS analysis. In addition, often the Al K-edge at 1559 eV interferes with the EXAFS signal, as Al is frequently present either as contaminant in biogenic samples or even within different components of the beamline. The limited *k*-range available for the Mg K-edge EXAFS analyses (Δk = 5–6) results in poor spatial resolution in the FT, on the order of 0.26–0.3 Å, as compared to the larger *k*-ranges that are so far been used for fitting EXAFS data of biominerals at the Ca K-edge (E_0 = 4,038 eV, Δk = 7 Å^{-1}, Δr = 0.22 Å), the Ga (E_0 = 10,367 eV, Δk = 8 Å^{-1}, Δr = 0.19 Å), and Sr K-edges (E_0 = 16,105 eV, Δk = 10 Å^{-1}, Δr = 0.15 Å). Thus, fitting various reference samples is required in order to evaluate the accuracy of bond length determination (Finch and Allison 2007). The short *k*-range allows refinement of only the first coordination shell; thus, also coordination number precision is reduced due to the high correlation between all of the parameters that contribute to the FT amplitude (S_0^2, N, σ^2) (Politi et al. 2010). For these reasons, many works have focused on the Mg K-edge XANES region and used fingerprint spectra to study Mg incorporation in various biominerals. Nonetheless, despite the technical difficulties,

EXAFS analysis has been applied for the study of various recent and fossil biominerals (Cusack et al. 2008; Finch and Allison 2008). The Mg environment in three biogenic Mg-ACCs was studied and compared with that of synthetic Mg-ACC and crystalline Mg carbonates (Politi et al. 2010). It has been found that the Mg–O distance in all amorphous minerals was shorter in relation to the anhydrous crystalline polymorphs, contrary to the Ca–O bond length, which is larger in the amorphous phases relative to crystalline $CaCO_3$ phases (Table 7.1). A comparison with hydrated forms of Mg carbonates and with the $Mg–O_{(water)}$ bonds in organic molecules suggests that water molecules are involved in magnesium coordination in ACC.

7.4.2.3 Magnesium and strontium ions in crystalline $CaCO_3$ minerals

An interesting example for the power of XAS as a probe of dopant elements is the study performed by Finch and Allison (Allison et al. 2005; Finch and Allison 2007, 2008), in which they investigated the Mg environment in recent and fossil aragonite minerals. While Mg is usually incorporated in calcite within lattice positions and can therefore be studied by XRD, it is incompatible with the aragonite lattice. In such circumstance as with amorphous minerals, magnesium in aragonite cannot be studied by means of XRD. The Mg environment in aragonite was found by XAS to be significantly disordered, and the element was suggested to reside in a disordered mineral phase or incorporated within the organic matrix. Interestingly, the Mg–O bond length determined by EXAFS fitting (Finch and Allison 2008) is similar to what was found for magnesium incorporated in biogenic amorphous Mg-ACC phases (Politi et al. 2010).

Incorporation of strontium as well as magnesium in carbonate skeletons is frequently used in paleoclimatic research to calculate past sea surface temperatures (SSTs). Importantly, these calculations assume homogenous substitutions of the dopant ions in the mineral. The way in which strontium is incorporated in aragonite biominerals was under a debate after a report suggested, based on Sr K-edge EXAFS, that up to 40% of strontium ions are present in coral aragonite as a separate $SrCO_3$ phase (strontianite) (Greegor et al. 1997). Recent EXAFS work, however, determined that Sr in coral aragonite resides as a single-phase solid solution (Allison et al. 2005). From the various reports using EXAFS analysis and XRD, it is now recognized that the validity of SST calculations depends on the type of incorporated ions (Mg or Sr), the carbonate polymorph (calcite or aragonite), and the biomineralization mechanism. While Mg in calcitic skeletons can be used to calculate past SST in some mineralizing species, it is not a valid proxy in the aragonitic skeletons of corals. In those skeletons, the Sr/Ca ratio instead is more appropriate.

7.4.2.4 Sr K-edge EXAFS in developing sea urchin spicules

Exploiting the advantages mentioned earlier of working at heavy-element K-edges rather than the soft Ca K-edge in XAS, Sr labeling has been recently employed to follow ACC to calcite transformation in sea urchin larva spicules (Tester et al. 2013). Using pulse-chase experiments, Tester et al. were able to extract the kinetics of the transformation by probing small labeled regions of the spicules at different times after the pulse; it was found that dehydration and formation of calcitic short-range order occur within the first 3 h and full crystallization is obtained in the next 24 h. The authors also suggested that the intermediate phase in the transformation is poorly crystalline calcite rather than an anhydrous ACC phase.

7.5 SUMMARY

The wide range of uses of EXAFS in biomineralization is directly related to the power of this technique in extracting high-resolution structural information out of a wide range of samples in different sample environments. Although the measurements and analysis are not simple, the technique is becoming increasingly accessible with the development of specialized beamlines and the advancement of dedicated analysis software. There are many questions in biomineralization that could benefit from the use of EXAFS analysis and teach us more on the variation in biomineral structures and their formation mechanisms. With the rapid advances of fast measurements (QEXAFS, ED-EXAFS), many processes can now be followed in real time to study the very early stages of mineralization, and samples can now also be manipulated *in situ* to follow their crystallization or phase transformation in real time. We expect that many interesting studies will be performed exploring these possibilities of EXAFS in biomineralization.

ACKNOWLEDGMENTS

The authors are grateful to Profs. Steve Weiner, Irit Sagi, and Alexei Erko for discussions and comments on the manuscript.

REFERENCES

Aaseth, J., Boivin, G., and Andersen, O. 2012. Osteoporosis and trace elements—An overview. *Journal of Trace Elements in Medicine and Biology*, 26: 149–152.

Addadi, L., Raz, S., and Weiner, S. 2003. Taking advantage of disorder: Amorphous calcium carbonate and its roles in biomineralization. *Advanced Materials*, 15: 959–970.

Allison, N., Finch, A.A. Newville, M., and Sutton, S.R. 2005. Strontium in coral aragonite: 3. Sr coordination and geochemistry in relation to skeletal architecture. *Geochimica et Cosmochimica Acta*, 69: 3801–3811.

Ankudinov, A.L., Ravel, B., Rehr, J.J., and Conradson, S.D. 1998. Real-space multiple-scattering calculation and interpretation of X-ray absorption near-edge structure. *Physical Review B*, 58: 7565–7576.

Batchelor, D.R., Follath, R., and Schmeisser, D. 2001. Commissioning results of the BTUC-PGM beamline. *Nuclear Instruments & Methods in Physics Research*, 468: 470–473.

Baumgartner, J., Morin, G., Menguy, N., Perez Gonzales, T., Widdrat, M., Cosmidis, J., and Faivre, D. 2013. Magnetotactic bacteria form magnetite from a phosphate-rich ferric hydroxide via nanometric ferric (oxyhydr)oxide intermediates. *Proceedings of the National Academy of Sciences*, 110: 14883–14888.

Becker, A., Bismayer, U., Epple, M. et al. 2003. Structural characterisation of X-ray amorphous calcium carbonate (ACC) in sternal deposits of the crustacea *Porcellio scaber*. *Dalton Transactions*, 4: 551–555.

Becker, A., Ziegler, A., and Epple, M. 2005. The mineral phase in the cuticles of two species of crustacea consists of magnesium calcite, amorphous calcium carbonate and amorphous calcium phosphate. *Dalton Transactions*, 10: 1814–1820.

Beniash, E., Aizenberg, J., Addadi, L., and Weiner, S. 1997. Amorphous calcium carbonate transforms into calcite during sea urchin larval spicule growth. *Proceedings of the Royal Society B: Biological Sciences*, 264: 461–465.

Berman, A., Hanson, J., Leiserowitz, L., Koetzle, T.F., Weiner, S., and Addadi, L. 1993. Biological control of crystal texture: A widespread strategy for adapting crystal properties to function. *Science*, 259: 776–779.

Bianconi, A. and Benfatto, M. 1988. XANES in condensed systems. In. *Synchrotron Radiation in Chemistry and Biology*, eds. A. Mosset and J. Galy, pp. 29–67. Berlin, Germany: Springer-Verlag.

Binsted, N., Hasnain, S.S., and Hukins, D.W.L. 1982. Developmental changes in bone mineral structure demonstrated by extended x-ray absorption fine structure (EXAFS) spectroscopy. *Biochemical and Biophysical Research Communications*, 107: 89–92.

Bunker, G. 2010. *Introduction to XAFS: A Practical Guide to X-ray Absorption Fine Structure Spectroscopy*. Cambridge, U.K.: Cambridge University Press.

Cuif, J., Dauphin, Y., Doucet, J., Salomé, M., and Susini, J. 2003. XANES mapping of organic sulfate in three scleractinian coral skeletons. *Geochimica et Cosmochimica Acta*, 67: 75–83.

Cusack, M., Dauphin, Y., Cuif, J., Salomé, M., Freer, A., and Yin, H. 2008a. Micro-XANES mapping of sulphur and its association with magnesium and phosphorus in the shell of the brachiopod, *Terebratulina retusa*. *Chemical Geology*, 253: 172–179.

Cusack, M., Pérez-Huerta, A., Janousch, M., and Finch, A. 2008b. Magnesium in the lattice of calcite-shelled brachiopods. *Chemical Geology*, 257: 59–64.

Dent, A.J. 2002. Development of time-resolved XAFS instrumentation for quick EXAFS and energy-dispersive EXAFS measurements on catalyst systems. *Topics in Catalysis*, 18: 27–35.

Eanes, E.D., Powers, L., and Costa, J.L. 1981. Extended x-ray absorption fine structure (EXAFS) studies on calcium in crystalline and amorphous solids of biological interest. *Cell Calcium*, 2: 251–262.

Favre-Nicolin, V., Bos, S., Lorenzo, J.E. et al. 2000. Integration procedure for the quantitative analysis of dispersive anomalous diffraction. *Journal of Applied Crystallography*, 33: 52–63.

Fdez-Gubieda, M. L., Muela, A., Alonso, J., Garcia-Prieto, A., Olivi, L., Fernandez-Pacheco, R., and Barandiaran, J.M. 2013. Magnetite Bbiomineralization in magnetospirillum gryphiswaldense: Time-resolved magnetic and structural studies. *American Chemical Society Nano*, 7: 3297–3305.

Filipponi, A. and Di Cicco, A. 1995. X-ray-absorption spectroscopy and n-body distribution functions in condensed matter. 2. Data analysis and applications. *Physical Review B: Condensed Matter and Materials Physics*, 52: 15135–15149.

Filipponi, A., Di Cicco, A., and Natoli, C.R. 1995. X-ray-absorption spectroscopy and n-body distribution functions in condensed matter. 1. Theory. *Physical Review B*, 52: 15122–15134.

Finch, A.A. and Allison, N. 2007. Coordination of Sr and Mg in calcite and aragonite. *Mineralogical Magazine*, 71: 539–552.

Finch, A.A. and Allison, N. 2008. Mg structural state in coral aragonite and implications for the paleoenvironmental proxy. *Geophysical Research Letters*, 35: 1–5.

Frenkel, A.I., Kleifeld, O., Wasserman, S.R., and Sagi, I. 2002. Phase speciation by extended x-ray absorption fine structure spectroscopy. *The Journal of Chemical Physics*, 116: 9449–9456.

Fricke, H. 1920. The K-characteristic absorption frequencies for the chemical elements magnesium to chromium. *Physical Review*, 16: 202–216.

Gebauer, D., Gunawidjaja, P.N., Ko, J.Y.P. et al. 2010. Proto-calcite and proto-vaterite in amorphous calcium carbonates. *Angewandte Chemie (International ed. in English)*, 49: 8889–8891.

Gong, Y.U.T., Killian, C.E., Olson, I.C. et al. 2012. Phase transitions in biogenic amorphous calcium carbonate. *Proceedings of the National Academy of Sciences of the United States of America*, 109: 6088–6093.

Goulon, J., Goulon-Ginet, C., Cortes, R., and Dubois, J.M. 1982. On experimental attenuation factors of the amplitude of the EXAFS oscillations in absorption, reflectivity and luminescence measurements. *Journal of Physique*, 43: 539–548.

Greegor, R.B., Pingitore, N.E., and Lytle, F.W. 1997. Strontianite in coral skeletal aragonite. *Science*, 275: 1452–1454.

Grossman, M., Born, B., Heyden, M. et al. 2011. Correlated structural kinetics and retarded solvent dynamics at the metalloprotease active site. *Nature Structural and Molecular Biology*, 18: 1102–1108.

Grossman, M. and Sagi, I. 2012. Application of stopped-flow and time-resolved X-ray absorption spectroscopy to the study of metalloproteins molecular mechanisms. In *X-Ray Spectroscopy*, ed. S. K. Sharma, pp. 245–264. Rijeka, Croatia: InTech. http://www.intechopen.com/books/x-ray-spectroscopy.

Günther, C., Becker, A., Wolf, G., and Epple, M. 2005. In vitro synthesis and structural characterization of amorphous calcium carbonate. *Zeitschrift für Anorganische und Allgemeine Chemie*, 631: 2830–2835.

Hasse, B., Ehrenberg, H., Marxen, J.C. et al. 2000. Calcium carbonate modifications in the mineralized shell of the freshwater snail *Biomphalaria glabrata*. *Chemistry—A European Journal*, 6: 3679–3685.

Kelly, S.D., Hesterberg, D., and Ravel, B. 2008. Analysis of soils and minerals using X-ray absorption spectroscopy. In *Methods in Soil Analysis*, eds. A.A. Ulery and L.R. Drees, p. 522. Madison, WI: Soil Science Society of America, Inc.

Kleifeld, O., Kotra, L.P., Gervasi, D.C. et al. 2001. X-ray absorption studies of human matrix metalloproteinase-2 (MMP-2) bound to a highly selective mechanism-based inhibitor. *The Journal of Biological Chemistry*, 276: 17125–17131.

Kleifeld, O., Van den Steen, P.E., Frenkel, A.I. et al. 2000. Structural characterization of the catalytic active site in the latent and active natural Gelatinase B from human neutrophils. *The Journal of Biological Chemistry*, 275: 34335–34343.

Koningsberger, D.C. and Prins, R. 1988. *X-ray Absorption Principles, Applications, Techniques of EXAFS, SEXAFS and XANES*. New York: John Willey & Sons.

Korbas, M., Rokita, E., Meyer-Klaucke, W., and Ryczek, J. 2004. Bone tissue incorporates in vitro gallium with a local structure similar to gallium-doped brushite. *Journal of Biological Inorganic Chemistry*. 9: 67–76.

Kronig, R.D.L. 1931. Zur Theorie der Feinstruktur in den Röntgenabsorptionsspektren. *Zeitschrift für Physik*, 70(5–6): 317–323.

Kronig, R.D.L. 1932a. Zur Theorie der Feinstruktur in den Röntgenabsorptionsspektren. II. *Zeitschrift für Physik*, 75(3–4): 191–210.

Kronig, R.D.L. 1932b. Zur Theorie der Feinstruktur in den Röntgenabsorptionsspektren. III. *Zeitschrift für Physik*, 75(7–8): 468–475.

Lam, R.S.K., Chernock, J.M., Lennie, A., and Meldrum, F.C. 2007. Synthesis-dependent structural variations in amorphous calcium carbonate. *Crystal Engineering Communication*, 9: 1226–1236.

Laurencin, D., Wong, A., Chrzanowski, W. et al. 2010. Probing the calcium and sodium local environment in bones and teeth using multinuclear solid state NMR and X-ray absorption spectroscopy. *Physical Chemistry Chemical Physics*, 12: 1081–1091.

Levi-Kalisman, Y., Raz, S., Weiner, S., Addadi, L., and Sagi, I. 2000. X-ray absorption spectroscopy studies on the structure of a biogenic "amorphous" calcium carbonate phase. *Journal of the Chemical Society, Dalton Transactions*, 21: 3977–3982.

Levi-Kalisman, Y., Raz, S., Weiner, S., Addadi, L., and Sagi, I. 2002. Structural differences between biogenic amorphous calcium carbonate phases using X-ray absorption spectroscopy. *Advanced Functional Materials*, 12: 43–48.

Lowenstam, H.A. and Weiner, S. 1989. *On Biomineralization*. Oxford, U.K.: Oxford University Press.

Manceau, A., Chateigner, D., and Gates, W.P. 1997. Polarized EXAFS, distance-valence least-squares modeling (DVLS), and quantitative texture analysis approaches to the structural refinement of Garfield nontronite. *Physics and Chemistry of Minerals,* 25: 347–365.

Manceau, A. and Schlegel, M.L. 2001. Texture effect on polarized EXAFS amplitude. *Physics and Chemistry of Minerals*, 28: 52–56.

Marxen, J.C., Becker, W., Finke, D., Hasse, B., and Epple, M. 2003. Early mineralization in *Biomphalaria glabrata* microstructure and structural results. *Journal of Molluscan Studies*, 69: 113–121.

Matsunaga, T., Ishizaki, H., Tanabe, S., and Hayashi, Y. 2009. Synchrotron radiation microbeam X-ray fluorescence analysis of zinc concentration in remineralized enamel in situ. *Archives of Oral Biology*, 54: 420–423.

Miller, R.M., Hukins, D.W.L., Hasnain, S.S., and Lagarde, P. 1981. Extended x-ray absorption fine structure (EXAFS) studies of the calcium ion environment in bone mineral and related calcium phosphates. *Biochemical and Biophysical Research Communications*, 99: 102–106.

Mustre De Leon, J., Rehr, J.J., and Zabinsky, S.I. 1991. Ab initio curved-wave X-ray-absorption fine-structure. *Physical Review B: Condensed Matter and Materials Physics*, 44: 4146–4156.

Newton, M.A. 2009. Applying dynamic and synchronous DRIFTS/EXAFS to the structural reactive behaviour of dilute (≤ 1 wt%) supported Rh/Al_2O_3 catalysts using quick and energy dispersive EXAFS. *Topics in Catalysis*, 52: 1410–1424.

Newton, M.A., Dent, A.J., and Evans, J. 2002. Bringing time resolution to EXAFS: Recent developments and application to chemical systems. *Chemical Society Reviews*, 31: 83–95.

Newville, M. 2001. IFEFFIT: Interactive XAFS analysis and Feff fitting. *Journal of Synchrotron Radiation*, 8: 322–324.

Newville, M., Kas, J.J., and Rehr, J.J. 2009. Improvements in modeling EXAFS with many-pole self-energy and FEFF 8.5. In *Proceedings of the 14th International Conference on X-ray Absorption Fine Structure (XAFS14)*, eds. A. DiCicco and A. Filipponi. *Journal of Physics Conference Series*, 190: 012023. Camerino, Italy: IOP Publishing Ltd.

Newville, M., Livins, P., Yacoby, Y., Stern, E.A., and Rehr, J.J. 1933. Near-edge x-ray-absorption fine structure of Pb: A comparison of theory and experiment. *Physical Review B*, 47: 14126–14131.

Pérez-Huerta, A., Cusack, M., Janousch, M., and Finch, A.A. 2008. Influence of crystallographic orientation of biogenic calcite on in situ Mg XANES analyses. *Journal of Synchrotron Radiation*, 15: 572–575.

Pokroy, B., Quintana, J.P., Caspi, E., Berner, A., and Zolotoyabko, E. 2004. Anisotropic lattice distortions in biogenic aragonite. *Nature Materials*, 3: 5–7.

Politi, Y., Batchelor, D.R., Zaslansky, P. et al. 2010. Role of magnesium ion in the stabilization of biogenic amorphous calcium carbonate: A structure–function investigation. *Chemistry of Materials*, 22: 161–166.

Politi, Y., Levi-Kalisman, Y., Raz, S. et al. 2006. Structural characterization of the transient amorphous calcium carbonate precursor phase in sea urchin embryos. *Advanced Functional Materials*, 16: 1289–1298.

Politi, Y., Metzler, R.A., Abrecht, M. et al. 2008. Transformation mechanism of amorphous calcium carbonate into calcite in the sea urchin larval spicule. *Proceedings of the National Academy of Sciences of the United States of America*, 105: 17362–17366.

Ravel, B. and Newville, M, 2005. ATHENA, ARTEMIS, HEPHAESTUS: Data analysis for X-ray absorption spectroscopy using IFEFFIT. *Journal of Synchrotron Radiation*, 12: 537–541.

Rehr, J.J. and Albers, R.C. 2000. Theoretical approaches to X-ray absorption fine structure. *Reviews of Modern Physics*, 72: 621–654.

Rehr, J.J., Kas, J.J., Prange, M.P. et al. 2009. Ab initio theory and calculations of X-ray spectra. *Comptes Rendus Physique*, 10: 548–559.

Rehr, J.J., Kas, J.J., Vila, D.F., Prange, M.P., and Jorissen, K. 2010. Parameter-free calculations of X-ray spectra with FEFF9. *Physical Chemistry Chemical Physics*, 12: 5503–5513.

Rehr, J.J., Stern, E.A., Martin, R.L., and Davidson, E.R. 1978. Extended X-ray-absorption fine-structure amplitudes—Wave-function relaxation and chemical effects. *Physical Review B*, 17: 560–565.

Rokita, E., Hermes, C., Nolting, H.F., and Ryczek, J. 2000. Substitution of calcium by strontium within selected calcium phosphates. *Journal of Crystal Growth*, 130: 543–552.

Rosenblum, G., Meroueh, S.O., Kleifeld, O. et al. 2003. Structural basis for potent slow binding inhibition of human matrix metalloproteinase-2 (MMP-2). *The Journal of Biological Chemistry*, 278: 27009–27015.

Sayers, D.E., Stern, E.A., and Lytle, F.W. 1971. New technique for investigating noncrystalline structures—Fourier analysis of extended X-ray-absorption fine structure. *Physical Review Letters*, 27: 1204–1207.

Shulman, R.G., Eisenberger, P., and Kincaid, B.M. 1978. X-ray absorption spectroscopy of biological molecules. *Annual Review of Biophysics and Bioenergy*, 7: 559–578.

Sickafus, K.E., Matzke, H., Hartmann, T. et al. 1999. Radiation damage effects in zirconia. *Journal of Nuclear Materials*, 274: 66–77.

Sorensen, L.B., Cross, J.O., Newville, M., Ravel, B., and Rehr, J.J. 1994. Diffraction anomalous fine structure: Unifying x-ray diffraction and x-ray absorption with DAFS. In *Resonant Anomalous X-Ray Scattering: Theory and Applications*, eds. G. Materlik, C.J. Sparks, and K. Fischer, pp. 389–420. Amsterdam, the Netherlands: Elsevier Science.

Sowrey, F.E., Skipper, L.J., Pickup, D.M. et al. 2004. Systematic empirical analysis of calcium–oxygen coordination environment by calcium K-edge XANES. *Physical Chemistry Chemical Physics*, 6: 188–192.

Stern, A.E. and Kim, K. 1981. Thickness effect on the extended-x-ray-fine-structure amplitude. *Physical Review B*, 23: 3781–3787.

Stern, E.A., Sayers, D.E., and Lytle, F.W. 1975. Extended X-ray-absorption fine-structure technique: 3. Determination of physical parameters. *Physical Review B*, 11: 4836–4846.

Stötzel, J., Lützenkirchen-Hecht, D., Fonda, E., De Oliveira, N., Briois, V., and Frahm, R. 2008. Novel angular encoder for a quick-extended x-ray absorption fine structure monochromator. *The Review of Scientific Instruments*, 79: 0831070–0831074.

Stragier, H., Cross, J.O., Rehr, J.J., and Sorensen, L.B. 1992. Diffraction anomalous fine-structure—A new X-ray structural technique. *Physical Review Letters*, 69: 3064–3067.

Taylor, M.G., Simkiss, K., Greaves, N.G., Okazaki, M., and Mann, S. 1993. An X-ray absorption spectroscopy study of the structure and transformation of amorphous calcium carbonate from plant cystoliths. *Proceedings: Biological Science*, 252: 75–80.

Teramura, K., Okuoka, S., Yamazoe, S., Kato, K., Shishido, T., and Tanaka, T. 2008. In situ time-resolved energy-dispersive XAFS study on photodeposition of Rh particles on a TiO_2 photocatalyst. *The Journal of Physical C*, 112: 8495–8498.

Terra, J., Dourado, E.R., Eon, J.-G., Ellis, D.E., Gonzalez, G., and Rossi, A.M. 2009. The structure of strontium-doped hydroxyapatite: An experimental and theoretical study. *Physical Chemistry Chemical Physics*, 11: 568–577.

Tester, C.C., Wu, C.-H., Krejci, M.R. et al. 2013. Time-resolved evolution of short- and long-range order during the transformation of amorphous calcium carbonate to calcite in the sea urchin embryo. *Advanced Functional Materials*, 23: 4185–4194.

Wang, R.Z., Addadi, L., and Weiner, S. 1997. Design strategies of sea urchin teeth: Structure, composition and micromechanical relations to function. *Philosophical transactions of the Royal Society of London. Series B*, 352: 469–480.

Zolotoyabko, E., Caspi, E.N., Fieramosca, J.S. et al. 2010. Differences between bond lengths in biogenic and geological calcite. *Crystal Growth & Design*, 10: 1207–1214.

8 Soft x-ray scanning transmission spectromicroscopy

Julie Cosmidis and Karim Benzerara

Contents

8.1 INTRODUCTION

It has been increasingly suggested that minerals formed by living organisms, that is, biominerals, have peculiar features compared to abiotically formed minerals. This results from the fact that biominerals are formed in highly supersaturated solutions and/or in the presence of a templating organic matrix that controls the nucleation and growth of the minerals (Weiner and Dove, 2003; Chan et al., 2011; Suzuki et al., 2011). As a nonexhaustive list of the particular features that can be observed sometimes, one can cite, for example, poor crystallinity in some cases, absence of lattice defects in other cases, mesocrystallinity (i.e., clustering of nanodomains with a common crystallographic orientation resulting in an apparent single crystal at a larger scale), association with organic molecules at the nanometer scale, major and/or trace element composition departing from what is expected in abiotically precipitated minerals, particular textural arrangement, and particular morphology. There are several reasons why it matters to precisely analyze these features. First, biominerals can sometimes have a major environmental impact depending on their structural and/or surface properties (Benzerara et al., 2011a). For example, Fe-oxides, oxyhydroxides, carbonates, phosphates, sulfates, and sulfides formed by microbes in various environments offer huge reactive surfaces and can sorb or release metal pollutants and also catalyze redox reactions modifying the speciation of inorganic pollutants (Stumm et al., 1994; Borch et al., 2010; Gadd, 2010). As a result, biominerals impact massively the mobility of pollutants in various environmental settings, and it is of high importance to better assess these interactions between microbes, minerals, and pollutants (e.g., Benzerara et al., 2008; Ona-Nguema et al., 2010; Jorand et al., 2011; Dong and Lu, 2012; Pantke et al., 2012). An additional motivation for deciphering potential differences between biominerals and abiotic minerals is the search for traces of life in ancient rocks (Golden et al., 2004; Ohmoto et al., 2008; Jimenez-Lopez et al., 2012). It is indeed often assumed that organisms can leave traces of their activity in the geological record in the form of biominerals (e.g., Benzerara and Menguy, 2009). Such traces may be more resistant to transformations during diagenesis and metamorphism than organic molecules and may thus provide more reliable targets for exobiology and paleobiology studies (Benzerara and Miot, 2010). Deciphering how much we can learn from these mineral fossils on past metabolic activities is currently an active quest. The importance of biominerals is not restricted to Earth sciences. The study of the origin and properties of biominerals concerns other scientific fields such as cultural heritage, in particular regarding the formation of biominerals as a route to stone monument restoration (e.g., Webster and May, 2006); material chemistry, with the possibility to synthesize new materials with interesting chemical and/or physical properties (e.g., Mann, 1993; Nassif and Livage, 2011; Andre et al., 2012); or medical science, for example, regarding pathogenic calcifications (e.g., Buckley et al., 1995; Benzerara et al., 2006).

Despite the importance of biominerals, we still have a limited knowledge of their formation processes. Understanding molecular mechanisms involved in their formation requires an advanced characterization of both organic molecules and minerals at the scale at which the processes operate, that is, the submicrometer scale. For a long time, this has been a major methodological challenge. Several analytical facilities providing such characterization capabilities are now available, including synchrotron-based soft x-ray scanning transmission x-ray microscopy (STXM). This microscopy technique provides chemical speciation-sensitive images at a spatial resolution down to ~15 nm coupled with x-ray absorption near-edge spectra (XANES) over a relatively extended range of energies (between 100 and 2000 eV). Soft x-ray STXM allows characterizing the speciation (i.e., coordination or redox state) of various elements including the major elements composing organic molecules (e.g., C, N, O, S, and P) and various metals forming biominerals (e.g., Ca, Fe). Although there is yet a limited access to this technique that requires specialized synchrotron radiation beamlines (less than a dozen of such beam lines worldwide are available currently or planned, Table 8.1), it has been proved over the last 10 years that soft x-ray STXM provides first-order information on biomineralization processes. Soft x-ray XANES has been used to study biominerals (here, calcium phosphates) as early as 1995 (Buckley et al., 1995). Hitchcock et al. (2002) reviewed some capabilities of soft x-ray STXM of interest to the study of biomaterials, focusing mostly on the organic part. Pecher et al. (2003) and Benzerara et al. (2004a) showed the first direct applications of soft x-ray STXM to the study of manganese-forming and calcium phosphate-forming bacteria, respectively. Since then, the number of studies using STXM for the study of biomineral interactions has increased year after year. The kinds of information that are retrieved and the types of signals that are measured have been widely diversified until recently from the acquisition of plain images and XANES spectra to the achievement of tomographs or the use of linear or circular polarization to map the crystallographic orientation of crystals or their magnetic moment.

Here, we review former studies using soft x-ray STXM in the field of biomineralization with the scope of presenting the basic principles, advantages, and limits of this technique and mentioning some future developments that are expected. Among the take-home messages, one has to remember that there is no unique technique able to characterize exhaustively any biomineral. We will thus also evoke how soft x-ray STXM is complementary to other techniques that allow obtaining information on different chemical, biological, and structural properties of a single sample at different length scales.

8.2 HOW DOES SOFT X-RAY STXM WORK?

An extensive documentation about the principles and functioning of STXM microscopes can be found on beamlines websites, especially the manual for beamline 5.3.2.2 at the Advanced Light Source (ALS) (http://www.lbl.gov/Science-Articles/Archive/sabl/2005/August/assets/docs/STXM_Beamline_5-3-2_Manual.pdf) or videos on the SM beamline at the Canadian Light Source (CLS) (http://exshare.lightsource.ca/sm/Pages/SM-Home.aspx), and in Howells et al. (2007). We provide a list of the STXM beamlines that are available or in the design stage with their principal technical characteristics and references for further information (Table 8.1). Here, we only provide a brief description of a STXM microscope, which will allow nonexperts to understand basically how STXM data are generated. A schematic sketch of a microscope is shown in Figure 8.1.

Some benchtop full-field x-ray microscopes (TXM) have been developed for 3D-imaging applications (Bertilson et al., 2009). In contrast, STXM crucially requires an intense source of x-rays provided by synchrotron radiation. STXM beamlines available over the world can be distinguished based on whether they are on a bending magnet or an insertion device (undulator). Indeed, this determines the capabilities of these beamlines. For example, some measurements involving the control of the polarization of the beam can only be conducted on an undulator beamline.

8.2.1 SOURCE OF HIGH-BRIGHTNESS MONOCHROMATIC X-RAYS

The photons produced by synchrotron radiation are monochromatized before being directed to the STXM. Monochromators available on existing beamlines allow working in the soft x-ray range (~80–2100 eV on 11.0.2 at the ALS, 130–2500 eV on the SM beamline at the CLS, more restricted range for other beamlines) with spectral resolutions better than 3000 ($E/\Delta E$) in general. In the end, a flux of ~10^8 photons/s is achieved in a 30 × 30 nm spot on the sample. On most existing STXM beamlines, moving from one edge energy to another takes several minutes including optimization of the beam.

The presence of harmonics, that is, photons with an energy that is a multiple of the selected photon energy, can alter significantly the quality of XANES spectra especially in the case of highly absorbing samples. These harmonics are produced by second- and third-order diffraction of photons on monochromator diffraction gratings. It has been noted that bending magnet STXMs are particularly sensitive to such contaminations (Kilcoyne et al., 2003).

8.2.2 MICROSCOPE

Very roughly, most of the currently available microscopes consist in ~1 m^3 sized aluminum chambers containing motor stages for moving the optics and the samples, a lens, an order sorting aperture

Table 8.1 List of soft x-ray STXM beamlines

SYNCHROTRON LIGHT SOURCE	LOCATION	BEAMLINE	STATUS	SOURCE	ENERGY RANGE (eV)	WEBSITE	REFERENCES
Advanced Light Source (ALS)	Berkeley, California	11.0.2	Operational, open for user proposals	EPU	200–1900	http://beamline1102.als.lbl.gov/	Bluhm et al. (2003)
Advanced Light Source (ALS)	Berkeley, California	5.3.2.2	Operational, open for user proposals	BM	250–800	http://ssg.als.lbl.gov/ssgbeamlines/beamline5-3-2	Ade et al. (2003), Kilcoyne et al. (2003)
Advanced Light Source (ALS)	Berkeley, California	5.3.2.1	Operational, not available for general users	BM	350–2500	http://ssg.als.lbl.gov/ssgbeamlines/beamline5-3-2	
Canadian Light Source (CLS)	Saskatoon, Canada	SM	Operational, open for user proposals	EPU	130–2500	http://exshare.lightsource.ca/sm/Pages/SM-Home.aspx	Kaznatcheev et al. (2007)
Swiss Light Source (SLS)	Villigen, Switzerland	PolLux	Operational, open for user proposals	EPU	250–1600	http://www.psi.ch/sls/pollux/pollux	Flechsig et al. (2007), Raabe et al. (2008, 2009), Frommherz et al. (2010)
Bessy II	Berlin, Germany	Maxymus	Operational, open for user proposals	EPU	120–1900	http://www.helmholtz-berlin.de/pubbin/igama_output?modus = einzel&sprache = en&gid = 1885&typoid = 35517	Nolle et al. (2012)
Soleil	Gif-sur-Yvette, France	HERMES	Expected to be operational by the end of 2014	EPU	70–2500	http://www.synchrotron-soleil.fr/portal/page/portal/Recherche/LignesLumiere/HERMES	
Diamond Light Source (DLS)	Didcot, United Kingdom	I08	In design phase	EPU	250–4000	http://www.diamond.ac.uk/Home/Beamlines/I08.html	
National Synchrotron Radiation Laboratory (NSRL)	Hefei, China	U12A	?	BM	250–1800	http://mems.ustc.edu.cn	Jiang et al. (2003)
Pohang Light Source (PLS)	Pohang, Korea	10A	Expected to be operational at the end of 2013	EPU	100–2000	http://pal.postech.ac.kr/paleng/bl/10A/	Lee and Shin (2001), Shin and Lee (2001)
Australian Synchrotron (AS)	Clayton, Australia		In design phase?	EPU	200–3000	http://www.synchrotron.org.au/index.php/about-us/australian-synchrotron-development-plan/asdp-beamlines-phase-3/soft-x-ray-microscopy-and-spectroscopy-cluster	

Note: BM, bending magnet; EPU, elliptically polarizing undulator.

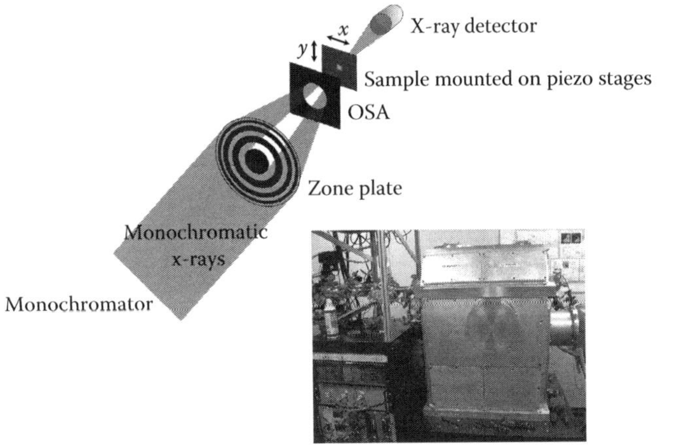

Figure 8.1 Schematic sketch of an STXM microscope. A picture of the STXM on beamline 5.3.2.2 at the ALS is provided.

(OSA), the sample holder, and an x-ray detector (Figure 8.1). A lock placed at the top or in front of the chamber allows exchange of the sample holders. The chamber can be evacuated to 10^{-6} torr. Some He is usually added to reduce interferometer drifts and increase thermal stability. In the case of *wet samples*, exchange of air by helium can be performed so that the sample stays permanently at 1 atm during the sample exchange.

8.2.3 LENS: USE OF A ZONE PLATE

The monochromatic x-ray beam is focused on the sample by a Fresnel zone plate. It consists of concentric alternating absorbing and transparent rings forming a circular diffraction grating. The sample is positioned at the focal point of the zone plate (focusing step, see the following text). The focal length depends on the energy of the incident x-ray beam. At 300 eV (i.e., around the K-edge of carbon), the distance between the zone plate and the sample (i.e., working distance) is in the order of few hundreds of micrometers. As a result, there is very little space upstream of the sample, which constrains the sample environment (see Section 8.3.6) and limits the use of fluorescence (which is feasible however, see the following text). Since the focal length depends on the energy of the incident x-ray beam, the distance between the zone plate and the sample (working distance) has to be changed each time the beam energy is varied (this can be the case during data acquisition, when an energy range of several tens of eV is scanned). If all the optics is not perfectly aligned along the optical path, there might be some apparent drift of the sample in the observation plane during an energy scan. Most of the time, however, beamline scientists have done a great job in having their beamline well aligned, and this drift can be easily corrected by data processing.

Theoretically, the spatial resolution achieved by STXM depends on the precision of stage movements and the size of the spot formed by the zone plate. Currently, zone plates (in particular, the outermost zone width) are the limiting devices for the spatial resolution. Standard commercial zone plates purchased by beamlines commonly achieve spatial resolutions between 20 and 40 nm. Using zone plates, record spatial resolutions of 10 nm were achieved at the ALS (Chao et al., 2012) and at the Swiss Light Source (SLS) (Vila-Comamala et al., 2009). In the future, the development of ptychography may help to improve spatial resolutions using standard zone plates. However, it should be noted that biomineralogists should not always want zone plates with the highest spatial resolution. Indeed, there can be several drawbacks: for example, the working distance of zone plates with the best spatial resolution may be inappropriate for studying samples at very low energy (e.g., around 130 eV for P $L_{2,3}$-edges); upon aging, a given zone plate may have a low transmission efficiency and might not be appropriate for measurements requiring high fluxes. In any case, this should be systematically discussed with the beamline scientist prior to beamtime based on the research goals in mind. Extensive information about Fresnel zone plate physics and their applications in microscopy can be found in Howells et al. (2007).

8.2.4 ORDER SORTING APERTURE

An OSA, which is a pinhole measuring ~50 μm in diameter, is placed between the zone plate and the sample, decreasing even further the space available upstream to the sample. As mentioned previously, zone plates are diffractive lenses. Direct light transmitted through the zone plate may thus combine with the different orders of diffraction generated by the zone plate, altering the quality of the images and spectra. A central stop on the zone plate and the OSA filter out zero-order light and higher-order focal points of the zone plate, respectively. The OSA has to be placed at the proper distance between the sample and the zone plate to eliminate these higher-order diffracted light and let only first-order diffracted light to the sample.

8.2.5 SCANNING AND CONTROLLING SPECIMEN POSITION

For most current STXM applications and on most of the current beamlines, the sample is scanned in order to obtain an image. Several samples can be placed on a single sample holder. This holder is mounted on two sets of stages, one on top of the other, enabling the sample holder to be moved in three dimensions: x–y (i.e., the plane perpendicular to the incident beam that is the sample plane) and z (i.e., in the direction parallel to the beam, which is perpendicular to the sample). One set of stages at the base of the whole device has a large movement range (e.g., several tens of millimeters) and is used to move from one sample to another and for coarse sample positioning and imaging. Using these stages, images can be obtained measuring up to several millimeters in length with a pixel size of few micrometers. Piezo stages located on top of coarse stages are used to scan samples in the x–y plane with higher precision in order to obtain well-resolved images. The position of the piezo stages is monitored by laser interferometers, and the precision is usually better than 10 nm (e.g., Kilcoyne et al., 2003). Piezo stages have a reduced range of movement of the order of a few tens of micrometers, the exact size of which depends on the microscope (presently ~50 μm on 11.0.2 at the ALS). This limits the maximum size of images that can be obtained at high spatial resolution. Finally, the position of the zone plate can be moved along the z position to adjust the focus. Development of new stages is in progress on some beamlines to reduce vibrations and increase scanning speed.

8.2.6 X-RAY DETECTION

Currently available STXM use point detectors for most of the analyses. There are different types of such detectors, either photomultiplier tubes (PMTs) with a scintillator or silicon photodiodes (Kilcoyne et al., 2003). Fast avalanche photodiodes have been implemented to provide time-resolved measurements. Other detection systems of potential interest for biomineralization studies have been implemented on some STXM. This includes segmented silicon and CCD detectors that offer the possibility of differential phase contrast imaging (e.g., Hornberger et al., 2007; Raabe et al., 2009), total electron yield detectors for surface-sensitive measurements under high vacuum conditions (Hub et al., 2010; Behyan et al., 2011), or fluorescence detectors allowing measurement of diluted samples (Hitchcock et al., 2010). To our knowledge, only the last one has been used on biomineralizing systems as detailed in the following discussion (Hitchcock et al., 2012).

8.3 PERFORMING THE EXPERIMENTS

Different types of measurements can be obtained by STXM: images at a single energy (Figure 8.2), XANES spectra measured at a single point (point scan, Figure 8.3) or over a segment (line scan), and stacks of images of a single area at different energies (Figure 8.4). From the latter, it is possible to extract XANES spectra from different subareas if the stacks are run over an appropriate energy range with relatively short energy increments. Here, we provide some recommendations for the acquisition of quality images and stacks.

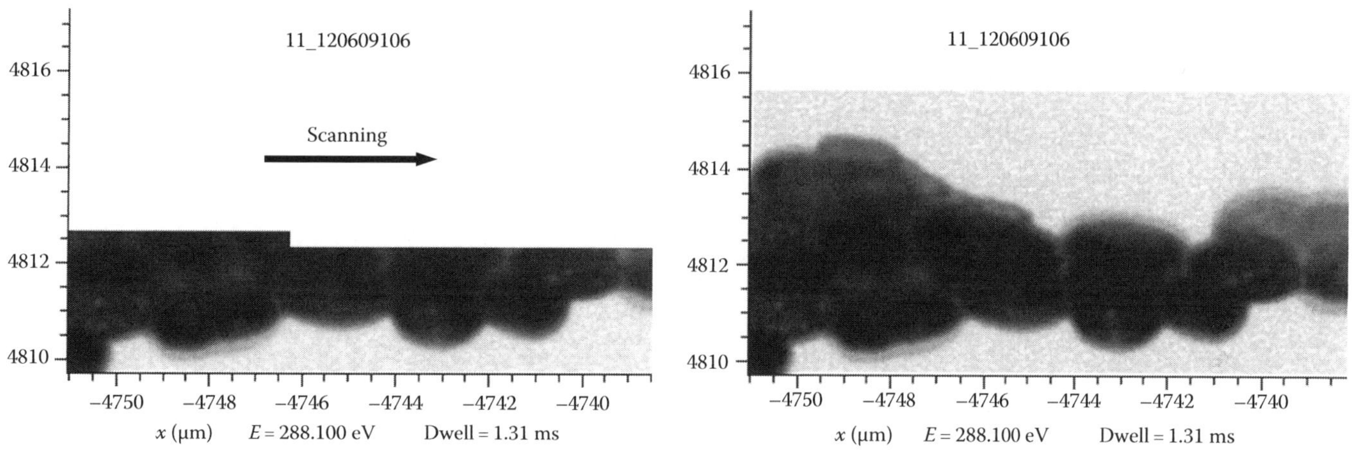

Figure 8.2 Acquisition of an STXM image at 288.1 eV. Sample is scanned pixel by pixel or line by line. This image measures 13 × 6 μm; the step size is 60 nm; the image thus measures 217 × 100 points. The dwell time is 1.3 ms per point. The whole image is obtained in 40 s.

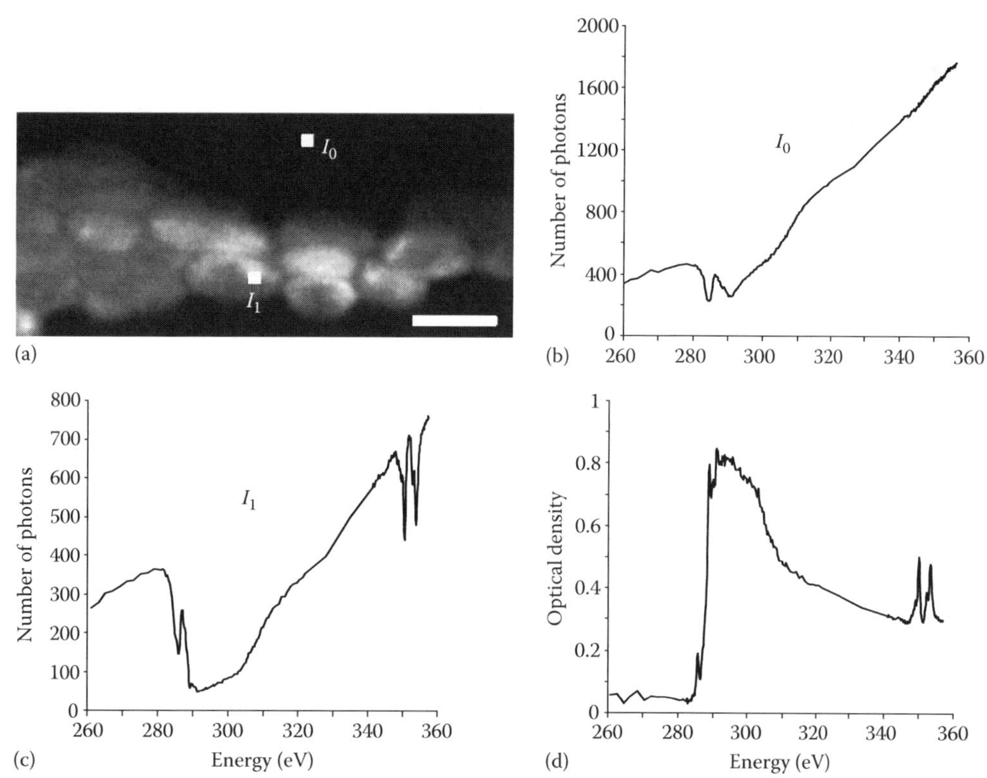

Figure 8.3 (a) Acquisition of an energy scan at a fixed sample point. (b) One pixel is picked up outside the sample to measure I_0. (c) One pixel is picked up on the sample to measure I_1. Energy is scanned and number of photons is measured as a function of energy. (d) The XANES spectrum is obtained as $-\log(I_1/I_0) = f(\text{energy})$. Scale bar of the image is 2 μm.

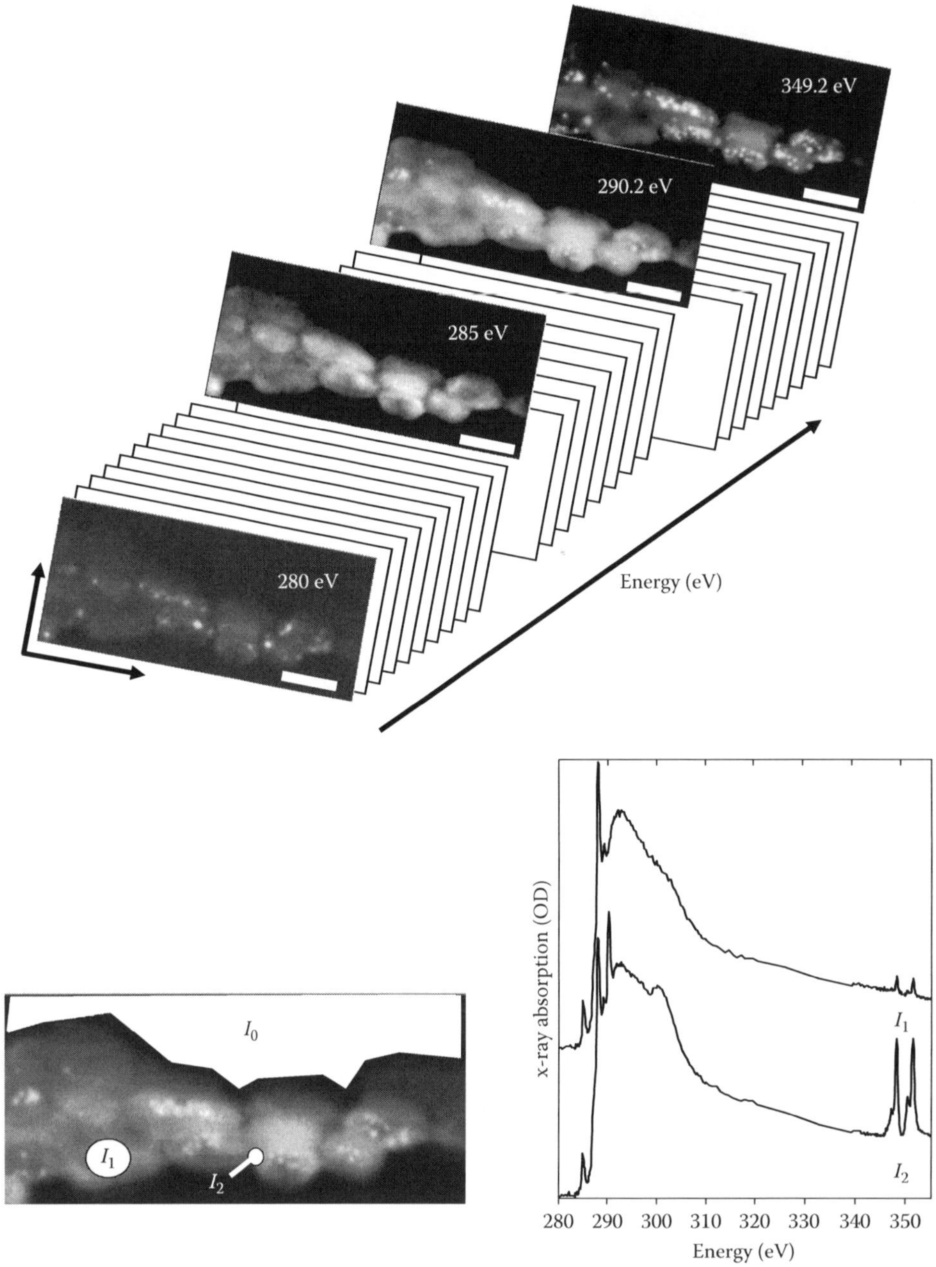

Figure 8.4 Principles of image stack acquisition. An image is scanned at 280 eV. Then energy is changed and a new image is scanned. Here, for example, 10 images are acquired between 280 and 283 eV, 89 images between 283 and 293 eV (0.11 eV step), 57 images between 293 and 310 eV, 13 images between 310 and 320 eV, 4 images between 320 and 340 eV, 18 images between 340 and 345 eV, 104 images between 345 and 356 eV (0.11 eV step), and 9 images between 356 and 370 eV. For an image that measures 113 × 75 pixels, the whole stack acquisition takes 1h33. With the aXis software, it is possible to realign all the images of the stack after acquisition. Then an I_0 area is defined where there is no sample. XANES spectra can be extracted from different areas of the image (here I_1 and I_2) allowing to check speciation variations in the sample.

8.3.1 SETTINGS OF THE EXPERIMENTS: SAMPLE LOCALIZATION, MICROSCOPE ALIGNMENT, AND ENERGY CALIBRATION

Samples of interest to biomineralogists are sometimes unique and of small size compared to the support frame on which they are deposited (e.g., a transmission electron microscopy [TEM] grid or a silicon nitride window). Finding the sample by scanning the whole sample holder can be tedious in STXM. This is why a light microscope was implemented on the former STXM beamline at NSLS. On more recent beamlines, the use of an offline visual light microscope (VLM) with an encoded *x*–*y* stage has proved to be very efficient in reducing the amount of time spent for sample localization once the VLM stage and the STXM stages have been calibrated so that the coordinates match together. Pictures of interesting areas of the sample with encoded coordinates can be loaded in the software used to run the STXM.

The complexity of the overall alignment process depends on the beamlines, in particular the efforts invested by the beamline scientists in aligning the whole beamline regularly and designing software procedures to facilitate that task. Basic operations that can be performed by the user consist in modifying the opening of the slits, adjusting the gap and the offset of the elliptically polarizing undulator (EPU) when the beamline is on an insertion device, and adjusting positions of the diverse optical components (e.g., M1 mirror, zone plate, OSA) in order to obtain noiseless images. In order to conduct these settings, it is good to have one hole on the sample holder with no sample unless the motor stages offer the possibility to have the beam illuminating a spot outside the sample holder.

Energy calibration can be performed at several edges. Electronic transitions in gaseous CO_2 (Ma et al., 1991), N_2 (Chen et al., 1989), and/or O_2 (Chen et al., 1989), which can be flushed in the STXM chamber, offer a convenient and efficient way to calibrate energy at the C, N, and O K-edges. Most of the beamlines are usually stable so that this calibration need only be done once over a whole beamtime.

8.3.2 IMAGE ACQUISITION

Focusing is achieved relatively easily by STXM. One sets a line scan running across a particle with a sharp edge. An energy around the edge of interest is chosen so that the particle is absorbing as much as possible. The distance between the zone plate and the sample is varied by small increments, and the absorption profile along the line is recorded. The best focus corresponds to the sharpest absorption profile. This procedure takes few minutes depending on how far the sample sits from the best focus. When correct focus is achieved, an image can be obtained by scanning the sample in the *x*–*y* plane. This is done at a single energy (Figure 8.2). Dwell time for each pixel can be chosen by the users. It can vary between a fraction of millisecond and several tens of millisecond depending on the signal to noise ratio that is required. Scanning can be performed line by line or for slower scans point by point. The intensity of transmitted x-rays is measured by the detector for each pixel. As an indication, a 5 × 5 μm image with a spatial resolution of 25 nm in the two dimensions (200 × 200 pixels) and a dwell time of 1.7 ms can be obtained in ~100 s on beamline 11.0.2 at the ALS. Several images at several energies can be recorded, for example, one below and one above the edge of interest (Figure 8.5). Background intensity (I_0) is extracted from one or several pixels where there is no sample (I_0 is not measured upstream as with some other techniques). Absorption by the support film on which the sample is lying can thus be subtracted. Images can be converted to optical density (OD) units that correspond to $-\log(I/I_0)$. Maps are processed by subtracting the pre-edge OD image from the post-edge one. Post-edge images can be measured at different pertinently chosen energies. This relatively quick and simple approach provides first hints on the localization of the element of interest and potential heterogeneities of its speciation (Figure 8.6).

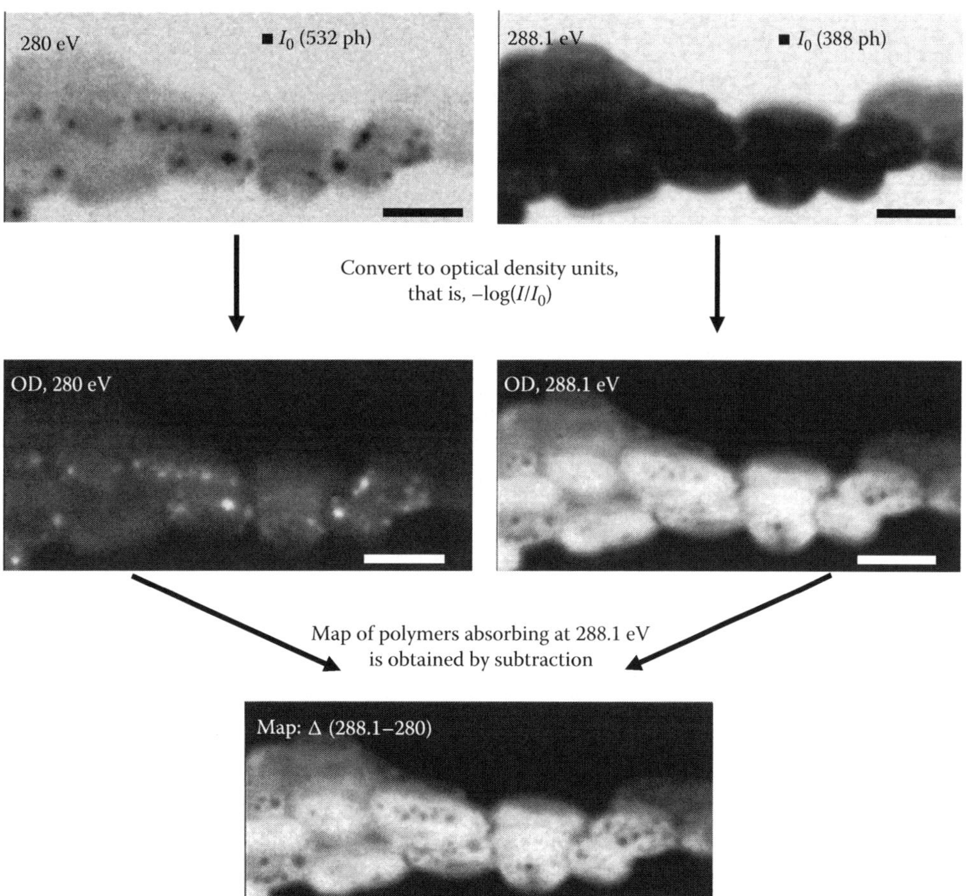

Figure 8.5 Conversion of an image to OD units. I_0 is measured where there is nothing but the support film (here Si_3N_4). The function $-\log(I/I_0)$ is applied to each pixel. Here, I_0 is lower at 288.1 eV (388 photons) than at 280 eV (532 photons). This is due to contaminating carbon on the optics of the microscope. This contribution is also removed in the I_0 from the signal measured on the sample. The map of C-compounds absorbing at 288.1 eV (in this example, it is mostly proteins) is obtained by subtracting the image in OD units below the edge from the one at 288.1 eV.

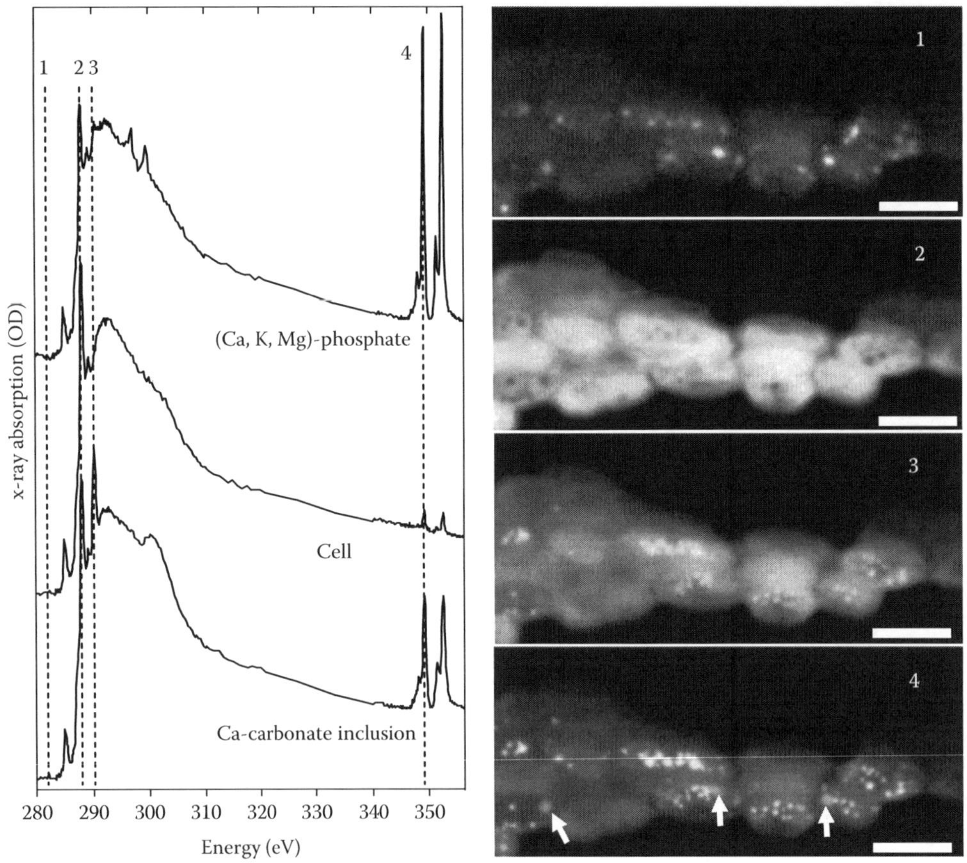

Figure 8.6 Energy-filtered imaging by STXM. On the left: Three different XANES spectra measured on three different areas from the sample at the C K-edge and Ca $L_{2,3}$-edges. One area corresponds to inclusions of phosphate of Ca, Mg, and K within the cells. One area corresponds to a cell without any inclusion (cell). The last area corresponds to inclusions of Ca carbonates within the cells. On the right: STXM images of the exact same zone converted to OD units. Scale bar is the same for all images: 2 µm. Image 1 was taken at 280 eV, an energy below the C K-edge. Carbon is not absorbing at this energy. The contrast is weak except for few bright spots corresponding mostly to P-containing inclusions (P $L_{2,3}$-edges are at ~130 eV, thus P-containing areas absorb photons at this energy). Image 2 was taken at 288.2 eV, an energy corresponding to 1s → π* electronic transition in amide groups. Bacterial cells contain a lot of proteins and appear very bright at this energy. Image 3 was taken at 290.2 eV, an energy corresponding to 1s → π* electronic transition in carbonate functional groups. Image 4 was taken at 349.2 eV, an energy corresponding to the Ca L_3-edge. In this image, both Ca-containing phosphate (see arrows) and carbonate inclusions can be observed. Obtaining such images at different energies allow to map different compounds. However, acquisition of spatially resolved spectra is usually necessary, especially when one does not know the sample yet.

8.3.3 POINT SCANS AND LINE SCANS FOR XANES SPECTRUM ACQUISITION

XANES spectra can be measured using different procedures. Point scans consist in measuring the intensity of the transmitted beam on a single pixel on the sample over the energy range of interest (Figure 8.3). The same scan is also measured outside the sample. The spectrum is obtained by plotting $-\log(I/I_0)$ as a function of the energy. The measurement is fast but there is no spatial information associated with it; hence, potential heterogeneities are overlooked. Moreover, potential drifts during the scan in energy cannot be corrected. Finally, the spot is illuminated for a long time and beam damages (see the following text) can become significant. Alternatively, line scans consist in measuring beam intensity on several pixels over a line that expands both over the sample (to measure I) and outside the sample (to measure I_0). This divides the exposure time by the number of pixels for the same quality of spectrum. The spatial information is very limited. This procedure, when the energy drift is not too important, is interesting to get a quick idea of how the average XANES spectrum of the sample looks.

8.3.4 STACK ACQUISITION

In order to capture the spatial distribution and variations of an element speciation in a sample, stacks of images of the same field are recorded over the energy range of interest and with the required energy step size (Figure 8.4). The images can be aligned afterward to correct the potential drift in the x–y plane due to the zone plate movement in the z direction. The definition of the energy scan is flexible. Smaller energy steps are usually chosen around the edge (e.g., a step of 0.1 eV in the 283–293 eV range at the C K-edge) in order to get higher spectral resolution over this range. Additional documentation on stacks can be found in Jacobsen et al. (2000). Typical duration of a stack is between 30 and 90 min. For example, on beamline 11.0.2, a stack at the C K-edge containing 164 images measuring 13 × 7 µm with a step size of 80 nm and obtained with a dwell time of ~1 ms is acquired in ~90 min, providing a spectral resolution of 0.1 eV over part of the spectrum.

8.3.5 BEAM DAMAGE

Although it has been assessed to be relatively low compared to other techniques in general (Braun et al., 2005; Hitchcock et al., 2008),

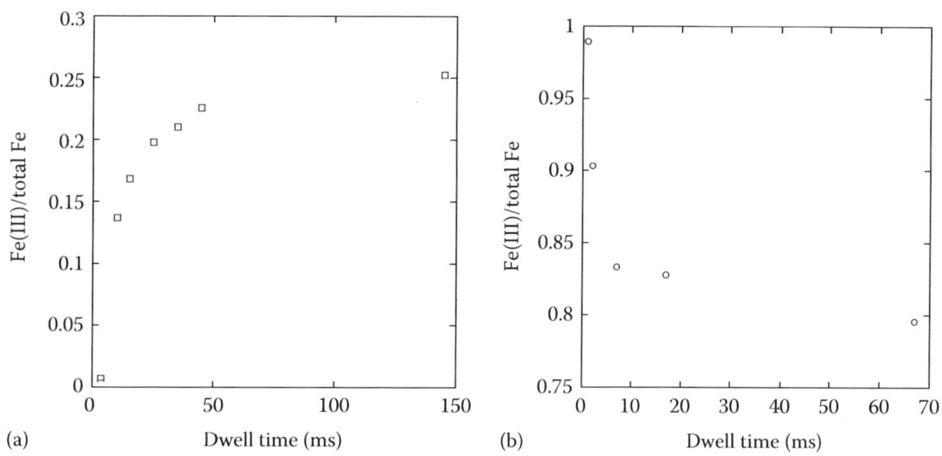

Figure 8.7 Beam radiation damages on a reference Fe(II)-phosphate (a) and Fe(III)-encrusted bacteria (b). Fe(III)/Fe(total) ratio changes with dwell time (which is related to dose). Photooxidation is observed in the Fe(II)-phosphate, while photoreduction is observed when organic carbon is present.

beam damage induced by STXM analyses can be detected as a mass loss (especially for carbon) and also a change in the element speciation. This has been particularly studied for carbon in organic polymers, of direct relevance to biomineralization studies (Beetz and Jacobsen, 2002; Coffey et al., 2002; Howells et al., 2009; Wang et al., 2009). Mass loss might be corrected by the use of cryo-conditions but not the break of chemical bonds (Beetz and Jacobsen, 2002). Moreover, some studies have shown that when using a dwell time less than few ms per energy point, which is typical for acquisition of stacks, there is very limited photoreduction or photooxidation of Fe at the Fe $L_{2,3}$-edges (Figure 8.7; Miot et al., 2009a).

8.3.6 SAMPLE PREPARATION

Samples for STXM analyses need to be thin enough so that at least 10% (depending on the linearity working range of the detector) of the soft x-ray photons go through. By knowing approximately the chemical composition of the sample, knowing the mass absorption coefficients of the element of interest, and assessing the amplitude of the resonances in the XANES spectra, it is possible to estimate the maximum thickness of the sample that is suitable with an STXM analysis. As an example, bacteria up to ~1 μm in thickness can be analyzed. Minerals have to be thinner. Absorption features at the Ca and Fe $L_{2,3}$-edges, for example, are particularly intense. Thus, samples have to be even thinner (<100 nm) when studying mineral phases rich in these elements at these edges. Otherwise, absorption saturation effects are created affecting the relative ratios of peaks in XANES spectra and leading possibly to wrong interpretations (Hanhan et al., 2009). Samples that can be observed by TEM can most often be analyzed by STXM. Different types of sample preparation are then possible. Some are common with TEM: powders that can be deposited on standard TEM Formvar- or anything else-coated grids. Alternatively, they can be deposited on Si_3N_4 thin (25–100 nm) windows to avoid any carbon contribution from the substrate (even if this contribution can be removed by I_0 correction). On most of currently available beamlines and for usual operation modes (except tomography), the sample holders are aluminum plates with six holes. Removable double-sided adhesive tape or glue can be used to attach the sample to the sample holder.

Thin sections prepared by ultramicrotomy can be observed (e.g., Hernández Cruz et al., 2006; Bernard et al., 2009). The signal from the epoxy resin used to embed the sample may be difficult to subtract from the signal of the area of interest, but the most prominent issue with ultramicrotomy samples is that apparently much of the organic content of the cells is lost when using standard protocols that makes STXM analyses relatively useless at the C K-edge (e.g., Lawrence et al., 2003). Ultrathin sections prepared by focused ion beam (FIB) milling have been repeatedly analyzed by STXM (Bernard et al., 2007; Lepot et al., 2008; Maclean et al., 2008; Benzerara et al., 2010b; Carlut et al., 2010; Galvez et al., 2012). This technique might produce artifacts such as local gallium implantation or amorphization of the sample (e.g., Bernard et al., 2009 and references therein). Interestingly, Bernard et al. (2009) and Bassim et al. (2012) have compared XANES spectra measured at the C K-edge on FIB-milled organic samples with XANES spectra of the same samples obtained by ultramicrotomy. It was concluded that the same chemical structures were preserved in both types of samples.

Finally, it is possible to perform STXM analyses under hydrated conditions: wet cells can be obtained by sandwiching a small drop of a sample (fraction of μL) between two silicon nitride windows (Hitchcock et al., 2002; Dynes et al., 2006a,b; Hunter et al., 2008; Chan et al., 2009; Christl et al., 2012). This can also be done under controlled atmosphere in a glove box when dealing with oxidation-sensitive samples (e.g., Miot et al., 2009a–c; Coker et al., 2012; Pantke et al., 2012).

8.3.7 DATA PROCESSING

Very conveniently, the same aXis2000 software (Hitchcock, 2013) can be used to process data obtained on most of currently accessible beamlines. It is a freeware and generates files with common extensions (.txt, .tiff, .jpeg) that can be easily used by other software in which users may have developed specific routines. Hyperspectral data generated by STXM contain a wealth of information that may need specific statistical procedures to be analyzed. Principal component analysis routines compatible with aXis2000 have been designed (e.g., Lerotic et al., 2004, 2005).

8.4 STXM IN STUDIES OF BIOMINERALIZATION

In the following, we review specific studies using STXM on biomineralizing systems to illustrate possible sample preparation strategies, the various types of measurements that can be achieved, and resulting advances and limits offered by soft x-ray STXM for biomineralogy.

8.4.1 FINDING TRACES OF MICROBIAL MINERALIZATION IN ANCIENT ROCKS

Biomineralization can lead to the preservation of bacterial structures in the geological record in the form of mineralized microfossils (Benzerara and Menguy, 2009; Benzerara and Miot, 2010). This is of high interest to geobiologists since it provides constraints on the history of life. These traces can be searched in ancient rocks, and there is a crucial need to characterize them exhaustively in order to be able to distinguish them from possible abiotic biomorphs (e.g., Lepot et al., 2008, 2009; De Gregorio et al., 2009; Fliegel et al., 2011). Here, we show how a combination of FIB, STXM, and TEM was used to characterize putative microfossils in a Paleocene phosphorite from Morocco at the nanometer scale (Cosmidis et al., 2013).

Phosphorites are large marine sedimentary formations containing high amounts of phosphate minerals. It is believed that they were formed by microbial biomineralization partly based on the scanning electron microscopy (SEM) observation of a great number of purported fossil traces of microorganisms in phosphorites of different ages and different locations (Soudry, 2000; Zanin and Zamirailova, 2011). However, the correct interpretation of these putative microfossils has long been debated, since it was first only based on images suggesting an ambiguous morphological resemblance with modern-type bacteria (Lamboy, 1994; Lundberg and McFarlane, 2011). Biomorphs similar to the ones described previously in phosphorites were observed by Cosmidis et al. (2013) in the Ouled Abdoun phosphorites (Morocco). They appear as micrometer-sized spheres that were cut by FIB milling (Figure 8.8). A crown of 20–40 nm in thickness delimiting the spheres was observed by TEM. This crown was interpreted as a fossilized bacterial cell wall based on the comparison with modern bacteria biomineralizing phosphate minerals in laboratory experiments. STXM and XANES analyses performed at the C K-edge on the same FIB foil (Figure 8.8) showed that these fossilized cell walls were relatively enriched in organic functional groups (aromatics, esters, and carboxyls) compared to the rest of the sample that was richer in inorganic carbon functional groups (i.e., carbonate groups). The preservation or organic matter within the bacterial cell wall is consistent with a model in which the fossilization (i.e., biomineralization) starts very rapidly when cells are alive and before the decay and oxidation of the microbial organic structures. Biomineralization initiates within the bacterial cell wall and then proceeds outside and inside (Benzerara et al., 2004b; Miot et al., 2011).

This example illustrates how STXM coupled with TEM on FIB foils can be used to characterize finely and identify microfossils but also to gain further insight into bacterial fossilization mechanisms. Compared with standard SEM-based observations, this approach provides improved spatial resolution, the possibility to characterize the internal structures of the

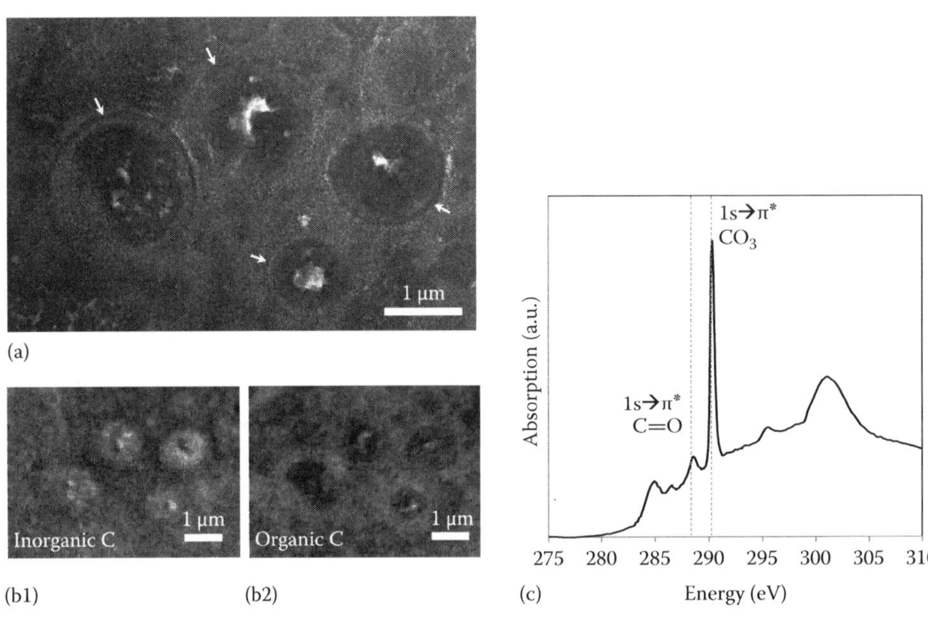

Figure 8.8 (a) TEM image of fossil bacteria in a phosphatic coprolite. The periplasm of the fossil bacteria is preserved as a dense 20–40 nm thick crown dense to electrons (arrows). (b1 and b2) STXM maps at the C K-edge of the same fossil bacteria. b1 is obtained by subtracting an image at 280 eV converted to OD to an image at 290.3 eV (C 1s → π^* transition peak of carbonates) converted to OD. B2 B1 is obtained by subtracting an image at 280 eV converted to OD to an image at 288.5 eV (C 1s → π^* transitions in carbonyl (C=O) bonds in carboxyl and/or esters) converted to OD. The intracellular and extracellular spaces are rich in inorganic C (the CO_3 ions present in the francolite structure), whereas the fossilized periplasm is relatively enriched in organic C. (c) XANES C K-edge spectrum of the sample. (From Cosmidis, J., Benzerara, K., Gheerbrant, E., Estève, I., Bouya, B., and Amaghaz, M.: Nanometer-scale characterization of exceptionally preserved bacterial fossils in Paleocene phosphorites from Ouled Abdoun (Morocco). *Geobiology*. 2013. 11. 139–153. Copyright Wiley-VCH Verlag GmbH & Co. KGaA. Reproduced with permission.)

bacterial fossils, and the speciation and distribution of organic carbon at the submicrometer scale. XANES at the C K-edge achieves higher spectral resolution than electron energy loss spectroscopy on standard TEM. Such a high spectral resolution, better than 0.1 eV at the C K-edge, is crucial in order to be able to retrieve meaningful data on organic carbon speciation. This is of crucial importance since the origin and burial history of fossils can be determined by the characterization of the functional groups composing organic matter within these fossils (Lepot et al., 2008; De Gregorio et al., 2009; Bernard et al., 2010; Cody et al., 2011).

8.4.2 GAINING INSIGHTS INTO THE MECHANISMS OF BACTERIAL CALCIFICATION BY STXM

The study of bacterial Ca-containing biominerals requires the capability to characterize and image both mineral phases and associated organic compounds at the nanometer scale. This is made possible by STXM analyses. STXM can be used to discriminate between the most common calcium carbonates and phosphates, whatever their crystallinity, involved in bacterial calcification as soon as they present distinctive XANES features at the Ca $L_{2,3}$-edge (Benzerara et al., 2004a). For example, Figure 8.9 illustrates how STXM analyses at the C K- and Ca $L_{2,3}$-edges have been used to map organic groups and amorphous Ca-containing carbonates in cyanobacterial cells of *Candidatus*

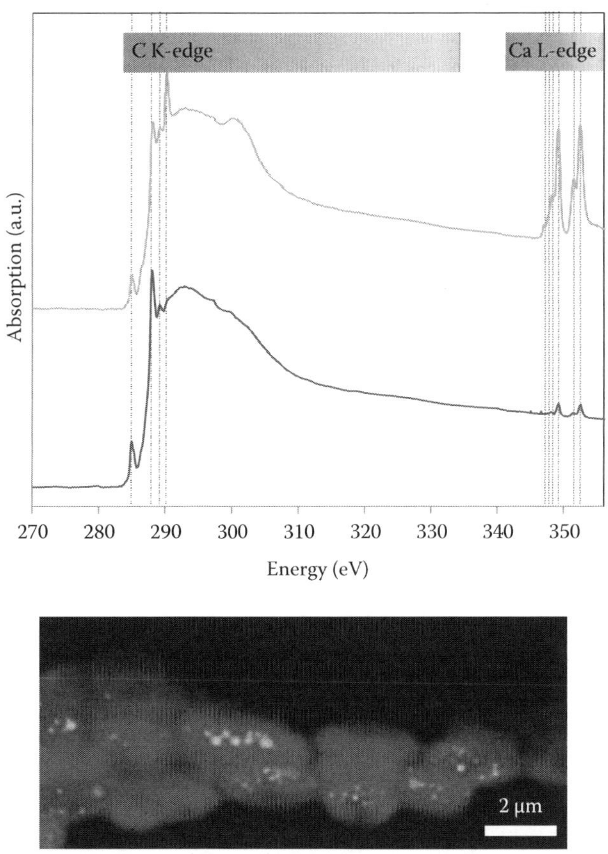

Figure 8.9 (See color insert.) STXM colored map of the cyanobacterium *Candidatus Gloeomargarita lithophora*. In red, areas exhibiting the bottom XANES spectrum (in red) that is typical of a bacterial cell spectrum. In blue, area showing the blue XANES spectrum that is a combination of the cell XANES spectrum plus a calcium carbonate contribution (see peak at 290.2 eV and Ca peaks).

Gloeomargarita lithophora. Until recently, calcium-carbonate formation by cyanobacteria was thought to occur exclusively outside of the cells, leading to their encrustation in extracellular precipitates. However, this general scheme was questioned recently by the discovery of a new cyanobacterium species, originating from a lacustrine microbialite in Mexico, which is able to form intracellular carbonate precipitates providing a new case of controlled biomineralization (Couradeau et al., 2012). A combination of several microscopies, including STXM, TEM, and confocal laser scanning microscopy (CLSM), has been used to characterize these intracellular precipitates. CLSM revealed the presence of chlorophyll-a and phycobiliproteins in the bacteria, confirming that they are cyanobacteria. The major element composition of the precipitates was measured by TEM- and SEM-coupled x-ray energy dispersive spectroscopy (XEDS) and showed that they contain calcium, magnesium, strontium, and barium. Electron diffraction in TEM demonstrated that the precipitates are poorly crystallized, that is, providing an example of amorphous calcium carbonate (ACC). By comparing the XANES signatures at the Ca $L_{2,3}$-edge of these poorly crystallized precipitates with those of various carbonate phases, Couradeau et al. (2012) suggested that intracellular precipitates display some local ordering consistent with the structure of benstonite, a Mg^-, Ca^-, Sr^-, and Ba^- bearing carbonate.

Poorly crystallized materials are common in biomineralization, notably as transient precursors to more stable crystalline phases. As XANES spectroscopy provides a very local atomic information, it is a powerful tool to characterize these low-crystallinity phases. For example, Figure 8.10 summarizes spectral features at the Ca $L_{2,3}$-edges of various Ca-containing

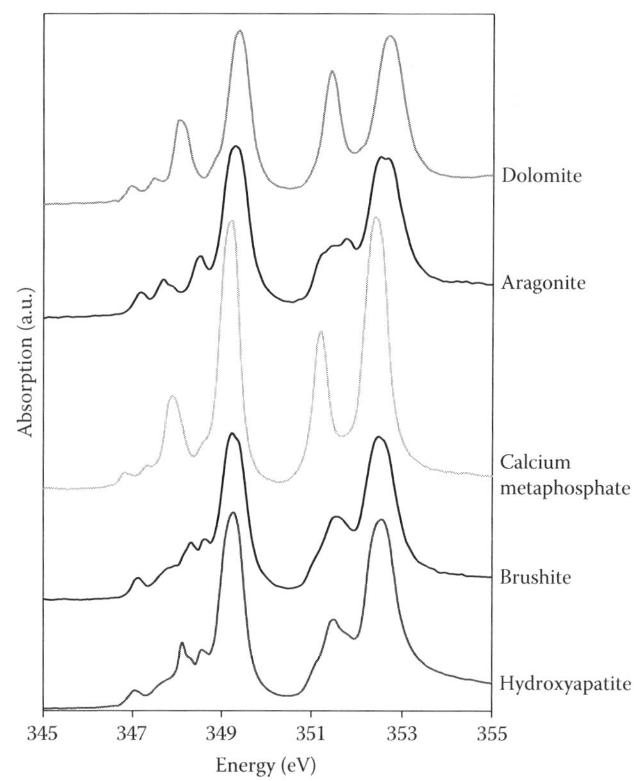

Figure 8.10 XANES spectra at the Ca $L_{2,3}$-edge of several Ca-bearing minerals involved in biomineralization. The energy positions of the different peaks are provided in Table 8.2.

Table 8.2 Names, formulas, and major XANES peak positions of some Ca-containing carbonate and phosphate minerals

	MINERAL	FORMULA	Ca L-EDGE XANES PEAK POSITIONS (eV)
Ca phosphates	Hydroxyapatite	$Ca_{10}(PO_4)_6(OH)_2$	347.1, 347.7, 348.2, 348.5, 349.3, 351.6, 352.5
	Brushite	$CaHPO_4 \cdot 2H_2O$	347.1, 347.7, 348.2, 348.5, 349.3, 351.6, 352.5
	Calcium metaphosphate	β-$Ca(PO_3)_2$	346.8, 347.3, 347.9, 348.5, 349.3, 351.2, 352.5
Ca carbonates	Aragonite	$CaCO_3$	347.2, 347.7, 348.5, 349.3, 351.2, 351.5, 351.8, 352.6
	Dolomite	$CaMg(CO_3)_2$	347, 347.5, 348.1, 349.4, 351.4, 352.7

phases of interest to biomineralogists. XANES spectroscopy at the Ca $L_{2,3}$-edge in x-ray photoelectron emission spectromicroscopy (XPEEM) has, for instance, been used to study the transformation of ACC into calcite in sea urchin spicules (Politi et al., 2008; Gong et al., 2012). Coupled with the high-resolution capabilities of STXM, this method has been also used to evidence the formation of an aragonite-like ACC phase by the cyanobacterial strain *Synechococcus leopoliensis* PCC 7942 (Obst et al., 2009a). It was suggested that this ACC phase may serve as a precursor in bacterially mediated calcite precipitation.

STXM provides images that result from an integrated signal coming from the whole thickness of the samples. In biomineralization, however, the question of the localization of the biominerals relative to the cells (intracellular, extracellular, or within the cell wall) is crucial and requires to determine the 3D structure of the bacteria–mineral assemblages. The use of ultramicrotomed sections of fixed, dehydrated cells embedded in an epoxy resin is not recommended for STXM observations since the preparation procedure alters the XANES chemical signatures at the C K-edge (Lawrence et al., 2003). Alternatively, STXM angle-scan tomography has been developed to permit 3D imaging and chemical mapping in a STXM, allowing to access directly the spatial configuration of bacteria–mineral interfaces (Johansson et al., 2007; Hitchcock et al., 2009). A special sample holder is used in order to allow the sample attached to a TEM grid to be rotated at ±70° relatively to the plane perpendicular to the incident photon beam. The orientation of the sample is changed by steps of 5° and images are recorded at each step. The images obtained are then carefully aligned and analyzed by the means of specific tomography algorithms that generate a 3D data set. This 3D speciation-sensitive imaging approach has been used, for example, by Obst et al. (2009b) to characterize the first steps of calcium carbonate nucleation at the surface of cyanobacterial cells.

8.4.3 STUDY OF THE REDOX STATE OF METALS AND METALLOIDS IN NATURAL AND EXPERIMENTAL BIOMINERALIZING SYSTEMS

Bacteria can impact strongly the mobility of metal and metalloid pollutants by changing their redox state and/or by inducing the formation of biomineralized solid phases that capture pollutants. These processes are of particularly critical environmental concern, and they offer an interesting bioremediation strategy. Biomineralization occurs naturally in very diverse environments but may be artificially stimulated for bioremediation purposes as proposed in several past studies (e.g., Beazley et al., 2007; Yabusaki et al., 2007). However, in order to optimize this approach, we need to better understand mineral nucleation and growth processes in the presence of microorganisms. STXM has proved to be useful for the study of this type of biomineralization: the sample preparation for bacteria cultures is relatively simple even when redox state preservation is necessary; the minerals that are formed are usually mixtures of small, disordered, and often poorly crystallized minerals containing metals with heterogeneous redox states.

Benzerara et al. (2008) used STXM to analyze the heterogeneities of arsenic redox states and the speciation of carbon at the submicrometer scale in an acid mine drainage (AMD) in the south of France (Carnoules, Gard). This AMD is strongly enriched in arsenic (80–350 mg L^{-1}) upstream. At 30 m downstream, 20%–60% of the As has been removed by the formation of As-rich (up to 22 wt% As) sediments. It has been shown that microbes have a prominent role in the formation of these sediments (e.g., Benzerara et al., 2011a and references herein). The use of STXM combined with TEM allowed to detail the microbial processes involved in the mediation of Fe- and As-containing mineral phases. Benzerara et al. (2008) first showed that it is possible to map As(III) and As(V) at the submicrometer scale by using As $L_{2,3}$-edges and to correlate this with the distribution of organic C species analyzed at the C K-edge. Then, this study showed that some mineral phases are associated directly with microbial cells showing a variety of biomineralization patterns, forming either extracellularly or within the periplasm. However, most of the mineral phases at Carnoulès were not associated with cells but with pervasive organic carbon interpreted as extracellular polymeric substance (EPS) produced by cells. Finally, abundant biomineralized organic vesicles resembling the outer membrane vesicles produced by Gram-negative bacteria were documented and may represent a significant biomineralization process that has been overlooked so far in AMD systems.

STXM has also been used to analyze samples produced by laboratory experiments. For example, Toner et al. (2005) used the model strain *Pseudomonas putida* MnB1 to quantify Mn oxidation in laboratory cultures over a 48 h period in order to progress on the determination of mechanisms of bacterial Mn oxidation and the characterization of resulting biominerals. Manganese is an environmentally abundant transition element with three redox states: +II, +III, and +IV. Mn(II) is soluble, but upon oxidation to Mn(III) or Mn(IV), manganese precipitates as a series of chemically and structurally diverse oxides. There is an increasing record of evidence that this redox transformation is catalyzed by microorganisms in nature (e.g., Spiro et al., 2010). Toner et al. (2005) performed STXM analyses at the Mn $L_{2,3}$-edges on hydrated samples collected at discrete times and sandwiched between two Si_3N_4 windows, sealed with epoxy. They fit measured XANES spectra by linear combinations of reference spectra to obtain abundances of Mn(II), Mn(III), and Mn(IV). They tested qualitatively the impact of beam damage and showed that the presence of organic molecules significantly enhances x-ray-induced photoreduction. They also suggested that the first-order spectral features at the Mn $L_{2,3}$-edges are dominated by the charge state rather than the structure, that is, errors are relatively low when using reference compounds that are not identical to the sample in structure.

Using similar approaches, several STXM studies were performed on Fe-oxidizing bacteria, including microaerophilic Fe-oxidizing bacteria such as *Gallionella* (Chan et al., 2004, 2009), anaerobic nitrate-reducing and Fe-oxidizing bacteria such as the BoFeN1 strain (Miot et al., 2009a,b, 2011; Pantke et al., 2012), and anoxygenic photosynthesizers (Miot et al., 2009c). Since Fe(III) is highly insoluble at neutral pH, it precipitates spontaneously once it gets oxidized metabolically by bacteria, typically in the form of an oxyhydroxide such as ferrihydrite or goethite. As a consequence, inorganic metal pollutants can be trapped by coprecipitation or adsorption at the surfaces of these Fe biominerals that have received great attention. Miot et al. (2009a,b) and Pantke et al. (2012) studied oxidation of Fe and resulting biominerals by strain BoFeN1 cultured in the laboratory. In this case, they worked on dried samples sandwiched between two Si_3N_4 windows sealed with epoxy under N_2 atmosphere in the glove box. Indeed, the presence of water can upon photooxidation induce artifactual Fe redox transformation during STXM analyses. Miot et al. (2009a,b) used phosphate-rich culture media and observed the formation of Fe(II) and/or Fe(III) phosphates. Alternatively, Pantke et al. (2012) used phosphate-poor culture media and observed the formation of mixed-valence Fe hydroxides and oxyhydroxides. It was shown that beam damage induces minimal redox changes (Figure 8.7; Miot et al., 2009a). Moreover, relative redox ratios can be retrieved by fitting the XANES spectra at the Fe $L_{2,3}$-edges by linear combinations of reference spectra (as shown by Dynes et al., 2006b as well). Miot et al. (2009b) showed that some parts of the Fe phases precipitate within the periplasm of BoFeN1 cells, that is, where Fe oxidation is catalyzed (Figure 8.11). They evidenced also some Fe oxidation occurring outside the cells and suggested that it might be mediated by nitrites produced by reduction of nitrates at the cell surface. By following the oxidation of Fe in a time course series both on the cellular and extracellular precipitates, they showed that oxidation of Fe is slower extracellularly than in the cell wall of BoFeN1 cells. This kind of information was impossible to achieve by bulk analyses. Other observations were uniquely provided by STXM: the existence of Fe redox state gradients forming around some cells along organic polymers of lipopolysaccharidic composition (Miot et al., 2009c) and correlations between Fe and organic C contents interpreted as the trapping of protein globules within Fe precipitates (Figure 8.11). The latter feature was confirmed by cryo-TEM (Miot et al., 2011) performed on some bacteria from the same cultures. Cryo-TEM is similar to STXM in the sense that it preserves the samples in their frozen-hydrated state, without any chemical pretreatment. On the one hand, cryo-TEM

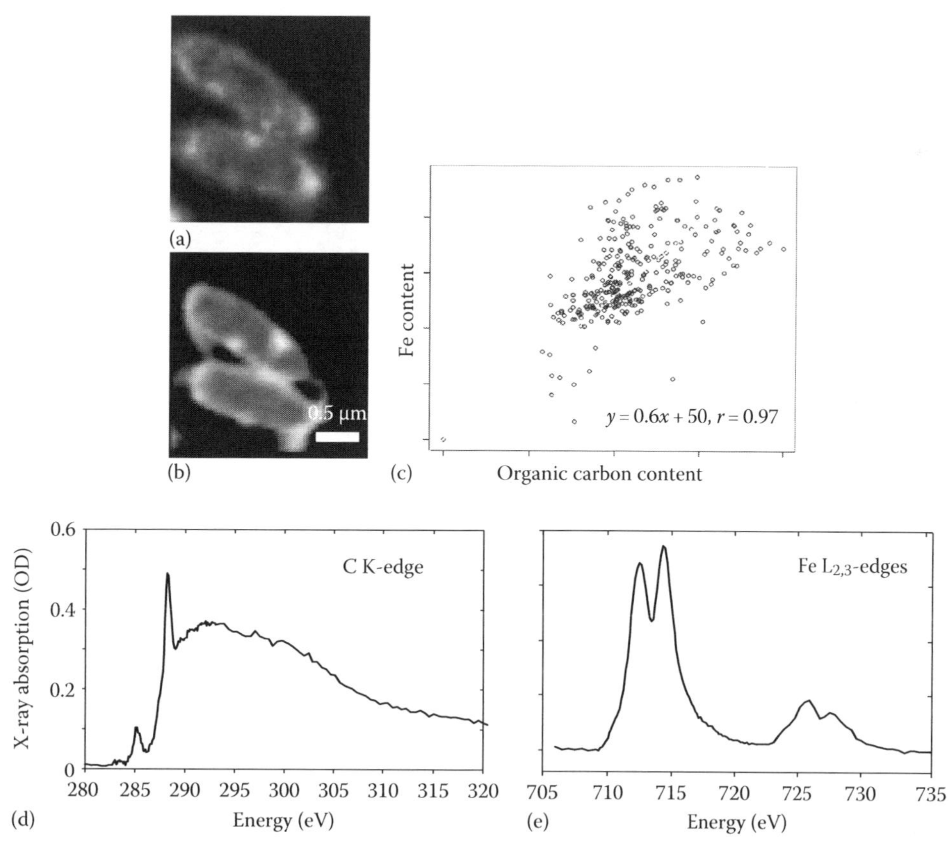

Figure 8.11 (a) and (b) are maps of organic carbon (here, difference between OD at 288.2 eV and OD at 280 eV) and Fe, respectively, obtained on two mineralized BoFeN1 cells. (c) Plot of organic carbon content and Fe content for each pixel showing a correlation between these two parameters. (d) XANES spectrum at the C K-edge of a mineralized BoFeN1 cell. (e) XANES spectrum at the Fe $L_{2,3}$-edges of a mineralized BoFeN1 cell after 6 h of culture. It indicates some mixture of Fe(II) and Fe(III) as shown in Miot et al. (2009b).

moreover offers high spatial resolution and a good contrast for cellular ultrastructures. On the other hand, STXM provides a high amount of chemical information. Observation of exactly the same sample by cryo-TEM and STXM is not yet possible, but implementation of cryo-conditions on STXM is in progress. Yet Miot et al. (2011) provide a direct illustration of how information provided by STXM and cryo-TEM can be combined to identify protein globules trapped in biominerals.

A few other studies have used STXM to analyze the speciation of transition elements such as Christl et al. (2012) on aerobic reduction of Cr(VI) to Cr(III) by the model strain *Pseudomonas corrugata* 28. Recently, Hitchcock et al. (2012) showed that the use of x-ray fluorescence coupled with STXM allows achieving much lower detection limits, opening broad perspectives for the study of mechanisms of trace element trapping by biominerals. The capabilities of XRF microscopy has already been shown on tender-hard x-ray beamlines (Kemner, 2008). The additional capability to correlate measured signals with the distribution and speciation of light elements such as carbon is crucial in biomineralization.

8.4.4 USE OF X-RAY BEAM POLARIZATION IN STXM BIOMINERALIZATION STUDIES

Several STXM beamlines (Table 8.1) are equipped with an EPU, which provides linearly and circularly polarized x-ray beam with variable orientations. The different polarization can be used to map heterogeneities in the crystallographic or magnetic properties of biomaterials at the spatial scale achieved by STXM, that is, <100 nm.

8.4.4.1 Use of x-ray magnetic circular dichroism for STXM study of magnetite biomineralization

Magnetite is a mixed-valence spinel with chemical formula Fe_3O_4 that has been an emblematic mineral for the search of traces of life, particularly because of the finding of magnetites of specific sizes and habits in the Martian meteorite ALH 84001, which have been first interpreted as ancient biominerals from Mars (McKay et al., 1996). Numerous papers have detailed the various abiotic (Vayssieres et al., 1998) and biotic (Komeili, 2012) mechanisms of magnetite formation and the chemistry and mineralogy of the end products (Devouard et al., 1998). Biomineralization of magnetites has also attracted the attention of people interested in sediment magnetization (Paasche et al., 2004) or condensed matter scientists with the aim of producing nanoparticles with particular magnetic properties that can be used in cancer therapy (e.g., Alphandery et al., 2011). Magnetotactic bacteria biomineralize intracellular chains of magnetites (in some cases, greigite, Fe_3S_4) within vesicles called magnetosomes. The chain is fixed within the cell so that the bacterium can passively align in and navigate along geomagnetic fields (Blakemore, 1975). Recently, there has been a great effort in characterizing systematically the control by the Fe(III)/Fe(II) behavior of magnetic properties in bacterially produced magnetites. For that purpose, the measurement of the Fe $L_{2,3}$ x-ray magnetic circular dichroism (XMCD) signal by STXM in combination with analyses at the C K-edge has proved to be an efficient route (e.g., Lam et al., 2010; Coker et al., 2012; Kalirai et al., 2012). XANES spectra are recorded at the Fe $L_{2,3}$-edges with right and left circularly polarized x-rays without an additional external magnetic field. The difference spectrum between these two spectra is the XMCD signal that provides information on the magnitude of the magnetic moment and the site occupancies of Fe ions in the three distinct sites of Fe in magnetite (octahedral Fe^{2+}, tetrahedral Fe^{3+}, and octahedral Fe^{3+} sites, Brice-Profeta et al., 2005). As a result, it is possible to obtain measurements of the magnetic properties in a spatially resolved manner at the ~50 nm scale. Lam et al. (2010) have performed the first measurement of the individual magnetic moment in magnetite nanocrystals produced by the marine magnetotactic bacterial strain MV-1. The method was then improved by Kalirai et al. (2012). They worked on air-dried cells. For higher statistical adequacy, Lam et al. (2010) used longer dwell times than in typical imaging/spectroscopy analyses (5–8 ms/pixel/energy point vs. 0.8–1.2 ms/pixel/energy point in Miot et al., 2009b). As a result, they could show that the magnetic moments of individual magnetites within a chain are all oriented in the same direction and that there is an excess Fe(II) in MV-1 magnetosomes. The latter feature was however reinterpreted secondarily as a beam damage artifact (Kalirai et al., 2012). Coker et al. (2012) studied extracellular magnetites formed by the Fe-reducing *Shewanella oneidensis* MR-1 bacterial strain. Interestingly, they observed heterogeneities in the amount of Fe(II) contained in magnetites depending on whether they were associated directly with the cells of extracellular polymeric matrix. The first ones showed an excess of Fe(II) compared to stoichiometric magnetites, while the second ones were relatively undersaturated with Fe(II). They suggested that the role of cell surface contact-mediated electron transfer might explain that observation. Since reduced forms of magnetite may be more effective in environmental remediation, better understanding of the parameters controlling their formation and spatial distribution is crucial.

8.4.4.2 Use of x-ray linear dichroism for STXM study of carbonate biomineralization

XANES is not only sensitive to the chemical composition of the sample that is analyzed, but the near-edge cross sections also depend strongly on the geometric orientation of the chemical bonds relative to the electric field vector. In materials with a strong anisotropy in the orientation of their molecular orbitals, this effect, known as x-ray linear dichroism (XLD), can be used to map the crystallographic orientation of these materials (Figure 8.12). For that purpose, the sample or more conveniently the direction of fully linear polarized x-rays is varied by increments of few degrees, and the absorption by the sample is measured at an appropriate energy. One obtains linear polarization-dependent images. A $\cos^2(\alpha)$ dependence of the absorbance is observed, where α is the angle difference between the polarization vector of the beam and the polarization angle of the absorbance maximum in the sample (Figure 8.12). It was shown previously by x-ray photoemission electron microscopy (XPEEM), which is a more surface-sensitive synchrotron-based method (~3 nm depth), that the absorption of x-rays at 290.3 eV, which corresponds to $1s \rightarrow \pi^*$ electronic transitions in carbonate groups, by an aragonite single crystal depends on the in-plane orientation of the aragonite *c*-axis relatively to the polarization vector of the x-ray beam (Metzler et al., 2007).

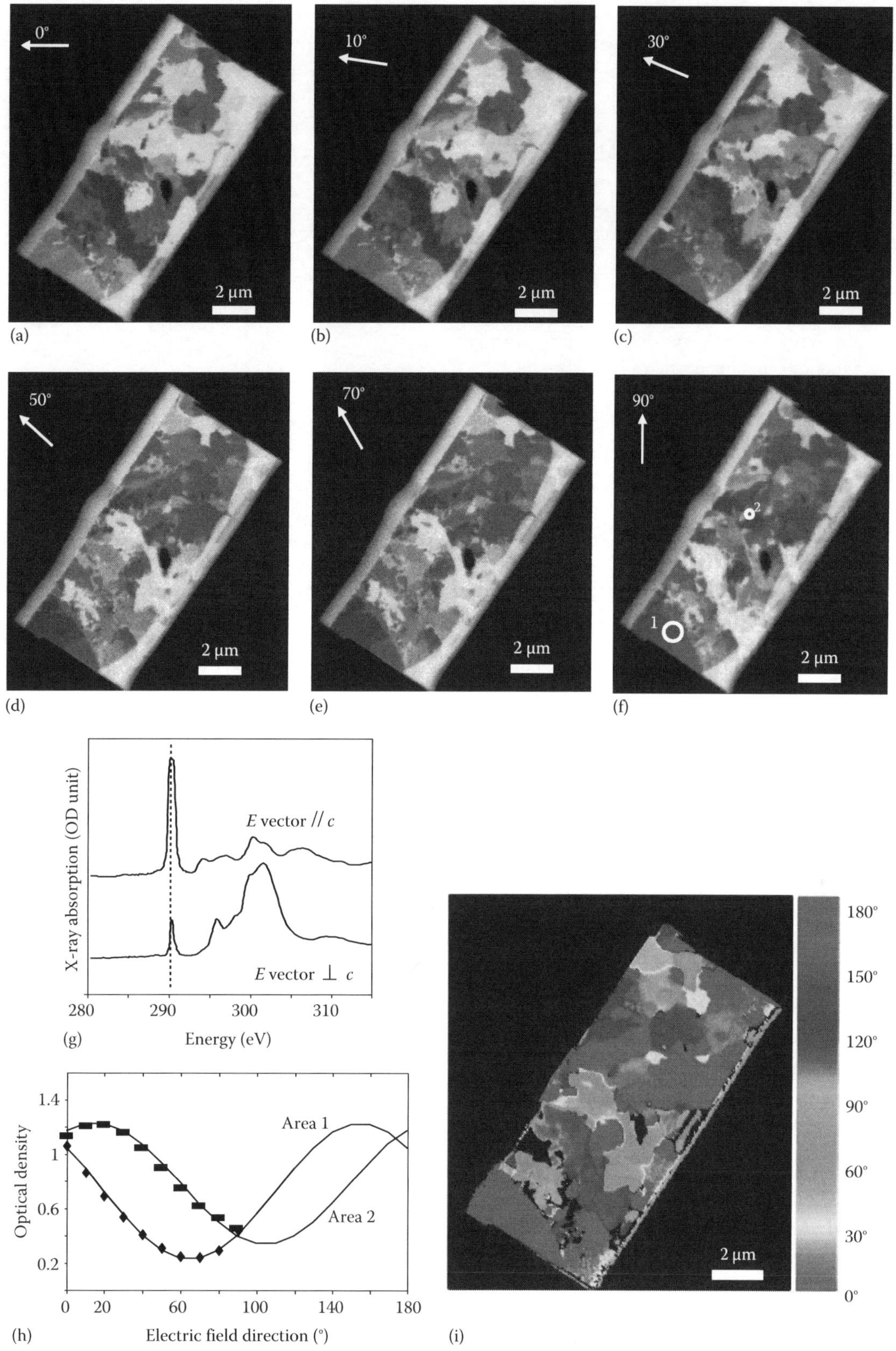

Figure 8.12 (See color insert.) (a) through (f) STXM map of the in-plane projection of the *c*-axis of aragonite in a FIB foil cut in stromatolites from Satonda (Indonesia). STXM images of the FIB foil at 290.3 eV (i.e., 1s → π* electron transition in carbonates) with varying directions of the polarization of the incident beam (0°, 10°, 30°, 50°, 70°, and 90°), indicated by arrows. For each pixel, the variation of OD is due to linear dichroism. (g) XANES spectra measured at the C K-edge on the same aragonite crystal with two different orientations of the polarization vector, parallel or perpendicular to the *c*-axis. In the first case, absorption is very high at 290.3 eV, while absorption is much weaker at the same energy in the second case. (h) These variations were plotted for area 1 and area 2. The curves can be fitted as $A\cos^2(\theta - \phi)+B$. The maximum of absorption gives the direction of the *c*-axis of aragonite (ϕ). Maximum absorption for areas 1 and 2 is at 155° and 15°, respectively. (i) The same procedure can be applied to each pixel in image (a). This provides a map of the orientation of the in-plane projection of the *c*-axis of aragonite as shown in (i).

XLD measurements in the STXM were applied to microbial aragonite by Benzerara et al. (2010) and corals by Benzerara et al. (2011b). The two studies were performed on FIB foils that were analyzed in parallel by TEM. Moreover, it was shown that amorphization of aragonite by FIB milling was very limited. Owing to the combined use of FIB milling, TEM, and STXM, they could find evidence for the existence of mesocrystals in the centers of calcification of a *Porites* sp. coral sample and in some lake stromatolites. Such mesocrystals in biomineralizing living organisms have been increasingly in evidence recently. Polarization-dependent STXM imaging allows mapping of the orientation of the in-plane projection of the *c*-axis of aragonite on the FIB foil at 40 nm resolution. Very few techniques actually provide crystallographic information at such a high spatial resolution except electron backscatter diffraction (EBSD) and TEM. For example, automated procedures used in mapping nanocrystalline phase orientations by TEM have been developed recently (Rauch et al., 2010) and provide a powerful way of characterizing crystallographic orientations with a spatial resolution of about 15 nm. Benzerara et al. (2011b) illustrated how the two techniques, STXM and TEM, can still be very complementary. Computation of orientation maps in TEM remains yet challenging for compounds with low symmetry that is usually the case for biominerals, while this is not a problem for XLD. Moreover, STXM polarization-dependent imaging contrast is much less sensitive to variations in the orientation of aragonite than electron diffraction contrast, in particular out-of-plane variations. As a result, STXM provides an interesting average view of the crystallographic orientations of aragonite in the whole FIB foil that can be used as a first step in the analysis of a biomineralized structure, while TEM provides a very precise (sometimes too precise) picture that can be used to refine the first picture offered by STXM. For example, in Benzerara et al. (2011b), the particular crystallographic arrangement of aragonite nanocrystals could not be evidenced by a first TEM inspection and only became visible after STXM analyses. Further high-resolution TEM analyses could then be performed. It can also be noticed that some organic polymers can also present a dichroic signal. This is the case of proteic β-sheets in *Bombyx mori* cocoon silk fibers, the orientations of which were mapped following the same polarization-dependent STXM imaging approach (Hernández Cruz et al., 2006).

8.5 CONCLUSIONS

In conclusion, STXM provides chemical speciation-sensitive images at a spatial resolution down to ~15 nm coupled with XANES over a relatively extended range of energies, that is, 100–2000 eV. It allows studying the speciation of light elements composing organic molecules such as C, N, O, S, and P as well as heavier elements that are contained by biominerals (e.g., Ca, Fe, Mn, and As). Samples can be studied under wet, dry, and/or anoxic conditions, which is a particularly interesting capability for biomineralogists. Bacteria are perfectly suited to STXM analyses regarding their size. Otherwise, the need for soft x-ray transparent samples requires either to slice the samples by FIB milling or ultramicrotomy (with potential resulting artifacts) or to grind it. Recent studies have shown the versatility of the signals that can be measured including 2D images and XANES spectra but also tomographs, x-ray fluorescence spectra, circular or linear polarization-dependent STXM images, and spectra. Several other developments are in progress and should bring valuable tools for biomineralogists. These include the development of cryo-STXM or the achievement of higher spatial resolution, for example.

REFERENCES

Ade, H., Kilcoyne, A.L.D., Tyliszczak, T. et al. 2003. Scanning transmission X-ray microscopy at a bending magnet beamline at the Advanced Light Source. *Journal De Physique IV* 104: 3–8.

Alphandery, E., Carvallo, C., Menguy, N., and Chebbi, I. 2011. Chains of cobalt doped magnetosomes extracted from AMB-1 magnetotactic bacteria for application in alternative magnetic field cancer therapy. *Journal of Physical Chemistry C* 115: 11920–11924.

Andre, R., Tahir, M.N., Natalio, F., and Tremel, W. 2012. Bioinspired synthesis of multifunctional inorganic and bio-organic hybrid materials. *FEBS Journal* 279: 1737–1749.

Bassim, N.D., De Gregorio, B.T., Kilcoyne, A.L.D. et al. 2012. Minimizing damage during FIB sample preparation of soft materials. *Journal of Microscopy* 245: 288–301.

Beazley, M.J., Martinez, R.J., Sobecky, P.A., Webb, S.M., and Taillefert, M. 2007. Uranium biomineralization as a result of bacterial phosphatase activity: Insights from bacterial isolates from a contaminated subsurface. *Environmental Science & Technology* 41: 5701–5707.

Beetz, T. and Jacobsen, C. 2002. Soft X-ray radiation-damage studies in PMMA using a cryo-STXM. *Journal of Synchrotron Radiation* 10: 280–283.

Behyan, S., Haines, B., Karanukaran, C. et al. 2011. Surface detection in a STXM microscope. *10th International Conference on X-Ray Microscopy* 1365: 184–187.

Benzerara, K., Meibom, A., Gautier, Q. et al. 2010. Nanotextures of aragonite in stromatolites from the quasi-marine Satonda crater lake, Indonesia. *Geological Society, London, Special Publications* 336: 211–224.

Benzerara, K. and Menguy, N. 2009. Looking for traces of life in minerals. *Comptes Rendus Palevol* 8: 617–628.

Benzerara, K., Menguy, N., Guyot, F. et al. 2004b. Biologically controlled precipitation of calcium phosphate by *Ramlibacter tataouinensis*. *Earth and Planetary Science Letters* 228: 439–449.

Benzerara, K., Menguy, N., Obst, M. et al. 2011b. Study of the crystallographic architecture of corals at the nanoscale by scanning transmission X-ray microscopy and transmission electron microscopy. *Ultramicroscopy* 111: 1268–1275.

Benzerara, K., Miller, V.M., Barell, G. et al. 2006. Search for microbial signatures within human and microbial calcifications using soft X-ray spectromicroscopy. *Journal of Investigative Medicine* 54: 367–379.

Benzerara, K. and Miot, J. 2010. Biomineralisation mechanisms. In *Origins and Evolution of Life: An Astrobiological Perspective*, eds. M. Gargaud, P. Lopez-Garcia, and H. Martin, pp. 450–468. Cambridge, U.K.: Cambridge University Press.

Benzerara, K., Miot, J., Morin, G. et al. 2011a. Significance, mechanisms and environmental implications of microbial biomineralization. *Comptes Rendus Geoscience* 343: 160–167.

Benzerara, K., Morin, G., Yoon, T.H. et al. 2008. Nanoscale study of As biomineralization in an acid mine drainage system. *Geochimica et Cosmochimica Acta* 72: 3949–3963.

Benzerara, K., Yoon, T.H., Tyliszczak, T. et al. 2004a. Scanning transmission X-ray microscopy study of microbial calcification. *Geobiology* 2: 249–259.

Bernard, S., Benzerara, K., Beyssac, O. et al. 2007. Exceptional preservation of fossil plant spores in high-pressure metamorphic rocks. *Earth and Planetary Science Letters* 262: 257–272.

Bernard, S., Benzerara, K., Beyssac, O. et al. 2009. Ultrastructural and chemical study of modern and fossil sporoderms by scanning transmission X-ray microscopy (STXM). *Review of Palaeobotany & Palynology* 156: 248–261.

Bernard, S., Benzerara, K., Beyssac, O., and Brown, G.E. 2010. Multiscale characterization of pyritized plant tissues in blueschist facies metamorphic rocks. *Geochimica et Cosmochimica Acta* 74: 5054–5068.

Bertilson, M., von Hofsten, O., Vogt, U., Holmberg, A., and Hertz, H.M. 2009. High-resolution computed tomography with a compact soft x-ray microscope. *Optics Express* 17: 11057–11065.

Blakemore, R. 1975. Magnetotactic bacteria. *Science* 190: 377–379.

Bluhm, H., Andersson, K., Araki, T. et al. 2006. Soft X-ray microscopy and spectroscopy at the molecular environmental science beamline at the advanced light source. *Journal of Electron Spectroscopy and Related Phenomena* 150: 86–104.

Borch, T., Kretzschmar, R., Kappler, A. et al. 2010. Biogeochemical redox processes and their impact on contaminant dynamics. *Environmental Science & Technology* 44: 15–23.

Braun, A., Huggins, F.E., Shah, N. et al. 2005. Advantages of soft X-ray absorption over TEM-EELS for solid carbon studies—A comparative study on diesel soot with EELS and NEXAFS. *Carbon* 43: 117–124.

Brice-Profeta, S., Arrio, M.A., Tronc, E. et al. 2005. XMCD investigation of spin disorder in gamma-Fe_2O_3 nanoparticles at the Fe L-2,L-3 edges. *Physica Scripta* T115: 626–628.

Buckley, C.J., Bellamy, S.J., Zhang, X., Dermody, G., and Hulbert, S. 1995. The NEXAFS of biological calcium phosphates. *Review of Scientific Instruments* 66: 1322–1324.

Carlut, J., Benzerara, K., Horen, H. et al. 2010. Microscopy study of biologically mediated alteration of natural mid-oceanic ridge basalts and magnetic implications. *Journal of Geophysical Research-Biogeosciences* 115: G00G11.

Chan, C.S., De Stasio, G., Welch, S.A. et al. 2004. Microbial polysaccharides template assembly of nanocrystal fibers. *Science* 303: 1656–1658.

Chan, C.S., Fakra, S.C., Edwards, D.C., Emerson, D., and Banfield, J.F. 2009. Iron oxyhydroxide mineralization on microbial extracellular polysaccharides. *Geochimica et Cosmochimica Acta* 73: 3807–3818.

Chan, C.S., Fakra, S.C., Emerson, D., Fleming, E.J., and Edwards, K.J. 2011. Lithotrophic iron-oxidizing bacteria produce organic stalks to control mineral growth: Implications for biosignature formation. *Isme Journal* 5: 717–727.

Chao, W., Fischer, P., Tyliszczak, T. et al. 2012. Real space soft x-ray imaging at 10 nm spatial resolution. *Optics Express* 20.

Chen, C.T., Ma, Y., and Sette, F. 1989. K-shell photoabsorption of the N_2 molecule. *Physical Review A* 40: 6737–6740.

Christl, I., Imseng, M., Tatti, E. et al. 2012. Aerobic reduction of chromium(VI) by *Pseudomonas corrugata* 28: Influence of metabolism and fate of reduced chromium. *Geomicrobiology Journal* 29: 173–185.

Cody, G.D., Gupta, N.S., Briggs, D.E.G. et al. 2011. Molecular signature of chitin-protein complex in Paleozoic arthropods. *Geology* 39: 255–258.

Coffey, T., Urquhart, S.G., and Ade, H. 2002. Characterization of the effects of soft X-ray irradiation on polymers. *Journal of Electron Spectroscopy and Related Phenomena* 122: 65–78.

Coker, V.S., Byrne, J.M., Telling, N.D. et al. 2012. Characterisation of the dissimilatory reduction of Fe(III)-oxyhydroxide at the microbe–Mineral interface: The application of STXM-XMCD. *Geobiology* 10: 347–354.

Cosmidis, J., Benzerara, K., Gheerbrant, E., Estève, I., Bouya, B., and Amaghaz, M. 2013. Nanometer-scale characterization of exceptionally preserved bacterial fossils in Paleocene phosphorites from Ouled Abdoun (Morocco). *Geobiology* 11: 139–153.

Couradeau, E., Benzerara, K., Gerard, E. et al. 2012. An early-branching microbialite cyanobacterium forms intracellular carbonates. *Science* 336: 459–462.

De Gregorio, B.T., Sharp, T.G., Flynn, G.J., Wirick, S., and Hervig, R.L. 2009. Biogenic origin for Earth's oldest putative microfossils. *Geology* 37: 631–634.

Devouard, B., Posfai, M., Hua, X. et al. 1998. Magnetite from magnetotactic bacteria: Size distributions and twinning. *American Mineralogist* 83: 1387–1398.

Dong, H.L. and Lu, A.H. 2012. Mineral-microbe interactions and implications for remediation. *Elements* 8: 95–100.

Dynes, J.J., Lawrence, J.R., Korber, D.R. et al. 2006a. Quantitative mapping of chlorhexidine in natural river biofilms. *Science of the Total Environment* 369: 369–383.

Dynes, J.J., Tyliszczak, T., Araki, T. et al. 2006b. Speciation and quantitative mapping of metal species in microbial biofilms using scanning transmission X-ray microscopy. *Environmental Science and Technology* 40: 1556–1565.

Flechsig, U., Quitmann, C., Raabe, J. et al. 2007. The PolLux microspectroscopy beamline at the Swiss light source. *Synchrotron Radiation Instrumentation* 879: 505–508.

Fliegel, D., Wirth, R., Simonetti, A. et al. 2011. Tubular textures in pillow lavas from a Caledonian west Norwegian ophiolite: A combined TEM, LA-ICP-MS, and STXM study. *Geochemistry Geophysics Geosystems* 12, Q02010.

Frommherz, U., Raabe, J., Watts, B., Stefani, R., and Ellenberger, U. 2010. Higher order suppressor (HOS) for the PolLux microspectroscope beamline at the Swiss light source SLS. *The 10th International Conference on Synchrotron Radiation Instrumentation* 1234: 429–432.

Gadd, G.M. 2010. Metals, minerals and microbes: Geomicrobiology and bioremediation. *Microbiology-Sgm* 156: 609–643.

Galvez, M.E., Beyssac, O., Benzerara, K. et al. 2012. Morphological preservation of carbonaceous plant fossils in blueschist metamorphic rocks from New Zealand. *Geobiology* 10: 118–129.

Golden, D.C., Ming, D.W., Morris, R.V. et al. 2004. Evidence for exclusively inorganic formation of magnetite in Martian meteorite ALH84001. *American Mineralogist* 89: 681–695.

Gong, Y.U.T., Killian, C.E., Olson, I.C. et al. 2012. Phase transitions in biogenic amorphous calcium carbonate. *Proceedings of the National Academy of Sciences of the United States of America* 109: 6088–6093.

Hanhan, S., Smith, A.M., Obst, M., and Hitchcock, A.P. 2009. Optimization of analysis of soft X-ray spectromicroscopy at the Ca 2p edge. *Journal of Electron Spectroscopy and Related Phenomena* 173: 44–49.

Hernandez Cruz, D., Rousseau, M.-E., West, M.M., Pezolet, M., and Hitchcock, A.P. 2006. Quantitative mapping of the orientation of fibroin beta-sheets in *B. mori* cocoon fibers by scanning transmission X-ray microscopy. *Biomacromolecules* 7: 836–843.

Hitchcock, A. 2013. aXis 2000—Analysis of x-ray images and spectra. Last modified September 3, 2013. http://unicorn.mcmaster.ca/aXis2000.html.

Hitchcock, A.P., Dynes, J.J., Johansson, G., Wang, J., and Botton, G. 2008. Comparison of NEXAFS microscopy and TEM-EELS for studies of soft matter. *Micron* 39: 311–319.

Hitchcock, A.P., Morin, C., Heng, Y.M., Cornelius, R.M., and Brash, J.L. 2002. Towards practical soft X-ray spectromicroscopy of biomaterials. *Journal of Biomaterials Science-Polymer Edition* 13: 919–937.

Hitchcock, A.P., Obst, M., Wang, J., Lu, Y.S., and Tyliszczak, T. 2012. Advances in the detection of As in environmental samples using low energy X-ray fluorescence in a scanning transmission X-ray microscope: Arsenic immobilization by an Fe(II)-oxidizing freshwater bacteria. *Environmental Science & Technology* 46: 2821–2829.

Hitchcock, A.P., Tyliszcak, T., Obst, M., Swerhone, G.D.W., and Lawrence, J.R. 2010. Improving sensitivity in soft X-ray STXM using low energy X-ray fluorescence. *Microscopy Microanalysis* 16: 924.

Hitchcock, A.P., Wang, J., and Obst, M. 2009. 3D chemical imaging with STXM spectro-tomography. *Microscopy and Microanalysis* 16: 850–851.

Hornberger, B., Feser, M., and Jacobsen, C. 2007. Quantitative amplitude and phase contrast imaging in a scanning transmission X-ray microscope. *Ultramicroscopy* 107: 644–655.

Howells, M., Jacobsen, C., and Warwick, T. 2007. Principles and applications of zone plate X-ray microscopes. In *Science of Microscopy*, eds. P.W. Hawkes and J.C.H. Spence, pp. 835–926. New York: Springer.

Howells, M.R., Beetz, T., Chapman, H.N. et al. 2009. An assessment of the resolution limitation due to radiation-damage in X-ray diffraction microscopy. *Journal of Electron Spectroscopy and Related Phenomena* 170: 4–12.

Hub, C., Wenzel, S., Raabe, J., Ade, H., and Fink, R.H. 2010. Surface sensitivity in scanning transmission x-ray microspectroscopy using secondary electron detection. *Review of Scientific Instruments* 81: 033704.

Hunter, R.C., Hitchcock, A.P., Dynes, J.J., Obst, M., and Beveridge, T.J. 2008. Mapping the speciation of iron in *Pseudomonas aeruginosa* biofilms using scanning transmission X-ray microscopy. *Environmental Science & Technology* 42: 8766–8772.

Jacobsen, C., Wirick, S., Flynn, G., and Zimba, C. 2000. Soft X-ray spectroscopy from image sequences with sub-100 nm spatial resolution. *Journal of Microscopy-Oxford* 197: 173–184.

Jiang, S., Chen, L., Xu, C.Y. et al. 2003. The scanning transmission X-ray microscope at NSRL. *Journal De Physique IV* 104: 81–84.

Jimenez-Lopez, C., Rodriguez-Navarro, C., Rodriguez-Navarro, A. et al. 2012. Signatures in magnetites formed by (Ca,Mg,Fe)CO_3 thermal decomposition: Terrestrial and extraterrestrial implications. *Geochimica et Cosmochimica Acta* 87: 69–80.

Johansson, G.A., Tyliszczak, T., Mitchell, G.E., Keefe, M.H., and Hitchcock, A.P. 2007. Three-dimensional chemical mapping by scanning transmission X-ray spectromicroscopy. *Journal of Synchrotron Radiation* 14: 395–402.

Jorand, F., Zegeye, A., Ghanbaja, J., and Abdelmoula, M. 2011. The formation of green rust induced by tropical river biofilm components. *Science of the Total Environment* 409: 2586–2596.

Kalirai, S.S., Lam, K.P., Bazylinski, D.A., Lins, U., and Hitchcock, A.P. 2012. Examining the chemistry and magnetism of magnetotactic bacterium *Candidatus* Magnetovibrio blakemorei strain MV-1 using scanning transmission X-ray microscopy. *Chemical Geology* 300: 14–23.

Kaznatcheev, K.V., Karunakaran, C., Lanke, U.D. et al. 2007. Soft X-ray spectromicroscopy beamline at the CLS: Commissioning results. *Nuclear Instruments and Methods in Physics Research Section A-Accelerators Spectrometers Detectors and Associated Equipment* 582: 96–99.

Kemner, K.M. 2008. Hard X-ray micro(spectro)scopy: A powerful tool for the geomicrobiologists. *Geobiology* 6: 270–277.

Kilcoyne, A.L.D., Tyliszczak, T., Steele, W.F. et al. 2003. Interferometer-controlled scanning transmission X-ray microscopes at the advanced light source. *Journal of Synchrotron Radiation* 10: 125–136.

Komeili, A. 2012. Molecular mechanisms of compartmentalization and biomineralization in magnetotactic bacteria. *FEMS Microbiology Reviews* 36: 232–255.

Lam, K.P., Hitchcock, A.P., Obst, M. et al. 2010. Characterizing magnetism of individual magnetosomes by X-ray magnetic circular dichroism in a scanning transmission X-ray microscope. *Chemical Geology* 270: 110–116.

Lamboy, M. 1994. Nanostructure and genesis of phosphorites from Odp Leg-112, the Peru margin. *Marine Geology* 118: 5–22.

Lawrence, J.R., Swerhone, G.D.W., Leppard, G.G. et al. 2003. Scanning transmission X-ray, laser scanning, and transmission electron microscopy mapping of the exopolymeric matrix of microbial biofilms. *Applied and Environmental Microbiology* 69: 5543–5554.

Lee, M.K. and Shin, H.J. 2001. Soft x-ray spectromicroscope at the Pohang light source. *Review of Scientific Instruments* 72: 2605–2609.

Lepot, K., Benzerara, K., Brown, Jr., G.E., and Philippot, P. 2008. Microbially influenced formation of 2,724-million-year-old stromatolites. *Nature Geoscience* 1: 118–121.

Lepot, K., Benzerara, K., Rividi, N. et al. 2009. Organic matter heterogeneities in 2.72 Ga stromatolites: Alteration versus preservation by sulfur incorporation. *Geochimica et Cosmochimica Acta* 73: 6579–6599.

Lerotic, M., Jacobsen, C., Gillow, J.B. et al. 2005. Cluster analysis in soft X-ray spectromicroscopy: Finding the patterns in complex specimens. *Journal of Electron Spectroscopy and Related Phenomena* 144: 1137–1143.

Lerotic, M., Jacobsen, C., Schafer, T., and Vogt, S. 2004. Cluster analysis of soft X-ray spectromicroscopy data. *Ultramicroscopy* 100: 35–57.

Lundberg, J. and McFarlane, D.A. 2011. Subaerial freshwater phosphatic stromatolites in Deer Cave, Sarawak—A unique geobiological cave formation. *Geomorphology* 128: 57–72.

Ma, Y., Chen, C.T., Meigs, G., Randall, K., and Sette, F. 1991. High-resolution K-shell photoabsorption measurements of simple molecules. *Physical Review A* 44: 1848–1858.

MaClean, L.C.W., Tyliszczak, T., Gilbert, P. et al. 2008. A high-resolution chemical and structural study of framboidal pyrite formed within a low-temperature bacterial biofilm. *Geobiology* 6: 471–480.

Mann, S. 1993. Molecular tectonics in biomineralization and biomimetic materials chemistry. *Nature* 365: 499–505.

McKay, D.S., Gibson, E.K., ThomasKeprta, K.L. et al. 1996. Search for past life on Mars: Possible relic biogenic activity in Martian meteorite ALH84001. *Science* 273: 924–930.

Metzler, R.A., Abrecht, M., Olabisi, R.M. et al. 2007. Architecture of columnar nacre, and implications for its formation mechanism. *Physical Review Letters* 98.

Miot, J., Benzerara, K., Morin, G. et al. 2009a. Transformation of vivianite by anaerobic nitrate-reducing iron-oxidizing bacteria. *Geobiology* 7: 373–384.

Miot, J., Benzerara, K., Morin, G. et al. 2009b. Iron biomineralization by anaerobic neutrophilic iron-oxidizing bacteria. *Geochimica et Cosmochimica Acta* 73: 696–711.

Miot, J., Benzerara, K., Obst, M. et al. 2009c. Extracellular iron biomineralization by photoautotrophic iron-oxidizing bacteria. *Applied and Environmental Microbiology* 75: 5586–5591.

Miot, J., Maclellan, K., Benzerara, K., and Boisset, N. 2011. Preservation of protein globules and peptidoglycan in the mineralized cell wall of nitrate-reducing, iron(II)-oxidizing bacteria: A cryo-electron microscopy study. *Geobiology* 9: 459–470.

Nassif, N. and Livage, J. 2011. From diatoms to silica-based biohybrids. *Chemical Society Reviews* 40: 849–859.

Nolle, D., Weigand, M., Audehm, P. et al. 2012. Note: Unique characterization possibilities in the ultra high vacuum scanning transmission x-ray microscope (UHV-STXM) "MAXYMUS" using a rotatable permanent magnetic field up to 0.22 T. *Review of Scientific Instruments* 83: 046112.

Obst, M., Dynes, J.J., Lawrence, J.R. et al. 2009a. Precipitation of amorphous $CaCO_3$ (aragonite-like) by cyanobacteria: A STXM study of the influence of EPS on the nucleation process. *Geochimica et Cosmochimica Acta* 73: 4180–4198.

Obst, M., Wang, J., and Hitchcock, A.P. 2009b. Soft X-ray spectrotomography study of cyanobacterial biomineral nucleation. *Geobiology* 7: 577–591.

Ohmoto, H., Runnegar, B., Kump, L.R. et al. 2008. Biosignatures in ancient rocks: A summary of discussions at a field workshop on biosignatures in ancient rocks. *Astrobiology* 8: 883–895.

Ona-Nguema, G., Morin, G., Wang, Y.H. et al. 2010. XANES evidence for rapid Arsenic(III) oxidation at magnetite and ferrihydrite surfaces by dissolved O-2 via Fe2+-mediated reactions. *Environmental Science & Technology* 44: 5416–5422.

Paasche, O., Lovlie, R., Dahl, S.O., Bakke, J., and Nesje, A. 2004. Bacterial magnetite in lake sediments: Late glacial to Holocene climate and sedimentary changes in northern Norway. *Earth and Planetary Science Letters* 223: 319–333.

Pantke, C., Obst, M., Benzerara, K. et al. 2012. Green rust formation during Fe(II) oxidation by the nitrate-reducing *Acidovorax* sp. strain BoFeN1. *Environmental Science & Technology* 46: 1439–1446.

Pecher, K., McCubbery, D., Kneedler, E. et al. 2003. Quantitative charge state analysis of manganese biominerals in aqueous suspension using scanning transmission X-ray microscopy (STXM). *Geochimica et Cosmochimica Acta* 67: 1089–1098.

Politi, Y., Metzler, R.A., Abrecht, M. et al. 2008. Transformation mechanism of amorphous calcium carbonate into calcite in the sea urchin larval spicule. *Proceedings of the National Academy of Sciences of the United States of America* 105: 17362–17366.

Raabe, J., Tzvetkov, G., Flechsig, U. et al. 2008. PolLux: A new facility for soft x-ray spectromicroscopy at the Swiss Light Source. *Review of Scientific Instruments* 79: 113704.

Raabe, J., Watts, B., Tzvetkov, G., Fink, R.H., and Quitmann, C. 2009. First differential phase contrast results from PolLux. *J. Phys. Conf. Ser.* 186: 012012.

Rauch, E.F., Portillo, J., Nicolopoulos, S. et al. 2010. Automated nanocrystal orientation and phase mapping in the transmission electron microscope on the basis of precession electron diffraction. *Zeitschrift Fur Kristallographie* 225: 103–109.

Soudry, D. 2000. Microbial phosphate sediments. In *Microbial Sediments*, eds. R.E. Riding and S.M. Awramik, pp. 127–136. Berlin, Germany: Springer.

Shin, H.J. and Lee, M.K. 2001. Scanning soft X-ray spectromicroscopy at the Pohang Light Source: Commissioning results. *Nuclear Instruments and Methods in Physics Research Section A—Accelerators Spectrometers Detectors and Associated Equipment* 467–468: 909–912.

Spiro, T.G., Bargar, J.R., Sposito, G., and Tebo, B.M. 2010. Bacteriogenic manganese oxides. *Accounts of Chemical Research* 43: 2–9.

Stumm, W., Sigg, L., and Sulzberger, B. 1994. General chemistry of aquatic systems. In *Chemical and Biological Regulation of Aquatic Systems*, eds. J. Buffle and R.R. DeVitre. Boca Raton, FL: Taylor & Francis.

Suzuki, T., Hashimoto, H., Matsumoto, N. et al. 2011. Nanometer-scale visualization and structural analysis of the inorganic/organic hybrid structure of *Gallionella ferruginea* twisted stalks. *Applied and Environmental Microbiology* 77: 2877–2881.

Toner, B., Fakra, S., Villalobos, M., Warwick, T., and Sposito, G. 2005. Spatially resolved characterization of biogenic manganese oxide production within a bacterial biofilm. *Applied and Environmental Microbiology* 71: 1300–1310.

Vayssieres, L., Chaneac, C., Tronc, E., and Jolivet, J.P. 1998. Size tailoring of magnetite particles formed by aqueous precipitation: An example of thermodynamic stability of nanometric oxide particles. *Journal of Colloid and Interface Science* 205: 205–212.

Vila-Comamala, J., Jefimovs, K., Raabe, J. et al. 2009. Advanced thin film technology for ultrahigh resolution X-ray microscopy. *Ultramicroscopy* 109: 1360–1364.

Wang, J., Morin, C., Li, L. et al. 2009. Radiation damage in soft X-ray microscopy. *Journal of Electron Spectroscopy and Related Phenomena* 170: 25–36.

Webster, A. and May, E. 2006. Bioremediation of weathered-building stone surfaces. *Trends in Biotechnology* 24: 255–260.

Weiner, S. and Dove, P.M. 2003. An overview of biomineralization processes and the problem of the vital effect. In *Biomineralization*, eds. P.M. Dove, J.J. De Yoreo, and S. Weiner, pp. 1–29. Washington, DC: Mineralogical Society of America.

Yabusaki, S.B., Fang, Y., Long, P.E. et al. 2007. Uranium removal from groundwater via in situ biostimulation: Field-scale modeling of transport and biological processes. *Journal of Contaminant Hydrology* 93: 216–235.

Zanin, Y.N. and Zamirailova, A.G. 2011. The history of the study of bacterial/cyanobacterial forms in phosphorites. *Russian Geology and Geophysics* 52: 1134–1139.

9 Photoemission spectromicroscopy for the biomineralogist

*Pupa U.P.A. Gilbert**

Contents

Photoelectron emission spectromicroscopy (PEEM) is highly informative for analyses of minerals and biominerals. In this review, we provide little information about the instruments or the samples themselves; rather, we focus on providing extensive, representative data obtained on various biominerals, with the hope of inspiring the uninitiated to devise new experiments with their own samples, and start using PEEM at a nearby synchrotron. Here, we present, for the first time, the complete details of how to prepare a biomineral sample, that is, how to make it flat and conductive and therefore compatible with any PEEM at any synchrotron. PEEM combined with x-ray absorption near-edge structure (XANES) spectroscopy is informative because it detects the oxidation state of elements, the amorphous or crystalline structure of biominerals, as well as distinguishing crystalline polymorphs. It is also able to identify different classes of organic molecules and the orientation of selected organic molecules. We briefly discuss polarization-dependent imaging contrast (PIC) mapping, which reveals the orientation of aragonite, calcite, or vaterite (all are $CaCO_3$) nanocrystals and the patterns they form in biominerals. We conclude by providing a list of synchrotrons and PEEMs around the world, where an interested biomineralogist will be most welcome to bring samples and ideas, and receive all the technical help necessary to produce publication-quality results. This review is, therefore, an instruction manual for the curious biomineralogist to start utilizing PEEM.

9.1 INTRODUCTION TO PEEM SPECTROMICROSCOPY

9.1.1 WHAT IS PEEM GOOD FOR?

Photoelectron emission microscopy is sometimes called photoemission electron microscopy or spectromicroscopy, and the acronym used is PEEM. This is a popular method at all synchrotrons because it provides both a magnified image of the sample surface and chemical analysis (Tonner and Harp 1988; Engel et al. 1991; Tromp and Reuter 1991; Bauer 1994; Anders et al. 1999; Politi et al. 2008). The year 2013 marks the 80th anniversary of the invention of PEEM (Brüche 1933), which has since been dramatically refined and perfected (Bauer 2012). The Tonner group was the first to build a PEEM at a synchrotron (Tonner and Harp 1988), and the author of this chapter did the first PEEM experiments on biological samples (De Stasio, Koranda, et al. 1992; De Stasio, Perfetti, et al. 1992; De Stasio, Dunham, et al. 1993; De Stasio, Hardcastle, et al. 1993) using Tonner's microscope; she designed and built a PEEM (De Stasio et al. 1998), and did the first experiment on bacterial biominerals

* Previously publishing as Gelsomina De Stasio.

(Labrenz et al. 2000; Chan et al. 2004) and eukaryotic biominerals (Metzler et al. 2007). In a nutshell, PEEM provides high-resolution imaging, down to a few tens of nanometers (De Stasio et al. 1999; Frazer et al. 2004). Combined with XANES, also known as near-edge x-ray absorption fine structure (NEXAFS) spectroscopy (Stöhr 2003), X-PEEM or XANES-PEEM detects the elemental composition of the sample, the crystal structure of minerals, the molecular structure of organic molecules, and, using linearly polarized x-rays, it also provides the orientation of birefringent crystals. This chapter will show ample data on biominerals for each one of these PEEM capabilities. Despite these observations, most PEEMs around the world are used to study metals, semiconductors, or magnetic materials, that is, only conductors, not insulators such as minerals or biominerals. Lack of conductivity often discouraged PEEM owners and users from trying experiments on insulators. We have devised a unique but simple coating method that makes any insulating material behave as a good conductor in a PEEM or any other photoemission experiment (De Stasio et al. 2003; Gilbert et al. 2005); hence, the main limit for the biomineralogist or the mineralogist has been removed.

9.1.2 WHAT IS A PEEM?

In a XANES-PEEM experiment, the sample is kept at a constant **high negative voltage** between −15 and −20 kV, and the sample surface is illuminated by x-rays from a synchrotron, and an **electron optics column** is mounted with its axis perpendicular to the sample surface (Tonner and Harp 1988; Engel et al. 1991; Tromp and Reuter 1991; Nettesheim et al. 1993; Bauer 1994). The first element of the electron optics column is an objective lens, with its entrance aperture typically only 2 mm away from the sample surface to collect as many electrons as possible. After the objective lens, the optics column has additional lenses and a variety of optical elements, including apertures, stigmators, and sometimes aberration correctors (Skoczylas et al. 1991; Feng et al. 2002; Schonhense and Spiecker 2002; Koshikawa et al. 2005; Renault et al. 2007; Tromp et al. 2010; Tromp 2011). The objective and all other lenses in the column may be either electrostatic or magnetic. However, this makes very little difference for the user. The salient point is that the electron optics column **produces a magnified image of the sample surface.** This image is an electron image or more accurately a photoelectron image. Because of photoelectric effect, under x-ray illumination, all elements emit electrons (Einstein 1905). PEEM is, therefore, a photon-in-electron-out experiment. It requires a **tunable source of x-rays**, that is, a **synchrotron**, so that the number of photoelectrons emitted by an element in the sample, for example, Ca, increases dramatically as the x-ray photon energy is scanned across the calcium L-edge absorption, from 344 to 356 eV. In a real-time PEEM image observed while scanning across the Ca L-edge, the observer notices the Ca-rich structures in the sample become brighter at the Ca L-edge absorption energies. Figure 9.1 shows Ca spectra from aragonite and calcite, taken from the mollusk shell *Pinctada fucata*. Figure 9.2 shows calcium at the forming part of a mouse incisor tooth.

9.1.3 WHO IS PEEM GOOD FOR?

PEEM is a powerful tool when combined with XANES for any scientist interested in investigating biominerals. Despite

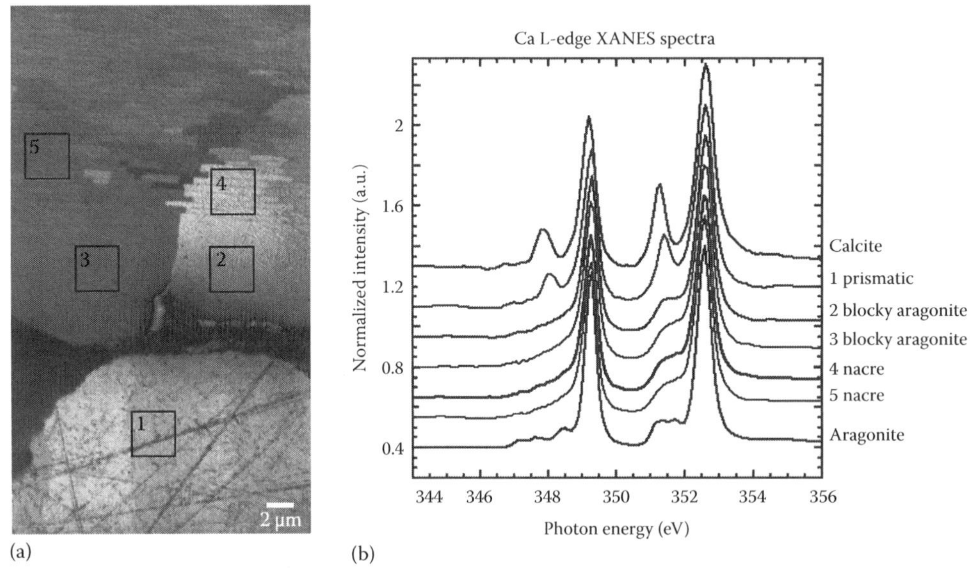

Figure 9.1 Polarization-dependent imaging contrast (PIC)-map and Ca spectra taken at the prismatic-nacre boundary in *Pinctada fucata*. (a) PIC-map showing one prism at the bottom, a dark organic wall of organic molecules separating the calcite from aragonite compartments. Immediately above this organic wall two large and blocky aragonite crystals are observed, and approximately 8 micron above the familiar lamellar nacre appearance begins, and continues for several millimeters (not shown). Notice that the gray levels of the two large aragonite crystals are different, indicating that their c-axes are not parallel. Interestingly, their crystal orientations propagate into the nacre lamellae, as these grow across the organic matrix layers. At a greater distance from the onset of lamellar nacre formation, much greater c-axis co-orientation is observed. (b) Ca spectra extracted from the correspondingly numbered regions in (a). For reference, spectra from geologic aragonite and calcite are also presented at the top and bottom, respectively. The spectrum from the prism (1) is calcite, those from the two large blocky aragonite crystals (2,3) are aragonite, as are the nacre spectra (4,5). (Reprinted with permission from Metzler, R.A., Evans, J.S., Killian, C.E., Zhou, D., Churchill, T.H., Appathurai, N.P., Coppersmith, S.N., and Gilbert, P.U.P.A., Nacre protein fragment templates lamellar aragonite growth, *J. Am. Chem. Soc.*, 132, 6329–6334. Copyright 2010 American Chemical Society.)

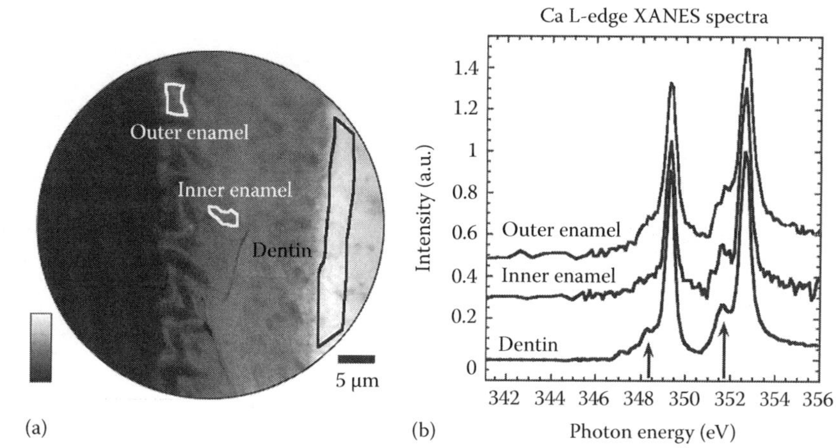

Figure 9.2 (a) Ca distribution map in a region of forming mouse incisor tooth at a distance of approximately 1 mm from the cervical loop. High Ca concentration is represented by lighter gray level. The Ca map was obtained by digital ratio of 349.3 and 344 eV X-PEEM images. (b) XANES spectra extracted from the outer secretory enamel, inner secretory enamel, and mantle dentin regions outlined in (a). Notice the significant differences in the spectral region usually affected by the crystal field (arrows). (Data from Beniash, E. et al., *Proc. R. Soc. Lond. B: Biol. Sci.*, 264(1380), 461, 2009. With permission.)

the considerable amount of hardware and technology that drives the PEEM instrument, most of the challenging technical aspects are handled by the beamline scientists. The beamline scientists are a most welcoming lot, eager to collaborate with researchers with interesting samples and novel ideas, regardless of their familiarity with the technical aspects. Once you locate a synchrotron near you—there will likely be a PEEM spectromicroscope there that could become the source of your future data. The only caveat is that, quite likely, the beamline scientist you contact will tell you that your mineral or biomineral sample cannot be analyzed! This chapter will help you prepare your samples for PEEM and make them perform as well the metallic samples most often analyzed by PEEM.

9.2 HOW TO USE PEEM SPECTROMICROSCOPY TO ANALYZE BIOMINERALS

9.2.1 SAMPLE PREPARATION

The ideal sample for a PEEM experiment is **flat**, **conductive**, and **vacuum compatible**. The sample should be *flat* for three reasons: (1) The depth of field, that is, the range of sample distances that can be imaged in sharp focus by the objective lens, has to be small to enable high magnification, just as in an optical microscope the highest magnification and highest numerical aperture objectives provide a very shallow depth of field (Hecht 2002). (2) If the sample has an uneven surface, for instance, protruding tips, asperities, crevices, or edges, once it is floated at high voltage, arcing occurs, the voltage trips off, and imaging is impossible. There is also undesirable loss of material from the sample and acceleration of debris through the electron optics column, which may render it unusable. (3) A flat sample gives higher-resolution images because the electric field is homogeneous for all electrons emitted from all locations in the imaged sample area; hence, all these electrons are accelerated consistently through the same electric field for the same distance to the objective lens and can therefore be imaged (Griffith et al. 1972).

The sample should be a good **conductor**, so when it photoemits electrons under x-ray illumination, it will not charge. The problem with charging is explained in the following discussion. Furthermore, an insulating sample cannot be floated at a constant and stable voltage and therefore cannot be imaged by PEEM.

The sample should be **vacuum compatible** because of the following reasons: (1) Photoemitted electrons would rapidly recombine with gas molecules. If this combination occurred near the sample, the sample would become undetectable. (2) PEEM experiments are always done in the soft x-ray energy range, that is, approximately between 100 and 2000 eV because this is where the binding energies of the electrons, and therefore their absorption edges, are located for most relevant elements. Electron binding energies of all elements are listed in reference "Electron Binding Energies". For the same reason, there are no transparent materials in this energy range; hence, there cannot be a window between the synchrotron ring in which the electrons are circulating in ultrahigh vacuum; the beamline through which the photons are transmitted, monochromatized, and focused; and the PEEM chamber. Hence, the PEEM experiment is done in ultrahigh vacuum (typically 10^{-9}–10^{-10} Torr). This is a major limit, which reduces the range of experiments that can be done on biologically relevant samples. No hydrated tissue sample can be analyzed. For minerals and biominerals however, this is not as severe a limitation, because at least the dry mineral components can be safely analyzed.

9.2.1.1 Making the sample flat by embedding and polishing

Making a mineral or biomineral sample flat is not difficult. Geologists are familiar with petrological thin sections, in which a rock, a mineral, a fossil, or a fresh biomineral is adhered to a glass microscope slide with a glue or epoxy and then polished to a thin section, typically 30 μm in thickness. Such samples are perfectly fine for PEEM experiments, except that the microscope slide will have to be cut to a slightly smaller size, typically on the order of 10–16 mm squares, with 14–23 mm diagonal. This is done easily with a diamond scribe. Other samples such as sea urchin biominerals, bone, teeth, sponge

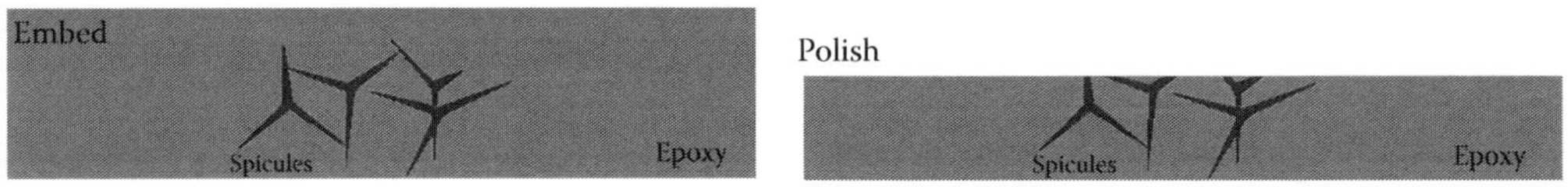

Figure 9.3 Schematic diagram of a 3D sample, sea urchin spicules, embedded into epoxy and then polished. The same approach is valid for any solid mineral or biomineral sample. (Reprinted from *Ultramicroscopy*, 98(1), De Stasio, G., Frazer, B.H., Gilbert, B., Richter, K.L., and Valley, J.W., Compensation of charging in X-PEEM: A successful test on mineral inclusions in 4.4 Ga old zircon, 57–62. Copyright 2003, with permission from Elsevier; *Mol. Geomicrobiol.*, Gilbert, P.U.P.A., Frazer, B.H., and Abrecht, M., The organic-mineral interface in biominerals, p. 1570185. Copyright 2005, with permission from Elsevier.)

or ascidian spicules, or mollusk shells should be embedded in epoxy and polished as shown in Figure 9.3. The best epoxy we have thus far identified on the market is EpoFix (Electron Microscopy Science, Hatfield, PA). Previously, we used EpoThin (Buehler, Lake Bluff, IL) but EpoFix gives better and more consistent results, although it must be replaced every ~10 months even if it was not used, since it degrades with time. One important trick is to mount the sample in its mold with super glue or double-stick tape in the desired orientation, then slowly pour EpoFix on and around it, and immediately place it into a desiccator that can be pumped down with a rotary pump (down to ~10^{-3} torr). Pump down the desiccator three consecutive times, and vent it to air pressure each time. This dramatically reduces the number of air bubbles in the EpoFix and at the interface of EpoFix with the biominerals.

Polishing should also be done with a special precaution for biominerals, as they often contain minor or major proportions of amorphous minerals. Therefore, one must prevent the dissolution of the amorphous phases during wet polishing. For example, if the sample is made of crystalline or amorphous calcium carbonate (ACC), we do not put it in contact with water at any stage of the preparation or cleaning, but polish it and clean it only with $-CO_3$-saturated solutions. Since the solubility is 3–4 orders of magnitude greater for Na_2CO_3 (216 g/L) than for other carbonates, calcite (15 mg/L), aragonite (16 mg/L), and ACC (200 mg/L), it is best to polish the sample using 22 g/L Na_2CO_3 solutions. This is 100× greater $-CO_3$ concentration compared with ACC solubility. If the sample contains ACC, this will not be dissolved during polishing.

We prepare all polishing solutions and suspensions in advance as follows. We use MicroPolish II and MasterPrep (Buehler, Lake Bluff, IL) polishing pastes that have 300 and 50 nm alumina (Al_2O_3) particles, respectively, suspended in water at pH 7. We dialyze these suspensions in dialysis tubing against 22 g/L Na_2CO_3 solution at room temperature and change the Na_2CO_3 solution at least three times in 72 h. With this precaution, polishing is conventionally done, using coarse grinding paper first and then decreasing size alumina grit until the surface is scratch-free and mirror flat.

Once polished, the sample should be sized to fit into the sample holder of the PEEM that will be used. For the Elmitec GmbH (http://www.elmitec.de/Components.php?Bereich=Cartridges) PEEM (Frazer et al. 2004), the maximum diameter of a round sample or diagonal of a square sample is 14 mm (Elmitec GmbH). For PEEM-3 at the advanced light source (ALS), it is 23 mm (http://xraysweb.lbl.gov/peem2/webpage/Home.shtml). A good sample thickness is usually 3 mm or less. To reduce the sample to these sizes, we use the rotating polisher plate, with a coarse silicon carbide grinding paper disc, for example, 300 grit (also from Buehler), and hold the sample with hands covered with gloves or use a low-speed sectioning saw with a diamond blade, for example, the IsoMet low-speed saw (Buehler, Lake Bluff, IL).

9.2.1.2 Making the sample conductive by coating

Making the sample conductive by coating is relatively simple, and every electron microscopy lab is equipped to do this. For a PEEM sample, there is an additional requirement, which not all labs can satisfy: the sample has to be coated with a differential-thickness coating approach, as described in reference De Stasio et al. (2003). This means that at the edges, the coating must be thicker, for example, 30–40 nm, but at the center, on the polished surface that one wishes to analyze, it should be exactly 1 nm. Not all coaters are equipped with a thickness monitor; hence, the requirement that the coating be exactly 1 nm may be prohibitive for certain groups. This is quite simple for us to do, so just send us your samples and we are happy to coat them for you. We use two high-resolution sputter coaters from Cressington Scientific, model 208HR (http://www.cressington.com/product_208hr.html), purchased from Ted Pella (http://www.tedpella.com/cressing_html/crs208hr.htm). The reason for having a 1 nm thick platinum coating on the area to be analyzed by PEEM is that the escape depth of the photoelectrons is small, on the order of a few nanometers, and depends on which element one wishes to analyze. For oxygen, the maximum probing depth (MPD) is 5 nm, for carbon and calcium 3 nm, for iron 8 nm, and for copper 13 nm. These probing depths depend on the kinetic energy of the Auger electrons emitted by each element: the greater the binding energy, the greater the kinetic energy of the primary Auger electron, hence, the greater the depth from which electrons originate when they escape the sample surface, after having undergone multiple inelastic collisions inside the sample. At each of these collisions, the kinetic energy of the electrons gets a little smaller; hence, the most energetic electrons can come from deeper, still have nonzero kinetic energy, and thus reach the sample surface. Extensive work was done by Bradley H. Frazer, a former graduate student in the Gilbert group, to measure the MPD of various elements, as shown in Figure 9.4.

9.2.1.3 Sample charging and the need for coating

Illuminating samples with x-rays provokes the photoemission of electrons (Einstein 1905). Because the electrons have negative charge, once they are extracted, the sample itself becomes positively charged. If the sample is a good conductor, the extracted electrons are rapidly replenished, and no charging

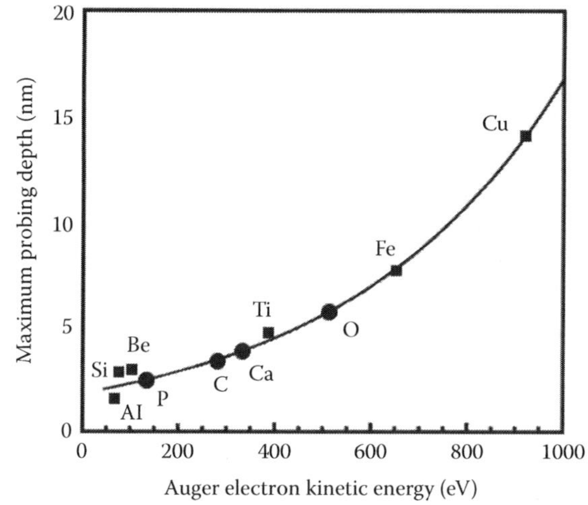

Figure 9.4 MPD as a function of Auger electron initial kinetic energy. The square symbols represent elements for which MPD data were acquired; the solid line is an exponential fit of these data. The circular symbols represent elements for which the MPD data were not acquired, but their MPD was deduced from the exponential fit. (Reprinted from *Surf. Sci.*, 537, Frazer, B.H., Gilbert, B., Sonderegger, B.R., and De Stasio, G., The probing depth of total electron yield in the sub keV range: TEY-XAS and X-PEEM, 161–167. Copyright 2003, with permission from Elsevier; *Phys. Rev. B*, 77, Metzler, R.A., Zhou, D., Abrecht, M., Chiou, J.-W., Guo, J., Ariosa, D., Coppersmith, S.N., and Gilbert, P.U.P.A., Polarization-dependent imaging contrast in abalone shells, 064110. Copyright 2008, with permission from Elsevier; *J. Electron Spectrosc. Relat. Phenom., Special Issue on Photoelectron Microscopy, Time-Resolved Pump-Probe PES*, Gilbert, P.U.P.A., Polarization-dependent imaging contrast (PIC) mapping reveals nanocrystal orientation patterns in carbonate biominerals, pp. 395–405. Copyright 2012, with permission from Elsevier.)

is observed. If the sample is an insulator, however, all sorts of interesting but undesirable charging phenomena occur (Gilbert et al. 2000, 2001). Biominerals are almost invariably insulators; hence, they do not lend themselves to PEEM or any other photoemission experiment, unless they are coated. As shown in Figure 9.4, the MPD for the elements of biomineral relevance is only 2–5 nm; hence, it is vital to coat the sample with a conducting layer thinner than 2–5 nm. The best results are obtained coating with 1 nm platinum. Gold is not recommended, as gold coatings invariably crack at the nanoscale and do not provide a continuous conducting layer. Chromium is a favorite for scanning electron microscopy because of its low yield of back-scattered electrons, but a Cr coating oxidizes within a few minutes and becomes insulating. Therefore, Cr-coated samples must be prepared and promptly analyzed, and they can only be studied once. Coating samples with platinum guarantees both a continuous conducting layer and stability over time. Thus, the same sample can be analyzed immediately or years later, and it remains just as viable, for single or repeated experiments.

9.2.1.4 How to coat a sample for PEEM

The Cressington sputter coater mentioned previously is well suited to produce a homogeneous 1 nm thick Pt coating layer. We usually use a taller glass jar (15 cm) to achieve lower coating rate and thereby higher precision, by placing the sputter source and Pt target farther away from the sample and thickness monitor. In addition, we tilt and rapidly rotate the sample with a rotary–planetary–tilting stage (http://www.tedpella.com/cressing_html/crs208hr.htm) during the 1 nm Pt coating, so any sample surface asperities, crevices, or scratches are coated evenly on all their sides. In addition, we need to measure the thickness of the coating layer very accurately. To achieve greater accuracy on the thickness monitor, which is a quartz balance, whose mass changes as it gets coated, hence its oscillator frequency varies and the deposited amount of Pt can be measured precisely (O'Sullivan and Guilbault 1999). We use an important trick: when programming into the thickness controller the thickness to be deposited, we enter 10 nm. However, instead of entering the correct density for Pt, which is 21.45 g/cm^3, we enter one-tenth of that density, 2.15 g/cm^3; thus, the thickness monitor reads in Å, not in nm, is far more accurate, and the deposited layer is exactly 10 Å = 1 nm.

Once coated with 1 nm Pt, the sample should, in theory, be ready for a PEEM experiment, but usually such samples do not work well, as 1 nm Pt is insufficient to make a very good electrical contact between the sample holder and the sample surface itself. Any rigid metallic surface in contact with such a thin Pt layer would just scratch it and thus break electrical conductivity of the sample surface. To make a good electrical contact, one needs to have at least 30 nm Pt. Hence, we usually coat the sample as shown in Figure 9.5. First, we place a mask typically 3–5 mm in diagonal or diameter at the center of the sample. The function of the mask is to cover the area of interest, while coating everything else with thicker Pt, typically 30–50 nm, and then remove the mask to make the

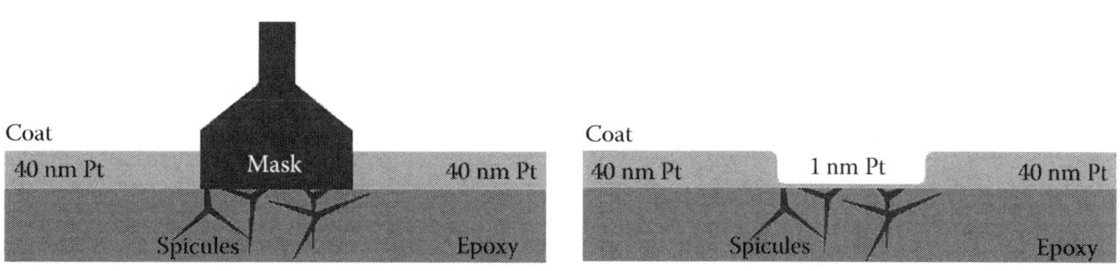

Figure 9.5 The biomineral sample could be a piece of shell or bone, tooth, etc. We first deposit a mask at the center of the sample, rapidly coat with 40 nm Pt on the non-moving sample, then remove the mask, and slowly coat the whole sample, tilted and rotating, with 1 nm Pt. (Adapted from De Stasio, G. et al., *Ultramicroscopy*, 98(1), 57, 2003; Gilbert, P.U.P.A. et al., The organic-mineral interface in biominerals, in *Molecular Geomicrobiology*, eds. Banfield, J.F, Nealson, K.H., and Cervini-Silva, J., *Reviews in Mineralogy and Geochemistry*, Mineralogical Society of America, Washington, DC, p. 1570185, 2005; Gilbert, P.U.P.A., Polarization-dependent imaging contrast (PIC) mapping reveals nanocrystal orientation patterns in carbonate biominerals, in *J. Electron Spectrosc. Relat. Phenom., Special Issue on Photoelectron Microscopy, Time-Resolved Pump-Probe PES*, eds. Kiskinova, M. and Scholl, A., Elsevier, pp. 395–405, 2012. With permission.)

1 nm coating. The mask could be a square or rectangular piece of Si wafer, with its polished surface down, in contact with the polished surface of the biomineral sample. It could also be a 3 mm diameter TEM grid with a sturdy, not holey nor laced, membrane coating. In both cases, however, it is easy to scratch the sample surface when removing such thin masks with tweezers. The best mask is one obtained by machining a small piece of any metal into two connected cylinders with different diameters, the larger one to be used as the mask and the smaller as a handle. The mask covers the area of interest, and the top cylinder provides a handle to deposit and pick up the mask with tweezers. A possible shape for this convenient mask is shown in Figure 9.5. One or more masks should completely cover the areas of interest for the PEEM experiment. The thicker Pt coating, for example, 40 nm thick as in Figure 9.5, should be deposited rapidly, and without moving the sample, to prevent undesirable motion of the mask. In order to obtain greater coating rate, we use a shorter glass jar (6.5 cm), enter the actual Pt density (21.45 g/cm^3) into the thickness controller, and program it to coat with 40 nm Pt. With the shorter glass jar, the Pt source and the sample are much closer; hence, the coating rate is greater. The deposition rates are on the order of 2 min for the 40 nm Pt coating (fast, short jar) and 30 s for the 1 nm Pt coating (slow, tall jar).

Once coated with this method, the biomineral sample is now ready for PEEM analysis. Notice that in all PEEMs, the electrical contact is done from above the sample surface, by placing a thin metal plate with a large circular aperture at the center (7 mm for the Elmitec PEEM, up to 15 mm for PEEM-3 at ALS), through which the sample surface is exposed for analysis. The plate will therefore make an excellent electrical contact with the 40 nm Pt coating all around the sides of the sample and well away from the region of interest for PEEM analysis, which should always be at the center of the sample, the plate, and the sample holder for best undistorted imaging in PEEM.

9.2.1.5 Sectioned samples

Other samples containing small amounts of hard minerals can be embedded in resin and sectioned with a microtome using a diamond knife as usually done for TEM analysis. The difference for preparation of a PEEM sample is that the section or ribbon of sections should be deposited on a clean silicon wafer, which is sufficiently conductive. Such samples do not require the thicker Pt coating, thus the 1 nm Pt coating is sufficient. Examples of sectioned samples prepared in this way are the forming part of the mouse incisor in Figure 9.2 and the rat-tail tendon shown in Figure 9.8. The latter is made of collagen and not at all mineralized (Lam et al. 2012), and the former is only partly mineralized, with the majority of the tooth tissues being hydrated and soft when extracted; hence, that sample lent itself to ultramicrotomy (Beniash et al. 2009).

9.2.2 CHEMICAL ANALYSIS OF BIOMINERALS WITH XANES SPECTROSCOPY

Once prepared as described in Section 9.2.1, mineral and biomineral samples are ideally suited for PEEM experiments. Different kinds of chemical analysis are possible with XANES spectroscopy. XANES distinguishes the **oxidation state of elements** as shown in Figure 9.6 for iron and Mn minerals, based on their electronic structures, and **organic molecules** as shown in Figure 9.7 for representative references from each group of organics (nucleic acids, lipids, proteins, polysaccharides) and bacterial biofilm filaments. Recently, it was also revealed that the orientation of parallel fibers in rat-tail collagen, thus the **orientation of selected organic molecules**, can be detected by XANES spectroscopy, as shown in Figure 9.8. It is important to notice, however, that organic molecules are extremely sensitive to radiation damage (Wang et al. 2009) and are therefore damaged during a PEEM experiment much more than the mineral components in biominerals. XANES spectra also provide information on

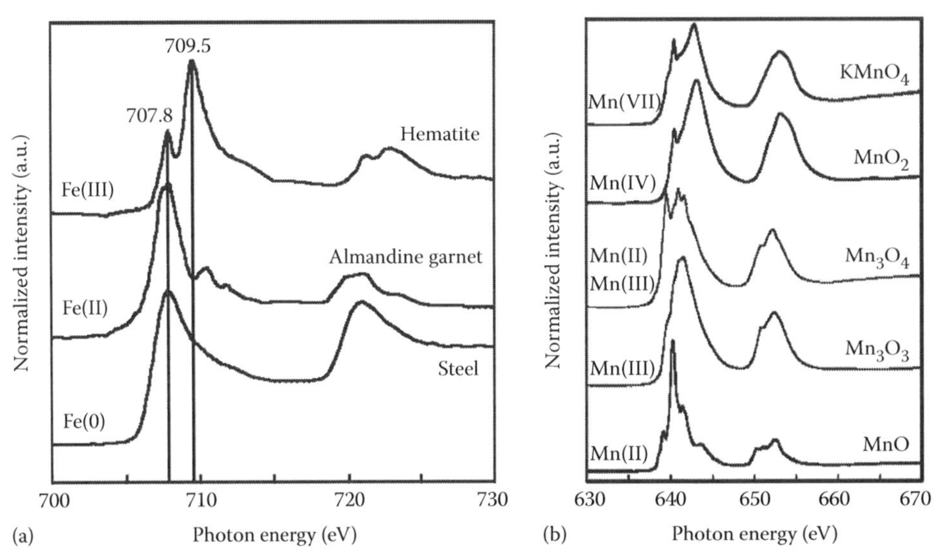

Figure 9.6 (a) Iron L-edge XANES spectra from ferric (III), ferrous (II), and metallic iron (0), in the minerals and metal indicated. (Data from Gilbert, P.U.P.A. et al., The organic-mineral interface in biominerals, in *Molecular Geomicrobiology*, eds. Banfield, J.F, Nealson, K.H., and Cervini-Silva, J., *Reviews in Mineralogy and Geochemistry*, Mineralogical Society of America, Washington, DC, p. 1570185, 2005. With permission.) (b) Manganese L-edge XANES spectra of manganese oxides. The formal Mn oxidation states are given on the left. (Reprinted with permission from Gilbert, B., Frazer, B.H., Belz, A., Conrad, P.G., Nealson, K.H., Haskel, D., Lang, J.C., Srajer, G., and De Stasio, G., Multiple scattering calculations of bonding and x-ray absorption spectroscopy of manganese oxides, *J. Phys. Chem. A*, 107(16), 2839–2847, 2003a. Copyright 2003 American Chemical Society.)

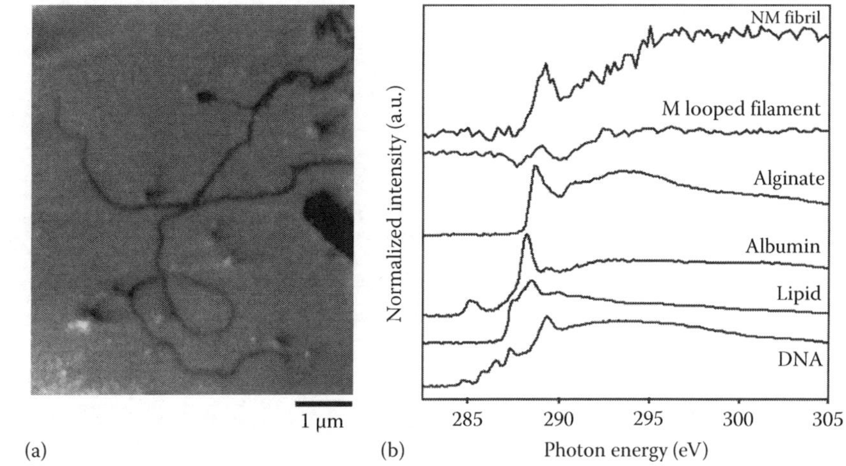

Figure 9.7 PEEM image and XANES spectra of the filaments produced by iron-oxidizing bacteria. (a) Mineralized filaments from the bacterial biofilm, which contain akaganeite (β-FeOOH) crystalline cores, templated by acidic polysaccharides. (b) Carbon K-edge XANES spectra from non-mineralized (NM) fibril not shown and the mineralized (M) filament in (B) and reference organic molecules: alginate (a representative acidic polysaccharide), albumin (a representative protein), a lipid, and DNA. Comparison with the bottom four reference standards enabled us to identify the chemical nature of the unknown filaments. Notice the similarity of the spectra from the NM fibrils and M filaments with the polysaccharide spectrum and the additional structure in the one from the M filament: the peak at 292.4 eV was assigned to the C–O bond in carboxyl groups. (Data from Chan, C.S., De Stasio, G., Welch, S.A., Girasole, M., Frazer, B.H., Nesterova, M.V., Fakra, S., and Banfield, J.F., Microbial polysaccharides template assembly of nanocrystal fibers, *Science*, 303(5664), 1656–1658, 2004. Reprinted with permission of AAAS.) The references standards in Chan et al. (2004) were courtesy of Adam Hitchcock, previously published in Lawrence et al. (2003).

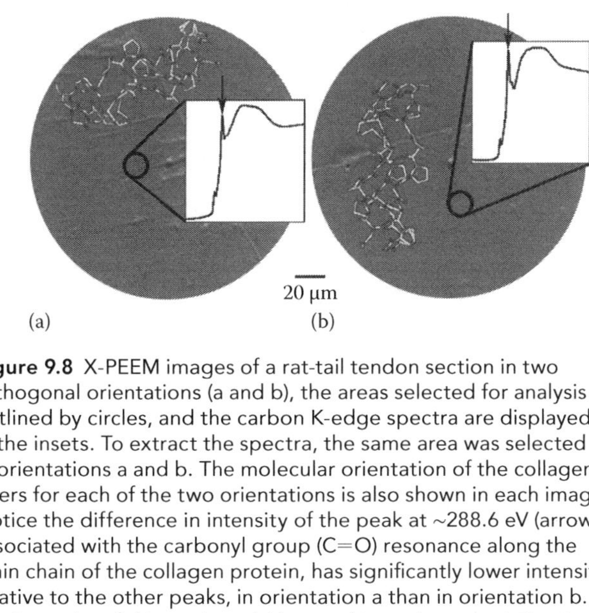

Figure 9.8 X-PEEM images of a rat-tail tendon section in two orthogonal orientations (a and b), the areas selected for analysis are outlined by circles, and the carbon K-edge spectra are displayed in the insets. To extract the spectra, the same area was selected in orientations a and b. The molecular orientation of the collagen fibers for each of the two orientations is also shown in each image. Notice the difference in intensity of the peak at ~288.6 eV (arrows), associated with the carbonyl group (C=O) resonance along the main chain of the collagen protein, has significantly lower intensity, relative to the other peaks, in orientation a than in orientation b. The intensity of this peak could be used to assign orientation to collagen fibers or individual microscopic fibrils in unknown samples. (Reprinted with permission from Lam, R.S.K., Metzler, R.A., Gilbert, P.U.P.A., and Beniash, E., Anisotropy of chemical bonds in collagen molecules studied by x-ray absorption near-edge structure (XANES) spectroscopy, *ACS Chem. Biol.*, 7(3), 476–480. Copyright 2012 American Chemical Society.)

crystal structures and **distinguish polymorphs** of the same compound, as shown in Figure 9.1 for calcite and aragonite (both $CaCO_3$) and in Figure 9.9 for sphalerite and wurtzite (both ZnS) minerals. Different minerals can be mapped by XANES-PEEM as shown in Figure 9.10 for calcite and aragonite, in Figures 9.11 and 9.12 for quartz inclusions in the oldest object ever found on Earth, a 220 μm zircon that is 4.4 billion years old, and in Figure 9.13 for Mg in the sea urchin tooth.

Since XANES spectroscopy is sensitive to the local environment of each atom, it provides distinct spectroscopic signatures of the **degree of crystallinity**, as shown in Figure 9.14 for Fe minerals and biominerals and in Figures 9.15 and 9.16 for **amorphous vs. crystalline** components in sea urchin spicules and teeth.

The capability of distinguishing amorphous from crystalline phases with spatial resolutions on the order of 20 nm (Rempfer and Griffith 1989; De Stasio et al. 1999; Frazer et al. 2004) is unique to XANES-PEEM instruments and particularly interesting for the biomineralogist, because ACC has been identified as a precursor phase in sea urchin spicules by Beniash

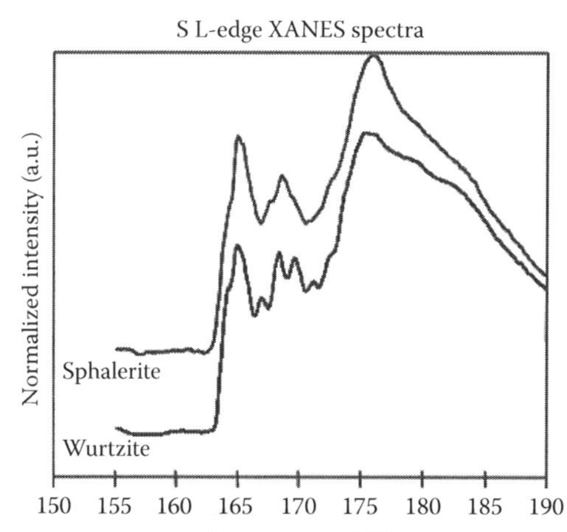

Figure 9.9 Sulfur L-edge spectra from two mineral polymorphs of zinc sulfide (ZnS): cubic sphalerite and hexagonal wurtzite. (Reprinted with permission from Gilbert, B., Frazer, H., Zhang, H., Huang, F., Banfield, J.F., Haskel, D., Lang, J.C., Srajer, G., and De Stasio, G., X-ray absorption spectroscopy of the cubic and hexagonal polytypes of zinc sulfide, *Phys. Rev. B*, 66(24), 245205. Copyright 2002, by the American Physical Society.)

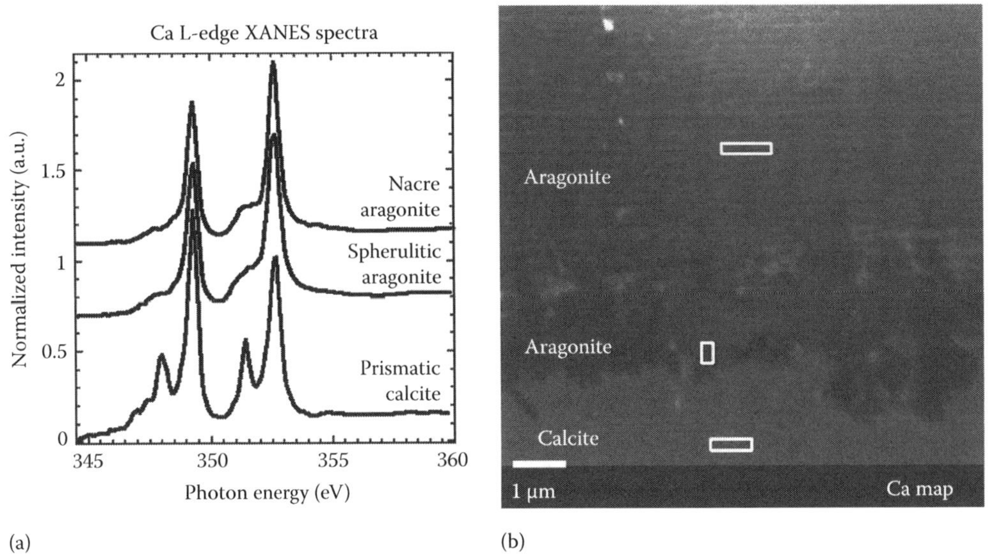

Figure 9.10 (a) Ca L-edge spectra from red abalone nacre layers, the spherulitic aragonite crystals near the boundary, and the prismatic crystals, extracted from the three regions shown in (b). The prismatic spectra are typical of calcite, while the spectra from nacre and spherulitic crystals are clearly aragonite as their spectra resemble those from geologic and biogenic aragonite shown in Figure 9.1. (b) Distribution map of calcite, obtained by ratio of images at 351.6 eV (a sharp calcite peak) and 345 eV. This selection of energies shows the highest contrast between calcite (lighter gray levels at the bottom) and aragonite (darker) and the sharp boundary between the two polymorphs. (Reprinted with permission from Gilbert, B., Metzler, R.A., Zhou, D., Scholl, A., Doran, A., Young, A., Kunz, M., Tamura, N., and Coppersmith, S.N., Gradual ordering in red abalone nacre, *J. Am. Chem. Soc.*, 130(51), 17519–17527. Copyright 2008 American Chemical Society.)

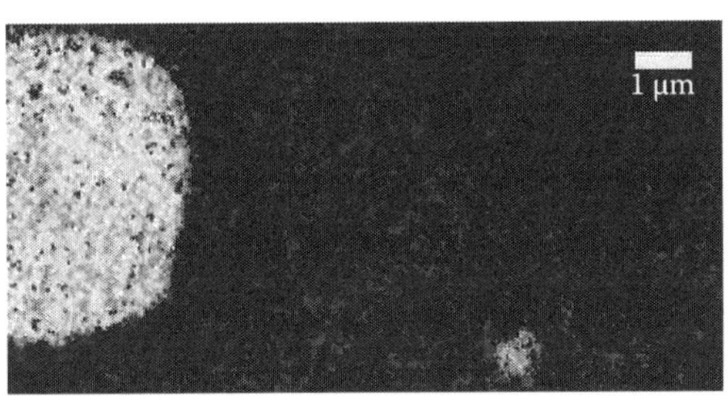

Figure 9.11 Distribution maps of quartz (lighter gray levels) and zircon (darker gray levels), obtained from digital ratio of XANES-PEEM images at specific energies, as described in Figure 9.12. A larger inclusion and a smaller (submicron) quartz inclusion are identified by their spectroscopic signatures shown in Figure 9.12 and mapped here. (Reprinted from *Ultramicroscopy*, 98(1), De Stasio, G., Frazer, B.H., Gilbert, B., Richter, K.L., and Valley, J.W., Compensation of charging in X-PEEM: A successful test on mineral inclusions in 4.4 Ga old zircon, 57–62. Copyright 2003, with permission from Elsevier.)

et al. (1997), in sea urchin spines by Politi et al. (Politi et al. 2004; Seto et al. 2012), and in sea urchin teeth by Killian et al. (2009). Furthermore, Politi et al. (2008) identified not one but two amorphous precursor phases, a hydrated phase, which here we term $ACCH_2O$, and another majority phase, presumably anhydrous ACC. These two phases coexisted with calcite in freshly extracted sea urchin spicules but transformed into calcite with time (Politi et al. 2008). Subsequently, Radha et al. (2010) showed that the ACC phase intermediate between synthetic $ACCH_2O$ and calcite has precisely the same enthalpy of transformation into calcite as the majority amorphous phase in fresh sea urchin spicules; hence, the tentative assignment of Politi et al. of the latter phase to ACC was likely correct. Finally, Gong et al. (2012) produced low-noise reference spectra from $ACCH_2O$, ACC, and calcite, as shown in Figure 9.17; mapped the three phases in sea urchin spicule cross sections; and showed that indeed the first phase deposited is $ACCH_2O$, the intermediate phase is ACC, and the final phase is calcite. The mapping of the three phases and their spatial sequence in sea urchin spicules is shown in Figure 9.18, whereas Figure 9.19 shows the relative abundances of the three phases in spicules at different developmental stages: 36, 48, and 72 h after fertilization of sea urchin eggs. More information on the developmental stages of spicules is reviewed in Gilbert and Wilt (2011).

An amorphous precursor phase has also been observed in zebrafish bone by Mahamid et al. (2008) and in mouse tooth enamel by Beniash et al. (2009). Hence, it may be possible in the future to map amorphous precursors in bone or teeth, as was done for sea urchin spicules in Figures 9.18 and 9.19.

9.2.3 CRYSTAL ORIENTATION ANALYSIS OF BIOMINERALS: PIC-MAPPING

PEEM microscopes are very sensitive to the orientation of birefringent crystals, as Metzler et al. first showed in nacre aragonite (Metzler et al. 2007). This sensitivity originates from x-ray linear dichroism, an effect previously well known in the condensed matter physics community and observed extensively in a variety of man-made systems, including magnetic materials (van der Laan and Thole 1991; Stohr et al. 1999; Scholl et al. 2000; Lüning et al. 2003; Holcomb et al. 2010), organic molecules (Ade and Hsiao 1993), molecular

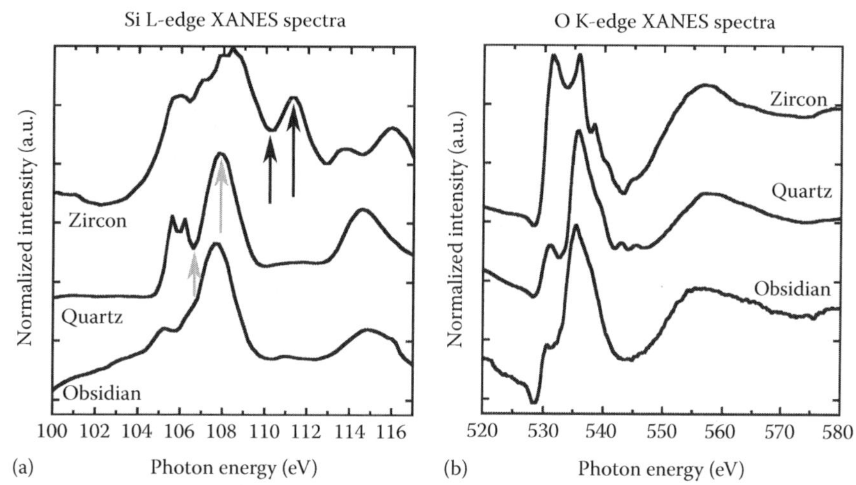

Figure 9.12 (a) Si and (b) O spectra acquired from three silicate minerals, two crystalline (zircon and quartz), and one glassy (obsidian). An unknown mineral or a microscopic mineral inclusion such as those in Figure 9.11 can be identified by their XANES spectra, as these are quite distinct. The combined distribution maps of Figure 9.11 were obtained by digital ratio of on-peak and off-peak images acquired at the energies indicated by the black arrows for zircon and the light gray arrows for quartz. At those black-arrow energies, the spectral intensities for zircon are quite different but nearly identical for quartz; the reverse is true for the light-gray-arrow energies. Such spectral differences make it possible to map these two minerals by simple digital ratio of two PEEM images, acquired at those energies. (Data adapted from Gilbert, B. et al., *Am. Mineral.*, 88(5–6), 763, 2003b. With permission.)

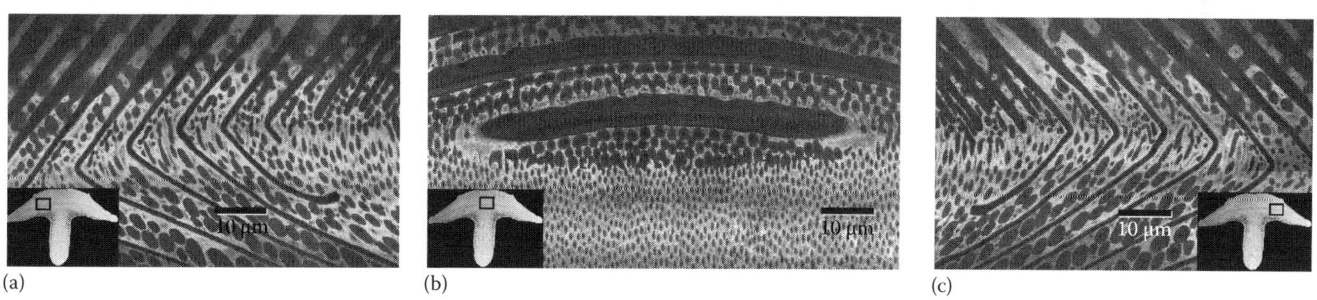

Figure 9.13 Magnesium maps of a sea urchin tooth from *Strongylocentrotus purpuratus*, obtained by X-PEEM. High Mg concentration (lighter gray level) is observed in the polycrystalline matrix surrounding plates and fibers. (a and c) Mg maps showing the areas 250 μm away from the center on either side, where the stone part is delimited laterally by plates that kink and extend toward the keel in a chevron shape. The black boxes in the insets indicate the position of the images in a, b, and c with respect to the whole tooth cross section. (b) Mg map at the center of the tooth cross section, including the stone part and the first two plates delimiting it at the top. (Killian, C.E., Metzler, R.A., Gong, Y.U.T., Churchill, T.H., Olson, I.C., Trubetskoy, V., Christensen, M.B., Fournelle, J.H., De Carlo, F., Cohen, S., Mahamid, J., Wilt, F.H., Scholl, A., Young, A., Doran, A., Coppersmith, S.N., and Gilbert, P.U.P.A.: Self-sharpening mechanism of the sea urchin tooth. *Adv. Funct. Mater.* 2011. 21. 682–690. Copyright Wiley-VCH Verlag GmbH & Co. KGaA. Reproduced with permission.)

monolayers (Stohr et al. 1981; Madix et al. 1988), and liquid crystals (Stohr et al. 2001), but it had never been observed in a natural system or a biological one. Since the first discovery in 2005 and publication in 2007 (Metzler et al. 2007; Gilbert 2012), we have consistently observed this effect in biomineral, synthetic, and geologic crystals of calcite and aragonite. Crystals birefringent in visible polarized light show linear dichroism in the soft x-ray regime. This is not a coincidence, as the physical basis of the polarization effects is the same in the visible light and x-ray energy ranges.

The observation of x-ray linear dichroism in biominerals is relevant because it makes it possible to display the orientation of individual biomineral component crystals at the nanoscale. The orientation patterns of nano- and microcrystals in pristine biominerals are now routinely revealed using the imaging method that our group introduced: PIC-mapping (Metzler et al. 2008; Gilbert 2012). This imaging modality uses x-ray linear dichroism and is done using XANES-PEEM and an elliptically polarizing undulator (EPU) as the source of x-rays, in which the linear polarization can be rotated at will in the plane perpendicular to the direction of propagation of the x-rays. PIC-mapping has been used for a variety of biominerals in the last 6 years, including the nacre (Metzler et al. 2007, 2008, 2010; Gilbert et al. 2008) and prismatic (Metzler et al. 2008, 2010; Gilbert et al. 2011) layers of mollusk shells and sea urchin teeth (Killian et al. 2009, 2011; Ma et al. 2009; Gilbert and Wilt 2011). Please see Gilbert (2012) for a detailed description of the PIC-mapping imaging modality and a recent review of the results. Here, we only present three examples in these three popular biominerals.

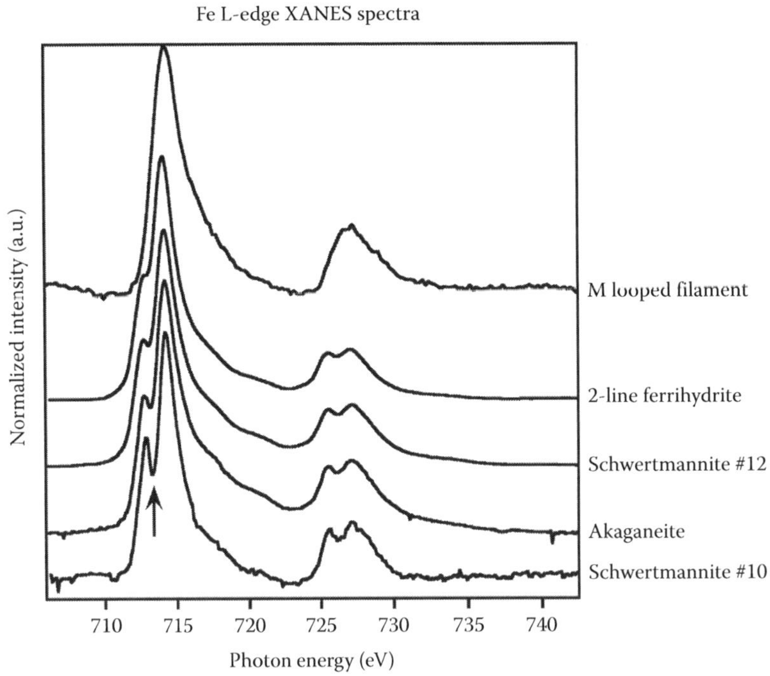

Figure 9.14 X-PEEM Fe L-edge XANES spectra of the FeOOH-mineralized looped filament shown in Figure 9.7, compared with iron oxyhydroxide standards, arranged (bottom to top) in order of decreasing crystallinity, as measured by x-ray diffraction peak broadening. Crystalline phases have well-resolved peaks at ~708.4 and ~709.8 eV. As crystallinity decreases, the dip separating peaks (arrow) decreases in depth. In the mineralized filament, the lack of dip can therefore be interpreted as amorphous or nanocrystalline iron oxyhydroxide. The Fe spectra from the NM fibrils of Figure 9.7 are flat (not shown), indicating that they are not mineralized. (Data from Chan, C.S., De Stasio, G., Welch, S.A., Girasole, M., Frazer, B.H., Nesterova, M.V., Fakra, S., and Banfield, J.F., Microbial polysaccharides template assembly of nanocrystal fibers, *Science*, 303(5664), 1656–1658, 2004. Reprinted with permission of AAAS.)

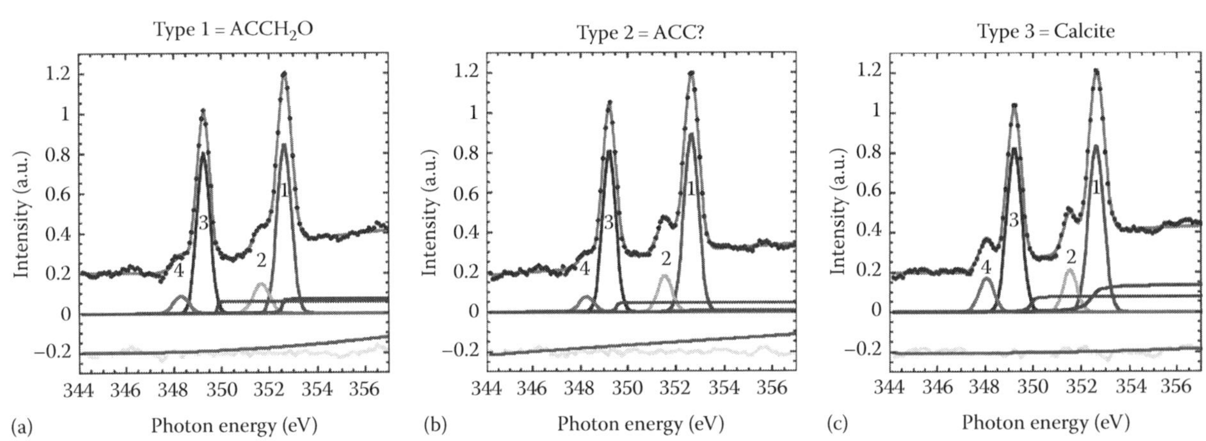

Figure 9.15 Peak fitting of XANES spectra acquired with PEEM from sea urchin 48 h spicules at the calcium L-edge. Each spectrum in (a), (b), and (c) was acquired from a single pixel, 200 nm in size, located on the spicule surface. Fit components are as follows. Two arctangents are the ionization potentials of the L_2- and L_3-edges. Two Gaussian highest components, labeled peaks 1 and 3, are the L_2 and L_3 main white-line peaks. Two smaller Gaussian components, labeled peaks 2 and 4, are the crystal field peaks. Notice that peak 4 is lower and broader in both amorphous phases (a) and (b) compared to calcite, whereas peak 3 is similar in calcite and the presumed ACC phase and lower and broader in $ACCH_2O$. These differences are easy to see considering that the position of the label numbers 2 and 4 is identical in all spectra. The dark gray line at the bottom is a third-order polynomial background. Black dots, experimental spectra; gray solid line, fit result. Light gray noisy line at the bottom: residue, that is, the difference between the experimental spectra and the fits. In each panel, the experimental spectrum and fit are displaced up; the polynomial background and the residue are displaced down, for clarity. Peak fitting was done using the GG macros, downloadable free of charge from the author's website, which are accompanied by an easy to use manual (GG–Macros 2013). (Data from Politi, Y., Metzler, R.A., Abrecht, M., Gilbert, B., Wilt, F.H., Sagi, I., Addadi, L., Weiner, S., and Gilbert, P.U.P.A., Transformation mechanism of amorphous calcium carbonate into calcite in the sea urchin larval spicule, *Proc. Natl. Acad. Sci. USA*, 105(45), 17362. Copyright 2008 National Academy of Sciences, U.S.A.)

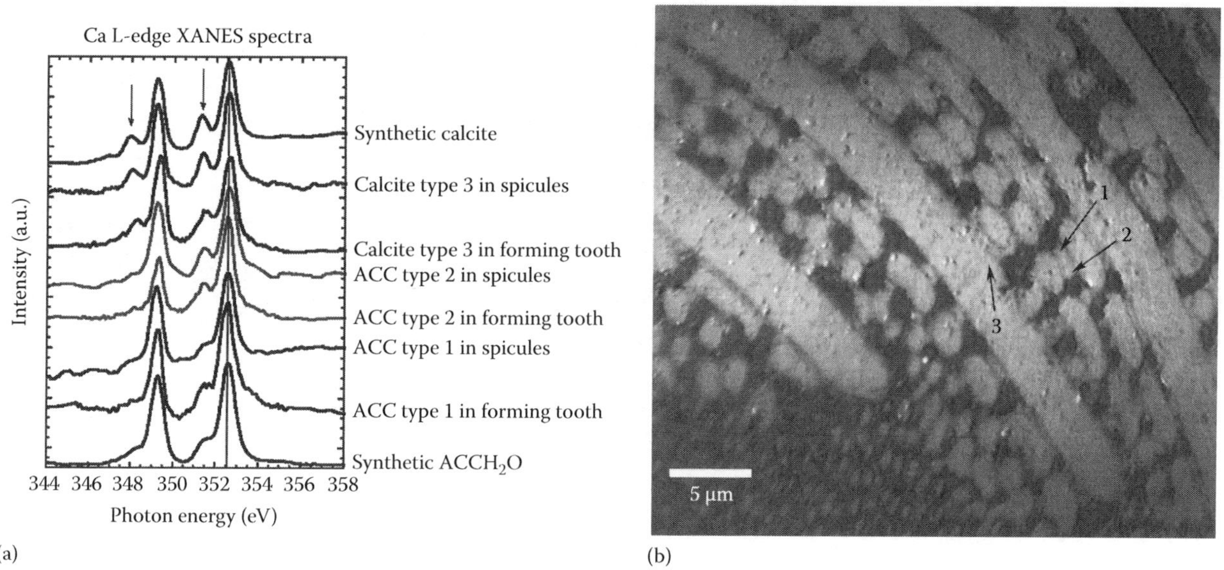

Figure 9.16 (a) Calcium spectra from the forming end of the sea urchin tooth. The calcium L-edge spectra were acquired from the 20 nm pixels shown in (b) and compared to those from sea urchin spicules from (Politi et al. 2008) and to synthetic calcite and ACCH$_2$O. The spectra are distinguished by the relative intensities of the two crystal field peaks (arrows in a). Spicule spectra from types 1, 2, and 3 $CaCO_3$ (same as in Figure 9.15) are also displayed for comparison. The spectra from 20 nm pixels in the forming tooth are in excellent agreement with those from spicules; thus, they are assigned to the same three types of $CaCO_3$. (b) Calcium spatial distribution map of a cross section of the sea urchin tooth, taken 3 mm away from the forming end of an *S. purpuratus* tooth. Notice that the plates and the fibers are fully formed and mineralized. They are Ca rich, as indicated by the light gray level in this Ca map. The matrix between plates and fibers is still forming, and many voids are present; these appear dark as they are filled by epoxy, which is Ca-free. The arrows indicate the single pixels (20 nm in size) from which the spectra in (a) were extracted, labeled 1, 2, and 3 as the types of $CaCO_3$ spectroscopically identified in those pixels and Figure 9.15. The amorphous phases occur most frequently in the forming matrix (arrows 1 and 2 in b). Notice that the Ca-rich regions shown in this cross section are fully aggregated and *solid*. Individual nanoparticles not yet aggregated, whether amorphous or crystalline, would have been washed away during polishing of this sample, which was done with 50 nm alumina grit suspended in water saturated with $CaCO_3$. This is important because it shows that amorphous phases aggregate first and are extremely stable and insoluble; hence, they are observed here. They will be converted into crystals at a later stage. Normalization and alignment of the spectra and distribution mapping were done using the GG macros (GG–Macros 2013). (Reprinted with permission from Killian, C.E., Metzler, R.A., Gong, Y.T., Olson, I.C., Aizenberg, J., Politi, Y., Addadi, L., Weiner, S., Wilt, F.H., Scholl, A., Young, A., Coppersmith, S.N., and Gilbert, P.U.P.A., The mechanism of calcite co-orientation in the sea urchin tooth, *J. Am. Chem. Soc.*, 131, 18404–18409. Copyright 2009 American Chemical Society.)

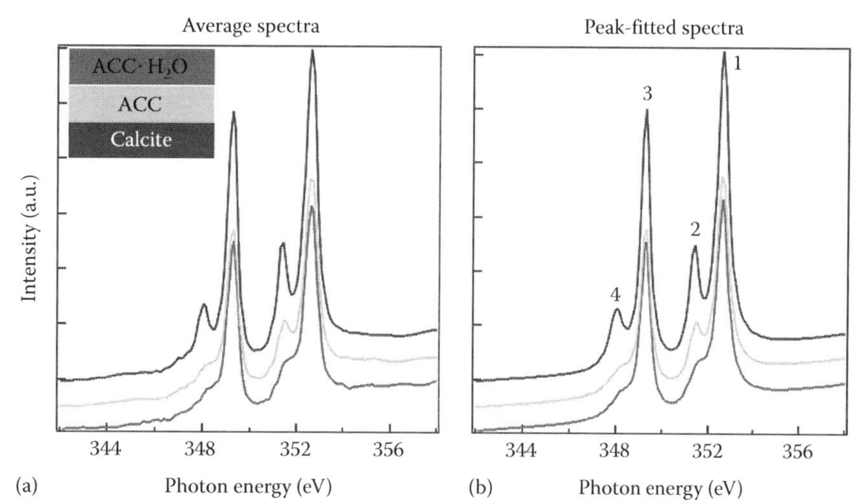

Figure 9.17 The three reference spectra used for component analysis in Figures 9.18 and 9.19. (a) XANES spectra across the calcium L-edge extracted from sea urchin spicules. To minimize experimental noise, 6–10 independently acquired single-pixel spectra, with pixel sizes of 20 nm, were averaged to give each reference spectrum. (b) Spectra resulting from peak fitting the spectra in (a), which completely eliminates experimental noise from these spectra; thus, they can be used as references for component analysis as described in reference Gong et al. (2012). Spectra are offset for clarity: ACC·H$_2$O at bottom and calcite at top. These spectra originate from single pixels in Ca stacks of XANES-PEEM images. Each spectrum was extracted using the GG macros and normalized; all groups of 6–10 similar spectra were averaged and then peak fitted. All of these steps are much simpler and faster to do using the GG macros (GG–Macros 2013) than in any other software package. (Data from Gong, Y.U.T. et al., *Proc. Natl. Acad. Sci. USA*, 109, 6088, 2012.)

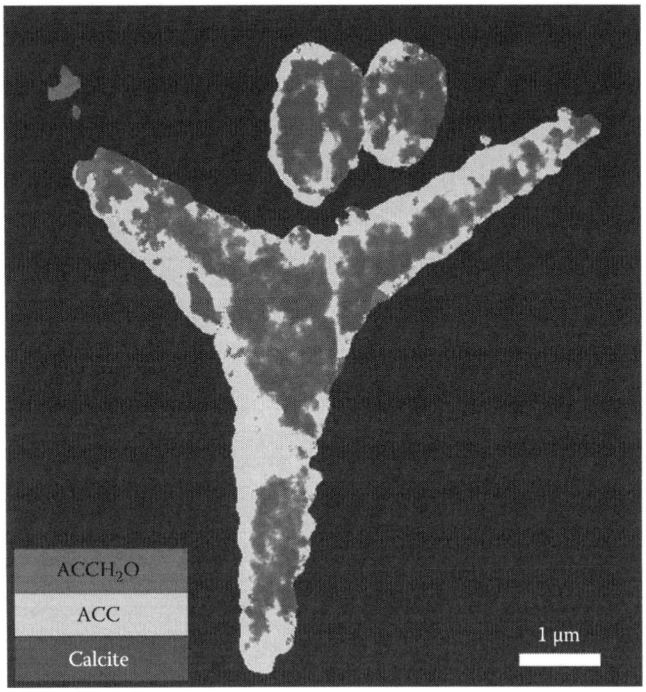

Figure 9.18 (See color insert.) Component mapping in 48 h spicules, at the prism developmental stage, analyzed within 24 h of extraction from the embryo. XANES-PEEM component analysis map of three spicules embedded in epoxy, polished to expose a cross section, and coated. The larger triradiate spicule at the center is polished in plane, whereas two other cylindrical spicules at the top have their long axes perpendicular to the plane of the image and appear elliptical. This red, green, and blue (RGB) map displays the results of *component analysis* mapping, in each component spectrum (Figure 9.17). Notice the greater concentration of red $ACCH_2O$ at the top left tip of the triradiate spicule and near its center, on the right-hand side. These are areas in which $ACCH_2O$ was freshly deposited at the outer rim of the spicule. Moving toward the inside of each spicule, most of the mineral detected is green ACC, and finally at the center, the main component is blue calcite. These data provided the first direct evidence for the sequence of transforming phases in sea urchin spicules: $ACCH_2O \rightarrow ACC \rightarrow$calcite. Component analysis was done using the GG macros (GG–Macros 2013). (Data from Gong, Y.U.T. et al., *Proc. Natl. Acad. Sci. USA*, 109, 6088, 2012.)

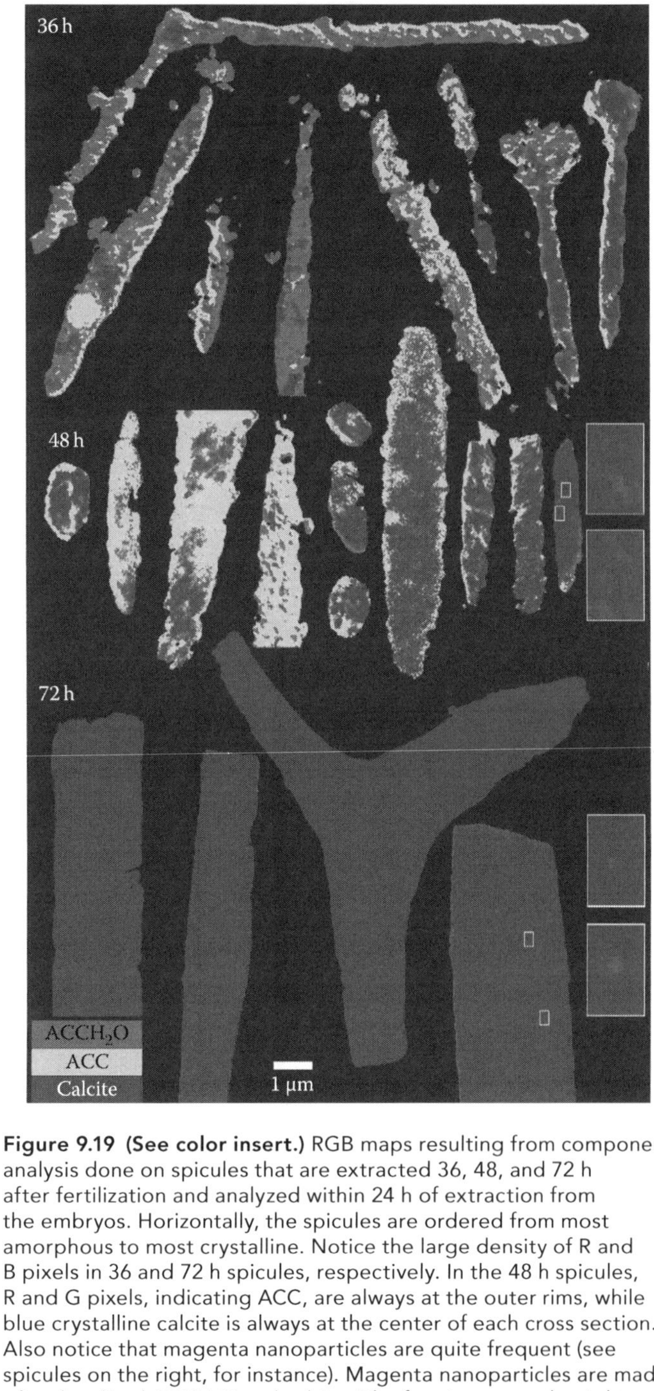

Figure 9.19 (See color insert.) RGB maps resulting from component analysis done on spicules that are extracted 36, 48, and 72 h after fertilization and analyzed within 24 h of extraction from the embryos. Horizontally, the spicules are ordered from most amorphous to most crystalline. Notice the large density of R and B pixels in 36 and 72 h spicules, respectively. In the 48 h spicules, R and G pixels, indicating ACC, are always at the outer rims, while blue crystalline calcite is always at the center of each cross section. Also notice that magenta nanoparticles are quite frequent (see spicules on the right, for instance). Magenta nanoparticles are made of co-localized $ACCH_2O$ and calcite. The four insets on the right show zoomed-in maps of the four regions in white boxes on the 48 and 72 h spicules on the right. In the insets, pixels are 20 nm, and the color balance has been adjusted to enhance the magenta nanoparticle, otherwise faint, because magenta nanoparticles contain a much greater proportion of calcite than $ACCH_2O$. These nanoparticles are 60–120 nm in size and are consistently surrounded by blue calcite. Component analysis was done using the GG macros (GG–Macros 2013). (Data from Gong, Y.U.T. et al., *Proc. Natl. Acad. Sci. USA*, 109, 6088, 2012.)

In Figure 9.20, we show PIC-mapping of *Atrina rigida* nacre, in Figure 9.21 a portion of a prism in *Pinctada fucata*, and in Figure 9.22 a sea urchin tooth from *Strongylocentrotus purpuratus*.

9.3 DATA PROCESSING FOR SPECTRA, IMAGES, MAPS, AND PIC-MAPS

All the data in Figures 9.15 through 9.21 were analyzed using the Gilbert Group (GG) macros, developed by our research group, and distributed free of charge on the author's research website (GG–Macros 2013). They run in Igor Pro® and can be used interchangeably on PC, Mac, or UNIX operating systems. Extractions of spectra from single pixels in a stack, from a line of pixels in any orientation, or from a region of interest are instantaneous. All sorts of spectral data processing such as averaging, smoothing, adding, subtracting, scaling, aligning, normalizing, and peak fitting are straightforward. Image subtraction or division, elemental distribution mapping, component analysis, and quantitative PIC-mapping are all included in the GG macros and are designed to be very user-friendly and simple to learn.

Figure 9.20 PIC-maps of nacre from an *Atrina rigida* shell. The sample is polished so the nacre layer planes are perpendicular to the plane of the image. The PIC-map quantitatively shows in grayscale the orientation of the aragonite crystal c-axis in each 20 nm pixel, according to the grayscale bar at the bottom (0° corresponds to alignment within the image plane). Notice that these tablets are not all co-oriented and do not have all their c-axes perpendicular to the nacre layers, as previously extensively assumed in the literature. If this were the case, all tablets would share the same gray level, but they clearly do not: stacks of 1–6 tablets have different gray levels, indicating an angle spread greater than 20°. These and seven other shell species showed clear trends in tablet width, thickness, and angle spread and also the first correlation of angle spread with water temperature (Olson and Gilbert 2012). (Reprinted with permission from Olson, I.C., Kozdon, R., Valley, J.W., and Gilbert, P.U.P.A., Mollusk shell nacre ultrastructure correlates with environmental temperature and pressure, *J. Am. Chem. Soc.*, 134, 7351–7358. Copyright 2012 American Chemical Society.)

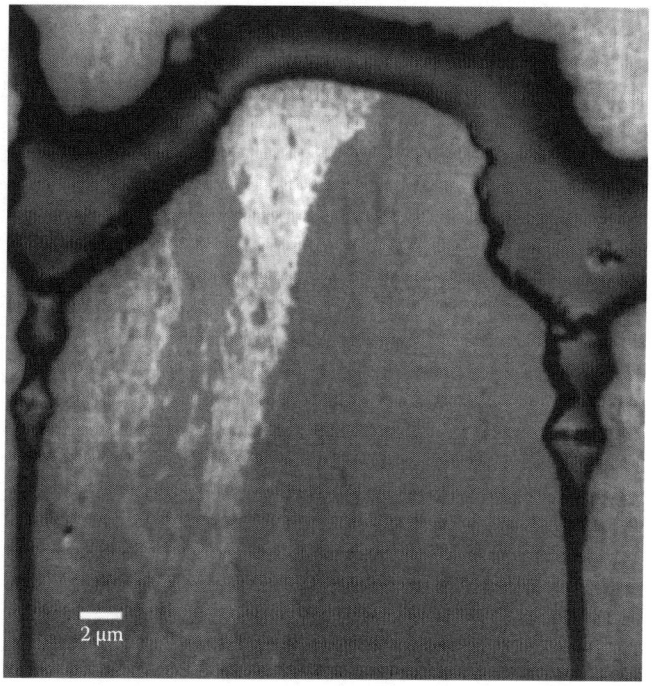

Figure 9.21 The nanocrystalline structure of the prismatic layer in *Pinctada fucata*. The PIC-map shows one prism near the prismatic-nacre boundary of the shell. The boundary itself is the darker, thick organic layer at the top, and two vertical periprismatic sheaths separate the central prism from the adjacent ones on the left- and the right-hand sides. The central prism extends for ~350 μm and is only partially displayed in this 30 μm field of view image. Different gray levels represent different orientations, and corresponding spectra as a function of polarization are displayed in the original publication. An animated version of this image as the polarization angle varies is displayed on the PNAS website: http://www.pnas.org/content/suppl/2011/06/21/1107917108.DCSupplemental/pnas.1107917108_SI.pdf#SM2. (Data from Gilbert, P.U.P.A. et al., *Proc. Natl. Acad. Sci. USA*, 108, 11350, 2011.)

9.4 OUTLOOK

PEEM microscopes are available at most synchrotrons around the world; some of them are designed and built by the local scientists, others are commercially available from four independent German companies (Elmitec: http://www.elmitec.de/, Omicron: http://www.omicron.de/en/products/focus-peem-/instrument-concept, Specs: http://www.specs.de/cms/front_content.php?idcat=84, Staib: http://www.staibinstruments.com/products/peem/peem.html). PEEMs are extremely useful for a diversity of scientific fields, and they are so widely available that they are not likely to become obsolete in the near future. All PEEM instruments continue to be perfected and optimized; hence, their performances are constantly improving. Aberration-corrected PEEMs have been developed and are being commissioned at Bessy II in Germany (Schmidt et al. 2002; Tromp et al. 2010) and the ALS in the United States (Feng et al. 2002; PEEM-3_webpage 2007). They are also beginning to be available commercially. Aberration correction will further improve the resolution and the transmission efficiency of the optics column, hence reducing the time the sample must be exposed to the x-ray beam and consequently reducing radiation damage.

Another significant and future feasible improvement is the use of a rotatable sample holder, which would provide completely quantitative, as opposed to the current semiquantitative, PIC-mapping (Gilbert 2012).

Because of the rapidly increasing availability and performances of all PEEMs, more biomineralization studies will likely rely on PEEM experiments. If you have an interesting biomineral or mineral, I hope this chapter has enticed you to explore using PEEM-XANES as a tool to address your scientific questions. With the recent advances in PEEM instrumentation and sample preparation along with the assistance of beamline scientists, even the scientist with little or no experience working at a synchrotron should be able to take advantage of this powerful tool.

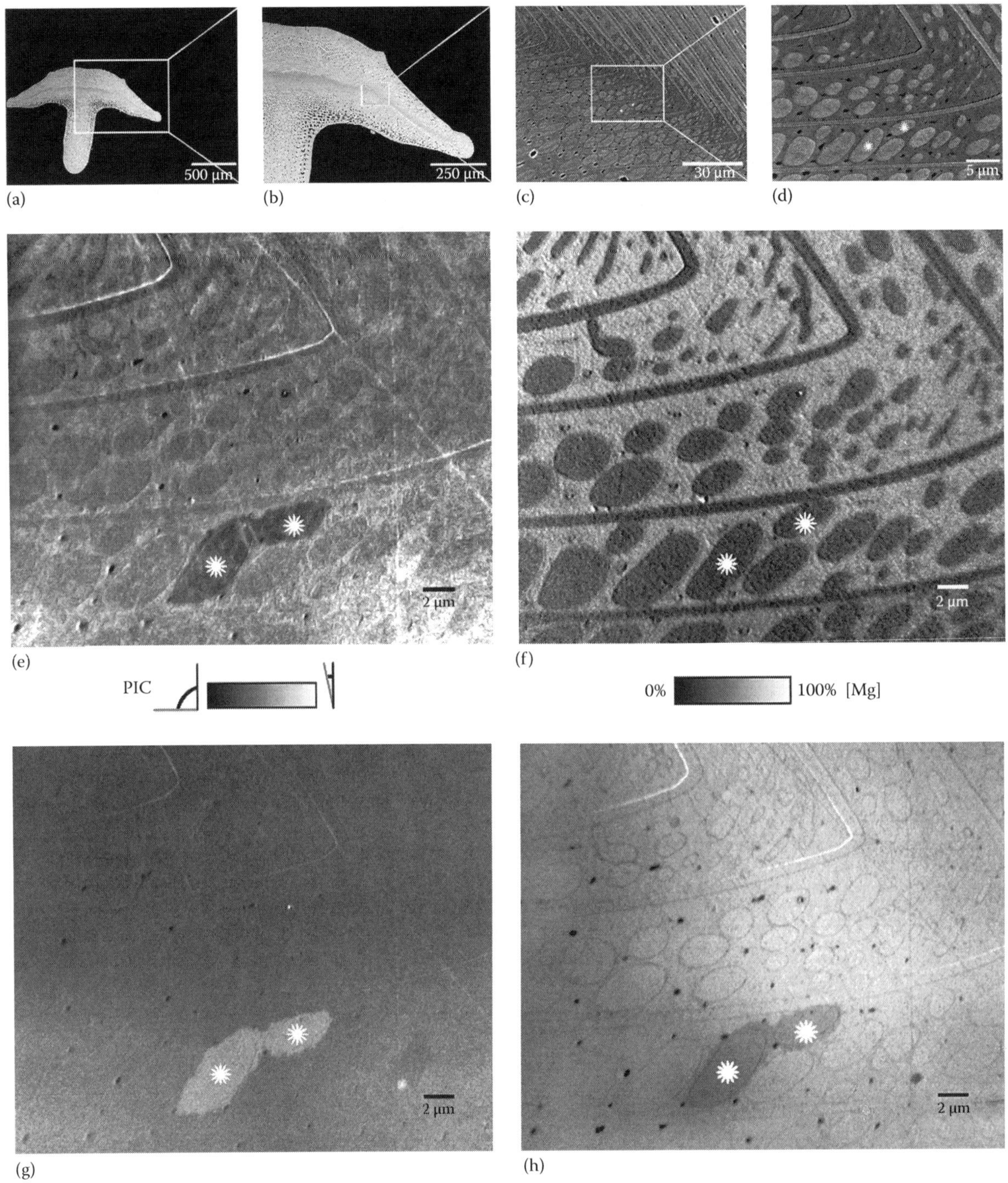

Figure 9.22 Two stray fibers reveal how co-orientation arises in sea urchin teeth. The same region of a sea urchin tooth was analyzed with SEM, PIC-mapping, and Mg mapping. In this region of a sea urchin tooth cross section, the primary and secondary plates join each other and have a chevron morphology, ~300 μm away from the center of the stone. (a–d) Increasing magnification SEM micrographs of the tooth cross section. The boxes in a, b, and c indicate the areas where the micrographs in b, c, and d were acquired. Notice that these images are acquired in SEM-back-scattered electron (SEM-BSE) mode; thus, the greater the atomic mass, the greater the brightness. In d, ^{40}Ca-rich plates and fibers appear brighter than the ^{24}Mg-rich polycrystalline matrix. Organic layers surround plates and fibers, which are rich in lighter elements ^{16}O, ^{14}N, ^{12}C, and ^{1}H and therefore appear darkest. The asterisks indicate the two fibers with strong misorientation in e, f, g, and h. (e) PIC-map showing two strongly misoriented fibers (white asterisks). This PIC-map was acquired with EPU° = 45° and was obtained by digital ratio of carbon π^* and pre-edge images. (f) Mg distribution map from the same region as d–h, showing well-defined and sharp elliptical fiber edges. The strongly misaligned fibers in E are extremely rare in the sea urchin tooth, in which all other fibers are co-oriented with each other and with the polycrystalline matrix. Despite the sharp edges of the fibers in the Mg map in f, the PIC-map in e shows that the polycrystalline matrix surrounding and between the two stray fibers is as misoriented as the fibers themselves. This indicates that the nanoparticles in the polycrystalline matrix are imparted in their orientation by the fibers. (g and h) PIC-maps of the same region in d–h, acquired with linearly polarized x-ray illumination with horizontal (EPU° = 90°, G) and vertical polarization (EPU° = 0°, H). Notice that at these polarization angles, the plates, the fibers, and the polycrystalline matrix all exhibit similar gray levels, while the two stray fibers (asterisks) are strongly different. This means that the crystal orientation difference between the plates and the surrounding matrix observed in e is small, on the order of a few degrees, while the misorientation of the stray fibers is large, on the order of tens of degrees. The crystal orientation of the stray fibers spills out of the fibers into the surrounding polycrystalline matrix nanoparticles. Data acquired using the PEEM-3 spectromicroscope (PEEM-3_webpage 2007) on beamline 11.0.1 at ALS and published in Killian et al. (2009).

9.5 FIND A PEEM AT A SYNCHROTRON NEAR YOU

Hereafter is a growing and incomplete list of countries, synchrotrons, and PEEMs:

Canada
CLS, Elmitec PEEM: http://www.lightsource.ca/experimental/sm.php

China
SSRF, PEEM, Elmitec PEEM: http://ssrf.sinap.ac.cn/english/3/BL08U1-A.htm

France
Soleil, Elmitec PEEM: http://www.synchrotron-soleil.fr/Recherche/LignesLumiere/HERMES

In addition, two other PEEMs use beamlines at Soleil; their home institutes should be contacted if you wish to do experiments:

Focus PEEM run by the Micro & Nano Magnetism Group of the Institute Néel:
http://neel.cnrs.fr/spip.php?rubrique52
periodically installed on the TEMPO beamline:
http://www.synchrotron-soleil.fr/Recherche/LignesLumiere/TEMPO
Omicron NanoESCA designed for high spectroscopic resolution run by the LENSIS laboratory of the CEA/IRAMIS institute:
http://iramis.cea.fr/spcsi/spectromicroscopy/

Germany
BESSY II, Elmitec PEEM: http://www.helmholtz-berlin.de/forschung/magma/m-dynamik/forschungsgruppen/x-peem/instrumente/x-peem_en.html
BESSY II, SMART project: http://www.helmholtz-berlin.de/pubbin/igama_output?modus=einzel&sprache=en&gid=1687
BESSY II, Aberration-corrected Specs PEEM: http://www.helmholtz-berlin.de/pubbin/igama_output?modus=einzel&sprache=en&gid=1889&typoid=37587

Italy
Elettra, SPELEEM: http://www.elettra.trieste.it/lightsources/elettra/elettra-beamlines/nanospectroscopy/beamline-description/all.html
Elettra, NanoESCA: http://www.elettra.trieste.it/it/lightsources/elettra/elettra-beamlines/nanoesca/nanoesca.html

Japan
SRL, PEEM: http://www.issp.u-tokyo.ac.jp/labs/sor/tsukuba-E/I-U-R_index.html
SPring 8, Elmitec PEEMSPECTOR, Omicron PEEM: http://www.spring8.or.jp/wkg/BL25SU/instrument/lang-en/INS-0000000344/instrument_summary_view

Jordan
SESAME, PEEM: http://www.sesame.org.jo/sesame/

Korea
PAL, find PEEM on beamline 2A: http://paleng.postech.ac.kr/

Singapore
SSLS, PEEM: http://ssls.nus.edu.sg/facility/sins.html

Spain
ALBA, PEEM: http://www.cells.es/Beamlines/CIRCE

Sweden
Max-Lab, PEEM: https://www.maxlab.lu.se/node/26

Switzerland
Swiss Light Source, Elmitec PEEM: http://www.psi.ch/sls/sim/endstations

Taiwan
National Synchrotron Radiation Research Center, Omicron PEEM: http://www.nsrrc.org.tw/english/research8_1_peem.aspx

Thailand
SLRI, PEEM: *http://www.slri.or.th/en/index.php?option=com_content&view=article&id=41&Itemid=96*

United Kingdom
Diamond, Elmitec PEEM: http://www.diamond.ac.uk/Home/Beamlines/I06/peem.html

United States
ALS, PEEM-2, PEEM-3: http://xraysweb.lbl.gov/peem2/webpage/Home.shtml
APS, Omicron PEEM, Elmitec PEEM: http://www.aps.anl.gov/Beamlines/Directory/beamline.php?beamline_id=7
NSLS, PEEM: http://beamlines.ps.bnl.gov/beamline.aspx?blid=U5UA

ACKNOWLEDGMENTS

I thank former students Brad H. Frazer, Rebecca A. Metzler and Ian C. Olson for their tireless work, which made PIC-mapping of biominerals possible. I also thank the expert technical support of past and present beamline scientists Bradley H. Frazer, Mike Abrecht, and Narayan Appathurai at SRC and Andreas Scholl, Anthony Young, Matthew Marcus, Andrew Doran, Nobumichi Tamura, Martin Kunz, and Jinghua Guo at ALS. This work was supported by NSF awards CHE-0613972 and DMR-1105167, and DOE Award DE-FG02-07ER15899 to PUPAG. The PEEM experiments were performed at the University of Wisconsin Synchrotron Radiation Center, supported by NSF awards DMR-0084402 and DMR-0537588, and at the Berkeley Advanced Light Source, supported by DOE under contract DE-AC02-05CH11231.

REFERENCES

Ade, H. and B. Hsiao. 1993. X-ray linear dichroism microscopy. *Science* 262(5138):1427–1429. doi: 10.1126/science.262.5138.1427.

Anders, S., H.A. Padmore, R.M. Duarte, T. Renner, T. Stammler, A. Scholl, M.R. Scheinfein, J. Stohr, L. Seve, and B. Sinkovic. 1999. Photoemission electron microscope for the study of magnetic materials. *Rev Sci Instrum* 70(10):3973–3981.

Bauer, E. 1994. Low-energy-electron microscopy. *Rep Prog Phys* 57(9):895–938. doi: 10.1088/0034-4885/57/9/002.

Bauer, E. 2012. A brief history of PEEM. *J Electron Spectrosc Relat Phenom* 185(10):314–322.

Beniash, E., J. Aizenberg, L. Addadi, and S. Wzeiner. 1997. Amorphous calcium carbonate transforms into calcite during sea urchin larval spicule growth. *Proc R Soc Lond B: Biological Sci* 264(1380):461–465.

Beniash, E., R.A. Metzler, R.S.K. Lam, and P.U.P.A. Gilbert. 2009. Transient amorphous calcium phosphate in forming enamel. *J Struct Biol* 166:133–143.

Brüche, E. 1933. Elektronenmikroskopische abbildung mit lichtelektrischen elektronen. *Z Phys A Hadron Nucl* 86(7):448–450.

Chan, C.S., G. De Stasio, S.A. Welch, M. Girasole, B.H. Frazer, M.V. Nesterova, S. Fakra, and J.F. Banfield. 2004. Microbial polysaccharides template assembly of nanocrystal fibers. *Science* 303(5664):1656–1658. doi: 10.1126/science.1092098.

Cressington coater. http://www.cressington.com/product_208hr.html.

De Stasio, G., M. Capozi, G.F. Lorusso, P.A. Baudat, T.C. Droubay, P. Perfetti, G. Margaritondo, and B.P. Tonner. 1998. MEPHISTO: Performance tests of a novel synchrotron imaging photoelectron spectromicroscope. *Rev Sci Instrum* 69(5):2062–2066.

De Stasio, G., B.H. Frazer, B. Gilbert, K.L. Richter, and J.W. Valley. 2003. Compensation of charging in X-PEEM: A successful test on mineral inclusions in 4.4 Ga old zircon. *Ultramicroscopy* 98(1):57–62. doi: 10.1016/S0304-3991(03)00088-3.

De Stasio, G., S.F. Koranda, B.P. Tonner, G.R. Harp, D. Mercanti, M.T. Ciotti, and G. Margaritondo. 1992. X-ray secondary-emission microscopy (XSEM) of neurons. *Europhys Lett* 19(7):655.

De Stasio, G., L. Perfetti, B. Gilbert, O. Fauchoux, M. Capozi, P. Perfetti, G. Margaritondo, and B.P. Tonner. 1999. MEPHISTO spectromicroscope reaches 20 nm lateral resolution. *Rev Sci Instrum* 70(3):1740–1742.

Einstein, A. 1905. The photoelectric effect. *Annalen der Physik*, Berlin, Germany: Wiley-VCH Verlag GmbH & Co. KgaA.

Electron binding energies. http://xdb.lbl.gov/Section1/Sec_1-1.html.

Elmitec GmbH. http://www.elmitec.de/Components.php?Bereich=Cartridges.

Engel, W., M.E. Kordesch, H.H. Rotermund, S. Kubala, and A. Vonoertzen. 1991. A UHV-compatible photoelectron emission microscope for applications in surface science. *Ultramicroscopy* 36:148–153.

Feng, J., H. Padmore, D.H. Wei, S. Anders, Y. Wu, A. Scholl, and D. Robin. 2002. Modeling the acceleration field and objective lens for an aberration corrected photoemission electron microscope. *Rev Sci Instrum* 73(3):1514–1517. doi: 10.1063/1.1423631.

Frazer, B.H., B. Gilbert, B.R. Sonderegger, and G. De Stasio. 2003. The probing depth of total electron yield in the sub keV range: TEY-XAS and X-PEEM. *Surf Sci* 537:161–167.

Frazer, B.H., M. Girasole, L.M. Wiese, T. Franz, and G. De Stasio. 2004. Spectromicroscope for the PHotoelectron Imaging of Nanostructures with x-rays (SPHINX): Performance in biology, medicine and geology. *Ultramicroscopy* 99(2–3):87–94. doi: 10.1016/j.ultramic.2003.10.001.

GG–Macros. 2013. http://home.physics.wisc.edu/gilbert/.

Gilbert, B., R. Andres, P. Perfetti, G. Margaritondo, G. Rempfer, and G. De Stasio. 2000. Charging phenomena in PEEM imaging and spectroscopy. *Ultramicroscopy* 83(1–2):129–139. doi: 10.1016/s0304-3991(99)00196-5.

Gilbert, B., B.H. Frazer, A. Belz, P.G. Conrad, K.H. Nealson, D. Haskel, J.C. Lang, G. Srajer, and G. De Stasio. 2003a. Multiple scattering calculations of bonding and x-ray absorption spectroscopy of manganese oxides. *J Phys Chem A* 107(16):2839–2847. doi: 10.1021/jp021493s.

Gilbert, B., B.H. Frazer, F. Naab, J. Fournelle, J.W. Valley, and G. De Stasio. 2003b. X-ray absorption spectroscopy of silicates for in situ, sub-micrometer mineral identification. *Am Mineral* 88(5–6):763–769.

Gilbert, B., B.H. Frazer, H. Zhang, F. Huang, J.F. Banfield, D. Haskel, J.C. Lang, G. Srajer, and G. De Stasio. 2002. X-ray absorption spectroscopy of the cubic and hexagonal polytypes of zinc sulfide. *Phys Rev B* 66(24):245205-1–245205-6. doi: 10.1103/PhysRevB.66.245205.

Gilbert, B., G. Margaritondo, S. Douglas, K.H. Nealson, R.F. Egerton, G.F. Rempfer, and G. De Stasio. 2001. XANES microspectroscopy of biominerals with photoconductive charge compensation. *J Electron Spectrosc Relat Phenom* 114:1005–1011. doi: 10.1016/s0368-2048(00)00342-x.

Gilbert, P.U.P.A. 2012. Polarization-dependent Imaging Contrast (PIC) mapping reveals nanocrystal orientation patterns in carbonate biominerals. *J Electron Spectrosc Relat Phenom, Special Issue on Photoelectron Microscopy, Time-Resolved Pump-Probe PES*, eds. M. Kiskinova and A. Scholl, pp. 395–405. http://dx.doi.org/10.1016/j.elspec.2012.06.001.

Gilbert, P.U.P.A., B.H. Frazer, and M. Abrecht. 2005. The organic-mineral interface in biominerals. In *Molecular Geomicrobiology*, eds. J.F. Banfield, K.H. Nealson, and J. Cervini-Silva, p. 1570185. Washington, DC: Reviews in Mineralogy and Geochemistry, Mineralogical Society of America.

Gilbert, P.U.P.A., R.A. Metzler, D. Zhou, A. Scholl, A. Doran, A. Young, M. Kunz, N. Tamura, and S.N. Coppersmith. 2008. Gradual ordering in red abalone nacre. *J Am Chem Soc* 130(51):17519–17527. doi: 10.1021/ja8065495.

Gilbert, P.U.P.A. and F.H. Wilt. 2011. Molecular aspects of biomineralization of the echinoderm endoskeleton. In *Molecular Biomineralization*, ed. W.E.G. Müller, pp. 199–223. Heidelberg, Germany: Springer.

Gilbert, P.U.P.A., A. Young, and S.N. Coppersmith. 2011. Measurement of c-axis angular orientation in calcite ($CaCO_3$) nanocrystals using x-ray absorption spectroscopy. *Proc Natl Acad Sci USA* 108:11350–11355.

Gong, Y.U.T., C.E. Killian, I.C. Olson, N.P. Appathurai, A.L. Amasino, M.C. Martin, L.J. Holt, F.H. Wilt, and P.U.P.A. Gilbert. 2012. Phase transitions in biogenic amorphous calcium carbonate. *Proc Natl Acad Sci USA* 109:6088–6093.

Griffith, O.H., G.B. Birrell, C.A. Burke, G.F. Rempfer, G.B. Lee, G.H. Lesch, Schlosse W et al. 1972. Photoelectron microscopy—New approach to mapping organic and biological surfaces. *Proc Natl Acad Sci USA* 69(3):561–565. doi: 10.1073/pnas.69.3.561.

Hecht, E. 2002. *Optics*, 4th edn., Reading, MA: Addison-Wesley.

Holcomb, M.B., L.W. Martin, A. Scholl, Q. He, P. Yu, C.-H. Yang, S.Y. Yang et al. 2010. Probing the evolution of antiferromagnetism in multiferroics. *Phys Rev B* 81:134406.

Killian, C.E., R.A. Metzler, Y.U.T. Gong, T.H. Churchill, I.C. Olson, V. Trubetskoy, M.B. Christensen et al. 2011. Self-sharpening mechanism of the sea urchin tooth. *Adv Funct Mater* 21:682–690. doi: 10.1002/adfm.201001546.

Killian, C.E., R.A. Metzler, Y.U.T. Gong, I.C. Olson, J. Aizenberg, Y. Politi, L. Addadi et al. 2009. The mechanism of calcite co-orientation in the sea urchin tooth. *J Am Chem Soc* 131:18404–18409.

Koshikawa, T., H. Shimizu, R. Amakawa, T. Ikuta, T. Yasue, and E. Bauer. 2005. A new aberration correction method for photoemission electron microscopy by means of moving focus. *J Phys Condens Matter* 17(16):S1371–S1380. doi: 10.1088/0953-8984/17/16/008.

Labrenz, M., G.K. Druschel, T. Thomsen-Ebert, B. Gilbert, S.A. Welch, K.M. Kemner, G.A. Logan et al. 2000. Formation of sphalerite (ZnS) deposits in natural biofilms of sulfate-reducing bacteria. *Science* 290(5497):1744–1747. doi: 10.1126/science.290.5497.1744.

Lam, R.S.K., R.A. Metzler, P.U.P.A. Gilbert, and E. Beniash. 2012. Anisotropy of chemical bonds in collagen molecules studied by x-ray absorption near-edge structure (XANES) spectroscopy. *ACS Chem Biol* 7(3):476–480. doi: 10.1021/cb200260d.

Lawrence, J.R., G.D.W. Swerhone, G.G. Leppard, T. Araki, X. Zhang, M.M. West, and A.P. Hitchcock. 2003. Scanning transmission x-ray, laser scanning, and transmission electron microscopy mapping of the exopolymeric matrix of microbial biofilms. *Appl Environ Microbiol* 69:5543–5554.

Lüning, J., F. Nolting, A. Scholl, H Ohldag, J.W. Seo, J. Fompeyrine, J.-P. Locquet, and J. Stöhr. 2003. Determination of the antiferromagnetic spin axis in epitaxial $LaFeO_3$ films by x-ray magnetic linear dichroism spectroscopy. *Phys Rev B* 67:214433.

Ma, Y.R., B. Aichmayer, O. Paris, P. Fratzl, A. Meibom, R.A. Metzler, Y. Politi, L. Addadi, P.U.P.A. Gilbert, and S. Weiner. 2009. The grinding tip of the sea urchin tooth exhibits exquisite control over calcite crystal orientation and Mg distribution. *Proc Natl Acad Sci USA* 106:6048–6053.

Madix, R.J., J.L. Solomon, and J. Stohr. 1988. The orientation of the carbonate anion on Ag(110). *Surf Sci* 197(3):L253–L259.

Mahamid, J., A. Sharir, L. Addadi, and S. Weiner. 2008. Amorphous calcium phosphate is a major component of the forming fin bones of zebrafish: Indications for an amorphous precursor phase. *Proc Natl Acad Sci USA* 105(35):12748–12753.

Metzler, R.A., M. Abrecht, R.M. Olabisi, D. Ariosa, C.J. Johnson, B.H. Frazer, S.N. Coppersmith, and P.U.P.A. Gilbert. 2007. Architecture of columnar nacre, and implications for its formation mechanism. *Phys Rev Lett* 98(26):268102.

Metzler, R.A. J.S. Evans, C.E. Killian, D. Zhou, T.H. Churchill, N.P. Appathurai, S.N. Coppersmith, and P.U.P.A. Gilbert. 2010. Nacre protein fragment templates lamellar aragonite growth. *J Am Chem Soc* 132:6329–6334.

Metzler, R.A., D. Zhou, M. Abrecht, J.-W. Chiou, J. Guo, D. Ariosa, S.N. Coppersmith, and P.U.P.A. Gilbert. 2008. Polarization-dependent imaging contrast in abalone shells. *Phys Rev B* 77:064110-1/9.

Nettesheim, S., A. Vonoertzen, H.H. Rotermund, and G. Ertl. 1993. Reaction-diffusion patterns in the catalytic cooxidation on Pt(110)—Front propagation and spiral waves. *J Chem Phys* 98(12):9977–9985. doi: 10.1063/1.464323.

Olson, I.C. and P.U.P.A. Gilbert. 2012. Aragonite crystal orientation in mollusk shell nacre may depend on temperature. The angle spread of crystalline aragonite tablets records the water temperature at which nacre was deposited by *Pinctada margaritifera*. *Faraday Discuss* 159:421–432. doi: 10.1039/c2fd20047c.

Olson, I.C., R. Kozdon, J.W. Valley, and P.U.P.A. Gilbert. 2012. Mollusk shell nacre ultrastructure correlates with environmental temperature and pressure. *J Am Chem Soc* 134:7351–7358. doi: dx.doi.org/10.1021/ja210808s.

O'Sullivan, C.K. and G.G. Guilbault. 1999. Commercial quartz crystal microbalances—Theory and applications. *Biosens Bioelectron* 14(8–9):663–670. doi: 10.1016/s0956-5663(99)00040-8.

PEEM-3_webpage. 2007. http://xraysweb.lbl.gov/peem2/webpage/Home.shtml.

Politi, Y., T. Arad, E. Klein, S. Weiner, and L. Addadi. 2004. Sea urchin spine calcite forms via a transient amorphous calcium carbonate phase. *Science* 306(5699):1161–1164.

Politi, Y., R.A. Metzler, M. Abrecht, B. Gilbert, F.H. Wilt, I. Sagi, L. Addadi, S. Weiner, and P.U.P.A. Gilbert. 2008. Transformation mechanism of amorphous calcium carbonate into calcite in the sea urchin larval spicule. *Proc Natl Acad Sci USA* 105(45):17362–17366.

Radha, A.V., T.Z. Forbes, C.E. Killian, P.U.P.A. Gilbert, and A. Navrotsky. 2010. Transformation and crystallization energetics of synthetic and biogenic amorphous calcium carbonate. *Proc Natl Acad Sci USA* 107:16438–16443.

Rempfer, G.F. and O.H. Griffith. 1989. The resolution of photoelectron microscopes with UV, x-ray, and synchrotron excitation sources. *Ultramicroscopy* 27(3):273–300. doi: 10.1016/0304-3991(89)90019-3.

Renault, O., N. Barrett, A. Bailly, L.F. Zagonel, D. Mariolle, J.C. Cezar, N.B. Brookes, K. Winkler, B. Kromker, and D. Funnemann. 2007. Energy-filtered XPEEM with NanoESCA using synchrotron and laboratory x-ray sources: Principles and first demonstrated results. *Surf Sci* 601(20):4727–4732. doi: 10.1016/j.susc.2007.05.061.

Schmidt, T., U. Groh, R. Fink, and E. Umbach. 2002. XPEEM with energy-filtering: Advantages and first results from the SMART project. *Surf Rev Lett* 9(01):223–232.

Scholl, A., J. Stohr, J. Luning, J.W. Seo, J. Fompeyrine, H. Siegwart, J.P. Locquet et al. 2000. Observation of antiferromagnetic domains in epitaxial thin films. *Science* 287(5455):1014–1016.

Schonhense, G. and H. Spiecker. 2002. Correction of chromatic and spherical aberration in electron microscopy utilizing the time structure of pulsed excitation sources. *J Vac Sci Technol B* 20(6):2526–2534. doi: 10.1116/1.1523373.

Seto, J., Y. Ma, S.A. Davis, F. Meldrum, A. Gourrier, Y.Y. Kim, U. Schilde, M. Sztucki, M. Burghammer, and S. Maltsev. 2012. Structure-property relationships of a biological mesocrystal in the adult sea urchin spine. *Proc Natl Acad Sci USA* 109(10):3699–3704.

Skoczylas, W.P., G.F. Rempfer, and O.H. Griffith. 1991. A proposed modular imaging-system for photoelectron and electron-probe microscopy with aberration correction, and for mirror microscopy and low-energy electron-microscopy. *Ultramicroscopy* 36(1–3):252–261. doi: 10.1016/0304-3991(91)90154-x.

Stöhr, J. 2003. *NEXAFS Spectroscopy*. Vol. 25, Berlin, Germany: Springer.

Stohr, J., K. Baberschke, R. Jaeger, R. Treichler, and S. Brennan. 1981. Orientation of chemisorbed molecules from surface absorption fine structure measurements -CO and NO on Ni(100). *Phys Rev Lett* 47(5):381–384.

Stohr, J., M.G. Samant, J. Luning, A.C. Callegari, P. Chaudhari, J.P. Doyle, J.A. Lacey, S.A. Lien, S. Purushothaman, and J.L. Speidell. 2001. Liquid crystal alignment on carbonaceous surfaces with orientational order. *Science* 292(5525):2299–2302.

Stohr, J., A. Scholl, T.J. Regan, S. Anders, J. Luning, M.R. Scheinfein, H.A. Padmore, and R.L. White. 1999. Images of the antiferromagnetic structure of a NiO(100) surface by means of x-ray magnetic linear dichroism spectromicroscopy. *Phys Rev Lett* 83(9):1862–1865.

Ted Pella coater. http://www.tedpella.com/cressing_html/crs208hr.htm.

Tonner, B.P. and G.R. Harp. 1988. Photoelectron microscopy with synchrotron radiation. *Rev Sci Instrum* 59(6):853–858.

Tromp, R.M. 2011. Measuring and correcting aberrations of a cathode objective lens. *Ultramicroscopy* 111(4):273–281. doi: 10.1016/j.ultramic.2010.11.029.

Tromp, R.M., J.B. Hannon, A.W. Ellis, W. Wan, A. Berghaus, and O. Schaff. 2010. A new aberration-corrected, energy-filtered LEEM/PEEM instrument. I. Principles and design. *Ultramicroscopy* 110(7):852–861. doi: 10.1016/j.ultramic.2010.03.005.

Tromp, R.M. and M.C. Reuter. 1991. Design of a new photo-emission/low-energy electron microscope for surface studies. *Ultramicroscopy* 36(1):99–106.

van der Laan, G. and B.T. Thole. 1991. Strong magnetic-x-ray dichroism in 2p absorption-spectra of 3d transition-metal ions. *Phys Rev B* 43(16):13401–13411.

Wang, J., C. Morin, L. Li, A.P. Hitchcock, A. Scholl, and A. Doran. 2009. Radiation damage in soft x-ray microscopy. *J Electron Spectrosc Relat Phenom* 170(1–3):25–36. doi: 10.1016/j.elspec.2008.01.002.

10 Solid-state NMR spectroscopy: A tool for molecular-level structure analysis and dynamics

Melinda J. Duer

Contents

10.1 NMR METHODOLOGY

10.1.1 WHAT IS NMR SPECTROSCOPY AND WHAT DOES IT MEASURE?

Nuclear magnetic resonance spectroscopy (NMR) is a technique that detects structural features and dynamics at the molecular level. The method can be applied to samples in the solid, as well as the more common solution phase, and is well suited to examining the heterogeneous systems that most biominerals consist of. Quantitative molecular structural measurements using solid-state NMR are feasible and are becoming increasingly common as equipment and methodologies improve. The essential tool of solid-state NMR spectroscopy is the NMR spectrometer, which consists of a very high-field magnet (10–20 T is typical) into which the sample is placed and exposed to a band of radiofrequencies (electromagnetic radiation). Under the magnetic field conditions, the nuclei in the sample absorb radio-frequency radiation at characteristic frequencies. The crucial points are that the frequency of radiation that a particular nucleus absorbs depends on (1) the identity of the nuclear species, that is, ^{13}C will absorb in a very different frequency range to ^{1}H, and (2) the detailed environment of the nucleus.

The frequency range that a particular nuclear species will absorb in is (almost always) distinctly separate from that of other nuclear species, and thus NMR spectroscopy can examine one nuclear species at a time. Figure 10.1 shows a ^{13}C solid-state NMR spectrum of bone as an example. The absorbance is plotted versus a scaled resonance frequency and yields distinct peaks at specific frequencies.* The most abundant organic species in bone is collagen, and this spectrum is largely that of collagen, although there are a few other components visible in this spectrum that are of some interest and that are discussed later in this chapter.

Assigning a spectrum such as that in Figure 10.1 is always the first challenge in solid-state NMR. The process is greatly eased by the fact that species in similar chemical functional groups and similar environments give signals at closely similar frequencies, so a complex spectrum such as that in Figure 10.1 may be assigned to a large extent by examining spectra of model compounds, as in Figure 10.1. For ^{13}C specifically (and to a lesser extent ^{15}N), there is such a wealth of data available that there are web-based databases of ^{13}C *chemical shifts*† to assist with assignments, as well as many very good programs, such as Collaborative Computing Project for NMR (CCPN) for proteins (CCPN software, www.ccpn.ac.uk), which suggest possible assignments using such databases.

The chemical shift depends not only on the chemical functional group that the (in this case) ^{13}C nucleus is part of; it depends on the detailed geometry and bonding around the nucleus. So for a protein, the ^{13}C and ^{15}N chemical shifts of nuclei in the back bone, that is, α-carbon, peptide carbonyl carbon, and nitrogen, will have different chemical shifts for the same amino acid residue species in an α-helix and a β-sheet structure, the latter giving ^{13}C C_α and CO chemical shifts that are of order 3–6 ppm lower than for a random coil structure; the collagen or proline helix has lower chemical shifts still for its backbone nuclei. This dependence of chemical shift on molecular geometry allows qualitative information on molecular structure to be obtained directly from the NMR spectrum. Hydrogen bonding differences are another cause of chemical shift differences between otherwise similar chemical functional groups and molecular geometries,

* By convention, the measured frequency shift relative to a reference is what is plotted on an NMR spectrum, where the frequency shift, δ, is labeled parts per million or "ppm" on the spectrum with δ given by $\delta = (\nu - \nu_{ref})/\nu_{ref} \cdot 10^6$, in which ν is the signal frequency and ν_{ref} is the reference frequency.

† Chemical shifts are signal frequencies relative to a defined reference frequency, measured in ppm.

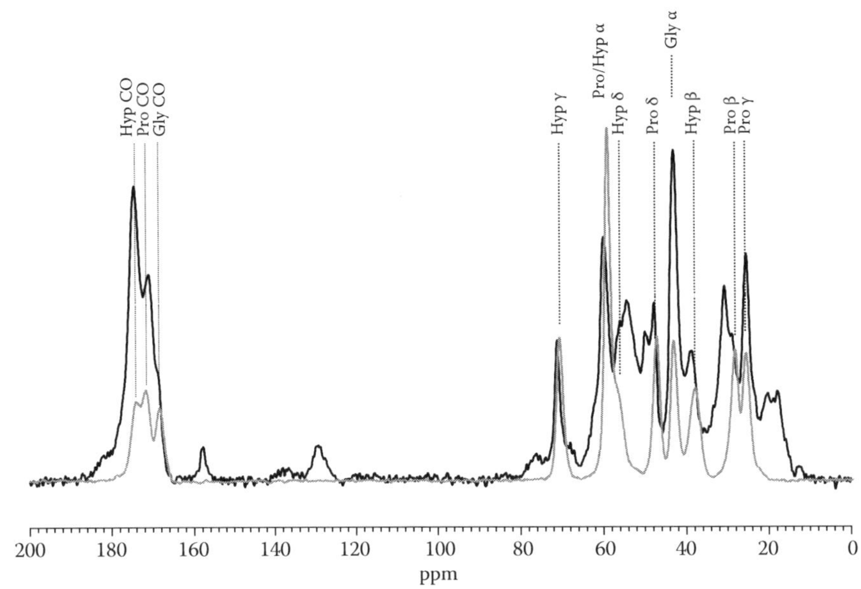

Figure 10.1 Black trace: ^{13}C (cross polarization, MAS) solid-state NMR spectrum of intact equine bone. The main component represented in this spectrum is collagen (type I) from the organic matrix of the bone. Grey trace: ^{13}C (cross polarization, MAS) solid-state NMR spectrum of $(Gly\text{-}Pro\text{-}Hyp)_{10}$, a model of the collagen triple helix. The assignment of this spectrum is shown and allows the assignment of much of the bone spectrum. Note, however, that other non-Pro/Hyp residues will contribute to the carbonyl (CO) signal between 174 and 179 ppm.

and again, this can provide useful qualitative information on the relative strength of hydrogen bonds.

Figure 10.2 compares the ^{13}C NMR spectra of two model peptides and makes a useful point about what causes the linewidths of NMR signals: as is seen in Figure 10.2, the widths of signals from the larger triple-helical $(Gly\text{-}Pro\text{-}Hyp)_{10}$ peptide are significantly larger than those for the simpler Gly-Pro-Gly-Gly peptide. The linewidth in an NMR spectrum is a reflection of heterogeneity or, to put it another way, is determined by the range of different environments for a particular chemical functional group within a sample. For the GPGG peptide, separate signals are resolved for every carbon atom in the molecule; thus, there are three Gly C_α signals, three Gly ^{13}CO signals, one Pro C_α signal, and so on, each of which is sharp as the signal represents a single environment. For the $(GPO)_{10}$ triple helix, we expect $3 \times 10 = 30$ Gly C_α signals, 30 Gly ^{13}CO signals, etc. The NMR spectrum is not able to resolve all these signals (which are likely to be very closely spaced in frequency, as the environments are very similar), and so the result is a broader band of unresolved signals.

This dependence of the chemical shift on the detailed molecular geometry and bonding means that NMR spectra act as detailed "fingerprints" of the underlying molecular structures in a material, even if the details of the structure, or even the composition, are not known. Thus, it is relatively easy to spot whether two materials have similar underlying molecular structures simply by measuring NMR spectra of relevant nuclear species; likewise, it is easy to spot if a process on the sample has *changed* underlying molecular structures. In the latter case, one may not be able to say what the structural change is, but one can say which functional groups it affects, providing the spectrum can be assigned sufficiently.

An important question is how different do two molecular structures have to be to show different NMR spectra? The qualitative answer is not much. Protein ^{13}C solution-state chemical shifts change by order of −0.2 ppm for C_α, CH, and CH_3 sites and by order of −0.1 ppm for CH_2 sites when the pressure is changed from 3 to 200 MPa, due to associated bond length and angle changes, for instance (Wilton et al., 2009). As a practical example, Figure 10.3 shows the ^{31}P solid-state NMR spectrum of octacalcium phosphate (OCP), a mineral of relevance in bone mineralization. In the OCP structure (Figure 10.3), there are two crystallographically distinct phosphate groups in the apatitic layers (labeled P1 and P4), and these give resolvably different ^{31}P chemical shifts, although both are nominally tetrahedral phosphate groups. The phosphate group that interacts with water in the hydrated layer of the OCP structure (P2) has a significantly different chemical shift again (~2 ppm), because of its hydrogen bonding interaction with a water molecule. The formal hydrogen phosphates, P5 and P6, have much lower chemical shifts (−1.2 and +0.06 ppm, respectively), characterizing their protonation state; again, there is a resolvable chemical shift difference between these ^{31}P sites because of the small structural differences between them.

The P3 site in OCP is interesting from an NMR point of view: formally an orthophosphate group, it is (from its position in the unit cell) very strongly hydrogen bonded to P5 and P6 and to water. A first-principles chemical shift calculation (Davies et al., 2012) puts the ^{31}P chemical shift of this species at −3 ppm, demonstrating that the effect of the hydrogen bonding is to make P3 appear as a partially protonated phosphate group in terms of electronic structure. In fact, at room temperature, the signal for P3 appears with that of P5 at −1.2 ppm, because of the other feature to which NMR is sensitive: *molecular dynamics*.

The reason that NMR is sensitive to molecular dynamics is simple: as shown earlier, each distinct chemical site is associated with a distinct chemical shift. If there is molecular motion such that a nucleus is moving between different sites, then its chemical shift changes with time. If the motion is on a timescale similar to that of the NMR measurement (i.e., correlation time of 10^{-3}–10^{-6} s), then the chemical shift for the nucleus is partially

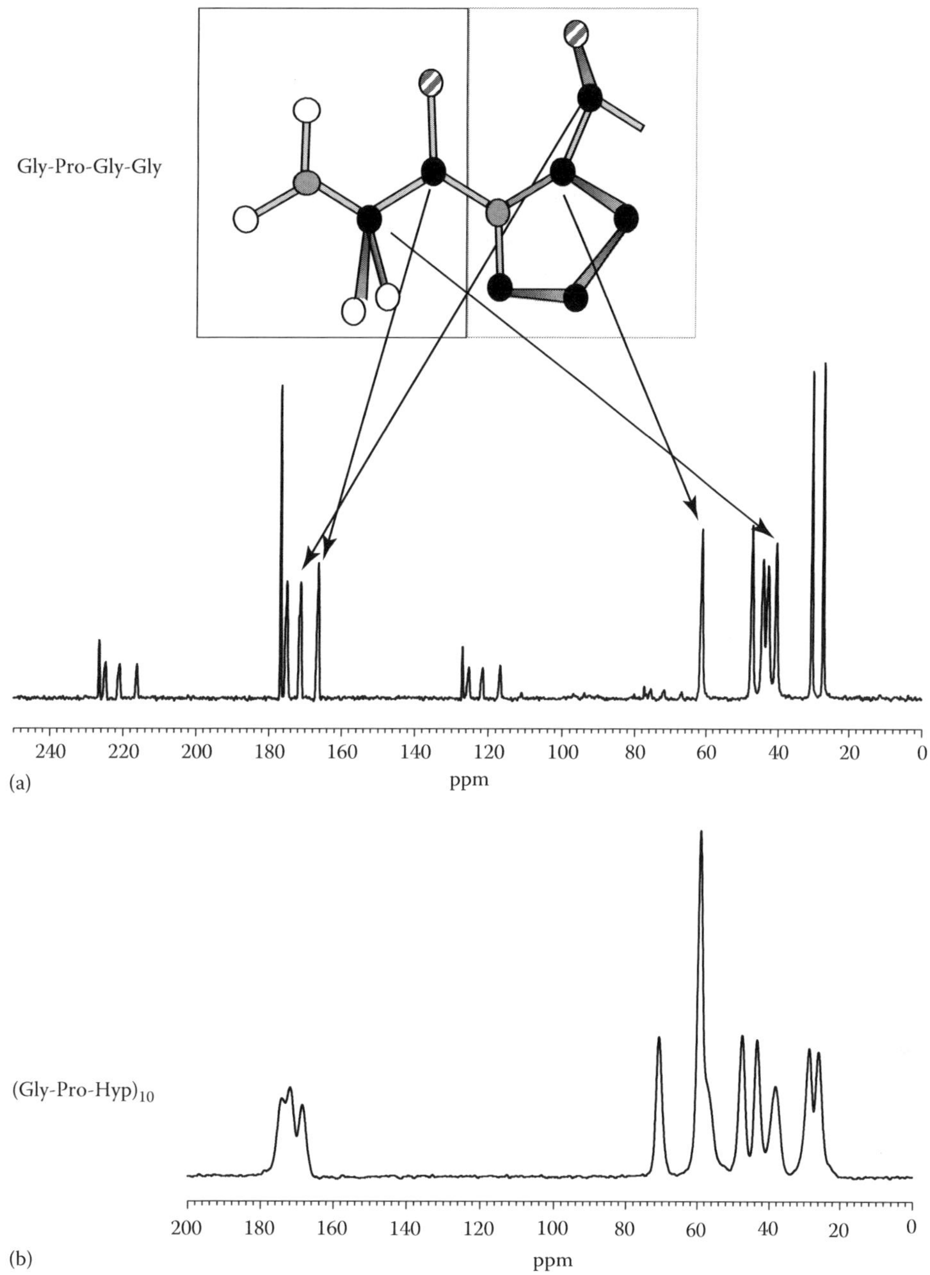

Figure 10.2 (a) ^{13}C (cross polarization, MAS) solid-state NMR spectrum of the peptide Gly-Pro-Gly-Gly (random coil/no particular secondary structure; black atoms, carbon; grey atoms, nitrogen; white atoms, hydrogen; hatched atoms, oxygen). The first two residues of the peptide are shown and the assignment of their α-carbon and carbonyl carbons indicated. (b) ^{13}C (cross polarization, MAS) solid-state NMR spectrum of the peptide $(Gly\text{-}Pro\text{-}Hyp)_{10}$ that is triple helical and a model for triple-helical collagens. The linewidths of the signals in this spectrum are larger than those of the top spectrum, simply because there are more residues in the peptide and thus a greater range of physical environments for the ^{13}C nuclei in each residue type. Note that the Pro α-carbon signal in the GPGG peptide (random) coil is at a higher chemical shift than that in the triple-helical $(GPO)_{10}$ peptide. It is also significant that the chemical shift of the Gly CO shown for the GPGG peptide is significantly lower in chemical shift than the other Gly CO (not shown) in the peptide. This is because this Gly CO is bonded to an *imino* rather than an amino acid (i.e., Pro) as the other Gly COs in the peptide are. This is discussed further in Section 10.2.2.

averaged between those associated with the different sites it moves between. The result is that the lines associated with the different sites involved in the motion first *coalesce* then collapse into a single broad line that sharpens as the motional timescale shortens. In OCP, the motion affecting P3 also involves P5 and is a hopping of the protons involved in the hydrogen bonding. Note that the motion in this case probably does not involve the observed ^{31}P nucleus itself moving, rather the environment around the ^{31}P nuclei moving or changing with time.

Although the ability to fingerprint complex molecular structures and qualitatively assess molecular geometry is highly useful and the everyday bread and butter of NMR, it is the ability to determine internuclear distances quantitatively and to assess which nuclear sites are close in space on a relative scale that makes NMR so powerful. Figure 10.4 shows schematically the principle behind this: 2D spectra. The 2D NMR spectrum in Figure 10.4 consists of a projection on each of the two spectral axes, consisting of the normal 1D NMR spectrum for the sample;

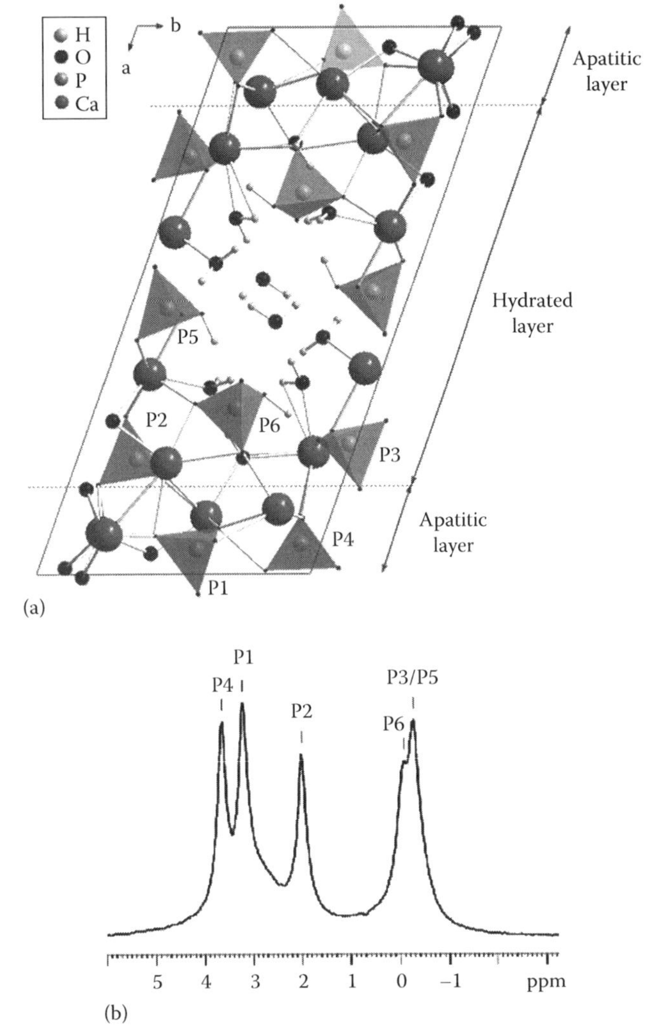

Figure 10.3 (a) The crystal structure of OCP showing the six different phosphorus environments in the unit cell. (b) The ^{31}P (MAS, direct polarization) NMR spectrum of OCP with the assignment of the signals according to the earlier structure. Note that separate signals are resolved for every phosphate site, except P3 and P5, whose signals are coalesced due to mobile protons in the structure that hop between these phosphate groups, thus varying the environment of these phosphorus sites with time. Even the two orthophosphate groups, P1 and P4, have well-resolved signals, despite having very similar structures for the two phosphate groups. (From Davies, E. et al., *J. Am. Chem. Soc.*, 134(30), 12508, 2012, doi:10.1021/ja3017544.)

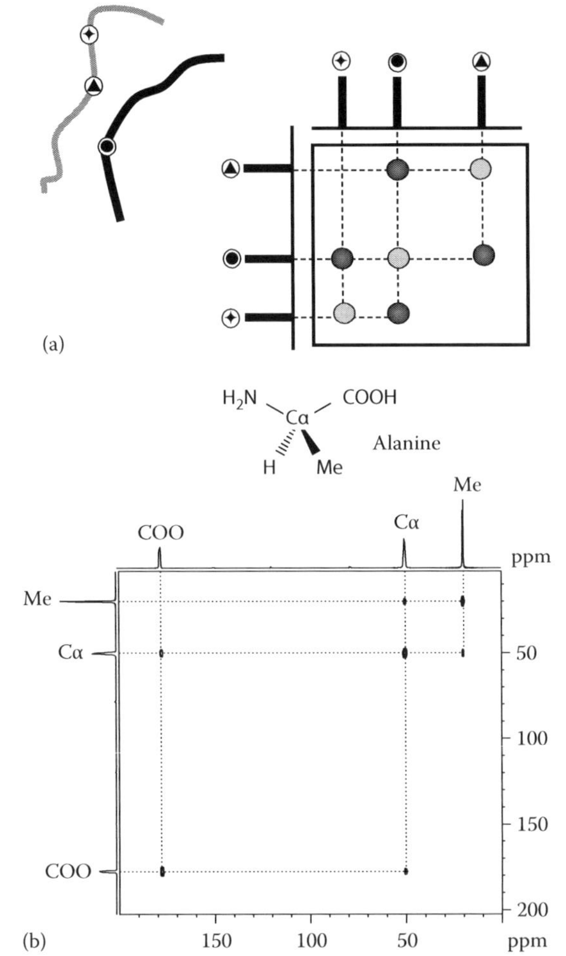

Figure 10.4 (a) Schematic of a 2D NMR correlation spectrum. The nuclear sites that give rise to signals are indicated with ♦, ●, and ▲ symbols. The projections on each of the two spectral axes shown are the normal 1D NMR spectra for the given nucleus type (homonuclear in this case, i.e., all the same nuclear species). Each signal gives rise to a so-called *autocorrelation* peak along the leading diagonal of the spectrum (pale grey). Signals from sites that are close in space also give rise to off-diagonal *correlation* peaks (dark grey). Thus, ▲ shows a correlation peak with ● and ● with ▲. (b) A simple experimental example of a ^{13}C-^{13}C correlation spectrum of U-^{13}C-alanine. The mixing time of the correlation experiment has been set so that only the closest ^{13}C sites give cross peaks, that is, C_α with $^{13}COOH$ and $^{13}CH_3$ (Me = methyl).

to illustrate the principles here, this sample is assumed to have three different sites for the observed nucleus, designated by the different symbols, as shown in the figure. In the 2D plane of the spectrum, signals appear as standard down the main diagonal of the spectrum, connecting the signals along each spectral axis from the same species. The interesting signals are the off-diagonal ones as these *connect signals from species that are close in space.* Thus, the ● and ♦ sites are close in space and show a cross peak between them, likewise the ● and ▲ sites.

This form of spectrum is known as a (2D) *correlation spectrum* because it correlates signals for nuclear species that are close in space. The particular correlation experiment type just described uses the *magnetic dipolar coupling* usually just called the *dipolar coupling* between nuclei to determine the nuclear correlations. The dipolar coupling is a through-space interaction that drops off as $1/r^3$, where r is the internuclear distance, and the experiments can be designed such that only correlations arising from dipolar couplings larger than a certain value can effectively be observed. Note that the ● and ▲ pair of sites in the example in Figure 10.4 are not in the same molecule but are nevertheless close in space because of the molecular packing; this is a particularly powerful feature as in many biomaterials, the aggregation geometry of molecules in the material is as important as the individual molecular structures.

The fact that the 2D correlation experiment is sensitive to internuclear distance gives a method for *measuring internuclear distances* via this type of experiment. This uses the dipolar coupling to investigate which nuclei are close in space. The effects of dipolar coupling are usually removed from solid-state NMR spectra by spinning the sample rapidly about an axis at the so-called magic angle with respect to the applied field (see later), but in a 2D correlation spectrum, the experiment is arranged so that the dipolar coupling between nuclei operates for a fixed period of time, the so-called mixing time, in the NMR experiment, determined by the experimenter. By increasing this mixing time in successive experiments, one can follow the buildup of cross-peak intensities and the buildup curves analyzed to determine *r*, the internuclear distance for the particular pair of nuclear sites involved in the cross-peak correlation (see, for instance, Middleton (2011) for a more detailed discussion of the area). Thus, one can in principle determine internuclear distances for any pair of nuclei, which give resolved signals and for which a sufficiently detailed buildup curve can be measured. This principle can also be used to edit the 2D correlation spectrum, for instance, by setting the mixing time so that only cross peaks between the closest sites are observed, as demonstrated in Figure 10.4b, where the mixing time has been set so that only cross peaks from ^{13}C nuclei within a normal C–C bond length of each other give rise to cross peaks.

The range of internuclear distances that can be measured depends on type of nucleus, as the dipolar coupling also depends on the nuclear gyromagnetic ratio, effectively a measure of the strength of magnetic field associated with the nucleus. ^{1}H has the largest gyromagnetic ratio (other than tritium), with ^{31}P being just under half the size and ^{13}C, a quarter. Roughly speaking, one can reasonably expect to measure ^{13}C–^{13}C distances of up to 4–8 Å via these methods.

Other 2D NMR experiments can distinguish between nuclei that are close because they are covalently bonded and those that are close because of intermolecular proximity. The so-called J-coupling interaction between nuclei is mediated by the bond electrons between the coupled nuclei, and experiments such as the beautifully named INEPT (Morris and Freeman, 1979) and INADEQUATE (Freeman et al., 1980) experiments utilize this to produce 2D correlation spectra in which correlations appear only between the signals of bonded nuclei (and both experiments can be arranged so that nuclei that are one, two, or even three bonds apart can be highlighted).

The principles of 2D NMR spectra can be used to design 3D, 4D, and higher-dimensional NMR spectra that correlate three, four, and more spectral frequencies, respectively. Thus, for proteins, ^{15}N–$^{13}C_{\alpha}$–^{13}CO 3D NMR spectra are commonly found where the fact that such spectra pick out nuclear species close in space is used to assign the highly complex ^{13}C and ^{15}N NMR spectra of proteins. Note that here, different nuclear types, that is, ^{15}N and ^{13}C, are being correlated according to their proximity in space. This is common and many more examples will appear in Section 10.3.

10.1.2 SAMPLE PREPARATION (OR THE LACK OF IT)

So far, we have discussed NMR spectroscopy in general terms without mentioning the *solid-state* part of it. *Solid-state* simply refers to the state of the material being studied—solid, in this case. Solution-state NMR is vastly more common but is restricted as its name suggests to materials that are soluble and for which useful information can be obtained in its dissolved state. For the majority of biominerals, the solid state is the important one, hence the focus on solid-state NMR here.

One of the beauties of solid-state NMR is the relatively mild sample preparation required compared to other nano-structural techniques such as electron microscopy. In order to remove the line broadening effects of the so-called anisotropic nuclear interactions (there are whole volumes devoted to anisotropic interactions, so they will only be described further here as necessary; for now, suffice to say, the internuclear dipolar coupling utilized in 2D spectroscopy is an example of one such interaction), most solid samples are spun at the "magic angle" while NMR spectra are recorded. Spinning rapidly at the magic angle (54.7° to the applied magnetic field in the NMR experiment) has the effect of removing these line broadening effects (Duer, 2004). The ^{13}C NMR spectrum of bone in Figure 10.1 is a magic-angle spinning (MAS) NMR spectrum for this reason. "Rapid" spinning is a spin rate several times in excess of the largest anisotropic nuclear interaction affecting the spectrum; in practice, spin rates of 5–45 kHz are typically used with samples contained in zirconia rotors 1.5–4 mm in diameter.

Obviously, to spin stably at such rates, the sample itself must by highly homogenously packed with respect to the rotor. This is most easily achieved by grinding the sample to a powder. The particle size within the powder is not critical, only that it is small enough to allow the sample to pack sufficiently to spin stably. Samples are commonly ground in cryo-ball mills, using balls made from an inert material, such as stainless steel. Milling at liquid nitrogen temperatures retains volatile components that might otherwise evaporate and prevents excessive sample heating that might damage the sample.

There are pitfalls associated with sample milling. One is that the sample dehydrates much more easily when milled to a fine powder, and dehydration is highly likely to affect the molecular structures that are being sought by NMR (Wilson et al., 2005; Zhu et al., 2009).

It may also disrupt the molecular and/or intermolecular structures to some extent. Recent work (Laurencin et al., 2010) has shown that milling bone, for example, leads to significant line broadening in the ^{13}C NMR spectra. Where possible, better alternatives include shaping the solid sample to fit exactly inside the NMR rotor (Laurencin et al., 2010) or taking slices of material, packed in the NMR rotor with an inert powder that gives no NMR spectrum in the frequency range of the sample.

As described in Section 10.1.1, the frequency ranges for different nuclei do not overlap and this means that one can record NMR spectra of, say, the mineral component of a biomineral *without having to first remove the organic matrix from the sample.* So, as will be shown in Section 10.2.1, one can record ^{31}P NMR spectra of bone mineral, with no need of physically isolating the mineral from the bone sample. As will be shown in Section 10.2.1, by judicious use of 2D correlation methods, it is also possible to record the ^{1}H NMR spectrum exclusively of the mineral phase, on an intact biomineral, without interference of the ^{1}H signals from the organic matrix in the sample.

By and large, then solid-state NMR requires minimal sample preparation compared to other techniques and therefore is capable of observing structures not able to be seen in other methods.

10.1.3 WHAT CAN GO WRONG?

As described earlier, samples can dehydrate rather rapidly when they are ground into a fine powder and this can affect the molecular structures being observed. This can be exacerbated by the NMR experiment itself that uses high-power pulses of radiofrequency, which can cause sample heating. Furthermore, spinning the sample at tens of kHz inevitably causes some frictional heating of the sample. Both heating effects can be mitigated against to some extent by actively cooling the sample during the NMR experiment. This may be needed in any case to prevent any decomposition of biological samples through enzyme or bacterial action, as well as through dehydration and protein denaturing.

In this respect, one of the big advantages of NMR, the minimal sample preparation required, can also be one of its biggest weaknesses. Whereas samples for electron microscopy, for instance, are dehydrated and fixed, which preserves the sample against most forms of decomposition, there is little to prevent a sample prepared for solid-state NMR undergoing enzymatic and/or bacterial decomposition. Even if a sample is prepared under sterile conditions, it is more or less impossible to maintain these during the NMR experiment—the NMR rotor must have some sort of vent in it to allow stable spinning. For structural methods examining longer lengths scales, this issue may not be so critical, but NMR is a very sensitive observer of atomic-level structure and the spectra will show the effects of decomposition, but without knowledge of what the spectrum of a fresh sample looks like, it is extremely difficult to know what spectral features are due to sample alteration postmortem. For example, Figure 10.5 shows some NMR spectra of bone samples prepared in different ways and stored for different lengths of time. The NMR experiment used here is the so-called $^{13}C\{^{31}P\}$ REDOR experiment (Gullion and Schaeffer, 1989). In this experiment, two ^{13}C NMR spectra are recorded, a reference spectrum and the REDOR spectrum. The reference spectrum is effectively a normal ^{13}C NMR spectrum of the sample; the REDOR spectrum is recorded after a train of radio-frequency pulses are applied to ^{31}P and results in a ^{13}C spectrum in which signals due to carbons that are close in space to phosphorus are reduced in intensity. Thus, the comparison of the two spectra allows immediate determination of the organic species in close proximity to bone mineral.

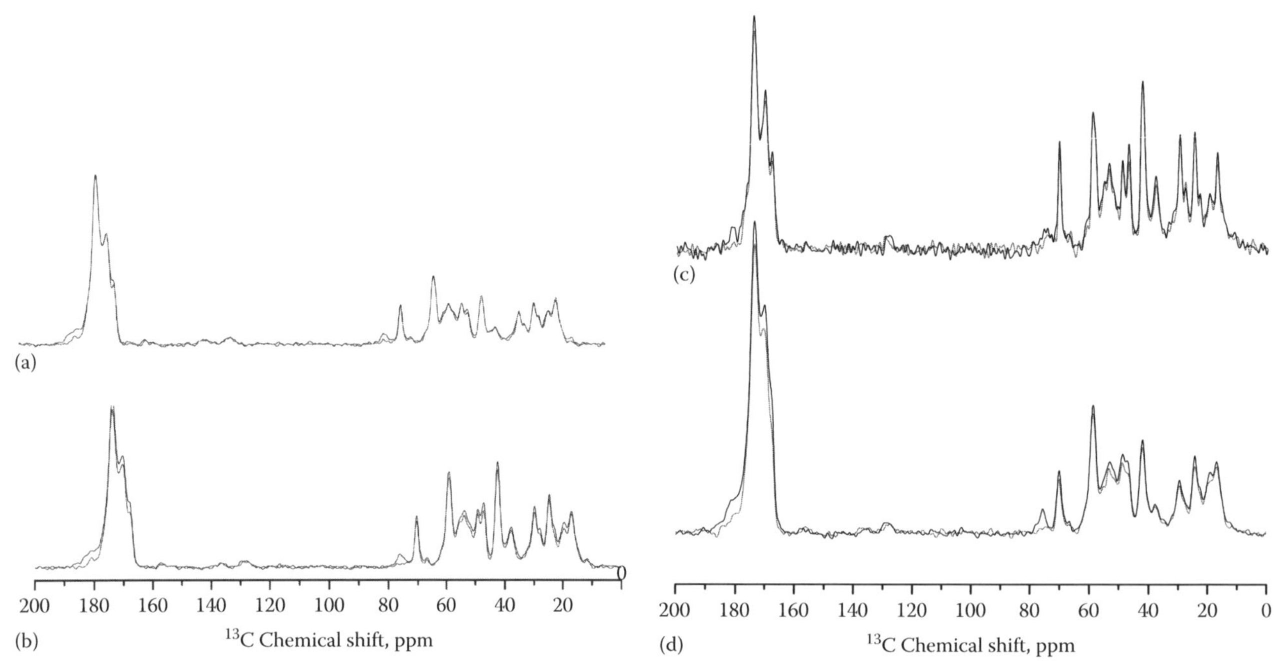

Figure 10.5 (a) $^{13}C\ \{^{31}P\}$ REDOR spectrum of fresh equine fetal bone (metacarpal, 6 months gestational age; stored at −80°C for less than 1 week, ground, and spectrum recorded immediately). The black line is the reference spectrum and comprises a normal ^{13}C NMR spectrum of bone. The gray line is the REDOR spectrum where signals dephase according to the proximity of the ^{13}C site to ^{31}P in the sample. In this spectrum, signals due to carboxylate groups (180–182 ppm), mineral carbonate (168 ppm), glycosylation/GAG sugar rings and citrate (74–78 ppm), and methyl groups (14 ppm, GAGs and/or lipids) are the main signals to dephase indicating that these organic groups are in close proximity (<6 Å) to mineral. (b) $^{13}C\ \{^{31}P\}$ REDOR spectrum of the same sample recorded under identical spectral conditions after storing the ground sample at −18°C for 6 weeks. Now most of the signals in the spectrum dephase, probably as a result of sample dehydration or decomposition (by enzymes or bacteria) and calcium and phosphate ions previously in solution in the sample precipitating out on the remaining organic matrix. (c) $^{13}C\ \{^{31}P\}$ REDOR spectrum of fresh, unground rabbit limb bone (stored for less than 1 week at −80°C). Only protein carboxylate signals (180–182 ppm), GAG carboxylate signals, and mineral carbonate (168 ppm) show any significant dephasing and are therefore known to be close to bone mineral. The rest of the GAG signals (sugar ring 74–78 ppm and 53–55 ppm; methyl groups 16–18 ppm) and any citrate in the sample (methylene CH_2, 44–49 ppm, quaternary carbon ~76 ppm) are too mobile to show any dephasing due to proximity to mineral, presumably because they are in an aqueous phase, so only the bound functional groups show proximity to mineral. (d) $^{13}C\ \{^{31}P\}$ REDOR spectrum of ground equine limb bone after storage for several weeks at −18°C. Now dephasing of characteristic GAG sugar ring and methyl side chains and citrate appears, as well as dephasing of other signals. This sample has dehydrated (which can be shown by ^{1}H NMR), with a result that species which in fresh bone would be sitting in an aqueous phase have now become solid-like and sit in close proximity to bone mineral. It is worth noting that it is the mobility of GAGs and citrate in the fresh bone sample in (a) that results in a lower intensity for the aqueous phase signals for these species (sugar ring 53–55 ppm and 74–78 ppm; side chain methyl groups 16–18 ppm). The spectra are all recorded using the so-called *cross polarization* from ^{1}H; the ^{13}C signal intensity is *all* derived from nearby ^{1}H via the mutual magnetic dipolar coupling between ^{13}C and ^{1}H. Any process, such as molecular motion that diminishes the dipolar coupling interaction, necessarily also diminishes the intensity of the resulting ^{13}C signal; indeed, signals from lipids (highly mobile species, giving the sharp signals in spectrum (a)) appear more intense than many of the sugar and protein side chain signals, indicating the degree of mobility of these latter components in natively hydrated samples.

Characterization of atomic and molecular structure: Spectroscopy and spectromicroscopy

These spectra show that the effects, primarily of sample dehydration, exacerbated by grinding the sample to a fine powder, are significant and could lead to unwarranted conclusions as to the molecular organization of the sample.

10.2 APPLICATIONS OF NMR SPECTROSCOPY

10.2.1 MINERALS

NMR spectroscopy can be used as a basic characterization tool to assess the composition and degree of crystallinity of the mineral phase in most biominerals. 2D methods can be used to give more detailed structural information at the molecular level and have been the source of some critical information on mineral structures. For example, Figure 10.6a shows the ^{1}H-^{31}P 2D HETCOR

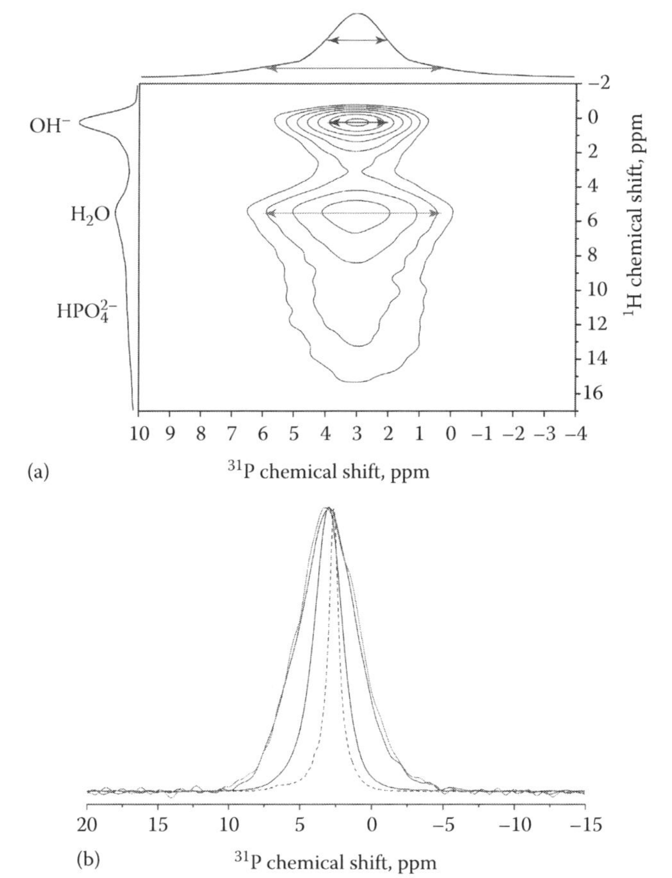

Figure 10.6 (a) ^{1}H-^{31}P 2D HETCOR spectrum of equine limb bone. The ^{1}H-^{31}P 2D correlation spectrum of bone has the normal ^{1}H spectrum of bone mineral on the vertical axis and the ^{31}P NMR spectrum of bone mineral on the horizontal axis. The ^{1}H signal from apatitic OH groups is correlated with a relatively sharp ^{31}P signal (indicated by the black arrow), which shows these sites are in a relatively ordered phase (shown more clearly in (b)), while ^{1}H in HPO_4^{2-} and H_2O are correlated with much broader ^{31}P signals (indicated by the gray arrow), showing that these species are associated with a much more disordered phase (again, shown more clearly in (b)). (b) ^{31}P spectra obtained by taking slices through the ^{1}H-^{31}P 2D HETCOR spectrum in (a). ^{31}P slices shown are those correlated with the OH^- (black, narrower signal), H_2O, and HPO_4^{2-} (both gray, broader signals) ^{1}H signals, showing the different linewidths for the phosphate species closest in space to these ^{1}H species. A ^{31}P spectrum of crystalline hydroxyapatite (dashed line) is shown for comparison.

spectrum of a sample of equine limb bone (utilizing the through-space, dipolar coupling). Such a spectrum was first recorded by Chuck Bronimann and coworkers in 1984 (Santos et al., 1994), a remarkable feature for the NMR equipment available at the time, then revisited in the light of new hypotheses in 2003 (Cho et al., 2003) giving excellent insight into the composition of bone mineral, not previously obtainable by other methods.

The ^{1}H-^{31}P 2D correlation spectrum of the bone has the normal ^{1}H spectrum of bone mineral on the vertical axis and the ^{31}P NMR spectrum of bone mineral on the horizontal axis. First, note that we know that we are only seeing ^{31}P signals from bone mineral even though the sample is intact bone; the selectivity of NMR by nucleus type ensures we only observe phosphorus species, and the vast majority of phosphorus in bone is in the mineral component, with only very small amounts in the organic component. Secondly note that, by using the 2D HETCOR, we are seeing ^{1}H signals also only from bone mineral as we are only seeing in this spectrum ^{1}H signals from ^{1}H nuclei that are close in space to ^{31}P nuclei, which in turn are largely confined to the bone mineral; thus, we have achieved a ^{1}H spectrum of bone mineral, without signals from the organic matrix interfering and crucially without having to first physically remove the organic matrix from the sample. As an aside, this spectrum was recorded on intact bone; there is no need to remove the organic component of the sample in order to observe a ^{31}P NMR spectrum of the mineral component. The organic component of bone contains very little phosphorus; the very vast majority is in the mineral phase and so that is what is observed in the ^{31}P NMR spectrum.

Clearly, the 1D ^{31}P NMR signal in Figure 10.6a (horizontal projection) overlaps with the ^{31}P signal from the orthophosphate ions in hydroxyapatite, as shown in Figure 10.6b, and so it is tempting to say that this spectrum shows that bone mineral consists of hydroxyapatite. But the ^{31}P spectrum contains further information that we have overlooked: the 1D bone ^{31}P signal is considerably broader than that of pure hydroxyapatite, indicating that there are a much larger variety of ^{31}P environments in bone mineral than in pure hydroxyapatite. We should not ignore this but seek to account for this variety of phosphorus environments in order to provide a better model of bone mineral. One possibility is that the bone mineral particles are very small and so surface effects may become visible in the NMR spectrum. The surface atomic structure of any particle will in general be different to that of the bulk. For an ionic crystal lattice, ions in the bulk are coordinatively saturated, in positions of equilibrium with respect to the electrostatic forces provided by the other ions around any particular one. Ions on the surface, on the other hand, are coordinatively unsaturated—they have "dangling" bonds and ions missing from their coordination sphere. Thus, they will in general not be in positions of equilibrium with respect to the electrostatic forces provided by the ions around them in the surface. Thus, ions at the surface will move, may become coordinated to other species, etc., in order to minimize their energy. They will, therefore, have quite different environments to those ions in the interior of the crystal and thus different NMR signal frequencies.

Surface effects of this nature may well be the main cause of the broad ^{31}P NMR signal for bone mineral, but without knowing more about the mineral particle surface structures, it is difficult to make this conclusion for sure. This is where the correlation details in the 2D plane of the HETCOR spectrum come into play.

The crystal structure of hydroxyapatite contains just one hydrogen site, namely, the hydroxyl group, that sits at the center of a triangular column of calcium ions. The ^{1}H signal from this hydroxyl site is indicated in Figure 10.6a. In the 2D plane of the spectrum of Figure 10.6a, this hydroxyl ^{1}H NMR signal is correlated with a relatively *sharp* ^{31}P signal, that is, the hydroxyl proton is close by to a relatively well-ordered, single ^{31}P environment. Such ^{1}H and ^{31}P environments have all the characteristics we would expect for the hydroxyapatite structure, and so we assign these $^{1}H/^{31}P$ signals to relatively crystalline hydroxyapatite.

But there are two more ^{1}H signals in the ^{1}H dimension of the 2D spectrum, and neither is expected for the hydroxyapatite structure. From their chemical shifts, the ^{1}H signal at ~5.5 ppm is due to water hydrogens, and the very broad signal at ~15–20 ppm is due to the hydrogens in hydrogen phosphate ions (Wilson et al., 2006). Both ^{1}H signals are themselves broad and correlated in the 2D plane with broad ^{31}P signals. The ^{31}P environments represented by these signals are the ones responsible for the broad ^{31}P signal for bone. The broadness of both ^{1}H and ^{31}P lines means that either there is a broad distribution of environments available to these sites or there is some molecular dynamics associated with these sites. Given the presence of water, molecular dynamics cannot be ruled out. The water component must be strongly bonded into the mineral, or there would be no correlation signal with ^{31}P (because the ^{1}H-^{31}P dipolar coupling that is responsible for the correlation signal would be averaged to zero by rapid, isotropic motion, resulting in no correlation signal and thus no ^{1}H water signal). The presence of HPO_4^{2-} ions in bone mineral is unequivocally established by this 2D correlation spectrum and was suggested to explain the apparent calcium deficiency of bone mineral when this finding was published in 2003.

For comparison, Figure 10.7 shows the ^{1}H-^{31}P 2D correlation spectrum from a barnacle shell mineralized with phosphatic mineral (*Ibla cumingi*) (Reid et al., 2012), showing that the mineral phase in this case is *not* related to hydroxyapatite; there is no ^{1}H O-^{1}H signal. Indeed, the only ^{1}H signal is a very broad signal in the hydrogen phosphate ^{1}H region (10–20 ppm), showing that this is the predominant phosphate species in this biomineral.

Whether the broad ^{1}H and ^{31}P lines are due to static disorder/heterogeneity or molecular dynamics is still to be established; to establish for certain, it is necessary to be absolutely sure that the hydration state of the bone sample is as in the native tissue, and this is very tricky to determine, as described in Section 10.1.3. But progress is being rapidly made, and it is highly likely that NMR will make yet further progress into establishing a clear model for bone mineral structure.

Other NMR-active nuclei that are useful for examining mineral composition and structure are ^{29}Si, which has been used to determine the ratio of Q^n species, that is, $SiO_n(OH)_{4-n}$ tetrahedral units (Bertermann et al., 2003), for instance, via MAS NMR and ^{13}C to study carbonate minerals, of which there are many interesting examples in the literature. One of particular note, in that it demonstrates what NMR can do over and above other characterization techniques, is the discovery

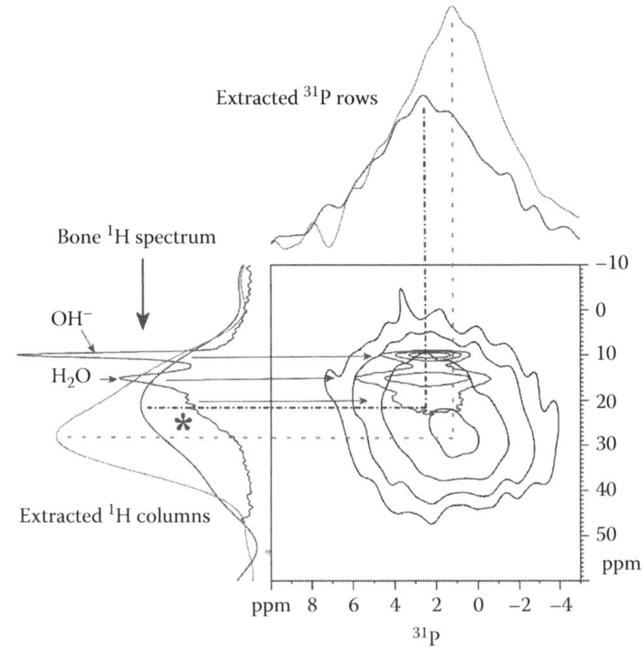

Figure 10.7 ^{1}H-^{31}P 2D correlation spectrum of barnacle (*Ibla cumingi*) shell, which is based on a calcium phosphate mineral phase (and chitin organic matrix). The 1D spectra on the spectra axes are projections of the 2D data, that is, the sum of the data in the given dimension. The broad lines in both the ^{1}H and ^{31}P spectral dimensions show that there is a wide variety of environments for both nuclear species. The ^{1}H signal is mainly associated with the hydrogen of hydrogen phosphate ions and the likewise ^{31}P signal. There is neither an indication of a hydroxyapatite-like hydroxyl group nor orthophosphate. Slices through the 2D spectrum show that the high-frequency ^{31}P sites are associated with lower-frequency ^{1}H sites, indicating that there is some order in the mineral phase. (From Reid, D.G. et al., *J. R. Soc., Interface/R. Soc.*, 9(72), 1510, 2012, doi:10.1098/rsif.2011.0895.)

that an amorphous calcium carbonate layer coats the aragonite particles in nacre (Nassif et al., 2005); solid-state NMR easily distinguishes amorphous or disordered phases from ordered ones on the basis of linewidth (much broader in disordered materials reflecting the larger range of physical and chemical environments) and relaxation times for observed signals, which are likely to be faster in a more mobile, disordered phase, for instance.

More recently, ^{43}Ca (Laurencin et al., 2010; Xu et al., 2010) and ^{17}O (Wu et al., 1997) have been target nuclear species for biominerals (for ^{43}Ca) and biomineral-like species, such as hydroxyapatite in the case of ^{17}O. Both these nuclei are so-called quadrupolar nuclei. This means that as well as interacting with the applied magnetic field of the NMR experiment via their intrinsic magnetic dipole moments, they also have an electric quadrupole moment that interacts with local electric field gradients. This results in broadening of the associated NMR signal, into structured *powder patterns* in the case of powder samples. The line broadening is not easily removed although there are 2D schemes (multiple-quantum MAS [MQMAS] [Frydman and Harwood, 1995] and satellite-transition MAS [STMAS] [Gan, 2000], for instance) that will remove the broadening in one of the spectral dimensions, but these are beyond the scope of this review. Becoming more useful at higher applied magnetic field strengths, double-rotation (DOR) NMR

(Samoson et al., 1988; Chmelka et al., 1989) involves spinning the sample simultaneously at both the magic angle (as in MAS) and a second angle (70.12°) that has the effect of averaging out both the normal anisotropic interactions that affect the sample and all components of the quadrupole interactions. Although it requires a sophisticated piece of equipment to achieve the double spinning, the resulting spectra are easy to interpret, whereas the 2D schemes require more knowledge of NMR if the raw data from the experiments are to be successfully processed. As an example of what can be achieved, Figure 10.8 shows an ^{17}O DOR NMR spectrum of bone (the first of its kind!) and compares it with a straightforward MAS spectrum. The main signals, between 90 and 120 ppm, are the oxygens of mineral phosphate groups; the low-frequency shoulder on this broad signal is due to those phosphate oxygens that are hydrogen bonded or covalently bonded to hydrogen, that is, ^{17}O–H groups in hydrogen phosphate groups. Water ^{17}O signals appear below 0 ppm, while the oxygens of the organic matrix are at high frequency.

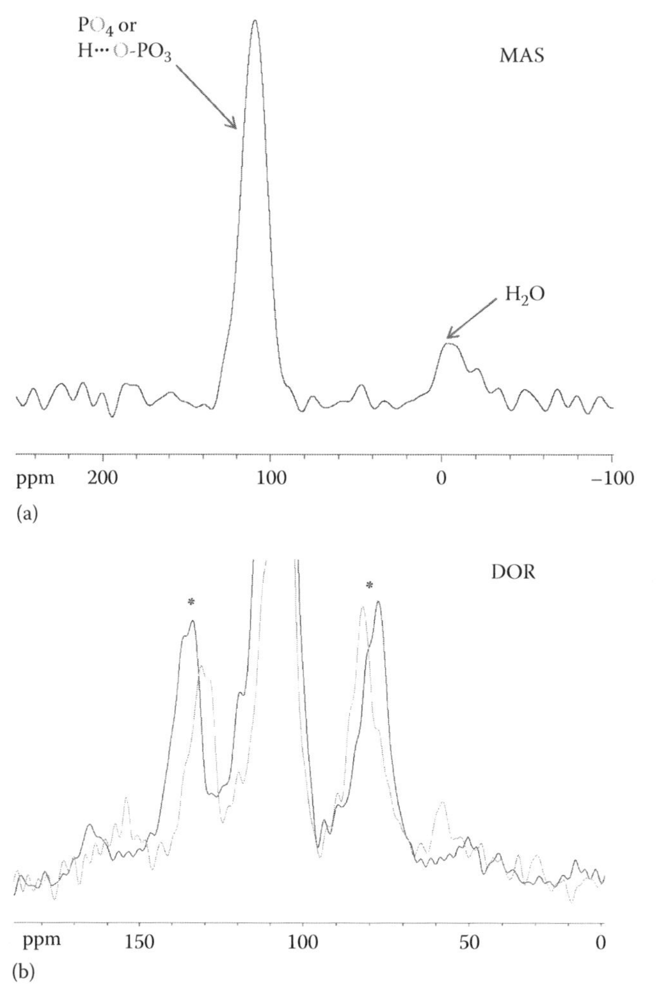

Figure 10.8 (a) ^{17}O MAS NMR spectrum of equine bone, with assignments indicated. The signals in this spectrum are subject to the influence of quadrupolar broadening. (b) ^{17}O DOR NMR spectrum of the same bone sample focusing on the orthophosphate region of the spectrum. Light line in the DOR spectrum corresponds to O indicated as gray letters in MAS labels. *Spinning sidebands (experimental artifacts due to slow sample spinning rate of the DOR outer rotor relative to the quadrupolar broadening and ^{1}H-^{17}O dipolar coupling).

10.2.2 ORGANIC MATRIX

As shown in Section 10.1.1, ^{13}C NMR readily observes the organic matrix of biominerals, with sensitivity to the underlying molecular structures present there. Polysaccharide and protein organic matrices are equally accessible to the methodology; Figure 10.9 shows the ^{13}C spectrum of a sample of a barnacle shell from *Ibla cumingi*, whose shell is known to contain calcium phosphate mineral. As noted in Section 10.1.1, the ^{13}C and ^{15}N chemical shifts in proteins are very sensitive to the secondary structure/molecular conformation of the protein. In particular, the ^{13}C chemical shift of the protein backbone α-carbon and carbonyl/peptide carbons is strongly indicative of secondary structure, with the chemical shifts for α-carbons in α-helices 2–4 ppm more than those in random coil structures and those in β-sheets 1–2 ppm less than for the random coil structure (Spera and Bax, 1991). We have shown that the ^{13}C α-carbon and peptide carbonyl carbon chemical shifts in a collagen triple helix are lower still. For instance, the Gly α-carbon in collagen is routinely found at 42 ppm compared with 45.1 ppm in random coil structures. The Gly ^{13}CO signal is indicative of whether the Gly residue precedes a Pro/Hyp residue (~168 ppm) or a non-imino acid residue (~170 ppm), both being significantly lower in chemical shift than the Gly ^{13}CO in random coil (174.9 ppm). Presumably, forming a peptide bond with an imino, rather than amino, acid forces a more

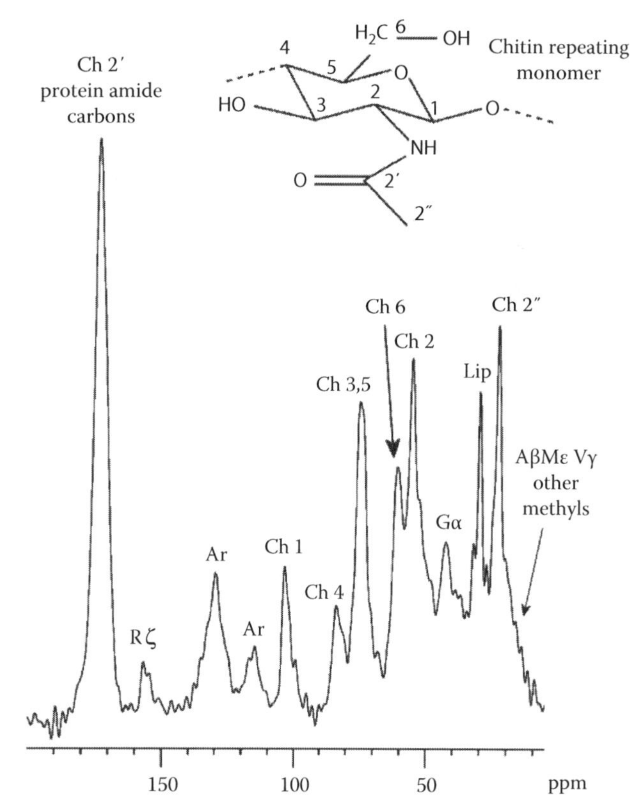

Figure 10.9 ^{13}C (cross polarization, MAS) NMR spectrum of barnacle (*Ibla cumingi*) shell, the organic component of which is primarily chitin, a polysaccharide with the repeating unit shown in the figure. The spectrum is assigned according to the chitin ^{13}C sites and other components of the organic matrix: G = glycine; A = alanine; V = valine; R = arginine; lip = lipid; Ar = aromatic species. (From Reid, D.G. et al., *J. R. Soc., Interface/R. Soc.*, 9(72), 1510, 2012, doi:10.1098/rsif.2011.0895.)

unusual geometry on the resulting peptide bond, and this is what is being reflected in the ^{13}C chemical shift.

In the collagen GXY triplet, the X, often Pro, ^{13}CO chemical shift is always lower than that for the Y, often Hyp, ^{13}CO, primarily because of differences in the hydrogen bonding that the respective CO groups participate in. The X residue participates in hydrogen bonding across the interior of the triple helix, with cross-helix Gly NH, while the Y carbonyl is oriented out of the triple helix and presumably participates in (weaker?) hydrogen bonding with surrounding water molecules and other residues in the helix where available. This marked difference in hydrogen bonding available to the X and Y members of the GXY triplet is what resolves the main peptide carbonyl spectral region into two distinct (broad) signals (plus the lower-frequency signals from Gly CO).

2D ^{1}H-^{13}C correlation spectra can assist in the assignment of ^{13}C signals and also yield the ^{1}H NMR spectrum of the organic matrix, without having to physically separate that matrix from the mineral component; compare ^{1}H-^{31}P correlation spectroscopy yielding the ^{1}H NMR spectrum of the mineral phase in phosphatic biominerals as described in Section 10.2.1.

For further molecular structural detail on the organic matrix in a biomineral sample, the ideal experiment would be ^{13}C-^{13}C 2D correlation experiments, so that the spatial juxtaposition of the different organic functional groups could be determined. However, at only 1% natural abundance for ^{13}C, the statistical likelihood of finding two ^{13}C nuclei in close proximity, whatever the molecular structure, is only $1/100 \times 1/100 = 0.01\%$, which means that the signal intensity of any ^{13}C-^{13}C correlation signal will be extremely weak.

The obvious way around this is to enrich the biomineral organic matrix in ^{13}C. We have very recently achieved this by feeding a mouse a diet in which the protein component was 30% enriched in ^{13}C. At this level of enrichment, 2D ^{13}C-^{13}C correlation NMR spectra become possible as shown in Figure 10.10. These 2D spectra can be compared with those for ^{13}C-enriched synthetic model peptides, for instance, in order to gain insight into the particular molecular structures present in the organic matrix. In the future, such experiments will probably form the basis of first-principles structure determination of crucial components of the organic matrix, such as sections of the collagen triple-helix structure in bone, in which structural constraints determined from NMR correlation experiments, computational modeling, and data from techniques such as high-resolution TEM are combined to produce accurate structure models.

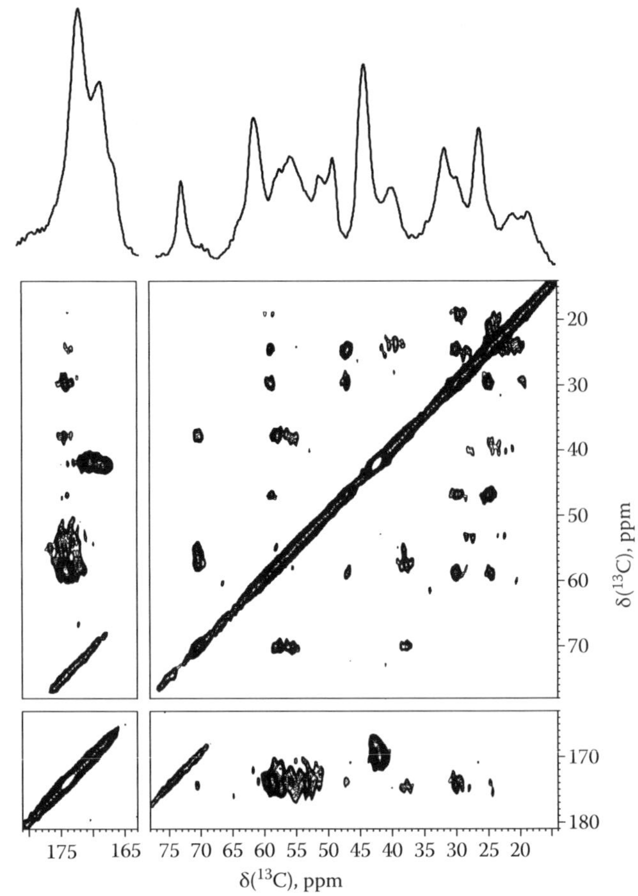

Figure 10.10 ^{13}C-^{13}C (proton-driven spin diffusion) correlation spectrum of ^{13}C-enriched mouse limb bone showing predominantly the signals from Pro, Hyp, and Gly ^{13}C. Note that the resolution between signals in the 2D plane is significantly better than that in the 1D ^{13}C NMR spectrum shown in the projections, meaning that many more signals can be assigned than in a normal 1D spectrum.

10.2.3 ORGANIC–INORGANIC INTERFACE

Arguably, one of the most important aspects of a biomineral is what holds its organic matrix and mineral components together. NMR is uniquely placed to examine this at the molecular level. Once again, it is the dipolar interaction between nuclei that are close in space that is used to distinguish the organic ^{13}C sites that are close in space to mineral species, but rather than a 2D correlation experiment, the experiment of choice is a 1D experiment. Introduced in Section 10.1.3, the so-called REDOR experiment (Gullion and Schaeffer, 1989) consists of recording a normal, 1D, in this case, ^{13}C NMR spectrum as a reference spectrum, then rerecording that same spectrum having first applied a series of radio-frequency pulses to the mineral nuclei of interest, for instance, ^{31}P in bone. The series of radio-frequency pulses has the effect of reintroducing the dipolar coupling between the ^{13}C nuclei and the mineral nuclei to which the pulses are applied, this dipolar coupling normally being averaged away to zero by the MAS routinely applied to collect solid-state NMR spectra. The resulting REDOR spectrum shows a reduction in signal intensity compared to the reference spectrum for signals associated with ^{13}C species close in space to the mineral nucleus selected.

$^{13}C\{^{31}P\}$ REDOR has shown that in mineralized collagenous tissues, that is, bone (Jaeger et al., 2005; Wise et al., 2007), tooth enamel (Reid et al., 2008), calcified cartilage (Duer et al., 2009), and pathological mineralized vascular plaques (Duer et al., 2008), it is the glycosylation components of the collagen, that is, sugar moieties covalently bound to collagen lysine and hydroxylysine residues, that are the closest components of the organic matrix to the mineral particles, possibly along with the sugar components of non-collagenous proteoglycans and along with some citrate (Hu et al., 2010). To this end, note that the main sugar signal, between 72 and 80 ppm (due to the non-anomeric ring carbons), is more intense in the calcified cartilage $^{13}C\{^{31}P\}$ REDOR spectrum than bone and tooth enamel (Figure 10.11). It seems likely that the collagen glycosylation presents a relatively

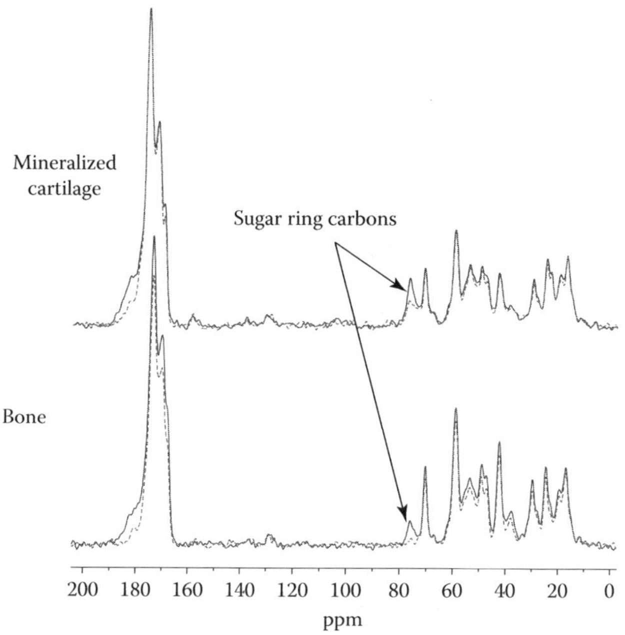

Figure 10.11 $^{13}C\{^{31}P\}$ REDOR spectra of equine mineralized joint cartilage (top) and bone (bottom). Both samples were dried so that the sugar (collagen glycosylation and GAGs) signals appear with maximum intensity (in a natively hydrated sample, these species are partially contained in a mobile aqueous phase, which reduced their signal intensity when recorded with cross polarization from ^{1}H as these spectra are in order to obtain sufficient signal-to-noise ratio for clear spectra). The broad signal between 74 and 78 ppm is due to the carbon atoms in sugar rings of collagen glycosylation and GAGs (and citrate quaternary carbon) and is more intense in the mineralized cartilage spectrum reflecting the greater GAG content of that tissue. The black solid line in each case is the reference spectrum and comprises a normal ^{13}C NMR spectrum of bone. The grey dashed line is the REDOR spectrum where signals dephase according to the proximity of the ^{13}C site to ^{31}P in the sample. Note that in these dehydrated samples, species that may normally be in an aqueous phase in the tissue under normal hydrated conditions bind or precipitate onto the tissue mineral component, leading to dephasing of the associated ^{13}C signals.

hydrophilic patch on the "exterior" of the collagen triple helix, allowing the binding/hydrogen bonding of mineral ions, where the remaining, relatively hydrophobic part of the collagen triple helix does not.

As discussed in Section 10.1.3, it is critical to ensure that the sample does not dehydrate *at all* when examining the organic–inorganic interface at this molecular level as even a small loss of water can cause ions that would normally be in the aqueous fluid of a biological sample to precipitate at any convenient nucleation site, very often on the organic matrix, giving REDOR spectra that appear to suggest that mineral is bound to parts of the organic matrix that natively it would not be.

10.3 CONCLUSIONS

Solid-state NMR spectra are highly sensitive observers of the molecular structure of not only the mineral and organic matrix components of biominerals but also of the interface between the organic and mineral components. The organic species most intimately associated with mineral are readily picked out from REDOR NMR spectra. 2D correlation NMR spectra yield valuable qualitative information on the spatial arrangement of molecular groups and have the potential to provide quantitative information. In the future, this will be used to provide structural constraints for quantitative structural modeling of key sections of the organic matrix in biominerals. Improved methodology for so-called quadrupolar nuclei such as ^{43}Ca and ^{17}O and higher magnetic fields for NMR spectrometers in the future may allow further detailed structural information on mineral phases not yet accessed.

As computational methodologies improve and resolution is gained in NMR spectra by working at higher magnetic fields, for instance, there are very real possibilities for determining accurate, quantitative molecular and crystal structures using a combined NMR/XRD/computational approach (commonly referred to as "NMR crystallography"). This approach also has the potential to deal with heterogeneous and partially disordered solids and so is likely to be used much more in future studies of biomineral atomic-level structure.

REFERENCES

Bertermann, R., Kröger, N., and Tacke, R. 2003. Solid-state ^{29}Si MAS NMR studies of diatoms: Structural characterization of biosilica deposits. *Analytical and Bioanalytical Chemistry, 375*(5), 630–634. doi:10.1007/s00216-003-1769-5.

Chmelka, B.F., Mueller, K.T., Pines, A., Stebbins, J., Wu, Y., and Zwanziger, J.W. 1989. Oxygen-17 NMR in solids by dynamic-angle spinning and double rotation. *Nature, 339*, 42–43.

Cho, G., Wu, Y., and Ackerman, J.L. 2003. Detection of hydroxyl ions in bone mineral by solid-state NMR spectroscopy. *Science* (New York), *300*(5622), 1123–1127. doi:10.1126/science.1078470.

Davies, E., Duer, M.J., Ashbrook, S.E., and Griffin, J.M. 2012. Applications of NMR crystallography to problems in biomineralization: Refinement of the crystal structure and ^{31}P solid-state NMR spectral assignment of octacalcium phosphate. *Journal of the American Chemical Society, 134*(30), 12508–12515. doi:10.1021/ja3017544.

Duer, M.J. 2004. *An Introduction to Solid-State Nuclear Magnetic Resonance Spectroscopy*. p. 400. Blackwell Science Ltd, London, U.K.

Duer, M.J., Friscić, T., Murray, R.C., Reid, D.G., and Wise, E.R. 2009. The mineral phase of calcified cartilage: Its molecular structure and interface with the organic matrix. *Biophysical Journal, 96*(8), 3372–3378. doi:10.1016/j.bpj.2008.12.3954.

Duer, M.J., Friscić, T., Proudfoot, D., Reid, D.G., Schoppet, M., Shanahan, C.M., Skepper, J.N. et al. 2008. Mineral surface in calcified plaque is like that of bone: Further evidence for regulated mineralization. *Arteriosclerosis, Thrombosis, and Vascular Biology, 28*(11), 2030–2034. doi:10.1161/ATVBAHA.108.172387.

Freeman, R., Bax, A., and Kempsell, S.P. 1980. Natural abundance ^{13}C-^{13}C coupling observed via double-quantum coherence. *Journal of the American Chemical Society, 102*(12), 4849–4851.

Frydman, L. and Harwood, J.S. 1995. Isotropic spectra of half-integer quadrupolar spins from bidimensional magic-angle spinning NMR. *Journal of the American Chemical Society, 117*(19), 5367–5368. doi:10.1021/ja00124a023.

Gan, Z. 2000. Isotropic NMR spectra of half-integer quadrupolar nuclei using satellite transitions and magic-angle spinning. *Journal of the American Chemical Society, 122*(13), 3242–3243. doi:10.1021/ja9939791.

Gullion, T. and Schaeffer, J. 1989. Rotational-echo double-resonance NMR. *Journal of Magnetic Resonance, 81*, 196–200.

Hu, Y.-Y., Rawal, A. and Schmidt-Rohr, K. 2010. Strongly bound citrate stabilizes the apatite nanocrystals in bone. *Proceedings of the National Academy of Sciences of the United States of America, 107*(52), 22425–22429. doi:10.1073/pnas.1009219107.

Jaeger, C., Groom, N.S., Bowe, E.A., Horner, A., Davies, M.E., Murray, R.C., and Duer, M.J. 2005. Investigation of the nature of the protein–mineral interface in bone by solid-state NMR. *Chemistry of Materials*, *17*(12), 3059–3061. doi:10.1021/cm050492k.

Laurencin, D., Wong, A., Chrzanowski, W., Knowles, J.C., Qiu, D., Pickup, D.M., Newport, R.J. et al. 2010. Probing the calcium and sodium local environment in bones and teeth using multinuclear solid state NMR and X-ray absorption spectroscopy. *Physical Chemistry Chemical Physics*, *12*(5), 1081–1091. doi:10.1039/b915708e.

Middleton, D.A. 2011. Crystallography and NMR: Applications to structural biology. *Encyclopedia of Magnetic Resonance*. John Wiley & Sons, Ltd., Chichester, U.K.

Morris, G.A. and Freeman, R. 1979. Enhancement of nuclear magnetic resonance signals by polarization transfer. *Journal of the American Chemical Society*, *101*(2), 760–762.

Nassif, N., Pinna, N., Gehrke, N., Antonietti, M., Jäger, C., and Cölfen, H. 2005. Amorphous layer around aragonite platelets in nacre. *Proceedings of the National Academy of Sciences of the United States of America*, *102*(36), 12653–12655. doi:10.1073/pnas.0502577102.

Reid, D.G., Duer, M.J., Murray, R.C., and Wise, E.R. 2008. The organic–mineral interface in teeth is like that in bone and dominated by polysaccharides: Universal mediators of normal calcium phosphate biomineralization in vertebrates? *Chemistry of Materials*, *20*(11), 3549–3550. doi:10.1021/cm800514u.

Reid, D.G., Mason, M.J., Chan, B.K.K., and Duer, M.J. 2012. Characterization of the phosphatic mineral of the barnacle *Ibla cumingi* at atomic level by solid-state nuclear magnetic resonance: Comparison with other phosphatic biominerals. *Journal of the Royal Society, Interface/the Royal Society*, *9*(72), 1510–1516. doi:10.1098/rsif.2011.0895.

Samoson, A., Lippmaa, E., and Pines, A. 1988. High resolution solid-state NMR: Averaging of second-order effects by means of a double rotor. *Molecular Physics*, *65*(4), 1013–1018.

Santos, R.A., Wind, R.A., and Bronnimann, C.E. 1994. 1H CRAMPS and ^{1}H-^{31}P HetCor experiments on bone, bone mineral, and model calcium phosphate phases. *Journal of Magnetic Resonance Series B*, *105*(2), 183. Retrieved from http://www.ncbi.nlm.nih.gov/pubmed/7952933.

Spera, S. and Bax, A. 1991. Empirical correlation between protein backbone conformation and Cα and Cβ ^{13}C NMR chemical shifts. *Journal of the American Chemical Society*, *113*, 5490–5492.

Wilson, E.E., Awonusi, A., Morris, M.D., Kohn, D.H., Tecklenburg, M.M., and Beck, L.W. 2005. Highly ordered interstitial water observed in bone by nuclear magnetic resonance. *Journal of Bone and Mineral Research: The Official Journal of the American Society for Bone and Mineral Research*, *20*(4), 625–634. doi:10.1359/JBMR.041217.

Wilson, E.E., Awonusi, A., Morris, M.D., Kohn, D.H., Tecklenburg, M.M.J., and Beck, L.W. 2006. Three structural roles for water in bone observed by solid-state NMR. *Biophysical Journal*, *90*(10), 3722–3731. doi:10.1529/biophysj.105.070243.

Wilton, D.J., Kitahara, R., Akasaka, K., and Williamson, M.P. 2009. Pressure-dependent ^{13}C chemical shifts in proteins: Origins and applications. *Journal of Biomolecular NMR*, *44*(1), 25–33. doi:10.1007/s10858-009-9312-4.

Wise, E.R., Maltsev, S., Davies, M.E., Duer, M.J., Jaeger, C., Loveridge, N., Murray, R.C. et al. 2007. The organic–mineral interface in bone is predominantly polysaccharide. *Chemistry of Materials*, *19*(21), 5055–5057. doi:10.1021/cm702054c.

Wu, G., Rovnyak, D., Huang, P.C., and Griffin, R.G. 1997. High-resolution oxygen-17 NMR spectroscopy of solids by multiple quantum magic-angle spinning. *Chemical Physics Letters*, *277*(October), 79–83.

Xu, J., Zhu, P., Gan, Z., Sahar, N., Tecklenburg, M., Morris, M.D., Kohn, D.H. et al. 2010. Natural-abundance ^{43}Ca solid-state NMR spectroscopy of bone. *Journal of the American Chemical Society*, *1*, 244–274. Retrieved from http://pubs.acs.org/doi/abs/10.1021/ja101961x.

Zhu, P., Xu, J., Sahar, N., Morris, M.D., Kohn, D.H., and Ramamoorthy, A. 2009. Time-resolved dehydration-induced structural changes in an intact bovine cortical bone revealed by solid-state NMR spectroscopy. *Journal of the American Chemical Society*, *131*(47), 17064–17065. doi:10.1021/ja9081028.

11 Realities of disordered or unfolded proteins: Relevance to biomineralization

John Spencer Evans

Contents

11.1 INTRODUCTION

Today, we recognize that mineralized tissues contain proteins that are important for mineral formation, tissue function, and integrity. Since the 1960s, there have been numerous studies of mineral-associated proteins and much has been learned regarding their functionalities. However, compared to other scientific fields, our understanding of biomineralization protein structure and how it contributes to function has not made sufficient progress. Although there are many reasons for this, the major difficulty lies with the unusual and "uncooperative" properties of these proteins, such as solution aggregation or the inability to crystallize from solution. Prior to 1995, biophysical studies were unable to document or detect the presence of tertiary or folded/globular structure in some biomineralization proteins (Renogupalakrishnan et al., 1986; Evans and Chan, 1994; Evans et al., 1994, 1995a,b; Matsushima et al., 1998), and thus the field of biomineralization proteomics, as it is now called, made little progress.

What the biomineralization community was not aware of at the time was that there was a seismic shift occurring in the protein folding community. By the early 1990s, the field of protein folding emerged as an important arena for understanding how nature deals with the folding and processing of globular proteins in cells (Dobson and Evans, 1984; Kim and Baldwin, 1990; O'Neil and DeGrado, 1990; Englander and Mayne, 1992). But by the mid-1990s, it was realized that there was a subset of proteins that did not fully fold yet were fully functional (Baum et al., 1989; van Mierlo et al., 1993; Wright and Dyson, 1999; Tompa, 2002; Uversky, 2002; Ward et al., 2004; Xie et al., 2007; Uversky and Dunker, 2010, 2012). These unfolded protein sequences were given the name "intrinsically disordered" or "intrinsically unfolded proteins" (IDP or IUP) (Wright and Dyson, 1999; Tompa, 2002; Uversky, 2002; Ward et al., 2004; Xie et al., 2007; Uversky and Dunker, 2010, 2012). Once the human genome was sequenced, it was found that 30% of the human proteome was indeed unfolded (Wright and Dyson, 1999; Uversky, 2002; Tompa, 2002; Ward et al., 2004; Xie et al., 2007; Uversky and Dunker, 2010, 2012). This indicates that IDPs are not a biological anomaly but a biological reality. By the year 2000, the field of protein intrinsic disorder had come into being and has now become a major scientific area in biology and chemistry (Baum et al., 1989; van Mierlo et al., 1993; Wright and Dyson, 1999; Tompa, 2002; Uversky, 2002; Ward et al., 2004; Xie et al., 2007; Uversky and Dunker, 2010, 2012).

With the recent revelation that 50% of the human extracellular matrix (ECM) proteome (Peysselon et al., 2011) and >90% of the known mollusk shell aragonite-associated proteins (Evans, 2012) are intrinsically disordered, it is now evident that IDPs could play a more significant role in the formation of biominerals than was previously believed. Hence, rather than concentrating our efforts to search for evidence of order in biomineralization proteins, it is clear that we must adopt a 180° approach and search for evidence of disorder, too. The purpose of this handbook chapter is to introduce the concept of intrinsic disorder in proteins and how this impacts protein function and biomineralization. We will conclude this chapter with a look into the future and what we expect to happen in the IDP field and how this will impact the biomineralization field.

11.2 INTRINSICALLY DISORDERED OR UNFOLDED PROTEINS AND THEIR FUNCTIONAL TRAITS

IDPs are a protein class that do not fold into a well-defined 3D structure under physiologic conditions (Baum et al., 1989; van Mierlo et al., 1993; Wright and Dyson, 1999; Tompa, 2002; Uversky, 2002; Ward et al., 2004; Xie et al., 2007; Uversky and

Dunker, 2010, 2012; Peysselon et al., 2011; Evans, 2012). They are usually defined as being nonglobular in global conformational terms with random coil content but can possess regions of defined secondary structure (Baum et al., 1989; van Mierlo et al., 1993; Wright and Dyson, 1999; Tompa, 2002; Uversky, 2002). In a sense, the disordered protein or protein domain exists as an ensemble of structures that are in equilibrium with one another. The disordered regions can exist either as an open, extended conformation or as a collapsed domain with poorly organized side chains, and transitions can occur between these extremes (Baum et al., 1989; van Mierlo et al., 1993; Wright and Dyson, 1999; Tompa, 2002; Uversky, 2002; Xie et al., 2007; Uversky and Dunker, 2010). These collapsed states were found to be very similar to the partially structured folding intermediates that have been identified in globular proteins, such as the premolten globule or molten globule (Baum et al., 1989; van Mierlo et al., 1993; Wright and Dyson, 1999; Tompa, 2002). This suggests that IDP sequences may have arisen as "castoffs" from the cellular folding process, that is, proteins that could not fold yet somehow maintained functionality for a different purpose than they were initially intended for (Baum et al., 1989).

Although the functions of all IDPs have not been completely identified, certain trends or themes have emerged and these functions are linked to their structural disorder (Tompa, 2002). The major advantage that intrinsic disorder conveys to a protein is the disorder–order transition or induced local folding that accompanies the binding of an appropriate target to the IDP (Baum et al., 1989; van Mierlo et al., 1993; Wright and Dyson, 1999; Tompa, 2002; Uversky, 2002; Ward et al., 2004; Xie et al., 2007; Uversky and Dunker, 2010, 2012; Peysselon et al., 2011; Evans, 2012). This transition has a significant conformational entropy component associated with this binding step, which uncouples binding affinity from specificity and makes for highly specific yet reversible interactions. This is important for regulatory systems where specific proteins interact with other macromolecules such as RNA or DNA (Tompa, 2002; Uversky, 2002; Xie et al., 2007; Uversky and Dunker, 2010) or, in the case of biomineralization, transient interactions with forming mineral clusters (Gebauer et al., 2008; Demichelis et al., 2011; Gebauer and Coelfen, 2011; Evans, 2012). Other functional advantages arise from IDP sequences. Their open or unfolded structures are often preserved after target binding that allows multiple contact points and large binding surfaces for other targets to bind to (Tompa, 2002). These open unfolded regions can adopt different structures upon binding to different targets, a term referred to as "binding promiscuity" (Baum et al., 1989; Tompa, 2002; Uversky, 2002; Delak et al., 2009; Bromley et al., 2011; Ndao et al., 2011). The open structure of IDPs allows for increased rates of interaction and is usually highly flexible, a trait that is important in the construction of large complexes or hierarchical assemblies. This open structure also imparts protease sensitivity, and it has been documented that IDPs are effectively controlled by rapid cellular turnover (Tompa, 2002; Uversky, 2002; Ward et al., 2004; Xie et al., 2007; Uversky and Dunker, 2010, 2012; Peysselon et al., 2011). Collectively, these types of traits are ideal for ECM construction, integrity, communication, and repair (Peysselon et al., 2011; Evans, 2012), and it is clear that the significant content of intrinsic disorder with biomineralization protein sequences may facilitate these important ECM features.

11.2.1 MOLECULAR BASIS FOR INTRINSIC DISORDER

Considerable effort has been expended to discover why some protein sequences are incapable of folding. It appears that one factor is the distinctive amino acid compositions that IDPs possess and the enrichment of P, E, K, S, Q disorder-promoting amino acids in IDP sequence regions relative to W, Y, F, C, I, L, N order-promoting amino acids, which are considered fold inducers or stabilizers (Tompa, 2002). The role of Pro in IDP sequence behavior cannot be overstated, since this amino acid increases backbone mobility, increases the likelihood of open structure such as the polyproline type II (PPII) extended helix, and facilitates conformational transformations (Baum et al., 1989; Delak et al., 2009; Bromley et al., 2011; Ndao et al., 2011). The characteristic low mean hydrophobicity and high net charge of IDP sequences prevents the formation of hydrophobic clusters and increases charge–charge repulsion, respectively, and leads to extended, open polypeptide conformations (Baum et al., 1989; van Mierlo et al., 1993; Wright and Dyson, 1999; Tompa, 2002; Uversky, 2002). Other factors that foster the unfolded state are the low compositional complexity of IDP sequences that arises from enrichment patterns at the genomic level. This leads to imbalance of certain functionally equivalent amino acid pairs, such as Q > N and S > T, which precludes beta strand formation and thus destabilizes protein folding to a certain degree (Tompa, 2002).

Structurally, IDPs are often considered to be random coil-like yet display detectable residual structure and as a result a number of structural classes have been proposed (Baum et al., 1989; van Mierlo et al., 1993; Wright and Dyson, 1999; Tompa, 2002; Uversky, 2002; Ward et al., 2004; Xie et al., 2007; Uversky and Dunker, 2010, 2012; Peysselon et al., 2011). Based on the terminology that developed from the globular protein folding field, these classes have been defined as coil-like, premolten globule-like, and molten globular states (Baum et al., 1989; van Mierlo et al., 1993; Wright and Dyson, 1999; Tompa, 2002; Uversky, 2002). However, IDPs exhibit different behavior compared to unfolded globular proteins and care should be taken not to confuse the two species. For example, IDPs retain their open conformation when they complex with their molecular target(s), that is, they undergo local disorder–order transition or folding but do not fold into a compact globular state (Baum et al., 1989; van Mierlo et al., 1993; Wright and Dyson, 1999; Tompa, 2002; Uversky, 2002; Ward et al., 2004; Xie et al., 2007; Uversky and Dunker, 2010, 2012; Peysselon et al., 2011). Further, their function is intimately linked with disorder, which shows evolutionary stability. Finally, the conformation they adopt is largely defined by their interacting partner, and not so much by their amino acid sequence as with globular proteins.

11.2.2 WHY IS INTRINSIC DISORDER IMPORTANT FOR THE BIOMINERALIZATION PROCESS?

Clearly, if 50% of human ECM proteins contain some percentage of intrinsic disorder (Peysselon et al., 2011), then this speaks of the necessity of unfolded structures for ECM function. This would be true of extracellular biomineralization processes as well and we foresee several potential benefits arising from disordered biomineral matrix proteins. From the foregoing, it is clear that unfolded regions are "plastic" in their molecular behavior and possess multiple capabilities (Baum et al., 1989; van Mierlo et al., 1993). Thus, inter-protein interactions could be "tuned" by the location, length, and extent of disorder within a series of proteins. In turn, this would affect which proteins interact with each other, their affinities, and ultimately dictate the dimensions, stability, and hierarchal nature of the matrix itself. However, the term "plasticity" has another meaning that is relevant here. Recent studies of disordered synthetic polymer systems known as polymer-induced liquid precursor phases (PILP) have demonstrated that disordered polymers and polypeptides can form aqueous phase suspensions with amorphous clusters, resulting in liquid-like yet stabilized amorphous mineral phases that can be transformed into specific polymorphs or crystalline forms (Gower, 2008; Wolf et al., 2011). These PILP systems integrate very well with the nonclassical nucleation scheme that involves the formation of prenucleation clusters from supersaturated solution (Gebauer et al., 2008; Demichelis et al., 2011; Gebauer and Coelfen, 2011; Evans, 2012). Hence, a disordered biomineralization protein or protein assembly may possess PILP-like molecular characteristics as a result of the labile, unstable sequence regions that exist in the primary sequence of these proteins. This would explain why proteins can form compatible phases with amorphous biominerals and subsequently allow polymorph transformations or specific crystal growth scenarios to occur (Samata et al., 1999; Michenfelder et al., 2003; Treccani et al., 2006; Mann et al., 2007; Amos and Evans, 2009; Keene et al., 2010a,b; Miyazaki et al., 2010; Suzuki et al., 2010, 2011; Amos et al., 2011a,b; Ponce and Evans, 2011).

11.3 EXPERIMENTAL METHODS FOR IDENTIFYING INTRINSIC DISORDER

Identifying the presence of intrinsic disorder can be conceptualized in two ways: either the glass is half empty or half full. By that, we mean either we use methods to either identify the existence of structure (and hence the absence of disorder) (Baum et al., 1989; van Mierlo et al., 1993; Wright and Dyson, 1999; Tompa, 2002; Uversky, 2002; Ward et al., 2004; Xie et al., 2007; Uversky and Dunker, 2010, 2012; Peysselon et al., 2011) or specifically look for disorder itself (Baum et al., 1989; Wright and Dyson, 1999; Delak et al., 2009; Bromley et al., 2011; Ndao et al., 2011; Uversky and Dunker, 2012). Experimentally, identifying the lack of structure is relatively straightforward and the most precise method for this would be x-ray crystallography. Here, one would look for residues with missing backbone coordinates within 3D structures as evidence of disordered regions (Baum et al., 1989; Tompa, 2002). However, the drawback with this approach is that x-ray methods cannot directly assign the structure of these amorphous regions. For this reason, the primary method that is well suited for identifying both order and disorder is nuclear magnetic resonance (NMR) (Baum et al., 1989; van Mierlo et al., 1993; Dyson and Wright, 1998; Wright and Dyson, 1999; Delak et al., 2009; Bromley et al., 2011; Ndao et al., 2011; Uversky and Dunker, 2012). Here, one can sequentially identify both ordered (secondary, tertiary) and partially or fully disordered (chemical shift dispersion, random coil, extended beta strand, beta turn, PPII) regions (Baum et al., 1989; van Mierlo et al., 1993; Dyson and Wright, 1998; Wright and Dyson, 1999; Delak et al., 2009; Bromley et al., 2011; Ndao et al., 2011; Uversky and Dunker, 2012). Disordered regions are also identifiable by high flexibility of the polypeptide chain, the presence of Pro cis–trans isomerization, and numerous sites of chemical or conformational exchange, which can be probed using relaxation measurements (Dyson and Wright, 1998; Wright and Dyson, 1999). A qualitative, rapid evaluation method that complements NMR is far-UV circular dichroism (CD) spectroscopy, which detects the amount (or lack) of secondary structure. The ellipticity spectrum of disordered region has a large negative peak at around 200 nm and a value close to zero at 220 nm, which allows the identification of partially or fully unstructured proteins (Sreerama and Woody, 2000, 2004; Lobley et al., 2002).

There are other techniques that can supplement the datasets obtained by NMR and CD. These include Fourier transform infrared spectroscopy (FTIR) (Renogupalakrishnan et al., 1986; Tompa, 2002; Kumar et al., 2008) and Raman (Tompa, 2002; Uversky, 2002; Uversky and Dunker, 2010; Sikirzhytski et al., 2012) spectroscopies, which probe information at the secondary structure level, and fluorescence spectrometry (Tompa, 2002; Uversky, 2002; Uversky and Dunker, 2010, 2012; Neyroz and Ciurti, 2012), which probe the environment of aromatic residues (i.e., the lack of a tightly packed hydrophobic core). Unfolded conformations can also be detected and characterized by hydrodynamic techniques such as size-exclusion chromatography (Tompa, 2002; Uversky, 2002, 2012; Uversky and Dunker, 2010), small-angle x-ray scattering (Tompa, 2002; Uversky, 2002; Uversky and Dunker, 2010, 2012; Bernado and Svergun, 2012), sedimentation analysis (Tompa, 2002; Salvay et al., 2012), and dynamic light scattering (Tompa, 2002; Amos and Evans, 2009; Amos et al., 2011a,b; Ponce and Evans, 2011). These methods provide hydrodynamic parameters such as the Stokes radius and the radius of gyration, which assume different values for globular and unfolded proteins (Tompa, 2002). Other complementary methods that have been used to detect the absence of protein folding (and hence the presence of disorder) include differential scanning calorimetry (Tompa, 2002; Uversky, 2002; Uversky and Dunker, 2010; Wang et al., 2010), surface plasmon resonance (SPR) (Tompa, 2002; Uversky, 2002; Dagkessamanskaia et al., 2010; Uversky and Dunker, 2010), proteolytic and thermal sensitivities (Tompa, 2002; Suskiewicz et al., 2011), and SDS-PAGE electrophoresis (Tompa, 2002; Uversky et al., 2009). The reader should note that new approaches and insights into the disorder are continually evolving. As a result, rather than giving specific examples into the application of a particular method, the reader is advised that (1) ensembles of methods will undoubtedly be the best choices to confirm or identify disordered regions in biomineralization proteins; (2) the optimum method we would use today may very well be obsolete by tomorrow.

11.4 THEORETICAL METHODS FOR IDENTIFYING INTRINSIC DISORDER

One of the major problems with experimental approaches is the time that it takes to acquire data. This is not a restriction for the study of an individual biomineralization protein, but it does become a major problem if one is attempting to identify disorder content and location for an entire biomineral proteome of a given organism or species. In that instance, the better course of action is to rely on theoretical approaches to screen the sequence database for evidence of disorder and then select specific and interesting protein examples for follow-up experimental studies.

The field of intrinsic disorder prediction is continuing to evolve in terms of accuracy and sophistication. As of this writing, there exist several benchmarked algorithms that identify unfolded or disordered sequences using three different approaches (Linding et al., 2003; Vucetic et al., 2003; Ward et al., 2004; Dosztányi et al., 2005; Prilusky, 2005; Ishida and Kinoshita, 2008). Ab initio approaches involve predictions based upon sequence information alone, and these methods usually utilize machine learning or patterning techniques such as scalar vector machines (SVMs), neural networks, and Bayesian classifiers. Template-based approaches involve the use of similar structures or nonstructures for sequence examination. Finally, meta-approaches are those that combine the predictions of several algorithms together, such as ab initio and template based. Since these methods are continuing to evolve and improve, we will defer the discussion of specific algorithms in this handbook and recommend an examination of the primary literature for predictive methods and their relative success/failure rates. One should remember that a predictive method is just that—predictive—and must be confirmed by experimental methods in order to ensure that a sequence is truly disordered, partially disordered, or folded.

11.5 "DISORDERED" FUTURE

We are at the crossroads in terms of biomineralization protein research. Although it is too early to tell how many mineral-associated proteins contain partial or total disorder, we speculate that the 50% disorder figure that is touted for the human ECM proteome will probably be close to the mark for many other biomineralization proteomes as well. If this proves to be correct, then many of the mysteries of protein-mediated biomineralization processes will finally become more clear and easier to solve as a result of advances in the IDP field itself and with the continued development of new analytical techniques. In turn, the advances in the biomineralization field will stimulate research and development in parallel fields of materials, chemistry, and biology that utilize the biomineralization process as inspiration.

ACKNOWLEDGMENTS

We acknowledge the support of the US Department of Energy, Office of Basic Energy Sciences, Division of Materials Sciences and Engineering under Award DE-FG02-03ER46099, and the US Army Research Laboratory and the US Army Research Office under grant number W911NF-12-1-0255. This review represents Contribution Number 72 by the Laboratory for Chemical Physics, New York University.

REFERENCES

Amos, F.F. and J.S. Evans. 2009. AP7, a partially disordered pseudo C-RING protein, is capable of forming stabilized aragonite in vitro. *Biochemistry* 48: 1332–1339.

Amos, F.F., Ndao, M., Ponce, C.B., and J.S. Evans. 2011a. A C-RING-like domain participates in protein self-assembly and mineral nucleation. *Biochemistry* 50: 8880–8887.

Amos, F.F., Ponce, C.B., and J.S. Evans. 2011b. Formation of framework nacre polypeptide supramolecular assemblies that nucleate polymorphs. *Biomacromolecules* 12: 1883–1890.

Baum, J., Dobson, C.M., Evans, P.A., and C. Hanley. 1989. Characterization of a partly folded protein by NMR methods: Studies on the molten globule state of guinea pig alpha-lactalbumin. *Biochemistry* 28: 7–13.

Bernado, P. and D.I. Svergun. 2012. Structural analysis of intrinsically disordered proteins by small angle X-ray scattering. *Mol. BioSyst.* 8: 151–167.

Bromley, K.M., Kiss, A.S., Lokappa, S.B., Lakshminarayanan, R., Fan, D., Ndao, M., Evans, J.S., and J. Moradian-Oldak. 2011. Dissecting amelogenin nanospheres: Characterization of metastable oligomers. *J. Biol. Chem.* 286: 34643–34653.

Dagkessamanskaia, A., Durand, F., Uversky, V.N., Binda, M., Lopez, F., El Azzouzi, K., Francois, J.M., and H. Martin-Yken. 2010. Functional dissection of an intrinsically disordered protein: Understanding the roles of different domains of Knr4 protein in protein-protein interactions. *Protein Sci.* 19: 1376–1385.

Delak, K., Harcup, C., Lakshminarayanan, R., Zhi, S., Fan, Y., Moradian-Oldak, J., and J.S. Evans. 2009. The tooth enamel protein, porcine amelogenin, is an intrinsically disordered protein with an extended molecular configuration in the monomeric form. *Biochemistry* 48: 2272–2281.

Demichelis, R., Raiteri, P., Gale, J.D., Quigley, D., and D. Gebauer. 2011. Stable prenucleation mineral clusters are liquid-like polymers. *Nat. Commun.* 2: 590–598. doi: 10.1038/ncomms1604.

Dobson, C.M. and P.A. Evans. 1984. Protein folding kinetics from magnetization transfer nuclear magnetic resonance. *Biochemistry* 23: 4267–4270.

Dosztányi, Z., Csizmók, V., Tompa, P., and I. Simon. 2005. IUPred: Web server for the prediction of intrinsically unstructured regions of proteins based on estimated energy content. *Bioinformatics* 21: 3433–3434.

Dyson, H.J. and P.E. Wright. 1998. Equilibrium NMR studies of unfolded and partially folded proteins. *Nat. Struct. Biol.* 7: 499–503.

Englander, S.W. and L. Mayne. 1992. Protein folding studies using hydrogen-exchange labeling and two-dimensional NMR. *Ann. Rev. Biophys. Biomol. Struct.* 21: 243–265.

Evans, J.S. 2012. Identification of intrinsically disordered and aggregation—Promoting sequences within the aragonite-associated nacre proteome. *Bioinformatics* 28: 3182–3185.

Evans, J.S. and S.I. Chan. 1994. Phosphophoryn, a biomineralization template protein: pH-dependent protein folding experiments. *Biopolymers* 34: 507–527.

Evans, J.S., Chan, S.I., and W.A. Goddard. 1995a. Prediction of polyelectrolyte polypeptide structures using Monte Carlo conformational search methods with implicit solvation modeling. *Protein Sci.* 4: 2019–2031.

Evans, J.S., Chiu, T., and S.I. Chan. 1994. Phosphophoryn, an acidic biomineralization regulatory protein: Conformational folding in the presence of Cd(II). *Biopolymers* 34: 1359–1375.

Evans, J.S., Mathiowetz, A.M., Chan, S.I., and W.A. Goddard. 1995b. De novo prediction of polypeptide conformations using dihedral probability grid Monte Carlo methodology. *Protein Sci.* 4: 1203–1216.

Gebauer, D. and H. Coelfen. 2011. Prenucleation clusters and non-classical nucleation. *Nano Today* 6: 564–584.

Gebauer, D., Volkel, A., and H. Coelfen. 2008. Stable prenucleation of calcium carbonate clusters. *Science* 322: 1819–1822.

Gower, L.B. 2008. Biomimetic model systems for investigating the amorphous precursor pathway and its role in biomineralization. *Chem. Rev.* 108: 4551–4627.

Ishida, T. and K. Kinoshita. 2008. Prediction of disordered regions in proteins based on the meta approach. *Bioinformatics* 24: 1344–1348.

Keene, E.C., Evans, J.S., and L.A. Estroff. 2010a. Silk fibroin hydrogels coupled with the n16N—Beta-chitin complex: An in vitro organic matrix for controlling calcium carbonate mineralization. *Cryst. Growth Des.* 10: 5169–5175.

Keene, E.C., Evans, J.S., and L.A. Estroff. 2010b. Matrix interactions in biomineralization: Aragonite nucleation by an intrinsically disordered nacre polypeptide, n16N, associated with a β-chitin substrate. *Cryst. Growth Des.* 10: 1383–1389.

Kim, P.S. and R. Baldwin. 1990. Intermediates in the folding reactions of small proteins. *Ann. Rev. Biochem.* 51: 459–489.

Kumar, N., Shukla, S., Kumar, S., Suryawanshi, A., Chaudhry, U., Ramachandran, S., and S. Maiti. 2008. Intrinsically disordered protein from a pathogenic mesophile *Mycobacterium tuberculosis* adopts structured conformation at high temperature. *Proteins* 71: 1123–1133.

Linding, R., Russell, R.B., Neduva, V., and T.J. Gibson. 2003. GlobPlot: Exploring protein sequences for globularity and disorder. *Nucleic Acids Res.* 31: 3701–3708.

Lobley, A., Whitmore, L., and B.A. Wallace. 2002. DICHROWEB: An interactive website for the analysis of protein secondary structure from circular dichroism spectra. *Bioinformatics* 18: 211–212.

Mann, K., Siedler, F., Treccani, L., Heinemann, F., and M. Fritz. 2007. Perlinhibin, a cysteine-, histidine-, and arginine-rich miniprotein from abalone (*Haliotis laevigata*) nacre, inhibits in vitro calcium carbonate crystallization. *Biophys. J.* 93: 1246–1254.

Matsushima, N., Izumi, Y., and T. Aoba. 1998. Small angle x-ray scattering and computer aided molecular modeling studies of 20 kDa fragment of porcine amelogenin: Does amelogenin adopt an elongated bundle structure? *J. Biochem.* 123: 150–156.

Michenfelder, M., Fu, G., Lawrence, C., Weaver, J.C., Wustman, B.A., Taranto, L., Evans, J.S., and D.E. Morse. 2003. Characterization of two molluscan crystal-modulating biomineralization proteins and identification of putative mineral binding domains. *Biopolymers* 70: 522–533.

Miyazaki, Y., Nishida, T., Aoki, H., and T. Samata. 2010. Expression of genes responsible for biomineralization of *Pinctada fucata* during development. *Comp. Biochem. Physiol. B* 155: 241–248.

Ndao, M., Dutta, K., Bromley, K., Sun, Z., Lakshminarayanan, R., Rewari, G., Moradian-Oldak, J., and J.S. Evans, 2011. Probing the self-association, intermolecular contacts, and folding propensity of amelogenin. *Protein Sci.* 20: 724–734.

Neyroz, P. and S. Ciurti. 2012. Intrinsic fluorescence of intrinsically disordered proteins. *Methods Mol. Biol.* 895: 435–440.

O'Neil, K.T. and W.F. DeGrado. 1990. A thermodynamic scale for the helix-forming tendencies of the commonly occurring amino acids. *Science* 250: 646–650.

Peysselon, F., Xue, B., Uversky, V.N., and S. Ricard-Blum. 2011. Intrinsic disorder of the extracellular matrix. *Mol. BioSyst.* 7: 3353–3365.

Ponce, C.B. and J.S. Evans. 2011. Polymorph crystal selection by n16, an intrinsically disordered nacre framework protein. *Cryst. Growth Des.* 11: 4690–4696.

Prilusky, J. 2005. FoldIndex(C): A simple tool to predict whether a given protein sequence is intrinsically unfolded. *Bioinformatics* 21: 3435–3438.

Renugupalakrishnan, V., Strawich, E.S., Horowitz, P.M., and M.J. Glimcher. 1986. Studies of the secondary structures of amelogenin. *Biochemistry* 25: 4879–4887.

Salvay, A.G., Communie, G., and C. Ebel. 2012. Sedimentation velocity analytical ultracentrifugation for intrinsically disordered proteins. *Methods Mol. Biol.* 896: 91–105.

Samata, T., Hayashi, N., Kono, M., Hasegawa, K., Horita, C., and S. Akera. 1999. A new matrix protein family related to the nacreous layer formation in *Pinctada fucata*. *FEBS Lett.* 462: 225–229.

Sikirzhytski, V., Topilina, N.I., Takor, G.A., Higashiya, S., Welch, J.T., Uversky, V.N., and I.K. Lednev. 2012. Fibrillation mechanism of a model intrinsically disordered protein revealed by 2D correlation deep UV resonance Raman spectroscopy. *Biomacromolecules* 13: 1503–1509.

Sreerama, N. and R.W. Woody. 2000. Estimation of protein secondary structure from circular dichroism spectra: Comparison of CONTIN, SELCON, and CDSSTR methods with an expanded reference set. *Anal. Biochem.* 287: 252–260.

Sreerama, N. and R.W. Woody. 2004. On the analysis of membrane protein circular dichroism spectra. *Protein Sci.* 13: 100–112.

Suskiewicz, M.J., Sussman, J.L., Silman, I., and Y. Shaul. 2011. Context-dependent resistance to proteolysis of intrinsically disordered proteins. *Protein Sci.* 20: 1285–1296.

Suzuki, M., Okumura, T., Nagasawa, H., and T. Kogure. 2011. Localization of intracrystalline organic macromolecules in mollusk shells. *J. Cryst. Growth* 337: 24–29.

Suzuki, M., Saruwatari, K., Kogure, T., Yamamoto, Y., Nishimura, T., Kato, T., and H. Nagasawa. 2010. An acidic matrix protein Pif is a key macromolecule for nacre formation. *Science* 325: 1388–1390.

Tompa, P. 2002. Intrinsically unstructured proteins. *Trends Biochem. Sci.* 27: 527–533.

Treccani, L., Mann, K., Heinemann, F., and M. Fritz. 2006. Perlwapin, an abalone nacre protein with three four-disulfide core (whey acidic protein) domains, inhibits the growth of calcium carbonate crystals. *Biophys. J.* 91: 2601–2608.

Uversky, V.N. 2002. Natively unfolded proteins: A point where biology waits for physics. *Protein Sci.* 11: 739–756.

Uversky, V.N. 2012. Size-exclusion chromatography in structural analysis of intrinsically disordered proteins. *Methods Mol. Biol.* 896: 179–194.

Uversky, V.N. and A.K. Dunker. 2010. Understanding protein non-folding. *Biochim. Biophys. Acta* 1804: 1231–1264.

Uversky, V.N. and A.K. Dunker. 2012. Multiparametric analysis of intrinsically disordered proteins: Looking at intrinsic disorder through compound eyes. *Anal. Chem.* 84: 2096–2104.

Uversky, V.N., Oldfield, C.J., Midic, U., Xie, H., Xue, B., Vucetic, S., Iakoucheva, L.M., Obradovic, Z., and A.K. Dunker. 2009. Unfoldomics of human diseases: Linking protein intrinsic disorder with diseases. *BMC Genomics* 10: S1–S7.

van Mierlo, C.P.M., Darby, N.J., Keeler, J., Neuhaus, D., and T.E. Creighton. 1993. Partially folded conformation of the (30–51) intermediate in the disulphide folding pathway of bovine pancreatic trypsin inhibitor. 1-H and 15-N resonance assignments and determination of backbone dynamics from 15-N relaxation measurements. *J. Mol. Biol.* 229: 1125–1146.

Vucetic, S., Brown, C., Dunker, A., and Z. Obradovic. 2003. Flavors of protein disorder. *Proteins Struct. Funct. Genet.* 52: 573–584.

Wang, X., Zhang, S., Zhang, J., Huang, X., Xu, C., Wang, W., Liu, Z., Wu, J., and Y. Shi. 2010. A large intrinsically disordered region in SKIP and its disorder-order transition induced by PPIL1 binding revealed by NMR. *J. Biol. Chem.* 285: 4951–4963.

Ward, J.J., Sodhi, J.S., McGuffin, L.J., Buxton, B.F., and D.T. Jones. 2004. Prediction and functional analysis of native disorder in proteins from the three kingdoms of life. *J. Mol. Biol.* 337: 635–645.

Wolf, S.E., Leiterer, J., Pipich, V., Barrea, R., Emmerling, F., and W. Tremel. 2011. Strong stabilization of amorphous calcium carbonate emulsion by ovalbumin: Gaining insight into the mechanism of 'polymer-induced liquid precursor' processes. *J. Am. Chem. Soc.* 133: 12642–12649.

Wright, P.E. and H.J. Dyson. 1999. Intrinsically unstructured proteins: Re-assessing the protein structure–function paradigm. *J. Mol. Biol.* 293: 321–331.

Xie, H., Vucetic, S., Iakoucheva, L.M., Oldfield, C.J., Dunker, A.K., Uversky, V.N., and Z. Obradovic. 2007. Functional anthology of intrinsic disorder. 1. Biological processes and functions of proteins with long disordered regions. *J. Proteome Res.* 6: 1882–1898.

Part III

Imaging morphology and interfaces

12 Exploring dynamics at the biomolecule–crystal interface using real-time *in situ* atomic force microscopy

S. Roger Qiu

Contents

12.1 INTRODUCTION

Biominerals are composite materials blending an organic matrix with inorganic entities, most of which are crystalline substances (Bouropoulos et al. 2001; Lowenstam and Weiner 1989). The organic molecules are viewed as impurities or modulators to alter the energy landscape of the inorganic molecular assembly during the formation of biominerals (De Yoreo and Vekilov 2003). The advancement of crystal growth theories combined with *in situ* analytical tools in the past two decades have enabled the understanding of the dynamics of biomineral formation from the macroscopic level to molecular level. One of the tools that has emerged to help establish a new paradigm for organic controlled mineralization (Orme et al. 2001) is *in situ* atomic force microscopy (AFM), which encompasses several unique advantages over other imaging tools used in the crystal growth community through the years, including the scanning tunneling microscopy (STM), scanning electron microscopy (SEM), transmission electron microscopy (TEM), and phase shift interferometry (Onuma et al. 1994).

AFM is one of the quickest and widely utilized analytical tools for academic research and industrial characterization (Balke et al. 2012; Bonnell 2000; Meyer et al. 2004) since its invention (Binnig et al. 1986). The basic principle of AFM has been discussed in several chapters in this book. For spatial resolution, AFM can examine features from about a few nm to over 100 μm, which span the length scales from single molecules to entire crystals (De Yoreo et al. 2001; Land and De Yoreo 2000). For temporal resolution, AFM can obtain images at a rate up to a couple tens of Hz without losing imaging authenticity. This scan rate range is compatible to that of the growing minerals as most of them are sparingly soluble crystals. Because AFM acquires topographic information through analysis of the interaction force between atoms on the AFM tip and the surface it scans across (instead of electron beams or tunneling current), it can probe the surface of any solid material including conductors, semiconductors, and insulators. Furthermore, because the interaction force can be tuned to very low values during scanning, the alteration to the probing object by the scanning probe can be minimized. Thus, this tool can be extended to investigate surfaces of *soft* materials including DNA, proteins, virus, cells, and ice (Bash et al. 2006; Cho et al. 2011; Lyubchenko et al. 1996; Orme and Giocondi 2007). For the experimental environment, an AFM imaging system can be operated in either *ex situ* or *in situ* mode, in dry or saturated vapor environment (Burnham et al. 2009), below zero degrees (°C) (Orme and Giocondi 2007), or above room temperature. More importantly, unlike many other that use the snapshot approach to record the growth and to deduce the growth kinetics, the *in situ* AFM is capable of monitoring growing surfaces in aqueous solution on-the-fly and thus capture the kinetics of growth events in real time.

Ever since the first utilization to study solution crystal growth and dissolution (Gratz et al. 1991), *in situ* AFM has become a vital tool to reveal the molecular scale interactions between biological additives and crystal surfaces that lead to the control of biomineral formation (Davis et al. 2000; De Yoreo and Dove 2004; De Yoreo et al. 2006, 2007; Elhadj et al. 2006a,b; Fu et al. 2005; Guo et al. 2002; Jung et al. 2004; Orme et al. 2001; Qiu et al. 2004, 2005; Teng 1998; Touryan et al. 2004; Weaver et al. 2006). These investigations have shown that understanding the stereochemical recognition between biomolecules and the edges of the atomic steps on crystals is essential to determine the origin of modification to growth dynamics and kinetics. Examples from some of these studies are adopted for this chapter to demonstrate how results derived from those *in situ* AFM studies are used to define the physical mechanisms that govern the shape modification and kinetic control in biominerals. Although most of the examples are taken from our own work, numerous high-quality investigations using *in situ* AFM on biomineral systems have also been conducted by many groups around the world. Most of these works are cited in this chapter.

The complete review of crystal growth fundamentals is beyond the scope of this chapter. The crystal growth field bears a long history of research activities and is of a great breadth and depth. There is a rich amount of literature already existing including textbooks, book chapters, and review articles. More specifically, in the past decades, several thorough reviews and book chapters are devoted to addressing biomineral formation within the framework of crystal growth science (De Yoreo and Vekilov 2003; De Yoreo et al. 2001, 2007; Dove et al. 2004; Orme and Giocondi 2007; Qiu and Orme 2008). Readers are referred to those references for crystal growth background.

The purpose of this chapter is to show the utilization of *in situ* AFM imaging to probe interactions between biomolecules and crystal faces and to provide a brief review of recent advances in defining the underlying physical principles that are responsible for the control of biomineral formation through real-time characterization. By no means is this chapter intended to be an operational manual for using an AFM instrument; rather we intend to share our insights into the technical advantages as well challenges, encountered using *in situ* AFM to monitor crystal growth in aqueous solution under the influence of biomolecules in real time. We hope that readers—especially for those who are new to the application of the tool—find the information useful in assisting to prevent or to solve those encountered technical issues. We also hope that the presented results show capabilities that would inspire readers to integrate the *in situ* AFM tool into their respective research activities.

Section 12.2 of this chapter provides a short overview of the existing impurity models that describe the behavior of additives in modifying the growth of inorganic crystals. The models are derived based on the studies using impurities of small molecules. As will be seen in Section 12.4, more realistic models can be developed from these simple ones provided additional physical factors are considered. Section 12.2 also serves the purpose of introducing nomenclatures used in this chapter to assist the reader. Section 12.3 gives a detailed description of the experimental setup for real-time molecular scale characterization of growing crystals. The functionality, possible technical issues, and solutions for each component are provided. In addition, the imaging artifacts and step velocity determination based on image distortion are also presented. Examples utilizing the *in situ* AFM tool to probe the interaction and to discover the mechanism are given in Section 12.4. The chapter is concluded by a perspective on future directions.

12.2 NOMENCLATURE AND IMPURITY MODELS

There are several established models in the crystal growth community describing the mechanism by which impurities modify crystal growth at the molecular scale (De Yoreo and Vekilov 2003; Dove et al. 2004). These models are developed mostly based on studies of small-molecule modulation of the thermodynamic and kinetic factors of growth occurring on the advancement of atomic steps. In this chapter, the four end-member models are summarized, and the relevance to the examples given in the later sections of the chapter will also be discussed. However, before introducing these models, the basic concept of crystal growth on dislocations and the nomenclature used in this chapter are described next with the intention of better readership.

12.2.1 NOMENCLATURE

A step is an important physical feature for crystals to grow, and it can be generated either by the presence of imperfections or screw dislocations on the existing or newly expressed faces or by 2D nucleation on terraces at relatively high solution supersaturations (Burton et al. 1951; Markov 1994). The geometry of a typical screw dislocation with a unit height is shown in Figure 12.1. Dislocations provide a continuous source of step generation, and if the length of the initial step segment is greater than that of critical length r_c, the step will grow. As the first step advances, a second step with the same step height (h) is generated behind and so on. The propagation of these steps forms a spiral on the surface leading to spiral growth. The distance between the edges of two adjacent steps is characterized as the terrace width (W). A typical step structure for a growing crystal face is shown in Figure 12.2. As illustrated, the step edge is not smooth. The rough nature of the step edge is originated from the thermal fluctuations associated with the growing surface. The incomplete region is referred to as the kink site (specified by black dots) where more unsaturated bonds are available relative to its neighbors. When molecules are attached to the kink sites, they form more bonds

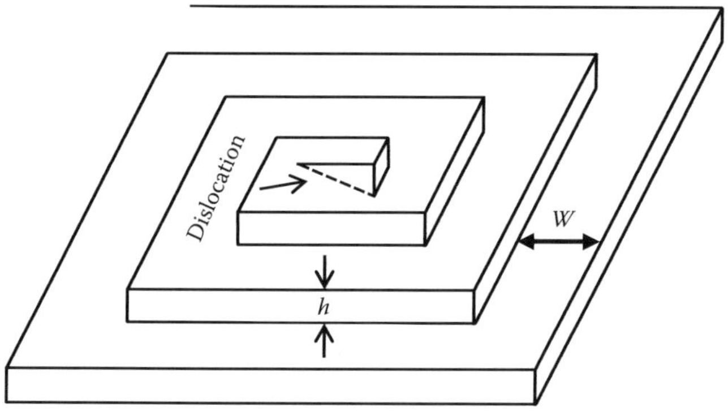

Figure 12.1 The schematic diagram illustrating a screw dislocation on a crystal surface that leads to a spiral growth hillock. The screw dislocation provides a continuous source of new steps. The distance between two adjacent step edges defines the terrace width *W* and *h* is the elementary step height.

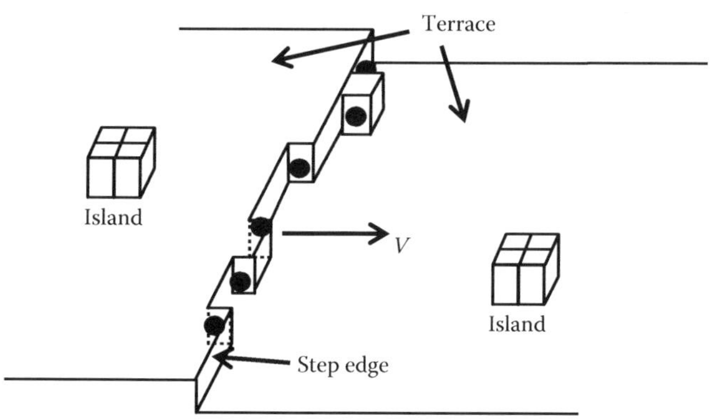

Figure 12.2 The schematic showing the characteristics of a growing step at velocity *V*. Kink sites on the step edge are specified by the black dots. Islands formed by 2D nucleation on two terraces are also demonstrated.

to neighboring molecules than the ones that are attached to the terraces or the step edges. On the other hand, molecules at kink sites can also break away from the step edge more easily because they have fewer bonds to break than those in the complete step structure. Thus, the activity of kink sites plays a critical role in step kinetics during crystal growth. The crystal grows when a flux of solute molecules is incorporated to kink sites along the step edge and dissolves when molecules detach from them.

The growth kinetics is characterized by the velocity of the growing step, which is controlled by the driving force of the system—or solution supersaturation. For a step of infinitely long length with high kink site density, the step velocity has a simple form

$$V = \beta \cdot \Omega(a_{(\mathrm{Ca}^{2+})} - a_{e(\mathrm{Ca}^{2+})}) \tag{12.1}$$

which is derived from the consideration that the step advancement is a consequence of the difference between molecules that are attached and detached at kink sites or by the mass preservation law (De Yoreo and Vekilov 2003). In this equation, β is the kinetic coefficient, Ω is the specific molecular volume of the growth unit, $a_{(\mathrm{Ca}^{2+})}$ is the calcium activity, and $a_{e(\mathrm{Ca}^{2+})}$ is the calcium activity at equilibrium. This formula, although having several exceptions (Orme and Giocondi 2007) including variations to step length, low kink density, multiple kink types, and the impurity interactions, has been shown to hold for many systems including biominerals (Chen and Vekilov 2002). The kinetic coefficient is an important parameter measuring the kinetics of adsorption, diffusion, desolvation, and incorporation of ions at the step edge and can be expressed as

$$\beta \propto n_{k,0} \exp\left(\frac{-E_a}{k_B T}\right) \tag{12.2}$$

where

$n_{k,0}$ is the kink site density in pure system
E_a is the net activation energy for the attachment of the growth unit to the kink site
k_B is the Boltzmann's constant
T is the absolute temperature

As shown in Equation 12.1, the kinetic coefficient is essentially the slope of the linear function between step velocity and solution concentration and thus can be determined by measuring the step velocity at different solution supersaturation.

12.2.2 IMPURITY MODELS

When present in the growth solution whether unwanted or intentionally added, impurities can modify the growth of crystal. The modification can be realized in several ways either in solution or on the growing surfaces, both thermodynamically and kinetically. In solution, interaction with solute molecules can lead to a modification of solution supersaturation or driving force for crystallization. This is accomplished thermodynamically by altering the solute ion activity through forming ion pairs or other chemical species with the solute molecules. This impurity–solute interaction can be quantified by solution speciation. In fact, there are several well-established thermodynamics databases available that can account for the impurity impact on solution supersaturation for several crystal systems including those commonly found biominerals such as calcium carbonate, calcium oxalate (CaOx), and calcium phosphate, especially for impurities of small molecules (Gustaffson 2004).

Most frequently observed growth modification by impurities cannot be explained by the changes in solution supersaturation alone, because the quantification through speciation does not account for the interactions occurring at the solid–liquid interface where adsorbed impurities are energetic. The growth modification through interactions at the solid–liquid interface, nevertheless, is rather complex owing to the diverse physical and chemical nature of the impurities as well as of the crystal with which they interact. As suggested by Equations 12.1 and 12.2, impurities can alter the growth by many ways; when adsorbed on the surface, they can either block or promote the adsorption and the movement of the solute on the surface, the attachment/detachment events at kink sites along the step edges, or the incorporation of the growth unit to the crystal lattice.

Based on investigations including *in situ* AFM analysis of step growth, the control mechanism by which small molecules control mineral growth can be categorized to four end-member models (De Yoreo and Vekilov 2003; Dove et al. 2004): (1) step pinning, (2) incorporation, (3) kink blocking, and (4) surfaction. Each of these models is summarized in the following. Advanced readers are referred to these books for details.

12.2.2.1 Step pinning

The step pinning model was first proposed by Cabrera and Vermilyea (C–V) (Cabrera and Vermilyea 1958). In the original treatment, the following assumptions are made: (1) the growing surface contains a uniform average concentration of steps, (2) there is constant current of impurity molecules deposited on the crystal face per unit surface area per unit time, (3) the average density of the adsorbed impurity (n_i) is proportional to the impurity solution concentration (C_i), and (4) the average distance (L_i) between two adjacent impurities is proportional to $1/C_i^{0.5}$. When impurity molecules adsorb on the crystal face or terraces, they can be either mobile or immobile. For the immobile adsorbed molecules, they remain at the location where they reach the lowest energy state on the surface separated by L_i (see Figure 12.3a). When a growing step encounters impurity molecules that act as blockers, the step front will try to squeeze through the space between any two neighboring impurities and result in a curved step edge with a radius of curvature of r. As pointed out by this classical model, there is a critical radius of curvature (r_c) associated with the step growth. If the average spacing (L_i) between a pair of impurities is greater than $2r_c$, the step will fence through the impurities and grow. On the other hand, if L_i is less than $2r_c$ apart, the step growth will be ceased. Since the steps are curved, by the Gibbs–Thomson effect that the effective supersaturation is reduced, the velocity of the step that is squeezing through will be less than that in the pure solution. For the classical C–V impurity model, the impurity modified step velocity has the following form:

$$V = V_0\sqrt{1 - 2r_c C_i^{0.5}} \quad (12.3)$$

where r_c is given by Burton et al. (1951)

$$r_c = \frac{\gamma\Omega}{k_B T\sigma} \quad (12.4)$$

where

γ is the step-edge free energy per unit step height
σ is the solution supersaturation

Because r_c is inversely proportional to σ, increasing solution supersaturation at a given impurity content can suppress the critical radius at the step edge, which leads to a recovery of the growth process. For some cases, a complete recovery is possible.

The supersaturation at which the step reaches its critical radius of curvature is usually referred as the width of the deadzone (σ_d), above which the step grows. Thus, σ_d defines a positive supersaturation region within which no growth occurs. Since at σ_d, the average impurity spacing is equal to $2r_c$, by setting this relation using Equation 12.4 and the formula for $L_i(=1/C_i^{0.5})$, the deadzone for a given impurity concentration can be expressed as

$$\sigma_d = \frac{2\gamma\Omega}{k_B T} C_i^{0.5} \quad (12.5)$$

It is apparent that the deadzone varies with impurity concentration. The dependence of deadzone on impurity content as well as the recovery process for high solution supersaturation is displayed in Figure 12.3a. Since the immobile impurity molecules on the surface pin the moving steps and cause the step front to bend, impurities following this model usually create a characteristic *scallop*-shaped step morphology around the growth hillocks. Examples of such morphology are given in Section 12.4.

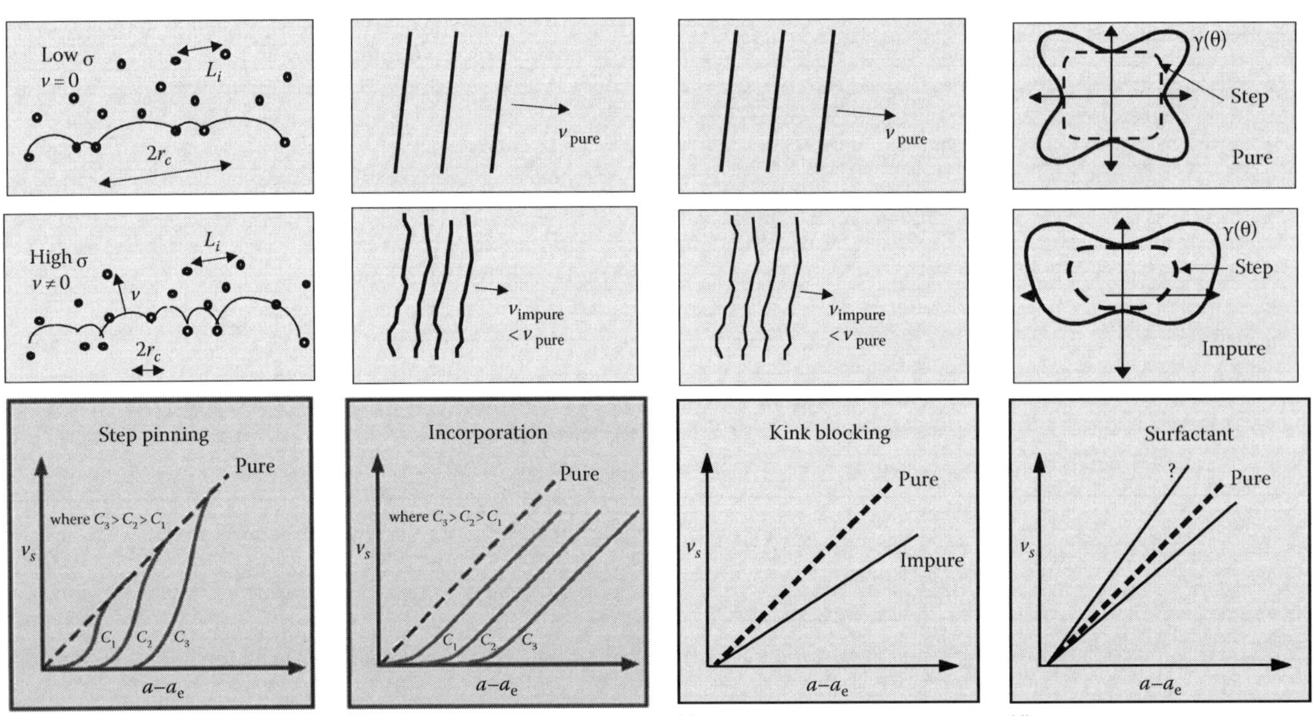

Figure 12.3 Four end-member models to describe the impact of impurity on step growth kinetics: (a) Step pinning, (b) incorporation, (c) kink blocking, and (d) surfaction. (After De Yoreo, J.J. and Vekilov, P., Principles of crystal nucleation and growth, in *Biomineraliztion*, ed. P.M. Dove, J.J. De Yoreo, and S. Weiner, Vol. 54, Mineralogical Society of America, Washington, DC, 2003. With permission.)

The mobile impurities on the other hand are not believed to impede step growth as moving steps can move the impurity along during growth. This is true for small impurity molecules. In fact, some large impurity molecules like proteins can act like mobile impurities and deposit on the crystal face without altering step kinetics (Qiu et al. 2004).

The simple model describing growth inhibition by Equations 12.3 and 12.5 has been shown to hold well in predicting the step kinetics modified by small impurities (De Yoreo and Vekilov 2003). For impurity at a larger size, the classical C–V can still be utilized to predict the modification of step kinetics provided that other more realistic physical parameters are included in the reformulation (Friddle et al. 2010; Weaver et al. 2006, 2007, 2010). Examples in Section 12.4 are given to specifically discuss the extensions applied to the classic model.

12.2.2.2 Incorporation

Certain types of impurity molecules can actually incorporate at the kink site or be seized by a growing step to become part of the crystal. Because the impurity molecules or ions usually carry different physical properties (e.g., charge, size, orientation) than that of the solute molecules, the existence of the impurity constituents distorts the crystal structure and introduces strain on the system. Consequently, the internal energy as well as the solubility of the system is changed (Davis et al. 2000; van Enckevort and van den Berg 1998). Incorporation of foreign molecules to crystal structure can lead to both inhibition and promotion of step growth. There are two common characteristics for the inhibition by incorporation: one is that the solution equilibrium activity is increased, which is exemplified by the non-growing region in the step velocity to supersaturation plot (Figure 12.3b), and the other is that the kinetic coefficient remains the same as that for the pure growth, which is shown by the same kinetic coefficient. The fact that the kinetic coefficient remains unchanged suggests that the impurity incorporation does not modify the kinetics of the attachment process (Davis et al. 2000).

12.2.2.3 Kink blocking

Step kinetics is closely related to the activity of kink site at the step edge including the generation of the new kink sites and the availability for attachment. Some impurities can adsorb at kink sites transiently to temporarily block the kink sites. This can result in reduction of available kink sites and blockage (Bliznakow 1958; Chernov 1961) of kink movement, both of which will lead to inhibition to step growth. Unlike the previous two cases, this type of step–impurity interaction does not create a *deadzone* within which the growth of step segment is nulled. The solubility is thus not changed. The kinetic coefficient however is greatly affected. The typical characteristics of the velocity vs. supersaturation affected by kink blocking are shown in Figure 12.3c.

12.2.2.4 Surfaction

Molecules or compounds that can lower interfacial energy are referred to as surfactants. At the solid–liquid interface where steps grow, many impurity molecules can act like surfactants to modify their growth by either increasing the step velocity, altering growth morphology in the absence of the kinetics effect, or a combination of both. The enhancement of the growth kinetics can be achieved either thermodynamically by lowering the step-edge free energy (Fu et al. 2005) or kinetically by reducing the activation energy of the attachment. Examples of each case are given in Section 12.4. The alteration to step morphology with the step velocity unchanged is attributed to a pure thermodynamic effect where the directional dependence of the step free energy is modified (Orme et al. 2001). While some impurity molecules may adsorb to the terraces or step edge, they do not pin steps. The step edge remains smooth and stable, that is, no serrated segments formed although they may be curved. The typical nature of the velocity to supersaturation plot is shown in Figure 12.3d. A 2D Wulff plot showing the changes of the step-edge free energy for growth hillocks that develop in a pure and impurity-bearing solution (De Yoreo and Vekilov 2003) is also included.

12.3 EXPERIMENTAL CONSIDERATIONS FOR REAL-TIME *IN SITU* AFM IMAGING

12.3.1 SYSTEM SETUP FOR REAL-TIME *IN SITU* AFM IMAGING

The *in situ* AFM imaging system for real-time investigation of growth and growth modification of biominerals by biomolecules is composed of several key modules including the probing module capable of imaging in aqueous environment, temperature-control module, flow-control module, vibration damper, and solution reservoir. A schematic showing the layout of the system is displayed in Figure 12.4, and a simplified

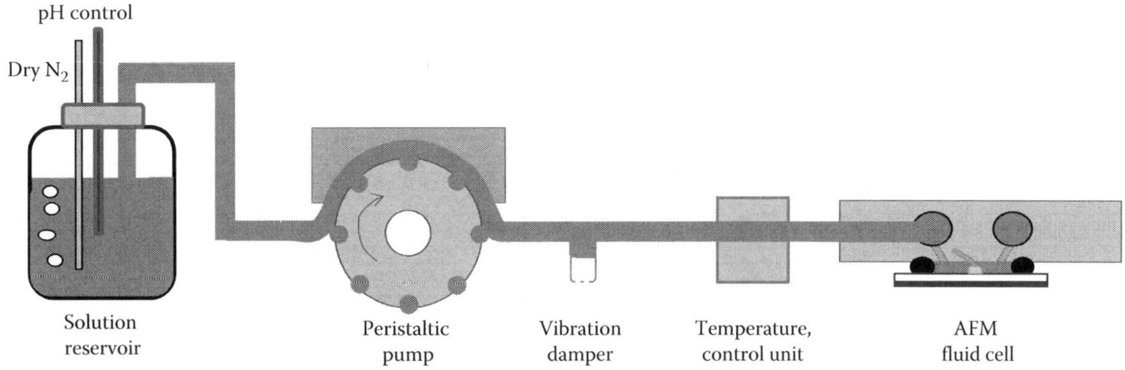

Figure 12.4 A schematic showing the system used for *in situ* imaging of a growing crystal surface in real time.

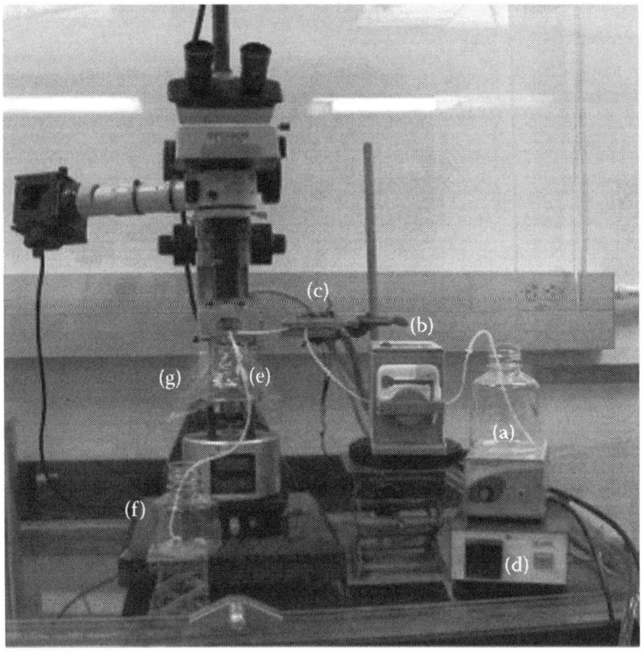

Figure 12.5 A simplified setup for *in situ* AFM imaging containing several control modules: (a) solution reservoir and stirrer, (b) peristaltic pump, (c) heating unit with chilled water circulation tubes, (d) temperature controller, (e) thermocouples, (f) waste container, and (g) AFM unit with a plastic wrap for liquid-leak protection. (Photo courtesy of Mathew L. Weaver.)

version of the experimental setup used in one of our laboratories is shown in Figure 12.5.

For the imaging module, besides the regular AFM imaging gadgets such as the base unit, scanner, and detection head, a liquid cell that can hold the AFM probe and create a reaction chamber when sealed with a sample holder is used to replace the regular tip holder. The function and application issues of the liquid cell will be discussed in the sample preparation section later.

The temperature-control module contains a heating/cooling unit (Peltier) where the solution-supplying tubing runs through a thermocouple, a temperature controller with a feedback loop, and the water-cooling unit (NESLAB). The thermocouple is usually placed inside the liquid cell through the outlet tubing to measure the solution temperature at the region where the reactions are taking place. The temperature reading is used for the feedback loop to stabilize the temperature inside the liquid cell to a desired value. Because the thermocouple is located downstream of the solution flow, it causes minimal disturbance to the solution flow or contamination to the solution inside the reaction chamber. With temperature-ramping ability, the experiment setup is capable of determining important thermodynamic and kinetic parameters such as the activation energy. There are alternative ways to adjust the solution temperature. For example, one could place the entire solution reservoir inside a thermal bath and tune the cell liquid temperature with a feedback loop. This approach is suitable for those studies at a fixed solution temperature, that is, the biomineral system in the physiological conditions at 37°C. The drawback of this, however, is that the flow system has a much longer path length. Thus, a larger thermal loss and temperature fluctuations are expected. Insulating layers over the plastic tubing are usually required to minimize the heat loss during solution transportation. For example, one can place the flowing tubing inside a plastic pipe that has a large diameter where the air in the gap can serve as thermal insulation.

The system starts with the reservoir where the solution of preset conditions is first introduced. The module usually contains several components enabling an easy solution switch, impurity precursor addition, pH adjustment, and dissolved gases removal. To minimize unwanted impurity introduction from the solution container, a glass-based vessel should be avoided. This is especially true for those investigations at low pH and elevated temperature where the leaching of silicate species is enhanced. To ensure accuracy of the quantitative experiment, pH fluctuation needs to be minimized. This can be achieved by an online titration (e.g., pH-Stat) of the solution throughout the experiment. In any open container with solution, if not purged, gas molecules such as N_2, oxygen, and CO_2 are usually dissolved in the liquid at equilibrium with that in the atmosphere. The dissolved gases can often create bubbles in the solution. At room temperature, although rare, bubbles can be generated in the flowing solution. When these bubbles make their way to the liquid cell, they can be easily trapped on the AFM tip and scatter the probing laser beam and smear the experiment. While these bubbles can sometimes be squeezed out by purging the solution at a higher rate or pressure, they are usually difficult to remove from inside the liquid cell. In most occasions, one will have to dismantle, clean, dry, and remount the unit. If this happens in the middle of an experiment, besides wasting valuable time, it can result in loss of precious experimental agents including some proteins that are difficult to prepare. Thus, the removal of bubbles in solution before they reach the imaging unit is very important. In the setup described here, a manifold connector can be placed between the transporting tubing right before the liquid cell. When a bubble-containing liquid approaches to the liquid cell, stop the flow momentarily, open another path in the manifold, redirect and discard the bubble-containing portion, then resume the normal flow. In the meantime, degas the reservoir with dry N_2 or Ar gas to remove dissolved air. The latter step becomes more necessary when the reservoir is placed in a thermal bath.

The solution in the reservoir is delivered to the imaging cell by a pumping module utilizing either a syringe pump or a peristaltic pump. In both delivery systems, the flowing solution inevitably bears vibrational modes induced by the pumps or other components in the system. The vibrations can couple with the imaging system and lead to imaging defects if not damped or filtered properly. The extent of the interference to imaging strongly depends on the flow rate and imaging scan speed. For a combination of high rate of scan and flow in the system, the noise induced in the system often prevents the collection of discernible images.

As outlined in the setup sketch, a damper is installed in-line after the pump to suppress the vibrational noise generated by the periodic movement of the pump. The vibrational damper is made of a tube with one end connecting to the flowing tube while the other end is exposed to air. The vibrational modes in the flowing solution are mostly absorbed by the solution in the tube and transferred to air. This vibration damper can also serve as a tool to clean up the bubbles flowing along the way. There are a couple of ways to minimize the coupling of the noise in the flow system to the imaging module, which is discussed in Section 12.3.6. An obvious one is to keep the overall tubing length at a minimum. The second is not to connect the tubing and the liquid cell too tight. The short transportation length can also minimize the heat dissipation in the tube and help stabilize the solution temperature.

There are several inherent issues in this setup that need to be mitigated to ensure the experimental efficiency, continuity, stability, data authenticity, and image quality. The nature of each issue and possible strategy to minimize each one of them is discussed in the following in detail.

12.3.2 SEED CRYSTAL PREPARATION

The *in situ* experiment is usually accomplished on a seed crystal housed in a liquid cell that can either be built in-house (Hillner et al. 1992; Schoenwald et al. 2010; Sulchek et al. 2000; Wade et al. 1998) or acquired commercially. The liquid cell serves multiple functionalities including holding the AFM cantilever in position and placing the tip at the surface to be imaged while forming a sealed enclosure through an O-ring gasket to retain a small volume of solution that can be pumped or heated to mimic physiological conditions. To ensure a laminar flow within the cell and to minimize solution leakage during scanning, the seed crystal geometry must be prepared carefully.

Seed crystals for AFM-based experiments can either be harvested in-house from a solution grow experiment (e.g., CaOx mono-, dihydrate crystals or KDP) or purchased commercially (e.g., hydroxyapatite, brushite, calcite, and fluorapatite). There are two ways to select the seed crystal size to facilitate the liquid cell. One is to make the seed small enough to fit inside the space created by the O-ring and the sample holder (option 1) and the other is to use the crystal itself as the base to seal the cell with the O-ring gasket (option 2). The former is used for those systems that only small (less than 1 mm) crystals can be harvested in preparation like the CaOx monohydrate (COM) or dihydrate crystals, which have a nominal size ranging from a few hundreds of microns to a couple of millimeters. Crystals with accessible cleavage planes such as calcite crystals or KDP are good candidates for the second approach. The latter is usually a preferred method whenever crystals with decent size are available because it provides an open space for solution to flow and thus reduces the solution Reynolds number. For this method, a thinner O-ring is suggested in order to reduce the stress in the system due to the large compression needed to form a sealed compartment.

Figure 12.6a shows a schematic of a commercial fluid cell along with a typical layout of an O-ring that is comparable to those used for the Bruker Multimode liquid cell (part number MTFML). A top-view schematic displaying the layout for a small sample (option 1) housed inside the liquid cell is shown in Figure 12.6b. In this configuration, the crystal geometry cannot exceed those outlined in the plot for all three dimensions for the type of O-ring specified here. If the sample is too big, the solution will not flow smoothly over the crystal surface. Furthermore, the O-ring will not sit on top of the sample holder in a relaxed state, thus resulting in either leakage or false engaging when imaging. An O-ring with a large vertical thickness may be used to alleviate the sealing problem, but this type of O-ring usually leads to a large drift while imaging. Thus, cautions must be taken when

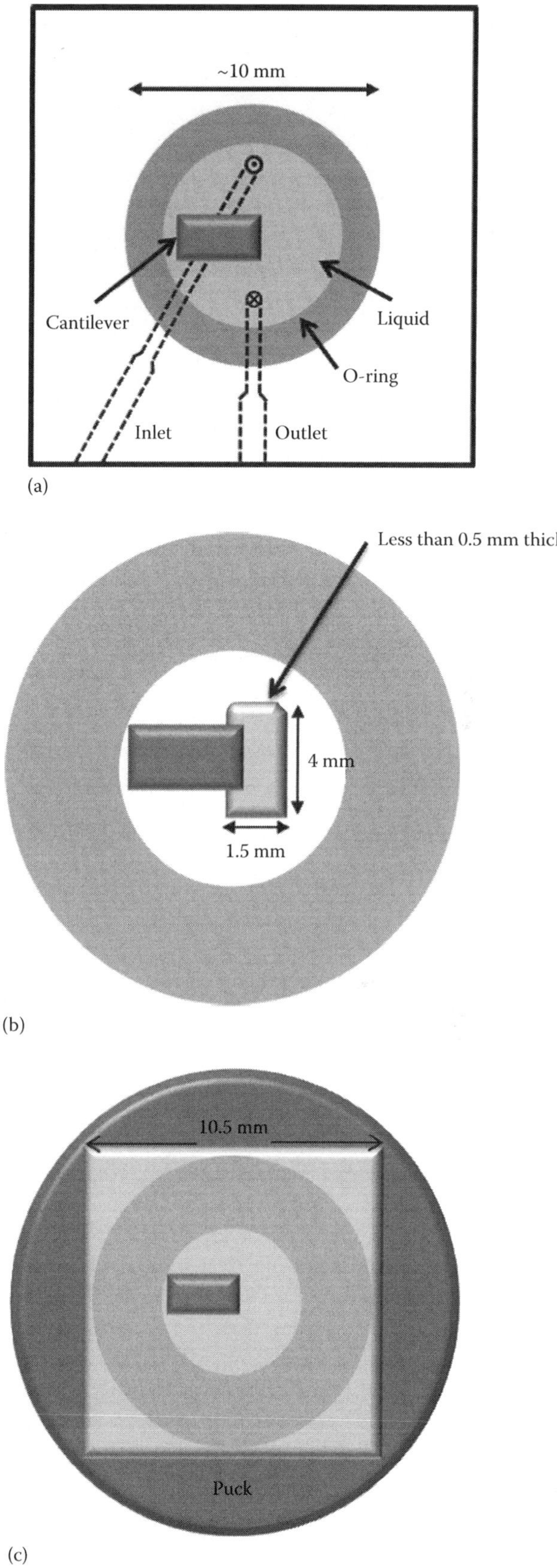

Figure 12.6 Plan views of (a) geometrical configuration of a typical commercially available AFM liquid cell, (b) estimated maximum sample dimensions that can be accommodated inside the liquid cell with minimal interference to imaging and solution flow, and (c) configuration using the sample itself as the substrate to seal the liquid cell.

such an O-ring is used. A similar sketch using a large sample is shown in Figure 12.6c. The space within the enclosure is large, and the solution can flow through the surface without much of the disturbances like that in option 1. In this configuration, the O-ring has to be pressed much harder to achieve a good seal with the substrate. To avoid crashing the tip during the operation, one may want to choose a thinner O-ring and level the relative height between the tip and the O-ring edge as close as possible before placing the liquid cell on top of the sample.

12.3.3 HYGIENE OF IMAGING COMPONENTS

Crystal growth kinetics is extremely sensitive to impurities including those contaminations that were introduced to the imaging system unintentionally. Residues from any components of the imaging system can smear the experimental data and lead to false conclusions. In many cases, these unwanted materials can halt an experiment even for those well-designed and known systems such as calcite and CaOx crystals. For example, dislocation hillocks on a calcite crystal (104) face can grow rather easily when under supersaturated solutions. On some occasions, it has been seen that steps grow for a short transient time under pure supersaturated calcium carbonate solution and then stop completely. A similar effect was also observed in the experiment with CaOx dehydrate crystals. It was found that the contaminants on the AFM tip were responsible for the inhibition. Thus, assurance on the cleanliness of experimental components is critical to the quality of the results especially for those investigations on the interaction between biomimetic additives and inorganic crystals. Methods discussed in the following have been shown to be effective in ensuring the cleanliness of the imaging system in our lab.

The AFM probes are in most cases stored in a container with a gel pad and inevitably organic residues often coat the tip as well as the cantilever. If not removed properly prior to an experiment, the loosely bound residues can easily break away from their host and enter the solution. Some of these will eventually land on the crystal surface and become active impurities to alter the growth kinetics. Fortunately, these organic residues can be removed effectively by a plasma cleaner. Exposure of the AFM tips to either an Ar or N_2 plasma source for 5–10 min can yield a clean tip.

The liquid cell (including the O-ring) should also be cleaned carefully before any experiment, and it can be accomplished through the flowing steps. First soak the entire unit in a diluted Micro-90 solution for at least 30 min to remove any grease or any other organic residues followed by a series of DI water rinsing. The next step is to wash the inner region (which will be in contact with the experimental solution and the crystal) including the inlet and outlet with mildly acidic water, then rinse it again multiple times by DI water. The same procedures are also applicable for cleaning the tubing and connectors. To enhance the removal efficiency and completeness, one may also incorporate an ultrasonic cleaner system in each of the cleaning steps especially during the final water rinse. Another possible impurity source is the metallic puck for anchoring the seed crystal when the sample is glued directly to it. An easy way to remove the problem is to cover the puck surface with a glass cover slip of the same diameter, which can be pre-sanitized by DI water and a plasma source readily.

12.3.4 EQUIPMENT PROTECTION FROM SOLUTION LEAKAGE

Solution leakage from the liquid cell is a commonly encountered problem in solution-based AFM experiments especially when solution is continuously pumped from a reservoir. Leakage can occur in any stage of the experiment. There are several components in the Bruker Multimode imaging system that are prone to electrolyte contamination should the aqueous solution leak out the liquid cell. The leakage can be detrimental to these components. The first one is the piezoelectric component in the scanner tube, which is right below the sample holder. The mechanical components of the scanner are next to the AFM base, which houses all the controlling electronic circuits of the system. Solution can penetrate to the scanner tube by letting it dwell too long under the sample puck, and consequently, crystallites form as water naturally evaporates. These randomly formed crystallites inside the scanner tube can skew the movement of piezoelectric components and eventually stop its functionality. Second, if the spilled liquid enters the base unit, it can shorten the circuitry and burn electrical components. Solution spills can also cause mechanical parts to rust and malfunction. Repairing or replacement of these units are time consuming and costly. However, large spills can be prevented if cautionary measures are implemented properly. Spills can be mitigated through several simple engineering and operational controls.

Although some have chosen the extreme by covering the gaps between the joints in the base system using sealant, a nonintrusive method would be preferred. One can install the following controls to minimize the negative impact of the leaking. First, use a large thin sheet of plastic wrap (like the Saran cling wrap, see Figure 12.5g) to cover the scanning tube and the scanner, then place the sample holder on top of the scanning tube to hold the plastic wrap in place. To eliminate the dragging resistance induced by the clinging plastic sheet on the piezoelectric components, pull the plastic up to leave enough space between the plastic and the tubing surface before setting the sample on top of it. Then press the plastic lightly around the side of the scanning tube to hold the plastic wrap in place (the noise and drift reduction due to this exercise is discussed in Section 12.3.6). For the base system, one can tailor a concentric ring-shaped water-resistant sheet to hold onto the scanner support ring on the AFM base. The plastic wrap and the cover sheet thus are acting as a shield to prevent any solution penetrating the imaging components should a leak occur.

To avoid large solution spill or leakage, one can also apply the following operational control by using a *leak indicator* to identify the small leaks at its earlier stage. The leak indicator can be any materials that absorb liquid and change their apparent color when absorbing. For example, the all-purpose Bluesorb 750 (Berkshire) is a good choice. One can cut two small paper strips from one sheet of it and place them under both sides of the O-ring. Any small leakage from the liquid cell will result in a noticeable stain on the strip(s). Sometimes, a small leak may stop on its own as the O-ring relaxes itself on the sample or the sample holder after the system is scanned repeatedly. In this case, the size of the stain will remain unchanged. If the water stain on the paper strip enlarges, the cell is leaking continuously and one would need to stop the experiment (or flow) and readjust the liquid cell system accordingly until a sealed system is achieved.

12.3.5 DETERMINATION OF PROPER SOLUTION FLOW RATE

The solution flow rate needs to be carefully determined for each mineral system investigated. The growth rate or step velocity on a crystal face depends on the crystal solubility, solution supersaturation, temperature, or surface kinetics. The flow rate must be adjusted such that the solute diffusion in the solution is not limited so as to cause artificial modifications to the growth kinetics. There is usually a flow rate window within which the step velocity is independent of the flowing rate through the reaction cell, that is, at the given supersaturation, the step velocity does not change when the flow rate is further increased (Gilmer et al. 1971). To define this benign flow rate window, one can first measure the step velocity as a function of the flow rate at the highest solution supersaturation that is to be investigated and then repeat the process at the minimum supersaturation. The latter is usually just above the equilibrium. In general, the step velocity increases with the flow rate initially and gradually reaches a plateau when the flow rate is further increased. Theoretically, operating at any flow rate within the plateau is acceptable; however, the flow rate at the lower end is often adopted for a reduced noise level. The flow rate determined from the former measurement is usually within the plateau region of the latter. For a crystal system that has multiple faces of interest, the same procures should be performed to determine the optimal flow rate for each face.

12.3.6 FINE TUNING THE IMAGING SYSTEM

As expected, the imaging system will be under a large stress after the liquid cell is sealed in both sample scenarios because of the strain imposed on the O-ring gasket after the sealing manipulations. To ensure the experimental continuity, the imaging authenticity, and a low chance of system leakage, the stress in the system must be minimized prior to massive data collection. There are several conservative but proven methods to release the system stress *in situ* without breaking the seal or the AFM tip. The general rationale is to pre-scan the system at a much higher scan rate and larger scan size than that one would normally apply in the regular experiment for a lengthy time. The first is to scan the tip without solution at a low scan rate (0.5–1.0 Hz) and a small scan area (~500 nm) with a minimum tip-surface force (near the attractive region) for about 1–5 min depending on how the system responds. Then scan the system by gradually changing the scan size only while monitoring the system response. Let the system scan at an area that is twice that of the experimental condition for at least 10 min and then gradually increase scan rate. If the system holds for the series of scans in air, repeat the scan while the solution is introduced for another 5–10 min. Then increase the scan rate to twice that of the experimental scan rate, and let it scan for another 10 min. This will usually tune the system to a stage where the stress-induced imaging defects or drift are minimized, at least for the system that scanned at half of that rate used in the tuning stage.

To ensure that the AFM images are authentic representations of the surface morphology and to minimize the tip interference on those large adsorbates (e.g., protein additives), several precautions need to be taken. First, the imaging force needs to be reduced to the minimum possible value that allows the tip to remain in contact with the surface but to impose no measurable effect of imaging on the growth kinetics or morphology. The tip-surface force meeting this criterion is usually in the attractive regime. There are a couple of ways to verify the effect of imaging force on the image appearance. The first approach is to gradually increase the imaging force until alteration on the morphology or step velocity is observed. The second way is to zoom out the image to a larger scan box and to look for the signature of the smaller scan area. When the force applied to the surface is kept minimal, the smaller scan area shows no difference to that displayed in the image at a large scan area. Otherwise, the subdivision within the larger image gives a different appearance. A consequence of imaging in the attractive regime is that sometimes the tip can transiently pull off the surface and lose contact with the imaging area. The resulting image often shows poor quality with missing scan lines. Although the quality of the image is somewhat reduced, the morphology and kinetics of surface features including steps and adsorbed additives are minimally affected.

12.3.7 STEP VELOCITY DETERMINATION

There is another apparent imaging artifact that is unique to the *in situ* AFM imaging of growing steps. Unlike any direct optical imaging or phase shift interferometry where every point in the area of interest is captured at the same time, AFM images are acquired through a series of scan lines in a finite time. For a stationary object, there would be no difference between the images collected from these two methods. However, for a moving object, like a step on a growing crystal face, images obtained using AFM will yield a step that is distorted from its true orientation. The images reported here are not corrected for this effect. In fact, this change in angle has been used to extract the step velocities at different growth conditions (Land et al. 1997).

The dynamic and kinetic factors that control crystal growth and the impact of impurities on these parameters can be extracted from the analysis of step morphology and kinetics. Thus, the ability to accurately determine step velocity from *in situ* AFM images collected at different conditions is critical to quantify these important physical parameters. There are three commonly accepted methods to evaluate the step velocity (De Yoreo et al. 2001; Land et al. 1997; Yau et al. 2000). The first method is to utilize the apparent angles of the selected step front measured in successive images of up and down scans. The second process is to quantify the displacement of the step front in a pseudo 2D image collected by disabling the slow-scanning axis. The third and most intuitive approach is to compare the step movement from a fixed reference point, which can be a stationary object on the surface or the origin of the dislocation in a series of sequential images. Detailed discussion of each method is given in the following.

As discussed in earlier sections, for active growing hillocks, the obtained apparent AFM image gives a distorted representation of the true morphology due to a finite scan rate. The extent of the deviation to the true structure depends on the scan direction as well as the contrast between the step velocity and the tip scan rate. For any step, the distortion is manifested as the changes in

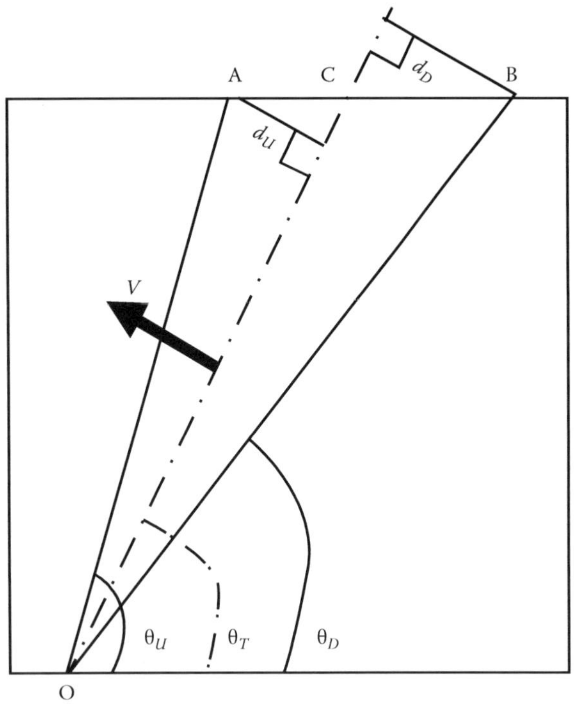

Figure 12.7 The schematic illustrating the apparent step orientation in an AFM image collected during up- and down-scans, respectively. OA represents the step edge in the up-scan image while OB is the edge in that of the down-scan. The estimated true step orientation is given by the dotted line OC.

the angle of step edge. Since the scan direction, scan rate, scan size, number of scan lines per frame, and the apparent angles of the step edge can be either preset or measured, the step velocity can be extracted by deconvoluting the sequential down- and up-scan images through the apparent angles. To demonstrate the change of the apparent step orientation between the down- and up-scan images, a schematic display of a single step edge in an apparent image relative to that of the true step front is shown in Figure 12.7. In this display, a straight step edge is assumed and is moving along the direction indicated by the solid arrow. While the line OA represents the step edge in the down-scan image and the line OB represents the step front in the up-scan image, the true location of the step front is assumed to be along the line OC. The step velocity is defined as the advancement of the step front along the direction that is perpendicular to the step edge in a unit time. From the down-scan image, the advancement of the step after one-frame time t is equal to d_D, the magnitude of the step velocity or step speed V can then be expressed as

$$V = \frac{d_D}{t} \tag{12.6}$$

Similarly, from the up-scan image, the step velocity can also be written as

$$V = \frac{d_U}{t} \tag{12.7}$$

where d_U is the displacement of the step in the up-scan image after one-frame time. Because the step velocity determined from any two adjacent AFM images for a given growth condition should be equal for the same step, the following condition holds:

$$\frac{d_D}{t} = \frac{d_U}{t} \tag{12.8}$$

This equation can be reformulated in terms of the step apparent angle in both down- (θ_D) and up- (θ_U) scan images and the true step angle (θ_T) through a simple trigonometric manipulation:

$$\frac{S \cdot \sin(\theta_T - \theta_D)}{\sin(\theta_D)} = \frac{S \cdot \sin(\theta_U - \theta_T)}{\sin(\theta_U)} \tag{12.9}$$

where S is the vertical dimension of the image frame. From the previous equation, the unknown parameter of θ_T can be solved analytically and expressed in terms of those measurable apparent angles

$$\theta_T = \cot^{-1}\left(\frac{\cot\theta_U + \cot\theta_D}{2}\right) \tag{12.10}$$

Using either Equation 12.6 or 12.7 and substituting Equation 12.10 for θ_T, the step velocity can be related to the three angles by

$$V \cdot t = \frac{S \cdot \sin(\theta_T - \theta_D)}{\sin(\theta_D)} = \frac{S}{2} \cdot \sin(\theta_T) \cdot (\cot\theta_D - \cot\theta_U) \tag{12.11}$$

Let N be the number of scan lines per frame and R be the scan rate, the time needed for capturing one frame of image is $t = N/R$. Solving Equation 12.11 for V, the step velocity thus carries the form

$$V = \frac{S \cdot R}{2 \cdot N} \cdot \sin\theta_T \cdot (\cot\theta_D - \cot\theta_U) \tag{12.12}$$

Thus, by measuring the apparent angles of a step in a sequential down- and up-scan AFM images and utilizing a united equation of (12.10) and (12.12), the velocity of a step is determined.

The second approach to obtaining the step velocity is based on images that collected with the slow-scan axis disabled. In this imaging mode, the AFM tip is only probing the surface along one single scan line. Nevertheless, the temporal evolution of one particular spot on the step edge is displayed in the pseudo 2D AFM image with the vertical dimension being the time elapsed for the frame. By measuring the horizontal displacement away from the initial step front location, the step velocity can be estimated readily. A schematic is displayed in Figure 12.8 to demonstrate this method. In this representation, a single straight step with a large terrace width (shaded area in Figure 12.8a) is assumed to be perpendicular to the fast-scan line and propagate along the arrow-pointed V direction. Suppose the slow scan is disabled when the tip is at location P where P is a single spot on the step edge, the subsequent image collected would capture the progression of the surface that covered the single scan line including point P. Because the step is advancing to the left, it is expected that point P would also move toward the left of the frame. In result, an image with morphology similar to that shown in Figure 12.8b is obtained. Please be cautious that the slanted

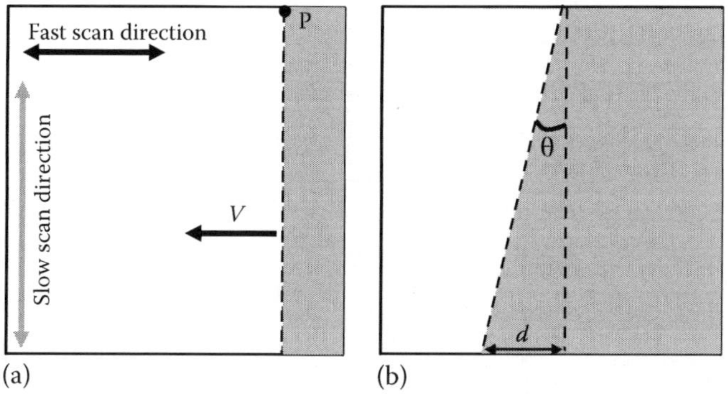

Figure 12.8 The schematic representation of an AFM image collected with the slow-scan axis disabled. (a) At time $t = 0$, a straight step is moving along the fast scan axis at velocity of V. (b) The appearance of point P in a slow-scan disabled AFM image where d is the distance that P is moved from its original location after one frame of scan.

line in the image is not the step edge rather it is the trace of the location P through time t, the time that elapsed after one frame of scan. The magnitude of the step velocity is essentially the same as the advancement rate of point P, and thus the step velocity can be evaluated using $V = d/t$ where d is the displacement of P from its original position. The quantity of d can be either measured directly from the image or calculated from the slant angle (θ) and the frame size S. The step velocity can be estimated from the following two equivalent formulas:

$$V = \frac{d \cdot R}{N} \tag{12.13a}$$

or

$$V = \frac{S \cdot R \cdot \tan\theta}{N} \tag{12.13b}$$

Since both θ and d are measured values and can be obtained readily, one may choose to use both measurements to estimate the step velocity and to increase the accuracy of the quantification.

The third method is to determine the step velocity by comparing the advancement of the step edge to a fixed reference point through a series of images that are collected under the same experimental conditions. This method is especially useful for those systems that have slow growth rate or for investigation at near equilibrium where the step velocity and the creation rate of new steps are usually small. Because of the results can be obtained by averaging calculations from a set of images in a reasonable lengthy time, the step velocity can be determined with good accuracy from this method.

In general, all these methods of calculation are valid to estimate the step velocity, and one can choose any of these depending on the image quality, the nature of the growth hillocks, and the dynamics of the system. One may want to use a combination of two methods to confirm the accuracy of the obtained results. The significance of the first method can also be reflected by the fact that one can use the true angle value and the step velocity to deconvolute the apparent images and to recover the true morphology of the growth hillocks and steps, by adding a post analysis tool in the imaging process program.

12.4 EXAMPLES OF USING *IN SITU* AFM TO PROBE BIOMOLECULE–MINERAL INTERACTION

12.4.1 NONEQUILIBRIUM MORPHOLOGY AND ITS IMPLICATIONS

In situ AFM can be used to reveal the morphology of a growing source on a crystal surface under nonequilibrium conditions. For some mineral systems, the dynamic morphology during growth resembles that at equilibrium while others exhibit a stark contrast. Examples are given in the following to demonstrate the two distinct cases.

The geometry of the microscopic growing hillocks in a nonequilibrium state for the COM crystal system is different from the shape of their equilibrium habit, and it is true for both faces that have been investigated. COM is the most thermodynamically stable form of CaOx crystals and is the primary constituent of the majority of human kidney stones (Coe et al. 1992). Figure 12.9a shows the equilibrium habit of COM with the commonly expressed faces {–101}, {010}, and {120}. Figure 12.9b shows the molecular structure of the (–101) face. The geometry of the growth hillocks on the (–101) and (010) face in pure solution is shown in Figure 12.9c and d, respectively. At equilibrium (Figure 12.9a), the (–101) face shows a hexagonal shape, and the (010) face displays a parallelogram geometry. At nonequilibrium, COM crystals grow on complex screw dislocation hillocks similar to other solution grown crystals such as KDP (Rashkovich 1991) or calcite (Teng et al. 2000). On the (–101) the face, the growth hillocks exhibit a triangular-shaped morphology bound by the [1–20], [120], and [–10–1] steps (Cho et al. 2012), which is very different from its equilibrium hexagonal habit (Figure 12.9a). On the (010) face, the geometry of growth hillocks exhibit a near rectangular-shaped morphology bound by <100> and <001> steps (Cho et al. 2012). Although this shape resembles that of the (010) face, the orientation is very different. Thus, the geometry during growth is in contrast to that at equilibrium for both faces of the COM crystal.

The triangular shape of growth hillocks on the (–101) face indicates that steps propagating toward the {010} faces and the (–1–20) and (–120) were not present. Instead, steps moving toward [–10–1] were observed that truncate the angle formed by the [120] and [1–20] directed steps. The implication is that steps directed toward these three unrepresented directions have considerably higher speeds than the others. Even so, for the growth hillock shown in Figure 12.9c, the [–10–1] directed steps move about 10 times faster than that of the other two. This means that there is an enormous anisotropy in step velocity or attachment/detachment kinetics between the [1–20]/[120] and [–1–20]/[–120] directions, despite the fact that the only significant difference is in the tilt angle of the oxalates (see Figure 1c in Qiu et al. 2005, and Figure 12.9b herein). The highly anisotropic step kinetics gives rise to closely spaced steps along the two slower directions.

In fact, this inference has been confirmed in a recent report (Cho et al. 2012). In this study, a racemic mixture containing equal amounts of 6-residue of L- and D-poly-Asp peptides is added to the growth solution. Because it is found that the L- and

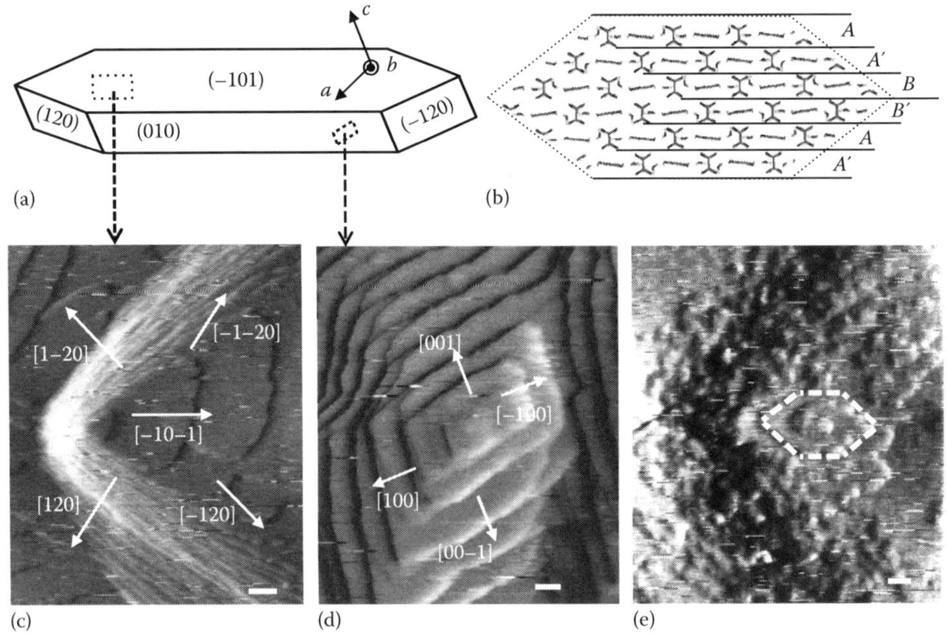

Figure 12.9 (a) Schematic of COM crystal equilibrium habit with three commonly expressed planes and the coordinate unit vectors. (b) Molecular structure of the COM (−101) face. The dotted hexagon shows the relationship between the crystal habit and the molecular structure. Solid lines represent the *AA′BB′* stacking sequence that is perpendicular to the [010] direction due to the different orientations of water molecules in successive rows. Representative AFM images show the growth hillocks (c) on the (−101) face in pure solution, (d) on the (010) face in pure solution, and (e) on the (−101) face under the influence of equal molar of L-and D-Asp_6. Scale bars are 200 nm in (c) and 100 nm in (d, e). (Reproduced from Cho, K.R. et al., Qiu, S.R. et al., *Cryst. Growth Des.*, 12(12), 5939, doi: 10.1021/cg3009317, 2012. *J. Am. Chem. Soc.*, 127(25), 9036, 2005. With permission.)

D-enantiomer only inhibits the growth of the upper and lower steps shown in Figure 12.9c, respectively, the growth of all four steps will be inhibited when both enantiomers are present. In fact, this is exactly what is observed on the (−101) face where steps along all directions are affected almost equally; the [120] and [1−20] steps exhibited nearly identical morphological and kinetic changes. Apparently, these two enantiomers acted in concert to inhibit the growth on the (−101) face by independently interacting with the steps. Furthermore, because of the inhibition along all step directions, those that previously were unexpressed in pure solution, such as the [010], [0−10], [−120], and [−1−20] steps, were expressed under the influence of the enantiomer peptides. As a result, steps in the first and second turns of the hillock exhibited a shape that resembled the bulk crystal morphology shown in Figure 12.9e.

The nonequilibrium growth hillock on the (010) face COM further provides an example that the microscopic growing morphology is directly related to the molecular structure of the crystal. As shown in Figure 12.9d, the growth source is rather complicated containing multiple screw dislocations and mostly of the Frank–Reed type (Frank and van der Merwe 1949). For the first couple turns of the growth spiral, the step heights range from 1 to 3 *h*, where *h* is the height of an elementary step, about 3.98 Å. However, at a sufficient distance away from the dislocation source, all steps are quadrupled with heights of ca. 16 Å. This quadrupling of steps is caused by step interlacing (Frank 1951; van Enckevort et al. 1981), which is most apparent at the hillock corners as shown in Figure 12.9d. The step interlacing and step bunching originate from molecular rows stacked in repeating sequences AA′BB′ within the (−101) face (Figure 12.9b), which has been discussed in detail in an earlier study (Qiu et al. 2005). In short, the unique packing structure in the (−101) face creates a screw–axis symmetry element perpendicular to the (010) face and leads to asymmetries in step velocity from layer to layer with a four-layer periodicity. The multiple step height also contributes to the much smaller step velocity for all the steps measured on the (010) face than that on the (−101) face. Judging by the shape of the growing hillock, the step velocity along each of the four expressed directions should be comparable.

In the calcite system, however, the geometry of the equilibrium crystal habit and growth hillocks during growth under pure supersaturated solution resembles each other. As shown in Figure 12.10a, the calcite crystal has a rhombohedral shape. Because of the structural symmetry, all four faces are equivalent. Growth on pyramidal hillocks from any of the freshly cleaved (104) faces exhibit the same rhombohedral shape. The growth hillocks contain two pairs of structurally distinct steps with a step height of 3.1 Å and related to each other by a *c*-glide plane as shown in Figure 12.10b (De Yoreo et al. 2001; Teng 1998). These steps are commonly labeled obtuse and acute or positive and negative, the former labels reflecting the angle formed between the step riser and the (104) plane. For the steps shown in the figure (Fu et al. 2005), the obtuse steps grew faster than the acute. This difference in step velocity causes the angle θ that bisects the two equivalent flanks of the growth hillock to deviate from 180°. Depending on the solution supersaturation, the obtuse and acute steps in the growth hillocks can be readily discerned from the appearance of the angle θ (De Yoreo et al. 2001; Teng 1998).

12.4.2 HOW CHIRALITY IS TRANSFERRED FROM MOLECULES TO CRYSTAL MORPHOLOGY

Many biogenic crystals exhibit chiral morphology, for example, the chiral calcite plates in coccoliths (Borman et al. 1982;

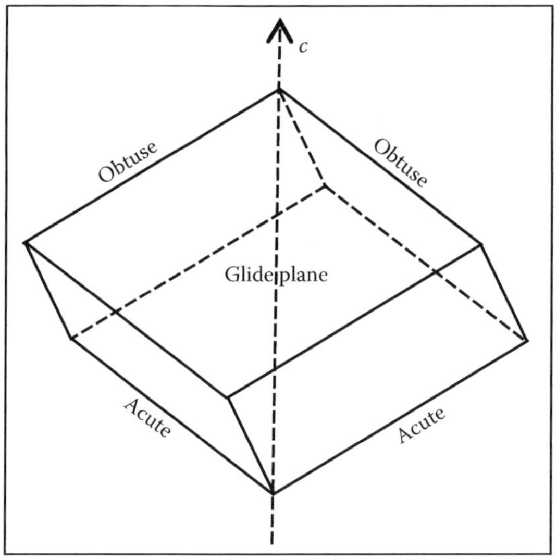

(a)

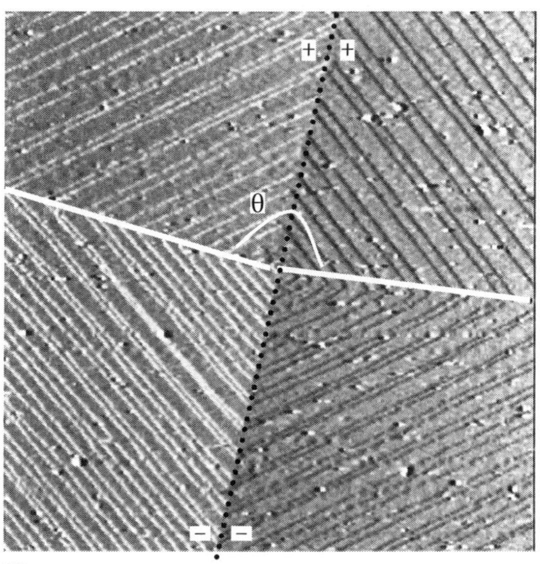

(b)

Figure 12.10 (a) Schematic showing the typical rhombohedral shape of a calcite crystal with six crystallographically equivalent (104) faces. The arrowed line indicates the glide plane. (b) Representative AFM image showing the growth hillock on the (104) calcite face in pure supersaturate calcium carbonate solution. The obtuse step is designated as the positive step and the acute step is designated as the negative step. The dotted line corresponds to the glide plane. Image horizontal dimension: 3.4 μm. (After Fu, G. et al., *Adv. Mater.*, 17(22), 2678, doi: 10.1002/adma.200500633, 2005. With permission.)

Didymus et al. 1994; Young et al. 1992) and the COM crystals in leaves of plants in the Solanaceae family (Bouropoulos et al. 2001; Levy-Lior et al. 2003). Biomolecules such as proteins, polysaccharides, and peptides are believed to play important roles in crystal symmetry reduction and the formation of the chiral morphology *in vivo* (Borman et al. 1982; Bouropoulos et al. 2001; Levy-Lior et al. 2003). *In situ* AFM can be used to observe the molecular scale growth modification of surface steps by chiral molecules and to reveal the mechanism by which the changes in small scale are translated into bulk crystal habit.

In an early study, Orme et al. (2001) showed that the addition of enantiomeric aspartic acid (Asp) leads to a growth hillock of broken symmetry with respect to the glide plane by modifying the geometry of the acute steps. The modification to the step geometry on the (104) face as shown in Figure 12.10b depends on the enantiomer introduced. At the presence of D-Asp, the acute step with the (01–4) riser is mostly affected and becomes curved. The presence of L-Asp, on the other hand, only modifies the acute step having the (1–1–4) riser. Neither enantiomer causes modification to the two obtuse steps. The resulting morphology of the growth hillocks is thus directly related to the chirality of the enantiomer, and they are mirror images of each other. Although the modified steps have curved geometry, the step edge remains smooth and the step velocity is unchanged. Moreover, no incorporation of the enantiomer is found in the growing calcite crystal. Theoretical calculations reveal that chiral modification is attributed to the stereochemical recognition between the enantiomer and specific acute step that leads to the reduction in step-edge free energy. Furthermore, the resulting chiral shape manifested by the atomic steps on the growth hillocks matches remarkably well with the bulk crystal habit that is harvested under the influence of the same Asp enantiomer. This study using *in situ* AFM provides the direct evidence that chiral modification can propagate from atomic to macroscopic length scales through enantiomer–step specific binding.

Similar biomolecular control in chiral formation has also been observed on the COM system (Cho et al. 2012). Figure 12.11a and b show the modification of growth hillocks on the (–101) face of COM under the influence of 6-residue linear Asp enantiomers (D-Asp_6, L-Asp_6). Similar to the calcite/Asp case, each enantiomer modifies the growth of only one specific set of steps and the resulting morphology with D-Asp_6 exhibited mirror symmetry to that with L-Asp_6. As shown in Figure 12.11a, under the influence of D-Asp_6, the growth hillock morphology changes from the triangular shape in pure solution (Figure 12.9c) to that of an ellipse, of which the major axis is tilted from the [–10–1] direction by approximately 20° counterclockwise (see inserted arrow in Figure 12.11a). In contrast, as displayed in Figure 12.11b, under the influence of L-Asp_6, the ellipse is tilted in the opposite direction with its major axis rotated clockwise about 20° from the [–10–1] direction.

The resulting enantiomorphic modification to the COM growth hillock under the D- and L-Asp_6 is attributed to the chiral-selective interaction between the enantiomers and the specific steps that leads to the asymmetric growth inhibition. While the D-Asp_6 strongly inhibits the [120] and [–120] steps, the L-Asp_6 preferentially interacts with the [1–20] and [–1–20] steps. This is reflected by the step velocity ratios $V_{[1-20]}/V_{[120]}$ and $V_{[-1-20]}/V_{[-120]}$, which are shown in Figure 12.11c and d, respectively, for both enantiomers. A ratio greater than unity indicates that the step velocity in the denominator is preferentially inhibited and vice versa. Furthermore, the step edges are heavily serrated, indicating that enantiomers are adsorbed to the terrace and pin the steps. Since the chiral modification is manifested through the reduction of step velocity and change in step shape, the observed effects must be related to kinetic factors, which are different from that of the calcite chiral modification by the simple Asp enantiomers.

As indicated in these examples, for chiral modification to occur, the following conditions must be satisfied. For one, the crystal itself must have mirror symmetry in the crystal structure;

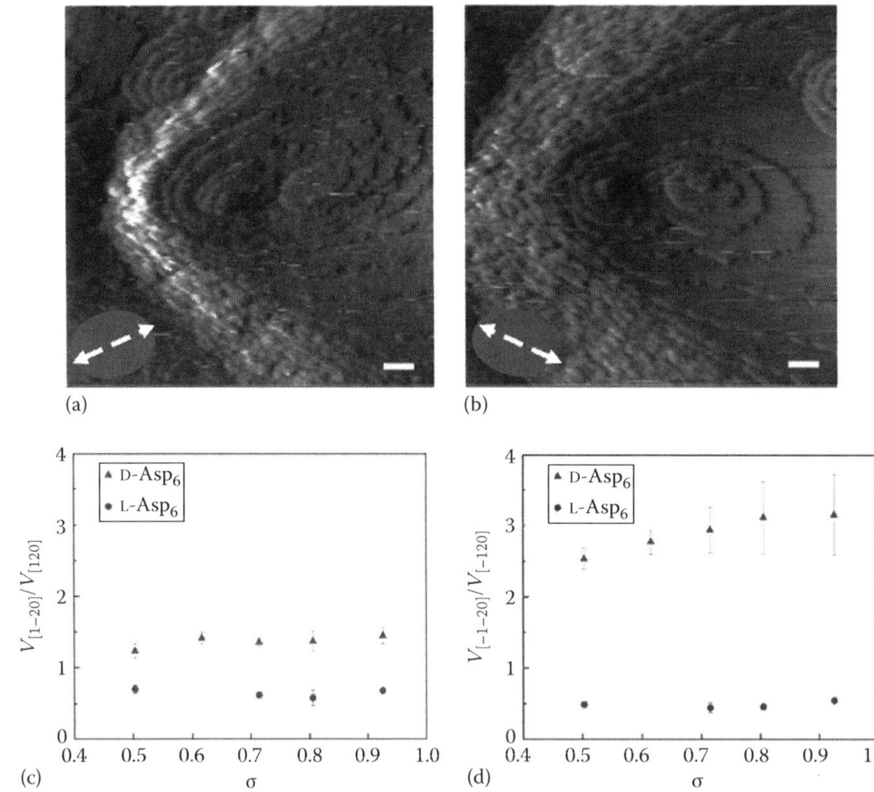

Figure 12.11 Representative AFM images showing the growth modification of the COM (−101) face by two enantiomers (a) D-Asp$_6$ and (b) L-Asp$_6$. (c) Plot of $V_{[1-20]}/V_{[120]}$ under the influence of either L-Asp6 (filled circle) or D-Asp$_6$ (filled triangle) at different solution supersaturation. (d) Plot of $V_{[-1-20]}/V_{[-120]}$ of either L-Asp$_6$ (filled circle) or D-Asp$_6$ (filled triangle) at different solution supersaturations. Scale bars are (a, b) 200 nm. (After Cho, K.R. et al., *Cryst. Growth Des.* 12(12), 5939, doi: 10.1021/cg3009317, 2012.)

and for the other, there must be a stereochemical match between the step structure and one of the enantiomers. The interaction between the enantiomer and the step can lead to the modification of step-edge free energy (i.e., thermodynamic control) (Orme et al. 2001) or of the rate at which the solute molecules attach or detach to the kinks at the step edge (i.e., kinetic control) (Maruyama et al. 2009). Regardless which factor dominates in the chiral modification, they all result from the preferential binding of specific chiral molecules to steps related by mirror symmetry. The studies from *in situ* AFM imaging suggest that formation of chiral morphologies in minerals may follow the general rule that molecular chirality of biomolecules is transferred to crystal morphology through preferential binding to mirror symmetry-related surface steps.

12.4.3 INHIBITION THROUGH STEP PINNING DESCRIBED BY A REFORMULATED C–V MODEL

Researchers have long used well-established impurity models in crystal growth theory to describe the growth modification of biomineral crystals. Recently, Weaver et al. have developed a mathematical model describing step kinetics in the COM system in the presence of biomolecules (Weaver et al. 2007). This model reformulates the classic Cabrera–Vermilyea model of step pinning (Cabrera and Vermilyea 1958) to more realistically account for impurity adsorption dynamics, utilize an exact expression for the Gibbs–Thomson effect, and include the reduction of kinetic coefficient through step pinning and kink blocking. The reformulated C–V model does an excellent job of predicting the effect of citrate (Weaver et al. 2006, 2007) as well as 6-residue linear Asp peptides (Cho et al. 2013) (both of which contains multiple carboxylic groups) on COM step kinetics. Detailed description can be found in the previous references; the key factors introduced in reformulating the C–V model, however, are worth revisiting and are summarized in the following.

Figure 12.12a shows the growth modification of the COM (−101) face by citrate, a naturally occurring molecule in the urinary system and a common therapeutic agent for treating kidney stone diseases. The final morphology of the growth hillocks is in stark contrast to that from the pure solution as shown in Figure 12.9c. The serrated step edge and the existence of a deadzone in step velocity indicate that citrate molecules modify the growth by pinning the step motion (Weaver et al. 2006, 2007). The normalized step velocity (to that in the pure solution) along the [−10−1] step as a function of the citrate concentration for different solution supersaturation is displayed in Figure 12.12b. Assuming the modified step kinetics by citrate follows the basic step-pinning model as described by Equation 12.3, the dependence of step velocity on the citrate level should follow the dotted line in Figure 12.12b. Clearly, the simple model does not reflect the experimental data well at all.

To accurately describe the inhibition kinetics, the following factors are considered to extend the original C–V model. The first factor introduced is the expression for the surface coverage of impurity. Instead of the $C_i^{1/2}$ assumption used in the original model, the Langmuir model of adsorption dynamics is adopted to truly reflect the impurity adsorption process and the

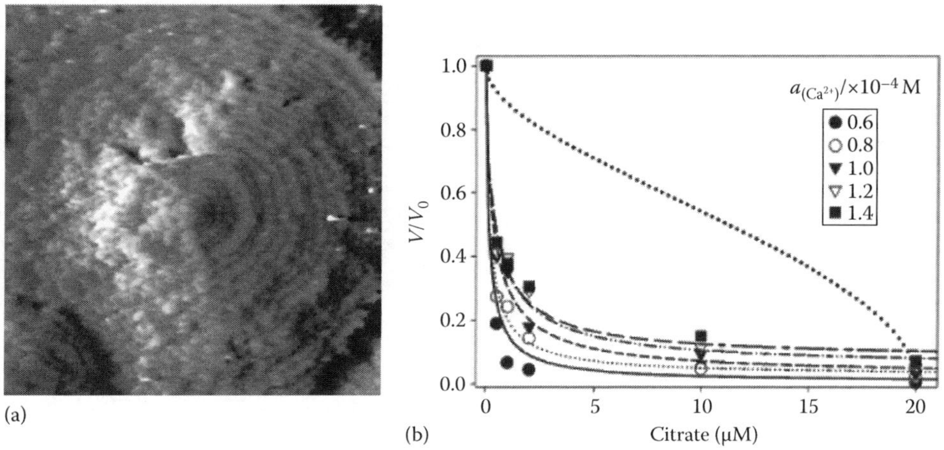

Figure 12.12 (a) AFM image showing the final shape of the growth hillock on the (−101) face of COM crystal under the influence of citrate. The growth hillock in the pure solution is shown in Figure 12.9c. (b) Dependence of relative step velocity on citrate concentration. Dotted line is fit to Equation 12.3. Dashed lines are fits to the data according to Equation 12.17. Horizontal dimension in a: 3 μm. (After Weaver, M.L. et al., *J. Cryst. Growth*, 306(1), 135, doi: 10.1016/j.jcrysgro.2007.04.053, 2007; Qiu, S.R. et al., *J. Am. Chem. Soc.*, 127(25), 9036, 2005.)

impurity surface coverage. This gives a fractional coverage of surface Θ in the following form:

$$\Theta = \frac{(k_A/k_D)C_i}{1+(k_A/k_D)C_i} \qquad (12.14)$$

where

k_A is the attachment rate coefficients
k_D is the detachment rate coefficients

The second factor considered is the impact of the Gibbs–Thomson effect on the solution activity. Following Teng et al. (Teng 1998), the Gibbs–Thomson effect on solution activity can be written in a general expression:

$$a_e = a_{e,\infty}\exp\left(\frac{\sigma r_c}{r}\right) \qquad (12.15)$$

where

a_e is the curvature-dependent equilibrium activity
$a_{e,\infty}$ is the equilibrium activity for a step with no curvature

This formula gives an exact expression that is corrected for all supersaturation due to the Gibbs–Thomson effect. The third addition is to include the kinetic coefficient in the C–V model's framework. In Weaver's report, for all citrate concentration, the step velocity does not recover to that in the pure solution, even for solution superstation that is much greater than that of the deadzone. The kinetic coefficients for all cases instead decrease, which is in contrast to the assumption used in the original C–V model. Based on the experimental observations, the reduction in kinetic coefficient is believed to arise from the diminishing of the active kink sites on the surface due to kink site poisoning upon citrate adsorption. Following this assumption, the kinetic coefficient (β) can be written in the following form with respect to that for the pure solution (β_0):

$$\beta = \beta_0\left\{1-\frac{a(h\Omega)^{0.5}}{n_{k,0}}\Theta^{0.5}\right\} \qquad (12.16)$$

In this form, a is the size of the growth unit along the step edge, which is typically the size of the solute molecule.

Combining the previous three expressions for Θ, a_e, and β with the expression for the step velocity given in Equation 12.1, a reformulated expression based on the original C–V model has the following form for the step velocity (V) of a crystal in the presence of an impurity relative to that for the pure solution (V_0):

$$\frac{V}{V_0} = \left\{1-A_3\left[\frac{A_2C_i}{1+A_2C_i}\right]^{0.5}\right\}\left\{1-\frac{e^{A_1(A_2C_i/1+A_2C_i)^{0.5}}-1}{e^{\sigma}-1}\right\} \qquad (12.17)$$

In this formula, $A_1 = \dfrac{2\gamma B(\Omega h)^{0.5}}{k_BT}$, $A_2 = \dfrac{k_A}{k_D}$, and $A_3 = \dfrac{aB(h)^{0.5}}{n_{k,0}(\Omega)^{0.5}}$. B is the product of the following proportionality constants (Weaver et al. 2007): the fraction of adsorbed surface impurities that stick to a step and pin it, the geometric factor relating linear spacing to aerial density, and the percolation threshold for a step to move through a field of blockers (Potapenko 1993). The curves shown in Figure 12.12b are best fits to Equation 12.17 for different solution superstation, and they are in excellent agreement with the experiment results. Similar agreement between the experiment and theory has also observed for the dependence of V on the solution supersaturation for different citrate concentrations.

The reformulated C–V type of impurity model is further validated by a recent study of successfully predicting the step kinetics under the influence of enantiomers on all steps in different faces of the COM crystal (Cho et al. 2013). Figure 12.13 shows the relative step velocity (V/V_0) as a function of Asp_6 concentration in COM solution at a fixed solution supersaturation for steps on both faces. The solid lines are best fit to the experimental data using Equation 12.17 at the fixed supersaturation, and it is clear that all fits match very well with the experimental data. Considering the structural similarity between citrate and poly-Asp, which both contain multiple carboxylic groups spaced along its backbone, it is expected that they would have similar effects in modulating the step kinetics of COM growth. And the reformulated C–V model prediction supports that.

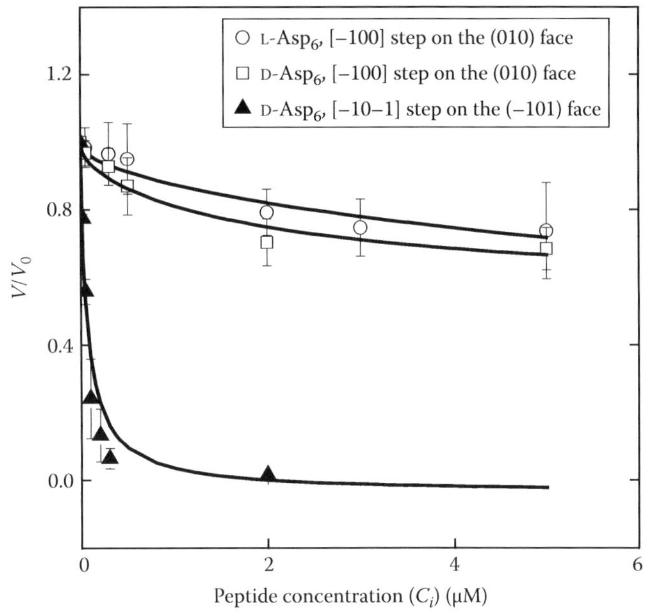

Figure 12.13 Relative step velocity (V/V_0) as a function of Asp_6 concentration (C_i) in COM solution at a given solution supersaturation. The unfilled black circle and the unfilled square represent the relative step velocity for the [−100] quadruple unit height on (010) face in the presence of L-Asp_6 and D-Asp_6, respectively. The filled triangle shows the relative velocity of the [−10−1] step on the (−101) face in the presence of D-Asp_6. The solid lines are fits of Equation 12.17 to the experimental data. (Cho, K.R., et al., *CrystEngComm* 15(1), 54, doi: 10.1039/c2ce25936b, 2013. Reproduced by permission of The Royal Society of Chemistry.)

The three fitting parameters A_1, A_2, and A_3 in the generalized formula for step velocity V are in fact controlled by three fundamental parameters of the system: the step-edge free energy, the kink site density, and the ratio of the adsorption/desorption rate coefficients. By fitting the experimental data using this equation, one could extract fundamental constants of the system (Cho et al. 2013; Weaver et al. 2007).

12.4.4 PROTEIN AS SURFACTANT TO ACCELERATE GROWTH

In addition to the inhibitive role that is commonly viewed for impurities on the growing crystals, additives can also serve as growth promoters by accelerating step kinetics. This has been observed by *in situ* AFM investigation for impurities of small peptides (Elhadj et al. 2006a,b) as well as full proteins on both calcite (Fu et al. 2005) and COM systems (Weaver et al. 2009). The argued view is that small amino acids, short or long chain of peptides, as well as proteins all act as surfactants to lower the step-edge free energy and to promote the ion attachment at the growing steps. Two examples are given in the following to demonstrate the role of proteins on promoting growth of calcite and COM crystals to show how the mechanism is revealed.

The growth modification of calcite by AP8 protein isolated and purified from abalone shell nacre (Fu et al. 2005) is exemplified in Figure 12.14 through the temporal evolution of the pyramidal growth hillocks on the (104) face. Similar to that discussed earlier on the modification of calcite by simple amino acids, only the acute step morphology is modified. After continuous exposure to AP8, both of the acute step edges of the hillock become progressively less straight, eventually producing a rounded morphology. Furthermore, all steps remain smooth and stable with no pinning features observed. The step velocity, on the other hand, for all four steps increases substantially with the obtuse steps amplified more. For example, at a protein concentration of 0.2 μM, the obtuse step velocity increases by a factor of 4.8 compared to that of 2.1 for the acute steps. As a result, the relative step velocity between the obtuse and acute steps is also increased by a factor of 2. This asymmetric modification has led to a reduced angle θ between the two obtuse steps as shown in Figure 12.14d.

As discussed in Section 12.2.2, there are four end-member models to describe growth modification by impurities: step pinning, incorporation, kink poisoning, and surfaction. The observation of an increase in step velocity accompanied by changes in average step spacing and hillock shape without step pinning (or incorporation) is consistent with only the fourth model. Therefore, it is postulated that, as with simple amino acids and poly-Asp (Elhadj et al. 2006a,b), AP8 proteins act as surfactants to modify the thermodynamics and kinetics of calcium and carbonate attachment at step edges without significant pinning, poisoning, or incorporation. The decrease in average terrace width indicates that the protein also reduces the step-edge free energy (De Yoreo and Vekilov 2003; Dove et al. 2004). This assumption is backed by the protein structural analysis, which shows that AP8 is amphiphilic (Lowik and van Hest 2004) and has an open conformation and an extended hydrodynamic radius when interacting with multiple ions or small $CaCO_3$ nuclei. The flexible structure enables the hydrophilic Asp domains to attract ions and water molecules and transport them to the step edges on the existing crystal without incorporating itself into the crystal lattice: a surfactant activity (Elhadj et al. 2006a).

The manifestation of Tamm–Horsfall glycoprotein (THP) accelerating the COM growth is somewhat different from that of AP8 protein. THP is a natural occurring macromolecule in urine and plays an important role in regulating kidney stone formation. Although THP does not incorporate into the COM crystal nor modify the edge stability, it adsorbs onto the crystal faces. Sequential AFM images showing the development of aggregates on the COM (−101) face after the immunoaffinity-purified protein (apTHP) is introduced in the growing solution are displayed in Figure 12.15. As shown in the images, the hillock shape remains unchanged even after a lengthy protein exposure time. THP proteins form aggregates and deposit on the (−101) face. In general, the aggregates have irregular filamentary shape with a dominant orientation parallel to the [−10−1] step edge. For sufficiently high protein concentration, the (−101) face becomes completely covered by the aggregates. Although the surface is covered by proteins, the triangular-shaped hillock and the steps can still be seen in the AFM image propagating underneath the proteins but with an increased velocity along all three step directions.

The [−10−1] step velocity as a function of the solution supersaturation is plotted in Figure 12.16 for two different protein concentrations and ionic strengths. Guided by the best-fit trend lines, the step velocity is clearly increased as the slope for both lines is larger than that of the pure solution. Furthermore, the

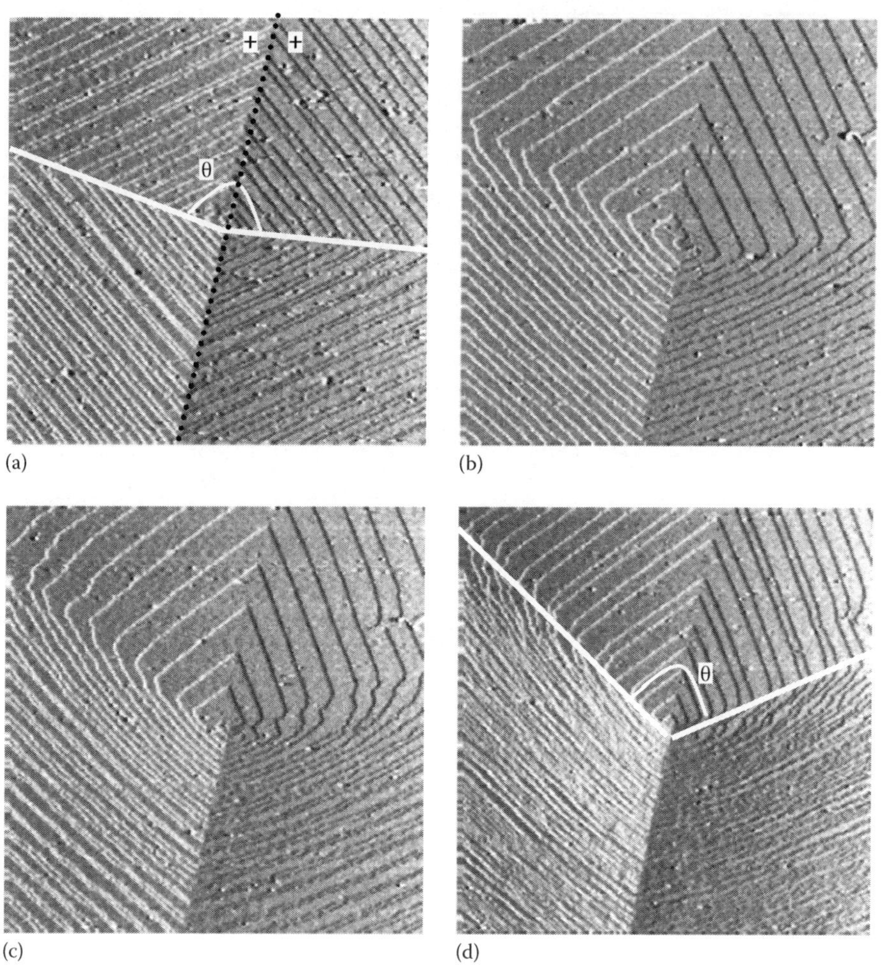

Figure 12.14 Representative AFM images showing the temporal evolution of morphological changes on the (104) face of calcite crystal induced by AP8 protein: (a) $t = 0$; in pure solution, (b) $t = 70$ min, (c) 90 min, and (d) 122 min. Horizontal scale: (a–d) 3.4 μm. (After Fu, G. et al., *Adv. Mater.*, 17(22), 2678, doi: 10.1002/adma.200500633, 2005. With permission.)

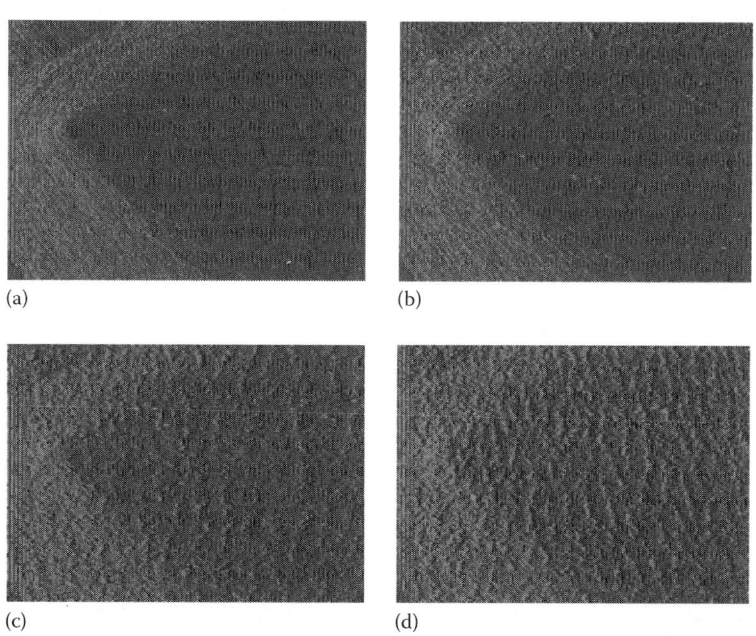

Figure 12.15 AFM images showing the growth hillock on the (−101) face after exposed to 100 nM apTHP for (a) 0 min, (b) 1.5 min, (c) 3 min, and (d) 4.5 min. The general morphology of the growth hillock does not change although proteins are deposited on the surface after a short exposure time. Images horizontal dimension: 3.5 μm. (After Weaver, M.L. et al., *Calcif. Tissue Int.*, 84(6), 462, doi: 10.1007/s00223-009-9223-0, 2009. With permission.)

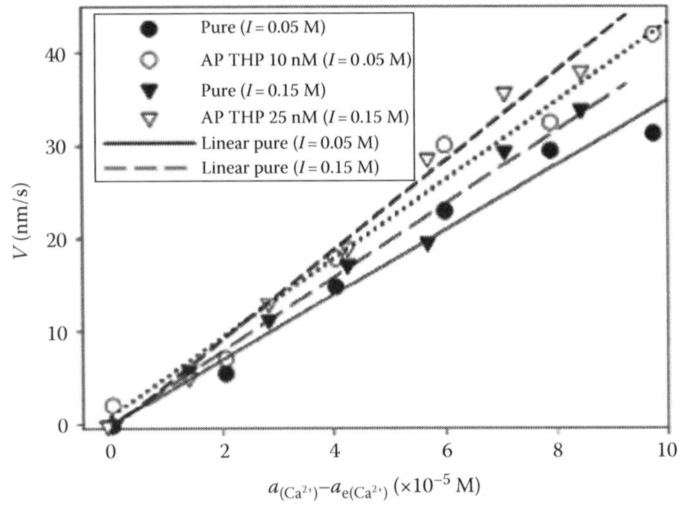

Figure 12.16 The effect of apTHP on the [−10−1] step of the (−101) face of COM at two different ionic strengths. In both cases, although in different extent, the presence of apTHP protein enhances the step velocity from that in the pure system. (After Weaver, M.L. et al., *Calcif. Tissue Int.*, 84(6), 462, doi: 10.1007/s00223-009-9223-0, 2009. With permission.)

equilibrium Ca^{2+} activity remains the same as that for the pure solution. The nonshift in the equilibrium activity further argues that the growth enhancement by the protein is not through incorporation of the impurity to the crystal lattice, rather by a surfactant effect.

12.5 SUMMARY

We have demonstrated the utility of combining *in situ* AFM imaging with crystal growth theory for understanding events underlying biomineral formation. Through the many systems that have been studied, *in situ* AFM has convincingly shown that the interactions between biomolecules and atomic steps at crystal surfaces are essential to the modulation of growth dynamics and kinetics that impact the rate and shape of crystal formation. While numerous mechanisms by which impurities control mineral formation have been defined, there are still many that remain undetermined. This is true especially for those interactions involving macromolecules at atomic steps as shown in many studies. An AFM that has a fast-scan rate matching atomic motion combined with a spatial resolution capable of routine true atomic scale imaging is the key to unlocking these mysteries. A high-speed/high-resolution AFM would enable direct real-time observation of kink activities, quantify the solute detachment/attachment rate, and reveal kink site-specific modification by macromolecules. In fact, the development of fast AFM has already been on the block (Ando 2012). The maturation of this technology and its utilization to appropriate dynamic growth systems combined with fast computational tools offer the promise to the next breakthrough in discovering new mechanisms by which biomolecules control crystal formation in biological systems.

ACKNOWLEDGMENTS

This work is performed under the auspices of the US Department of Energy by Lawrence Livermore National Laboratory under Contract DE-AC52-07NA27344. I would like to thank Dr. Raymond W. Friddle, Mr. Anthony Monterrosa, Dr. Mathew L. Weaver, Dr. Kang Rae Cho, and Dr. James J. De Yoreo for their assistance during the preparation of this chapter. I would also like to thank Dr. Elaine DiMasi and Prof. Laurie B. Gower for their patience and support to accomplish the manuscript.

REFERENCES

Ando, T. 2012. High-speed atomic force microscopy coming of age. *Nanotechnology* 23: 062001.

Balke, N., D. Bonnell, D.S. Ginger, and M. Kemerink. 2012. Scanning probes for new energy materials: Probing local structure and function. *MRS Bulletin* 37(7): 633–637. doi: 10.1557/mrs.2012.141.

Bash, R., H. Wang, C. Anderson, J. Yodh, G. Hager, S.M. Lindsay, and D. Lohr. 2006. AFM imaging of protein movements: Histone H2A-H2B release during nucleosome remodeling. *FEBS Letters* 580(19): 4757–4761. doi: 10.1016/j.febslet.2006.06.101.

Binnig, G., C.F. Quate, and C. Gerber. 1986. Atomic force microscope. *Physical Review Letters* 56(9): 930–933. doi: 10.1103/PhysRevLett.56.930.

Bliznakow, G. 1958. Crystal habit and adsorption of cosolutes. *Fortschritte Der Mineralogie* 36: 149–154.

Bonnell, D.A. 2000. *Scanning Probe Microscopy and Spectroscopy: Theory, Techniques, and Applications*. Wiley, New York.

Borman, A.H., E.W. Dejong, M. Huizinga, D.J. Kok, P. Westbroek, and L. Bosch. 1982. The role in $CaCO_3$ crystallization of an acid Ca^{2+}-binding polysaccharide associated with coccoliths of *Emiliania huxleyi*. *European Journal of Biochemistry* 129(1): 179–183.

Bouropoulos, N., S. Weiner, and L. Addadi. 2001. Calcium oxalate crystals in tomato and tobacco plants: Morphology and in vitro interactions of crystal-associated macromolecules. *Chemistry: A European Journal* 7(9): 1881–1888.

Burnham, A.K., S.R. Qiu, R. Pitchimani, and B.L. Weeks. 2009. Comparison of kinetic and thermodynamic parameters of single crystal pentaerythritol tetranitrate using atomic force microscopy and thermogravimetric analysis: Implications on coarsening mechanisms. *Journal of Applied Physics* 105(10): 104312. doi: 10431210.1063/1.3129504.

Burton, W.K., N. Cabrera, and F.C. Frank. 1951. The growth of crystals and the equilibrium structure of their surfaces. *Philosophical Transactions of the Royal Society of London, Series A. Mathematical and Physical Sciences* 243(866): 299–358. doi: 10.1098/rsta.1951.0006.

Cabrera, N. and D.A. Vermilyea. 1958. The growth of crystals from solution. *Growth and Perfection of Crystals*. ed. R.H. Doremus, B.W. Roberts, and D. Turnbul. pp. 393–410. Wiley, New York.

Chen, K. and P.G. Vekilov. 2002. Evidence for the surface-diffusion mechanism of solution crystallization from molecular-level observations with ferritin. *Physical Review E* 66(2): 021606. doi: 10.1103/PhysRevE.66.021606.

Chernov, A.A. 1961. The spiral growth of crystals. *Soviet Physics Uspekhi* 1: 116–148.

Cho, K.R., Y. Huang, S.L. Yu, S.M. Yin, M. Plomp, S.R. Qiu, R. Lakshminarayanan, J. Moradian-Oldak, M.S. Sy, and J.J. De Yoreo. 2011. A multistage pathway for human prion protein aggregation in vitro: From multimeric seeds to beta-oligomers and nonfibrillar structures. *Journal of the American Chemical Society* 133(22): 8586–8593. doi: 10.1021/ja1117446.

Cho, K.R., E.A. Salter, J.J. De Yoreo, A. Wierzbicki, S. Elhadj, Y. Huang, and S.R. Qiu. 2012. Impact of chiral molecules on the formation of biominerals: A calcium oxalate monohydrate example. *Crystal Growth & Design* 12(12): 5939–5947. doi: 10.1021/cg3009317.

Cho, K.R., E.A. Salter, J.J. De Yoreo, A. Wierzbicki, S. Elhadj, Y. Huang, and S.R. Qiu. 2013. Growth inhibition of calcium oxalate monohydrate crystal by linear aspartic acid enantiomers investigated by in situ atomic force microscopy. *CrystEngComm* 15(1): 54–64. doi: 10.1039/c2ce25936b.

Coe, F.L., J.H. Parks, and J.R. Asplin. 1992. The pathogenesis and treatment of kidney stones. *The New England Journal of Medicine* 327: 1141–1152.

Davis, K.J., P.M. Dove, and J.J. De Yoreo. 2000. The role of Mg^{2+} as an impurity in calcite growth. *Science* 290: 1134–1137.

De Yoreo, J.J. and P.M. Dove. 2004. Shaping crystals with biomolecules. *Science* 306: 1301.

De Yoreo, J.J., C.A. Orme, and T.A. Land. 2001. Using atomic force microscopy to investigate solution crystal growth. In *Advances in Crystal Growth Research*, ed. K. Nakajima. Elsevier Science, Amsterdam, the Netherlands.

De Yoreo, J.J., S.R. Qiu, and J.R. Hoyer. 2006. Molecular modulation of calcium oxalate crystallization. *American Journal of Physiology-Renal Physiology* 291(6): F1123–F1131. doi: 10.1152/ajprenal.00136.2006.

De Yoreo, J.J. and P. Vekilov. 2003. Principles of crystal nucleation and growth. In *Biomineraliztion*, ed. P.M. Dove, J.J. De Yoreo, and S. Weiner. Vol. 54. Mineralogical Society of America, Washington, DC.

De Yoreo, J.J., A. Wierzbicki, and P.M. Dove. 2007. New insights into mechanisms of biomolecular control on growth of inorganic crystals. *CrystEngComm* 9(12): 1144–1152. doi: 10.1039/b713006f.

Didymus, J.M., J.R. Young, and S. Mann. 1994. Construction and morphogenesis of the chiral ultrastructure of coccoliths from the marine alga *Emiliania huxleyi*. *Proceedings of the Royal Society of London. Series B* 258(1353): 237–245.

Dove, P.M., J.J. De Yoreo, and K.J. Davis. 2004. Inhibition of CaCO(3) crystallization by small molecules: The magnesium example. *Nanoscale Structure and Assembly at Solid-Fluid Interfaces*, Vol. II, pp. 55–82. Springer, London, U.K.

Elhadj, S., J.J. De Yoreo, J.R. Hoyer, and P.M. Dove. 2006a. Role of molecular charge and hydrophilicity in regulating the kinetics of crystal growth. *Proceedings of the National Academy of Sciences of the United States of America* 103(51): 19237–19242. doi: 10.1073/pnas.0605748103.

Elhadj, S., E.A. Salter, A. Wierzbicki, J.J. De Yoreo, N. Han, and P.M. Dove. 2006b. Peptide controls on calcite mineralization: Polyaspartate chain length affects growth kinetics and acts as a stereochemical switch on morphology. *Crystal Growth & Design* 6(1): 197–201. doi: 10.1021/cg050288+.

Frank, F.C. 1951. The growth of carborundum: Dislocation and polytypism. *Philosophical Magazine* 42: 1014.

Frank, F.C. and J.H. van der Merwe. 1949. One dimensional dislocations. 1. Static theory. *Proceedings of the Royal Society of London* A198: 205.

Friddle, R.W., M.L. Weaver, S.R. Qiu, A. Wierzbicki, W.H. Casey, and J.J. De Yoreo. 2010. Subnanometer atomic force microscopy of peptide-mineral interactions links clustering and competition to acceleration and catastrophe. *Proceedings of the National Academy of Sciences of the United States of America* 107(1): 11–15. doi: 10.1073/pnas.0908205107.

Fu, G., S.R. Qiu, C.A. Orme, D.E. Morse, and J.J. De Yoreo. 2005. Acceleration of calcite kinetics by abalone nacre proteins. *Advanced Materials* 17(22): 2678–2683. doi: 10.1002/adma.200500633.

Gilmer, G.H., R. Ghez, and N. Cabrera. 1971. An analysis of combined surface and volume diffusion processes in crystal growth. *Journal of Crystal Growth* 8(1): 79.

Gratz, A.J., S. Manne, and P.K. Hansma. 1991. Atomic force microscopy of atomic-scale ledges and etch pits formed during dissolution of quartz. *Science* 251(4999): 1343–1346. doi: 10.1126/science.251.4999.1343.

Guo, S., M.D. Ward, and J.A. Wesson. 2002. Direct visualization of calcium oxalate monohydrate crystallization and dissolution with atomic force microscopy and the role of polymeric additives. *Langmuir* 18(11): 4284–4291.

Gustaffson, J.P. 2004. Visual Minteq 2.30. Department of Land and Water Resources Engineering, KTH (or Royal Institute of Technology), Stockholm, Sweden.

Hillner, P.E., A.J. Gratz, S. Manne, and P.K. Hansma. 1992. Atomic-scale imaging of calcite growth and dissolution in real-time. *Geology* 20: 359.

Jung, T., X. Sheng, C.K. Choi, W. Kim, J.A. Wesson, and M.D. Ward. 2004. Probing crystallization of calcium oxalate monohydrate and the role of macromolecule additives with in situ atomic force microscopy. *Langmuir* 20: 8587–8596.

Land, T.A. and J.J. De Yoreo. 2000. The evolution of growth modes and activity of growth sources on canavalin investigated by in situ atomic force microscopy. *Journal of Crystal Growth* 208: 623–637.

Land, T.A., J.J. De Yoreo, and J.D. Lee. 1997. An in-situ AFM investigation of canavalin crystallization kinetics. *Surface Science* 384: 136–155.

Levy-Lior, A., S. Weiner, and L. Addadi. 2003. Achiral calcium-oxalate crystals with chiral morphology from the leaves of some Solanaceae plants. *Helvetica Chimica Acta* 86(12): 4007–4017.

Lowenstam, H.A. and S. Weiner. 1989. *On Biomineraliztion*. Oxford University Press, Oxford, U.K.

Lowik, D.W., P.M. and J.C.M. van Hest. 2004. Peptide based amphiphiles. *Chemical Society Reviews* 33(4): 234–245. doi: 10.1039/B212638A.

Lyubchenko, Y.L., A.A. Gall, L.S. Shlyakhtenko, and S.M. Lindsay. 1996. AFM imaging of DNA and other biological molecules: Use of silylated mica. *Biophysical Journal* 70(2): WAMG6–WAMG6.

Markov, I.V. 1994. *Crystal Growth for Beginners: Fundamentals of Nucleation, Crystal Growth, and Epitaxy*. World Scientific, Singapore.

Maruyama, M., K. Tsukamoto, G. Sazaki, Y. Nishimura, and P.G. Vekilov. 2009. Chiral and achiral mechanisms of regulation of calcite crystallization. *Crystal Growth & Design* 9(1): 127–135. doi: 10.1021/cg701219h.

Meyer, E., H.J. Hug, and R. Bennewitz. 2004. *Scanning Probe Microscopy: The Lab on a Tip*: Springer, Berlin, Germany.

Onuma, K., T. Kameyama, and K. Tsukamoto. 1994. In-situ study of surface phenomena by real-time phase-shift interferometry. *Journal of Crystal Growth* 137(3–4): 610–622. doi: 10.1016/0022-0248(94)91006-5.

Orme, C.A. and J.L. Giocondi. 2007. The use of scanning probe microscopy to investigate crystal-fluid interfaces. In *Perspectives on Inorganic, Organic, and Biological Crystal Growth: From Fundamentals to Applications*, ed. M. Skowronski, J.J. DeYoreo, and C.A. Wang, pp. 342–362. American Institute of Physics, Melville, New York.

Orme, C.A., A. Noy, A. Wierzbicki, M.T. McBride, M. Grantham, H.H. Teng, P.M. Dove, and J.J. DeYoreo. 2001. Formation of chiral morphologies through selective binding of amino acids to calcite surface steps. *Nature* 411(6839): 775–779.

Potapenko, S.Y. 1993. The threshold for percolation through impurity fence. *Journal of Crystal Growth* 133(1–2): 141–146.

Qiu, S.R. and C.A. Orme. 2008. Dynamics of biomineral formation at the near-molecular level. *Chemical Reviews* 108(11): 4784–4822. doi: 10.1021/cr800322u.

Qiu, S.R., A. Wierzbicki, C.A. Orme, A.M. Cody, J.R. Hoyer, G.H. Nancollas, S. Zepeda, and J.J. De Yoreo. 2004. Molecular modulation of calcium oxalate crystallization by osteopontin and citrate. *Proceedings of the National Academy of Sciences of the United States of America* 101(7): 1811–1815. doi: 10.1073/pnas.0307900100.

Qiu, S.R., A. Wierzbicki, E.A. Salter, S. Zepeda, C.A. Orme, J.R. Hoyer, G.H. Nancollas, A.M. Cody, and J.J. De Yoreo. 2005. Modulation of calcium oxalate monohydrate crystallization by citrate through selective binding to atomic steps. *Journal of American Chemical Society* 127(25): 9036–9044.

Rashkovich, L.N. 1991. KDP-family single crystal. In *The Adam Hilger Series on Optics and Optoelectronics*, ed. B.E.A. Saleh, E.R. Pike, and W.T. Welford. Adam Hilger, New York.

Schoenwald, K., Z.C. Peng, D. Noga, S.R. Qiu, and T. Sulchek. 2010. Integration of atomic force microscopy and a microfluidic liquid cell for aqueous imaging and force spectroscopy. *Review of Scientific Instruments* 81(5): 053704. doi: 10.1063/1.3395879.

Sulchek, T., R. Hsieh, J.D. Adams, S.C. Minne, C.F. Quate, and D.M. Adderton. 2000. High-speed atomic force microscopy in liquid. *Review of Scientific Instrument* 71: 2097.

Teng, H.H. 1998. Thermodynamics of calcite growth: Baseline for understanding biomineral formation. *Science* 282: 4.

Teng, H.H., P.M. Dove, and J.J. De Yoreo. 2000. Kinetics of calcite growth: Surface processes and relationships to macroscopic rate laws. *Geochimica et Cosmochimica Acta* 64(13): 2255–2266.

Touryan, L.A., M.J. Lochhead, B.J. Marquardt, and V. Vogel. 2004. Sequential switch of biomineral crystal morphology using trivalent ions. *Nature Materials* 3(4): 239–243. doi: 10.1038/nmat1096.

van Enckevort, W.J.P., P. Bennema, and W.H. van der Linden. 1981. On the observation of growth spirals with very low step heights on potash alum single crystals. *Zeitschrift für Physikalische Chemie Neue Folge* 124: 171.

van Enckevort, W.J.P. and A.C.J.F. van den Berg. 1998. Impurity blocking of crystal growth: A Monte Carlo study. *Journal of Crystal Growth* 183(3): 441–455. doi: 10.1016/s0022-0248(97)00432-6.

Wade, T., J.F. Garst, and J.L. Stickney. 1998. A simple modification of a commercial atomic force microscopy liquid cell for in situ imaging in organic, reactive, or air sensitive environments. *Review of Scientific Instruments* 70: 121.

Weaver, M.L., S.R. Qiu, R.W. Friddle, W.H. Casey, and J.J. De Yoreo. 2010. How the overlapping time scales for peptide binding and terrace exposure lead to nonlinear step dynamics during growth of calcium oxalate monohydrate. *Crystal Growth & Design* 10(7): 2954–2959.

Weaver, M.L., S.R. Qiu, J.R. Hoyer, W.H. Casey, G.H. Nancollas, and J.J. De Yoreo. 2006. Improved model for inhibition of pathological mineralization based on citrate-calcium oxalate monohydrate interaction. *ChemPhysChem* 7(10): 2081–2084. doi: 10.1002/cphc.200600371.

Weaver, M.L., S.R. Qiu, J.R. Hoyer, W.H. Casey, G.H. Nancollas, and J.J. De Yoreo. 2007. Inhibition of calcium oxalate monohydrate growth by citrate and the effect of the background electrolyte. *Journal of Crystal Growth* 306(1): 135–145. doi: 10.1016/j.jcrysgro.2007.04.053.

Weaver, M.L., S.R. Qiu, J.R. Hoyer, W.H. Casey, G.H. Nancollas, and J.J. De Yoreo. 2009. Surface aggregation of urinary proteins and aspartic acid-rich peptides on the faces of calcium oxalate monohydrate investigated by in situ force microscopy. *Calcified Tissue International* 84(6): 462–473. doi: 10.1007/s00223-009-9223-0.

Yau, S.T., B.R. Thomas, and P.G. Vekilov. 2000. Molecular mechanisms of crystallization and defect formation. *Physical Review Letters* 85(2): 353–356. doi: 10.1103/PhysRevLett.85.353.

Young, J.R., J.M. Didymus, P.R. Bown, B. Prins, and S. Mann. 1992. Crystal assembly and phylogenetic evolution in heterococcoliths. *Nature* 356(6369): 516–518.

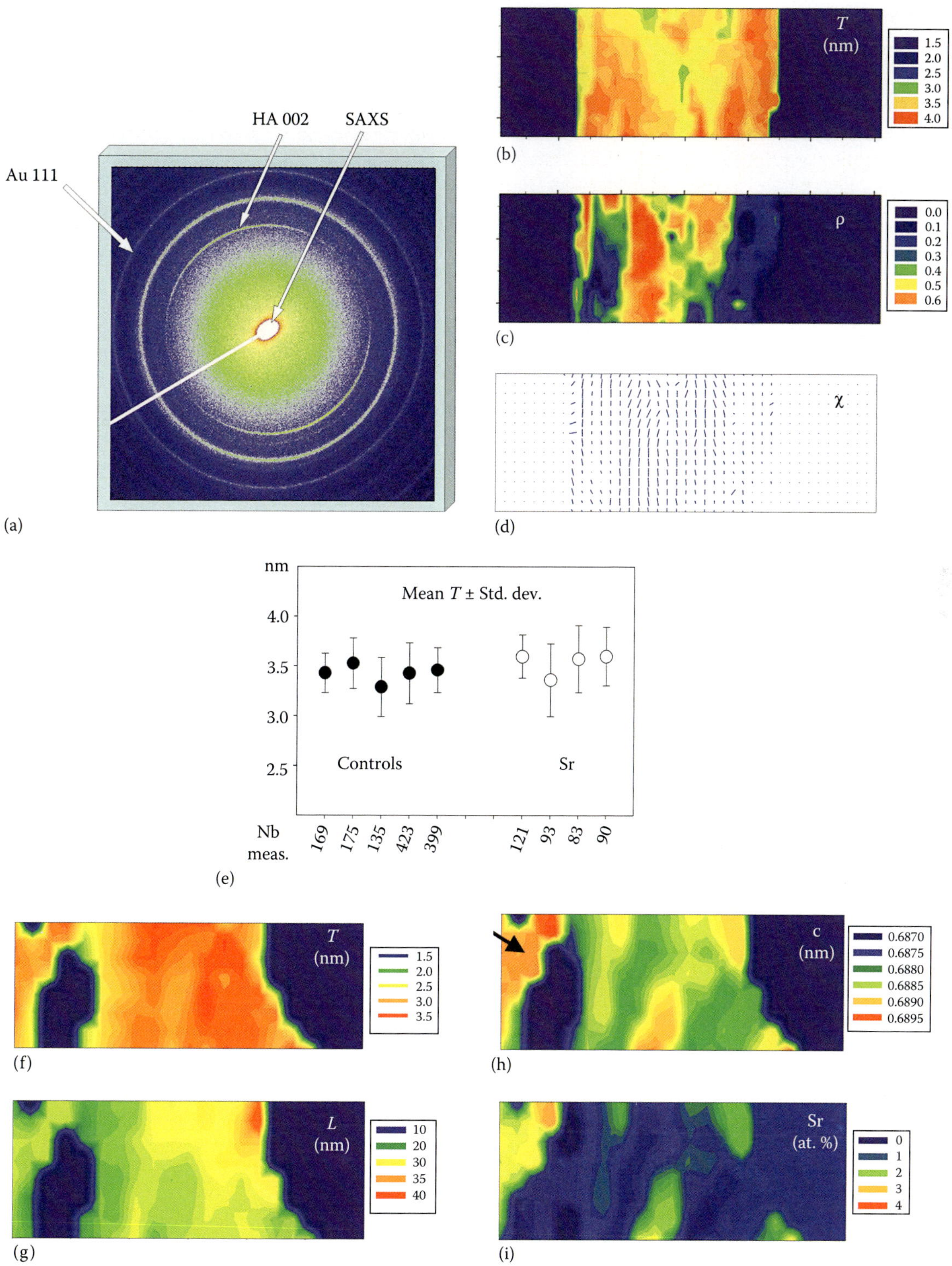

Figure 1.7 (a) Exemplary SAXS/WAXS pattern from a bone section acquired at the µSpot beamline at BESSY II, Berlin. The HAP peak (HA 002) appears in arcs that provide mineral orientation information. The Au 111 ring arises from an Au film sputtered onto the bone for calibration. (b–d) Parameters determined from SAXS data, mapped over a rectangular region of bone sampled from a placebo-treated patient: (b) typical mineral crystal thickness *T*, which is fairly constant across the sample; (c) degree of alignment ρ; and (d) orientations of the elongated crystals show how crystals are grouped into *packets*. (e) Dependence of the *T* parameter on sample and on Sr content. (f–i) Parameter maps of a sample from a Sr-treated patient: (f) the similar mineral thickness *T* from SAXS; (g) mineral crystal length *L* from WAXS; (h) the HA *c* lattice parameter obtained from WAXS, where higher values indicate Sr incorporation; (i) Sr concentration derived from analysis of the *c* lattice parameter. (Adapted from Li, C. et al., *J. Bone Miner. Res.*, 25, 968, 2010. With permission.)

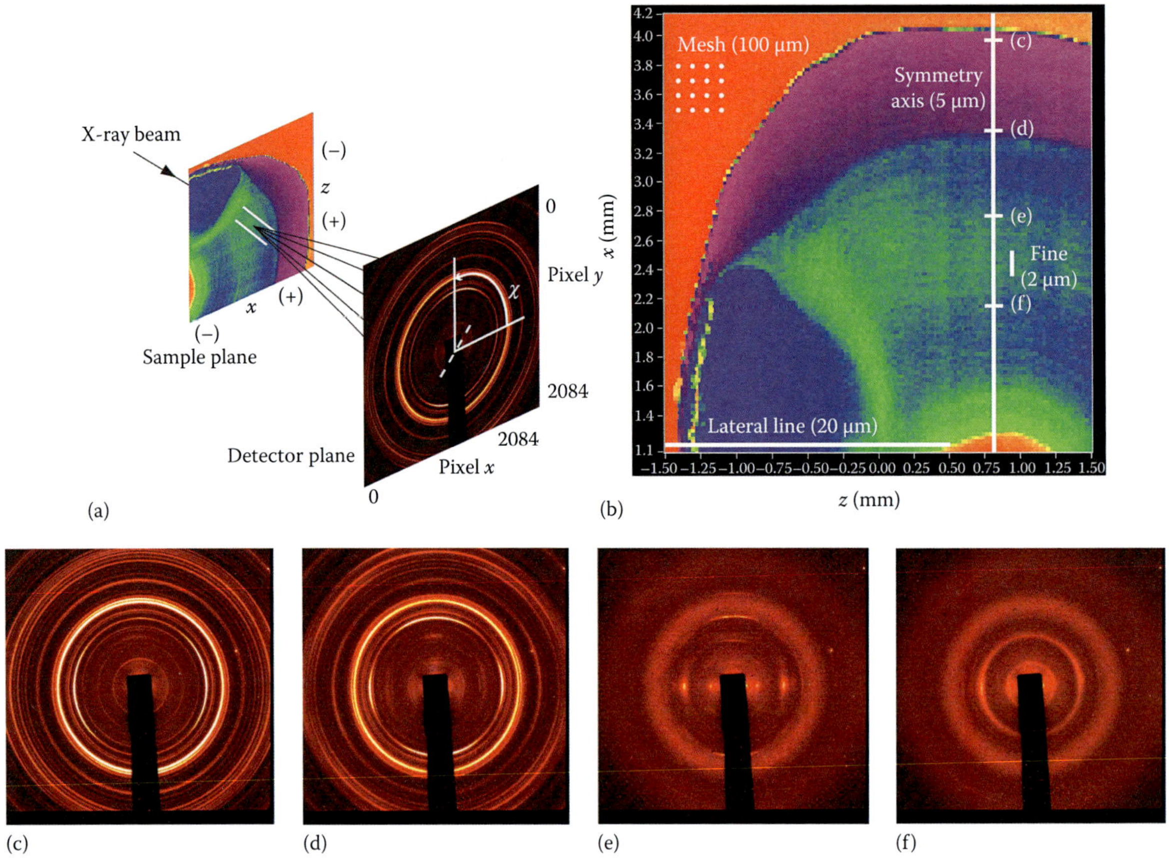

Figure 1.9 (a) Representation of the transmission mapping experiment. Aligned crystal planes (white lines in the sample plane) create diffraction peaks along a particular line in detector azimuth χ (dashed line in detector plane). (b) X-ray transmission map of sectioned stomatopod dactyl club. Regions of decreasing density lead from purple to blue to green (red color is the surrounding epoxy region). White lines and points indicate areas and spacings of diffraction maps taken. (c) WAXS pattern in mineralized impact region. (d) Oriented mineral at the boundary region. (e and f) WAXS patterns dominated by amorphous mineral and chitin.

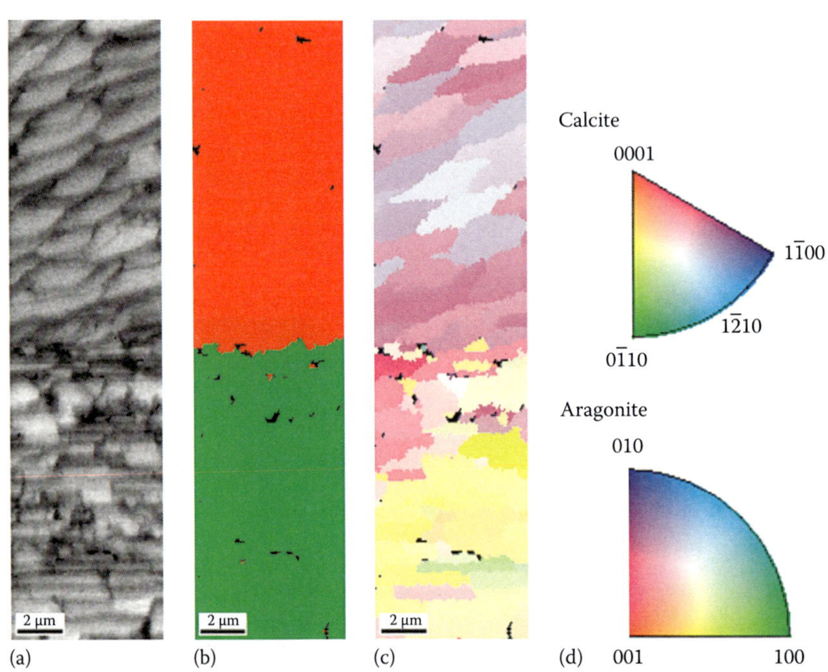

Figure 3.4 Output from EBSD analyses. EBSD analysis of the interface between calcite (top) and aragonite (bottom) in the shell of the common blue mussel, *M. edulis*. Data obtained using accelerating voltage of 20 kV, working distance of 10 mm, aperture of 50 nm, and step size of 0.25 μm The main aspects of data output from EBSD analyses are (a) diffraction intensity with brighter regions indicating more diffraction, (b) phase, with each assigned a separate color, which in this case is calcite (red) and aragonite (green) and (c) crystallographic orientation with a color key (d) for calcite (top) and aragonite (bottom) indicating which crystallographic planes are normal to the view. Scale bar = 2 μm.

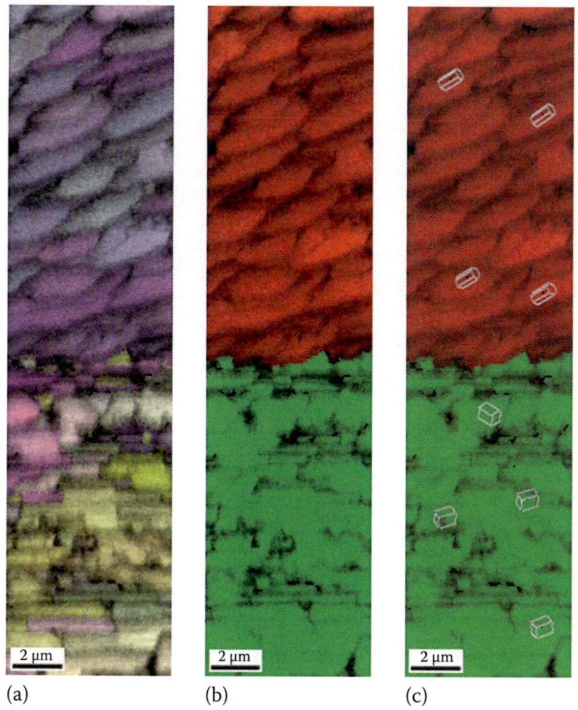

Figure 3.5 Some options for EBSD data presentation. These examples depict some of the options available for presenting EBSD analyses using the data from *M. edulis* from Figure 3.4. (a) Diffraction intensity map sits behind the crystallographic orientation map. (b) Phase map from 4B with diffraction intensity behind and then in (c) with wire frames depicting crystallographic orientation. Scale bar = 2 μm.

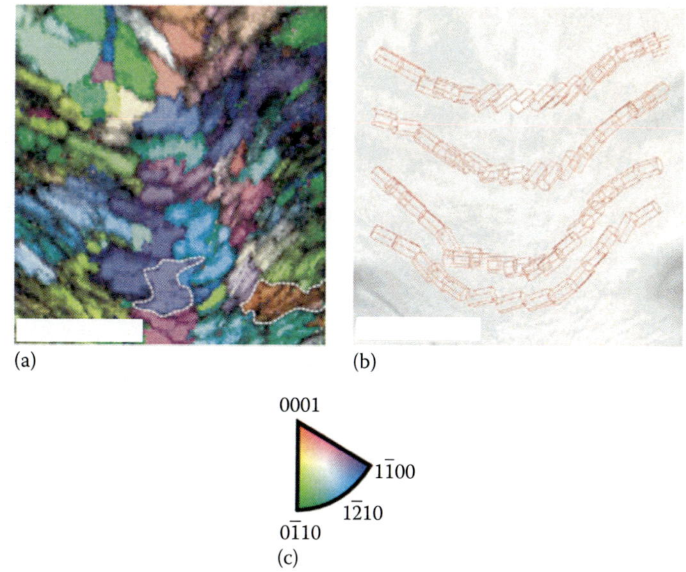

Figure 3.7 EBSD mapping of calcite semi-nacre *of the shell of the brachiopod, N. anomala*. (a) Crystallographic orientation map of the secondary layer of the shell of *Nysius huttoni*. (b) Wire frame unit cells superimposed on SEM image of *N. huttoni* laminae indicating that the *c*-axis of calcite is coincident with the undulations of the laminae and more or less parallel with the shell exterior. (c) Color key used in (a). (Reprinted with permission from England, J., Cusack, M., Dalbeck, P., and Perez-Huerta, A., Comparison of the crystallographic structure of semi nacre and nacre by electron backscatter diffraction, *Cryst. Growth Des.*, 7, 307–310. Copyright 2007 American Chemical Society.)

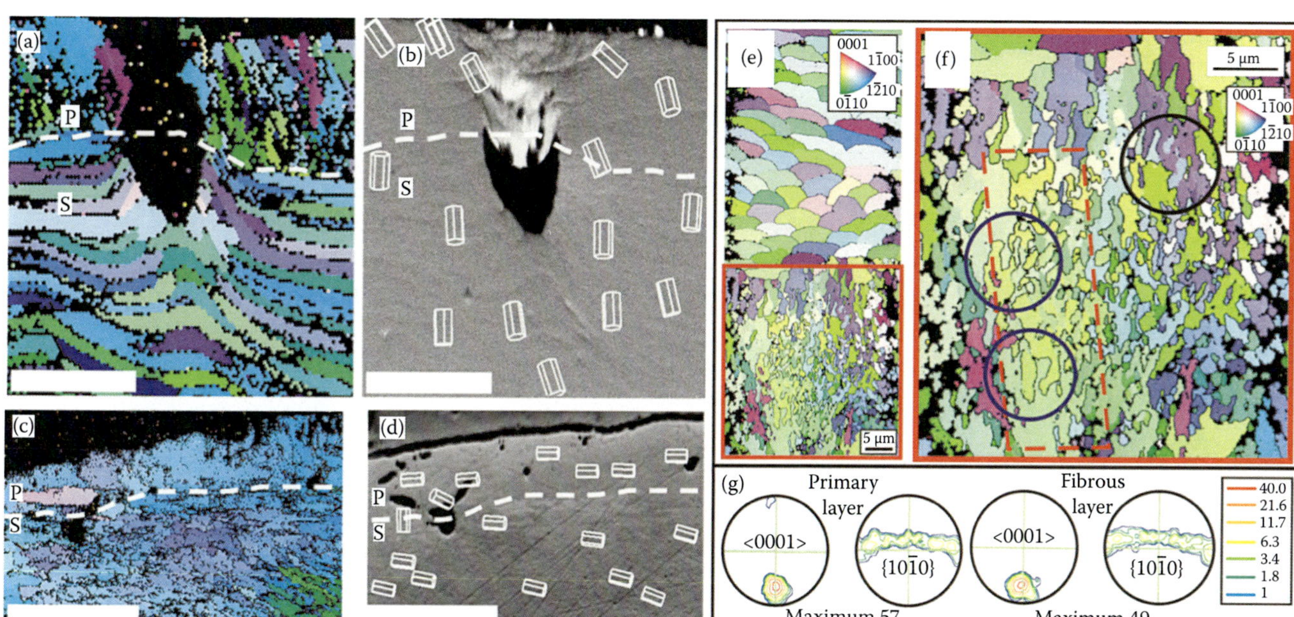

Figure 3.8 EBSD analyses of primary (outer) layer of calcite-shelled brachiopods. (a) Crystallographic orientation of primary (P) and secondary (S) layers of (a) *Terebratalia transversa* and (c) *N. anomala* with corresponding secondary electron images in (b) and (d), respectively, with wire frames indicating crystallographic orientation. Scale bars = 50 μm, 50 μm, 100 μm, and 100 μm for (a–d), respectively. (e) Crystallographic orientation map of *Gryphus vitreus*, with area of primary layer having interdigitating grains highlighted in the red box and enlarged in (f). (g) Pole figures of the primary and secondary layers of *G. vitreus*. (a–d: From Cusack, M., Chung, P., Dauphin, Y., and Perez-Huerta, A.: Brachiopod primary layer crystallography and nanostructure. *Evolution and Development of the Brachiopod Shell*. Special Papers in Palaeontology, 2010, 84, 99–105. Copyright Wiley-VCH Verlag GmbH & Co. KGaA. Reproduced with permission from the Palaeontological Association; e–g: From Goetz, A.J., Steinmetz, D.R., Griesshaber, E., Zaefferer, S., Raabe, D., Kelm, K., Irsen, S., Sehrbrock, A., and Schmahl, W.W.: Interdigitating biocalcite dendrites form a 3-D jigsaw structure in brachiopod shells. *Acta Biomater.* 2011. 7. 2237–2243. Copyright Wiley-VCH Verlag GmbH & Co. KGaA. Reproduced with permission.)

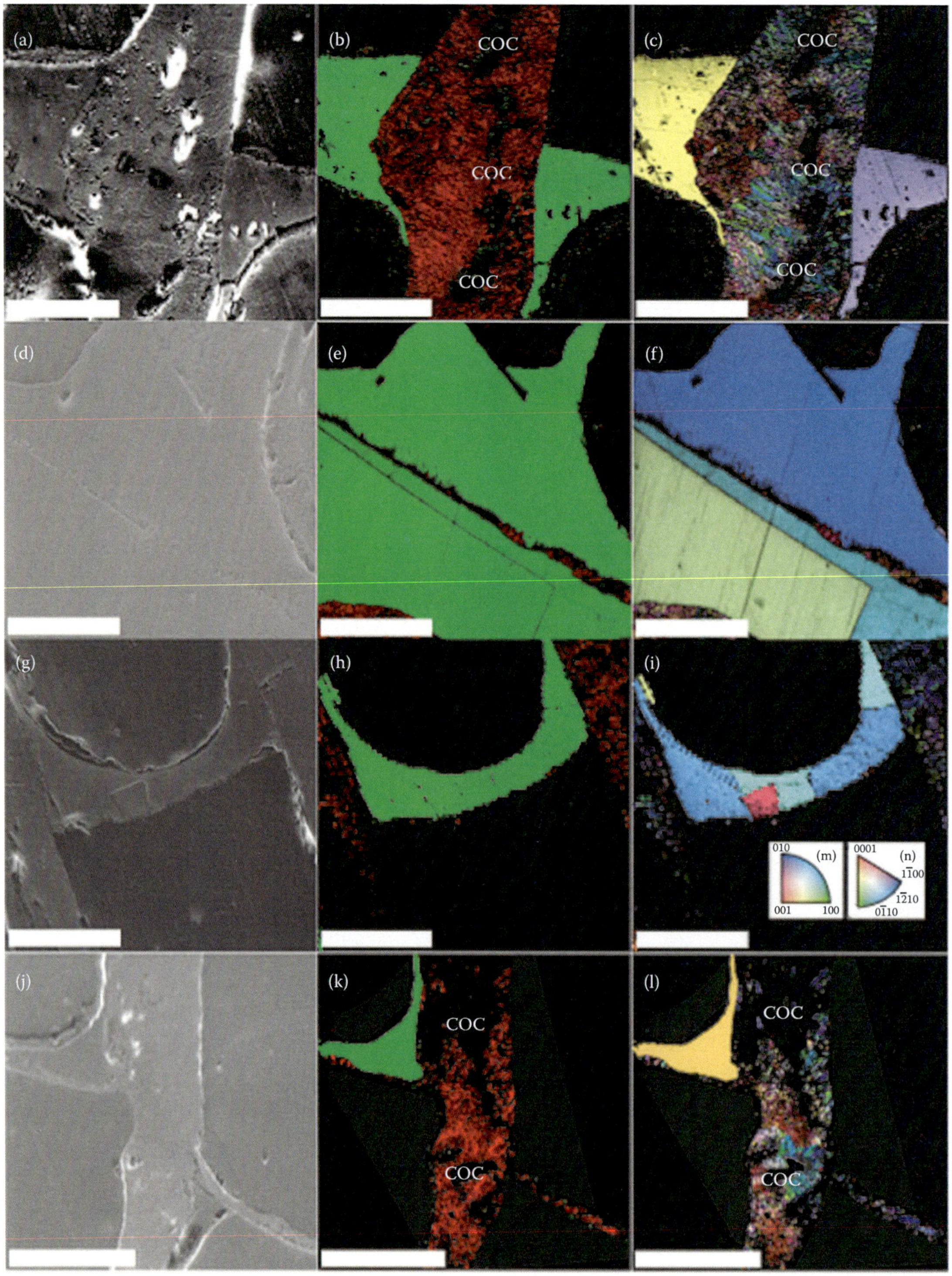

Figure 3.9 EBSD analysis of fossil coral revealing relatively large crystals of calcite resembling horizontal structures in pristine corals. Secondary electron images (a, d, g, j) of the areas analyzed by EBSD. Phase maps of these areas (b, e, h, k) where aragonite is indicated in red and calcite in green. Crystallographic orientation maps (c, f, i, l) of these same regions with crystallographic orientation indicated by color key (m and n) insets in (i) for aragonite and calcite, respectively. Scale bars for a–c = 50 μm, d–f = 25 μm, g–i = 50 μm, and j–l = 80 μm. (Reprinted from *Chem. Geol.*, 280, Dalbeck, P., Cusack, M., Dobson, P.S., Allison, N., Fallick, A.E., Tudhope, A.W., and EIMF, Identification and composition of secondary meniscus calcite in fossil coral and the effect on predicted sea surface temperature, 314–322, Copyright 2011 with permission from Elsevier.)

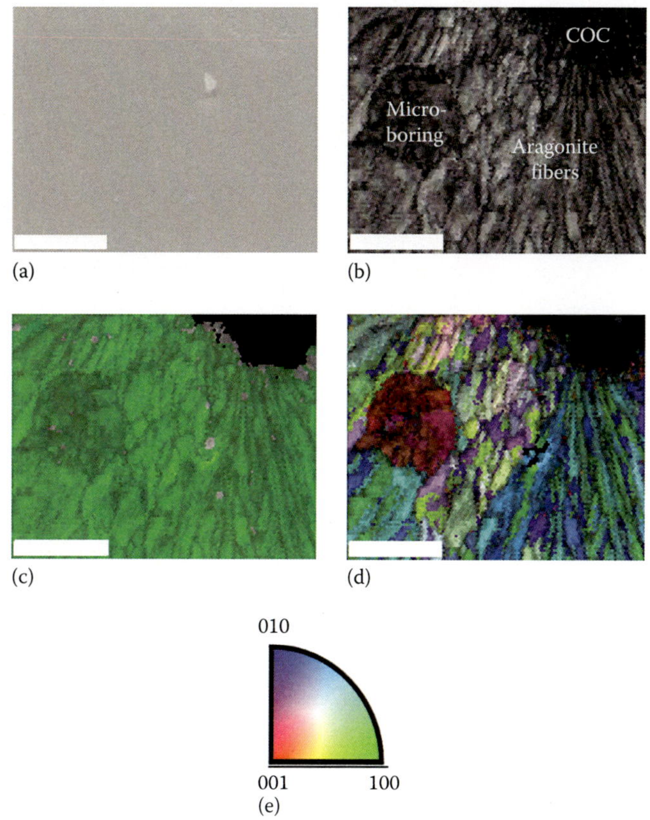

Figure 3.11 EBSD analysis of Porites coral revealing secondary aragonite within original primary aragonite. Images (a–d) depict the same area within the coral. Scale bar = 10 μm. (a) Secondary electron image of the coral region that is analyzed by EBSD in (c–d). (b) Diffraction intensity map. Dark area at top right indicates the poor diffraction of the COC. (c) Combined diffraction intensity and phase map with green indicating aragonite. (d) Map of combined diffraction intensity and crystallographic orientation of the same area (obtained simultaneously), with mainly blue and green color coding with reference to the key (e), indicating that the {010} and {100} planes of aragonite are normal to the plane of view indicating that the {001} is concurrent with the fiber axis. Secondary aragonite in what is likely to be a microboring has a different crystallographic orientation to the original aragonite fibers with the 001 plane normal to the plane of view, which is why the secondary aragonite appears red in the image. (Reprinted from *Coral Reefs*, 27, Cusack, M., England, J., Dalbeck, P., Tudhope, A.W., Fallick, A.E., and Allison, N., Electron backscatter diffraction (EBSD) as a tool for detection of coral diagenesis, 905–911, Copyright 2008, with permission from Elsevier.)

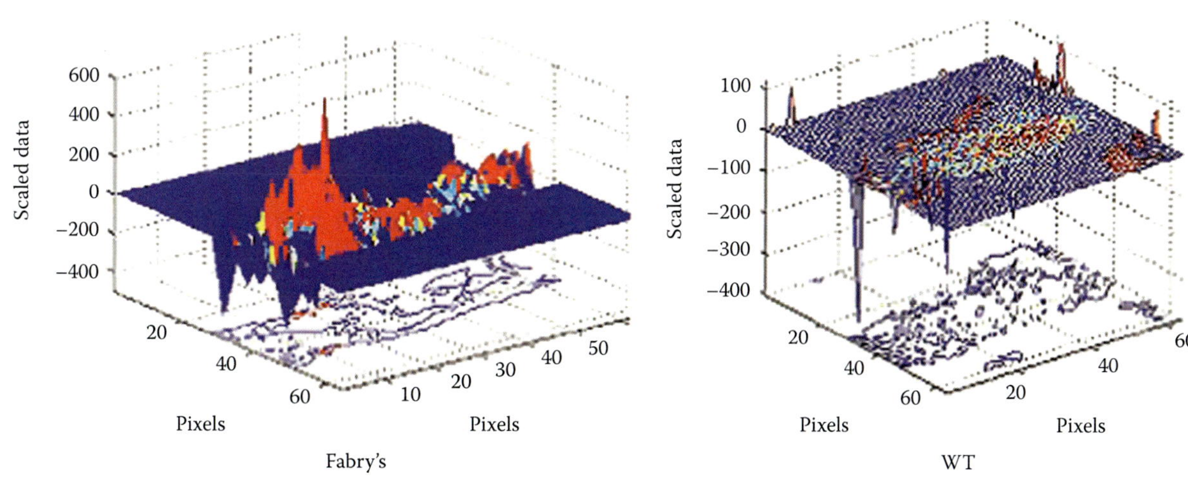

Figure 4.4 IR images can be used to illustrate the lipid distribution in healthy and diseased bone. A 3D projection of the lipid/protein ratio in the long bone of a mouse with Fabry's disease (a lipid storage disease that results in the accumulation of globotriaosylceramide in tissues) compared to a WT mouse. Notice the increased accumulation of lipid in Fabry's mouse (note that the scale for Fabry's mouse is 6× that of the WT).

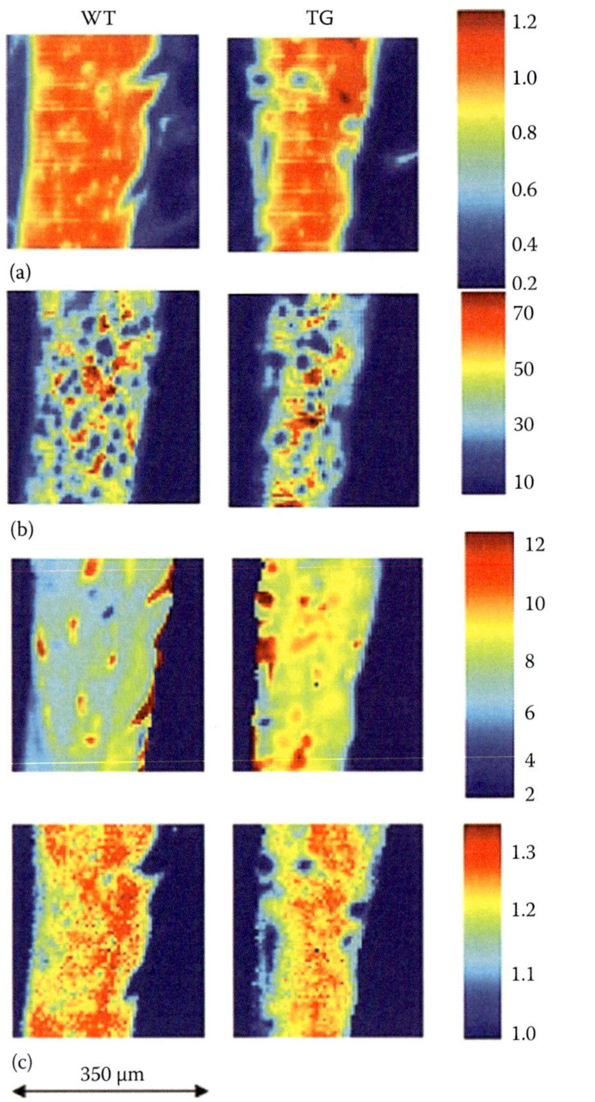

Figure 4.5 Imaging processing comparing a 2-month-old transgenic (TG) mouse overexpressing a mutant collagen (TG) and its background-matched WT control. (a) Unprocessed image of a TG and age- and background-matched WT mouse bone as seen on the spectrometer. The spectrum in the center corresponds to the point indicated by the arrowhead. (b) Corrected image of the data shown in (a). These data were corrected for background, PMMA, and water vapor, indicated in the adjacent spectrum by spectral subtraction, resulting in the processed spectrum and image. (c) Mineral-to-matrix ratio and crystallinity distribution in the same bone. The scale bar indicates the x and y dimensions of the figure.

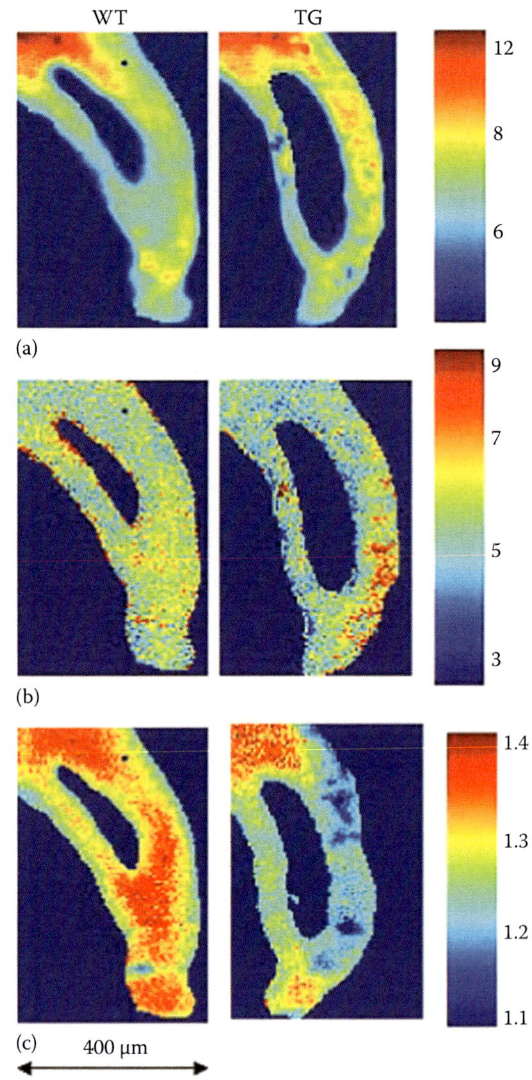

Figure 4.6 Images of the dental root in the same animals whose bones are shown in Figure 4.5 representing (a) mineral/matrix ratio, (b) carbonate/phosphate ratio, and (c) crystallinity.

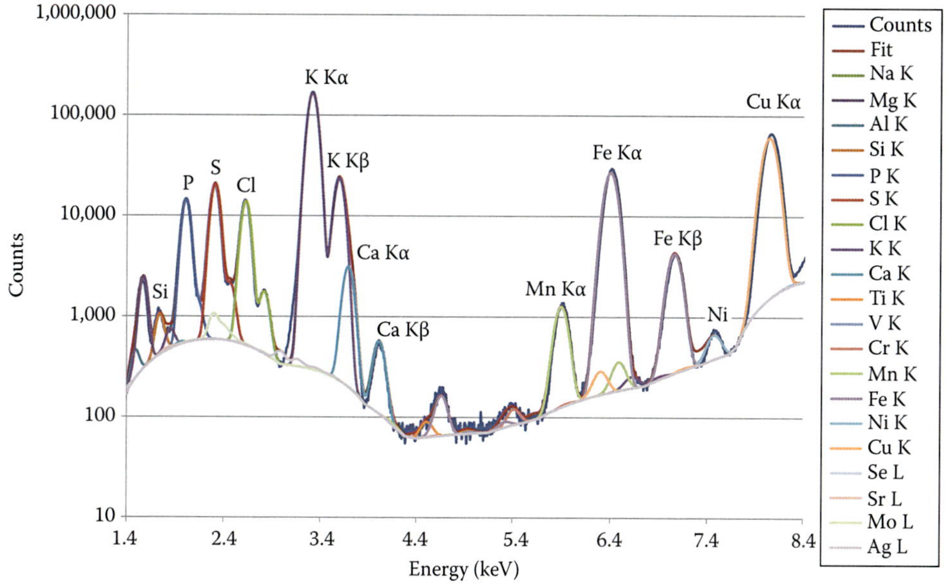

Figure 6.2 Example of fitting of a fluorescence spectrum using PyMca (Solé et al. 2007).

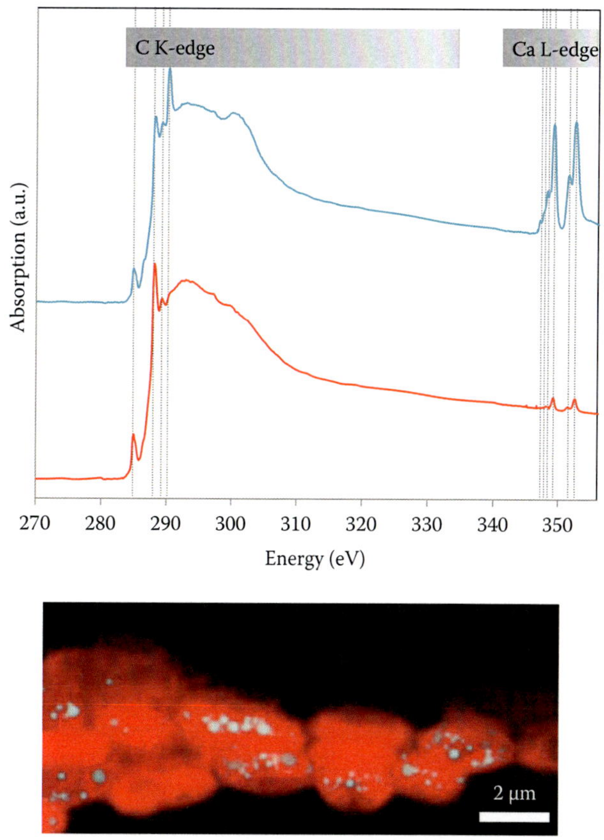

Figure 8.9 STXM colored map of the cyanobacterium *Candidatus Gloeomargarita lithophora*. In red, areas exhibiting the bottom XANES spectrum (in red) that is typical of a bacterial cell spectrum. In blue, area showing the blue XANES spectrum that is a combination of the cell XANES spectrum plus a calcium carbonate contribution (see peak at 290.2 eV and Ca peaks).

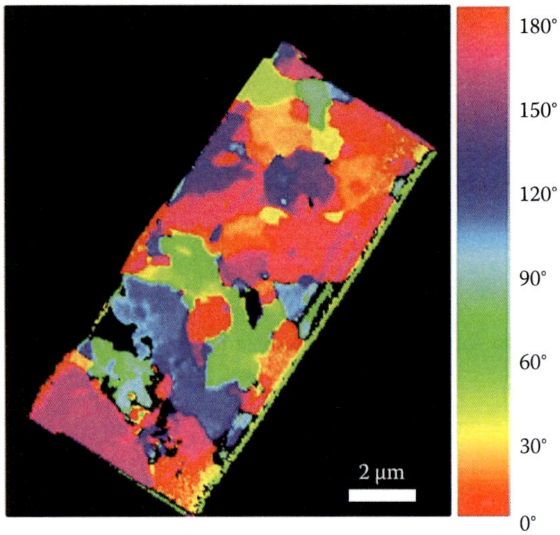

(i)

Figure 8.12 (i) The same procedure can be applied to each pixel in image (a). This provides a map of the orientation of the in-plane projection of the *c*-axis of aragonite as shown in (i).

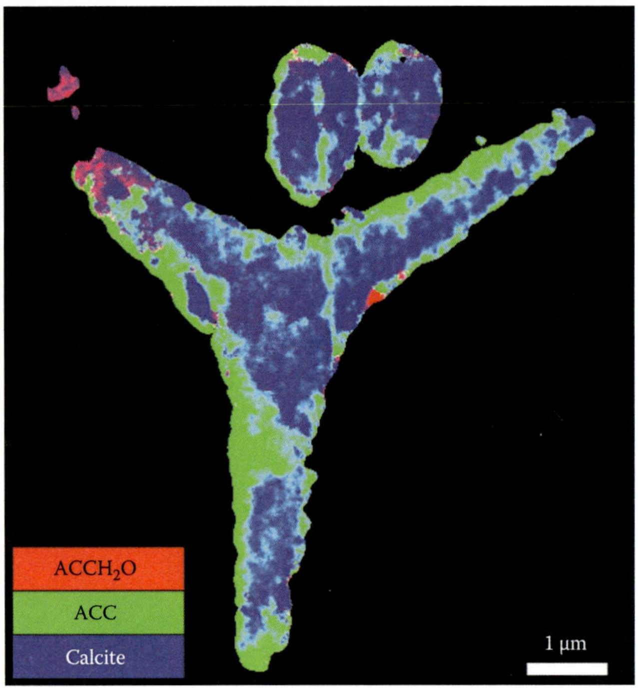

Figure 9.18 Component mapping in 48 h spicules, at the prism developmental stage, analyzed within 24 h of extraction from the embryo. XANES-PEEM component analysis map of three spicules embedded in epoxy, polished to expose a cross section, and coated. The larger triradiate spicule at the center is polished in plane, whereas two other cylindrical spicules at the top have their long axes perpendicular to the plane of the image and appear elliptical. This red, green, and blue (RGB) map displays the results of *component analysis* mapping, in each component spectrum (Figure 9.17). Notice the greater concentration of red $ACCH_2O$ at the top left tip of the triradiate spicule and near its center, on the right-hand side. These are areas in which $ACCH_2O$ was freshly deposited at the outer rim of the spicule. Moving toward the inside of each spicule, most of the mineral detected is green ACC, and finally at the center, the main component is blue calcite. These data provided the first direct evidence for the sequence of transforming phases in sea urchin spicules: $ACCH_2O \rightarrow ACC \rightarrow$ calcite. Component analysis was done using the GG macros (GG–Macros 2013). (Data from Gong, Y.U.T. et al., *Proc. Natl. Acad. Sci. USA*, 109, 6088, 2012.)

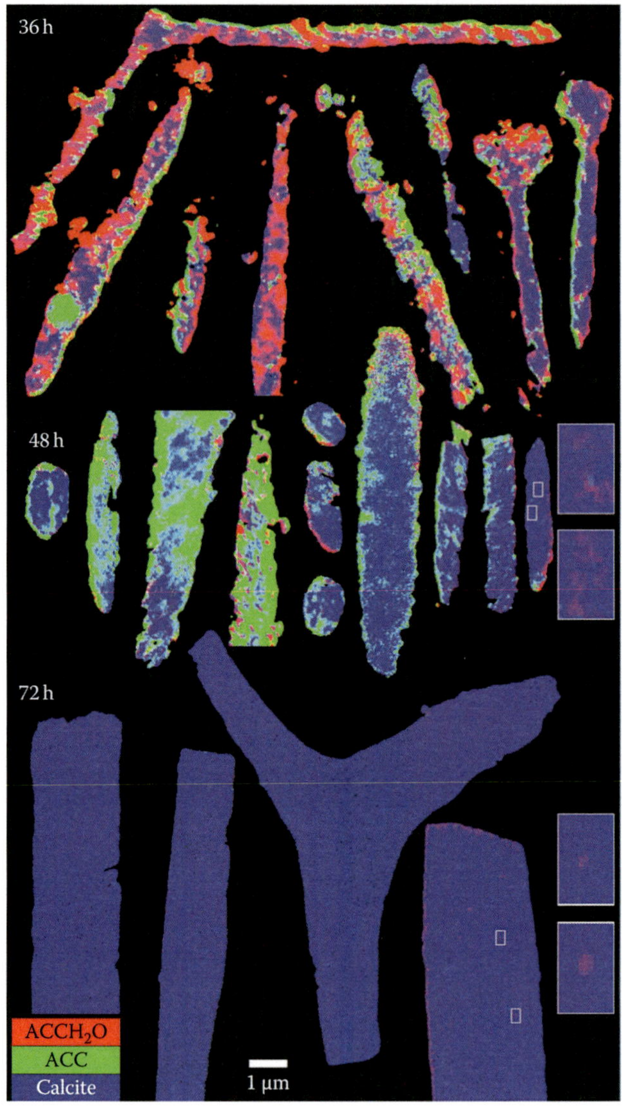

Figure 9.19 RGB maps resulting from component analysis done on spicules that are extracted 36, 48, and 72 h after fertilization and analyzed within 24 h of extraction from the embryos. Horizontally, the spicules are ordered from most amorphous to most crystalline. Notice the large density of R and B pixels in 36 and 72 h spicules, respectively. In the 48 h spicules, R and G pixels, indicating ACC, are always at the outer rims, while blue crystalline calcite is always at the center of each cross section. Also notice that magenta nanoparticles are quite frequent (see spicules on the right, for instance). Magenta nanoparticles are made of co-localized $ACCH_2O$ and calcite. The four insets on the right show zoomed-in maps of the four regions in white boxes on the 48 and 72 h spicules on the right. In the insets, pixels are 20 nm, and the color balance has been adjusted to enhance the magenta nanoparticle, otherwise faint, because magenta nanoparticles contain a much greater proportion of calcite than $ACCH_2O$. These nanoparticles are 60–120 nm in size and are consistently surrounded by blue calcite. Component analysis was done using the GG macros (GG–Macros 2013). (Data from Gong, Y.U.T. et al., *Proc. Natl. Acad. Sci. USA*, 109, 6088, 2012.)

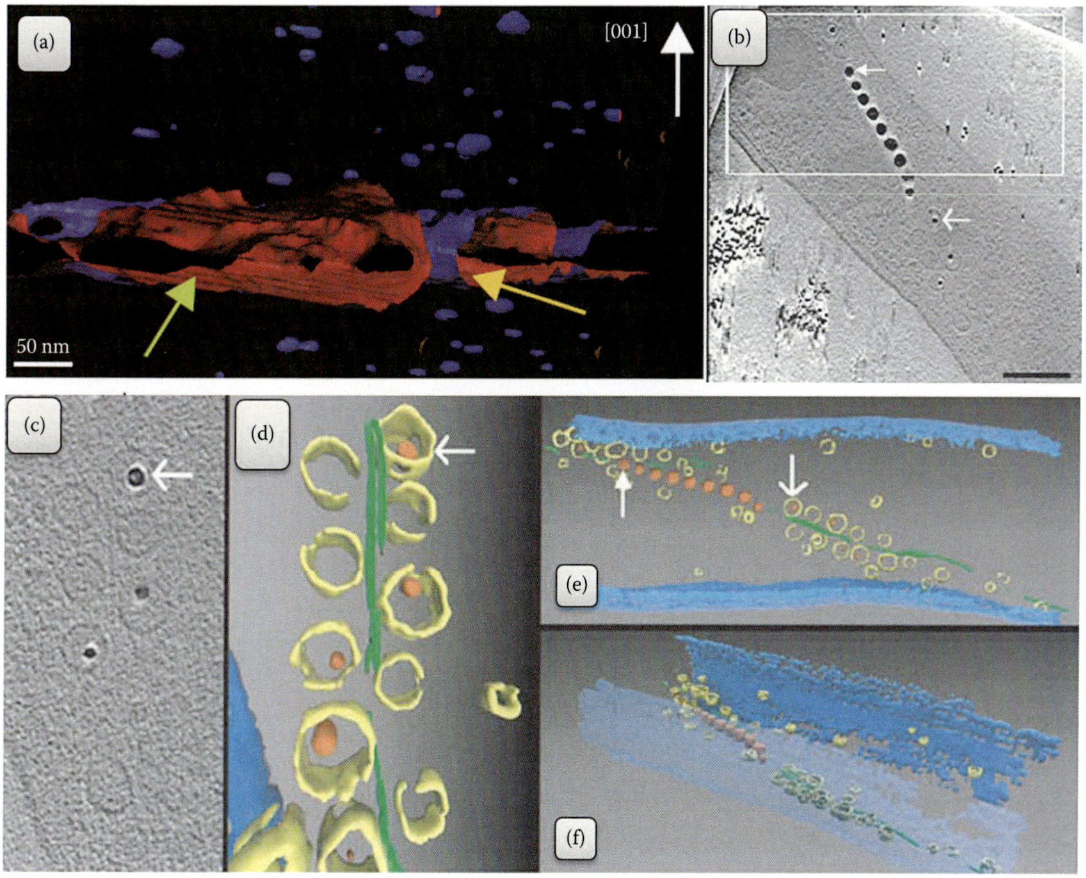

Figure 14.4 (a) Tomogram of aragonite platelets in nacre, showing connectivity via mineral bridges. Blue: aragonite; Red: organic matrix; Yellow arrow: pointing at a mineral bridge. (Reprinted from *Ultramicroscopy*, 109(3), Gries, K., Kröger, R., Kübel, C., Schowalter, M., Fritz, M., and Rosenauer, R., Correlation of the orientation of stacked aragonite platelets in nacre and their connection via mineral bridges, 230–236. Copyright 2009, with permission from Elsevier.) (b–f) Cryo-ET of wild-type cells. (b) 3D reconstruction: superimposed x–y slices along the z-axis through a typical tomogram. (c) Magnified x–y slice of vesicles, which are either empty or contain growing crystals (arrow) connected to the filamentous structure. (d) Surface-rendered representation of a segment of the cell showing vesicles (yellow), magnetite crystals (red), and a filamentous structure (green). (e and f) Complete 3D visualization of the whole cell in different views (membrane in blue). (Reprinted by permission from Macmillan Publishers Ltd. *Nature*, Scheffel, A., Gruska, M., Faivre, D., Linaroudis, A., Plitzko, J.M., and Schüler, D., An acidic protein aligns magnetosomes along a filamentous structure in magnetotactic bacteria, 440(7080), 110–114, Copyright 2006.)

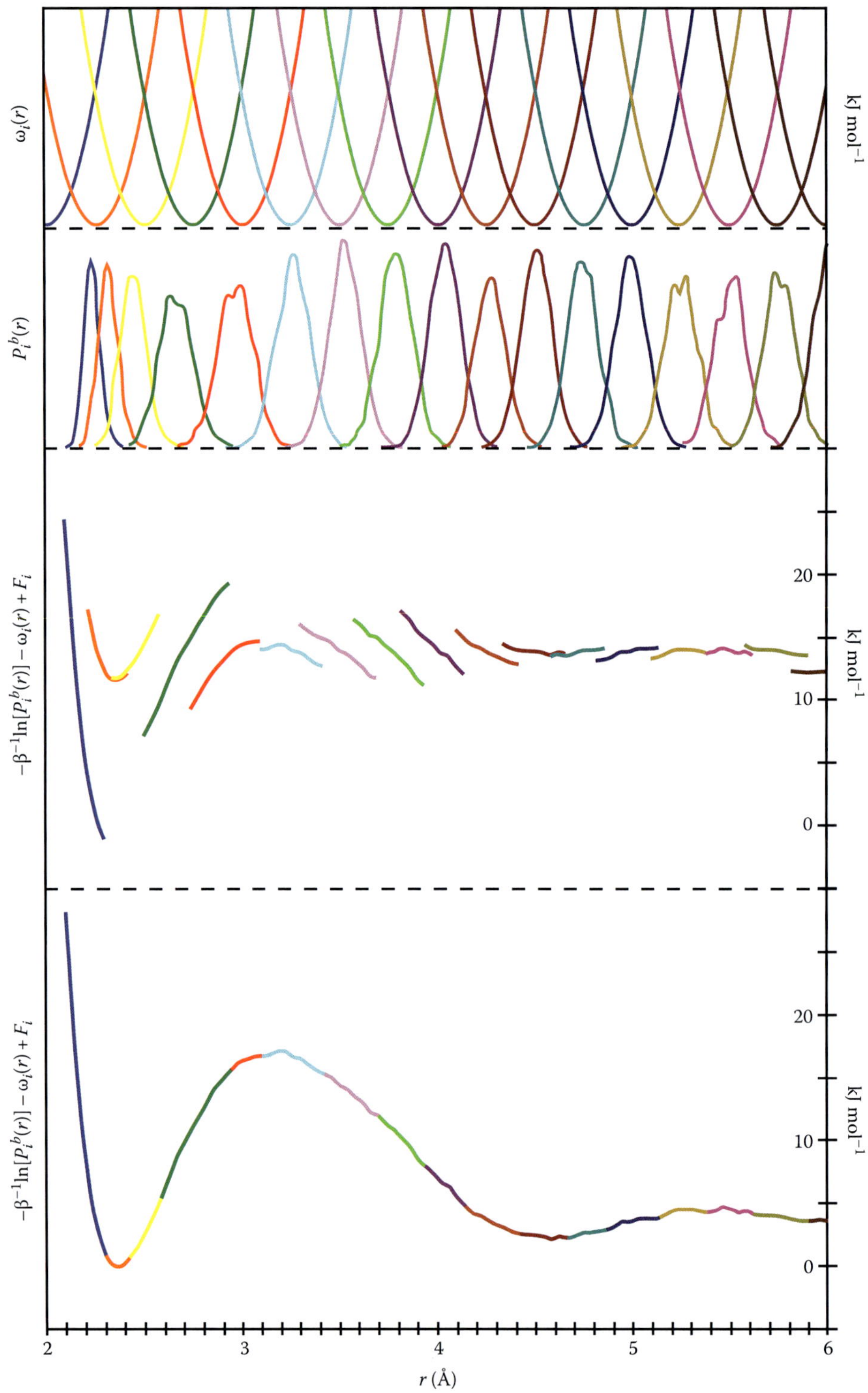

Figure 18.5 Construction of the free energy landscape for solvent exchange about calcium as determined by umbrella sampling without the Jacobian correction. The topmost portion shows the individual restraining potentials $\omega_i(r)$ and the biased probability distributions $P_i^b(r)$ that result from running an 0.5 ns MD simulation in each of the windows. Respectively, the bottom two panels show the individual free energy segments that are computed for each window and the assembly of those segments into the full landscape using the WHAM.

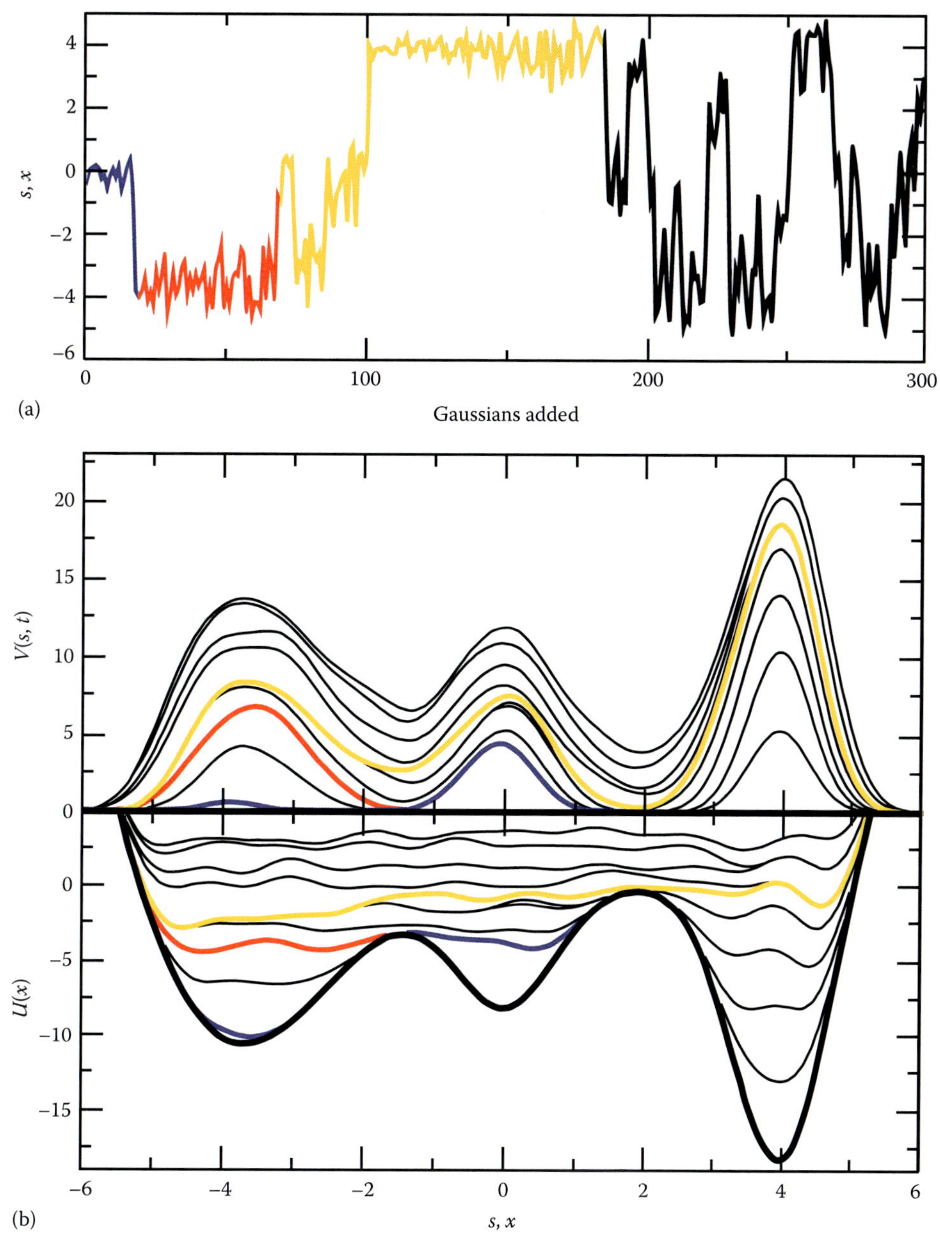

Figure 18.7 Compensation of the energy landscape, $U(x)$, by the history-dependent metadynamics bias potential $V(s, t)$. In the direct variant of metadynamics, $s = x$. (a) Evolution of the collective variable, s, throughout the simulation. (b) The upper panel shows the bias potential at several intermediate states during the simulation. The lower panel shows the current estimate of the energy landscape at each intermediate time. (Reproduced from Laio, A. and Gervasio, F.L., *Rep. Prog. Phys.*, 71(12), 126601, doi:10.1088/0034-4885/71/12/126601, 2008. With permission from IOP.)

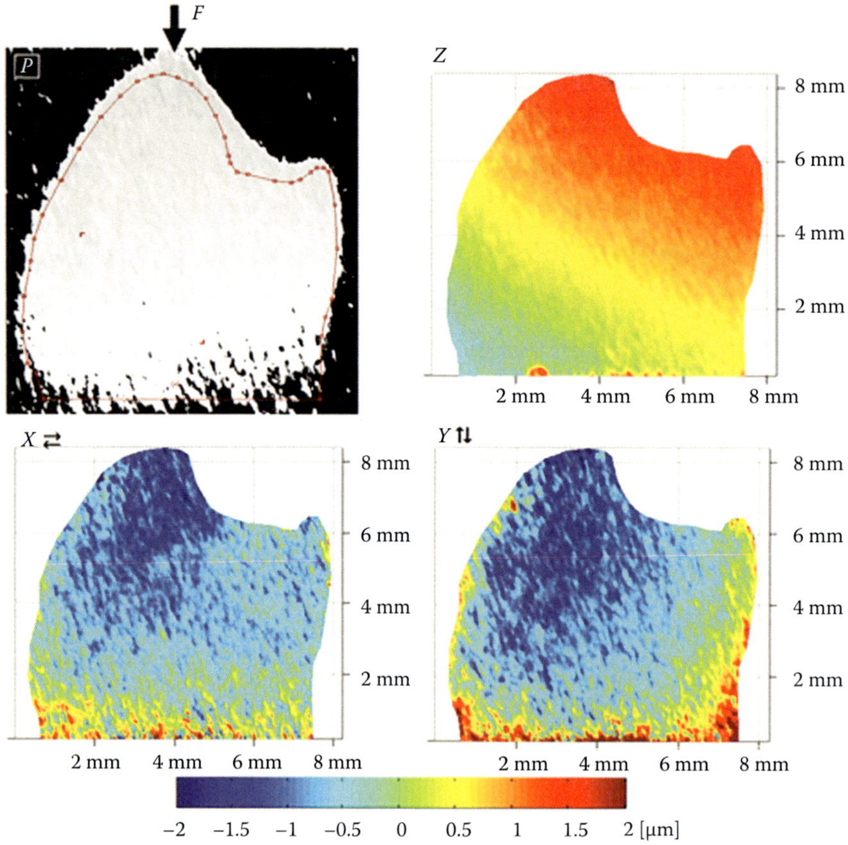

Figure 22.6 Typical phase-shift and displacement data obtained using ESPI. A tooth under compression (*F*) will deform, and the relative distance changes in the path of the light across the sample surface will change the intensity detected by the interferometer. A phase-map reveals small changes to the source-object-surface-camera distances (*P*). Upon phase unwrapping, displacement maps along the *z*- (out-of-plane) and *x*- and *y*- (in-plane) directions are retrieved and can be quantified. The micrometer-scale displacements across the surface are color-coded.

Figure 23.9 Using a highly reduced color palette, the illustrator can dramatically draw attention to the species of interest, while simultaneously maintaining a high level of detail in the image background. Following the printing of each metal engraving in black ink, a small army of artists would individually hand-color each illustration. (a) European Mantis; (b) The Solan Goose (young and old plumage). (Images adapted from Jardine, W., *The Naturalist's Library. Introduction to Entomology I: 1840; Birds of Great Britain and Ireland IV: 1860*, H.W. Lizars, Edinburgh, Scotland.)

13 *In situ* atomic force microscopy as a tool for investigating assembly of protein matrices

Sungwook Chung and James J. De Yoreo

Contents

13.1 INTRODUCTION

The development of atomic force microscopy (AFM) as a molecular-scale imaging tool that can probe surface processes *in situ* has enabled researchers to explore some of the physical mechanisms that underlie the control of organic matrices and molecules on mineral nucleation and growth. In this current chapter, our goal is to cover the topics related to using the tool of *in situ* AFM to investigate organization of protein matrices that will eventually act as primary templates for subsequent biomineralization processes. With this objective, our approach is to first overview the examples of the roles of protein matrices on biomineralization, the principles of operation, and the advantages of *in situ* AFM techniques and then finally to discuss examples of the studies by focusing on how the analysis of measured quantities from *in situ* AFM techniques can be used to extract parameters that provide a quantitative understanding of the underlying mechanisms of matrix assembly and mineralization.

13.1.1 ROLE OF PROTEIN MATRICES ON BIOMINERALIZATION

Macromolecular matrices play a key role in directing the formation of mineralized tissues, determining both the location and orientation of mineral components (Addadi et al. 2006; Braissant et al. 2003; Weiner et al. 1999). The architecture of the matrix is itself often complex. For example, collagen matrices, which constitute the organic scaffolds of bones and teeth in all higher organisms, are constructed from triple helices of the individual collagen monomers (Shoulders and Raines 2009). These helices further assemble into highly organized, twisted fibrils exhibiting a pseudohexagonal symmetry. The ability of the assembled matrix to direct mineral formation is certainly affected by its structural relationship with the incipient nucleus (Nudelman et al. 2010). However, because mineral formation is, at its most fundamental level, a phase transition in which solute ions make their way from a state of high free energy in a fully or partially solvated condition to a state of low free energy within the growing mineral phase, it is also controlled by the changes to the energy landscape the matrix imposes upon the mineralizing constituents (De Yoreo and Vekilov 2003; De Yoreo et al. 2013; Giuffre et al. 2013; Habraken et al. 2013; Hu et al. 2012). This effect on the energetics of nucleation has been demonstrated for both natural protein matrices (De Yoreo et al. 2013; Habraken et al. 2013) and synthetic analogues (Giuffre et al. 2013; Hu et al. 2012).

13.1.2 EXPERIMENTAL TOOLS FOR STUDYING ASSEMBLY OF PROTEIN MATRICES

While detailed studies on the structure of organic matrices have been performed for many systems (Lowenstam and Weiner 1989; Orgel et al. 2006; Shoulders and Raines 2009; Weiner et al. 1999), only a few have considered the mechanisms and pathways underlying matrix assembly (Chung et al. 2010; Lopez et al. 2010; Shin et al. 2012). Similarly, many experimental studies have explored the structural relationships between matrix and mineral (Addadi et al. 2006; Bilan and Usov 2001; Braissant et al. 2003; Fang et al. 2011; Jager and Fratzl 2000; Nudelman et al. 2010; Weiner et al. 1999), but only a handful ones have investigated the energetic controls on mineral formation (Giuffre et al. 2013; Habraken et al. 2013; Hu et al. 2012). The reason for this paucity of experimental studies is that probing the dynamics of matrix assembly and the energetics of matrix-directed mineralization requires *in situ* methods. To quantify the energetic controls on mineral formation, one must be able to measure nucleation rates as a function of temperature and/or supersaturation (Giuffre et al. 2013; Habraken et al. 2013; Hu et al. 2012). In the case of matrix assembly, a true understanding of the dynamics demands an ability to observe molecular-scale events (Chung et al. 2010; Lopez et al. 2010; Shin et al. 2012). While some of

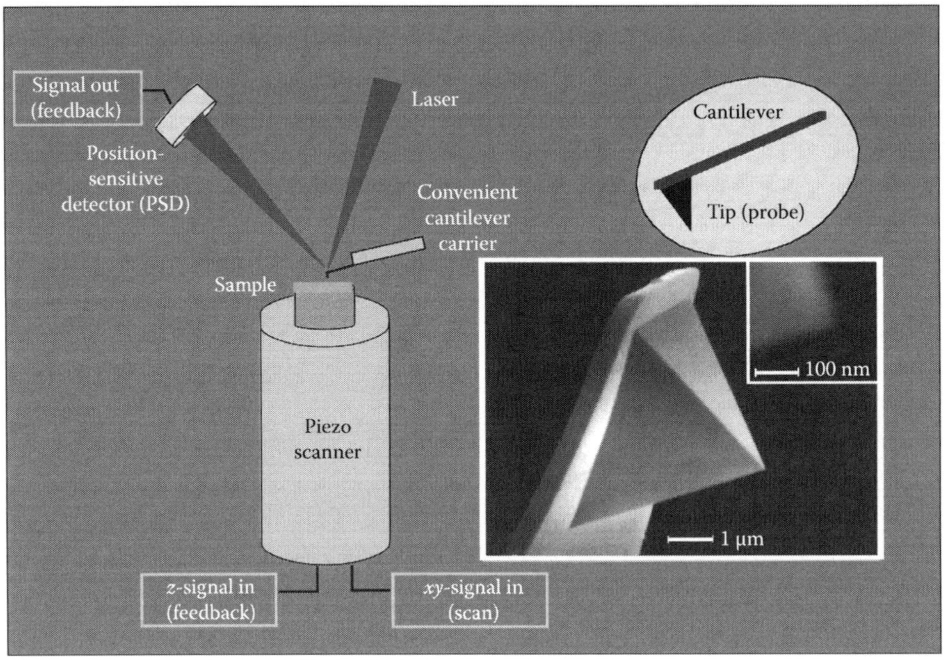

Figure 13.1 Schematic diagram of an atomic force microscope. The system consists of a probe that measures very tiny forces (inset SEM image shows a typical AFM probe made out of silicon nitride), an optical detection setup where a laser is focused on the back of the cantilever and reflected back to a PSD to measure the deflection of the cantilever, and a piezoelectric scanner that moves the sample stage along the *x*, *y*, and *z* direction when voltage is applied.

these measurements can be performed using optical microscopy (Giuffre et al. 2013; Hu et al. 2012), the atomic force microscope offers the unique ability to observe these processes *in situ* at molecular length scales (Chung et al. 2010; Habraken et al. 2013; Lopez et al. 2010; Shin et al. 2012).

13.1.3 OPERATION PRINCIPLES AND ADVANTAGES OF *IN SITU* AFM TECHNIQUES

Scanning probe microscopy (SPM) is one of the microscopy techniques that generate images of surfaces using a physical probe that scans across the specimen. An image of the surface is acquired by mechanically moving the probe, typically in a raster scan, line by line, on the specimen while recording the probe–surface interaction as a function of probe position. AFM is a type of SPM with a very high resolution on the order of fractions of a nanometer, which is more than 1000 times better than the optical diffraction limit. The granddaddy of AFM is the scanning tunneling microscope developed by Gerd Binnig and Heinrich Rohrer in the early 1980s at International Business Machines (IBM) Research at Zurich, and this development eventually earned them the Nobel Prize for Physics in 1986 (Rohrer et al. 1982). Binning et al. (1986) invented AFM in 1986. The first commercially available AFM was introduced in 1989. Since then, AFM has been one of the principal tools for imaging, measuring, and manipulating matters at the nanoscale, particularly in the areas of physical, analytical, biological, and materials science. Furthermore, since early in its development, AFM has been especially useful for imaging of biomolecular structures under ambient conditions or in solutions because it can deliver topographic images with submolecular resolution in aqueous environments (Horber and Miles 2003). However, obtaining high resolution on protein structures has been a challenge because the inherent fluctuations of the solution due to Brownian motion interfere with imaging through mechanisms that will become clear in what follows.

Figure 13.1 shows the schematic diagram of an AFM. The AFM consists of a cantilever with a sharp probe (or tip) at its end that is used to scan the surface of the specimen. The cantilever is typically made out of silicon (Si) or silicon nitride (Si_3N_4) with a tip radius of curvature on the order of a few nanometers. When the tip of the cantilever is brought into proximity of a specimen surface, forces interacting between the tip and the specimen lead to the mechanical deflection of the cantilever according to Hook's law. The forces that are measured have either an attractive or repulsive nature and include mechanical contact and friction forces, capillary forces, van der Waals forces, electrostatic force, magnetic force, forces due to chemical bonding, solvation forces, and Casimir forces. Along with the forces, additional material properties may be simultaneously measured through the use of specialized probes. In the case of protein structures imaged in solution, capillary forces and magnetic forces play little or no role, but the response of the cantilever involves some combination of the others. Typically, the cantilever deflection is measured using a laser reflected from the top surface of the cantilever into an array of photodiodes known as a position-sensitive detector (PSD). Other methods of detecting cantilever deflection that have been used include optical interferometry (Erlandsson et al. 1988), capacitive sensing (Kopanski 2007), and sensing of piezoelectric AFM cantilevers fabricated with piezoelectric elements that act as strain sensors (Minne et al. 1995).

In most cases, in order to prevent the tip from colliding with high points on the surface while scanning at a constant height, a feedback mechanism is employed to adjust the tip-to-sample distance and instead maintain a constant force between the tip and the sample. Conventionally, the sample is mounted on a piezoelectric tube known as a scanner that can move the sample in the x and y directions and in the z direction for maintaining a constant force or height (see Figure 13.1). A *tripod* configuration of three piezo crystals may be employed, with each responsible for scanning in the x, y, and z directions independently. This can potentially eliminate some of the distortion effects observed with the tube piezo scanner. There is an alternative design of the tip being mounted on a piezo tube scanning either in the x, y, and z direction over the sample or in the z direction while the sample is being scanned in the x and y direction.

Because AFM relies on the forces between the tip and the surface of the sample, understanding and knowing these forces are very important for proper AFM operation. Figure 13.2 shows a simplified force curve of an AFM. The force is not measured directly; rather it is calculated by measuring the deflection of the cantilever while knowing the stiffness of the cantilever. Hook's law gives $F = -k \times x$ where F is the force, k is the stiffness of the cantilever (i.e., spring constant), and x is the distance over which the cantilever is being bent.

The AFM can be operated in a number of different imaging modes depending on the applications. In general, these imaging modes are categorized into static modes (also called contact mode based on the constant height or contact force between tip and sample) and dynamic modes (also called noncontact, *tapping mode*, or other imaging modes) in which the cantilever or the scanner is oscillated. These imaging modes can be also applied to different environments such as vacuum, air, gas, and liquid. When imaging proteins, noncontact or tapping mode is almost always the most useful, because it minimizes the shear deformation of the sample that occurs when a tip used in contact mode is dragged across the surface.

In the static mode of operation, the cantilever is dragged across the surface of the sample while the tip is either in contact or near contact with the surface and the topography of the surface is measured directly using the deflection of the cantilever. This static cantilever deflection is used as a feedback signal for the piezo scanner. When the tip of the cantilever is close to the surface of the sample in the ambient conditions, attractive forces (such as that due to the formation of water meniscus layer) can be quite strong, causing the tip and cantilever to snap into the surface. Furthermore, the measurement of a static deflection signal is prone to noise and drift. Typically, low-stiffness cantilevers are used for the static-mode AFM (i.e., most contact-mode AFM cantilevers have a spring constant (k) less than 1 N/m), and the tip is always in contact with the surface so that the overall force is repulsive. This repulsive force between the tip and the

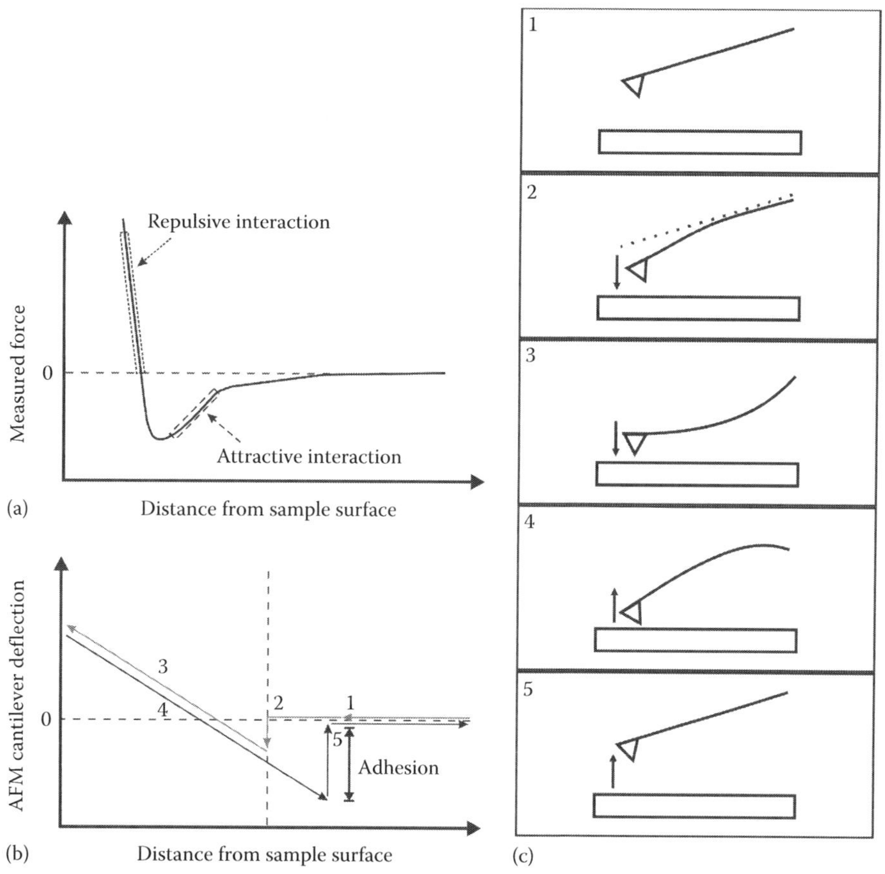

Figure 13.2 (a) The relationship between the force measured by an AFM probe and the distance of the probe from the sample surface. Because AFM relies on the forces between the AFM probe and the sample surface, knowing these forces is important for proper AFM imaging. When the probe is far from the surface, it does not sense any interaction. As the distance from the sample surface decreases, the probe senses an attractive interaction (i.e., probe is pulled toward the surface). As the probe comes in contact with and pushes on the surface, it starts to sense a repulsive interaction and the cantilever of the probe is bent. (b and c) AFM can be used to determine force by measuring the deflection of the cantilever and calculating the actual force based on the stiffness of the cantilever using Hooke's law. A typical force–distance curve (b) obtained from AFM shows the deflection of the cantilever while the AFM probe is moving toward the sample (c1, 2, and 3) and away from the sample (c4 and 5).

surface is kept constant during contact-mode AFM operation by maintaining a constant deflection signal. As a consequence of the lateral motion of the tip under application of a repulsive force, static-mode imaging generates frictional forces. Although these forces can be measured and quantified—giving rise to the term lateral force microscopy (LFM)—to reveal areas of relatively high and low friction, the high shear force is generally deleterious to imaging soft materials like proteins. Hence, imaging of protein matrices is best done with a dynamic mode.

In the dynamic mode of operation, the tip of the cantilever either is not in contact or is in intermittent contact with the surface of the sample. In the case of noncontact imaging, the cantilever is oscillated at a frequency slightly above its resonance frequency where the amplitude of oscillation is a few nanometers (<10 nm). The van der Waals forces, which are attractive and strongest from 1 to 10 nm above the surface, or any other long-ranged forces applied from the surface, decrease either the resonance frequency or the amplitude of oscillation. The feedback loop acts to counter changes in the frequency or amplitude by adjusting the tip-to-sample distance. This adjustment then provides the information for constructing a topographic image of the sample surface. In the case of intermittent contact (i.e., also known as *AC mode* or *tapping mode*), the cantilever is driven to oscillate near its resonance frequency by a small piezoelectric crystal mount in the AFM tip holder as in the case of noncontact-mode imaging. However, the amplitude of the oscillation typically is greater than that (<10 nm) used in noncontact mode. The interaction of forces (i.e., as van der Waals forces, electrostatic forces) acting on the cantilever causes the amplitude of the oscillation to decrease as the tip comes close to the surface. The feedback loop uses the change of the amplitude of the oscillation to adjust the distance between the tip and surface and maintain a predetermined cantilever oscillation (known as *a set point*) as the cantilever is scanned over the surface of the sample. Therefore, a *tapping-mode* AFM image is produced by displaying the force of the intermittent contact of the tip with the sample surface as a function of tip position (Geisse 2009). Because tapping mode largely eliminates the shear forces present in contact mode, it is the most useful imaging mode for investigating protein systems.

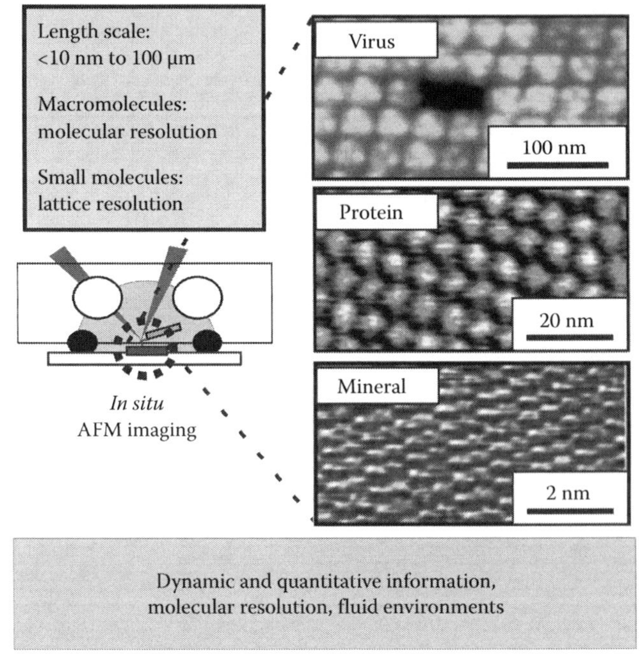

Figure 13.3 The attributes of *in situ* AFM showing its powerful capability for investigating biochemical processes in both ambient and solution conditions. AFM height images of (top) 2D lattice of cowpea mosaic virus, (middle) surface of canavalin crystal, and (bottom) surface of calcium oxalate monohydrate crystal.

Figure 13.3 illustrates the attributes of *in situ* AFM techniques that have been used to investigate organization of materials in biological and environmental systems. Whether through assembly of protein complexes or biologically directed mineral nucleation and growth, their formation process typically occurs in aqueous media. This aqueous environment presents many challenges to both theoretical and experimental investigations. At the same time, however, it presents a great opportunity for application of AFM. Both imaging and force spectroscopy via AFM can be carried out in fluidic environments. Therefore, the interactions of biomolecules including proteins with other biomolecules plus or minus additional surfaces, aggregation, and self-assembly processes and biologically directed mineral nucleation and growth templated by the protein matrices can be measured *in situ* with spatial resolution at the level of single protein complexes and atomic step heights. This unique *in situ* capability has many advantages over other analytical tools and impacts on the study of protein matrix assembly in a number of ways. First, an *in situ* AFM capability allows one to capture the structural and morphological evolution during assembly under physiologically relevant environmental and temperature conditions. Second, the dynamics of these processes from the initial to final state provide information about the energy landscape experienced by the constituent ions, molecules, and macromolecular building blocks. Third, *in situ* AFM imaging has the ability both to precisely control the solution conditions such as composition, temperature, pH, and flow rate and to change the conditions or solutions during the course of an experiment. Fourth, *in situ* AFM imaging removes the uncertainty associated with inferring the pathways of formation processes associated with observing just the initial and final structure or morphology.

There are many technological challenges associated with applying *in situ* AFM imaging to protein and biomineral systems. Some are associated with producing a suitable sample. For example, because AFM requires a substrate, investigation of proteins by AFM can only be performed once a set of solution conditions and substrate composition is identified that results in adsorption of the proteins in a state that is relevant. Generally this means that binding has to be strong enough to keep the proteins in place during imaging but weak enough to allow their inherent intra- and interprotein interactions to determine their conformation and assembled architecture. Other challenges are associated with imaging. The most important are the time scale of the processes of interest versus the rate at which one can image and the distortion of protein structure due to the force applied by the tip. However, these challenges are rarely insurmountable. In Section 13.2, we review some recent examples of investigation of self-assembly by different protein systems, illustrate the unique information obtained through *in situ* AFM imaging, and explain how the analysis of measured quantities is converted into the parameters of scientific interest.

13.2 EXAMPLES OF ASSEMBLY OF PROTEIN MATRICES

Self-assembled protein architectures exhibit a range of structural motifs (Mann 2008) including particles (Johnson 2008), fibers (Rambaran and Serpell 2008), ribbons (Du et al. 2005), and sheets (Tanaka et al. 2008). Their functions include selective transport (Tanaka et al. 2008), structural scaffolding (Engelhardt 2007), mineral templating (Du et al. 2005; Schultzelam and Beveridge 1994), and propagation of or protection from pathogenesis (Cherny and Gazit 2008; Rambaran and Serpell 2008). Although the molecular structures of the isolated proteins dictate their governing interactions, these functions emerge from the nanoscale organization that arises out of self-assembly. Typically, proteins that naturally self-assemble into extended ordered structures adopt conformations that are distinct from those of the individual monomeric proteins (Chiti and Dobson 2009; Salgado et al. 2008; Schoen et al. 2011; Shoulders and Raines 2009). For example, as discussed previously, collagen matrices, which constitute monomeric scaffolds of bones and teeth in all higher organisms, are constructed from triple helices of the individual collagen monomers (Shoulders and Raines 2009). These helices further assemble into highly organized twisted fibrils exhibiting a pseudohexagonal symmetry. In some cases, assembly is inexorably linked to folding transformations, as in the case of prion or amyloid fibrils, where misfolding of the monomers triggers assembly, which in turn drives misfolding of new monomers (Chiti and Dobson 2009). Nonetheless, while the phenomenon of folding by individual proteins has been well explained both experimentally and theoretically, much less attention has been given to protein assembly, and the role of folding transformations in defining the assembly pathway is largely unexplored.

13.3 S-LAYER PROTEIN ASSEMBLY ON BIOMIMETIC AND INORGANIC SURFACES

Recently, *in situ* AFM was used to investigate the dynamics and underlying mechanisms of S-layer protein assembly. The cell surface layers known as S-layers are composed of a monomolecular layer of identical proteins of glycoproteins. S-layer proteins form into compact 2D crystalline sheet arrays on the outermost cell envelopes and membranes of many strains of bacteria and archaea (Engelhardt 2007; Sleytr et al. 1999). The crystalline lattice of S-layer proteins is typically built from multimeric growth units of individual monomers, which in some cases can embrace more than one structural conformation within a membrane (Stewart et al. 1986). The individual proteins (molecular weight [MW] ~ hundreds of kDa) are produced within the cell, transformed from inside, and assembled on the outside of the cell. They overlie a lipid membrane or polymeric cell wall and exhibit crystal symmetries including p1, p2, p4, and p6. They play a role in all the functions such as selective transport (Tanaka et al. 2008), structural scaffolding (Engelhardt 2007), propagation of or protection from pathogenesis (Cherny and Gazit 2008; Rambaran and Serpell 2008), and even mineral templating (Du et al. 2005; Schultzelam and Beveridge 1994) and can be isolated and reconstituted *in vitro* into 2D arrays both in bulk solutions and at surfaces. It has been reported that they were among the first self-assembled protein structures to be exploited as nanoscale scaffolds for organizing nanomaterials via a bottom-up strategy because of their large-scale order and a periodicity commensurate with the dimensions of nanomaterials (Moll et al. 2002; Shenton et al. 1997).

Chung et al. (Chung et al. 2010) used *in situ* AFM to investigate the nucleation and growth of crystalline S-layer in 2D on supported lipid bilayers (SLBs). S-layer SbpA was isolated from *Lysinibacillus sphaericus* (ATCC 4525, MW ~132 kDa). The SLB of 1-palmitoyl-2-oleoyl-sn-glycero-3 phosphocholine (POPC) was formed on freshly peeled mica surface glued on metal disks. Monomeric SbpA dissolved in pure water was mixed with 10 mM tris(hydroxymethyl)aminomethane (Tris), pH 7.1, 100 mM NaCl, and 50 mM $CaCl_2$ and injected in the fluid cell of an AFM containing an SLB deposited mica. The AFM probe consisted of a sharp silicon tip on a soft silicon nitride cantilever (nominal tip diameter <15 nm, spring constant *k*—ca. 0.035 N/m). *In situ* AFM imaging was employed to investigate the dynamics of S-layer assembly on the SBL for a range of protein concentrations.

For typical imaging conditions, contact-mode AFM images were collected at 0.5–2 Hz scan speed (tip speed = 0.25–2 μm/s) while applying imaging force of ~150 pN or less using optimized feedback and set point parameters for minimal tip perturbation and therefore stable imaging conditions. The requirement for low imaging forces in order to successfully image the S-layer system highlights a general challenge and approach to imaging biological materials. Most biological structures are relatively soft and delicate, and typical imaging forces of tens of nanonewtons can easily distort or even permanently destroy such structures. Therefore, a two-pronged approach to imposing lower imaging forces (down to piconewton range) has to be taken. First, the cantilever properties need to be chosen carefully. As indicated previously, one of the most successful strategies is to use silicon nitride cantilevers, which have force constants of less than 0.1 N/m, with Si tips, which are much sharper than silicon nitride and thus give higher resolution. However, the exact choice of cantilever stiffness that gives the best results is system dependent, and we have found that even a factor of two in stiffness—both larger and smaller—can lead to large differences in image quality. Optimization of cantilever stiffness is generally a matter of trial and error. Once the AFM tip approached and engaged in either normal contact or tapping mode, the set point must be carefully decreased until the microscope can barely sense the interactions between the tip and surface. By thusly minimizing the average force exerted during *in situ* AFM imaging, high resolution is generally possible.

Typical results of *in situ* AFM imaging of S-layer assembly were shown in Figure 13.4, presenting a set of *in situ* AFM images and the corresponding height profiles selected from a time sequence. Figure 13.5 shows zoomed, sequential *in situ* AFM height images revealing a multistage assembly process of S-layer adsorption followed by condensation and phase transition of a single cluster as well as its subsequent growth

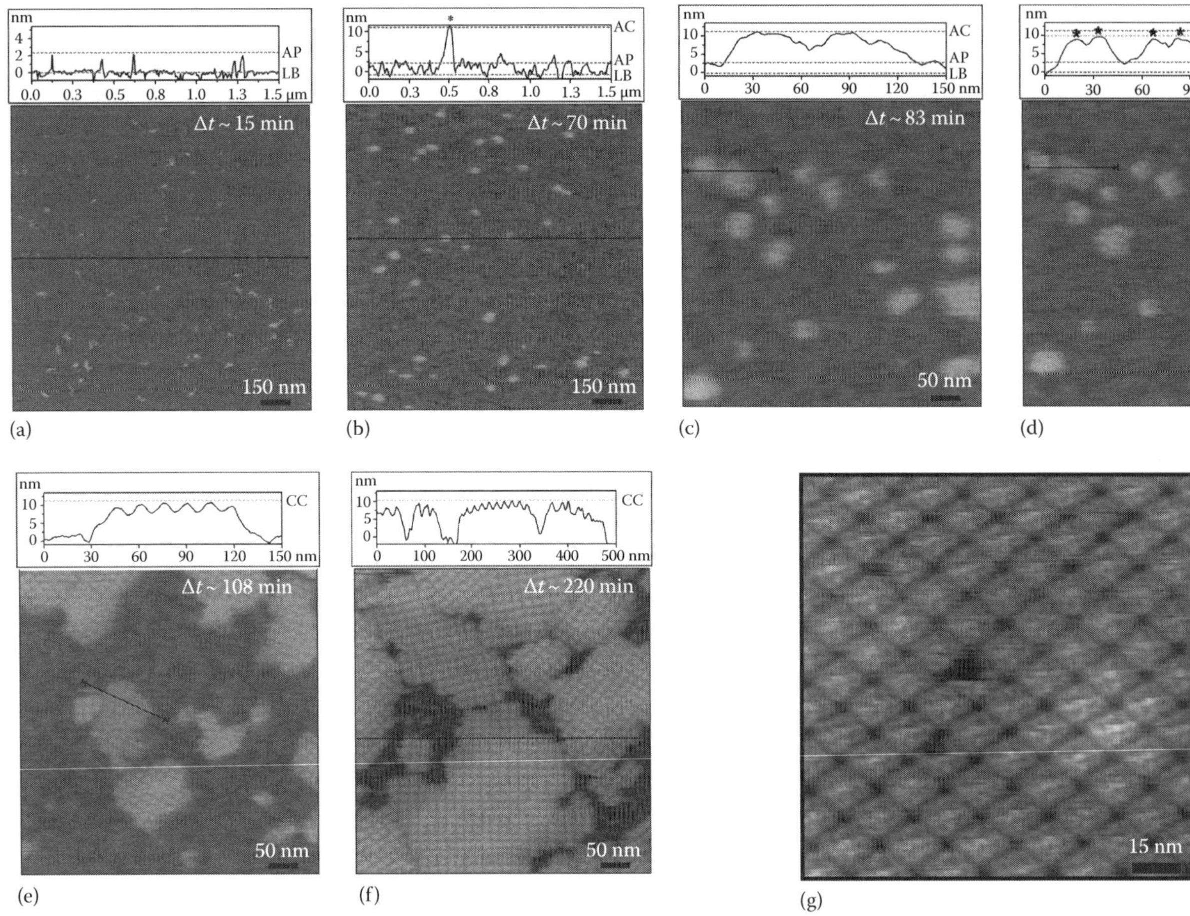

Figure 13.4 (a–f) Time series of *in situ* AFM images and height profiles showing the dynamics of S-layer assembly into 2D crystals on SLBs. Height profiles were measured along the horizontal black lines in each image and are labeled to denote the heights of the LB, adsorbed proteins (APs), amorphous clusters (ACs), and crystalline clusters (CCs). Δt indicates time elapsed since introduction of the solution containing SbpA protein into the AFM fluid cell. By $\Delta t = 15$ min (a), SbpA monolayers had started to adsorb onto the SLBs. During 15 min $\leq \Delta t \leq$ 70 min, the AP coverage increased, and by $\Delta t \cong 70$ min (b), not only was the surface well covered with APs but ACs of nearly uniform heights had formed. By $\Delta t \cong$ 83 min (c), some of the ACs had begun to show the emergence of internal structure and order, and by $\Delta t \cong 87$ min (d), nearly all of ACs in (c) has transformed into CCs with a clearly visible tetragonal lattice (P4 symmetry). The height profiles before and after the transformation ((c) vs. (d)) demonstrate that the ACs were slightly taller (i.e., about 2 nm) than the CCs and reflect the resulting periodic crystalline structures of CCs. Detailed volume measurements of ACs and CCs show that CCs were also structurally more compact. At later time (e.g., $\Delta t \cong 108$ min, (e) and $\Delta t \cong 220$ min, (f)), each CC continued to grow by consuming available APs near the clusters until growth was physically hindered by neighboring CCs. (g) A highly resolved AFM image from mature CCs reveals the tetrameric arrangement and subdomain details of a look-alike structure of the four S-layer monomers that comprise each lattice unit. (Reproduced with permission from Chung, S., Shin, S.H., Bertozzi, C.R., and De Yoreo, J.J., Self-catalyzed growth of S layers via an amorphous-to-crystalline transition limited by folding kinetics. *Proc. Natl. Acad. Sci. USA*, 107(38), 16536–16541, 2010, doi: 10.1073/Pnas.1008280107. Copyright 2010 National Academy of Sciences, U.S.A.)

by addition of new tetramers. The results by Chung et al. (2010) revealed a multistage assembly pathway of S-layers, which comprises four distinct processes (see Figure 13.6): (1) adsorption of S-layer monomers in extended conformations on the SLB, (2) condensation of amorphous or liquid-like clusters, (3) relaxation (via rearrangement and folding) into crystalline clusters (CCs) composed of a square lattice of tetramers, and (4) self-catalyzed growth by the formation of new tetramers at edge sites of the CCs (a movie of the growth process can be found in the supporting information [SI] of reference [Chung et al. 2010]). Not only did the *in situ* AFM imaging capability provide time sequential AFM images revealing structural information about protein assembly (i.e., heights and domain sizes of S-layer monomers, amorphous, and CCs), but it also enabled Chung et al. (2010) to measure cluster growth rates. Figure 13.7 shows the growth kinetics for the CCs.

The data suggest a simple model for growth that leads to a specific prediction for the dependence of CC size on time. The rate of increase in the number of tetramers in a cluster (i.e., the size of CC can be determined by counting the number of tetramers in a cluster) dN_T/dt should be proportional to the number of lattice sites around its perimeter ($4L/a$), which is the number of potentially available sites for the formation of new tetramers, the surface concentration of adsorbed monomeric proteins $n(t)$, and the rate coefficient for tetramer formation β. For a square island (or cluster), this is

$$\frac{dN_T}{dt} = \left(\frac{4L}{a}\right) \cdot \left(\frac{n(t)}{4}\right) \cdot \beta \qquad (13.1)$$

where
L is the island width
a is the lattice parameter

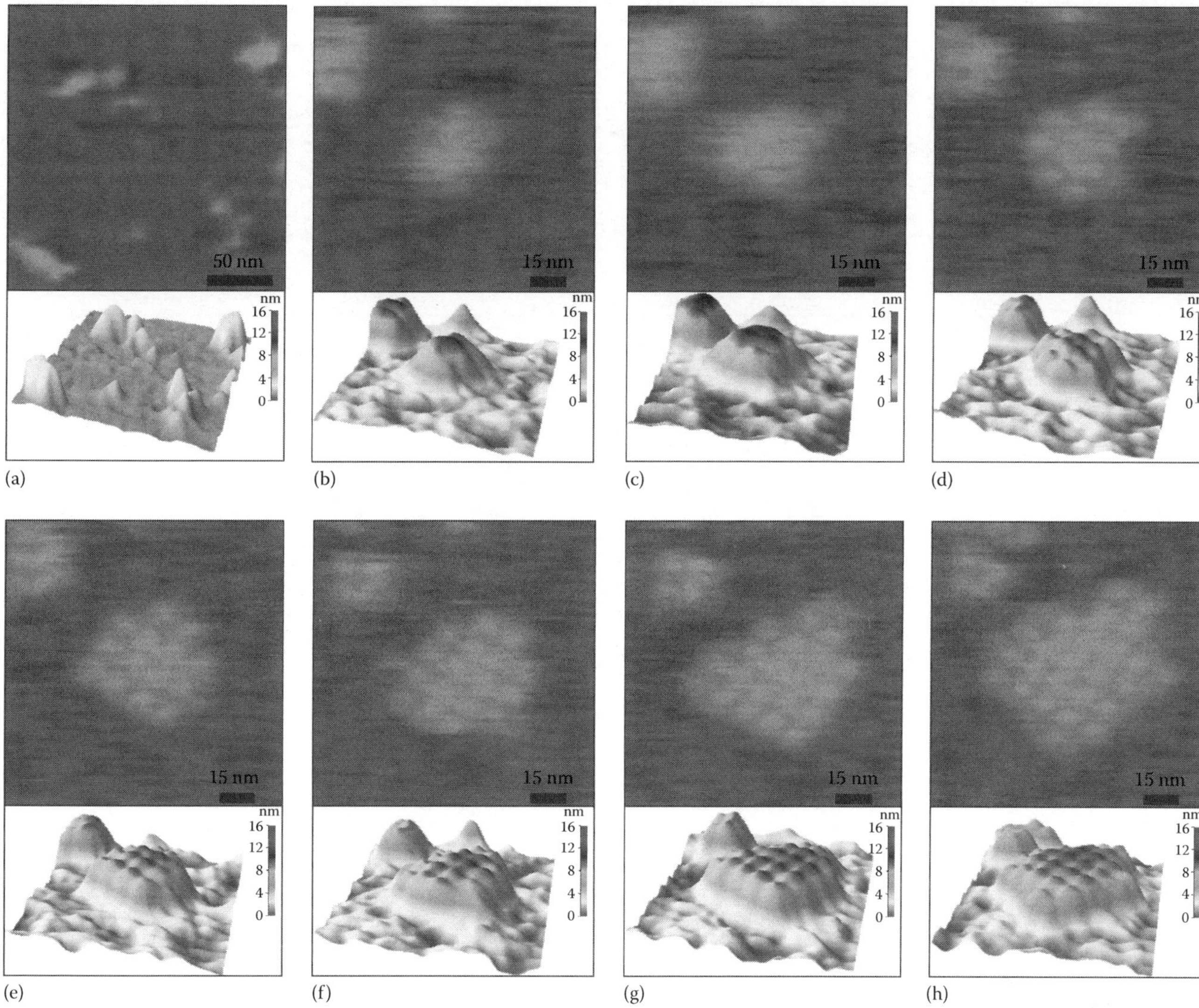

Figure 13.5 Sequential *in situ* AFM height images and surface plots showing S-layer adsorption followed by condensation and amorphous to crystalline transformation of a single cluster as well as subsequent domain growth. Δt indicates time elapsed since collection of the image in (a). (a) At $\Delta t = 0$, SbpA monomers or small oligomers in an extended conformation have adsorbed onto the SLBs. (b) By $\Delta t \cong 62$ min, the amorphous nucleus has formed but shows signs of emerging order. (c) By $\Delta t \cong 65.8$ min and (d) $\Delta t \cong 70.3$ min, they show the amorphous to crystalline transition, after which time each individual lattice unit is clearly discernible. After the transformation ($\Delta t \cong 70.3$ min (d) to 116.8 min [h]), the crystalline domain grows by formation of new lattice units at—and only at—unpopulated lattice sites along the perimeter of the crystalline domain. Times are (a) 0, (b) 62, (c) 65.8, (d) 70.3, (e) 88.5, (f) 95.5, (g) 102.6, and (h) 116.8 min. (Reproduced with permission from Chung, S., Shin, S.H., Bertozzi, C.R., and De Yoreo, J.J., Self-catalyzed growth of S layers via an amorphous-to-crystalline transition limited by folding kinetics. *Proc. Natl. Acad. Sci. USA*, 107(38), 16536–16541, 2010, doi: 10.1073/Pnas.1008280107. Copyright 2010 National Academy of Sciences, U.S.A.)

Equation 13.1 implies that the CC growth rate should be greater for larger initial cluster size ($N_T(t = 0)$) and increase as the CCs grow. However, as the adsorbed monomeric proteins are consumed (i.e., $n(t)$ decreases), the growth rate should approach zero. Note that $n(t)$ can be expressed in terms of the average island area $\langle L(t)^2 \rangle$ versus time, the surface number density of cluster m, and the initial surface coverage (i.e., number density) of monomeric proteins n_0, all of which can be determined from the experiments. By substituting the expression for $n(t)$ in terms of $\langle L(t)^2 \rangle$, Equation 13.1 can be solved exactly and gives a simple expression for the number of tetramers $N_T(t)$ in a cluster as a function of time:

$$N_T(t) = \left[\beta n_0 f(t) + \sqrt{N_{T,0}}\right]^2 \qquad (13.2)$$

where

n_0 is the initial surface coverage (i.e., number density) of monomeric proteins

$N_{T,0}$ is the number of tetramers in the cluster at the time of nucleation

$f(t)$ is a function of time (t) that is linear at small t and approaches a constant at large t (the detailed expression is in the SI of the reference [Chung et al. 2010])

The rate coefficient β was expressed as the product of the diffusive collision rate (aD/d), where a is the S-layer lattice parameter, D is the diffusivity, and d is the typical jump distance of a few water molecules. In the Boltzmann factor ($\exp[-E_A/kT]$), the activation energy E_A is associated with the creation of new tetramers. The dotted line in Figure 13.7 gives the best fit to

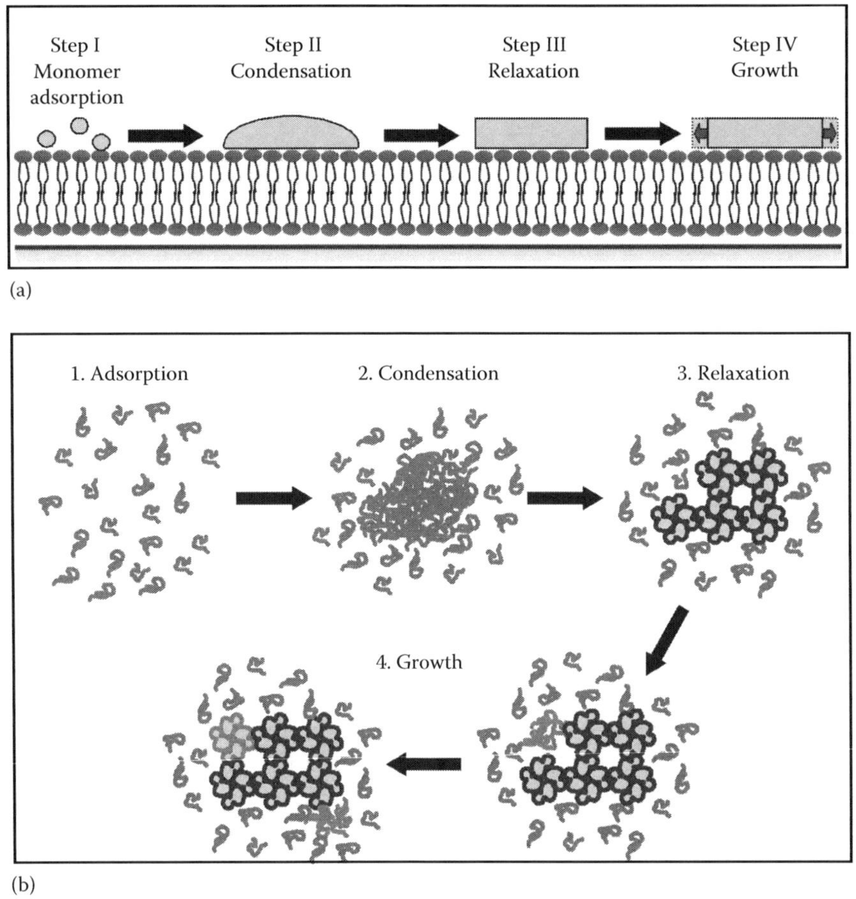

Figure 13.6 Proposed scheme of 2D crystallization of S-layer from solution on SLBs. (a) Side view of membrane and (b) top view of proteins on surface. Step I, adsorption of Sbpa S-layer monomers in extended conformations on the surface; Step II, condensation of the ACs; Step III, relaxation to the CCs; and Step IV, self-catalyzed crystal growth. (Reproduced with permission from Chung, S., Shin, S.H., Bertozzi, C.R., and De Yoreo, J.J., Self-catalyzed growth of S layers via an amorphous-to-crystalline transition limited by folding kinetics. *Proc. Natl. Acad. Sci. USA*, 107(38), 16536–16541, 2010, doi: 10.1073/Pnas.1008280107. Copyright 2010 National Academy of Sciences, U.S.A.)

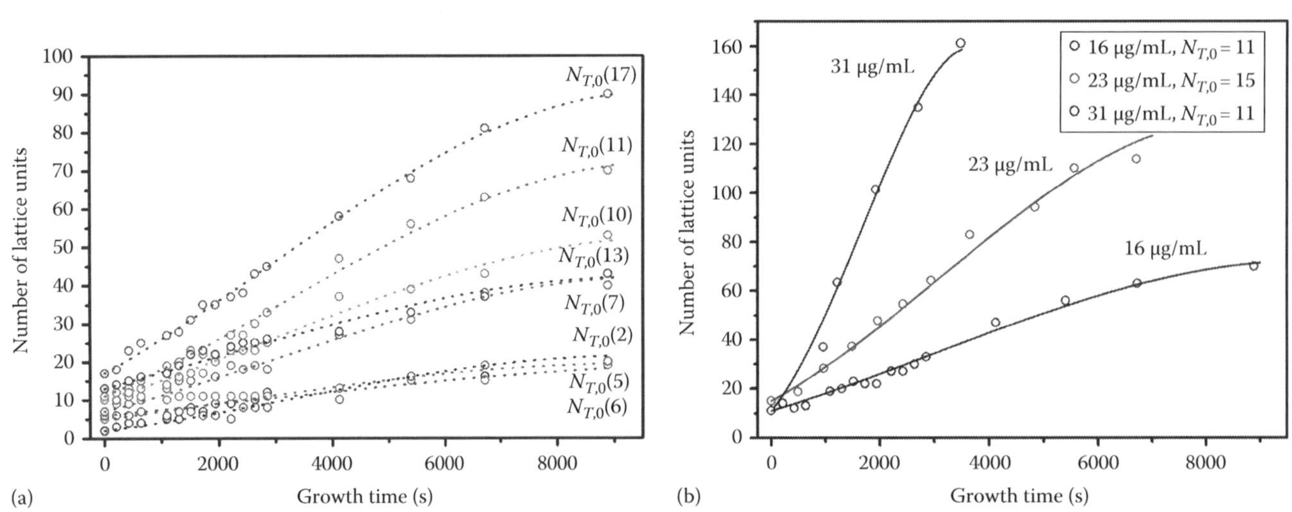

Figure 13.7 Growth kinetics of CCs: (a) The main plot shows growth of individual clusters for a range of initial sizes at protein concentration in solution (C_p) ≅ 16 μg/mL. The plot shows that the growth rate for an individual cluster depends on its size at $t = 0$ when the crystalline phase emerged, with larger clusters exhibiting faster growth rates. Dotted line gives best fits to experimental data where the fitting is based on the model described in the text (Equation 13.2). The legend relates the data for each curve to the initial number of lattice units in the crystalline domain. (b) Plot of three growth curves and the theoretical fits obtained from independent experiments at three different C_p concentrations. Three similar sizes of crystalline nuclei were compared to emphasize the phenomena of accelerated growth rate as C_p was increased. (Reproduced with permission from Chung, S., Shin, S.H., Bertozzi, C.R., and De Yoreo, J.J., Self-catalyzed growth of S layers via an amorphous-to-crystalline transition limited by folding kinetics. *Proc. Natl. Acad. Sci. USA*, 107(38), 16536–16541, 2010, doi: 10.1073/Pnas.1008280107. Copyright 2010 National Academy of Sciences, U.S.A.)

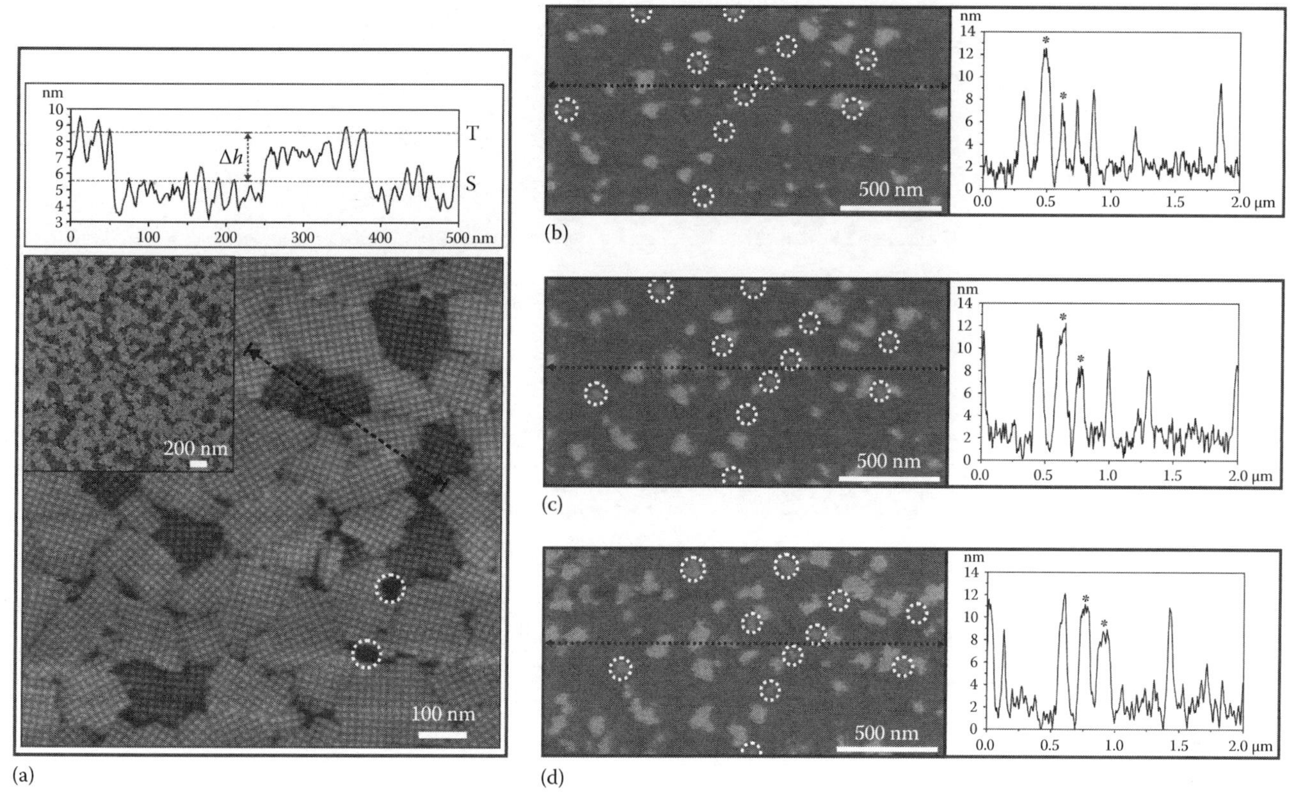

Figure 13.8 (a) AFM height profile and image of 2D S-layer organization on mica showing two types of domains. The height profile (top) measured along the black dotted line in (a) shows that the average difference between the domain height (h) of the *tall phase* (denoted as T) and the *short phase* (denoted as S) is approximately 3 nm. The average height of the tall domains measured from the mica surface (highlighted by white dotted circles) corresponds to ca. 8–9 nm, which is consistent with measurements of S-layers on SLBs. (b–d) Time sequence of *in situ* AFM images and height profiles showing the early state of S-layer assembly. White circles highlight initial nuclei formed from the APs. Height profiles were measured along the horizontal dotted black lines in each image. Two dotted circles on the dotted black line highlight a pair of clusters, one *tall* and the other *short*, which maintain a consistent height difference of approximately 3 nm during the observed growth period. The nonzero baselines in each height profile denote the level of the adsorbed monomers, which have an average height of approximately 2 nm. The time AFM images were captured at (b) 37, (c) 48, and (d) 59 min after the start of the growth experiment. (Reproduced with permission from Shin, S.H., Chung, S., Sanii, B., Comolli, L.R., Bertozzi, C.R., and De Yoreo, J.J., Direct observation of kinetic traps associated with structural transformations leading to multiple pathways of S-layer assembly. *Proc. Natl. Acad. Sci. USA*, 109(32), 12968–12973, 2012, doi: 10.1073/Pnas.1201504109. Copyright 2012 National Academy of Sciences, U.S.A.)

the experimental data. From the analysis, the value of E_A was estimated to be about 52 kJ/mol or $20kT$. This value is about half of the free energy barrier expected for folding of a single protein of the size of monomeric SbpA based on the literature study of folding barrier (Naganathan and Munoz 2005). Therefore, this reduction of folding barrier suggests that the presence of folded tetramers has a catalytic effect of lowering the activation barrier to the formation of new tetramers. The importance of this catalytic effect on tetramer formation at the edge sites around existing crystalline S-layer domains on SLBs is highlighted by the lack of any such tetramer formation events in regions where there exist no crystalline domains. The findings show that the capability of imaging the assembly process *in situ* enabled the discovery of a two-step assembly process in which initial formation of amorphous or liquid-like clusters is required to get to the ordered state, as well as an autocatalytic process in which the emergence of order in S-layer crystals assists the formation of the next ordered growth unit by lowering the energy barrier, thus providing an energetic rationale for the accelerated rate of the tetramer growth.

Shin et al. (2012) extended the study of S-layer assembly by examining the importance of the underlying surface. They followed S-layer assembly on atomically flat mica surfaces and discovered that, in the absence of the lipid bilayers (LBs), two structurally distinct types of S-layer domains are formed. One is identical to that observed on the SLB. The other, however, is about 2–3 nm shorter in height, although it possesses nearly identical crystal symmetry and lattice constant as the ones formed on the SLB (Figure 13.8a). In order to understand the origin of the two crystalline S-layer domains, Shin et al. (2012) used *in situ* AFM to investigate the assembly process. Figure 13.8b through d shows the early growth process. It reveals a simultaneous emergence of two types of clusters, once nucleated. Both types of clusters continuously grew laterally while maintaining their height difference. They discontinued their growth phase when expansion became physically hindered by neighboring domains. After the end of growth phase, the short domains gradually transformed into tall domains by *standing upright*, shown in Figure 13.9. The capability to investigate the system *in situ* enabled the quantitative measurement of the

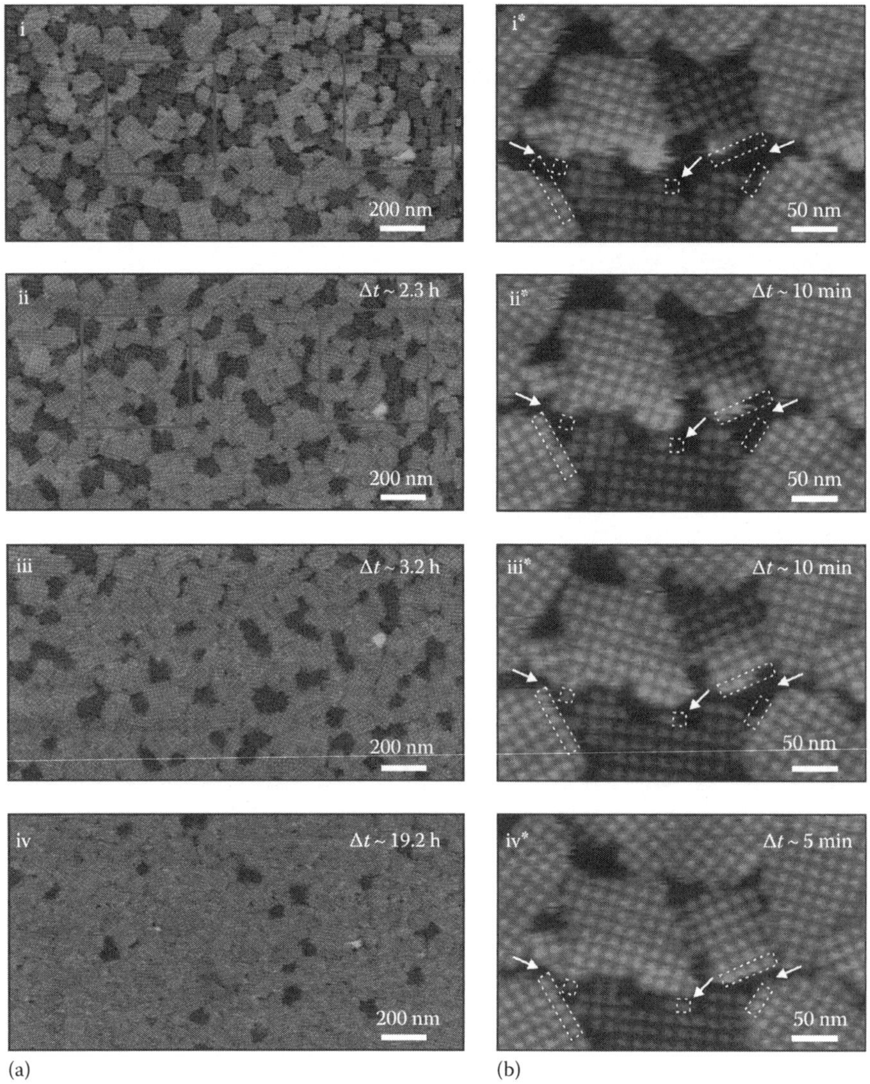

Figure 13.9 (a) Time sequence of *in situ* AFM images of 2D S-layer assembly for the two types of domains (short and tall), as well as the transformation from short to tall domains. Here, Δt indicates time elapsed since collection of images i and i*. S-layers were initially grown in solution at 25°C and further development of the domains was monitored in the same solution condition. Series in (a) shows evolution in distribution of domains in protein solution following incubation time of 4.5 h. (a, i) Initial ratio of tall to short domains by area was 1.3. (a, ii) By Δt = 2.3 h, about 25% of the short domains in (a, i) had transformed into tall domains without any dissolution of S-layer proteins from the domains. (a, iii) and (a, iv) At Δt = 3.2 and 19.2 h, most of the short domains had transformed into tall domains. However, a few short domains still remained, indicating that the transformation does not go to 100% completion until much larger times. Series in (b) shows transformation of a single domain in protein solution following an incubation time of 2.5 h. The transition began at the free edge of the short domain. In between (b, iii*) and (b, iv*), the short domain continued to grow by adding new tetramers at the bottom edge. Images were collected at time (b, i*) 2.5, (b, ii*) 2.7, (b, iii*) 2.8, and (b, iv*) 2.9 h. (Reproduced with permission from Shin, S.H., Chung, S., Sanii, B., Comolli, L.R., Bertozzi, C.R., and De Yoreo, J.J., Direct observation of kinetic traps associated with structural transformations leading to multiple pathways of S-layer assembly. *Proc. Natl. Acad. Sci. USA*, 109(32), 12968–12973, 2012, doi: 10.1073/Pnas.1201504109. Copyright 2012 National Academy of Sciences, U.S.A.)

number density of nuclei of the tall and short phases, from which one could extract the relative magnitude of the energy barriers to nucleation of each phase. The relative number of short and tall nuclei is given by

$$\frac{N_S}{N_T} = \exp\left[-\frac{(E_S - E_T)}{kT}\right] \quad (13.3)$$

where N_i and E_i are the fractions of nuclei clusters and activation energy barriers, respectively subscripts S and T refer to short and tall phase.

From the analysis of the data in Figure 13.8b through d, the difference between the two energy barriers ($\Delta = E_S - E_T$) was found to be about 1.6 kJ/mol or $0.7kT$ (Shin et al. 2012). This difference is much smaller than the energy barrier to creating new tetramers either through new tetramer formation at the edge sites around existing crystalline domains or isolated S-layer folding (Chung et al. 2010).

Figure 13.9 shows the transformation process from short to tall domains. Analysis of the kinetics of transformation from the short to the tall phase revealed that, in contrast to the small energy difference in the probability of nucleating the short and tall phases, the energy barrier to transformation is quite large

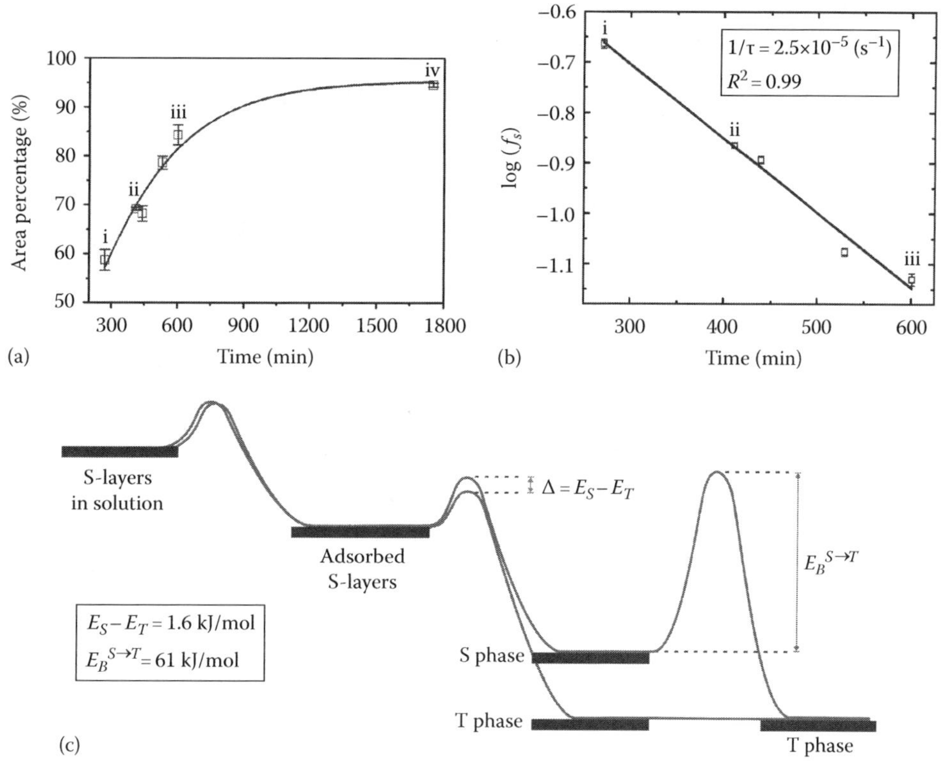

Figure 13.10 (a and b) Kinetics of short-to-tall S-layer domain transformation on bare mica and (c) proposed energy diagram. (b) The plot shows that the ratio of short-to-tall domains (f_s) versus time exhibits a simple exponential decay. The best fit is shown as a solid line in (b). From these data, the energy barrier to transformation of a domain from short to tall was found to be about ≅61 kJ/mol. Labeling of the data point (i–iv) corresponds to the image in Figure 13.9 on which the analysis was performed. Times at which images were captured are (i) 4.5, (ii) 6.8, (iii) 10.0, and (iv) 29.2 h. (c) Proposed energy diagram of S-layer assembly from the monomer in solution to the crystal on mica. Energy barriers to formation of the two types of domains only differ by 1.6 kJ/mol. However, the energy barrier to transformation from one to the other is ≅38 times larger. These energy barriers are qualitatively presented along with two pathways during assembly. (Reproduced with permission from Shin, S.H., Chung, S., Sanii, B., Comolli, L.R., Bertozzi, C.R., and De Yoreo, J.J., Direct observation of kinetic traps associated with structural transformations leading to multiple pathways of S-layer assembly. *Proc. Natl. Acad. Sci. USA*, 109(32), 12968–12973, 2012, doi: 10.1073/Pnas.1201504109. Copyright 2012 National Academy of Sciences, U.S.A.)

(Figure 13.10c). The time dependence of the relative number of short and tall nuclei should be given by

$$\frac{N_S(t)}{N_S(t=0)} = \exp\left[-\frac{t}{\tau}\right] \tag{13.4}$$

$$\frac{1}{\tau} = f \exp\left[-\frac{E_B}{kT}\right] \tag{13.5}$$

The number of short domains at any given time is $N_S(t)$ where $N_S(t = 0)$ is initial number of short domains before transformation. $1/\tau$ is a characteristic rate constant, which is related to f the characteristic attempt frequency associated with conformational fluctuation of S-layer proteins and E_B is the energy barrier to transformation. Figure 13.10b shows that the fraction of short domains indeed follows a simple exponential dependence. Assuming that the transformation of a single domain is initiated by a single tetramer (see Figure 13.9b; a movie of the transformation process can be found in the SI of the reference (Shin et al. 2012)) and using a physically appropriate value of f, the energy barrier to transformation of a domain from short to tall was found to be approximately 61 kJ/mol or $25kT$ (Shin et al. 2012). In the context of an activation energy barrier, the barrier of relaxation from high-energy domains into low-energy domains is almost as big as the folding barrier of individual S-layer proteins.

High-resolution AFM height images (Figure 13.8a) of the short and tall domains revealed nearly identical lattice spacing. However, image analysis in Figure 13.11 showed clear topological differences in the submolecular details of the tetramers in two domains. For example, the tall domain (Figure 13.11a and c) consists of tetrameric units that are more compact within the lateral plane of the domain and exhibit loop-like regions that are not apparent in the tetramers of the short domain (Figure 13.11b and d), which expose some portions of the proteins in interstitial regions that are not evident in the tall domains. Therefore, these combined results from high-resolution *in situ* AFM suggest that the transformation from high-energy, less stable short to low-energy, tall domains is due to a slight conformational transition that is equivalent to folding of under portions of the S-layer proteins in order to produce a less extended and more compact structure.

The findings of these two studies of S-layer assembly on biomimetic and inorganic surfaces highlight the benefit of *in situ* AFM imaging, in probing the assembly of protein matrices. These experiments reveal the complex nature of the nucleation and growth pathway of protein assembly, exhibit the importance of kinetic traps in deciding that pathway, and

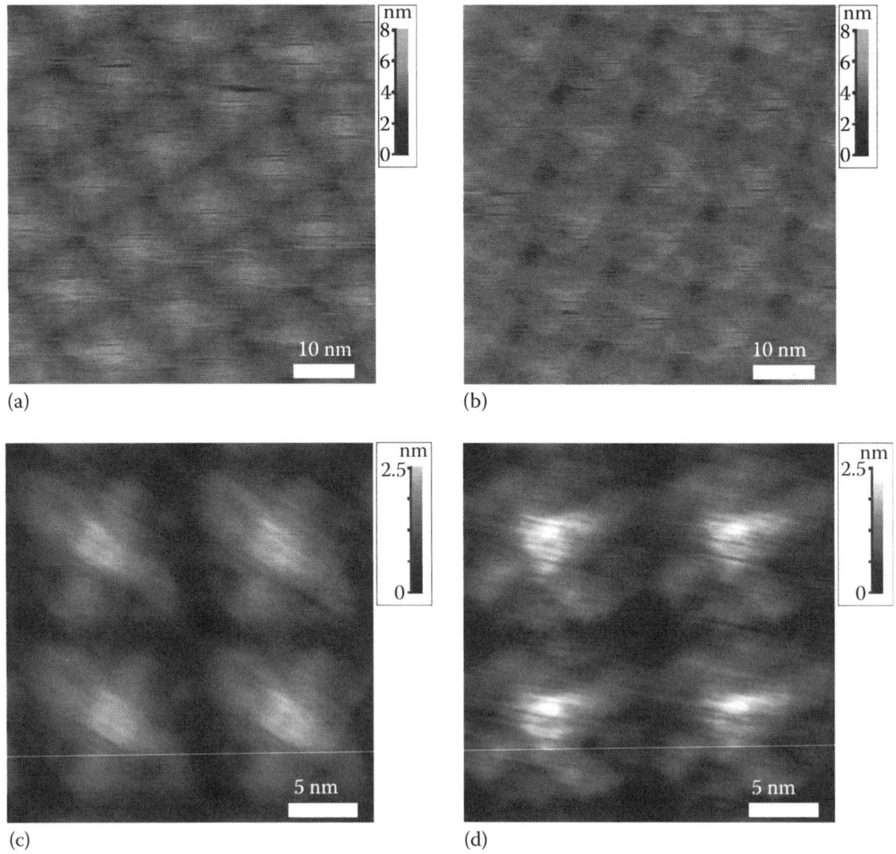

Figure 13.11 High-resolution AFM images of tall (a, c) and short (b, d) domains: (a) and (b) show unprocessed AFM images. (c) and (d) show high-resolution height images of four tetramers from tall and short domains that reveal distinct topological differences ((c) and (d) were processed through correlation averaging). Both (a) and (b) were imaged in the same solution with low imaging force (≅85 pN). Full *Z* scale is 8 nm for (a) and (b) and 2.5 nm for (c) and (d). (Reproduced with permission from Shin, S.H., Chung, S., Sanii, B., Comolli, L.R., Bertozzi, C.R., and De Yoreo, J.J., Direct observation of kinetic traps associated with structural transformations leading to multiple pathways of S-layer assembly. *Proc. Natl. Acad. Sci. USA*, 109(32), 12968–12973, 2012, doi: 10.1073/Pnas.1201504109. Copyright 2012 National Academy of Sciences, U.S.A.)

provide quantitative values of energy barriers that influence the rate of assembly shown in Figure 13.10c. They also allow us to recognize the resemblance between the dynamics of individual protein folding and the assembly of proteins into extended ordered structures that require the individual components to undergo conformational changes. When individual proteins transform from an unstructured state into the final equilibrium state, the concept of a folding funnel characterized by a large number of configurations associated with potential states higher in energy than the final state is often invoked (Figure 13.12) (Dill and Chan 1997; Gruebele 2009). The potential walls of the folding funnel are not smooth, so the resulting bumps and valleys define the kinetic traps where the proteins unveil non-equilibrium structures for extended periods of time. These transient structures can be disordered *molten globules* or partially ordered intermediates as illustrated in Figure 13.12 (Radford 2000). These *in situ* AFM studies of the S-layer assembly process suggest that the concept of the folding funnel for individual proteins can be equally applied to assembly of extended ordered protein structures. Starting with rapid collapse into clusters of monomers, the S-layer system explores many different configurations before transforming into the final ordered states. However, a large percentage of the configurational trajectories drive the system temporarily into a kinetic trap consisting of a high-energy, less-ordered structure from which it slowly relaxes

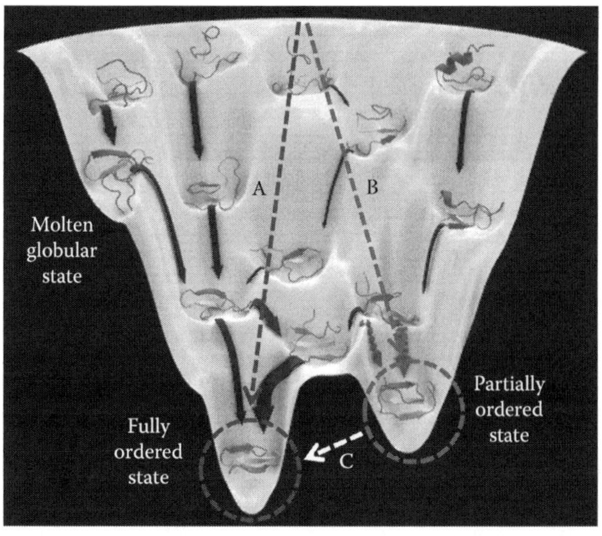

Figure 13.12 Protein folding funnel (Gruebele 2009; Noe et al. 2009) showing unfolded, molten globule, partially ordered state and fully ordered state. Barriers to forming partially ordered (passage B) and fully ordered state (passage A) from molten globule are small; however, barrier to transformation from partially ordered to fully ordered states is large (passage C). (De Yoreo, J.J., Chung, S., and Friddle, R.W.: In situ atomic force microscopy as a tool for investigating interactions and assembly dynamics in biomolecular and biomineral systems. Advanced Functional Materials. 2013. 23. 2525–2538. Copyright 2013 Wiley-VCH Verlag GmbH & Co. KGaA. Reproduced with permission.)

into the low-energy equilibrium structure. Indeed, the language used to describe the protein folding funnel is well suited to this picture of S-layer assembly, with the amorphous liquid-like clusters investigated in the study by Chung et al. (2010) representing the multiprotein equivalent of the *molten globules* and the high-energy domains constituting the *partially ordered, folding intermediates*. Because extended protein architectures with long-range order normally require conformational changes from the monomeric structures, this picture obtained from *in situ* AFM study may be generally applicable to the phenomenon of protein assembly.

13.4 CONCLUSION AND FUTURE OUTLOOK

We have discussed the importance of protein matrix formation in biomineralization processes and the advantages of using *in situ* AFM techniques to investigate the dynamics and underlying mechanisms of matrix assembly. The example of S-layer assembly presented in this chapter provides a concrete demonstration of this approach. This system highlights the experimental challenges associated with imaging protein systems with molecular resolution and provides a guide for overcoming those challenges. In particular, creating a substrate upon which the matrix will assemble, finding solution conditions for which assembly occurs on a time scale compatible with the microscope, minimizing the loading force while maximizing tip sharpness using silicon tips on silicon nitride cantilevers, and optimizing the set point parameters of the AFM were necessary to observe assembly of this soft system at molecular resolutions.

The results show how one can obtain information about both assembly pathways and energetics. In the case of the S-layer, the results give a molecular-scale picture of a multistage assembly pathway that begins with condensation of monomers into liquid-like clusters. These clusters eventually transform into CCs, after which they grow into an extended 2D lattice through formation of new tetramers at the cluster edges. Moreover, the presence of the ordered lattice of tetramers reduces the energy barrier to the formation of new tetramers. Because of the conformational changes required to assemble S-layer matrices from monomeric species, assembly is autocatalytic and the emergence of order promotes the growth of the ordered matrix. This kind of *in situ* study will be essential for probing both the dynamics of matrix assembly and the energetics of matrix-directed mineralization.

Looking toward the future, in order to expand the ability of *in situ* AFM to provide a deep level of understanding about the dynamics and energetics of assembly in biological systems, there are two main features that must be improved. The first one is imaging speed. A typical image collection time for tapping-mode AFM is on order of 10^2 s due to the low resonance frequency of cantilevers in aqueous environments. In the past few years, there have been significant developments in this direction, and we are now seeing the advent of an era of high-speed AFM, in which imaging rates will routinely be 100–1000 times faster (i.e., ~ a few tens of frames/s) than for conventional AFM (Ando 2012; Ando et al. 2001; Kodera et al. 2010). This capability will enable investigation of important dynamic processes in matrix assembly and matrix-directed mineralization. For example, conformational changes associated with protein folding and matrix assembly occur too fast for conventional AFM to resolve, but these new high-speed machines should be able to capture these processes. This capability will also open up new possibilities for studying nucleation events during matrix-directed mineralization.

The second limitation of conventional AFM that must be overcome to expand its impact is the magnitude of the minimum force that must be applied to obtain an image. This is particularly critical for advancing our ability to image cell membranes at molecular resolution. This has not been possible to date, because the large amplitude of AFM cantilever's Brownian vibrations in fluid environments requires the use of large drive amplitudes in order to obtain a signal that is above the noise level (Sanii and Ashby 2010). Furthermore, oscillation of the cantilever in fluid incurs some disadvantages associated with damping of the cantilever motion by the fluid, resulting in decreased quality factor (Q) of the cantilever and a consequent reduction in the resolution of the image. New developments that remove the fluid-induced vibrations of the cantilever are needed to improve the spatial resolution and provide us with a molecular-scale view of cell membrane structure and dynamics.

REFERENCES

Addadi, L., D. Joester, F. Nudelman, and S. Weiner. 2006. Mollusk shell formation: A source of new concepts for understanding biomineralization processes. *Chemistry—A European Journal* 12(4): 981–987. doi: 10.1002/chem.200500980.

Ando, T. 2012. High-speed atomic force microscopy coming of age. *Nanotechnology* 23(6): 062001–062027. doi: 10.1088/0957-4484/23/6/062001.

Ando, T., N. Kodera, E. Takai, D. Maruyama, K. Saito, and A. Toda. 2001. A high-speed atomic force microscope for studying biological macromolecules. *Proceedings of the National Academy of Sciences of the United States of America* 98(22): 12468–12472.

Bilan, M.I. and A.I. Usov. 2001. Polysaccharides of calcareous algae and their effect on the calcification process. *Russian Journal of Bioorganic Chemistry* 27(1): 2–16. doi: 10.1023/A:1009584516443.

Binnig, G., C.F. Quate, and C. Gerber. 1986. Atomic force microscope. *Physical Review Letters* 56(9): 930–933. doi: 10.1103/Physrevlett.56.930.

Braissant, O., G. Cailleau, C. Dupraz, and A.P. Verrecchia. 2003. Bacterially induced mineralization of calcium carbonate in terrestrial environments: The role of exopolysaccharides and amino acids. *Journal of Sedimentary Research* 73(3): 485–490. doi: 10.1306/111302730485.

Cherny, I. and E. Gazit. 2008. Amyloids: Not only pathological agents but also ordered nanomaterials. *Angewandte Chemie-International Edition* 47(22): 4062–4069.

Chiti, F. and C.M. Dobson. 2009. Amyloid formation by globular proteins under native conditions. *Nature Chemical Biology* 5(1): 15–22. doi: 10.1038/Nchembio.131.

Chung, S., S.H. Shin, C.R. Bertozzi, and J.J. De Yoreo. 2010. Self-catalyzed growth of S layers via an amorphous-to-crystalline transition limited by folding kinetics. *Proceedings of the National Academy of Sciences of the United States of America* 107(38): 16536–16541. doi: 10.1073/Pnas.1008280107.

De Yoreo, J.J., S. Chung, and M.H. Nielsen. 2013. The dynamics and energetics of matrix assembly and mineralization. *Calcified Tissue International* 93(4): 316–328. doi: 10.1007/s00223-013-9707-9.

De Yoreo, J.J. and P.G. Vekilov. 2003. Principles of crystal nucleation and growth. *Biomineralization* 54: 57–93. doi: 10.2113/0540057.

Dill, K.A. and H.S. Chan. 1997. From Levinthal to pathways to funnels. *Nature Structural Biology* 4(1): 10–19. doi: 10.1038/Nsb0197-10.

Du, C., G. Falini, S. Fermani, C. Abbott, and J. Moradian-Oldak. 2005. Supramolecular assembly of amelogenin nanospheres into birefringent microribbons. *Science* 307(5714): 1450–1454.

Engelhardt, H. 2007. Are S-layers exoskeletons? The basic function of protein surface layers revisited. *Journal of Structural Biology* 160(2): 115–124.

Erlandsson, R., G.M. Mcclelland, C.M. Mate, and S. Chiang. 1988. Atomic force microscopy using optical interferometry. *Journal of Vacuum Science & Technology A-Vacuum Surfaces and Films* 6(2): 266–270. doi: 10.1116/1.575440.

Fang, P.A., J.F. Conway, H.C. Margolis, J.P. Simmer, and E. Beniash. 2011. Hierarchical self-assembly of amelogenin and the regulation of biomineralization at the nanoscale. *Proceedings of the National Academy of Sciences of the United States of America* 108(34): 14097–14102. doi: 10.1073/Pnas.1106228108.

Geisse, N.A. 2009. AFM and combined optical techniques. *Materials Today* 12(7–8): 40–45.

Giuffre, A.J., L.M. Hamm, N. Han, J.J. De Yoreo, and P.M. Dove. 2013. Polysaccharide chemistry regulates kinetics of calcite nucleation through competition of interfacial energies. *Proceedings of the National Academy of Sciences of the United States of America* 110(23): 9261–9266. doi:10.1073/Pnas.1222162110.

Gruebele, M. 2009. How to mark off paths on the protein energy landscape. *Proceedings of the National Academy of Sciences of the United States of America* 106(45): 18879–18880. doi: 10.1073/Pnas.0910764106.

Habraken, W.J., J. Tao, L.J. Brylka, H. Friedrich, L. Bertinetti, A.S. Schenk, A. Verch et al. 2013. Ion-association complexes unite classical and non-classical theories for the biomimetic nucleation of calcium phosphate. *Nature Communications* 4: 1507. doi: 10.1038/ncomms2490.

Horber, J.K.H. and M.J. Miles. 2003. Scanning probe evolution in biology. *Science* 302(5647): 1002–1005. doi: 10.1126/Science.1067410.

Hu, Q., M.H. Nielsen, C.L. Freeman, L.M. Hamm, J. Tao, J.R.I. Lee, T.Y.J. Han, U. Becker, J.H. Harding, P.M. Dove, and J.J. De Yoreo. 2012. The thermodynamics of calcite nucleation at organic interfaces: Classical vs. non-classical pathways. *Faraday Discussions* 159: 509–523. doi: 10.1039/C2fd20124k.

Jager, I. and P. Fratzl. 2000. Mineralized collagen fibrils: A mechanical model with a staggered arrangement of mineral particles. *Biophysical Journal* 79(4): 1737–1746.

Johnson, J.E. 2008. Multi-disciplinary studies of viruses: The role of structure in shaping the questions and answers. *Journal of Structural Biology* 163(3): 246–253.

Kodera, N., D. Yamamoto, R. Ishikawa, and T. Ando. 2010. Video imaging of walking myosin V by high-speed atomic force microscopy. *Nature* 468(7320): 72–76. doi: 10.1038/Nature09450.

Kopanski, J.J. 2007. Scanning capacitance microscopy for electrical characterization of semiconductors and dielectrics. In *Scanning Probe Microscopy—Electrical and Electromechanical Phenomena at the Nanoscale*, eds. S. Kalinin and A. Gruverman, pp. 88–112. Springer Science, New York.

Lopez, A.E., S. Moreno-Flores, D. Pum, U.B. Sleytr, and J.L. Toca-Herrera. 2010. Surface dependence of protein nanocrystal formation. *Small* 6(3): 396–403. doi: 10.1002/Smll.200901169.

Lowenstam, H.A. and S. Weiner. 1989. *On Biomineralization*. Oxford University Press, Oxford, U.K.

Mann, S. 2008. Life as a nanoscale phenomenon. *Angewandte Chemie-International Edition* 47(29): 5306–5320.

Minne, S.C., S.R. Manalis, and C.F. Quate. 1995. Parallel atomic force microscopy using cantilevers with integrated piezoresistive sensors and integrated piezoelectric actuators. *Applied Physics Letters* 67(26): 3918–3920. doi: 10.1063/1.115317.

Moll, D., C. Huber, B. Schlegel, D. Pum, U.B. Sleytr, and M. Sara. 2002. S-layer-streptavidin fusion proteins as template for nanopatterned molecular arrays. *Proceedings of the National Academy of Sciences of the United States of America* 99(23): 14646–14651.

Naganathan, A.N. and V. Munoz. 2005. Scaling of folding times with protein size. *Journal of the American Chemical Society* 127(2): 480–481.

Noe, F., C. Schutte, E. Vanden-Eijnden, L. Reich, and T.R. Weikl. 2009. Constructing the equilibrium ensemble of folding pathways from short off-equilibrium simulations. *Proceedings of the National Academy of Sciences of the United States of America* 106(45): 19011–19016. doi: 10.1073/Pnas.0905466106.

Nudelman, F., K. Pieterse, A. George, P.H.H. Bomans, H. Friedrich, L.J. Brylka, P.A.J. Hilbers, G. de With, and N.A.J.M. Sommerdijk. 2010. The role of collagen in bone apatite formation in the presence of hydroxyapatite nucleation inhibitors. *Nature Materials* 9(12): 1004–1009. doi: 10.1038/Nmat2875.

Orgel, J.P.R.O., T.C. Irving, A. Miller, and T.J. Wess. 2006. Microfibrillar structure of type I collagen in situ. *Proceedings of the National Academy of Sciences of the United States of America* 103(24): 9001–9005. doi: 10.1073/Pnas.0502718103.

Radford, S.E. 2000. Protein folding: Progress made and promises ahead. *Trends in Biochemical Sciences* 25(12): 611–618. doi: 10.1016/S0968-0004(00)01707-2.

Rambaran, R.N. and L.C. Serpell. 2008. Amyloid fibrils abnormal protein assembly. *Prion* 2: 112–117.

Rohrer, H., C. Gerber, and E. Weibel. 1982. Surface studies by scanning tunneling microscopy. *Physical Review Letters* 49(1): 57–61.

Salgado, E.N., R.A. Lewis, J. Faraone-Mennella, and F.A. Tezcan. 2008. Metal-mediated self-assembly of protein superstructures: Influence of secondary interactions on protein oligomerization and aggregation. *Journal of the American Chemical Society* 130(19): 6082–6084. doi: 10.1021/Ja8012177.

Sanii, B. and P.D. Ashby. 2010. High sensitivity deflection detection of nanowires. *Physical Review Letters* 104(14). doi: Artn 147203 Doi 10.1103/Physrevlett.104.147203.

Schoen, A.P., D.T. Schoen, K.N.L. Huggins, M.A. Arunagirinathan, and S.C. Heilshorn. 2011. Template engineering through epitope recognition: A modular, biomimetic strategy for inorganic nanomaterial synthesis. *Journal of the American Chemical Society* 133(45): 18202–18207. doi: 10.1021/Ja204732n.

Schultzelam, S. and T.J. Beveridge. 1994. Nucleation of celestite and strontianite on a cyanobacterial S-layer. *Applied and Environmental Microbiology* 60(2): 447–453.

Shenton, W., D. Pum, U.B. Sleytr, and S. Mann. 1997. Synthesis of cadmium sulphide superlattices using self-assembled bacterial S-layers. *Nature* 389(6651): 585–587.

Shin, S.H., S. Chung, B. Sanii, L.R. Comolli, C.R. Bertozzi, and J.J. De Yoreo. 2012. Direct observation of kinetic traps associated with structural transformations leading to multiple pathways of S-layer assembly. *Proceedings of the National Academy of Sciences of the United States of America* 109(32): 12968–12973. doi: 10.1073/Pnas.1201504109.

Shoulders, M.D. and R.T. Raines. 2009. Collagen structure and stability. *Annual Review of Biochemistry* 78: 929–958. doi: 10.1146/Annurev.Biochem.77.032207.120833.

Sleytr, U.B., P. Messner, D. Pum, and M. Sara. 1999. Crystalline bacterial cell surface layers (S layers): From supramolecular cell structure to biomimetics and nanotechnology. *Angewandte Chemie-International Edition* 38(8): 1035–1054.

Stewart, M., T.J. Beveridge, and T.J. Trust. 1986. 2 patterns in the *Aeromonas-salmonicida* a-layer may reflect a structural transformation that alters permeability. *Journal of Bacteriology* 166(1): 120–127.

Tanaka, S., C.A. Kerfeld, M.R. Sawaya, F. Cai, S. Heinhorst, G.C. Cannon, and T.O. Yeates. 2008. Atomic-level models of the bacterial carboxysome shell. *Science* 319(5866): 1083–1086.

Weiner, S., W. Traub, and H.D. Wagner. 1999. Lamellar bone: Structure-function relations. *Journal of Structural Biology* 126(3): 241–255. doi: 10.1006/Jsbi.1999.4107.

14 Transmission electron microscopy in biomineralization research: Advances and challenges

Elia Beniash, Archan Dey, and Nico A.J.M. Sommerdijk

Contents

14.1 INTRODUCTION: BIOMINERALIZATION AND BIOMIMETIC HYBRID MATERIALS

Biominerals are a diverse group of biological materials that include inorganic solids and hierarchical organo-inorganic nanocomposites produced by organisms for a variety of purposes, such as navigation, mechanical support, and defense (Lowenstam and Weiner 1989). Biominerals are produced by organisms in every phylum of every kingdom of life; they vary widely in terms of the chemical composition and structural organization (Lowenstam 1981). According to the most recent inventory, 56 different mineral phases have been described in biominerals so far, and this list is likely to grow (Weiner and Dove 2003). Twenty-five percent of these minerals are amorphous and more than half contain calcium. The organic composition of biominerals is also very diverse, with biomacromolecules, such as proteins and polysaccharides comprising the major fraction of the organic matrix. Despite this diversity, a number of common strategies of biomineralization have been identified that distinguish them from geological mineral forms and synthetic minerals (Lowenstam 1981; Mann 1995; Weiner and Addadi 1997). One of them is the use of bottom-up approaches, in which the material synthesis starts at the atomic and molecular levels, leading to the formation of nanostructured building blocks, which organize into complex hierarchical structures (Weiner and Addadi 1997; Aizenberg et al. 2005; Fratzl and Weinkamer 2007). Interactions between macromolecular assemblies and forming mineral are essential elements of biomineralization processes (Cölfen and Mann 2003; Beniash et al. 2005; Fang et al. 2011a).

These biological strategies to mineralization lead to the formation of materials with intricate morphologies (Aizenberg et al. 2001) and often hierarchical structures (Weiner and Addadi 1997; Aizenberg et al. 2005). While modern man-made materials are often synthesized under extreme temperatures and pressures, biominerals are formed under ambient conditions from abundant, although relatively weak, individual components, yet they are remarkably resilient, due to their unique structural organization and robust molecular interfaces between the mineral and organic phases. As a result, the functional properties of biominerals are often superior to those of geological and man-made counterparts (Figure 14.1) (Mann 1995; Kamat et al. 2000; White et al. 2001; Davis 2004; Munch et al. 2008; Nudelman et al. 2008; Lee et al. 2012).

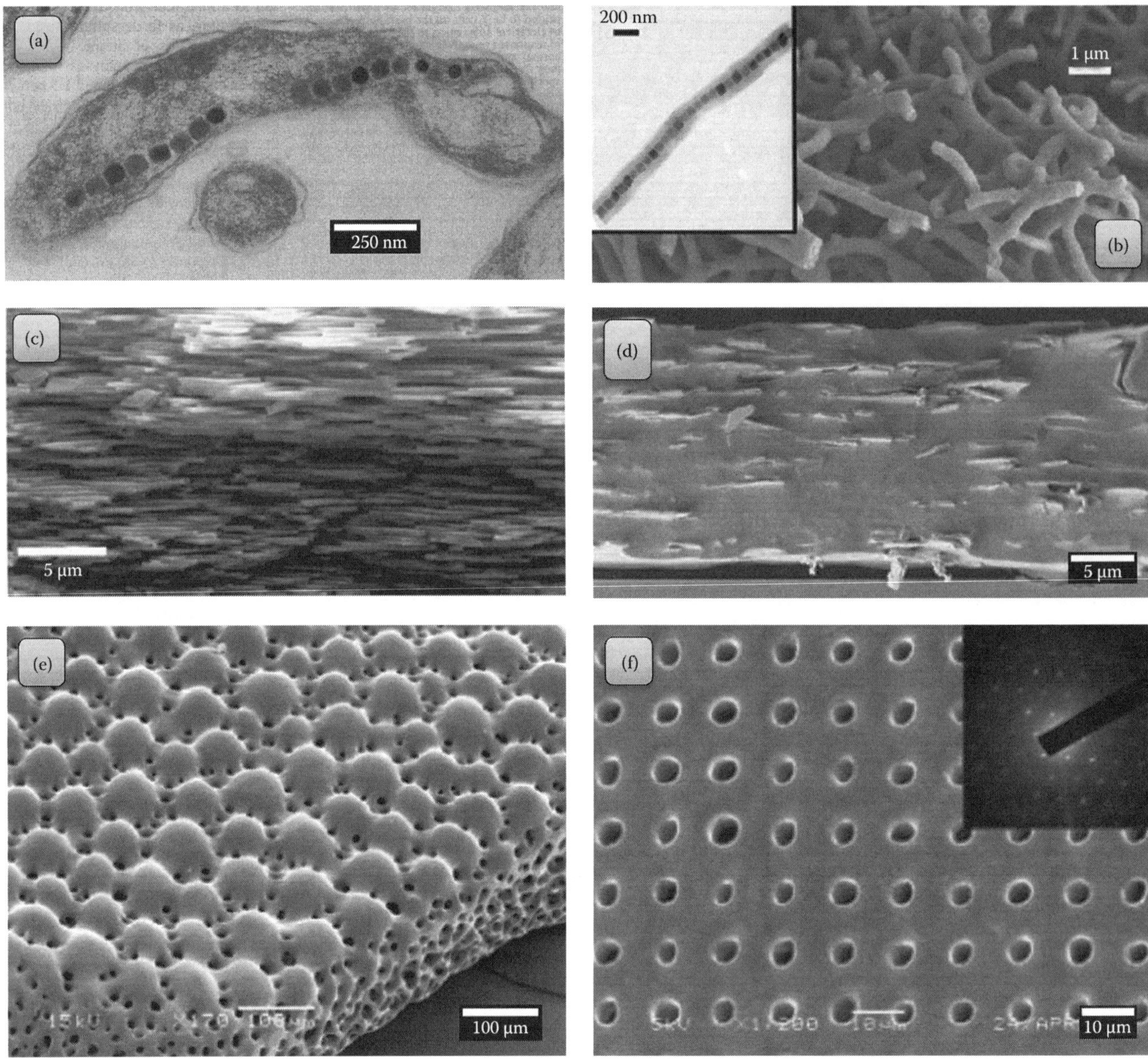

Figure 14.1 Structural comparison of biomineralized materials and their corresponding bioinspired counterparts: (a) TEM micrograph of magnetotactic bacteria, showing chains of magnetite nanocrystals within the cell. (From Frankel, R.B., Blakemore, R.P., and Wolfe, R.S., Magnetite in freshwater magnetotactic bacteria, *Science*, 203(4387), 1355–1356, doi: 10.1126/science.203.4387.1355. Copyright 1979. Reprinted with permission of AAAS.) (b) FESEM and TEM (inset) images of synthetic magnetite chains coated with amorphous carbon. (From Liu, Y. and Qianwang, C., Synthesis of magnetosome chain-like structures, *Nanotechnology*, 19(47), 475603. Copyright 2008. Reprinted by permission from IOP publishing.) (c) SEM image of a cross section of the shell of *Atrina rigida*, showing the nacreous layer, which is composed of aragonite tiles cemented with organic matrices (β-chitin, silklike proteins, and acidic glycoproteins). (Reprinted from *J. Struct. Biol.*, 162(2), Nudelman, F., Shimoni, E., Klein, E., Rousseau, M., Bourrat, X., Lopez, E., Addadi, L., and Weiner, S., Forming nacreous layer of the shells of the bivalves *Atrina rigida* and *Pinctada margaritifera*: An environmental- and cryo-scanning electron microscopy study, 290–300, doi: 10.1016/j.jsb.2008.01.008. Copyright 2008, with permission from Elsevier.) (d) SEM image of a multilayered alumina platelet-reinforced chitosan nanocomposite, inspired by the inner nacreous layer. (From Bonderer, L.J., Studart, A.R., and Gauckler, L.J., Bioinspired design and assembly of platelet reinforced polymer films, *Science*, 319(5866), 1069–1073, doi: 10.1126/science.1148726. Copyright 2008. Reprinted with permission AAAS.) (e) SEM micrograph of a part of the skeleton of a brittle star *Ophiocoma wendtii* (Ophiuroidea, Echinodermata). The entire structure (the mesh and the array of microlenses) is composed of a single calcite crystal used by the organism for mechanical and optical functions, and (f) SEM image of a sample micropatterned single calcite crystal fabricated using self-assembled monolayers. (From Aizenberg, J., Muller, D.A., Grazul, J.L., and Hamann, D.R., Direct fabrication of large micropatterned single crystals, *Science*, 299(5610), 1205–1208, doi: 10.1126/science.1079204. Copyright 2003. Reprinted with permission of AAAS.)

It is no surprise that these unique characteristics of biominerals as well as their formation processes attracted attention of researchers from different fields, interested in exploring the principles of biomineralization for the design of novel materials for a variety of applications (Heuer et al. 1992; Huo et al. 1994; Mann and Ozin 1996; Stupp and Braun 1997; Antonietti et al. 1998; Sarikaya 1999; Ball 2001; Hartgerink et al. 2001; Green et al. 2002; Sanchez et al. 2005; Palmer et al. 2008; Xia and Jiang 2008; Bhushan 2009; Aizenberg 2010; Wong et al. 2011). Biominerals became a source of inspiration for the development of materials (Bonderer et al. 2008; Kim et al. 2011) and structures (Kroger et al. 2002; Aizenberg et al. 2003; Liu and Qianwang 2008) with highly controllable and specialized properties. The exquisite control over polymorph selection, orientation, morphology, and organization of the inorganic phases in biominerals can be attributed to their intimate interaction with the directing organic phase (Sommerdijk and de With 2008). Hence, mimicking biominerals and their remarkable properties requires us to tune the interplay between organic (macro)molecules and mineral to orchestrate the effective coassembly of the organic and inorganic components (Aizenberg and Fratzl 2009; Sommerdijk and Cölfen 2010).

Our ability to develop bioinspired materials requires a detailed understanding of the biomineralization processes with a special focus on the dynamic interplay between biomacromolecules and developing mineral phases. However, due to the complexity and the hybrid nature of the biomineralization systems, it is often difficult to study these processes *in vivo* at the atomic, molecular, and nanometer scales (Dey et al. 2010b). Hence, a combination of *in vivo* studies and a reductionist approach relying on simplified *in vitro* model systems (Addadi et al. 1987; Mann et al. 1988) that replicate certain aspects of biomineralization is often employed to gain further mechanistic insights into these processes and is invaluable to devise protocols for bioinspired material synthesis.

As has been mentioned in the preceding paragraphs, biomineralization is a bottom-up process, and the interactions between organic and mineral phases at the atomic and molecular levels are key elements of any biomineralization system. Hence, the studies of the biomineralization processes rely on characterization techniques that can provide information on the protein–mineral interactions at atomic and molecular levels, that is, at nanometer and angstrom scales. Among the characterization techniques used in biomineralization studies, transmission electron microscopy (TEM) plays a critical role. This chapter provides an overview of the current state of the art of the application of TEM in biomineralization research with the focus on emerging approaches.

14.2 MECHANISMS OF BIOMINERALIZATION

Although structural, compositional, and mechanical properties of biominerals vary dramatically, there are a number of key principles that are common in a majority of biomineralization systems. One of these principles is compartmentalization that states that biomineralization processes generally take place in spatially confined environments in which chemical and physical conditions can be controlled by the cells (Weiner and Addadi 2011). Such compartments can be created by cell membranes, both intracellular as in foraminifera (Bentov et al. 2009) and diatoms (Sumper 2002), membrane-delineated extracellular domains, as in echinoderms (Märkel et al. 1989; Beniash et al. 1999) and sponges (Ledger and Jones 1977), and by combination of cell membranes and structural components of extracellular matrix as in mollusks (Wilt 2005; Nudelman et al. 2007) and vertebrates (Beniash 2011). Another important feature found in a variety of biomineralization systems is the presence of transient metastable mineral phases (Weiss et al. 2002; Beniash et al. 2009; Mahamid et al. 2010; Weiner and Addadi 2011; Gong et al. 2012). Metastable mineral phases are often associated with vesicular intracellular transport of mineral ions (Beniash et al. 1999; Mahamid et al. 2011b), providing a highly effective way to sequester and transfer large quantities of mineral ions in a solid concentrated form to the mineralization sites. Another important function of this process is to prevent cytotoxicity by isolating and maintaining physiological concentrations of free ions in the cytosol. Another advantage of the metastable phases in biomineralization is their stepwise kinetic transformation into more stable crystalline phases that allows for more precise regulation of the process in comparison with thermodynamic pathway (Cölfen and Mann 2003). Third and probably the most important principle of biomineralization is that all aspects of this process are tightly controlled by the cells, and studies of various biomineralizing systems, including research on coccolithophores (Marsh 2004; Ozaki et al. 2007), sea urchins (Beniash et al. 1997; Politi et al. 2004; Ma et al. 2009), diatoms (Kröger et al. 1999; Kröger and Sumper 2005), limpets (Towe and Lowenstam 1967; Lowenstam and Weiner 1985), zebrafish (Mahamid et al. 2008, 2010), and magnetotactic bacteria (Frankel et al. 1979; Dunin-Borkowski et al. 1998; Scheffel et al. 2006; Faivre et al. 2007), revealed that in all cases, complex cellular mechanisms are involved. This cellular control is exerted either directly by physically defining the mineralization site, ion sequestration, and transport or indirectly via specialized macromolecules that control all aspects of the mineralization process including mineral phases formed (Aizenberg et al. 1996; Belcher et al. 1996; Falini et al. 1996; Kwak et al. 2009), shape (Albeck et al. 1993, 1996; Moradian-Oldak et al. 1998), and organization (Fantner et al. 2005; Deshpande and Beniash 2008; Nudelman et al. 2010) of the mineral particles. Furthermore, the macromolecules are extremely important for maintaining unique mechanical properties of mature mineralized tissues (Wang and Weiner 1998; Baldassarri et al. 2008; Bajaj and Arola 2009).

14.3 EXPERIMENTAL APPROACHES IN BIOMINERALIZATION STUDIES

Due to the unique characteristics of biominerals that combine biological organic and inorganic components, studies of these systems are technically extremely challenging. Historically, methodologies for studies of biological systems and minerals have developed separately, and often classical characterization techniques used in these fields are not directly applicable to studies of biomineralization systems. Studies of soft highly hydrated biological organic phases and hard mineral phases typically require very different preparation and analytical techniques that are hard or impossible to combine. These challenges are quite significant when studying mature biominerals, but they pale in comparison to the challenges of studying dynamic processes of biomineral formation (Weiner and Addadi 2011). A number of methodological strategies have been developed over the years.

Artifacts are common issue in experimental research, regardless of the subject; however, due to the unique composite nature of biominerals, this problem is especially prominent in this field. Hence, the research question has to be carefully considered, and the research methods have to be chosen to minimize possible artifacts while answering this particular research question.

It is also important if possible to use a combination of analytical techniques to address the same question, corroborate the results, and reduce the number of potential artifacts.

Finally, where it is often impossible to gain a true understanding of biomineralization processes from *in vivo* experiments alone, the combination with *in vitro* approaches can often help the interpretation of the *in vivo* results.

14.4 TRANSMISSION ELECTRON MICROSCOPY

TEM is one of the most utilized techniques in the study of biomaterials; it is a versatile platform that can be used for a variety of applications. It allows to traverse almost seamlessly several scales from angstroms to tens of microns in one sample, which is a major advantage for studies of hierarchical nanostructured systems such as biominerals. It can be operated in several optical configurations such as bright- and dark-field mode as well as diffraction and can be used in combination with other analytical tools such as energy dispersive x-ray spectroscopy (EDS) and electron energy loss spectroscopy (EELS) (Plate et al. 1992; Gilis et al. 2011), which provide chemical information at the nanoscale (Figure 14.2).

14.4.1 PRINCIPLES OF OPERATION AND LIMITATIONS OF TEM

TEM is similar in terms of its optical configuration to a light microscope; however, instead of a light source generating a beam of photons that passes through a series of glass lenses, the electron microscope is equipped with an electron gun that generates an electron beam and electromagnetic lenses that focus it. The image is formed by the electron beam passing through the sample. The electrons pass through, get scattered, or absorbed by the sample, generating a projection on a fluorescent screen, a photographic film, or a charge-coupled device (CCD) chip, which represents a 2D electron density map of the sample. The first electron microscope was built in 1931 by the German scientist and engineer Ernst Ruska, who received a Nobel Prize for his revolutionary invention in 1986. Since then, TEM has become a major research tool, playing a critical role in our understanding of living cells, materials, and supramolecular assemblies; it will be fair to say that the field of nanotechnology would not exist without TEM. As the performance of microscopes is bound to the diffraction limit, the smaller wavelength of electrons allows us in the future to get thousand times better resolution than with a light microscope. As the magnetic lenses presently used in electron microscopes do not yet approach diffraction-limited performance, we are not yet able to take full advantage of the shorter wavelength of the electron (a few picometers). Nevertheless, through the recent implementation of aberration correction, a sub-angstrom point resolution has been demonstrated (Batson et al. 2002). In addition, it has been demonstrated that TEM can be used for time-resolved investigations (Browning et al. 2012; Yuk et al. 2012) (vide infra). As TEM provides a combination of temporal and spatial resolution that is not achievable with other techniques, it opens a unique window into bio(mimetic) mineralization studies. Furthermore,

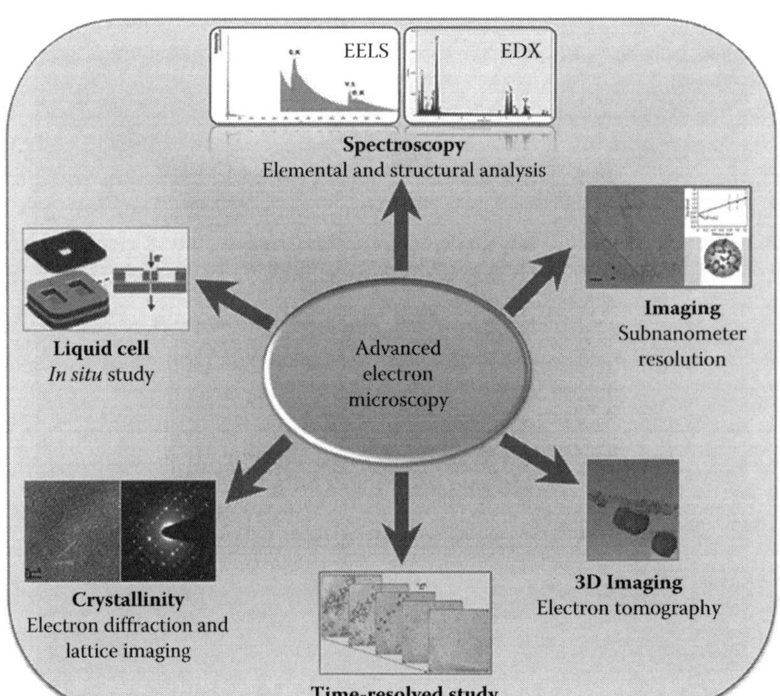

Figure 14.2 Schematic of various components in advanced EM toolbox.

as it combines imaging with other techniques such as diffraction, EDS measurements, and EELS, it allows to simultaneously obtain morphological together with structural and chemical information on the same part of the sample. For a comprehensive description of the principles of TEM, a curious reader can refer to textbooks (Williams and Carter 1996; Rieimer and Kohl 2008). This chapter will focus on specific aspects of TEM pertaining to biomineralization.

It is important to note that unlike photons that can penetrate through samples that are tens and even hundreds of microns thick without a significant loss, the penetration depth of electrons is much smaller and the thickness of a TEM sample is typically limited to 200–300 nm, unless an ultrahigh-voltage microscope is used (Landis and Song 1991). Since mineralized samples are denser than typical biological samples made of organics, it poses further limitation on their maximum thickness. A number of preparation techniques are used to produce thin electron-transparent samples, including ultramicrotomy (Landis et al. 1993; Beniash et al. 2006, 2009) and ion beam milling (Benzerara et al. 2011), which both can cause significant artifacts. Ultramicrotomy of mineralized tissues is technically difficult, especially in highly mineralized tissues such as dental enamel or single-crystalline skeletal elements. Furthermore, artifacts due to compression and fracture of the sample can occur. Ion beam milling is used for highly mineralized tissues such as mollusk shells (Su et al. 2002), yet an ion beam can affect unstable mineral phases and organics and hence is not applicable for specimens of forming tissues and tissues with high organic content.

Another major limitation of TEM is related to the fact that unlike photons, electrons strongly interact with gases, leading to beam scattering, ionization, and other undesirable effects that lead to a decrease in image quality and can damage the microscope. Therefore, the TEM column is kept under high vacuum and typical TEM samples are dry; however, drying can introduce major artifacts. It may cause changes in structural organization of biomacromolecules and their assemblies and can lead to the transformation of metastable mineral phases. One problem specific to biominerals is that drying causes shrinkage of the organic fraction, while the mineral volume remains unchanged, which can lead to major morphological changes in the samples. There are ways to minimize drying artifacts or eliminate them all together, such as fast freezing or freeze-drying of the samples, cryo-TEM, or use of liquid cells, which will be discussed in the following sections.

14.4.2 BRIGHT-FIELD ROOM TEMPERATURE TEM

Bright-field imaging is the most popular mode of TEM, and as mentioned in the preceding paragraphs, it is similar in terms of its optical configuration to bright-field light microscopy. It has been used for studies of biomineralization both *in vivo* (Robinson and Bishop 1950; Robinson 1952; Robinson and Watson 1952, 1955) and *in vitro* (Glimcher et al. 1957) since the 1950s. These very early studies have had a dramatic impact on our understanding of mineralized tissues, specifically bone, by characterizing mineral crystallites, identifying mineralized collagen fibrils as the building blocks of bone and dentin, and providing the first glimpse into the structural organization of these fibrils. During the 1950s, the major types of mineralized tissue preparation, such as fine particle suspensions, shadow casting, and ultramicrotomy of fully mineralized and demineralized samples, were established. Variations of these methods are still used in biomineralization studies today (Weiner et al. 1997, 1999; Beniash et al. 1999, 2009). At the same time, the limitations of electron microscopy (EM), related to dehydration and fixation of the samples and to beam damage, have been also recognized in those early years. One of these major limitations of routine TEM, which has led to heated debate over several decades, is concerned with the preservation of transient mineral phases. The pioneering studies of collagen mineralization conducted in the 1950s suggested that, although the main mineral phase in bone is carbonated apatite (Robinson and Bishop 1950), there are other mineral phases such as octacalcium phosphate (OCP) (Glimcher et al. 1957) and amorphous calcium phosphate (ACP) (Robinson and Watson 1955). Yet the presence of these transient mineral phases in bone has not been clearly established until recently (Weiss et al. 2002; Crane et al. 2006; Mahamid et al. 2008, 2010) and was a subject of heated debate in the 1960s–1980s (Boskey 1997). The main reason for this long-lasting controversy was the lack of methodologies that can minimize drying artifacts and beam damage. Recent progress in cryo-TEM, low-dose TEM, and other methods has enabled researchers to look at the bone mineral in its unaltered hydrated state, which led to a definite identification of transient ACP in bones. It is worth noticing that although cryo-EM is a gold standard, it is not always possible to apply this method to mineralized tissues. Other methods, such as freeze-drying of the samples, using cooled holders, and low-dose microscopy, are used to minimize mineral phase transition artifacts (Beniash et al. 1999, 2009).

In 1950 and 1960s, TEM was also applied to studies of dental enamel (Glimcher et al. 1965) and mollusk shells (Gregoire 1957), and with the increasing access of the scientific community to this technology, it became a routine technique in biomineralization research by the early 1970s.

Over the years, researchers have developed *in vitro* experimental approaches to better understand roles of individual biomacromolecules and small organelles such as matrix vesicles in mineralization (Heywood and Eanes 1992; Xu et al. 1998; Hunter et al. 2001). Although in earlier works, the products of mineralization reactions typically were transferred onto the TEM grids, recently new methods have been developed that allow mineralization reactions directly on the TEM grids (Hartgerink et al. 2001; Deshpande and Beniash 2008; Pichon et al. 2008; Pouget et al. 2009; Deshpande et al. 2010, 2011; Dey et al. 2010a; Fang et al. 2011a). This approach minimizes possible artifacts associated with sample transfer and allows to capture very fine details of the mineralization process.

14.4.3 HIGH-RESOLUTION ELECTRON MICROSCOPY, LATTICE IMAGING

High-resolution electron microscopy (HREM), a variation of bright-field TEM, is generally conducted at magnifications of 250 K or higher and with subnanometer resolution. The main application of HRTEM is the generation of so-called lattice images from crystalline materials. In lattice imaging, contrast is generated by interference of electron waves with the crystal lattices, allowing direct observation of the internal crystal structure. It can be used for structural studies of individual crystals, providing information regarding mineral phases, strains, and impurities. Cuisinier and colleagues have used this technique for studies of crystals in forming enamel. They have observed for the first time structural differences at the atomic level between the inner core of the crystals and the outer layers (Cuisinier et al. 1992a), which might reflect differences between the initial enamel mineral formed via assembly of prenucleation clusters (Fang et al. 2011a) and subsequent crystal growth via classical crystallization pathway. They have also observed very small nanometer-sized crystalline particles in proximity to the ribbonlike crystals in forming enamel (Cuisinier et al. 1992b, 1993). The authors have proposed that these particles were stabilized by self-assembled protein cages and acted as building blocks of the larger enamel crystals. Although at the time of discovery, the exact nature of these particles remained unclear and the possibility that these particles were simply preparation artifacts could not be ruled out, recent studies indicating that enamel crystals can form via assembly of protein-stabilized prenucleation clusters provide support for these earlier observations (Fang et al. 2011a). Another interesting application of this technique was presented in the article by Xu et al. (1998). In this work, the authors synthesized a macroscopic calcite film via amorphous calcium carbonate (ACC) precursor. Using HREM, they were able to clearly distinguish between amorphous and crystalline areas and to show that calcite was formed from ACC via phase transformation. Furthermore, they demonstrated that calcite crystals were all oriented with their (001) axes parallel to the porphyrin monolayer, suggesting that the monolayer regulated orientation of the forming crystals. In recent years, this method, in combination with electron diffraction, has been used to provide support to the idea that biominerals, behaving as single crystals in x-ray diffraction (XRD), might be highly aligned aggregates of nanocrystals (Figure 14.3a through d) (Sethmann et al. 2005, 2006; Oaki et al. 2006; Sethmann and Worheide 2008; Li and Huang 2009; Seto et al. 2012).

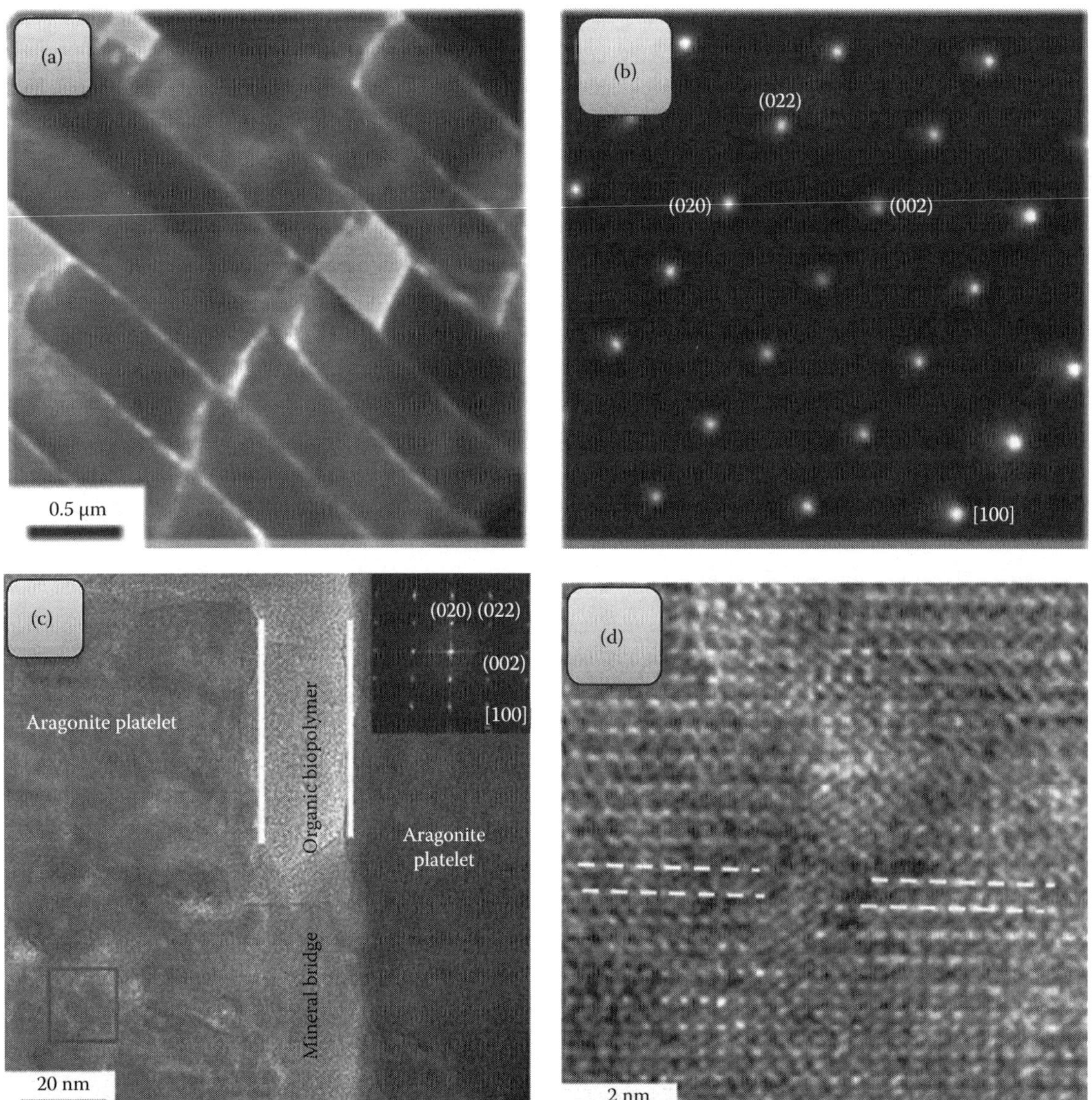

Figure 14.3 (a) TEM image of nacre's cross section, showing a brick wall–like architecture with aragonite platelets sandwiched with organic biopolymer interlayers. (b) Electron diffraction pattern of aragonite platelets, exhibiting single-crystal diffraction characteristics. (c) Magnified TEM image of two aragonite platelets with an organic biopolymer interlayer and a mineral bridge. (d) HRTEM image of the boxed area in (c), showing a screw dislocation lining two adjacent particles. (Reprinted with permission from Li, X. and Huang, Z., Unveiling the formation mechanism of pseudo-single-crystal aragonite platelets in nacre, *Phys. Rev. Lett.*, 102(7), 075502. Copyright 2009 by the American Physical Society.)

14.4.4 ELECTRON TOMOGRAPHY

An inherent limitation to TEM is that it produces 2D projections of 3D objects, resulting in the overlapping of multiple features that cannot be discerned individually. Electron tomography (ET) allows TEM to overcome this limitation (Friedrich et al. 2009). In this technique, images of the specimen are acquired at different tilt angles (often ~100 images between −65° and +65°) that are subsequently aligned in a stack and then reconstructed to the 3D object, revealing detailed information on structure, morphology, and 3D spatial organization within the specimen (Nudelman et al. 2010). Recently, fast computation methods have helped to make ET a powerful tool for the 3D structural determination of both synthetic and biological materials and assemblies. ET studies of mineralized tissues were pioneered by Landis and coworkers in the early 1990s, who published a series of studies on normal and pathological collagen mineralization in bones and mineralized tendons (Landis and Song 1991; Landis et al. 1993, 1996a,b; Landis 1995). These studies provided very detailed 3D information on the structure and mineralization dynamics of collagen fibrils. Furthermore, ET played a crucial role in determining that the mineral organization in peritubular dentin, which lacks collagen, is similar to that in mineralized collagenous fibrils, with platelike crystals organized into parallel arrays with their *c*-axes co-aligned (Weiner et al. 1999). ET in combination with HREM has also been used to study relationships between dentin and enamel crystals at the dentino-enamel boundary (Diekwisch et al. 1995; Fang et al. 2011b). In the last decade, this methodology has matured and is used widely for studies of biominerals and biomimetic mineralized materials. A number of groups have applied ET for studies of formation and structural organization of magnetosomes in several species of magnetotactic bacteria (Kasama et al. 2006; Byrne et al. 2010; Katzmann et al. 2011). Gries et al. (2009) successfully utilized ET to study the co-orientation of stacked aragonite platelets in nacre. They observed mineral bridges, which connect the stacked platelets and enable a transfer of the crystallographic orientation between adjacent plates during growth (Figure 14.4a). ET has been also used for structural studies of calcium phosphate polycrystalline-mineralized nanostructures grown in the presence of biomineralization proteins (Deshpande et al. 2010; Nudelman et al. 2010; Beniash et al. 2011) and their mimics (Dey et al. 2010a; Nudelman et al. 2010) and calcite single crystals grown

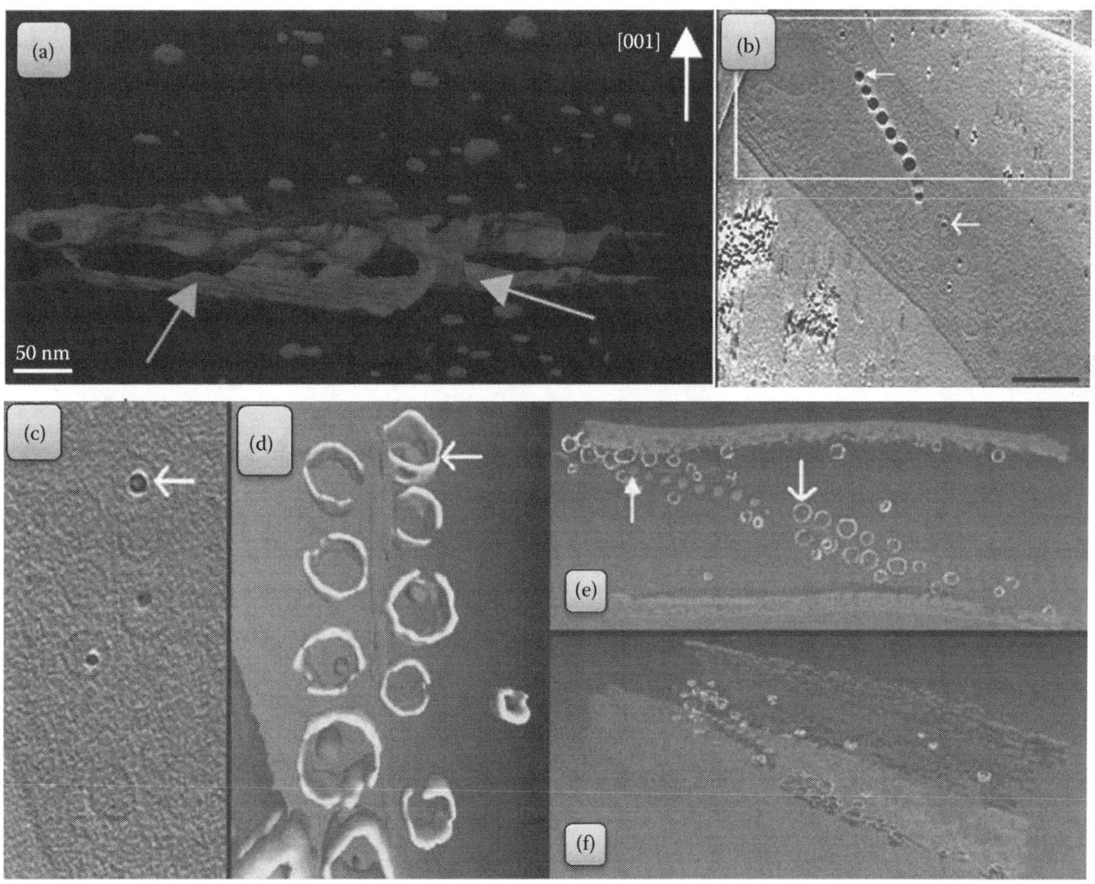

Figure 14.4 (See color insert.) (a) Tomogram of aragonite platelets in nacre, showing connectivity via mineral bridges. Blue: aragonite; Red: organic matrix; Yellow arrow: pointing at a mineral bridge. (Reprinted from *Ultramicroscopy*, 109(3), Gries, K., Kröger, R., Kübel, C., Schowalter, M., Fritz, M., and Rosenauer, R., Correlation of the orientation of stacked aragonite platelets in nacre and their connection via mineral bridges, 230–236. Copyright 2009, with permission from Elsevier.) (b–f) Cryo-ET of wild-type cells. (b) 3D reconstruction: superimposed x–y slices along the z-axis through a typical tomogram. (c) Magnified x–y slice of vesicles, which are either empty or contain growing crystals (arrow) connected to the filamentous structure. (d) Surface-rendered representation of a segment of the cell showing vesicles (yellow), magnetite crystals (red), and a filamentous structure (green). (e and f) Complete 3D visualization of the whole cell in different views (membrane in blue). (Reprinted by permission from Macmillan Publishers Ltd. *Nature*, Scheffel, A., Gruska, M., Faivre, D., Linaroudis, A., Plitzko, J.M., and Schüler, D., An acidic protein aligns magnetosomes along a filamentous structure in magnetotactic bacteria, 440(7080), 110–114, copyright 2006.)

in polysaccharide gels (Li et al. 2009). One of the major recent advances in ET is its application in cryo-EM of vitrified samples, which will be discussed in the following sections.

14.4.5 SELECTED AREA ELECTRON DIFFRACTION

Selected area electron diffraction (SAED) is a unique technique, which can be used to obtain crystallographic information from particles that are only a few nanometers across (Suvorova and Buffat 1999; Dwyer et al. 2007), two orders of magnitude better than the spatial resolution of microbeam XRD. The transition from the bright field to diffraction mode and back is achieved simply by changing the electron beam configuration in the microscope column. It is technically very easy to select an area of interest in the sample either by means of selective (diffraction) aperture or by concentrating the beam and obtain the diffraction pattern. Therefore, electron diffraction is a convenient and extremely powerful tool for identification of crystalline phases (Belcher et al. 1996; Ma and Liu 2009), for differentiation of crystalline from amorphous materials (Politi et al. 2004), for determination of crystal orientation and organization at the nanoscale (Gries et al. 2009), as well as for assessing structural relationships between organic matrices and mineral crystals (Weiner et al. 1983; Plate et al. 1992, 1994). It is widely used in studies of biominerals as well as *in vitro* biomineralization systems. One of the first applications of SAED was to study the organization of crystallites in mineralized collagen fibrils (Glimcher 1959) and enamel rods (Glimcher et al. 1965). Although the mineral phases of these tissues had been identified in the 1940s by means of XRD, structural studies of the organization of the crystallites at the nanoscale and their relations with the organic matrix only became possible using a combination of bright-field TEM and electron diffraction. Our current understanding of biominerals as hierarchical nanocomposites has its origins in these early studies. Electron diffraction is still a very important tool in biomineralization studies. It is widely used for determination of mineral phases as well as for the crystallographic texture of natural and synthetic biominerals at the nano- and microscales. Electron diffraction is capable of extracting structural information to much higher resolution (subpicometer regime) than imaging techniques because it is virtually immune to the resolution-limiting aberrations of the objective lens. However, as with any technique, SAED has its limitations and use of complimentary analytical approaches can help to better understand structural properties of biominerals. For example, using HRTEM, Li and Huang (Li and Huang 2009) have obtained new data, challenging the long-standing paradigm, based on the diffraction data, that each mineral tablet in nacre is a single crystal of aragonite. The results of their studies suggest that the aragonite platelets in nacre are highly organized assemblies of aragonite nanoparticles and hence are so-called mesocrystals (Figure 14.3a and b) (Cölfen and Antonietti 2005, 2008).

14.4.6 CRYOGENIC TEM

Biomineralization, as with the majority of biological processes, takes place in hydrated environments, while the dehydration and electron beam exposure required for routine TEM may cause major changes in both organic and mineral phases—especially metastable ones. This limitation of routine room temperature (RT) TEM has been on the minds of the researchers for many years, and a number of approaches to minimize the effects of dehydration such as anhydrous processing and freeze-drying of the samples and use of low-dose microscopy were employed (Landis et al. 1977; Landis and Glimcher 1978; Beniash et al. 1999). Nevertheless, the possibility of major artifacts caused by dehydration was hard to rule out. These limitations of the conventional RT TEM are especially detrimental for the studies of initial stages of biomineral formation due to the very delicate nature of macromolecular assemblies and their interactions with developing mineral phases.

The introduction of cryogenic TEM (cryo-TEM), which involves vitrification of aqueous samples and studying them at liquid nitrogen temperatures, has helped to overcome the limitations of conventional RT TEM and has led to major advances in our understanding of biomineralization processes. This technique, which has been around since the early 1980s, has emerged as a major tool for structural molecular biology (Dubochet et al. 1988; Frank et al. 1995; Medalia et al. 2002) and studies of surfactants, complex fluids, and supramolecular assemblies (Chretien and Wade 1991; Hartgerink et al. 1996; Talmon 1996; Dong et al. 2007) since it enables researchers to observe samples in their near-native hydrated state (Adrian et al. 1984; Al-Amoudi et al. 2004). In the 1990s and early 2000s, a handful of cryo-TEM studies of organic matrices of mineralized tissues have been published (Ziv et al. 1996; Weiner et al. 1997; Beniash et al. 2000; Levi-Kalisman et al. 2001). However, cryo-TEM in the mid-2000s has become a method of choice for studies of biomineralization *in vitro*, allowing for the first time a glimpse at the very early stages of the mineralization process (Pichon et al. 2008; Dey et al. 2010a; Nudelman et al. 2010; Pouget et al. 2010; Yuwono et al. 2010; Fang et al. 2011a). Recent developments in instrumentation and automation made cryo-TEM a reliable and accessible analytical tool leading to its expansion into biomineralization and other fields (Frederik and Sommerdijk 2005; Dey et al. 2010b; Nudelman et al. 2010).

Cryo-TEM relies on the ultrafast vitrification of thin, generally, aqueous films by plunge freezing in an appropriate coolant (generally melting ethane, $T_m = -183°C$). This yields a solid amorphous film in which all processes are arrested and hence provides a snapshot of the specimen in its solution state. The specimens containing vitreous ice film are electron transparent; therefore, the embedded nanostructures can be imaged at low dose to minimize beam damage. Reaction kinetics can be monitored by vitrification of samples at different time points, and reactions can even be performed on cryo-TEM grids to access short reaction times (Frederik and Sommerdijk 2005; Fang et al. 2011a). Thus, cryo-TEM provides unique possibilities for quasi *in situ*, time-resolved analysis of bio(mimetic) mineralization.

Recently, the combination of tomography and cryo-TEM (cryo-electron tomography [cryo-ET]) has emerged as a very powerful technique to investigate nanoscale biological structures in 3D. This was applied to the study of biomineralization by Scheffel et al. (2006) who investigated the subcellular structures of magnetotactic bacteria (Figure 14.4b through f). These authors demonstrated how an acidic protein aligns magnetosomes—vesicles containing a small magnetite crystal—along a

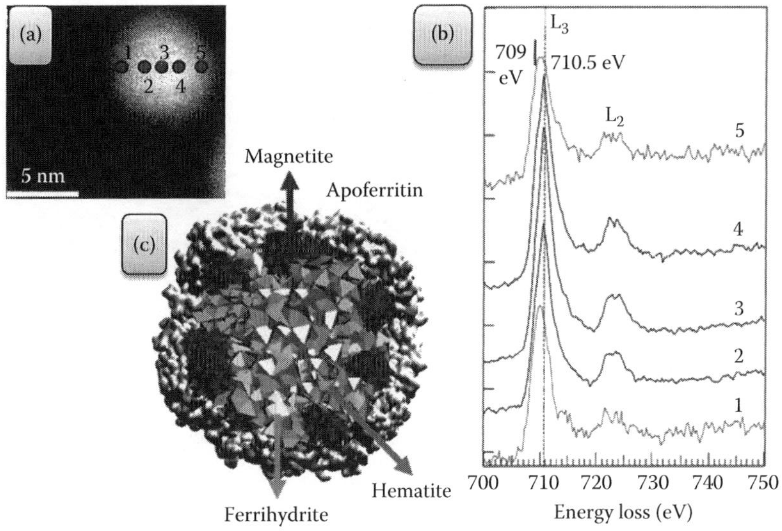

Figure 14.5 (a) A representative image of horse spleen ferritin particle, containing 500 iron atoms (points mark the locations where the electron energy loss spectra were acquired). (b) EEL spectra in the energy loss region of $FeL_{2,3}$ edges. The surface of the particle corresponds to 1 and 5 (the thinner regions), and 2, 3, and 4 to inner bulk sites. (c) Schematic core-shelled ferritin iron core structure. (Reprinted with permission from Galvez, N., Fernalndez, B., Salnchez, P., Cuesta, R., Ceolil n, M., Clemente-Leon, M., Trasobares, S., Lolpez-Haro, M., Calvino, J.J., Stephan, O., and Dominguez-Vera, J.M., Comparative structural and chemical studies of ferritin cores with gradual removal of their iron contents, *J. Am. Chem. Soc.*, 130(25), 8062–8068. Copyright 2008 American Chemical Society.)

filamentous structure, thereby creating the magnetic moment that the bacteria use to orient themselves in the earth magnetic field.

In principle, cryo-TEM can be combined with other analytical techniques such as EDS and EELS, although these techniques often require higher beam intensity that can damage the vitrified samples. Nevertheless, the use of cryo-EELS (Kim et al. 2006), low-dose selected area diffraction (LDSAED), and cryo-EDS was already demonstrated in soft nanoparticle literature (Pichon et al. 2008; Pouget et al. 2009; Dey et al. 2010a) (see the next section for more details). It is our conviction that the combination of cryo-TEM with such analytical techniques for structural and elemental analysis holds great promise for the study of the early stages of bio(mimetic) mineralization.

14.4.7 SPECTROSCOPY AND ELEMENTAL ANALYSIS

The toolbox of TEM can be further extended by incorporating spectroscopic techniques such as EDS and EELS by which we can analyze the elemental composition of the sample under investigation. For example, Lenders et al. (2012) recently presented the use of EDS to confirm a quantitative Mg incorporation in high-magnesian amorphous $CaCO_3$ and in high-magnesian calcite. EELS—most sensitive for the lighter elements—is generally considered being complementary to EDS, which is more sensitive for heavy elements (Leapman and Hunt 1991), but EELS is more efficient (Thomas 2009). Furthermore, using EELS, it is in principle possible to determine not only the composition and thickness of the examined specimens but also the atomic environment (local structure), oxidation state, and bond distances of elements, information that can be derived from the fine structure (near edge and extended edge) of the loss peak (Calvert et al. 2005; Gregori et al. 2006; Griesshaber et al. 2009). This was demonstrated for ferritin, an iron storage protein consisting of a spherical polypeptide shell (apoferritin) surrounding a 6 nm inorganic core of the hydrated iron oxide ferrihydrite. A magnetic mineral was synthesized within the nanodimensional cavity of horse spleen ferritin by the use of controlled reconstitution conditions (Meldrum et al. 1992). Spatially resolved EELS was used to probe the iron oxidation state distribution across individual nanoparticles of horse spleen ferritins from which iron had been gradually removed (Figure 14.5a). In particular, a qualitative analysis of the $FeL_{2,3}$ edge fine structures (Figure 14.5b) allowed to distinguish between iron(II) and iron(III), inferring that the ferritin iron core consisted of a polyphasic structure of ferrihydrite, magnetite, and hematite (Figure 14.5c) (Galvez et al. 2008).

14.5 CURRENT FRONTIERS IN TEM

14.5.1 HIGH-RESOLUTION LATTICE IMAGING IN CRYOGENIC TEM

Although high-resolution lattice imaging is often used in materials science, for 2D cryo-TEM, the resolution has long been limited to the nanometer regime. The main reason for this is that up to recently, cryo-TEM was predominantly used for the investigation of beam-sensitive soft matter and life science specimens, where the electron dose restrictions—with the concomitant low signal/noise ratios—and also the absence of crystallinity prohibited the recording of high-resolution information. So far, in soft matter and life science cryo-TEM, subnanometer resolution was only achieved when a large number of copies of biomolecules or biomolecular complexes could be averaged. However, in the more recent application of cryo-TEM for the investigation of (biomimetic) mineralization, there are several factors that allowed the recording of images with subnanometer resolution (Pouget et al. 2009; Baumgartner et al. 2013; Habraken et al. 2013). The first reports on the application of high-resolution cryo-TEM (cryo-HRTEM) appeared in the mid-2000s (Wang et al. 2004; Balmes et al. 2006), and this technique has now become a powerful tool in studies of early nucleation events. It is now possible to achieve lattice resolution in cryo-TEM, which was used in visualization

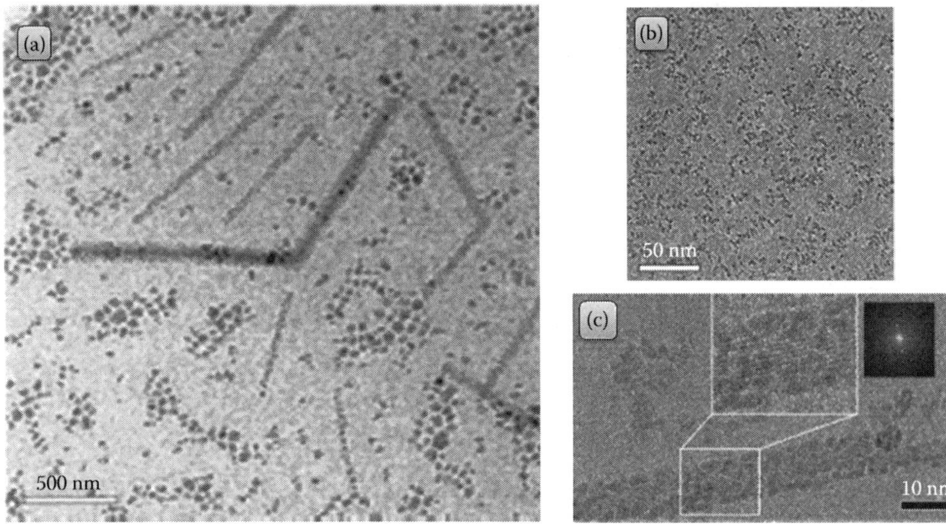

Figure 14.6 Cryo-TEM study of goethite mesocrystal formation from ferrihydrite: (a) Goethite mesocrystals after 10 days. (b) Freshly prepared suspension of dialyzed ferrihydrite. (c) Cryo-HRTEM images of mesocrystal intermediates showing lattice fringes that correspond to (110) planes (inset). (Reprinted with permission from Yuwono, V.M., Burrows, N.D., Soltis, J.A., and Penn, R.L., Oriented aggregation: Formation and transformation of mesocrystal intermediates revealed, *J. Am. Chem. Soc.*, 132(7), 2163–2165, doi: 10.1021/ja909769a. Copyright 2010 American Chemical Society.)

of multistep crystallization mechanisms for iron oxides. Penn and coworkers reported a time-resolved cryo-HRTEM study on the formation of goethite mesocrystals from ferrihydrite in aqueous suspension of goethite (Figure 14.6) (Yuwono et al. 2010), while Baumgarner et al. very recently demonstrated a new primary particle–based mechanism for the formation of magnetite (Baumgartner et al. 2013).

14.5.2 LIQUID CELL TEM

The most recent development in TEM that will likely revolutionize the field and provide a wealth of information on the most basic mechanisms of biomineralization is the advent of *in situ* TEM liquid cell. Although a detailed understanding of mineralization kinetics is best achieved by real-time observations, until recently, the only reliable way to carry out such studies was the liquid cell AFM (Teng et al. 2000; Qiu et al. 2004; Elhadj et al. 2005). However, this technique is limited to studies of surfaces. After the development of modern liquid cell TEM by Ross et al. (de Jonge and Ross 2011; Williamson et al. 2003), this technique was more recently used to visualize nanoparticle assembly *in situ* with nanoscale spatial resolution and subsecond temporal resolution (Li et al. 2012; Yuk et al. 2012). Alivisatos and coworkers (Zheng et al. 2009) observed the growth trajectories of individual colloidal platinum nanocrystals in solution by using a liquid cell. Although the nucleation of the nanoparticles that form through the reduction of Pt^{2+} precursor solution by the electron beam could not be recorded, the growth process was successfully imaged (Williamson et al. 2003). Here, the electrochemical formation of the nanoparticle was of great advantage as it allowed the process to be initiated when the liquid cell is already in the microscope, ready for imaging. The study revealed that the formation of these Pt nanocrystals was more complex than was previously envisioned. The nanocrystals were found to grow both by monomer attachment and by particle coalescence, combining both classical and nonclassical pathways to form monodisperse nanoparticles.

Very recently, Li et al. (2012) have published a study, which is an absolute tour de force in liquid cell *in situ* TEM. They used liquid cell TEM to study growth of iron oxyhydroxide (six-line ferrihydrite) nanoparticles by oriented attachment with atomic resolution (Li et al. 2012). The video-rate imaging at high resolution enabled them to track dynamics of the particle attachment and orientation. These experiments revealed that the nanoparticles randomly interact with each other due to the Brownian motion and electrostatic interactions until a certain level of co-alignment is achieved, triggering oriented attachment of the particles with complete lattice alignment (Figure 14.7).

It is clear that in the near future, such *in situ* TEM studies will also help us to understand the dynamics of bio(mimetic) mineral formation and the details of nonclassical crystal growth mechanisms that involve self-assembly of mineral clusters and primary nanocrystals, the crystallographic reorganization within assemblies, and the transformation of precursor phases.

14.6 APPLICATION OF ADVANCED TEM TECHNIQUES AND PROGRESS IN BIOMINERALIZATION RESEARCH

Studies of biomineralization processes as well as many other processes that occur at the nanoscale are technology driven, and in the first decade of the twenty-first century, the progress in TEM methodology has led to major discoveries and significant advances in our understanding of the basic biomineralization mechanisms. In this section, we provide an overview of these recent TEM investigations.

14.6.1 MONITORING THE BIOMIMETIC FORMATION OF $CaCO_3$

For many years, Langmuir monolayers have been used as model systems for template-directed $CaCO_3$ mineralization

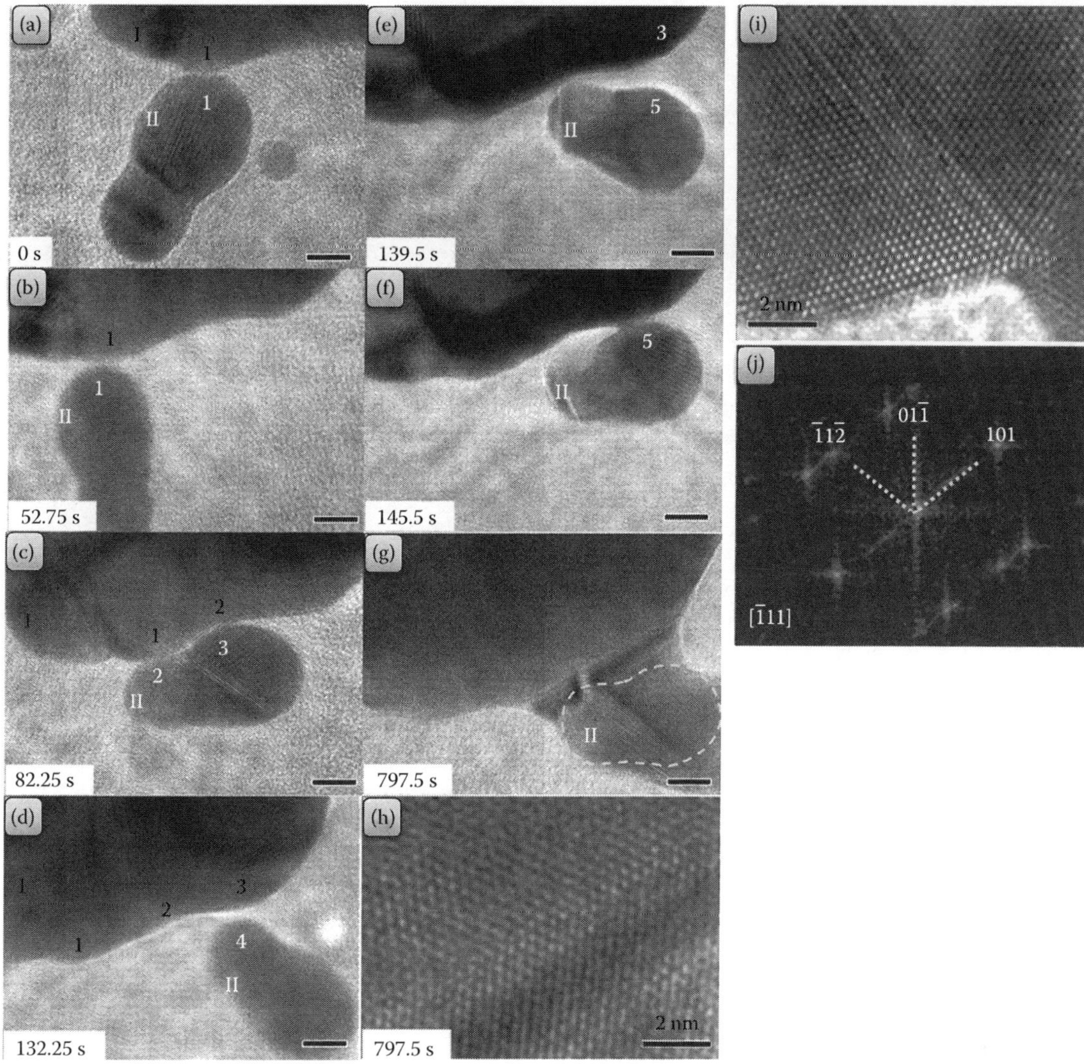

Figure 14.7 (a–g) Sequence of images showing typical dynamics of the attachment process. The surfaces of particles I and II made transient contact at many points and orientations (points 1–1, 1–2, 2–3, and 3–4) before finally attaching and growing together (points 3–5). (h) High-resolution image of interface in (g) showing twin structure (an inclined twin plane). The white dashed line in (g) shows the original boundary of the attached particle. (i and j) High-resolution *in situ* TEM image (i) and fast Fourier transform (FFT) (j) of an interface from another OA event demonstrating formation of a (101) twin interface after attachment. The grain boundary is delineated by a dashed line in (i). Scale bars are 5 nm for (a–g). (From Li, D.S., Nielsen, M.H., Lee, J.R.I., Frandsen, C., Banfield, J.F., and De Yoreo, J.J., Direction-specific interactions control crystal growth by oriented attachment, *Science*, 336(6084), 1014–1018, doi: 10.1126/science.1219643. Copyright 2012. Reprinted with permission of AAAS.)

(Mann et al. 1988). Despite the fact that several time-resolved *in situ* studies have addressed this topic (Berman et al. 1995; Pontoni et al. 2003; Popescu et al. 2007; Kewalramani et al. 2008), none of them was able to reveal unambiguously the interplay of template and growing mineral phase or to show how the template directs the formation of oriented crystals. Recently, we have used cryo-TEM in combination with other techniques such as LDSAED, dark-field imaging, and cryo-ET to address these issues. In these studies, an experimental protocol (Figure 14.8a) was devised to follow calcium carbonate mineralization under a Langmuir monolayer, making use of a glove box extension of the vitrification robot (Pichon et al. 2008). In this glove box, in which temperature and humidity are controlled by the software of the Vitrobot, samples are taken directly from the interface and subsequently vitrified. In this way, the growth process can be arrested at the very early stages of crystal formation. This glove box (Vos et al. 2008), which provides an extension of the workspace for the manipulation of instruments and specimens, has been instrumental in the study of processes at the air–water interface that otherwise would not be accessible to cryo-TEM.

Inside the glove box, a monolayer was prepared from a negatively charged valine-derived surfactant on a 10 mM $CaCl_2$ solution. This monolayer with a suspended solution droplet was deposited on a TEM grid and transferred to the Vitrobot. Here, the excess of liquid was removed by blotting, and the resulting thin film was exposed for short periods of 5–10 min to ammonium carbonate vapor, still at 100% humidity, thereby initiating the $CaCO_3$ formation in the thin $CaCl_2$ film. After the desired reaction time (30 s to 30 min), the sample was plunged into liquid ethane to vitrify the water and to quench the reaction. Cryo-TEM imaging in combination with LDSAED allowed us to observe ACC particles growing from a few tens of nanometers to hundreds of nanometers and then crystallizing to form [00.1]-oriented vaterite. The vaterite in turn transformed to yield

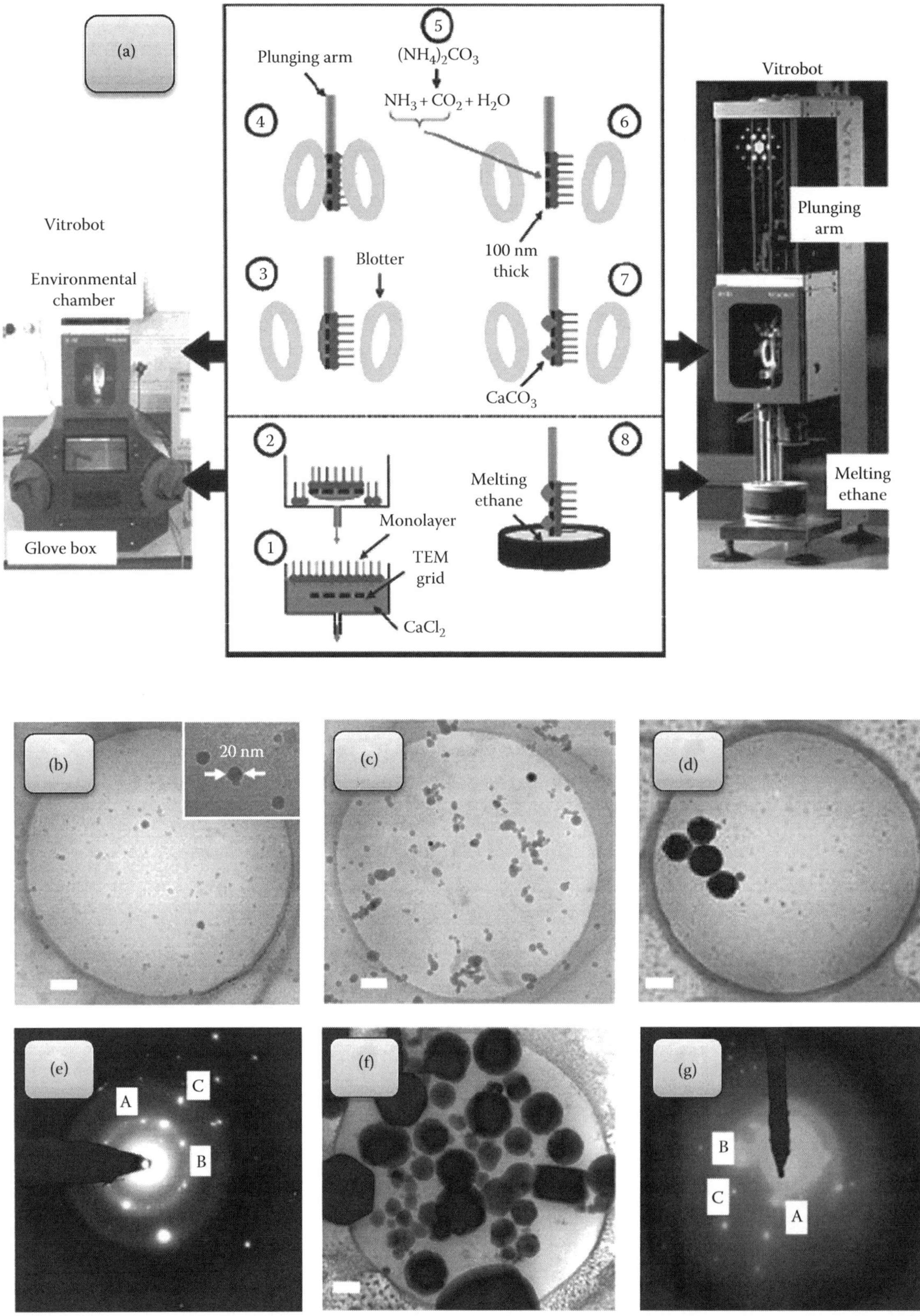

Figure 14.8 (a) Schematic of experimental design for cryo-TEM sample preparation of $CaCO_3$ mineralization under Langmuir monolayers. Step 1: Deposition of the monolayer on a $CaCl_2$ solution. Step 2: Draining of the solution to deposit the monolayer on the TEM grid. Step 3: Grid mounted on the plunging arm and introduction in the environmental chamber of the Vitrobot. Step 4: Blotting of the sample to remove the excess $CaCl_2$ solution. Step 5: Introduction of $(NH_4)_2CO_3$ and its decomposition into NH_3 and CO_2. Step 6: Diffusion of NH_3 and CO_2 into the solution. Step 7: Formation of $CaCO_3$. Step 8: Quenching of the reaction. (b–g) Micrographs and SAED of $CaCO_3$ particles (crystals) at different reaction times: (b) 1 min, (c) 2 min, (d) 3 min, and (f) 10 min. Scale bars are 200 nm. (e) SAED of [00.1]-oriented vaterite. (g) SAED of [10.0] calcite. A, B, C indicating indexed (*hkl*)s of vaterite and calcite crystal structure. (Reprinted with permission from Pichon, B.P., Bomans, P.H.H., Frederik, P.M., and Sommerdijk, N.A.J.M., A quasi-time-resolved CryoTEM study of the nucleation of $CaCO_3$ under Langmuir monolayers, *J. Am. Chem. Soc.*, 130(12), 4034–4040. Copyright 2008 American Chemical Society.)

the final product, [10.0]-oriented calcite. Control experiments with a structurally related neutral oligo(ethylene glycol)-derived surfactant yielded randomly oriented calcite, demonstrating the role of the negatively charged monolayer in the nucleation of the calcite [00.1] face. This role was further confirmed by the observation that ACC formed at higher calcium concentrations could be directly converted to [00.1] calcite—under cryogenic conditions—upon irradiation with the electron beam.

Arresting mineral formation in the very early stages also allowed us to study some interesting details concerning the amorphous phase. First, it was found that the very early 20–50 nm amorphous nanoparticles consisted of two populations of which one disintegrated upon exposure to the electron beam, whereas the other population survived. This observation is in line with other observations that ACC initially forms as a hydrated phase that dehydrates before it crystallizes. In later stages, nanoparticles of ~100 nm in diameter were observed that were largely amorphous but already contained regions that had crystallized into (00.1)-oriented vaterite, as was demonstrated with dark-field imaging (Pouget et al. 2010). This result demonstrates that at the single particle level, the transformation of ACC to vaterite occurs through a direct solid state transformation (Figure 14.8b through i).

Although it is likely that the monolayer directs the transformation of ACC into vaterite, the previously mentioned results do not provide evidence for this, as from the TEM projection images, one cannot locate and assess the location of the monolayer. Moreover, these experiments also do not show whether the monolayer plays a role in the nucleation of the amorphous phase. Therefore, in a subsequent study, we employed cryo-ET to distinguish between the mineralization events taking place in the bulk and the ones occurring at the monolayer surface (Pouget et al. 2009). Monitoring the $CaCO_3$ mineralization from a 9 mM $Ca(HCO_3)_2$ solution under a stearic acid monolayer revealed that the first precipitate consisted of amorphous nanoparticles with a size distribution around 30 nm forming in bulk solution rather than at the monolayer. At later stages, also larger particles were formed but exclusively in contact with the monolayer. At larger sizes, these particles developed randomly oriented vaterite domains that eventually evolved into [00.1]-oriented single-crystalline vaterite. It was concluded that the contact with the stabilizing monolayer allowed the amorphous particles to grow beyond their stable size, reaching dimensions (>70 nm) above which crystalline domains could develop. Also in this stage, apparently the interaction with the monolayer provides a stabilizing effect that promotes the growth of the [00.1]-oriented domains over the other orientations (Figure 14.9d).

Very similar results were obtained in a study where $CaCO_3$ was grown in the presence of ammonium ions (Pouget et al. 2010). Also, these were found to stabilize the (00.1) plane of vaterite, giving rise to hexagonal tablet-like crystals. Again, the first step in the nucleation process was the formation of ~35 nm ACC nanoparticles that represent the first metastable form of solid $CaCO_3$. These subsequently transform into a structurally probably more advanced form of ACC with a particle size of ~70 nm. The latter ones grow further to form particles with diameters >200 nm, for which

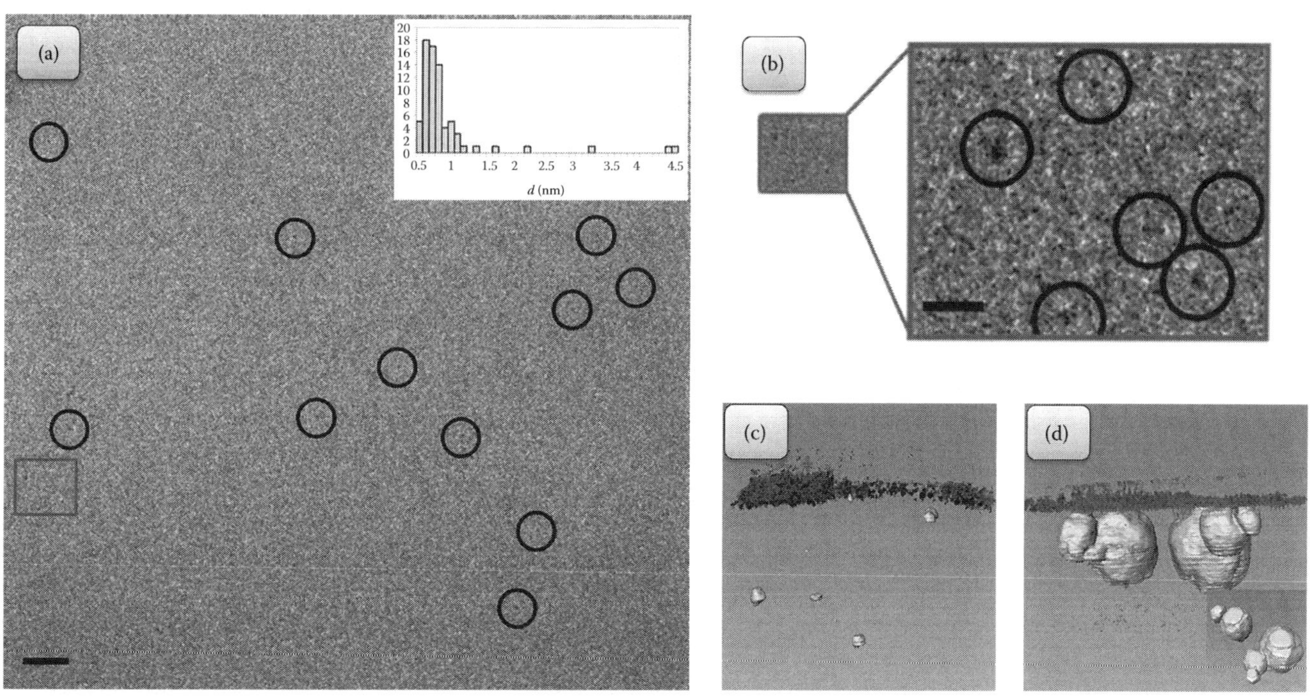

Figure 14.9 (a) Cryo-HRTEM imaging of a fresh 9 mM $Ca(HCO_3)_2$ solution after image processing in which prenucleation clusters are observed. An arbitrary number of clusters are highlighted by circles. Inset: Particle diameter (d) distribution of the prenucleation clusters observed in the cryo-TEM images. Scale bar, 20 nm. (b) Nonfiltered images representing the zone delimited by the square in (a). In the high-magnification image, all particles present are highlighted by circles. Particle sizes below the detection limit of 0.45 nm (three times the pixel size) are considered noise. Scale bar, 5 nm. (c and d) Computer-aided visualization of the tomograms after reaction time of 6 and 11 min, respectively. Please note that CdS hydrophobic nanorods were used to mark the monolayer and the arrow in (c) shows the nanorods marking the monolayer. (From Pouget, E.M., Bomans, P.H.H., Goos, J.A.C.M., Frederik, P.M., de With, G., and Sommerdijk, N.A.J.M., The initial stages of template-controlled $CaCO_3$ formation revealed by cryo-TEM, *Science*, 323(5920), 1455–1458. Copyright 2009. Reprinted with permission of AAAS.)

dark-field imaging demonstrated that they developed crystalline domains within the amorphous matrix. This suggests that also in this case, the development of single crystals occurs through the growth of domains that are stabilized by the interaction with NH_4^+ ions at the expense of the randomly oriented crystalline domains.

Interestingly, cryo-ET in combination with electron diffraction showed that the expression of the vaterite (00.1) face preceded the development of the single crystalline character of the vaterite particles. In fact, the stabilized crystal face appeared to act as a template from which crystallinity develops throughout the particle, in a similar manner as was described earlier for the crystallization under the stearic acid monolayer. This supports the proposal that within the amorphous matrix, crystallization occurs via a solid state dissociation–recrystallization process. Such a transition would require a reorganization and therefore mass transport within the solid phase. Hence, this suggests a solution-like character of the amorphous phase that allows separate nucleation events to take place. It is possible that it is the high degree of hydration in ACC that facilitates ion transport from one domain to the other.

However, the most intriguing result by far was the visualization of the recently discovered prenucleation clusters in a freshly prepared 9 mM $Ca(HCO_3)_2$ solution (Figure 14.9a and b). Previously, evidence from analytical ultracentrifuge experiments had indicated the existence of such clusters as solution species prior to the nucleation of ACC (Gebauer et al. 2008; Meldrum and Sear 2008). Using cryo-HRTEM, clusters with diameters of ~0.7 nm were indeed observed in solution. The existence of these clusters was again confirmed by analytical ultracentrifugation, which also confirmed the presence of aggregates that also had been observed in the cryo-TEM images. These aggregates were proposed as the first step towards ACC formation, and indeed, they could no longer be observed beyond the nucleation point. Hence, the visualization of prenucleation clusters confirmed a crystallization pathway that had not been predicted by classical nucleation theory.

14.6.2 MINERALIZATION PATHWAYS IN CALCIUM PHOSPHATE

Already in the 1960s, Posner, on the basis of XRD studies, postulated that ACP would consist of subnanometer $Ca_9(PO_4)_6$ clusters, the basic unit of the final apatite crystals (Posner and Betts 1975). Later light scattering studies indicated the existence of such clusters also in simulated body fluid (SBF), a solution containing near-physiological concentrations of the most important inorganic components in serum. These were related to 100–1000 times slower-than-expected rate of calcium phosphate mineralization (Onuma and Ito 1998), and accordingly, a cluster-growth model was proposed (Onuma et al. 1996). Using cryo-HRTEM, we confirmed the existence of such nanometer-sized entities in SBF solution, predominantly present as loosely aggregated networks (Figure 14.10a). Based on a defocus series—in which with decreasing underfocus values, the size of the particles becomes more accurate at the expense of their contrast—an average diameter (0.87 ± 0.2 nm) was estimated. This value agreed well with the previously reported cluster sizes (0.70–1.0 nm) and the theoretical size of Posner's clusters (0.95 nm, Figure 14.10b and c) (Dey et al. 2010a). However, more recently, it was demonstrated that these (sub)nanometer-sized building blocks were $Ca(HPO_4)_3^{4-}$ complexes rather than calcium phosphate clusters (Habraken et al. 2013). These take up a calcium ion at the nucleation point forming postnucleation clusters that precipitate as ACP.

SBF also was employed as a mineral source to study template-directed bio(mimetic) calcium phosphate mineralization under a Langmuir monolayer of arachidic acid (Dey et al. 2010a). Through its composition, SBF is stabilized against precipitation under physiological conditions. However, mineral formation was induced by the presence of the templating arachidic acid monolayer. Using the different components from the cryo-TEM toolbox, that is, cryo-TEM, cryo-ET, cryo-EDS, and low-dose SAED, we were able to demonstrate all stages of mineral formation from prenucleation complex aggregation to the formation of oriented crystals.

Cryo-ET demonstrated that the loose aggregation of prenucleation clusters in SBF (stage 1, Figure 14.10e) becomes denser in the presence of monolayer within a few hours (stage 2, Figure 14.10f). Moreover, whereas the network-like structures in stage 1 (Figure 14.10j) are solution species, in stage 2 (Figure 14.10k) and stage 3 (Figure 14.10g and l), these aggregates become more and more associated to the monolayer. Eventually, but exclusively at the interface, spherical particles with diameters of 40–80 nm nucleate from the densified amorphous aggregates (stage 4, Figure 14.10h and m). In time, the amorphous nanoparticles developed into spheroidal crystals with an average diameter of 120 nm (stage 5, Figure 14.10i and n). LDSAED identified these crystals as carbonated apatite, preferentially oriented with their *c*-axis parallel to the surface of the monolayer.

The earlier scenario indicates a clear role for the monolayer in the nucleation of calcium phosphate through the stabilization, densification, and arrangement of prenucleation species. For the first time, we were able to visualize the different steps leading up to the nucleation process, starting from prenucleation clusters and confirming the validity of the previously proposed two-step nucleation model (Erdemir et al. 2009). The fact that here mineral formation occurs exclusively through heterogeneous nucleation at the surface of the monolayer is in clear contrast to the formation of $CaCO_3$ in the section earlier (Figure 14.9c). This discrepancy was attributed to the higher driving force for homogeneous nucleation in calcium carbonate and the stabilization of calcium phosphate prenucleation clusters by foreign ions such as Na^+ in SBF (Xilin and Malcolm 2003).

Where the previously mentioned system is an *in vitro* model for pathological mineralization (Westenfeld et al. 2009), the recent reports that the formation of zebrafish bone (Mahamid et al. 2010) and also tooth enamel (Beniash et al. 2009) involves ACP precursor phases prompted us to reinvestigate collagen mineralization in an *in vitro* system (Nudelman et al. 2010). So far, it was believed that in bone biomineralization, collagen acts as inactive scaffold and crystal nucleation is guided by non-collagenous proteins (George and Veis 2008). Nudelman et al. (2010) deposited collagen on a cryo-TEM grid and confirmed that their reconstitution method yielded the native band structure. When the collagen was mineralized with calcium phosphate in the presence of polyaspartate (pAsp) as a

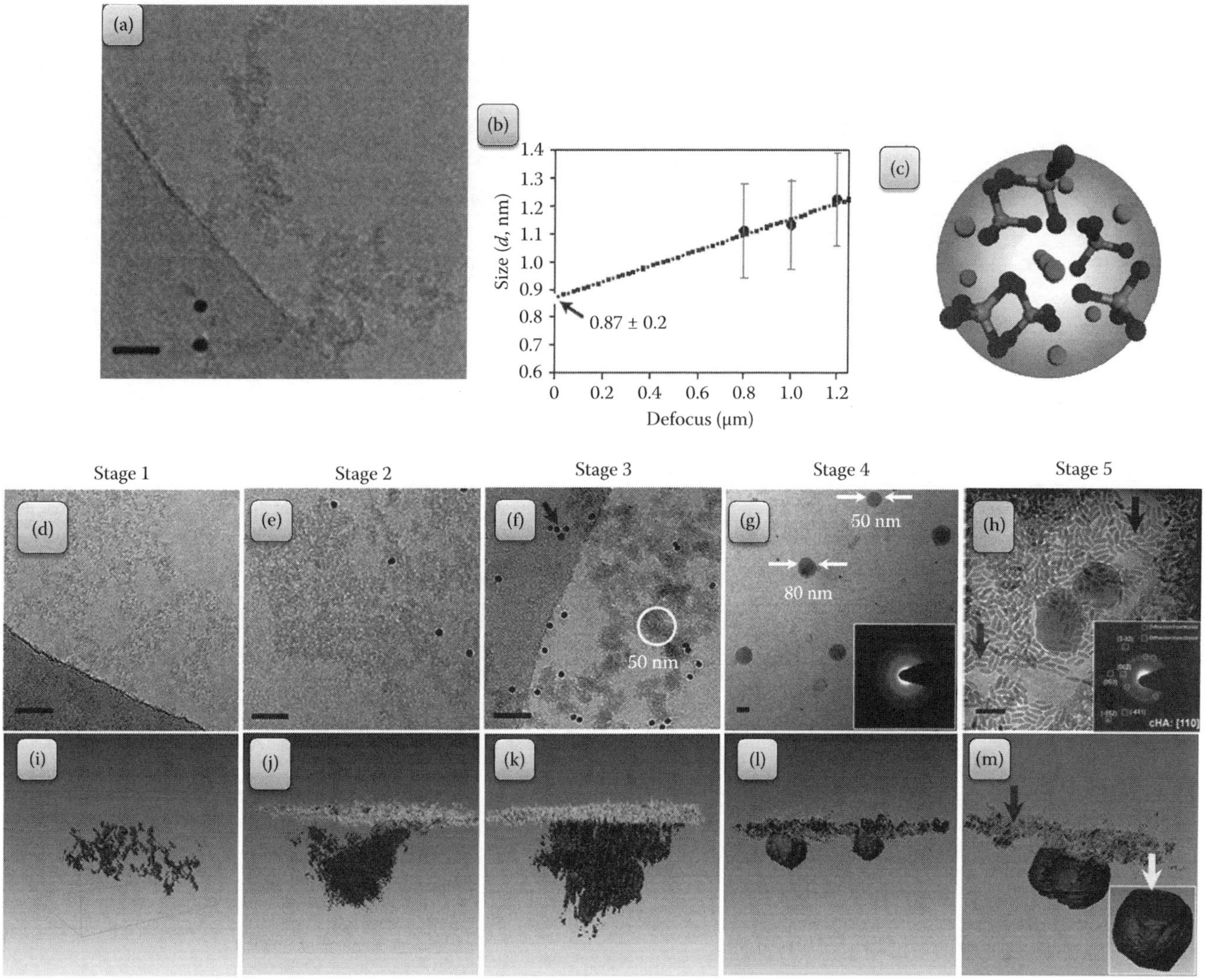

Figure 14.10 Cryo-TEM study of calcium phosphate mineralization under Langmuir monolayer: (a) Cryo-HRTEM image of assemblies of stable clusters in SBF kept at 37°C for 24 h. (b) Determination of diameter of prenucleation clusters by extrapolation of the diameter measured in cryo-HRTEM images recorded at different defocus values; error bars indicate the standard deviation. (c) Computer representation of the Posner cluster with composition $Ca_9(PO_4)_6$ and a diameter of 0.95 nm. (d–m) Different stages of the mineralization process. (d–h) 3D projection images; (i–m) computer-aided 3D visualizations of tomograms; (d and i) stage 1, the control experiment in the absence of a monolayer; (e and j) stage 2; (f and k) stage 3; (g and l) stage 4; the inset SAED in (g) shows that the spherical particles attached to the monolayer are amorphous; (h and m) stage 5; the inset SAED in (h) can be indexed as cHA with a [110] zone axis. The preferred nucleating face is (110), indicated by the white arrow in the inset of (m). Please note that for clarity (markers obscuring the image), we have not used the same area for 2D and 3D images except in (h and m). Markers (h and m) and gold beads (f) are indicated by arrows. (Reprinted by permission from Macmillan Publishers Ltd. *Nat. Mater.*, Dey, A., Bomans, P.H.H., Muller, F.A., Will, J., Frederik, P.M., de With, G., and Sommerdijk, N., The role of prenucleation clusters in surface-induced calcium phosphate crystallization, 9(12), 1010–1014, doi: 10.1038/nmat2900, Copyright 2010.)

mineralization inhibitor, as previously reported by Olszta et al. (2007), assemblies of nanometer-sized calcium phosphate entities were the first sign of mineral formation. These formed loosely packed, diffuse assemblies stabilized by pAsp, and zeta potential measurements showed that these consisted of negatively charged polymer–mineral complexes (Figure 14.11a). These assemblies were found to interact with specific sites in the collagen from where they infiltrated and deposited as interfibrillar material. The combination of LDSAED and cryo-EDS established that the deposited material was indeed ACP. By applying positive staining before vitrification of the samples, the details of the banding pattern were visualized, which revealed that the mineral was infiltrating through the so-called "a"-bands at the border of the gap and overlap zones. Modeling of collagen fibrils based on the collagen crystal structure showed that a positively charged region exists in a collagen fibril in this "a"-band region (Figure 14.11d). The charge complementarity between the mineral–polymer complex and the "a"-band explained the attraction of the mineral to this specific region of the collagen (Figure 14.11b). Finally, the ACP transformed to oriented crystalline hydroxyapatite, and cryo-ET demonstrated that these crystals were indeed inside the fibrils aligned along the long axis of collagen. A more detailed investigation on positively stained collagen demonstrated that the nucleation occurred predominantly on the charged bands but that there was no discernible selectivity between the bands (Figure 14.11c and e). Clearly, collagen can play an active role in the mineralization process being able to direct mineral infiltration as well as the nucleation of the apatite crystals.

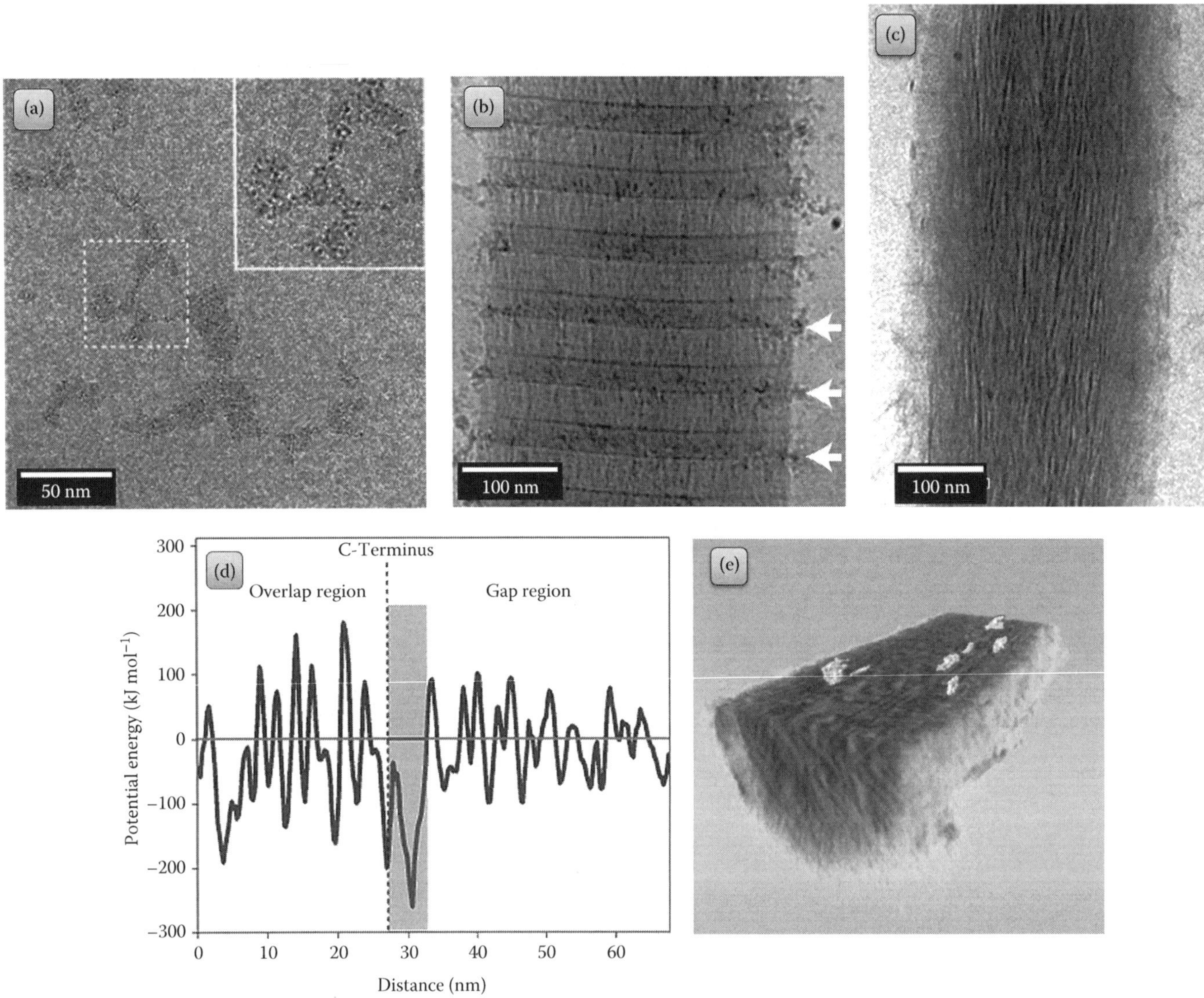

Figure 14.11 Cryo-TEM study of bone apatite formation: (a) Polymer induced liquid precursor (PILP)-like calcium phosphate aggregates formed after 10 min of reaction in the presence of pAsp. Inset: Higher magnification of the marked area. (b) Mineralization after 24 h, ACP infiltrating into the fibril (white arrows). (c) Fully mineralized collagen after 72 h reaction time. (d) Computer modeling of electrostatic potential energy of a collagen microfibril. The shaded area indicates the region where the potential energy is lowest, meaning that it is the most favorable for interaction with negative charges. This region is close to the C-terminus (dashed line) and corresponds to the mineral infiltration site, that is, the "a"-bands. (e) Computer-generated 3D visualization of mineralized collagen. The fibril is sectioned through the *xy* plane, revealing plate-shaped apatite crystals (colored in pink) embedded in the collagen matrix. (Reprinted by permission from Macmillan Publishers Ltd. *Nat. Mater.*, Nudelman, F., Pieterse, K., George, A., Bomans, P.H.H., Friedrich, H., Brylka, L.J., Hilbers, P.A.J., de With, G., and Sommerdijk, N., The role of collagen in bone apatite formation in the presence of hydroxyapatite nucleation inhibitors, 9(12), 1004–1009, doi: 10.1038/nmat2875, Copyright 2010.)

A recent *in vivo* study (Mahamid et al. 2011a) on bone formation in zebrafish reinforced the previously mentioned mechanism of mineral infiltration via infiltration of amorphous mineral droplets. Here, mineral globules were observed that appeared deformable and able to infiltrate first between the collagen fibers and then into collagen fibrils. Further, TEM imaging on cryo-sections showed the detailed structure of globules having a laminated arrangement composed of high electron-dense concentric rings with a low-dense material separating them. Most significantly, it was demonstrated that these mineralized globules were not confined by membranes.

Enamel is the hardest tissue in the vertebrate body and is made up from highly arranged hydroxyapatite crystals. It is well known that amelogenin assemblies play an important role in the formation and organization of the apatitic crystals. However, the details of the interactions between the amelogenin and the mineral phase during enamel formation are still not well understood. Fang et al. (2011a) recently presented a convincing *in vitro* cryo-TEM study on enamel formation, visualizing the regulation of biomimetic calcium phosphate by amelogenin at the nanoscale. The authors also, in this case, observed individual calcium phosphate prenucleation clusters that are stabilized by amelogenin. Eventually, highly organized mineral structures formed by the fusion of prenucleation clusters into chains that then fused together leading to the formation of bundles of elongated mineral particles (Figure 14.12). The densification and fusion of the clusters observed here for the formation of enamel strongly resemble the process we found for the formation of apatite from SBF (Figure 14.10) (Dey et al. 2010a).

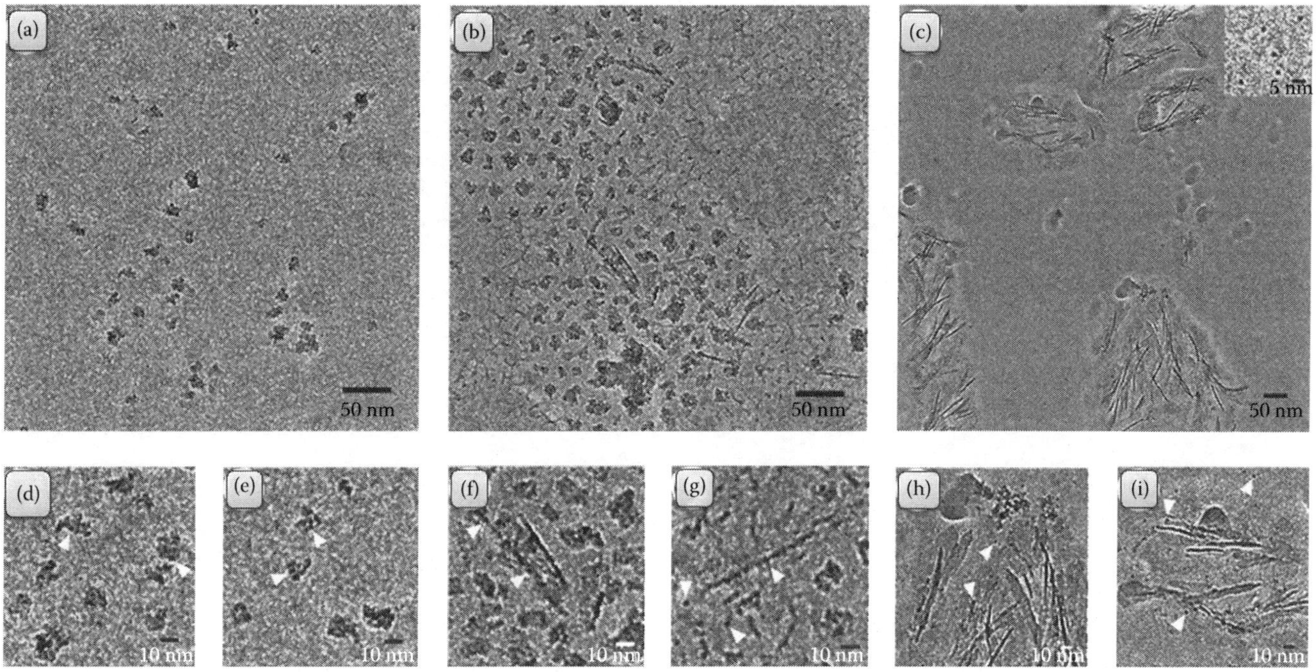

Figure 14.12 Cryo-TEM micrographs of calcium phosphate mineralization in the presence of full-length amelogenin rM179. (a, d, and e) 10 min in the reaction, (b, f, and g) 30 min in the reaction, and (c, h, and i) 120 min in the reaction. (c, Inset) Isolated mineral prenucleation clusters. Arrowheads in (d–i) point to individual prenucleation clusters. (From Fang, P.A., Conway, J.F., Margolis, H.C., Simmer, J.P., and Beniash, E., Hierarchical self-assembly of amelogenin and the regulation of biomineralization at the nanoscale, *Proc. Natl. Acad. Sci. USA*, 108(34), 14097–14102, doi: 10.1073/pnas.1106228108. Copyright 2011 National Academy of Sciences, U.S.A.)

14.7 CONCLUSIONS

The current state of affairs for TEM allows us to analyze structures with sub-angstrom resolution. However, the specifics of bio(mimetic) mineralization—which occurs in aqueous medium, through transient intermediates, and which involves sensitive hybrid materials with complicated hierarchical structures—for a long time have complicated the application of the full potential of TEM and its satellite techniques to resolve the mechanisms involved. Recently, cryo-TEM has manifested itself to be a relevant and powerful technique for the examination of the early stages of biomimetic mineralization. State-of-the-art instrumentation and automation in sampling have made it possible to achieve nanometer resolution in quasi *in situ*, time-resolved experiments. Studies applying cryo-TEM to investigate the biomimetic mineralization of calcium carbonate, calcium phosphate, and iron oxides revealed and visualized new pathways in the early stages of crystal formation. The aggregation-based mechanism for nucleation of the amorphous phase and crystal growth via oriented attachment were not foreseen by classical theories; however, recent studies have shown that these nonclassical mechanisms can be explained by classical theories (Baumgartner et al. 2013; Habraken et al. 2013). Moreover, the application of high-resolution 2D and 3D imaging along with electron diffraction has allowed insight into template and matrix-controlled mineralization processes down to the nanometer scale. A next step should be the development of cryo-EELS, which would allow to obtain information of the chemical evolution during the development of the mineral phase. Although this technique has been successfully demonstrated for polymeric nanoparticles, the marriage of the contrasting dose requirements for cryo-imaging and spectroscopy will require the application of absolute top of the range infrastructure. However, we expect that the implementation of recent developments such as the use of monochromatic electron sources and phase plates and aberration correctors in cryo-TEM would make it possible to achieve its goal.

Even more exciting are the possibilities of studying these processes in real time using liquid cell TEM, which has been demonstrated to combine lattice resolution with video-rate imaging. This technique will be of paramount importance for studying the role of the organic in the dynamic assembly of nanoparticles and the transformation of metastable phases.

In addition, we would like to draw attention to some new developments in materials science TEM. Femtosecond laser-induced electron sources have been used for the development of ultrafast 2D EM in solid samples and even time-resolved tomography (4D EM) for repetitive processes (Kwon and Zewail 2010; Zewail 2010). Of particular interest are the combinations with EELS (Carbone et al. 2009) and electron diffraction (Kwon et al. 2008) that have been applied for monitoring the development of chemistry and structure in solids, respectively. In addition, the development of fluctuation TEM (FTEM) allows the detection of subcritical nuclei in amorphous solids (Chen et al. 2008). These examples illustrate that advances in electron optics and recording systems for TEM enable rapid imaging of the crystallization process, although their application to biomimetic processes will require further development and adaptation to the presence of organic materials and/or aqueous environments. Finally, we anticipate that a combination of our ability to design new experiments and to interpret their results in a quantitative manner will enable us to obtain a deeper understanding of the processes involved in the formation of biominerals.

ACKNOWLEDGMENTS

The authors would like to thank Netherlands Science Foundation, NWO, and NIH/NIDCR for their financial support.

REFERENCES

Addadi, L., J. Moradian, E. Shay, N.G. Maroudas, and S. Weiner. 1987. A chemical model for the cooperation of sulfates and carboxylates in calcite crystal nucleation: Relevance to biomineralization. *Proceedings of the National Academy of Sciences of the United States of America* 84(9):2732–2736.

Adrian, M., J. Dubochet, J. Lepault, and A.W. McDowall. 1984. Cryo-electron microscopy of viruses. *Nature* 308(5954):32–36.

Aizenberg, J. 2010. New nanofabrication strategies: Inspired by biomineralization. *MRS Bulletin* 35(4):323–330. doi: 10.1557/mrs2010.555.

Aizenberg, J. and P. Fratzl. 2009. Biological and biomimetic materials. *Advanced Materials* 21(4):387–388.

Aizenberg, J., G. Lambert, L. Addadi, and S. Weiner. 1996. Stabilization of amorphous calcium carbonate by specialized macromolecules in biological and synthetic precipitates. *Advanced Materials* 8(3):222–226.

Aizenberg, J., D.A. Muller, J.L. Grazul, and D.R. Hamann. 2003. Direct fabrication of large micropatterned single crystals. *Science* 299(5610):1205–1208. doi: 10.1126/science.1079204.

Aizenberg, J., A. Tkachenko, S. Weiner, L. Addadi, and G. Hendler. 2001. Calcitic microlenses as part of the photoreceptor system in brittlestars. *Nature* 412(6849):819–822. doi: 10.1038/35090573.

Aizenberg, J., J.C. Weaver, M.S. Thanawala, V.C. Sundar, D.E. Morse, and P. Fratzl. 2005. Skeleton of *Euplectella* sp.: Structural hierarchy from the nanoscale to the macroscale. *Science* 309(5732):275–278. doi: 10.1126/science.1112255.

Al-Amoudi, A., J.-J. Chang, A. Leforestier, A. McDowall, L.M. Salamin, L.P.O. Norlen, K. Richter, N.S. Blanc, D. Studer, and J. Dubochet. 2004. Cryo-electron microscopy of vitreous sections. *EMBO Journal* 23(18):3583–3588.

Albeck, S., L. Addadi, and S. Weiner. 1996. Regulation of calcite crystal morphology by intracrystalline acidic proteins and glycoproteins. *Connective Tissue Research* 35(1–4):365–370.

Albeck, S., J. Aizenberg, L. Addadi, and S. Weiner. 1993. Interactions of various skeletal intracrystalline components with calcite crystals. *Journal of the American Chemical Society* 115(25):11691–11697.

Antonietti, M., M. Breulmann, C.G. Goltner, H. Colfen, K.K.W. Wong, D. Walsh, and S. Mann. 1998. Inorganic/organic mesostructures with complex architectures: Precipitation of calcium phosphate in the presence of double-hydrophilic block copolymers. *Chemistry—A European Journal* 4(12):2493–2500.

Bajaj, D. and D.D. Arola. 2009. On the R-curve behavior of human tooth enamel. *Biomaterials* 30(23–24):4037–4046. doi: 10.1016/j.biomaterials.2009.04.017.

Baldassarri, M., H.C. Margolis, and E. Beniash. 2008. Compositional determinants of mechanical properties of enamel. *Journal of Dental Research* 87(7):645–649.

Ball, P. 2001. Life's lessons in design. *Nature* 409(6818):413–416.

Balmes, O., J.O. Malm, N. Pettersson, G. Karlsson, and J.O. Bovin. 2006. Imaging atomic structure in metal nanoparticles using high-resolution cryo-TEM. *Microscopy and Microanalysis* 12(2):145–150. doi: 10.1017/s1431927606060119.

Batson, P.E., N. Dellby, and O.L. Krivanek. 2002. Sub-angstrom resolution using aberration corrected electron optics. *Nature* 418(6898):617–620.

Baumgartner, J., A. Dey, P.H.H. Bomans, C.L. Coadou, P. Fratzl, N.A.J.M. Sommerdijk, and D. Faivre. 2013. Nucleation and growth of magnetite from solution. *Nature Materials* 12(4):310–314. doi: http://www.nature.com/nmat/journal/v12/n4/abs/nmat3558.html#supplementary-information.

Belcher, A.M., X.H. Wu, R.J. Christensen, P.K. Hansma, G.D. Stucky, and D.E. Morse. 1996. Control of crystal phase switching and orientation by soluble mollusc-shell proteins. *Nature* 381(6577):56–58.

Beniash, E. 2011. Biominerals-hierarchical nanocomposites: The example of bone. *Wiley Interdisciplinary Reviews—Nanomedicine and Nanobiotechnology* 3(1):47–69. doi: 10.1002/wnan.105.

Beniash, E., L. Addadi, and S. Weiner. 1999. Cellular control over spicule formation in sea urchin embryos: A structural approach. *Journal of Structural Biology* 125(1):50–62.

Beniash, E., J. Aizenberg, L. Addadi, and S. Weiner. 1997. Amorphous calcium carbonate transforms into calcite during sea urchin larval spicule growth. *Proceedings of the Royal Society of London Series B—Biological Sciences* 264(1380):461–465.

Beniash, E., A.S. Deshpande, P.A. Fang, N.S. Lieb, X. Zhang, and C.S. Sfeir. 2011. Possible role of DMP1 in dentin mineralization. *Journal of Structural Biology* 174(1):100–106. doi: 10.1016/j.jsb.2010.11.013.

Beniash, E., R.A. Metzler, R.S.K. Lam, and P.U.P.A. Gilbert. 2009. Transient amorphous calcium phosphate in forming enamel. *Journal of Structural Biology* 166(2):133–143.

Beniash, E., J.P. Simmer, and H.C. Margolis. 2005. The effect of recombinant mouse amelogenins on the formation and organization of hydroxyapatite crystals in vitro. *Journal of Structural Biology* 149(2):182–190. doi: 10.1016/j.jsb.2004.11.001.

Beniash, E., Z. Skobe, and J.D. Bartlett. 2006. Formation of the dentino-enamel interface in enamelysin (MMP-20)-deficient mouse incisors. *European Journal of Oral Sciences* 114:24–29.

Beniash, E., W. Traub, A. Veis, and S. Weiner. 2000. A transmission electron microscope study using vitrified ice sections of predentin: Structural changes in the dentin collagenous matrix prior to mineralization. *Journal of Structural Biology* 132(3):212–225.

Bentov, S., C. Brownlee, and J. Erez. 2009. The role of seawater endocytosis in the biomineralization process in calcareous foraminifera. *Proceedings of the National Academy of Sciences of the United States of America* 106(51):21500–21504. doi: 10.1073/pnas.0906636106.

Benzerara, K., N. Menguy, M. Obst, J. Stolarski, M. Mazur, T. Tylisczak, G.E. Brown, and A. Meibom. 2011. Study of the crystallographic architecture of corals at the nanoscale by scanning transmission X-ray microscopy and transmission electron microscopy. *Ultramicroscopy* 111(8):1268–1275. doi: 10.1016/j.ultramic.2011.03.023.

Berman, A., D.J. Ahn, A. Lio, M. Salmeron, A. Reichert, and D. Charych. 1995. Total alignment of calcite at acidic polydiacetylene films: Cooperativity at the organic-inorganic interface. *Science* 269:515–518.

Bhushan, B. 2009. Biomimetics: Lessons from nature—An overview. *Philosophical Transactions of the Royal Society of London, Series A* 367(1893):1445–1486. doi: 10.1098/rsta.2009.0011.

Bonderer, L.J., A.R. Studart, and L.J. Gauckler. 2008. Bioinspired design and assembly of platelet reinforced polymer films. *Science* 319(5866):1069–1073. doi: 10.1126/science.1148726.

Boskey, A.L. 1997. Amorphous calcium phosphate: The contention of bone. *Journal of Dental Research* 76(8):1433–1436. doi: 10.1177/00220345970760080501.

Browning, N.D., M.A. Bonds, G.H. Campbell, J.E. Evans, T. LaGrange, K.L. Jungjohann, D.J. Masiel et al. 2012. Recent developments in dynamic transmission electron microscopy. *Current Opinion in Solid State and Materials Science* 16(1):23–30. doi: 10.1016/j.cossms.2011.07.001.

Byrne, M.E., D.A. Ball, J.L. Guerquin-Kern, I. Rouiller, T.D. Wu, K.H. Downing, H. Vali, and A. Komeili. 2010. Desulfovibrio magneticus RS-1 contains an iron- and phosphorus-rich organelle distinct from its bullet-shaped magnetosomes. *Proceedings of the National Academy of Sciences of the United States of America* 107(27):12263–12268. doi: 10.1073/pnas.1001290107.

Calvert, C.C., A. Brown, and R. Brydson. 2005. Determination of the local chemistry of iron in inorganic and organic materials. *Journal of Electron Spectroscopy and Related Phenomena* 143(2–3):173–187.

Carbone, F., B. Barwick, O.-H. Kwon, H.S. Park, J. Spencer Baskin, and A.H. Zewail. 2009. EELS femtosecond resolved in 4D ultrafast electron microscopy. *Chemical Physics Letters* 468(4–6):107–111.

Chen, Y., J. Xiao, Z. Wang, and S. Yang. 2008. Observation of an amorphous calcium carbonate precursor on a stearic acid monolayer formed during the biomimetic mineralization of $CaCO_3$. *Langmuir* 25(2):1054–1059.

Chretien, D. and R.H. Wade. 1991. New data on the microtubule surface lattice. *Biology of the Cell* 71(1–2):161–174. doi: 10.1016/0248-4900(91)90062-r.

Cölfen, H. and M. Antonietti. 2005. Mesocrystals: Inorganic superstructures made by highly parallel crystallization and controlled alignment. *Angewandte Chemie International Edition* 44(35):5576–5591. doi: 10.1002/anie.200500496.

Cölfen, H. and M. Antonietti. 2008. Mesocrystal systems. In *Mesocrystals and Nonclassical Crystallization*, pp. 113–177. Chichester, U.K.: John Wiley & Sons, Ltd.

Cölfen, H. and S. Mann. 2003. Higher-order organization by mesoscale self-assembly and transformation of hybrid nanostructures. *Angewandte Chemie International Edition* 42(21):2350–2365.

Crane, N.J., V. Popescu, M.D. Morris, P. Steenhuis, and M.A. Ignelzi. 2006. Raman spectroscopic evidence for octacalcium phosphate and other transient mineral species deposited during intramembranous mineralization. *Bone* 39(3):434–442. doi: 10.1016/j.bone.2006.02.059.

Cuisinier, F.J.G., P. Steuer, B. Senger, J.C. Voegel, and R.M. Frank. 1992a. Human amelogenesis. 1. High-resolution electron-microscopy study of ribbon-like crystals. *Calcified Tissue International* 51(4):259–268.

Cuisinier, F.J.G., P. Steuer, B. Senger, J.C. Voegel, and R.M. Frank. 1993. Human amelogenesis—High-resolution electron-microscopy of nanometer-sized particles. *Cell and Tissue Research* 273(1):175–182. doi: 10.1007/bf00304624.

Cuisinier, F.J.G., J.C. Voegel, J. Yacaman, and R.M. Frank. 1992b. Structure of initial crystals formed during human amelogenesis. *Journal of Crystal Growth* 116(3–4):314–318. doi: 10.1016/0022-0248(92)90638-y.

Davis, M.E. 2004. How life makes hard stuff. *Science* 305(5683):480. doi: 10.1126/science.1099773.

de Jonge, N. and F.M. Ross. 2011. Electron microscopy of specimens in liquid. *Nature Nanotechnology* 6(11):695–704. doi: 10.1038/nnano.2011.161.

Deshpande, A.S. and E. Beniash. 2008. Bioinspired synthesis of mineralized collagen fibrils. *Crystal Growth & Design* 8(8):3084–3090.

Deshpande, A.S., P.-A. Fang, J.P. Simmer, H.C. Margolis, and E. Beniash. 2010. Amelogenin-collagen interactions regulate calcium phosphate mineralization in vitro. *Journal of Biological Chemistry* 285(25):19277–19287. doi: 10.1074/jbc.M109.079939.

Deshpande, A.S., P.-A. Fang, X. Zhang, T. Jayaraman, C. Sfeir, and E. Beniash. 2011. Primary structure and phosphorylation of dentin matrix protein 1(DMP1) and dentin phosphophoryn (DPP) uniquely determine their role in biomineralization. *Biomacromolecules* 12(8):2933–2945. doi: 10.1021/bm2005214.

Dey, A., P.H.H. Bomans, F.A. Muller, J. Will, P.M. Frederik, G. de With, and N. Sommerdijk. 2010a. The role of prenucleation clusters in surface-induced calcium phosphate crystallization. *Nature Materials* 9(12):1010–1014. doi: 10.1038/nmat2900.

Dey, A., G. de With, and N.A.J.M. Sommerdijk. 2010b. In situ techniques in biomimetic mineralization studies of calcium carbonate. *Chemical Society Reviews* 39(2):397–409.

Diekwisch, T.G.H., B.J. Berman, S. Gentner, and H.C. Slavkin. 1995. Initial enamel crystals are not spatially associated with mineralized dentin. *Cell and Tissue Research* 279(1):149–167.

Dong, H., S.E. Paramonov, L. Aulisa, E.L. Bakota, and J.D. Hartgerink. 2007. Self-assembly of multidomain peptides: Balancing molecular frustration controls conformation and nanostructure. *Journal of the American Chemical Society* 129(41):12468–12472. doi: 10.1021/ja072536r.

Dubochet, J., M. Adrian, J.J. Chang, J.C. Homo, J. Lepault, A.W. McDowall, and P. Schultz. 1988. Cryo-electron microscopy of vitrified specimens. *Quarterly Reviews of Biophysics* 21(2):129–228.

Dunin-Borkowski, R.E., M.R. McCartney, R.B. Frankel, D.A. Bazylinski, M. Pósfai, and P.R. Buseck. 1998. Magnetic microstructure of magnetotactic bacteria by electron holography. *Science* 282(5395):1868–1870.

Dwyer, C., A.I. Kirkland, P. Hartel, H. Mueller, and M. Haider. 2007. Electron nanodiffraction using sharply focused parallel probes. *Applied Physics Letters* 90(15):151104–151106. doi: 151104 10.1063/1.2721120.

Elhadj, S., E.A. Salter, A. Wierzbicki, J.J. De Yoreo, N. Han, and P.M. Dove. 2005. Peptide controls on calcite mineralization: Polyaspartate chain length affects growth kinetics and acts as a stereochemical switch on morphology. *Crystal Growth & Design* 6(1):197–201. doi: 10.1021/cg050288+.

Erdemir, D., A.Y. Lee, and A.S. Myerson. 2009. Nucleation of crystals from solution: Classical and two-step models. *Accounts of Chemical Research* 42(5):621–629.

Faivre, D., L. Böttger, B. Matzanke, and D. Schüler. 2007. Intracellular magnetite biomineralization in bacteria proceeds by a distinct pathway involving membrane-bound ferritin and an iron(II) species. *Angewandte Chemie International Edition* 46(44):8495–8499.

Falini, G., S. Albeck, S. Weiner, and L. Addadi. 1996. Control of aragonite or calcite polymorphism by mollusk shell macromolecules. *Science* 271(5245):67–69.

Fang, P.A., J.F. Conway, H.C. Margolis, J.P. Simmer, and E. Beniash. 2011a. Hierarchical self-assembly of amelogenin and the regulation of biomineralization at the nanoscale. *Proceedings of the National Academy of Sciences of the United States of America* 108(34):14097–14102. doi: 10.1073/pnas.1106228108.

Fang, P.A., R.S.K. Lam, and E. Beniash. 2011b. Relationships between dentin and enamel mineral at the dentino-enamel boundary: Electron tomography and high-resolution transmission electron microscopy study. *European Journal of Oral Sciences* 119:120–124. doi: 10.1111/j.1600-0722.2011.00876.x.

Fantner, G.E., T. Hassenkam, J.H. Kindt, J.C. Weaver, H. Birkedal, L. Pechenik, J.A. Cutroni et al. 2005. Sacrificial bonds and hidden length dissipate energy as mineralized fibrils separate during bone fracture. *Nature Materials* 4(8):612–616.

Frank, J., J. Zhu, P. Penczek, Y.H. Li, S. Srivastava, A. Verschoor, M. Radermacher, R. Grassucci, R.K. Lata, and R.K. Agrawal. 1995. A model of protein-synthesis based on cryo-electron microscopy of the *E. coli* ribosome. *Nature* 376(6539):441–444. doi: 10.1038/376441a0.

Frankel, R.B., R.P. Blakemore, and R.S. Wolfe. 1979. Magnetite in freshwater magnetotactic bacteria. *Science* 203(4387):1355–1356. doi: 10.1126/science.203.4387.1355.

Fratzl, P. and R. Weinkamer. 2007. Nature's hierarchical materials. *Progress in Materials Science* 52(8):1263–1334. doi: 10.1016/j.pmatsci.2007.06.001.

Frederik, P.M. and N. Sommerdijk. 2005. Spatial and temporal resolution in cryo-electron microscopy—A scope for nano-chemistry. *Current Opinion in Colloid & Interface Science* 10(5–6):245–249.

Friedrich, H., P.E. de Jongh, A.J. Verkleij, and K.P. de Jong. 2009. Electron tomography for heterogeneous catalysts and related nanostructured materials. *Chemical Reviews* 109(5):1613–1629. doi: 10.1021/cr800434t.

Galvez, N., B. Fernalndez, P. Salnchez, R. Cuesta, M. Ceolіln, M. Clemente-Leon, S. Trasobares et al. 2008. Comparative structural and chemical studies of ferritin cores with gradual removal of their iron contents. *Journal of the American Chemical Society* 130(25):8062–8068.

Gebauer, D., A. Volkel, and H. Colfen. 2008. Stable prenucleation calcium carbonate clusters. *Science* 322(5909):1819–1822. doi: 10.1126/science.1164271.

George, A. and A. Veis. 2008. Phosphorylated proteins and control over apatite nucleation, crystal growth, and inhibition. *Chemical Reviews* 108(11):4670–4693. doi: 10.1021/cr0782729.

Gilis, M., O. Grauby, P. Willenz, P. Dubois, L. Legras, V. Heresanu, and A. Baronnet. 2011. Multi-scale mineralogical characterization of the hypercalcified sponge *Petrobiona massiliana* (Calcarea, Calcaronea). *Journal of Structural Biology* 176(3):315–329. doi: 10.1016/j.jsb.2011.08.008.

Glimcher, M.J. 1959. Molecular biology of mineralized tissues with particular reference to bone. *Reviews of Modern Physics* 31(2):359–393. doi: 10.1103/RevModPhys.31.359.

Glimcher, M.J., E.J. Daniel, D.F. Travis, and S. Kamhi. 1965. Electron optical and X-ray diffraction studies of the organization of the inorganic crystals in embryonic bovine enamel. *Journal of Ultrastructure Research* 12(Suppl 1):1–77. doi: 10.1016/s0022-5320(65)90001-8.

Glimcher, M.J., A.J. Hodge, and F.O. Schmitt. 1957. Macromolecular aggregation states in relation to mineralization—The collagen-hydroxyapatite system as studied in vitro. *Proceedings of the National Academy of Sciences of the United States of America* 43(10):860–867. doi: 10.1073/pnas.43.10.860.

Gong, Y.U.T., C.E. Killian, I.C. Olson, N.P. Appathurai, A.L. Amasino, M.C. Martin, L.J. Holt, F.H. Wilt, and P. Gilbert. 2012. Phase transitions in biogenic amorphous calcium carbonate. *Proceedings of the National Academy of Sciences of the United States of America* 109(16):6088–6093. doi: 10.1073/pnas.1118085109.

Green, D., D. Walsh, S. Mann, and R.O.C. Oreffo. 2002. The potential of biomimesis in bone tissue engineering: Lessons from the design and synthesis of invertebrate skeletons. *Bone* 30(6):810–815.

Gregoire, C. 1957. Topography of the organic components in mother-of-pearl. *Journal of Biophysical and Biochemical Cytology* 3(5):797–808. doi: 10.1083/jcb.3.5.797.

Gregori, G., H.-J. Kleebe, H. Mayr, and G. Ziegler. 2006. EELS characterisation of Î²-tricalcium phosphate and hydroxyapatite. *Journal of the European Ceramic Society* 26(8):1473–1479.

Gries, K., R. Kröger, C. Kübel, M. Schowalter, M. Fritz, and A. Rosenauer. 2009. Correlation of the orientation of stacked aragonite platelets in nacre and their connection via mineral bridges. *Ultramicroscopy* 109(3):230–236.

Griesshaber, E., K. Kelm, A. Sehrbrock, W. Mader, J. Mutterlose, U. Brand, and W.W. Schmahl. 2009. Amorphous calcium carbonate in the shell material of the brachiopod *Megerlia truncata*. *European Journal of Mineralogy* 21(4):715–723. doi: 10.1127/0935-1221/2009/0021-1950.

Habraken, W.J., J. Tao, L.J. Brylka, H. Friedrich, L. Bertinetti, A.S. Schenk, A. Verch et al. 2013. Ion-association complexes unite classical and non-classical theories for the biomimetic nucleation of calcium phosphate. *Nature Communications* 4:1507. doi: http://www.nature.com/ncomms/journal/v4/n2/suppinfo/ncomms2490_S1.html.

Hartgerink, J.D., E. Beniash, and S.I. Stupp. 2001. Self-assembly and mineralization of peptide-amphiphile nanofibers. *Science* 294(5547):1684–1688.

Hartgerink, J.D., J.R. Granja, R.A. Milligan, and M.R. Ghadiri. 1996. Self-assembling peptide nanotubes. *Journal of the American Chemical Society* 118(1):43–50. doi: 10.1021/ja953070s.

Heuer, A.H., D.J. Fink, V.J. Laraia, J.L. Arias, P.D. Calvert, K. Kendall, G.L. Messing et al. 1992. Innovative materials processing strategies—A biomimetic approach. *Science* 255(5048):1098–1105.

Heywood, B.R. and E.D. Eanes. 1992. An ultrastructural-study of the effects of acidic phospholipid substitutions on calcium-phosphate precipitation in anionic liposomes. *Calcified Tissue International* 50(2):149–156. doi: 10.1007/bf00298793.

Hunter, G.K., M.S. Poitras, T.M. Underhill, M.D. Grynpas, and H.A. Goldberg. 2001. Induction of collagen mineralization by a bone sialoprotein–decorin chimeric protein. *Journal of Biomedical Materials Research* 55(4):496–502.

Huo, Q.S., D.I. Margolese, U. Ciesla, P.Y. Feng, T.E. Gier, P. Sieger, R. Leon, P.M. Petroff, F. Schuth, and G.D. Stucky. 1994. Generalized synthesis of periodic surfactant inorganic composite-materials. *Nature* 368(6469):317–321.

Kamat, S., X. Su, R. Ballarini, and A.H. Heuer. 2000. Structural basis for the fracture toughness of the shell of the conch *Strombus gigas*. *Nature* 405(6790):1036–1040.

Kasama, T., M. Posfai, R.K.K. Chong, A.P. Finlayson, P.R. Buseck, R.B. Frankel, and R.E. Dunin-Borkowski. 2006. Magnetic properties, microstructure, composition, and morphology of greigite nanocrystals in magnetotactic bacteria from electron holography and tomography. *American Mineralogist* 91(8–9):1216–1229. doi: 10.2138/am.2006.2227.

Katzmann, E., F.D. Muller, C. Lang, M. Messerer, M. Winklhofer, J.M. Plitzko, and D. Schuler. 2011. Magnetosome chains are recruited to cellular division sites and split by asymmetric septation. *Molecular Microbiology* 82(6):1316–1329. doi: 10.1111/j.1365-2958.2011.07874.x.

Kewalramani, S., K. Kim, B. Stripe, G. Evmenenko, G.H.B. Dommett, and P. Dutta. 2008. Observation of an organic–inorganic lattice match during biomimetic growth of (001)-oriented calcite crystals under floating sulfate monolayers. *Langmuir* 24(19):10579–10582.

Kim, G., A. Sousa, D. Meyers, M. Shope, and M. Libera. 2006. Diffuse polymer interfaces in lobed nanoemulsions preserved in aqueous media. *Journal of the American Chemical Society* 128(20):6570–6571.

Kim, Y.-Y., K. Ganesan, P. Yang, A.N. Kulak, S. Borukhin, S. Pechook, L. Ribeiro et al. 2011. An artificial biomineral formed by incorporation of copolymer micelles in calcite crystals. *Nature Materials* 10(11):890–896.

Kröger, N., R. Deutzmann, and M. Sumper. 1999. Polycationic peptides from diatom biosilica that direct silica nanosphere formation. *Science* 286(5442):1129–1132. doi: 10.1126/science.286.5442.1129.

Kröger, N., S. Lorenz, E. Brunner, and M. Sumper. 2002. Self-assembly of highly phosphorylated silaffins and their function in biosilica morphogenesis. *Science* 298(5593):584–586. doi: 10.1126/science.1076221.

Kröger, N. and M. Sumper. 2005. The molecular basis of diatom biosilica formation. In *Biomineralization*, Bäuerlein, E. ed. pp. 135–158. Weinheim, Germany: Wiley-VCH Verlag GmbH & Co. KGaA.

Kwak, S.Y., F.B. Wiedemann-Bidlack, E. Beniash, Y. Yamakoshi, J.P. Simmer, A. Litman, and H.C. Margolis. 2009. Role of 20-kDa amelogenin (P148) phosphorylation in calcium phosphate formation in vitro. *Journal of Biological Chemistry* 284(28):18972–18979. doi: 10.1074/jbc.M109.020370.

Kwon, O.-H., B. Barwick, H.S. Park, J. Spencer Baskin, and A.H. Zewail. 2008. 4D visualization of embryonic, structural crystallization by single-pulse microscopy. *Proceedings of the National Academy of Sciences of the United States of America* 105(25):8519–8524. doi: 10.1073/pnas.0803344105.

Kwon, O.-H. and A.H. Zewail. 2010. 4D electron tomography. *Science* 328(5986):1668–1673. doi: 10.1126/science.1190470.

Landis, W.J. 1995. The strength of a calcified tissue depends in part on the molecular structure and organization of its constituent mineral crystals in their organic matrix. *Bone* 16(5):533–544.

Landis, W.J. and M.J. Glimcher. 1978. Electron-diffraction and electron-probe microanalysis of mineral phase of bone tissue prepared by anhydrous techniques. *Journal of Ultrastructure Research* 63(2):188–223. doi: 10.1016/s0022-5320(78)80074-4.

Landis, W.J., B.T. Hauschka, C.A. Rogerson, and M.J. Glimcher. 1977. Electron-microscopic observations of bone tissue prepared by ultra-cryomicrotomy. *Journal of Ultrastructure Research* 59(2):185–206. doi: 10.1016/s0022-5320(77)80079-8.

Landis, W.J., K.J. Hodgens, J. Arena, M.J. Song, and B.F. McEwen. 1996a. Structural relations between collagen and mineral in bone as determined by high voltage electron microscopic tomography. *Microscopy Research and Technique* 33(2):192–202.

Landis, W.J., K.J. Hodgens, M.J. Song, J. Arena, S. Kiyonaga, M. Marko, C. Owen, and B.F. McEwen. 1996b. Mineralization of collagen may occur on fibril surfaces: Evidence from conventional and high-voltage electron microscopy and three-dimensional imaging. *Journal of Structural Biology* 117(1):24–35.

Landis, W.J. and M.J. Song. 1991. Early mineral deposition in calcifying tendon characterized by high voltage electron microscopy and three-dimensional graphic imaging. *Journal of Structural Biology* 107(2):116–127. doi: 10.1016/1047-8477(91)90015-o.

Landis, W.J., M.J. Song, A. Leith, L. McEwen, and B.F. McEwen. 1993. Mineral and organic matrix interaction in normally calcifying tendon visualized in three dimensions by high-voltage electron microscopic tomography and graphic image reconstruction. *Journal of Structural Biology* 110(1):39–54.

Leapman, R.D. and J.A. Hunt. 1991. Comparison of detection limits for EELS and EDXS. *Microscopy Microanalysis and Microstructures* 2(2–3):231–244.

Ledger, P.W. and W.C. Jones. 1977. Spicule formation in the calcareous sponge *Sycon ciliatum*. *Cell and Tissue Research* 181(4):553–567. doi: 10.1007/bf00221776.

Lee, K., W. Wagermaier, A. Masic, K.P. Kommareddy, M. Bennet, I. Manjubala, S.-W. Lee, S.B. Park, H. Cofen, and P. Fratzl. 2012. Self-assembly of amorphous calcium carbonate microlens arrays. *Nature Communications* 3:725.

Lenders, J.J.M., A. Dey, P.H.H. Bomans, J. Spielmann, M.M.R.M. Hendrix, G. de With, F.C. Meldrum, S. Harder, and N.A.J.M. Sommerdijk. 2012. High-magnesian calcite mesocrystals: A coordination chemistry approach. *Journal of the American Chemical Society* 134(2):1367–1373.

Levi-Kalisman, Y., G. Falini, L. Addadi, and S. Weiner. 2001. Structure of the nacreous organic matrix of a bivalve mollusk shell examined in the hydrated state using cryo-TEM. *Journal of Structural Biology* 135(1):8–17. doi: 10.1006/jsbi.2001.4372.

Li, D.S., M.H. Nielsen, J.R.I. Lee, C. Frandsen, J.F. Banfield, and J.J. De Yoreo. 2012. Direction-specific interactions control crystal growth by oriented attachment. *Science* 336(6084):1014–1018. doi: 10.1126/science.1219643.

Li, H.Y., H.L. Xin, D.A. Muller, and L.A. Estroff. 2009. Visualizing the 3D internal structure of calcite single crystals grown in agarose hydrogels. *Science* 326(5957):1244–1247. doi: 10.1126/science.1178583.

Li, X. and Z. Huang. 2009. Unveiling the formation mechanism of pseudo-single-crystal aragonite platelets in nacre. *Physical Review Letters* 102(7):075502.

Liu, Y. and C. Qianwang. 2008. Synthesis of magnetosome chain-like structures. *Nanotechnology* 19(47):475603.

Lowenstam, H.A. 1981. Minerals formed by organisms. *Science* 211(4487):1126–1131.

Lowenstam, H.A. and S. Weiner. 1985. Transformation of amorphous calcium phosphate to crystalline dahillite in the radular teeth of chitons. *Science* 227(4682):51–53.

Lowenstam, H.A. and S. Weiner. 1989. *On Biomineralization*. New York: Oxford University Press.

Ma, G. and X.Y. Liu. 2009. Hydroxyapatite: Hexagonal or monoclinic? *Crystal Growth & Design* 9(7):2991–2994.

Ma, Y., B. Aichmayer, O. Paris, P. Fratzl, A. Meibom, R.A. Metzler, Y. Politi, L. Addadi, P.U. P. A. Gilbert, and S. Weiner. 2009. The grinding tip of the sea urchin tooth exhibits exquisite control over calcite crystal orientation and Mg distribution. *Proceedings of the National Academy of Sciences of the United States of America* 106(15):6048–6053. doi: 10.1073/pnas.0810300106.

Mahamid, J., L. Addadi, and S. Weiner. 2011a. Crystallization pathways in bone. *Cells Tissues Organs* 194:92–97.

Mahamid, J., B. Aichmayer, E. Shimoni, R. Ziblat, C. Li, S. Siegel, O. Paris, P. Fratzl, S. Weiner, and L. Addadi. 2010. Mapping amorphous calcium phosphate transformation into crystalline mineral from the cell to the bone in zebrafish fin rays. *Proceedings of the National Academy of Sciences of the United States of America* 107(14):6316–6321.

Mahamid, J., A. Sharir, L. Addadi, and S. Weiner. 2008. Amorphous calcium phosphate is a major component of the forming fin bones of zebrafish: Indications for an amorphous precursor phase. *Proceedings of the National Academy of Sciences of the United States of America* 105(35):12748–12753. doi: 10.1073/pnas.0803354105.

Mahamid, J., A. Sharir, D. Gur, E. Zelzer, L. Addadi, and S. Weiner. 2011b. Bone mineralization proceeds through intracellular calcium phosphate loaded vesicles: A cryo-electron microscopy study. *Journal of Structural Biology* 174(3):527–535. doi: 10.1016/j.jsb.2011.03.014.

Mann, S. 1995. Biomineralization and biomimetic materials chemistry. *Journal of Materials Chemistry* 5(7):935–946.

Mann, S., B.R. Heywood, S. Rajam, and J.D. Birchall. 1988. Controlled crystallization of calcium carbonate under stearic acid monolayers. *Nature* 334:692–695.

Mann, S. and G.A. Ozin. 1996. Synthesis of inorganic materials with complex form. *Nature* 382(6589):313–318.

Märkel, K., U. Röser, and M. Stauber. 1989. On the ultrastructure and the supposed function of the mineralizing matrix coat of sea urchins (Echinodermata, Echinoida). *Zoomorphology* 109(2):79–87. doi: 10.1007/bf00312313.

Marsh, M.E. 2004. Biomineralization in Coccolithophores. In *Biomineralization*, Bäuerlein, E. ed. pp. 195–215. Weinheim, Germany: Wiley-VCH Verlag GmbH & Co. KGaA.

Medalia, O., I. Weber, A.S. Frangakis, D. Nicastro, G. Gerisch, and W. Baumeister. 2002. Macromolecular architecture in eukaryotic cells visualized by cryoelectron tomography. *Science* 298(5596):1209–1213. doi: 10.1126/science.1076184.

Meldrum, F.C., B.R. Heywood, and S. Mann. 1992. Magnetoferritin: In vitro synthesis of a novel magnetic protein. *Science* 257(5069):522–523. doi: 10.1126/science.1636086.

Meldrum, F.C. and R.P. Sear. 2008. Now you see them. *Science* 322(5909):1802–1803. doi: 10.1126/science.1167221.

Moradian-Oldak, J., J. Tan, and A.G. Fincham. 1998. Interaction of amelogenin with hydroxyapatite crystals: An adherence effect through amelogenin molecular self-association. *Biopolymers* 46(4):225–238.

Munch, E., M.E. Launey, D.H. Alsem, E. Saiz, A.P. Tomsia, and R.O. Ritchie. 2008. Tough, bio-inspired hybrid materials. *Science* 322(5907):1516–1520. doi: 10.1126/science.1164865.

Nudelman, F., H.H. Chen, H.A. Goldberg, S. Weiner, and L. Addadi. 2007. Spiers Memorial Lecture. Lessons from biomineralization: Comparing the growth strategies of mollusc shell prismatic and nacreous layers in *Atrina rigida*. *Faraday Discussions* 136:9–25.

Nudelman, F., K. Pieterse, A. George, P.H.H. Bomans, H. Friedrich, L.J. Brylka, P.A.J. Hilbers, G. de With, and N. Sommerdijk. 2010. The role of collagen in bone apatite formation in the presence of hydroxyapatite nucleation inhibitors. *Nature Materials* 9(12):1004–1009. doi: 10.1038/nmat2875.

Nudelman, F., E. Shimoni, E. Klein, M. Rousseau, X. Bourrat, E. Lopez, L. Addadi, and S. Weiner. 2008. Forming nacreous layer of the shells of the bivalves *Atrina rigida* and *Pinctada margaritifera*: An environmental- and cryo-scanning electron microscopy study. *Journal of Structural Biology* 162(2):290–300. doi: 10.1016/j.jsb.2008.01.008.

Oaki, Y., A. Kotachi, T. Miura, and H. Imai. 2006. Bridged nanocrystals in biominerals and their biomimetics: Classical yet modern crystal growth on the nanoscale. *Advanced Functional Materials* 16(12):1633–1639. doi: 10.1002/adfm.200600262.

Olszta, M.J., X. Cheng, S.S. Jee, R. Kumar, Y.-Y. Kim, M.J. Kaufman, E.P. Douglas, and L.B. Gower. 2007. Bone structure and formation: A new perspective. *Materials Science and Engineering: R: Reports* 58(3–5):77–116.

Onuma, K. and A. Ito. 1998. Cluster growth model for hydroxyapatite. *Chemistry of Materials* 10(11):3346–3351.

Onuma, K., A. Ito, and T. Tateishi. 1996. Investigation of a growth unit of hydroxyapatite crystal from the measurements of step kinetics. *Journal of Crystal Growth* 167(3–4):773–776.

Ozaki, N., S. Sakuda, and H. Nagasawa. 2007. A novel highly acidic polysaccharide with inhibitory activity on calcification from the calcified scale "coccolith" of a coccolithophorid alga, *Pleurochrysis haptonemofera*. *Biochemical and Biophysical Research Communications* 357(4):1172–1176.

Palmer, L.C., C.J. Newcomb, S.R. Kaltz, E.D. Spoerke, and S.I. Stupp. 2008. Biomimetic systems for hydroxyapatite mineralization inspired by bone and enamel. *Chemical Reviews* 108(11):4754–4783. doi: 10.1021/cr8004422.

Pichon, B.P., P.H.H. Bomans, P.M. Frederik, and N.A.J.M. Sommerdijk. 2008. A quasi-time-resolved CryoTEM study of the nucleation of $CaCO_3$ under Langmuir monolayers. *Journal of the American Chemical Society* 130(12):4034–4040.

Plate, U., S. Arnold, L. Reimer, H.J. Hohling, and A. Boyde. 1994. Investigation of the early mineralization on collagen in dentin of rat incisors by quantitative electron spectroscopic diffraction (ESD). *Cell and Tissue Research* 278(3):543–547.

Plate, U., H.J. Hohling, L. Reimer, R.H. Barckhaus, R. Wienecke, H.P. Wiesmann, and A. Boyde. 1992. Analysis of the calcium distribution in predentine by EELS and of the early crystal formation in dentine by ESI and ESD. *Journal of Microscopy—Oxford* 166:329–341.

Politi, Y., T. Arad, E. Klein, S. Weiner, and L. Addadi. 2004. Sea urchin spine calcite forms via a transient amorphous calcium carbonate phase. *Science* 306(5699):1161–1164.

Pontoni, D., J. Bolze, N. Dingenouts, T. Narayanan, and M. Ballauff. 2003. Crystallization of calcium carbonate observed in-situ by combined small- and wide-angle x-ray scattering. *Journal of Physical Chemistry B* 107(22):5123–5125.

Popescu, D.C., M.M.J. Smulders, B.P. Pichon, N. Chebotareva, S.-Y. Kwak, O.L.J. van Asselen, R.P. Sijbesma, E. DiMasi, and N.A.J.M. Sommerdijk. 2007. Template adaptability is key in the oriented crystallization of $CaCO_3$. *Journal of the American Chemical Society* 129(45):14058–14067.

Posner, A.S. and F. Betts. 1975. Synthetic amorphous calcium phosphate and its relation to bone mineral structure. *Accounts of Chemical Research* 8(8):273–281.

Pouget, E.M., P.H.H. Bomans, A. Dey, P.M. Frederik, G. de With, and N.A.J.M. Sommerdijk. 2010. The development of morphology and structure in hexagonal vaterite. *Journal of the American Chemical Society* 132(33):11560–11565.

Pouget, E.M., P.H.H. Bomans, J.A.C.M. Goos, P.M. Frederik, G. de With, and N.A.J.M. Sommerdijk. 2009. The initial stages of template-controlled $CaCO_3$ formation revealed by cryo-TEM. *Science* 323(5920):1455–1458.

Qiu, S.R., A. Wierzbicki, C.A. Orme, A.M. Cody, J.R. Hoyer, G.H. Nancollas, S. Zepeda, and J.J. De Yoreo. 2004. Molecular modulation of calcium oxalate crystallization by osteopontin and citrate. *Proceedings of the National Academy of Sciences of the United States of America* 101(7):1811–1815. doi: 10.1073/pnas.0307900100.

Rieimer, L. and H. Kohl. 2008. *Transmission Electron Microscopy: Physics of Image Formation*, 5th ed. New York: Springer.

Robinson, R.A. 1952. An electron-microscopic study of the crystalline inorganic component of bone and its relationship to the organic matrix. *Journal of Bone and Joint Surgery—American Volume* 34(2):389–476.

Robinson, R.A. and F.W. Bishop. 1950. Methods of preparing bone and tooth samples for viewing in the electron microscope. *Science* 111(2894):655–657. doi: 10.1126/science.111.2894.655.

Robinson, R.A. and M.L. Watson. 1952. Collagen-crystal relationships in bone as seen in the electron microscope. *Anatomical Record* 114(3):383–409. doi: 10.1002/ar.1091140302.

Robinson, R.A. and M.L. Watson. 1955. Crystal-collagen relationships in bone as observed in the electron microscope. 3. Crystal and collagen morphology as a function of age. *Annals of the New York Academy of Sciences* 60(5):596–628. doi: 10.1111/j.1749-6632.1955.tb40054.x.

Sanchez, C., H. Arribart, and M.M.G. Guille. 2005. Biomimetism and bioinspiration as tools for the design of innovative materials and systems. *Nature Materials* 4(4):277–288.

Sarikaya, M. 1999. Biomimetics: Materials fabrication through biology. *Proceedings of the National Academy of Sciences of the United States of America* 96(25):14183–14185. doi: 10.1073/pnas.96.25.14183.

Scheffel, A., M. Gruska, D. Faivre, A. Linaroudis, J.M. Plitzko, and D. Schüler. 2006. An acidic protein aligns magnetosomes along a filamentous structure in magnetotactic bacteria. *Nature* 440(7080):110–114.

Sethmann, I., R. Hinrichs, G. Worheide, and A. Putnis. 2006. Nano-cluster composite structure of calcitic sponge spicules—A case study of basic characteristics of biominerals. *Journal of Inorganic Biochemistry* 100(1):88–96. doi: 10.1016/j.jinorgbio.2005.10.005.

Sethmann, I., A. Putnis, O. Grassmann, and P. Lobmann. 2005. Observation of nano-clustered calcite growth via a transient phase mediated by organic polyanions: A close match for biomineralization. *American Mineralogist* 90(7):1213–1217. doi: 10.2138/am.2005.1833.

Sethmann, I. and G. Worheide. 2008. Structure and composition of calcareous sponge spicules: A review and comparison to structurally related biominerals. *Micron* 39(3):209–228. doi: 10.1016/j.micron.2007.01.006.

Seto, J., Y. Ma, S.A. Davis, F. Meldrum, A. Gourrier, Y.-Y. Kim, U. Schilde et al. 2012. Structure-property relationships of a biological mesocrystal in the adult sea urchin spine. *Proceedings of the National Academy of Sciences of the United States of America* 109(10):3699–3704. doi: 10.1073/pnas.1109243109.

Sommerdijk, N. and H. Cölfen. 2010. Lessons from nature—Biomimetic approaches to minerals with complex structures. *MRS Bulletin* 35(2):116–119.

Sommerdijk, N.A.J.M. and G. de With. 2008. Biomimetic $CaCO_3$ mineralization using designer molecules and interfaces. *Chemical Reviews* 108(11):4499–4550.

Stupp, S.I. and P.V. Braun. 1997. Molecular manipulation of microstructures: Biomaterials, ceramics, and semiconductors. *Science* 277(5330):1242–1248. doi: 10.1126/science.277.5330.1242.

Su, X.W., A.M. Belcher, C.M. Zaremba, D.E. Morse, G.D. Stucky, and A.H. Heuer. 2002. Structural and microstructural characterization of the growth lines and prismatic microarchitecture in red abalone shell and the microstructures of abalone "flat pearls". *Chemistry of Materials* 14(7):3106–3117. doi: 10.1021/cm011739q.

Sumper, M. 2002. A phase separation model for the nanopatterning of diatom biosilica. *Science* 295(5564):2430–2433. doi: 10.1126/science.1070026.

Suvorova, E.I. and P.A. Buffat. 1999. Electron diffraction from micro- and nanoparticles of hydroxyapatite. *Journal of Microscopy—Oxford* 196:46–58.

Talmon, Y. 1996. Transmission electron microscopy of complex fluids: The state of the art. *Berichte Der Bunsen-Gesellschaft—Physical Chemistry Chemical Physics* 100(3):364–372.

Teng, H.H., P.M. Dove, and J.J. De Yoreo. 2000. Kinetics of calcite growth: Surface processes and relationships to macroscopic rate laws. *Geochimica et Cosmochimica Acta* 64(13):2255–2266. doi: 10.1016/s0016-7037(00)00341-0.

Thomas, S.J. 2009. The renaissance and promise of electron energy-loss spectroscopy. *Angewandte Chemie International Edition* 48(47):8824–8826.

Towe, K.M. and H.A. Lowenstam. 1967. Ultrastructure and development of iron mineralization in the radular teeth of *Cryptochiton stelleri* (mollusca). *Journal of Ultrastructure Research* 17(1–2):1–13.

Vos, M.R., P.H.H. Bomans, P.M. Frederik, and N. Sommerdijk. 2008. The development of a glove-box/Vitrobot combination: Air-water interface events visualized by cryo-TEM. *Ultramicroscopy* 108(11):1478–1483. doi: 10.1016/j.ultramic.2008.03.014.

Wang, C.Y., C. Bottcher, D.W. Bahnemann, and J.K. Dohrmann. 2004. In situ electron microscopy investigation of Fe(III)-doped TiO_2 nanoparticles in an aqueous environment. *Journal of Nanoparticle Research* 6(1):119–122. doi: 10.1023/b:nano.0000023222.85864.78.

Wang, R.Z. and S. Weiner. 1998. Strain-structure relations in human teeth using Moire fringes. *Journal of Biomechanics* 31(2):135–141.

Weiner, S. and L. Addadi. 1997. Design strategies in mineralized biological materials. *Journal of Materials Chemistry* 7(5):689–702.

Weiner, S. and L. Addadi. 2011. Crystallization pathways in biomineralization. *Annual Review of Materials Research* 41(1):21–40. doi: doi:10.1146/annurev-matsci-062910-095803.

Weiner, S., T. Arad, I. Sabanay, and W. Traub. 1997. Rotated plywood structure of primary lamellar bone in the rat: Orientations of the collagen fibril arrays. *Bone* 20(6):509–514.

Weiner, S. and P.M. Dove. 2003. An overview of biomineralization processes and the problem of the vital effect. In *Biomineralization*, eds. P.M. Dove, J.J. DeYoreo, and S. Weiner, pp. 1–29. Washington, DC: Mineralogical Society of America/Geochemical Society.

Weiner, S., Y. Talmon, and W. Traub. 1983. Electron-diffraction of mollusk shell organic matrices and their relationship to the mineral phase. *International Journal of Biological Macromolecules* 5(6):325–328. doi: 10.1016/0141-8130(83)90055-7.

Weiner, S., A. Veis, E. Beniash, T. Arad, J.W. Dillon, B. Sabsay, and F. Siddiqui. 1999. Peritubular dentin formation: Crystal organization and the macromolecular constituents in human teeth. *Journal of Structural Biology* 126(1):27–41.

Weiss, I.M., N. Tuross, L. Addadi, and S. Weiner. 2002. Mollusc larval shell formation: Amorphous calcium carbonate is a precursor phase for aragonite. *Journal of Experimental Zoology* 293(5):478–491.

Westenfeld, R., C. Schafer, T. Kruger, C. Haarmann, L.J. Schurgers, C. Reutelingsperger, O. Ivanovski et al. 2009. Fetuin—A protects against atherosclerotic calcification in CKD. *Journal of the American Society Nephrology* 20(6):1264–1274.

White, S.N., W. Luo, M.L. Paine, H. Fong, M. Sarikaya, and M.L. Snead. 2001. Biological organization of hydroxyapatite crystallites into a fibrous continuum toughens and controls anisotropy in human enamel. *Journal of Dental Research* 80(1):321–326.

Williams, D.B. and C.B. Carter. 1996. *Transmission Electron Microscopy 4 Vol Set: A Textbook for Materials Science*, 4 vols. New York: Plenum Press.

Williamson, M.J., R.M. Tromp, P.M. Vereecken, R. Hull, and F.M. Ross. 2003. Dynamic microscopy of nanoscale cluster growth at the solid-liquid interface. *Nature Materials* 2(8):532–536.

Wilt, F.H. 2005. Developmental biology meets materials science: Morphogenesis of biomineralized structures. *Developmental Biology* 280(1):15–25. doi: 10.1016/j.ydbio.2005.01.019.

Wong, T.-S., S.H. Kang, S.K.Y. Tang, E.J. Smythe, B.D. Hatton, A. Grinthal, and J. Aizenberg. 2011. Bioinspired self-repairing slippery surfaces with pressure-stable omniphobicity. *Nature* 477(7365):443–447.

Xia, F. and L. Jiang. 2008. Bio-inspired, smart, multiscale interfacial materials. *Advanced Materials* 20(15):2842–2858.

Xilin, Y. and J.S. Malcolm. 2003. Biological calcium phosphates and Posner's cluster. *Journal of Chemical Physics* 118(8):3717–3723.

Xu, G.F., N. Yao, I.A. Aksay, and J.T. Groves. 1998. Biomimetic synthesis of macroscopic-scale calcium carbonate thin films. Evidence for a multistep assembly process. *Journal of the American Chemical Society* 120(46):11977–11985. doi: 10.1021/ja9819108.

Yuk, J.M., J. Park, P. Ercius, K. Kim, D.J. Hellebusch, M.F. Crommie, J.Y. Lee, A. Zettl, and A.P. Alivisatos. 2012. High-resolution EM of colloidal nanocrystal growth using graphene liquid cells. *Science* 336(6077):61–64. doi: 10.1126/science.1217654.

Yuwono, V.M., N.D. Burrows, J.A. Soltis, and R.L. Penn. 2010. Oriented aggregation: Formation and transformation of mesocrystal intermediates revealed. *Journal of the American Chemical Society* 132(7):2163–2165. doi: 10.1021/ja909769a.

Zewail, A. 2010. The new age of structural dynamics. *Acta Crystallographica Section A* 66(2):135–136. doi: 10.1107/S0108767309047801.

Zheng, H.M., R.K. Smith, Y.W. Jun, C. Kisielowski, U. Dahmen, and A.P. Alivisatos. 2009. Observation of single colloidal platinum nanocrystal growth trajectories. *Science* 324(5932):1309–1312. doi: 10.1126/science.1172104.

Ziv, V., I. Sabanay, T. Arad, W. Traub, and S. Weiner. 1996. Transitional structures in lamellar bone. *Microscopy Research and Technique* 33(2):203–213. doi: 10.1002/(sici)1097–0029(19960201)33:2<203::aid-jemt10>3.0.co;2-y.

15 X-ray computed tomography

Xianghui Xiao and Stuart R. Stock

Contents

15.1 INTRODUCTION

Computed tomography (CT) is a volumetric measurement technique widely used to study mineralized tissue, mostly in the clinical environment. In this chapter, much of the clinical CT activity is of only peripheral interest, and the treatment here will focus on x-ray-based techniques as they are applied to study biomineralization. It is, however, important to remember that CT can also be performed with sound waves or with visible light, not just x-radiation. MicroCT and nanoCT are higher resolution versions of clinical CT and comprise the bulk of applications described in the following sections.

The technique will be introduced in Section 15.2; biomineralization applications follow in Section 15.3, and the outlook for the future is discussed in Section 15.4. There are many more significant details not covered in this chapter, and the reader can find amplification elsewhere (Stock 1999, 2008a,b, 2012a; Neues and Epple 2008) or in any of the several review papers published each year.

15.2 COMPUTED TOMOGRAPHY

Biomineralization can be studied with CT at many different size scales ranging from clinical (*in vivo* in human patients) to nanoscopic (in micrometer-sized specimens), and the instrumentation varies accordingly. Section 15.2.1 briefly outlines interactions of the x-ray beam with specimens. The principles of CT are described in Section 15.2.2 for two reconstruction algorithms. In Section 15.2.3, tomography systems are outlined, and Section 15.2.4 focuses on phase contrast CT. Section 15.2.5 treats practical aspects of CT including brief mention of some common reconstruction artifacts.

15.2.1 X-RAY INTERACTIONS WITH SPECIMENS

X-ray imaging has been explored since shortly after Röntgen discovered x-rays in 1895 (Röntgen 1898). In the early era of x-ray imaging, 2D radiography was applied. Because x-rays are highly penetrating, they are suitable for volume imaging. Driven by the high demands from medical community, development of 3D x-ray imaging eventually resulted in 1972 in the first CT system for medical imaging (Hounsfield 1973). Soon after its invention, the applications of CT extended to fields other than medical diagnosis including areas as diverse as industrial inspection and scientific research on inorganic materials.

When x-rays pass through a sample, the x-ray beam interacts with the sample and, upon exiting the specimen, carries information about the sample. In the low x-ray energy range, the interactions between x-ray and matter are dominated by the photoelectric process and by elastic scattering, both of which can modify the incident x-ray beam intensity and direction. Tomographic reconstruction of the sample structure can be achieved by extracting information from the exiting x-ray beam. Depending on the type of detected signal, it is possible to obtain CT reconstruction of the spatial distribution of the linear attenuation coefficient (absorption contrast), refractive index (phase contrast), crystallographic structure (diffraction contrast), elemental distribution (fluorescence contrast), particle size and shape (scattering contrast), and local bonding structure (spectroscopy contrast). In this section, the focus will only be on absorption contrast and on phase contrast.

15.2.2 PRINCIPLES OF COMPUTED TOMOGRAPHY

CT reconstructs slices of a specimen digitally from the projections of the specimen along different directions. One CT reconstruction algorithm is called the algebraic reconstruction technique (ART) and is based on a model of CT problem consisting of a system of linear equations (Kak and Slaney 2001). Intuitively, it is easy to understand CT principles from the ART point of view. In the case of absorption contrast imaging, shadow images of a specimen are recorded. The intensity fluctuation in a shadow image is due to the specimen attenuation of the incident x-ray, which can be expressed by Beer's law:

$$I = I_0 \exp[\int_l \mu(\vec{r})\,dt], \tag{15.1}$$

where

I and I_0 are the transmitted intensity at the image plane and the incident x-ray intensity, respectively

$\mu(\vec{r})$ is the position-dependent linear attenuation coefficient of the sample

The integral is along the x-ray path. Equation 15.1 can be approximated by a discrete model illustrated in Figure 15.1. In the discrete model, Equation 15.1 is rewritten as

$$-\ln\left(\frac{I}{I_0}\right)[s,\theta] = \sum_{i,j} t[s,\theta;i,j]\cdot\mu[i,j], \tag{15.2}$$

where

$\mu[i, j]$ is the sample attenuation coefficient associated with voxel (i, j)

$t[s, \theta; i, j]$ is the intersection length between the voxel (i, j) and the ray along the θ direction, which is received by the detector's sth pixel

This discrete model represents the tomographic reconstruction problem as a system of linear equations. If the data from the measurements is sufficient, the system is determined. Therefore, the sample structure, in terms of the linear attenuation coefficient map, can be reconstructed uniquely from the projection measurements. The rank of the linear equation system in Equation 15.2 is usually very high. The number of unknown variables is of the order of i^*j, and the independent projection measurements in terms of rays should exceed i^*j. For instance, to solve a sample structure of 128 × 128 voxels, the rank of the linear system has to be at least $128^2 \times 128^2$. It is impractical to solve a linear equation system of such size directly. ART solves such linear equations in an iterative manner by starting with a guessed or random initial sample structure $u[i, j]$ and modifying $u[i, j]$

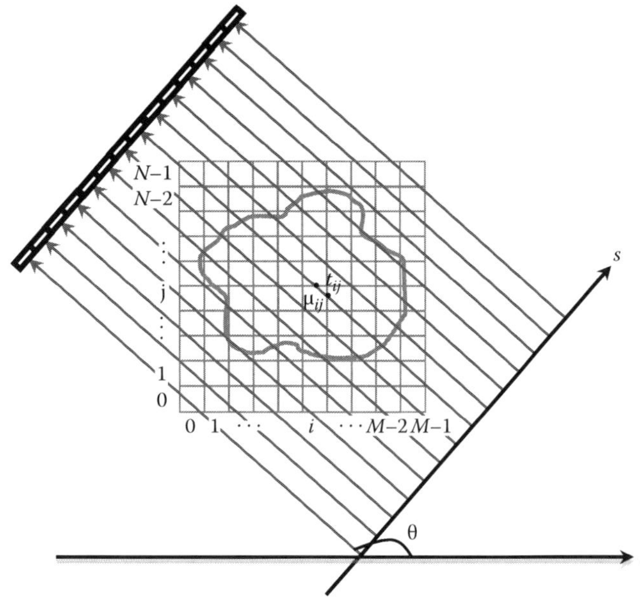

Figure 15.1 Illustration of geometry in ART. A sample is meshed in discrete Cartesian grid, and the linear attenuation coefficient at each pixel is represented as μ_{ij}, where (i, j) is the pixel coordinate in the Cartesian grid. Parallel x-ray beam illuminates on the sample. The ray shooting to detector pixel s intersects the pixel (i, j), and the intersection length is $t(i, j;\theta, s)$. The contribution to the total attenuation from pixel (i, j) is therefore $\mu(i, j)^*t(i, j; \theta, s)$.

according to the difference between the calculated projection values and the measured projection values at each detector pixel and each angle. The correct sample structure can be reached gradually after a few iterations (Natterer 2001).

ART is computation intensive, but there are other tomographic reconstruction algorithms that require less computer resources. Filtered back-projection (FBP) is one of the most popular approaches, because of its easy implementation and high computational efficiency. For the parallel-beam geometry shown in Figure 15.1, the Fourier transform of a sample projection along a line perpendicular to the x-ray direction is a line sliced from the 2D Fourier transform of the sample in the x–y plane. This is called the Fourier slice theorem (Kak and Slaney 2001). By imaging the sample along different projection directions, the 2D Fourier transform of the sample slice is obtained in polar coordinates. The inverse Fourier transform can then reconstruct the sample structure in real space. The inverse Fourier transform from a distribution in polar coordinates (in frequency space) to the corresponding distribution in Cartesian coordinates (in real space) involves filtering along the radial directions in the Fourier space and back-projecting along all angles in the real space. Thus, the name of the algorithm.

15.2.3 TOMOGRAPHY SYSTEMS

There are different geometrical arrangements practical for tomography (Stock 2008b). The first generation CT scanner that was developed during 1970s is based on pencil-beam raster scanning (i.e., translation–rotation scanning where the pencil beam travels across the sample, the specimen is rotated, etc.). This is parallel-beam CT geometry and is still in use in fluorescence, scattering, and diffraction-based tomography at synchrotron facilities. The parallel-beam CT geometry is also used in synchrotron-radiation-based microCT (SR microCT) with full field data acquisition. In this case, the sample is illuminated with an extended parallel x-ray beam, and the projection images of the sample are recorded with a 2D imaging detector, usually CCD or CMOS camera coupled to an x-ray scintillator. With parallel-beam FBP as described earlier, accurate sample structure can be reconstructed slice by slice, given suitable angular sampling. In second generation CT scanners, a fan-beam geometry is employed. An x-ray beam from a point source is limited to a thin ribbon by a slit, and the sample is positioned in the fan of radiation and rotated around an axis that is perpendicular to the fan plane. The parallel-beam FBP algorithm can be modified to accommodate the difference between parallel-beam and fan-beam geometries. In third generation CT scanners, a cone-beam geometry is employed and can be found in many commercial tube-based microCT systems. As suggested by the name, a cone of x-rays illuminates the sample. Unlike the fan-beam geometry, volumes can be reconstructed from a single sample scan. An algorithm developed by Feldkamp, Davis, and Kress, known as FDK algorithm (Feldkamp et al. 1984), is widely used for cone-beam tomography reconstruction. The only slice that is "accurately" reconstructed is that of the midplane (the plane perpendicular to the rotation axis and containing the source point), and distortion within reconstructed slices increases with increasing angle (distance from the midplane). The efficiency and simplicity of implementation outweigh the (mostly) minor inaccuracies, and many such commercial systems are available.

Tomographic resolution depends on several factors, including the intrinsic resolving power of the imaging system, the contrast between different sample constituents, and the tomography scan geometry. In tomography experiments, some kind of imaging detector is placed behind a sample to record the transmitted x-ray intensity as altered by the sample. In the pencil-beam raster scan approach, a detector receives the signals from the volume traversed by the x-ray beam, and the detector is not position sensitive. In this case, spatial resolution is limited to the pencil-beam size. In other tomography geometries, 1D or 2D x-ray imaging detectors are utilized. These detectors can be either directly sensitive to x-rays or can operate using photons from the visible spectrum produced from an x-ray scintillator. With point detection, the spatial resolution of the detector is limited to twice pixel size. In the second case, the spatial resolution of the detector system is limited by both blooming within the scintillation material and the detector pixel size. It is also possible to use an optical microscope objective lens to magnify the x-ray image on the scintillation screen and to project the image onto the detector. In this case, spatial resolution also depends on the detector size and the numerical aperture and aberrations of the lens. This is the approach adopted in SR microCT and some tube-source microCT systems. Alternatively, x-ray images of the sample can be magnified either by the natural divergence in cone-beam geometry or by x-ray optics. Although the magnification in cone-beam geometry can be very large if the detector is placed far from the sample and source, the spatial resolution in this case is limited by the source size since the source projection in the images is also proportionally increased along with magnification, blurring the resulting images. It is therefore critical to employ a very small source size to obtain high spatial resolution in cone-beam systems. This is challenging with a tube source because of the inverse correlation between the source size and source brightness, highlighting the trade-off between spatial resolution and temporal resolution. With a synchrotron source, the cone-beam geometry is achieved by focusing the near-parallel x-ray beam to a small spot with an x-ray mirror or lens. Because this case does not involve an inverse correlation between focused spot size and brightness, it is possible to achieve high spatial resolution and high temporal resolution simultaneously. The challenge in this case is how to make the focused spot as uniform as possible. Currently, 50 nm spatial resolution with synchrotron radiation and 100 nm resolution with a tube source can be achieved with cone-beam geometry (Withers 2007), but with very different temporal resolution. Imaging with voxel sizes below ~0.5 μm is often termed nanoCT.

15.2.4 X-RAY PHASE CONTRAST

In x-ray CT, contrast is also a concern, and it is generated by the differences in interaction of the x-rays with the regions of the specimen. In most circumstances, but certainly not all, the contrast in the radiograph can be attributed to absorption variation in the sample, as expressed by Beer's law earlier (Equation 15.1). This is the case of most commercial CT and microCT systems. In the case of materials like bone and soft tissues, absorption contrast allows excellent visualization of the bone constituent. In the relatively low x-ray energy regime (<50 keV), absorption contrast between different materials is small if the compositions of the materials are close (Carlson 2006). This is a problem, for example, when soft tissue samples are imaged at high resolution, because the x-rays are not absorbed significantly over a few micrometers length; so, features of that size do not have a signature in the measured images. Such features, therefore, will not be visible in CT reconstructions dominated by absorption contrast.

Absorption is not the only interaction of x-rays with matter. X-rays, like radio waves and visible light, are electromagnetic waves that propagate with a given amplitude and phase. X-ray wave fronts are distorted when propagating through regions of a specimen with different electron densities, and, phase varies across the wave front when it emerges from the specimen. Several methods exist for converting this x-ray phase modulation into variations in intensity at the detector, a subject that is beyond the scope of this brief treatment. Coherent or partially coherent x-ray sources (analogous to optical lasers) allow phase contrast to be seen, and incoherent sources (analogous to light bulbs) obscure the effects of phase contrast (see following paragraph). In the hard x-ray regime, phase contrast can be three orders of magnitude stronger than absorption contrast. Thus, it is possible to see a small feature that has no absorption contrast compared to its surroundings. Phase contrast is important in high-resolution CT imaging.

Source coherence is inversely proportional to the source size seen at the sample position, that is, a given size source appears more coherent when viewed from farther away. The x-ray sources in most commercial CT systems are almost incoherent and are, therefore, not suitable for direct phase contrast imaging. The microfocus x-ray sources used in the latest high-resolution CT systems, however, exhibit a limited degree of coherence and allow phase-contrast imaging. X-ray sources based on SEM systems possess coherence similar to synchrotron radiation sources. However lab-based sources are low in flux; so, the imaging speed is slower than that with synchrotron sources. Synchrotron sources on the other hand are very bright and highly collimated. Usually, synchrotron-imaging experiments are performed in some experimental station that is far away from the source (20 m or more). This guarantees high-flux, highly coherent x-ray illumination for rapid, high-resolution phase-contrast imaging. Phase-contrast images can be directly used for tomographic reconstructions. Edge-enhanced 3D structures result, but the contrast between the different material phases' interior regions is not very different from absorption contrast. Single-distance, single-shot phase-contrast imaging is more suitable for studying *in situ* dynamical processes. There are multiple approaches for quantitative phase-contrast imaging; the in-line approaches are widely used (Mayo et al. 2012), and descriptions of different experimental approaches for phase CT can be found elsewhere (Stock 2008b).

15.2.5 PRACTICAL ASPECTS OF TOMOGRAPHIC IMAGING

Before turning to applications of CT in studying biomineralization, it is useful to mention several practical topics very briefly. These are tradeoffs between noise in the data, counting time, spatial and angular resolutions; sample diameter versus voxel size and approximations including local tomography and other truncated data acquisition; reconstruction artifacts and partial voxel effects on segmentation.

Noise within reconstructions limits contrast sensitivity, and noise levels depend on the dynamic range of the detector as well as the sample characteristics. If the beam passing through the sample consists of N counts, then counting statistics dictate that this signal's standard deviation is $\sqrt{N}$. Increasing the signal-to-noise within the reconstruction by a factor of 2 requires counting for four times as long (but this is not the only factor that is important). If one decreases the size of the voxel in the reconstructed volume by a factor of q, one needs to collect more projections (see following paragraph) and count q^4 times as long.

Typically, specimens are rotated by small, constant angular increments, until the required views are recorded (180° for all but the cone-beam systems which require 360°). As dictated by the Nyquist sampling theorem, the number of views should be close to the number of samples of the radiograph, N_{pix} (Kak and Slaney 2001). The size of the voxels within a reconstructed slice relates resolution with which the radiograph is sampled. Practically, the projection of the specimen diameter *dia* onto the detector array of N_{pix} pixels allows reconstruction with voxel size $vox \sim dia/N_{pix}$, that is, a 2000 detector element array used with a 2 mm diameter specimen permits reconstruction with 1 μm voxels. This relationship can be circumvented by allowing some of the sample to lie outside the field-of-view (FOV) of the detector, an approach termed local or region-of-interest tomography. Local tomography generally produces geometrically accurate reconstructions, but the linear attenuation coefficients may be inaccurate. Samples can be reconstructed successfully with missing angular ranges of 10° or more, but often strong streaks artifacts result.

Artifacts include ring, beam hardening, phase contrast, motions, angular undersampling, range of missing angles, streaks from long, flat surfaces, or from opaque domains or cross talk between adjacent detector rows or columns (Stock 2008b). Ring artifacts result from abnormal response or incorrect normalization of individual detector elements, appear as arcs of high or low contrast concentric with the reconstruction center, and are easily recognized but sometimes interfere with segmentation. Beam hardening occurs with polychromatic radiation: sample interiors will appear to be less attenuating than the outer portions, that is, "cupping" is observed. Phase contrast may produce sharp light/dark fringes at interfaces; this edge-sharpening effect is helpful in detecting interfaces but can hinder segmentation. Any sample motion during data acquisition can spoil reconstructions; bubble formation in liquids in the FOV or motions of different portions of soft samples are examples. Angular undersampling tends to produce streaks or mottling within reconstructions, and the various streak artifacts reflect undersampling of a different sort.

Partial volume effects often confound analyses and complicate segmentation. If the specimen consists of one solid phase and air-filled porosity, the reconstructed voxels span the solid–air interface to greatly different extents, and this broadens the histogram of linear attenuation coefficients. If only air and solid were present, voxel values intermediate between air and dense solid simply reflect different volume fractions of solid.

15.3 APPLICATIONS

With >10^3 microCT and nanoCT papers appearing annually (Stock 2012b), there is no shortage of examples that could be used to illustrate how CT has been used to study biomineralization. A recent work (Neues and Epple 2008), for example, covers a wide variety of microCT studies of biomineralization but, by reason of length, only touches lightly on the subject. The examples of biomineralization studies could be organized in many different ways, but this section, in general terms, will proceed from larger to smaller "samples." As it happens, this dictates that the first portion of this section will center on bone and tooth (mineral phase of carbonated hydroxyapatite, cAp) and only later cover nonapatitic calcium-based and noncalcium-based mineral systems. Section 15.3.1 describes clinical CT and how it is used to study mineralization. Section 15.3.2 covers *ex vivo* CT imaging, mostly at the micro- but also at the nanoscale, and Section 15.3.3 covers *in vivo* microCT such as is applied to small animal models. Analysis of structures within mineralized tissue is the focus of Section 15.3.4. Finally, Section 15.3.5 discusses studies of scaffolds for mineralized tissue.

15.3.1 BIOMINERALIZATION IN THE CLINICAL SETTING

In the clinical *in vivo* setting, minimizing the radiation dose received by the patient dominates the study of biomineralization. This significantly impacts the achievable spatial resolution and contrast sensitivity. In the context of biomineralization, the scale of abdominal or peripheral CT dictates that pathologies of mineralization are most often studied, that is, where bone is absent or where mineralization occurs where it is not supposed to be. Vessel calcification is one example. Figure 15.2 shows

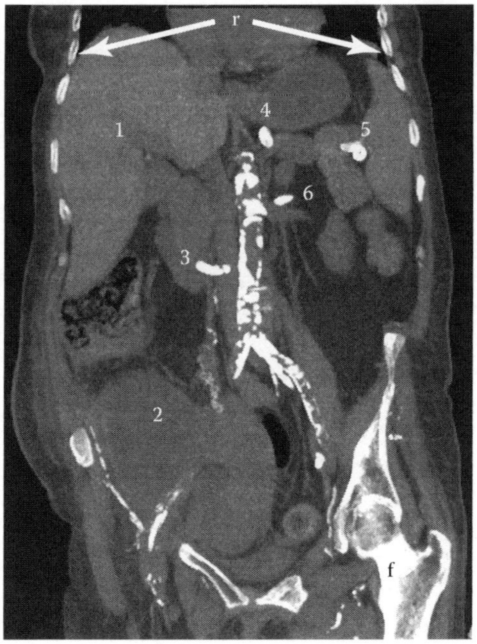

Figure 15.2 Patient with extensive abdominal vascular calcification. The lighter the pixel, the greater the x-ray attenuation; unless noted otherwise, the contrast is the same in all of the figures. Features identified in this coronal section are as follows: r, ribs; f, femur; 1, liver; 2, fluid collection right lower quadrant next to renal transplant; 3, right renal artery; 4, portion of proximal splenic artery; 5, distal splenic artery; 6, left renal artery. (Image provided by L.A. Goodhartz, Feinberg School of Medicine, Northwestern University, Chicago, IL.)

a coronal section (section parallel to the plane dividing the body into anterior and posterior halves) of a patient who had had a kidney transplant. In Figure 15.2, soft tissue "1" and fluid "2" are virtually indistinguishable (light gray), but have very different contrast from bone and calcified vessels (white). Ribs "r" and femur "f" on the one hand and calcified vessels "3–6" on the other are also difficult to differentiate based on the contrast in this image. Kidney stone diagnosis is also performed with CT. In Thoroughbred horse radia and tibiae, bone strength was significantly increased in exercise-conditioned versus unconditioned animals, a result that CT showed could be attributed to changes in bone size but not differences in bone mineral content or density (Nicholson and Firth 2010). Changes in bone density, geometry, and biomechanical properties of the proximal femur can be followed longitudinally with clinical CT (Ito et al. 2011). Clinical CT is capable of providing accurate thickness density and mass measurements of the femoral cortex (Treece et al. 2012), but probably not accurate values of quantities like trabecular thickness. Obtaining the requisite spatial resolution and biochemical information needed to illuminate the biomineralization processes generally requires the mineral products to be excised and studied *ex vivo*.

One exception is high-resolution peripheral quantitative CT (HR-pQCT) designed for studying trabecular bone in the limbs (typical FOV is ~125 mm with voxel sizes of ~120 μm); resolution suffices for accurate 3D histomorphometry of trabecular bone (see the following subsection); contrast can be related to apparent bone mineral density and tissue level mineral density if suitable calibration standards and beam hardening corrections are employed (Burghardt et al. 2010). Another arena where CT is employed to study biomineralization *in vivo* is in maxillofacial imaging. In preoperative planning for implanting prostheses, it is important to insure that the local bone environment is suitable; if not, the procedure must be delayed until osseous topography and/or bone volume excesses/deficiencies can be corrected. In order to evaluate the implant site, use of a maxillofacial CT system is recommended (Tyndall 2012).

The response of jaw (and other bones) to applied stress can be evaluated through CT-based finite element modeling. The biomechanical analysis of stresses and strains in the ostrich mandible is one example (Rayfield 2011). New clinical vertebral fractures can be predicted in elderly men using finite element analysis of CT scans (Wang et al. 2012).

15.3.2 *EX VIVO* ANALYSES

Osteoporosis is a skeletal disease characterized by low bone mass, microarchitectural deterioration, and altered nanoscopic properties of bone (i.e., lowered material quality). The result is bone fragility and increased susceptibility to fracture. In 2004, an estimated 10 million Americans older than 50 years had osteoporosis, with ~1.5 million fragility fractures occurring in these patients each year, and an additional 34 million Americans were at risk (US Surgeon General 2004). Osteoporosis is a significant biomineralization problem in that it may relate to an imbalanced bone removal versus bone addition during remodeling (Raisz 2008) or to changes in the mineralized tissue material properties (Ritchie et al. 2006).

Dual-energy x-ray absorptometry (DXA) is a radiography-based approach for diagnosing osteopenia (lower than normal bone mineral density) and osteoporosis (Chun 2011). DXA measures the transmission of x-rays (at two tube potentials) through selected bone sites. Clinical CT is problematic for *in vivo* osteoporosis studies, because sites of greatest concern, the femoral neck and vertebrae (Harvey et al. 2010), are dominated by trabecular bone with relatively thin cortices and large amounts of bone partial volumes, making it difficult to quantify architectural parameters (volume fraction of mineralized tissue BV/TV, mean trabecular thickness <Tb.Th.>, connectivity, structure model index SMI (parameter quantifying how rod- or plate-like the structure is on average), etc. (Hildebrand and Rüegsegger 1997b; Odgaard 2001; see following section) or cortical thicknesses. Some progress is being made through advanced image analysis (Treece et al. 2012). Femoral head failure during sideways falls is reasonably well predicted by CT-based FEM, which also correlates with DXA-based measures of bone mineralization (Koivumaki et al. 2012). Quality of implant fixation in the hip has also been studied with CT-based FEM (Shim et al. 2012).

An *ex vivo* scan of a human femoral head is shown in Figure 15.3. Here, the trabeculae and thin cortex are clearly visible, and quantification of microarchitectural parameters would be straightforward. This particular specimen exhibits fracture, with the neck of the femur being partially pushed into the head and locally crushing the preexisting trabeculae. Formation of a fracture callus shows where healing is progressing. With voxel size of 26 μm, the data in Figure 15.3 falls into the realm of microCT.

The entire skeleton of small animals can be imaged and used to establish the phenotype of different knockout models. Examples

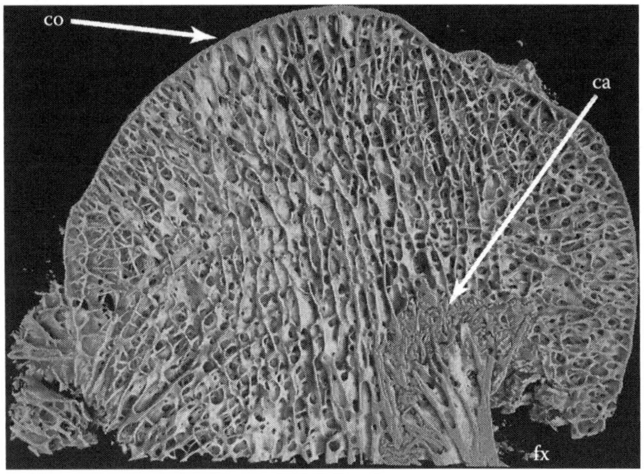

Figure 15.3 Human femoral head imaged *ex vivo* in a time-delay integration scanner (Davis and Elliott 1997). The image is a 3D rendering numerically sectioned to expose the interior of the bone, and the femoral neck lies below this volume. The bone's cortex is labeled "co," a fracture site where the femoral neck was pushed into the femoral head is "fx," and the healing callus at the end of the fracture is labeled "ca." The volume was scanned in four blocks of 2801 projections over 360°. The data required ~132 h to image, and was immersed in glycerol to prevent dehydration. Each block was reconstructed with 1888 × 1888 × 1306 voxels, 26 μm on a side. More details of the study can be found elsewhere (Ahmed 2011). (Image courtesy of F. Ahmed and G.R. Davis, Queen Mary, University of London, London, U.K.)

include the vertebrae and ribs of wild type and deformed medaka fish (Neues et al. 2007), and a mouse model of skeletal dysplasia (Wang et al. 2011). Since the mid-1990s, the most popular use of microCT has been to quantify microarchitecture of bone biopsies or of the entire distal or proximal ends of long bones of rat and mouse models of disease (Kinney et al. 1995; Bonse et al. 1996; Hildebrand and Rüeggeseger 1997a;); *ex vivo* microCT imaging of whole human vertebrae (Fields et al. 2012), for example, formed the basis of FEM analysis of whether age-related vertebral fragility was inversely related to the amount of structural redundancy present, that is, the ability to safely redistribute stress internally after local trabecular failure. Most studies of trabecular microarchitecture, however, focus on the differences between controls, disease models, and treatments where accurate comparisons must be made not only of changes in BV/TV but also in <Tb.Th.> and SMI.

The first step in quantifying mineralized tissue* microarchitectural parameters in microCT data sets is to segment the data into bone and other tissue. Typically, there are two peaks in the histogram of voxel values, one of bone and the other of soft tissue plus fluid (or medium in which the bone is cast or air), and the threshold for segmentation is typically chosen to be midway between the flanks of the two peaks. Most reasonable definitions for the segmentation do not appear to affect comparisons of populations, but the imaging conditions must remain the same. If the voxel dimensions are of the same size as the minimum dimensions of the structure of interest, however, then partial volume effects dominate and much more care must be taken to avoid unintended bias. Measurement of thicknesses requires 3D data sets and techniques such as the distance transform algorithm (Hildebrand and Rüegsegger 1997a), and not simple measurements of apparent thickness in 2D sections.

Segmented data are often presented as 3D renderings of trabecular structures and provide immediately intelligible comparison between populations, but numerical comparisons are definitive. For example, sites of osteoblastic metastases, degenerative osteosclerosis, and matching normal human lumbar vertebrae had very different microarchitectural parameters (Sone et al. 2004). Treatment with raloxifene produced trabecular bone (beagle femoral neck), which had significantly greater compressive ultimate stress, modulus, and toughness than bone from animals treated with vehicle, despite there being no difference in the microarchitectural parameters (Allen et al. 2007). Several parameters express more or less the same microarchitectural measures, and one group has used principal component analysis to determine the variables responsible for the variance in mechanical testing of cylindrical samples cut from human mandibular condyles (Giesen et al. 2003).†

Finite element analysis of microCT data frequently appears in the literature. In one study, the contributions of individual trabecular rods to elastic properties of human vertebral bone were examined virtually (Liu et al. 2009), something that cannot be done any other way. A second example is a study of different structures within diadematoid sea urchin spines (Yuan 2012), performed by virtual alteration of the structures. Finite elements were also used to study the mechanics of keratin sheaths over bone cores in dinosaur claws (Manning et al. 2009).

In many vertebrates, bone is replaced over the life of the animal (remodeling, discussed in Section 15.3.4). Some adult amphibians such as newts can completely regenerate amputated limbs, and microCT has been used to explore the timescale of bone replacement (Stock et al. 2003a). Mammals repair fractured bones by forming a mineralized callus around the break, a rapidly formed, highly trabecular, and relatively lightly mineralized tissue (Figure 15.4). MicroCT parameters reflecting callus structure and size (polar moment of inertia, bending moment of inertia, and total callus volume) were more sensitive to changes in the callus (mouse femur, 12 weeks) than measures assessing callus substance (tissue mineral density, BV/TV, and bone mineral density) (O'Neill et al. 2012). Mesenchymal stem cell treatment restored both fracture callus volume (measured with microCT) and biomechanical strength ($p < 0.05$) in animals with alcohol-impaired healing (Obermeyer et al. 2012).

Sea urchins and other echinoderms make their mineralized tissue from calcite. Each echinoderm ossicle, a skeletal organ that is an analog of a bone in vertebrates, is a single crystal that has a fine, fenestrated structure reminiscent of trabecular bone but with dimensions ~1–10 μm instead of bone's ~ 100 μm. MicroCT with 1–2 μm voxels can resolve stereom and allow measurement

* Although almost all microarchitecture quantification involves bone, the same approach has been applied to the part of the sea urchin jaw structure, a fenestrated, single crystal of calcite (Stock et al. 2003b).

† Principal component analysis has also been used to establish phylogenic relationships between claw-bearing and nail-bearing digits in primates (Maiolino et al. 2012) and between sea urchins based on the macrostructure of their tooth cross-sections (Ziegler et al. 2012).

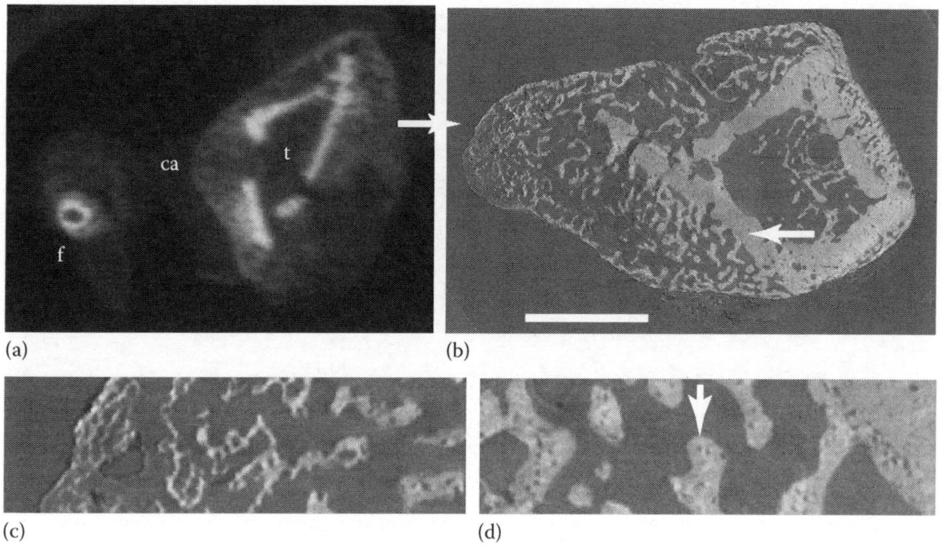

Figure 15.4 Murine fracture calluses compared in lab and synchrotron microCT slices. The lighter the voxel, the greater the attenuation is: (a) Lab microCT with 30 μm voxels with fibula "f," tibia "t," and fracture callus "ca" labeled (Obermeyer et al. 2012). The original cortical bone appears white, and the finer callus bone is light gray and appears to have lower mineral density, an artifact of partial volume sampling. (b) Synchrotron microCT performed at 2-BM, APS, 2.9 μm voxels. The arrow pointing right indicates the area enlarged in (c) and that pointing left indicates the area in (d). The enlargements do not show differences in contrast between the cortical and callus bone, but different positions within the callus show quite different microarchitecture and internal structure. The vertical arrow in (d) points to osteocyte lacunae. Scales bars in (a, b) equal 1 mm and those in (c, d) equal 0.20 mm. (From 2-BM, APS, S.R. Stock, X. Xiao, T. Obermeyer, J. Callaci, 2010.)

of quantities such as <Tb.Th.> (Stock et al. 2004a) as well as allow characterization of the structure within the protective spines (Stock et al. 2009). Figure 15.5 shows the stereom of a pyramid (one of five making up the urchin's oral apparatus) and its tooth, and the three orthogonal views through the suture between two demipyramids reveal two types of stereom, a galleried stereom to either side of the suture and a coarse, irregular imperforate stereom (see Smith 1980 for definition of these terms). In microCT data sets where the resolution is too coarse to allow the stereom to be resolved, the values of the voxels can be interpreted as partial volumes of calcite, and much can be inferred about the structure (Stock et al. 2003b).

The fact that different stereom types appear in the same positions and the ossicles of a given type always have the same crystallographic orientation means that there is significant control of mineralization in the sea urchin and other echinoderms. The structure and growth of sea urchin teeth not only differ from that of stereom but are also single crystals* and, if anything, mineralization in the continuously growing tooth is more tightly controlled than in stereom. Plates, needles, and prisms grow in a specific sequence with a well-defined Mg content and are in separate syncytial spaces until they are cemented together by a much higher Mg calcite (Cavey and Märkel 1994). The adjacent, unjoined structures can be resolved with synchrotron microCT (Stock et al. 2003c) and higher resolution tube-based microCT, except in the stone part, the hardest portion of the tooth (Märkel and Gorny 1973) that forms the tooth's cutting edge. This entails imaging with voxels in the range 1–2 μm, and the resulting 2–4 mm FOV (with 2 K detector elements) suffices to cover the entire cross-section of a typical sea urchin tooth. Unfortunately, the structure becomes more difficult to resolve once significantly very high Mg calcite has formed, because the absorption contrast of high and very high Mg calcite differs little (see Figure 15.6). The tooth cross-sectional shape (macrostructure) measured with microCT has been used as the basis for numerical phylogenic determinations (Ziegler et al. 2012), and similar numerical investigations based on tooth microstructure (plate orientations) are underway (Stock et al. 2012c). The carinar process plates in the keel of "T"-shaped teeth may provide mechanical advantage for camarodonts compared to stirodonts (which also possess "T"-shaped teeth, but without these plates) (Stock et al. 2003c).

One cannot appreciate the 3D complexity of sea urchin teeth in slices such as that in Figure 15.6. Figure 15.7 compares a slice through the tooth of *Micropyga tuberculata* with a 3D rendering of a stack of slices approximately encompassing the area in (Figure 15.7b). The curvature of the secondary plates is clear in the rendering; paging through the stack of slices also can help one appreciate such curvature.

Proteins are intimately involved in biomineralization, both in normal mineralization, for example, calcite of sea urchins (Veis et al. 1986), and in pathological mineralization such as that occurs in kidney stones, in heart valve calcification, in ectopic calcification such as that occurs in juvenile dermatomyositis, or in the complex osteoblastic/osteolytic lesions occurring in bone cancer. In human calcium phosphate (apatite and brushite) kidney stones, significant correlation was found between the stones' void volume (measured via microCT) and protein content (Pramanik et al. 2008). In human stones containing both calcium oxalate monohydrate and dehydrate, there is enough absorption contrast for the distribution of each crystallographic phase to be mapped with synchrotron microCT (Kaiser et al. 2011). Figure 15.8 shows examples of tube-based microCT of a human bladder stone and of a heavily calcified human heart valve. The layered structure in the bladder stone provides a temporal record of mineralization. The heart valve mineralized as a result of damage

* Actually, teeth are bicrystals with each half of the tooth slightly misoriented with respect to the other.

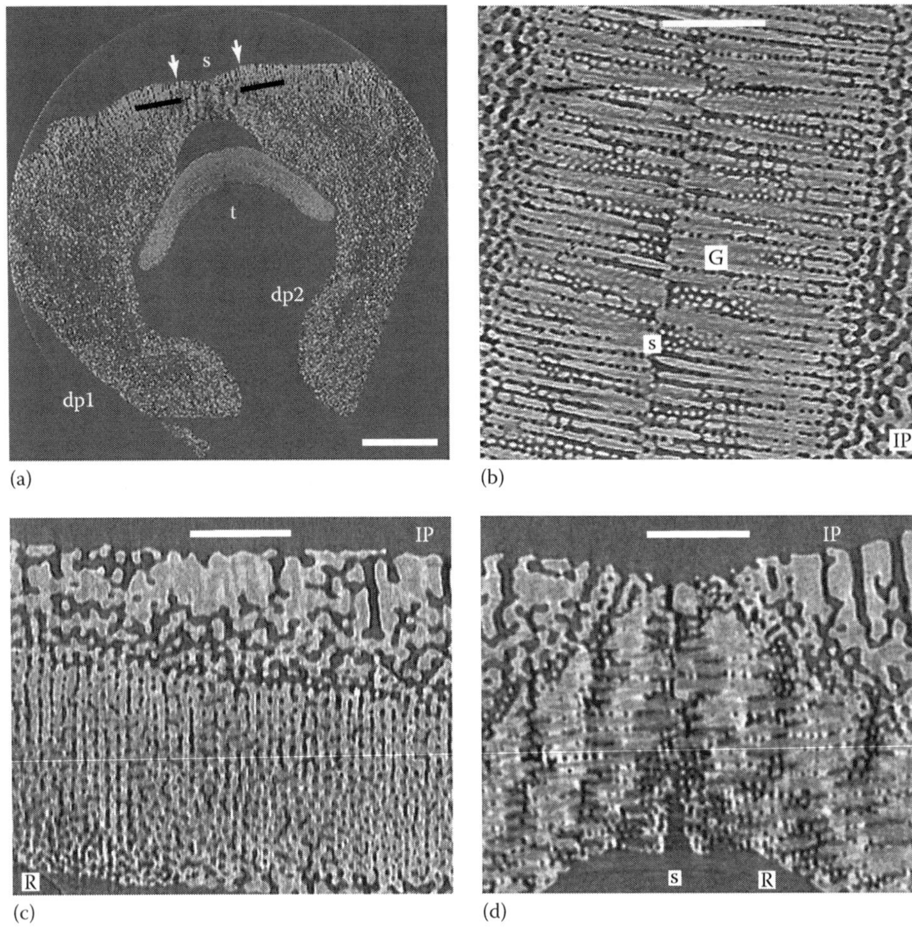

Figure 15.5 Synchrotron microCT of one of the five pyramids making up the jaw (Aristotle's lantern) of *Histocidaris elegans*: (a) In this slice (x–y section), the pyramid's two demipyramids "dp" join at the suture "s," and the grooved tooth "t" is attached to both demipyramids. The arrows mark the limits of the x–y enlargement in (d). (b) Enlargement of the x–z plane between the two black lines in (a). (c) Enlargement of the y–z plane at the position between "s" and the right-hand arrow in (a). (d) Enlargement of the x–y plane; the left and right sides are at the positions of the arrows in (a). Scale bar in (a) is 1 mm. In (b–d), the scale bar is 200 μm, and two types of stereom, galleried "G" and irregular perforate "IP," are visible. 20.7 keV, 0.12° rotation steps over 180°, 2 K × 2 K reconstructions, 2.9 μm voxels. (From 2-BM, APS, S.R. Stock, X. Xiao, A. Ziegler, 2011.)

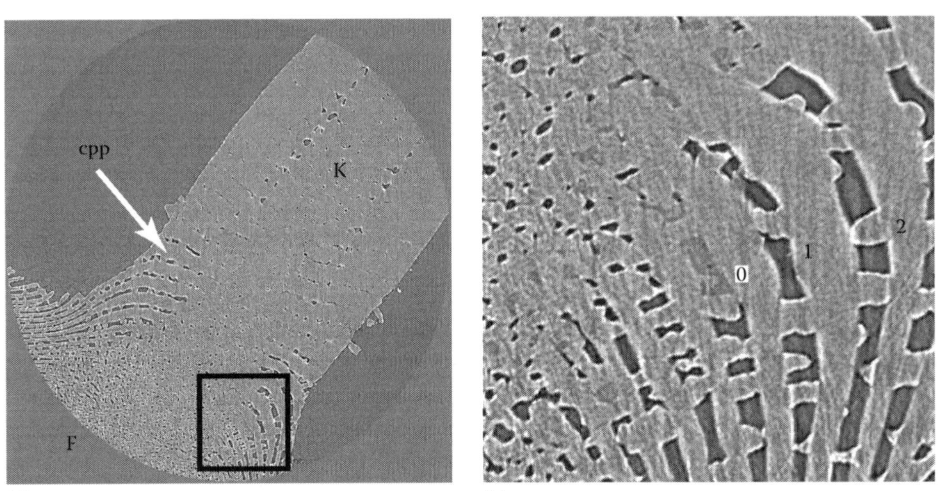

Figure 15.6 Synchrotron microCT of a tooth of the sea urchin *Lytechinus variegatus* whose soft tissue no longer remains in the spaces between calcite structural elements. The keel "K" and flange "F," leg and cross-bar, of the "T"-shaped tooth are labeled as well as the carinar process plates (cpp) within the keel: (a) Entire slice, horizontal FOV is 715 μm, and most of the flange and part of the keel are out of the FOV. (b) The area inside the box is enlarged and its horizontal FOV is 145 μm. Slight "hot edges," that is, light/dark fringes, appear at the air/calcite interfaces. Three curved plates of high Mg calcite (~0.13 mol fraction Mg (Stock et al. 2002)) are labeled "0–2." To the right of label "0" is open space; to the left is dried salt which has sufficient contrast to be distinguished from calcite. Between labels "1" and "2" and connecting these plates is a column of very high Mg calcite (~0.33 mol fraction Mg (Stock et al. 2002)) whose contrast does not differ from the lower Mg plates. 19 keV, 2000 projections over 180°, 2 K × 2 K reconstructions, 0.35 μm voxels. (From ID19, ESRF, A. Rack S.R. Stock, 2012.)

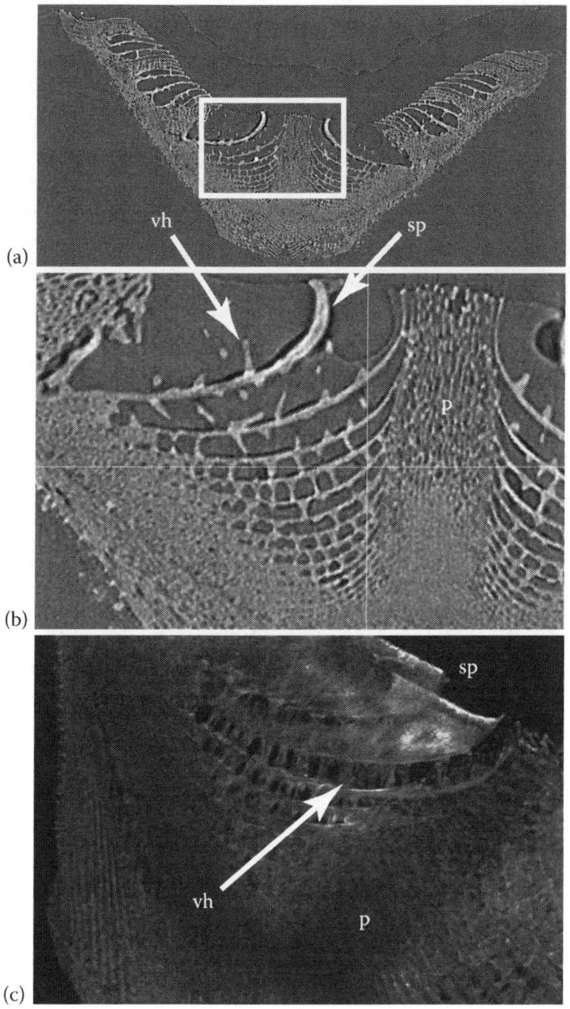

Figure 15.7 Slice and 3D rendering of the tooth of the sea urchin *Micropyga tuberculata*: (a) Slice of entire tooth, horizontal FOV = 1.81 mm. (b) Box shows approximate location of enlargement. p, prisms; sp, secondary plates; vh, very high Mg calcite columns linking adjacent secondary plates. (c) 3D rendering showing calcite structures as solid. 17 keV, 0.12° rotation steps over 180°, 2 K × 2 K reconstructions,1.45 μm voxels. (From 2-BM, APS, S.R. Stock, X. Xiao, A. Ziegler, 2012.)

from rheumatic fever, and the pattern of mineralization would be difficult to study through sectioning. MicroCT has also been useful in studying mineral distribution in biopsied tissue from patients suffering from juvenile dermatomyositis (Stock et al. 2004b), in osteoblastic/osteolytic lesions in bone cancer (Hu et al. 2011), or in arthritis (Chiba et al. 2012).

Synchrotron microCT has been used to study mineralized ectosymbionts anchored to sea urchin spines (David et al. 2009). Skeletal density and porosity characteristics are key parameters for investigations into scleractinian coral growth and for assessing the effects of a range of anthropogenic influences on coral reefs and have been studied by microCT (Roche et al. 2010). In sponges studied with synchrotron microCT, the distribution of silica skeletal elements, the canal system, and sponge tissue followed characteristic spatiotemporal morphological patterns (Hammel 2009). Calcareous plates in ammonite jaws were imaged by microCT to illuminate this animal's role in the Mesozoic marine food web (Kruta et al. 2011). The spatial variation of thickness and geometrical arrangement of Chiton plates, an organic matrix plus aragonite, have been quantified by microCT (Conners et al. 2012). Woodlice have a mineralized exoskeleton which is shed periodically; storage of the mineral (as amorphous calcium carbonate) within the body prior to molting has been studied with microCT (Neues et al. 2007). Development of the aragonite shells of embryonic and neonatal snails was also quantified with microCT (Marxen et al. 2008) as was the spatial distribution of different-sized statoliths (crystals of calcium sulfate hemihydrate) within the gravity sensing organ of medusae (Becker et al. 2005). Mineral distribution within plants does not appear to have been studied extensively with microCT, but the distribution of silica distribution in horsetail stalks is one application in the literature (Sapei et al. 2007).

Presence of very differently absorbing materials within a specimen can produce problems in microCT reconstructions because of the finite dynamic range in the projections. Small differences in absorptivity within both the higher absorption and the lower absorption phases often cannot be seen. One example

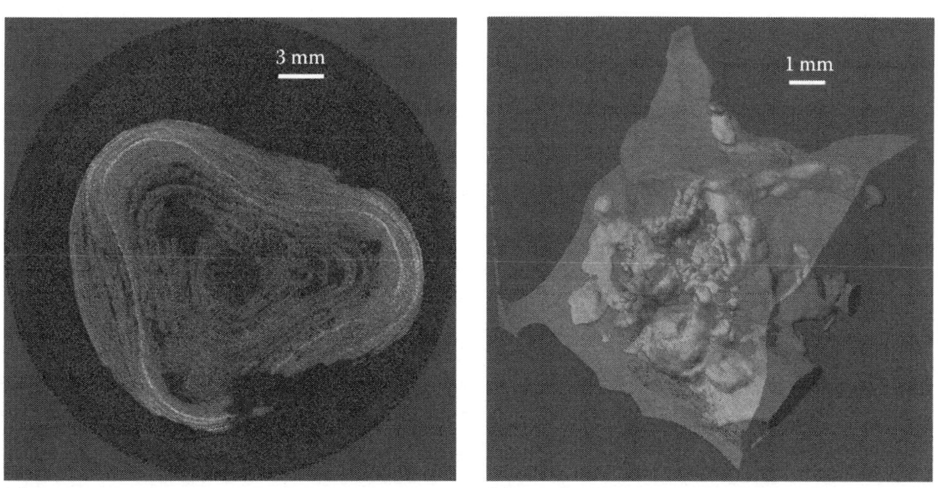

Figure 15.8 Tube-based microCT: (a) Slice of a human bladder stone imaged with 15 μm voxels and 70 kVp. S.R. Stock, E.R. Brooks, C.B. Langman. (b) 3D rendering of a heart valve damage by rheumatic fever with mineral render solid and soft tissue semitransparent. See Rajamannan et al. (2005) for further details.

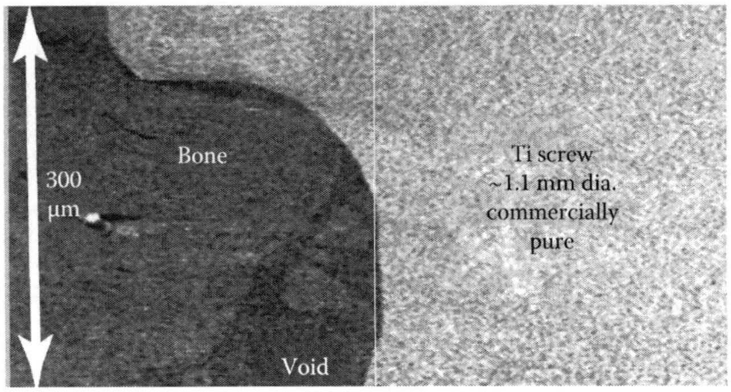

Figure 15.9 Numerical section perpendicular to a stack of 201 slices of Ti screw fixed in a block of dry bovine femoral bone. The screw was parallel to the tomography rotation axis, and some bone and Ti were out of the FOV (local tomography reconstruction). 63.35 keV, 1.45 μm voxels. (From 1-ID, APS, S.R. Stock, J.D. Almer, 2010.)

is seen in Figure 15.9, a Ti screw in bone. Bone can be barely discriminated from void space although some bone fragments produced by the screw rotation can be seen.

15.3.3 *IN VIVO* MICROCT

The ability to observe changes in a specimen with time is a powerful advantage of CT, allowing the specimen itself to serve as a control. *In vivo* microCT is being increasingly applied to biomineralization problems. One of the first such longitudinal studies used synchrotron microCT and examined the effect of parathyroid hormone (PTH) treatment on trabecular bone microarchitecture in osteopenic rats (Lane et al. 1995). Other examples of *in vivo* studies of bone include 3D mapping of changes in trabeculae (Willems et al. 2007; Schulte et al. 2011), quantitative analysis of tumor-induced bone destruction (Johnson et al. 2011), and quantification of dose–response behavior in neonatal murine calvariae in organ culture and treated with interleukin-1 (IL-1) or PTH (Stock et al. 2004c).

15.3.4 STRUCTURES WITHIN MINERALIZED TISSUES

MicroCT is very valuable for studying structures *within* mineralized tissues. This can be geometrically related or related to the composition of the tissue. The latter is particularly important in bone which is a composite material of cAp and collagen whose composition often varies through remodeling, the process whereby old, damaged bone surrounding Haversian canals is removed by osteoclasts, and new bone matrix is deposited by osteoblasts and subsequently remineralized. Precise quantification of differences in bone mineralization, however, generally requires much longer data acquisition times than in studies quantifying geometry such as trabecular architecture (Stock 2012a). Synchrotron microCT of heavily remodeling human cortical bone has been compared with scanning acoustic microscopy (Raum et al. 2006a), and local elastic constants have been derived from these data (Raum et al. 2006b). Differences in bone mineralization and porosity were quantified as a function of anatomical position in the human femoral cortex using synchrotron microCT (Rohrbach et al. 2012). In arthritis-affected tissue, differences in mineral level have also been reported for different positions (Chiba et al. 2012). Ravosa et al. (2007) compared mandible mineralization levels in wild-type mice and myostatin-deficient mice, which possess greatly increased masticatory muscle mass.

Structures such as osteocyte lacunae and microvessels can be mapped, and the distribution of their dimensions can be quantified with microCT performed with 1–2 μm voxels. One example is the measurement of a bimodal distribution of osteocyte lacunae volumes in human cortices (Hannah et al. 2010). A second is microvasculature mapping of murine cortical bone (Langer et al. 2012). Prolonged unloading in growing rats reduced cortical osteocyte lacunar density and volume in the distal tibia (Britz et al. 2012). Discussion of nanoCT imaging of osteocyte lacunae and their canaliculi is postponed until Section 15.4.

A characteristic feature of dentin, the tough mineralized tissue comprising the bulk of the volume of mammalian tooth, is the tubules that run from near the dentino-enamel junction (DEJ) or dentin–cementum junction into the pulp chamber within the tooth. MicroCT studies of tooth include high-resolution imaging of the tubule structure (Zaslansky et al. 2010) and of the DEJ (Stock et al. 2008a). Quantitative patterns of tooth mineralization were studied with synchrotron microCT of the intact maxilla and mandible of a Neanderthal child, and neither deciduous nor permanent tooth sequences are precisely found in modern comparative files (Bayle et al. 2009). Demineralization (carious lesion formation) of tooth is yet another application where microCT is very much valuable. Dowker et al. (2004) studied sound and carious human enamel, and others have quantified enamel mineral removal in a model demineralization/remineralization system (Delbem et al. 2009). MicroCT has also been used to study well-demarcated developmental defects of low mineral density dentin below the DEJ (Kühnisch et al. 2012).

15.3.5 SCAFFOLDS AND MODEL BIOMINERALIZATION SYSTEMS

Critical sized bone defects are those which are too large to heal naturally, and autografts (bones from another site in the patient) have been the best option for closing the defect. Complications associated with the donor site are significant enough, however, to motivate considerable research with artificial replacements. Scaffolds that are infiltrated by bone or that resorb with bone replacing the implant have been investigated by many groups. One microCT study investigated *in vitro* remodeling of a calcium phosphate scaffold in an osteoblast–osteoclast coculture system (Ruggiu et al. 2012). Others used microCT to compare *in vitro* mineralization of collagen and polymer scaffolds seeded with human stem cells (Kruger et al. 2011). The effectiveness of a calcium phosphate coated fiber mesh tube (mediating delivery of BMP-7) was evaluated for in critical size defects in rat long bones with microCT (Berner et al. 2012); a similar approach was used for studying the effect of nanofibers on critical defects (Mata et al. 2010) and for vertebral fusion (Hsu et al. 2011). *In vitro* seeding of bone marrow stromal cells into hydroxyapatite scaffold before implantation was another approach studied by microCT (Komlev et al. 2006), and the kinetics of such similarly seeded scaffolds was determined by scanning each scaffold before and after implantation (Peyrin 2011). Implanting bioceramic particles is another strategy for bone regeneration, and such material was examined *ex vivo* after bone in-growth (Rack et al. 2011).

15.4 OUTLOOK

Although the application of CT-based imaging and quantification of mineralized tissues will continue to increase, there are several areas where advances in the next few years are apt to be particularly great. These include phase tomography; real-time or longitudinal observation of specimens, tissues *in vivo*, and scaffolds for tissue growth; nanoCT; diffraction or scattering tomography; fluorescence tomography; and combinations of tomography with modalities such as position-resolved x-ray diffraction and scattering.

Phase microCT of mineralized tissues does not appear to have been widely applied, probably because absorption contrast provides adequate discrimination between mineralized and soft tissues, but examples do exist. The nondestructive in ovo study of fossil embryos is one application where phase microCT is extremely valuable (Fernandez et al. 2012). Phase microCT has been used to study the fossilized oral skeletal elements of conodonts, extinct jawless chordates, although, to be fair, there is still some debate about whether these animals possessed mineralized denticles (Goudemand et al. 2011). Cartilage associated with bone has also been studied with phase microCT and other techniques (Zehbe et al. 2012). Phase contrast can lead to significantly greater visibility of cracks than would be expected of purely absorption-based microCT (Ignatiev et al. 2006; Larrue et al. 2011) but this phenomenon has not been widely exploited in studies of cracks in mineralized tissues with investigators focusing on crack penetrants for enhancing crack visibility (Landrigan et al. 2011).

Real-time or near-real-time observation of mineralized tissues also does not appear to have received much attention although studies of cracks in mineralized tissue is an area where recent advances in synchrotron x-ray imaging may pay real dividends. In materials like bone that are tough, cracks are stable and real-time imaging is not needed because imaging can be done over minutes instead of fractions of seconds (Voide et al. 2009; Barth et al. 2010; Larrue et al. 2011), much like in early, lower resolution microCT studies in metals (Breunig et al. 1992). In mineralized tissues that contain collagen or other biopolymers, however, radiation damage can be significant enough to alter the materials' mechanical properties during data collection (Barth et al. 2010). In specimens that are too brittle to support open cracks without fracturing, that is, most shells, sea urchin spines, and ossicles, postfracture tomography following high-speed radiography of the propagating crack may add to understanding of these mineralized structures' resistance to fracture. Some mineralized tissues such as certain tendons deform viscoelastically and thus also present a challenge for *in situ* tomographic imaging under load.

NanoCT systems, both tube-based and synchrotron-radiation-based, are becoming more available, and studies of mineralized tissues have appeared in the literature. One obvious application of tomography with 50–100 nm voxels is the osteocyte–canaliculi system in bone: osteocyte lacunae are ellipsoidal with diameters on the order of 10 μm (Hannah et al. 2010; Tommasini et al. 2012), and the canaliculi radiating from the lacunae have diameters on the order of 100–300 nm (Cooper et al. 1966). Several reports have appeared where nanoCT imaged

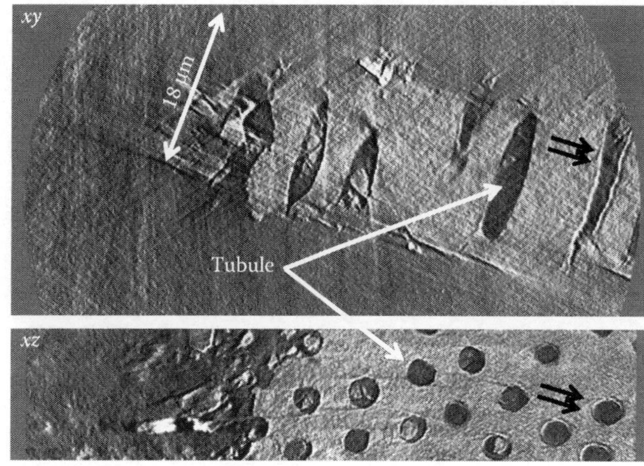

Figure 15.10 Tubules in equine crown dentin imaged with nanoCT. The *x–y* and *x–z* sections through a stack of 200 slices are shown. The double black arrows identify the same 8 keV, angular range −60° to 72°, angular step 0.25°, number of projections 529, detector binning 2 × 2. (From 32-ID, APS, S.R. Stock, S. Wang, 2012.)

osteocyte–canaliculi systems (Andrews et al. 2010; Dierolf et al. 2010; Schneider et al. 2010; Langer et al. 2012). Tubules in dentin are quite long and have diameters in the range of 1–3 μm, making the study of their 3D structure eminently suitable for nanoCT. Figure 15.10 shows two orthogonal sections through a plate-like section of equine coronal dentin reconstructed with nanoCT. Projections could be collected over only 132° and not 180°, meaning that the data was rather significantly truncated. Nonetheless, the microstructure of the tubules can be seen including what one interprets as polishing debris produced in thinning the specimen to 18 μm thickness.

X-ray diffraction tomography uses intensities diffracted by different crystallographic phases to reconstruct the distribution of these phases, and this can be particularly valuable in mineralized tissues where different polymorphs of calcium carbonate exist or where crystallographic texture varies with position. Crystallographic phases are, of course, different from x-ray phase contrast described elsewhere in this chapter. Reconstruction may also be performed with scattered radiation, for example, in the small-angle regime (Jensen et al. 2011). Stock et al. (2008b) performed diffraction microCT on a bone sample, and the technique has advanced recently in applications other than mineralized tissues (Alvarez-Murga et al. 2012; Stock et al. 2012a). Synchrotron x-radiation, a pinhole beam, and translation/rotation data collection are typically used, and Figure 15.11 shows reconstructions of a specimen of cortical and trabecular bone cut from a porcine spinous process. Reconstructions made with the cAp 22.2 and 00.4 reflections and the beam transmitted through specimen are compared with a matching slice from a tube-based commercial microCT system. The data for the diffraction reconstruction studies cited earlier were recorded with an area detector centered on the incident beam. In Figure 15.11, two area detectors (panels 3 and 4) were offset from the incident beam position, and two sectors of the diffraction rings rotated 90° from each other were covered. Cortical bone occupies positions at the left and right of the specimen in the absorption-based reconstructions (Figure 15.11a and b), with the trabecular bone filling the area between them. The cAp 22.2 reconstructions

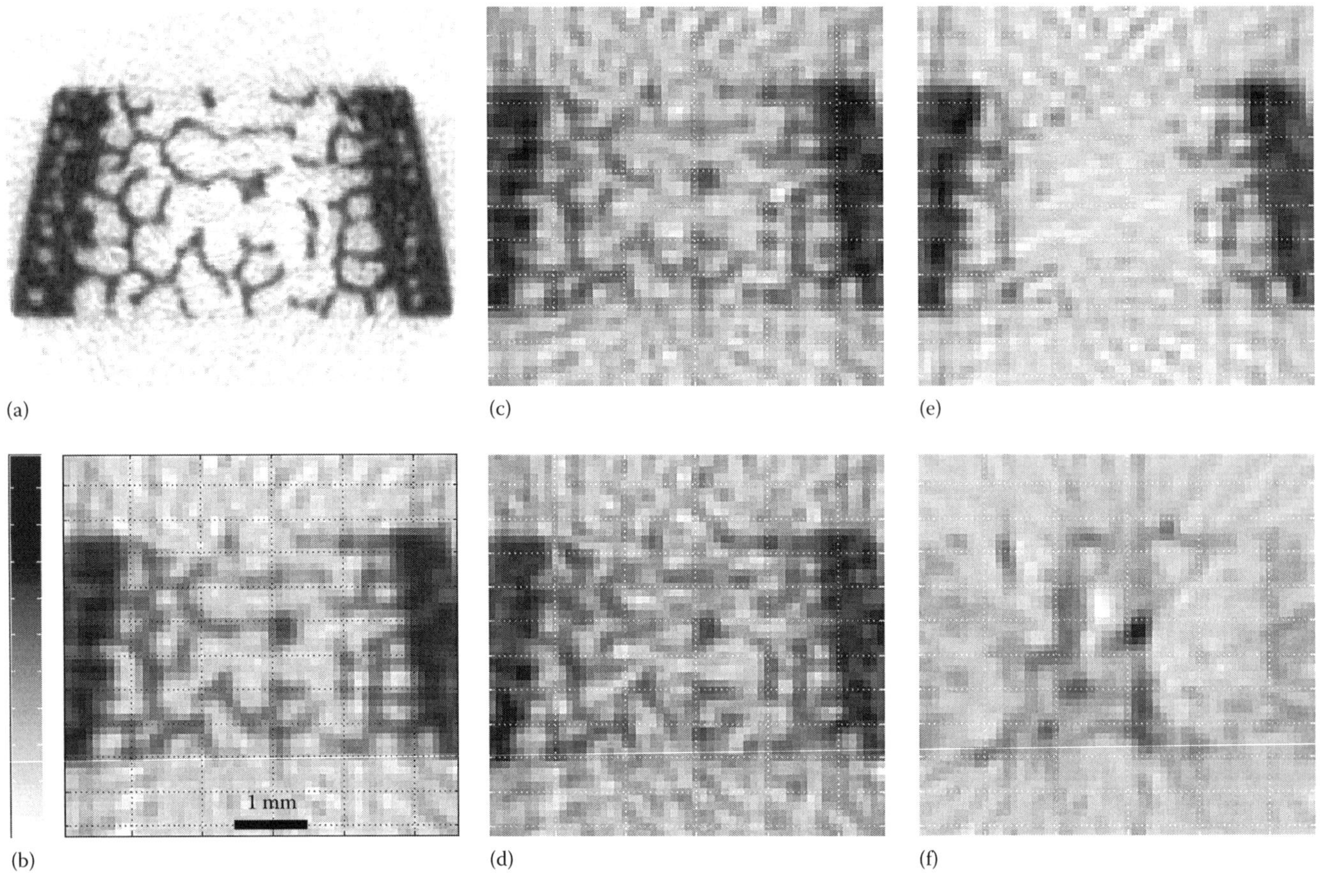

Figure 15.11 Reconstructed slices of a specimen of porcine spinous process with contrast reversed from that in the other figures (the darker the pixels, the greater the mineral content): (a) Absorption contrast from commercial, tube-based microCT system. 9.23 mm FOV, 45 kVp, 37 μm voxels. (b) Slice reconstructed with transmitted beam. Each reconstruction is scaled to its [minimum, maximum] values given in the following, and the linear gray scale is shown on the left-hand side of (b). (c, d) Slice reconstructed with cAp 22.2 diffracted intensity with ranges [−100, 500] and [−100, 360] recorded on panels 3 and 4, respectively. (e, f) Slice reconstructed with cAp 00.4 diffracted intensity with ranges [−500, 4000] and [−75, 450] recorded on panels 3 and 4, respectively. The very textured contrast in air outside the specimen (top and bottom of the diffraction reconstructions) is a consequence of the significant angular undersampling. 70 keV, beam 0.1 mm (in plane of reconstruction) and focused to ~0.05 mm (perpendicular), 80 positions per projection (0.1 mm translation increment), 31 projections (6° rotation steps), 5 s integration/pattern. (b–f: From 1-ID, APS, S.R. Stock, A.C. Deymier-Black, J.D. Almer.)

from panels 3 and 4 (Figure 15.11c and d) are very similar to each other and to both of the absorption reconstructions, whereas the panels 3 and 4 cAp 00.4 reconstructions differ enormously (see the cortical bone areas in Figure 15.11e and f), reflecting the strong *c*-axis texture in that portion of the bone. This result demonstrates the potential of diffraction tomography for accomplishing more than identifying what crystallographic phases are present where.

In general, microCT based on fluorescence signal from the specimen has not been used for many applications, in part because pencil beam data acquisition is so slow. The authors know of no examples where fluorescence tomography has been employed with mineralized tissue. This may be because typical elements associated with biominerals (Ca, Fe) fluoresce at relatively low energies, and samples are relatively thick, allowing little of this radiation to escape from the specimen.

Absorption microCT has long been compared with radiographs, histological sections, and other methods (Stiller et al. 2009; Zehbe et al. 2010). Absorption microCT combined with position-resolved x-ray diffraction of a crayfish mandible, for example, revealed an enamel-like apatite crown coexisting with other crystalline and amorphous calcium carbonate and calcium phosphate phases (Bentov et al. 2012). In a second example, x-ray diffraction of a living sea urchin tooth showed calcite was present very early in the mineralization process at a position where synchrotron microCT showed the primary plates were newly formed and quite small (Stock et al. 2012b).

ACKNOWLEDGMENT

SRS acknowledges support from NICDR grant DE001374 (to Arthur Veis). Use of the Advanced Photon Source was supported by the US Department of Energy, Office of Science, Office of Basic Energy Sciences, under Contract No. DE-AC02-06CH11357.

REFERENCES

Ahmed, F. 2011. Multiscale quantitative imaging of human femoral heads using x-ray microtomography. PhD thesis. Queen Mary, University of London.

Allen, M.R., H.A. Hogan, W.A. Hobbs, A.S. Koivuniemi, M.C. Koivuniemi, and D.B. Burr. 2007. Raloxifene enhances material-level mechanical properties of femoral cortical and trabecular bone. *Endocrinology* **148**:3908–3913.

Alvarez-Murga, M., P. Bleuet, and J.-L. Hodeau. 2012. Diffraction/scattering computed tomography for three-dimensional characterization of multi-phase crystalline and amorphous materials. *J Appl Cryst* **47**:1109–1124.

Andrews, J.C., E. Almeida, M.C. van der Meulen, J.S. Alwood, C. Lee, Y. Liu, J. Chen et al. 2010. Nanoscale x-ray microscopic imaging of mammalian mineralized tissue. *Microsc Microanal* **16**:327–336.

Barth, H.D., M.E. Launey, A.A. MacDowell, J.W. Ager, and R.O. Ritchie. 2010. On the effect of x-ray irradiation on the deformation and fracture behavior of human cortical bone. *Bone* **46**:1475–1485.

Bayle, P., J. Braga, A. Mazurier, and R. Macchiarelli. 2009. Dental developmental pattern of a Neanderthal child from Roc de Marsal: A high-resolution 3D analysis. *J Human Evol* **56**:66–75.

Becker, A., I. Sötje, C. Paulmann, F. Beckmann, T. Donath, R. Boese, O. Prymak, H. Tiemann, and M. Epple. 2005. Calcium sulfate hemihydrate is the inorganic mineral in statoliths of *Scyphozoan medusae* (Cnidaria). *DaltonTrans* (8):1545–1550.

Bentov, S., P. Zaslansky, A. Al-Sawalmih, A. Masic, P. Fratzl, A. Sagi, A. Berman, and B. Aichmayer. 2012. Enamel-like apatite crown covering amorphous mineral in a crayfish mandible. *Nature Comm* **3**(839):1–7.

Berner, A., J.D. Boerckel, S. Saifzadeh, R. Steck, J. Ren, C. Vaquette, J.Q. Zhang, M. Nerlich, R.E. Guldberg, D.W. Hutmacher, and M.A. Woodruff. 2012. Biomimetic tubular nanofiber mesh and platelet rich plasma-mediated delivery of BMP-7 for large bone defect regeneration. *Cell Tiss Res* **347**:603–612.

Bonse, U., F. Busch, O. Günnewig, F. Beckmann, G. Delling, M. Hahn, and A. Kvick. 1996. Microtomography (μCT) applied to structure analysis of human bone biopsies, *ESRF Newslett* March, 21–23.

Breunig, T.M., S.R. Stock, S.D. Antolovich, J.H. Kinney, W.N. Massey, and M.C. Nichols. 1992. A framework relating macroscopic measures and physical processes of crack closure of Al-Li alloy 2090. In *Fracture Mechanics: Twenty-Second Symposium Vol. 1*, Philadelphia, PA: ASTM STP, Vol. 1131, pp. 749–761.

Britz, H.M., Y. Carter, J. Jokihaara, O.V. Leppänen, T.L.N. Järvinen, G. Belev, and D.M.L. Cooper. 2012. Prolonged unloading in growing rats reduces cortical osteocyte lacunar density and volume in the distal tibia. *Bone* **51**:913–919.

Burghardt, A.J.,T.M. Link, and S. Majumdar. 2011. High-resolution computed tomography for clinical imaging of bone microarchitecture. *Clin Orthop Relat Res* **469**:2179–2193.

Carlson, W. 2006. Three-dimensional imaging of earth and planetary materials. *Earth Planetary Sci Lett* **249**:133–149.

Cavey, M.J. and K. Märkel. 1994. Echinoidea. In *Microscopic Anatomy of Invertebrates, Vol. 14, Echinodermata*, eds. E.W. Harrison and F.-S. Chia, New York, NY: Wiley, pp. 345–400.

Chiba, K., N. Nango, S. Kubota, N. Okazaki, K. Taguchi, M. Osaki, and M. Ito. 2012. Relationship between microstructure and degree of mineralization in subchondral bone of osteoarthritis: Asynchrotron radiation μCT study. *J Bone Miner Res* **27**:1511–1517.

Chun, K.J. 2011. Bone densitometry. *Semin Nucl Med* **41**:220–228.

Connors, M.J., H. Ehrlich, M. Hog, C. Godeffroy, S. Araya, I. Kallai, D. Gazit, M. Boyce, and C.J. Ortiz. 2012. Three-dimensional structure of the shell plate assembly of the chiton *Tonicella marmorea* and its biomechanical consequences. *J Struct Biol* **177**:314–328.

Cooper, R.R., J.W. Milgram, and R.A. Robinson. 1966. Morphology of the osteon. An electron microscopic study. *J Bone Joint Surg Am* **48**:1239–1271.

David, B., S.R. Stock, F. De Carlo, V. Hétérier, and C. De Ridder. 2009. Microtomographic survey of Antarctic echinoid spines: Diversity of microstructures and ectosymbionts anchorage. *Marine Biol* **156**:1559–1572.

Davis, G.R. and J.C. Elliott. 1997. X-ray microtomography scanner using time-delay integration for elimination of ring artefacts in the reconstructed image. *Nucl Instrum Meth A* **394**:157–162.

Delbem, A.C.B., K.T. Sassaki, A.E.M. Vieira, E. Rodrigues, M. Bergamaschi, S.R. Stock, M.L. Cannon, X. Xiao, F. De Carlo, and A.C.B. Delbem. 2009. In vitro validation of methodology to assess the effectiveness of fluoridated dentifrices on demineralization. *Caries Res* **43**:359–365.

Dierolf, M., A. Menzel, P. Thibault, P. Schneider, C.M. Kewish, R. Wepf, O. Bunk, and F. Pfeiffer. 2010. Ptychographic x-ray computed tomography at the nanoscale. *Nature* **467**:436–439.

Dowker S.E., J.C. Elliott, G.R. Davis, R.M. Wilson, and P. Cloetens. 2004. Synchrotron x-ray microtomographic investigation of mineral concentrations at micrometer scale in sound and carious enamel. *Caries Res* **38**:514–522.

Feldkamp, L.A., L.C. Davis, and J.W. Kress. 1984. Practical cone-beam algorithm, *J Opt Soc Am* **A1**:612–619.

Fernandez, V., E. Buffetaut, E. Maire, J. Adrien, V. Suteethorn, and P. Tafforeau. 2012. Phase contrast synchrotron microtomography: Improving noninvasive investigations of fossil embryos inovo. *Microsc Microanal* **18**:179–185.

Fields, A.J., S. Nawathe, S.K. Eswaran, M.G. Jekir, M.F. Adams, P. Papadopoulos, and T.M. Keaveny. 2012. Vertebral fragility and structural redundancy. *J Bone Miner Res* **27**:2152–2158.

Giesen, E.B.W., M. Ding, M. Dalstra, and T.M.G.J. van Ejiden. 2003. Architectural measures of the cancellous bone of the mandibular condyle identified by principal components analysis. *Calcif Tiss Int* **73**:225–231.

Goudemand, N., M.J. Orchard, S. Urdy, H. Bucher, and P. Tafforeau. 2011. Synchrotron-aided reconstruction of the conodont feeding apparatus and implications for the mouth of the first vertebrates. *Proc Natl Acad Sci USA* **108**:8720–8724.

Hammel, J.U., J. Herzen, F. Beckmann, and M. Nickel. 2009. Sponge budding is as patiotemporal morphological patterning process: Insights from synchrotron radiation-based x-ray microtomography into the asexual reproduction of *Tethya wilhelma. Front Zool* **6**:19.

Hannah, K.M., C.D. Thomas, J.G. Clement, F. DeCarlo, and A.G. Peele. 2010. Bimodal distribution of osteocyte lacunar size in the human femoral cortex as revealed by micro-CT. *Bone* **47**:866–871.

Harvey, N., E. Dennison, and C. Cooper. 2010. Osteoporosis: Impact on health and economics. *Nature Rev Rheum* **6**:99–105.

Hildebrand, T. and P. Rüeggeseger. 1997a. A new method for the model-independent assessment of thickness in three-dimensional images. *J Microsc* **185**:67–75.

Hildebrand, T. and P. Rüegsegger. 1997b. Quantification of bone architecture with the structure model index. *Comp Meth Biomech Biomed Eng* **1**:15–23.

Hounsfield, G.N. 1973. Computerized transverse axial scanning (tomography): Part I description of system. *Br J Radiol* **46**:1016–1022.

Hsu, W.K., R. Riaz, M. Polavarapu, G.C. Roc, S.R. Stock, J. Ghodasra, Z. Glicksman, and E.L. Hsu. 2011. A nanocomposite therapy as a more efficacious and less inflammatory alternative to bone morphogenetic protein-2 in a rodent arthrodesis model. *J Orthop Res* **29**:1812–1819.

Hu, Z., H. Gerseny, Z. Zhang, Y.-J. Chen, A. Berg, Z. Zhang, S. Stock, and P. Seth. 2011. Oncolytic adenovirus expressing soluble TGFβ receptor II-Fc-mediated inhibition of established bone metastases: A safe and effective systemic therapeutic approach for breast cancer. *Mol Ther* **19**:1609–1618.

Ignatiev, K.I., G.R. Davis, J.C. Elliott, and S.R. Stock. 2006. MicroCT (microtomography) quantification of microstructure related to macroscopic behavior. Part 1—Fatigue crack closure measured in situ in AA2090 compact tension samples. *Mater Sci Technol* **22**:1025–1037.

Ito, M., T. Nakata, A. Nishida, and M. Uetani. 2011. Age-related changes in bone density, geometry and biomechanical properties of the proximal femur: CT-based 3D hip structure analysis in normal postmenopausal women. *Bone* **48**:627–630.

Jensen, T.H., M. Bech, O. Bunk, A. Menzel, A. Bouchet, G. Le Duc, R. Feidenhans'l, and F. Pfeiffer. 2011. Molecular x-ray computed tomography of myelin in a rat brain. *Neuroimage* **57**:124–129.

Johnson, L.C., R.W. Johnson, S.A. Munoz, G.R. Mundy, T.E. Peterson, and J.A. Sterling. 2011. Longitudinal live animal micro-CT allows for quantitative analysis of tumor-induced bone destruction. *Bone* **48**:141–151.

Kaiser, J., M. Hola, M. Galiova, K. Novotny, V. Kanicky, P. Martinec, J. Scucka et al. 2011. Investigation of the microstructure and mineralogical composition of urinary calculi fragments by synchrotron radiation x-ray microtomography: A feasibility study. *Urol Res* **39**:259–267.

Kak, A.C. and M. Slaney. 2001. *Principles of Computerized Tomographic Imaging*. Philadelphia, PA: SIAM.

Kinney, J.H., N.E. Lane, and D.L. Haupt. 1995. In vivo three-dimensional microscopy of trabecular bone. *J Bone Miner Res* **10**:264–270.

Koivumaki, J.E.M., J. Thevenot, P. Pulkkinen, V. Kuhn, T.M. Link, F. Eckstein, and T. Jamsa. 2012. Cortical bone finite element models in the estimation of experimentally measured failure loads in the proximal femur. *Bone* **51**:737–740.

Komlev, V.S., F. Peyrin, M. Mastrogiacomo, A. Cedola, A. Papadimitropoulos, F. Rustichelli, and R. Cancedda. 2006. Kinetics of in vivo bone deposition by bone marrow stromal cells into porous calcium phosphate scaffolds: An x-ray computed microtomography study. *Tiss Eng* **12**:3449–3458.

Kruger, E.A., D.D. Im, D.S. Bischoff, C.T. Pereira, W. Huang, G.H. Rudkin, D.T. Yamaguchi, and T.A. Miller. 2011. In vitro mineralization of human mesenchymal stem cells on three-dimensional type I collagen versus PLGA scaffolds: A comparative analysis. *Plast Reconst Surg* **127**:2301–2311.

Kruta, I., N. Landman, I. Rouget, F. Cecca, and P. Tafforeau. 2011. The role of ammonites in the Mesozoic marine food web revealed by jaw preservation. *Science* **331**:70–72.

Kühnisch, J., M. Galler, M. Seitz, H. Stich, A. Lussi, R. Hickel, K.H. Kunzelmann, and K. Bücher. 2012. Irregularities below the enamel-dentin junction may predispose for fissure caries. *J Dent Res* **91**:1066–1070.

Landrigan, M.D., J. Li, T.L. Turnbull, D.B. Burr, G.L. Niebur, and R.K. Roeder. 2011. Contrast-enhanced micro-computed tomography of fatigue microdamage accumulation in human cortical bone. *Bone* **48**:443–450.

Lane, N.E., J.M. Thompson, G.J. Strewler, and J.H. Kinney. 1995. Intermittent treatment with human parathyroid hormone (hPTH[1-34]) increased trabecular bone volume but not connectivity in osteopenic rats. *J Bone Miner Res* **10**:1470–1477.

Langer, M., A. Pacureanu, H. Suhonen, Q. Grimal, P. Cloetens, and F. Peyrin. 2012. X-ray phase nanotomography resolves the 3D human bone ultrastructure. *PLoS One* 7:e35691.

Larrue, A., A. Rattner, Z.A. Peter, C. Olivier, N. Laroche, L. Vico, and F. Peyrin. 2011. Synchrotron radiation micro-CT at the micrometer scale for the analysis of the three-dimensional morphology of microcracks in human trabecular bone. *PLoS One* **6**:e21297.

Liu, X.S., X.H. Zhang, and X.E. Guo. 2009. Contributions of trabecular rods of various orientations in determining the elastic properties of human vertebral trabecular bone. *Bone* **45**:158–163.

Maiolino, S., D.M. Boyer, J.I. Bloch, C.C. Gilbert, and J. Groenke. 2012. Evidence for a grooming claw in a North American adapiform primate: Implications for anthropoid origins. *PLoS One* 7:e29135.

Manning, P.L., L. Margetts, M.R. Johnson, P.J. Withers, W.I. Sellers, P.L. Falkingham, P.M. Mummery, P.M. Barrett, and D.R. Raymont. 2009. Biomechanics of dromaeosaurid dinosaur claws: Application of x-ray microtomography, nanoindentation and finite element analysis. *Anat Rec* **292**:1397–1405.

Märkel, K. and P. Gorny. 1973. Zur funktionellen Anatomie der Seeigelzähne (Echinodermata, Echinoidea). *Z Morph Tiere* **75**:223–242.

Marxen, J.C., O. Prymak, F. Beckmann, F. Neues, and M. Epple. 2008. Embryonic shell formation in the snail *Biomphalaria glabrata*: A comparison between scanning electron microscopy (SEM) and synchrotron radiation microcomputer tomography (SRμCT). *J Mollusc Studies* **74**:19–25.

Mata, A., Y. Geng, K. Henrikson, C. Aparicio, S. Stock, R.L. Satcher, and S.I. Stupp. 2010. Bone regeneration mediated by biomimetic mineralization of a nanofiber matrix. *Biomaterials* **31**:6004–6012.

Mayo, S.C., A.W. Stevenson, and S.W. Wilkins. 2012. In-line phase contrast x-ray imaging and tomography for materials science. *Materials* **5**:937–965.

Natterer, F. 2001. *The Mathematics of Computerized Tomography*. Philadelphia, PA: SIAM.

Neues, F., F. Beckmann, A. Ziegler, and M. Epple. 2007. The application of synchrotron radiation-based micro-computer tomography in biomineralization. In *Biomineralisation: Biological Aspects and Structure Formation*, ed. E. Beuerlein. Weinheim, Germany: Wiley-VCH, pp. 369–380.

Neues, F. and M. Epple. 2008. X-ray microcomputer tomography for the study of biomineralized endo- and exoskeletons of animals. *Chem Rev* **108**:4734–4741.

Neues, F., R. Goerlich, J. Renn, F. Beckmann, and M. Epple. 2007. Skeletal deformations in medaka (*Oryzias latipes*) visualized by synchrotron radiation micro-computer tomography (SRμCT). *J Struct Biol* **160**:236–240.

Nicholson, C.J. and E.C. Firth. 2010. Assessment of bone response to conditioning exercise in the radius and tibia of young Thoroughbred horses using pQCT. *J Musculoskel Neuronal Interact* **10**:199–206.

Obermeyer, T., D. Yonick, K. Lauing, S. Stock, R. Nauer, P. Strotman, R. Shankar, R. Gamelli, M. Stover, and J.J. Callaci. 2012. Mesenchymal stem cells facilitate fracture repair in an alcohol-induced impaired healing model. *J Orthop Trauma* **26**: 712–718.

Odgaard, A. 2001. Quantification of cancellous bone architecture. In *Bone Mechanics Handbook,* 2nd edn., ed. S.C. Cowin. Boca Raton, FL: CRC Press, pp. 14-1-19.

O'Neill, K.R., C.M. Stutz, N.A. Mignemi, M.C. Burns, M.R. Murry, J.S. Nyman, and J.G. Schoenecker. 2012. Micro-computed tomography assessment of the progression of fracture healing in mice. *Bone* **50**:1357–67.

Peyrin, F. 2011. Evaluation of bone scaffolds by micro-CT. *Osteopor Inter* **22**:2043–2048.

Pramanik, R., J.R. Asplin, M.E. Jackson, and J.C. Williams. 2008. Protein content of human apatite and brushite kidney stones: Significant correlation with morphologic measures. *Urol Res* **36**:251–258.

Rack, A., M. Stiller, O. Dalügge, A.T. Rack, H. Riesemeier, and C. Knabe. 2011. Developments in high-resolution CT: Studying bioregeneration by hard x-ray synchrotron-based microtomography. In *Comprehensive Biomaterials*, Vol. 3., eds. P. Ducheyne, K.E. Healy, D.W. Hutmacher, D.W. Grainger, and C.J. Kirkpatrick. Amsterdam, the Netherlands: Elsevier, pp. 47–62.

Raisz, L.G. 2008. Overview of pathogenesis. In *Primer on the Metabolic Bone Diseases and Disorders of Mineral Metabolism,* 7th edn., ed. C.J. Rosen. Washington, DC: American Society for Bone and Mineral Research, pp. 203–206.

Rajamannan, N.M., T.B. Nealis, M. Subramaniam, S.R. Stock, C.I. Ignatiev, T.J. Sebo, J.W. Fredericksen et al. 2005. Calcified rheumatic valve neoangiogenesis is associated with VEGF expression and osteoblast-like bone formation. *Circulation* **111**:3296–3301.

Raum, K., R.O. Cleveland, F. Peyrin, and P. Laugier. 2006b. Derivation of elastic stiffness from site-matched mineral density and acoustic impedance maps. *Phys Med Biol* **51**:747–758.

Raum, K., I. Leguerney, F. Chandelier, M. Talmant, A. Saïed, F. Peyrin, and P. Laugier. 2006a. Site-matched assessment of structural and tissue properties of cortical bone using scanning acoustic microscopy and synchrotron radiation μCT. *Phys Med Biol* **51**:733–746.

Ravosa, M.J., E.P. Kloop, J. Pinchoff, S.R. Stock, and M.W. Hamrick. 2007. Plasticity of mandibular biomineralization in myostatin-deficient mice. *J Morphol* **268**:275–282.

Rayfield, E.J. 2011. Strain in the ostrich mandible during simulated pecking and validation of specimen-specific finite element models. *J Anat* **218**:47–58.

Ritchie, R.O., R.K. Nalla, J.J. Kruzic, J.W. Ager III, G. Balooch, and J.H. Kinney. 2006. Fracture and ageing in bone: Toughness and structural characterization. *Strain* **42**:225–232.

Roche, R.C., R.A. Abel, K.G. Johnson, and C.T. Perry. 2010. Quantification of porosity in *Acropora pulchra* (Brook1891) using x-ray micro-computed tomography techniques. *J Exp Marine Biol Ecol* **396**:1–9.

Rohrbach, D., S. Lakshmanan, F. Peyrin, M. Langer, A. Gerisch, Q. Grimal, P. Laugier, and K. Raum. 2012. Spatial distribution of tissue level properties in a human femoral cortical bone. *J Biomech* **45**:2264–2270.

Röntgen, W. 1898. Über eine neue Art von Strahlen (Concerning a new type of radiation). *Ann Phys Chem*, New Ser. **64**:1–37.

Ruggiu, A., F. Tortelli, V.S. Komlev, F. Peyrin, and R. Cancedda. 2012. Extracellular matrix deposition and scaffold biodegradation in an in vitro three-dimensional model of bone by x-ray computed microtomography. *J Tiss Eng Regen Med.* DOI: 10.1002/term.1559.

Sapei, L., N. Gierlinger, J. Hartmann, R. Nöske, P. Strauch, and O. Paris. 2007. Structural and analytical studies of silica accumulations in *Equisetum hyemale. Anal Bioanal Chem* **389**:1249–1257.

Schneider, P., M. Meier, R. Wepf, and R. Müller. 2010. Towards quantitative 3D imaging of the osteocyte lacuno-canalicular network. *Bone* **47**:848–858.

Schulte, F.A., F.M. Lambers, G. Kuhn, and R. Müller. 2011. In vivo micro-computed tomography allows direct three-dimensional quantification of both bone formation and bone resorption parameters using time-lapsed imaging. *Bone* **48**:433–442.

Shim, V.B., R.P. Pitto, and I.A. Anderson. 2012. Quantitative CT with finite elements analysis: Towards a predictive tool for bone remodeling around an uncemented tapered stem. *Inter Orthop* **36**:1363–1369.

Smith, A.B. 1980. Stereom microstructure of the echinoid test. In *Special Papers in Palaeontology*. No. 25. London, U.K.: The Palaeontology Association.

Sone, T., T. Tamada, Y. Jo, H. Miyoshi, and M. Fukunaga. 2004. Analysis of three-dimensional microarchitecture and degree of mineralization in bone metastases from prostate cancer using synchrotron microcomputed tomography. *Bone* **35**:432–438.

Stiller, M., A. Rack, S. Zabler, J. Goebbels, O. Dalügge, S. Jonscher, and C. Knabe. 2009. Quantification of bone tissue regeneration employing beta-tricalcium phosphate by three-dimensional non-invasive synchrotron micro-tomography—A comparative examination with histomorphometry. *Bone* **44**:619–628.

Stock, S.R. 1999. Microtomography of materials. *Inter Mater Rev* **44**:141–164.

Stock, S.R. 2008a. Recent advances in x-ray microtomography applied to materials. *Inter Mater Rev* **58**:129–181.

Stock, S.R. 2008b. *MicroComputed Tomography: Methodology and Applications*. Boca Raton, FL: Taylor & Francis.

Stock, S.R. 2012a. X-ray computed tomography. In *Characterization of Materials,* 2nd edn., ed. E.N. Kaufmann. New York: Wiley, pp. 1624–1641.

Stock, S.R. 2012b. Trends in micro- and nanoComputed Tomography 2010–2012. In *Developments in X-ray Tomography VIII,* SPIE Vol. 8506, ed. S.R. Stock. Bellingham, WA: SPIE, p. 850602.

Stock, S.R. and J.D. Almer. 2012a. Diffraction microComputed Tomography of an Al-matrix SiC-monofilament composite. *J Appl Cryst* **47**:1077–1083.

Stock, S.R., J. Barss, T. Dahl, A. Veis, and J.D. Almer. 2002. X-ray absorption microtomography (microCT) and small beam diffraction mapping of sea urchin teeth. *J Struct Biol* **139**:1–12.

Stock, S.R., D. Blackburn, M. Gradassi, and H.-G. Simon. 2003a. Bone formation during forelimb regeneration: A microtomography (microCT) analysis. *Dev Dyn* **226**:410–417.

Stock, S.R., F. DeCarlo, J.D. Almer. 2008b. High energy x-ray scattering tomography applied to bone. *J Struct Biol* **161**:144–150.

Stock, S.R., F. De Carlo, X. Xiao, and T.A. Ebert. 2009. Bridges between radial wedges (septs) in two diadematid spine types. In *Echinoderms: New Hampshire,* ed. L. Harris. London, U.K.: Taylor & Francis, pp. 263–267.

Stock, S.R., K.I. Ignatiev, T. Dahl, A. Veis, and F. De Carlo. 2003c. Three-dimensional microarchitecture of the plates (primary, secondary and carinar process) in the developing tooth of *Lytechinus variegatus* revealed by synchrotron x-ray absorption microtomography (microCT). *J Struct Biol* **144**:282–300.

Stock, S.R., K. Ignatiev, and F. De Carlo. 2004a. Very high resolution synchrotron microCT of sea urchin ossicle structure. In *Echinoderms: München*, eds. T. Heinzeller and J.H. Nebelsick. Leiden, South Holland: Balkema, pp. 353–358.

Stock, S.R., K.I. Ignatiev, S.A. Foster, L.A. Forman, and P.H. Stern. 2004c. MicroCT quantification of in vitro bone resorption of neonatal murine calvaria exposed to IL-1 or PTH. *J Struct Biol* **147**:185–199.

Stock S.R., K. Ignatiev, P.L. Lee, K. Abbott, and L.M. Pachman. 2004b. Pathological calcification in juvenile dermatomyositis (JDM): MicroCT and synchrotron x-ray diffraction reveal hydroxyapatite with varied microstructures, *Conn Tiss Res* **45**:248–256.

Stock, S.R., S. Nagaraja, J. Barss, T. Dahl, and A. Veis. 2003b. X-ray microCT study of pyramids of the sea urchin *Lytechinus variegatus. J Struct Biol* **141**:9–21.

Stock, S.R., A. Veis, X. Xiao, J.D. Almer, and J.R. Dorvee. 2012b. Sea urchin tooth mineralization: Calcite present early in the aboral plumula. *J Struct Biol* **180**:280–289.

Stock, S.R., A.E. Vieira, A.C. Delbem, M.L. Cannon, X. Xiao, and F. De Carlo. 2008a. Synchrotron microComputed Tomography of the mature bovine dentinoenamel junction. *J Struct Biol* **161**:162–171.

Stock, S.R., X. Xiao, S.R. Stock, and A. Ziegler. 2012c. Quantification of carinar process plate orientation in camarodont sea urchin teeth. *Cah Mar Biol* (in press).

Tommasini, S.M., A. Trinward, A.S. Acerbo, F. De Carlo, L.M. Miller, and S. Judex. 2012. Changes in intracortical microporosities induced by pharmaceutical treatment of osteoporosis as detected by high resolution micro-CT. *Bone* **50**:596–604.

Treece, G.M., K.E.S. Poole, and A.H. Gee. 2012. Imaging the femoral cortex: Thickness, density and mass from clinical CT. *Med Image Anal* **16**:952–965.

Tyndall, D.A., J.B. Price, S. Tetradis, S.D. Ganz, C. Hildebolt, and W.C. Scarfe. 2012. Position statement of the American Academy of Oral and Maxillofacial Radiology on selection criteria for the use of radiology in dental implantology with emphasis on cone beam computed tomography. *Oral Surg, Oral Med, Oral Path Oral Radiol* **113**:817–826.

US Surgeon General. 2004. *Bone Health and Osteoporosis: A Report of the Surgeon General.* Rockville, MD: US Department of Health and Human Services.

Veis D.J., T.M. Alberger, J. Clohisy, M. Rahima, B. Sabsay, and A. Veis. 1986. Matrix proteins of the teeth of the sea urchin *Lytechinus variegatus*. *J Exp Zool* **240**:35–46.

Voide, R., P. Schneider, M. Stauber, P. Wyss, M. Stampanoni, U. Sennhauser, G.H. van Lenthe, and R. Müller. 2009. Time-lapsed assessment of microcrack initiation and propagation in murine cortical bone at submicrometer resolution. *Bone* **45**:164–173.

Wang, W., J.S. Nyman, K. Ono, D.A Stevenson, X. Yang, and F. Elefteriou. 2011. Mice lacking Nf1 in osteochondroprogenitor cells display skeletal dysplasia similar to patients with neurofibromatosis type I. *Human Mol Gen* **20**:3910–3924.

Wang, X,. A. Sanyal, P.M. Cawthon, L. Palermo, M. Jekir, J. Christensen, K.E. Ensrud, S.R. Cummings, E. Orwoll, D.M. Black, and T.M. Keaveny. 2012. Prediction of new clinical vertebral fractures in elderly men using finite element analysis of CT scans. *J Bone Miner Res* **27**:808–816.

Willems, N.M.B.K., L. Mulder, G.E.J. Langenbach, T. Grünheid, A. Zentner, and T.M.G.J. vanEijden. 2007. Age-related changes in microarchitecture and mineralization of cancellous bone in the porcine mandibular condyle. *J Struct Biol* **158**:421–427.

Withers, P.J. 2007. X-ray nanotomography. *Mater Today* **10**:26–34.

Yuan, F. 2012. Finite element simulation of the mechanical properties of mineralized biomaterials. PhD Thesis. Northwestern University, Evanston.

Zaslansky, P., S. Zabler, and P. Fratzl. 2010. 3D variations in human crown dentin tubule orientation: A phase-contrast microtomography study. *Dent Mater* **26**:e1–10.

Zehbe, R., A. Haibel, H. Riesemeier, U. Gross, C.J. Kirkpatrick, H. Schubert, and C. Brochhausen. 2010. Going beyond histology. Synchrotron micro-computed tomography as a methodology for biological tissue characterization: From tissue morphology to individual cells. *J Roy Soc Interf* **7**:49–59.

Zehbe, R., H. Riesemeier, C.J. Kirkpatrick, and C. Brochhausen. 2012. Imaging of articular cartilage—Data matching using x-ray tomography, SEM, FIB slicing and conventional histology. *Micron* **43**:1060–1067.

Ziegler, A., S.R. Stock, B.H. Menze, and A.B. Smith. 2012. Macro- and microstructural diversity among sea urchin teeth revealed by large-scale micro-computed tomography survey, In *Developments in X-Ray Tomography VIII,* SPIE Vol. 8506, ed. S.R. Stock. Bellingham, WA: SPIE, p. 85061G.

16 SIMS method and examples of applications in coral biomineralization

Claire Rollion-Bard and Dominique Blamart

Contents

16.1 USE OF PROXIES

Increasing human activities and their impacts on the environment require an accurate estimate of what could be the environmental conditions for the next decades or centuries. Thus, major efforts have been made to predict future climate at different timescales and space resolution from models of various complexities. These currently developed models use physical parameters of the different compartments of climate system (e.g., ocean, clouds, and atmospheric particles). The collection of these data started 200 years ago, with the beginning of the instrumental measurement period. However, the last two centuries are not representative of climate variability encountered on Earth during the last tens or hundreds of thousands years at "astronomical" timescales (see the astronomical theory of Milankovitch). To fill this gap, paleoclimatologists used past climatic archives, also called proxies, that is, physical and chemical parameters of the environment are recorded indirectly in material mainly derived from biological activity. Because the ocean is one of the most important compartments of the climate system, it is fundamental to characterize as precisely as possible its major parameters. Therefore, knowledge of the evolution of the physical state of the world ocean in the past is essential to assess the environmental conditions in the near future. To obtain this information from the past, proxies that record chemically, isotopically, or genetically the signature of these parameters are used. This includes among them, temperature, salinity, or pH of the water masses.

In Earth sciences, especially in paleoceanography, biocarbonates (carbonates that are built by organisms, foraminifera, corals, bivalves, etc.) are widely used, because calcium carbonates are the most abundant, widespread, and are generally found continuously over geological time. Due to their chemical nature ($CaCO_3$), biocarbonates are easy to analyze for their trace element (Mg, Sr, B, Li, etc.) concentration, or for their isotopic composition (C, O, Ca, B, Li, etc.) (Figure 16.1). This explains why a large number of paleoceanographic studies are derived from chemical or isotopic data obtained on biocarbonates (Emiliani, 1954; Shackleton, 1974; Hays et al., 1976). Moreover, they also offer the opportunity to be dated accurately, on different timescales or window times, using radiogenic techniques such as U/Th (Cheng et al., 2000; Adkins et al., 2002), ^{14}C (Druffel and Linick, 1978; Bard, 1988, and references therein), or ^{210}Pb (Moore and Krishnaswami, 1972; Dodge and Thomson, 1974; Andrews et al., 2002; Sabatier et al., 2012).

Since more than 50 years, the use of geochemical data as climatic or environmental proxies is mostly based on chemical and thermodynamic considerations (Urey, 1947). However, this approach does not take into account the intimate relationship between the calcium carbonate crystals with the protein matrix and the terms and conditions of crystallization processes

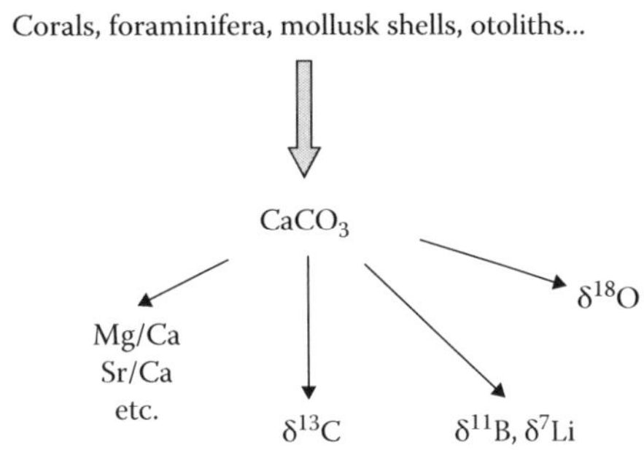

Figure 16.1 Cartoon representing some examples of biocarbonates used in paleoreconstructions, and some of the different proxies that can be measured in the carbonate structure.

during the growth of the organisms. Experiments conducted under controlled conditions have demonstrated, following the species investigated, a deviation of the data obtained from the thermodynamically expected values. This difference observed between the thermodynamically expected values of an inorganic carbonate and a biocarbonate was called "vital effect" (Urey et al., 1951) and is still a convenient "black box" containing all the impacts due to biological processes. For a given species, this effect is considered constant over geological time and also during the ontogenic cycle of the organism, which is not necessarily the case. Finally, one of the major points of interest is related to the biological material itself. The organisms develop skeletons that exhibit different morphology, size of crystal, and microstructures. These crystallographic heterogeneities develop some microstructures that have the particularity to be significantly different, from a biogeochemical point of view, from those of the surrounding crystals (Cohen and McConnaughey, 2003; Erez, 2003; Weiner and Dove, 2003).

To have an access to these biomineralization processes, we have to use a technique that does not average all the impact of these mechanisms, but in contrary, that can give some functionally resolved information. In that way, the secondary ion mass spectrometry (SIMS) is the most adapted instrument for that goal because it records these biomineralization processes via the heterogeneity at micrometer scale.

In this chapter, we show, based on some specific examples, how *in situ* measurements using ion microprobe (SIMS) have provided crucial key answers in understanding the mechanisms of calcification of living organisms. The results also demonstrate that the spatial heterogeneity in chemical and isotopic compositions of the skeleton is intimately related to microstructures. To illustrate our purpose, we focus mainly on the stable isotope geochemistry (C-O-B) of two deep-sea corals (DSCs), *Lophelia pertusa* and *Madrepora oculata*. The results obtained demonstrate clearly the complex relationship between the high variability of isotopic compositions and (micro) structures of the coral skeleton.

Nevertheless, even if it is sometimes not so clear, the medium of precipitation is related to the surrounding seawater. One example of this relation is the link that exists between the pH of seawater and pH of the fluid of calcification in foraminifera. Indeed, it was shown that foraminifera modified the pH of the seawater before calcification, but this modification is always related to the surrounding pH (Rollion-Bard and Erez, 2010). This link implies that biocarbonates obviously remain the material of choice in paleoceanographic studies however, it becomes increasingly clear that a careful sampling strategy will be the key to the accuracy and success of these studies (Lutringer et al., 2005).

16.2 ANALYTICAL METHOD

The first characteristic of SIMS is to be an instrument able to measure elements or their isotopes, at points resolved on a micrometer scale. This allows a precise relationship between the data and the sample studied, leading to a new approach in the interpretation of the results compared to bulk analysis, including mass balance calculation. Basically, this equipment is a mass spectrometer constituted of three main parts (Figure 16.2): the primary ions source system, the mass spectrometer itself (electrostatic sector and magnet), and the counting system, comprising Faraday cups and electron multipliers, all different parts being under high vacuum in the range of 10^{-9} mbar. Most of the SIMS have two different primary sources (generally cesium and oxygen) that can be used for the measurements. Depending on the element to be analyzed (see Section 16.2.2.1 for more details), one of the sources is selected and a focalized ion beam is produced on a sample (typically ≈10–20 μm). The primary ion beam sputters the sample, and some of the elements constitutive of the sample at the location of the spot are ionized and ejected toward the mass spectrometer, where they are sorted in energy with the electrostatic sector and in mass/charge with the magnet. The produced ions, called secondary ions, are then counted into a Faraday cup or into an electron multiplier for intensities lower than 10^6 cps (counts per second).

Various applications of ion probe measurements have been described in detail by Castaing et al. (1978), Shimizu et al. (1978), Shimizu and Hart (1982), Reed (1984), Ireland (1995), and Hinton (1995). In this chapter, we do not go through all the technical aspects and details of the measurements, but we just focus on some important points to take into account for SIMS applications, that is, sample preparation, sputtering process, mass resolution, and instrumental mass fractionation (IMF).

16.2.1 SAMPLE PREPARATION

Whatever the type of sample to be measured, certain rules must be observed to optimize the quality of analytical conditions and thus the quality of the measurements. The sample must be a polished section that can be contained in a 2.5 cm mount (1 in.), knowing that about 0.5 cm cannot be used for the analysis because it is too far from the center of the mount, and then some geometrical effects, called "X–Y effects," can be produced (Kita et al., 2009).

Geological samples are often mounted in epoxy or in indium (even if thin sections can be analyzed), and must be coated with C or Au for conductivity. The sample mount must be ground to produce a flat surface before final polishing. The grinding must be done progressively, generally with the use of 600, 1200, and 2400 grades. The final step of polishing is with diamond paste of 3 and 1 μm diameter. This last step must remove all the remaining scratches. It must be noted that in case of carbon analyses (isotope or content), the diamond paste is avoided due to some possible contamination, and it is then replaced by cerium oxide (CeO_2).

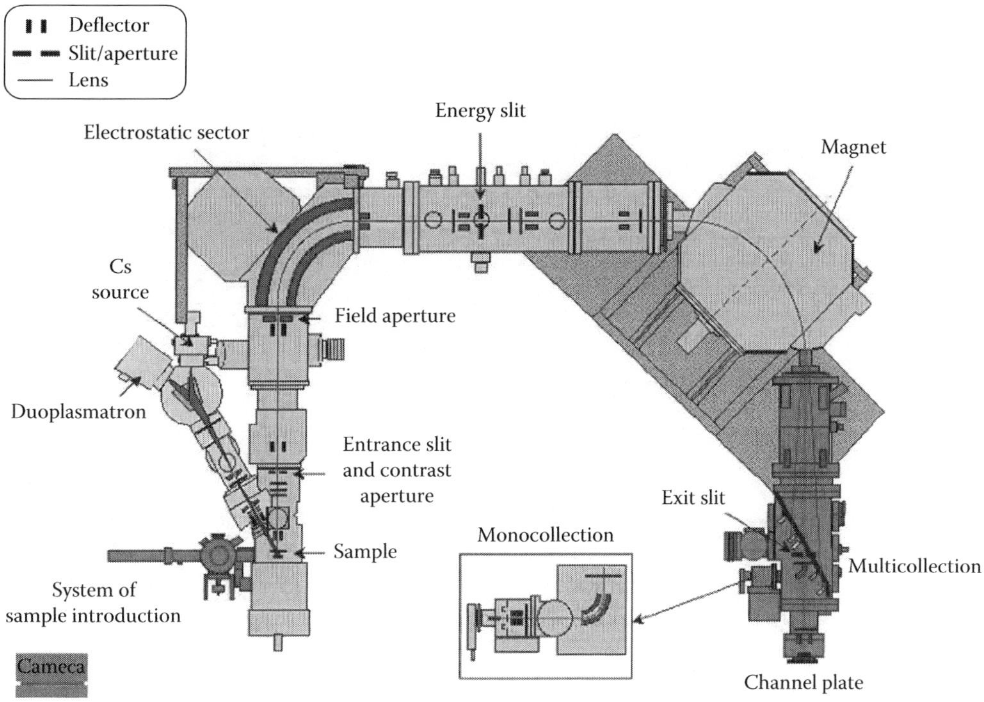

Figure 16.2 Schematic diagram of the Cameca IMS 1270. The primary column can produce a focused beam of either oxygen-negative ions (duoplasmatron source) or cesium-positive ions (Cs source) (at lower left). With the sputtering of the sample surface, some secondary ions are ejected and transferred toward the mass spectrometer with an electrostatic sector (upper left) and a magnet (upper right). Secondary ions can then be detected in (1) monocollection mode by peak jumping, (2) multicollection mode with simultaneous detection of up to five elements, or (3) image mode on the channel plate (lower right). (After a document by Cameca.)

16.2.2 DESCRIPTION OF THE TECHNIQUE

16.2.2.1 Sputtering

During the analysis by ion microprobe, the surface of the sample is sputtered by a focused primary ion beam of high energy (typically 10–13 keV). The impact of each ion leads to the displacement of some atoms in the sample, creating thus a collision in cascade that allows some atoms to be ejected from the surface of the sample when the given energy is higher than their binding force (Figure 16.3). All sputtered atoms come from only a few atomic layers, with the majority coming from the surface itself (Williams, 1983). Most of the ejected particles are neutral, but a small fraction is ionized and can be accelerated toward the mass spectrometer. The yield of ionization is about 0.1%.

So, in its simplest form, the ion probe consists of a primary ion beam source, an optical column that generates a focused primary ion beam, an extraction system that transfers the secondary ions from the surface of the sample to the entrance slit of the mass spectrometer, and the mass spectrometer itself (Figure 16.2). The choice of the primary ions depends of the potentiality of one element to be more ionized as negative or as positive (Storms et al., 1977). For example, if carbon isotopes analysis needs to be performed, it is preferable to use Cs^+ as the primary ion beam, because the ionization is more efficient in the negative form (M^-, the negative ion form of mass M) than in the positive one (M^+, the positive ion form of mass M). In that case, it will be preferable to use Cs^+ as the primary ion beam. In the case of bismuth, the M^-/M^+ ratio is close to 1; so, both primary ion beams can be used.

16.2.2.2 Mass resolution

Accurate abundance or isotopic analysis requires a clear identification of the masses of interest to be measured. The most important problem comes from the interferences on the masses that are either very close to each other or with a similar mass to charge ratio M/C, C being the charge of the ion. The simplest

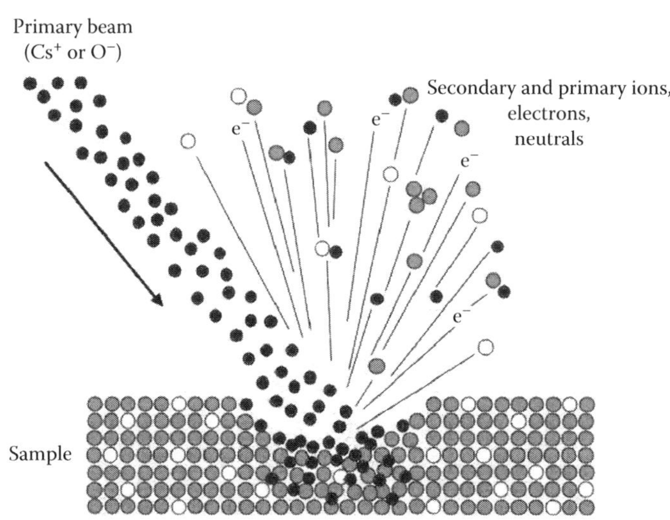

Figure 16.3 Sputtering of the surface of a sample by a primary beam of oxygen (O^-) or cesium (Cs^+) ions. By the impact of this high-energy beam (typically 10–13 keV), elements constitutive of the sample are ejected as mono- or pluriatomic secondary ions, electrons, and neutrals. The secondary ions are then accelerated toward the mass spectrometer of the ion probe. (Modified after Ireland, T.R., *Adv. Anal. Geochem.*, 2, 1–118, 1995.)

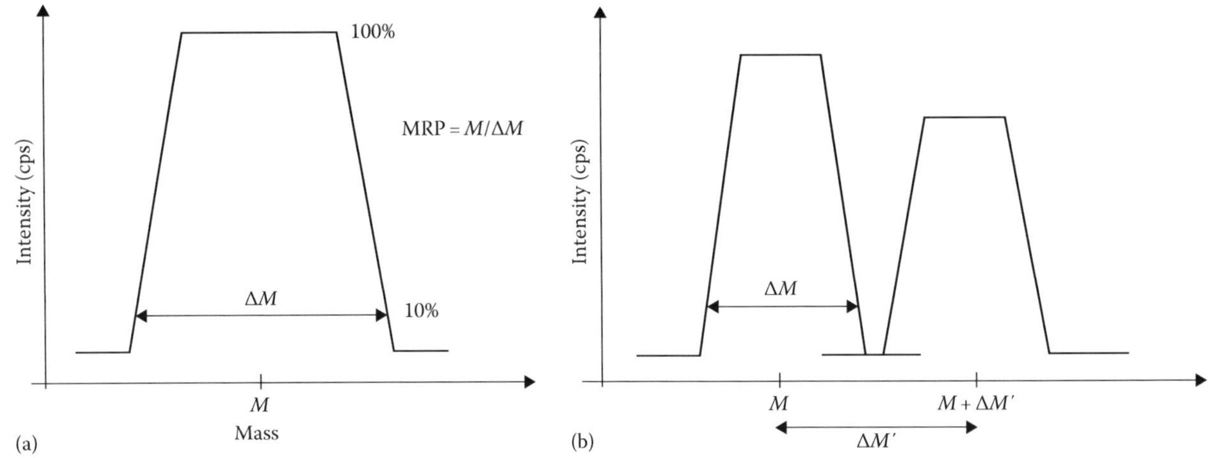

Figure 16.4 (a) Schematic diagram of a peak obtained by ion microprobe. The intensity of secondary ions is in cps. Mass resolution (MRP) is generally defined as the ratio $M/\Delta M$, M being the mass of the element, and ΔM, the width at 10% of the peak height. (b) Schematic peaks of two elements at masses M and $M + \Delta M'$. The two peaks are considered as separated when ΔM is lower than $\Delta M'$.

way to remove any interference is to have a mass resolution sufficient to separate the peaks. The mass resolution of the analytical settings is defined as the ratio between the mass of interest (M) and the width of the peak at 10% of the intensity (ΔM) (Figure 16.4). Two peaks are separated if ΔM is lower than $\Delta M'$, that is, the mass difference between the two peaks to be separated (Figure 16.4b). For example, in the case of the measurement of oxygen isotopes (^{16}O and ^{18}O), there is an isobaric interference of ^{17}OH (18.00695 amu) on ^{18}O (17.99916 amu). The difference $\Delta M'$ between these two masses is 0.00779 amu; then, a tuning with a mass resolution of at least 2310 (i.e., 18.00695/0.00779) is needed to separate these two peaks. The easiest way to tune the ion probe in order to get an optimal mass resolution is to diminish the width of the entrance and exit slits (see Figure 16.2). Indeed, the mass resolution for a given instrument is given by

$$\frac{M}{\Delta M} = k \times \frac{R}{s},$$

with

R is the radius of the ion path in the magnet
s is the width of the entrance slits
k is a constant defined by the design of the magnet (Long, 1995)

At the moment, the higher mass resolution that can be achieved by SIMS is around 35,000 with the Cameca IMS 1280 HR2. With this high mass resolution, most of the interferences can be removed. Nevertheless, in some cases, the resolution of two peaks is not possible. However, a correction can be applied by measuring another isotope (when possible) of the interfering element. For example, in the case of iron isotopes, a mass resolution of 73,000 is needed to remove the interference of mass ^{54}Cr on the peak of ^{54}Fe. The ^{54}Cr isobaric interference cannot be resolved; so, ^{52}Cr can be measured to estimate the intensity of ^{54}Cr (Marin-Carbonne et al., 2011).

16.2.2.3 Instrumental mass fractionation

Isotopic ratios of elements determined by ion microprobe are often shifted from their true ratios, with the measured ratios usually depleted in the heavy isotope relative to the true isotopic abundances. This difference is called IMF (also noted α_{inst}) and is defined as the ratio between the measured and the real values:

$$\alpha_{inst} = \frac{R_{measured}}{R_{true}},$$

where

$R_{measured}$ is the isotopic ratio measured by the ion probe
R_{true} is the true isotopic ratio of the reference material (RM)

The IMF is always less than 1, except for the lithium isotopes and is determined by measuring some RM with the same matrix as the sample and with a known isotopic and chemical composition. For convenience, the IMF is most of the time expressed in ‰ and corresponds to the difference between the isotopic composition (δ) measured with the ion probe and the true isotopic composition of the RM and is noted as Δ_{inst} (Figure 16.5). Different factors can affect the IMF; among them the more effective are the sputtering process and the analytical settings, the isotopic system to analyze, and finally the chemical composition of the matrix.

The IMF is related to the fact that, during the sputtering process, light isotopes are preferentially emitted relative to heavy isotopes of a given element, because light isotopes have lower bond energy (Slodzian et al., 1980; Shimizu and Hart, 1982). In the following, we do not describe the influence of the analytical settings that could include the pressure in the chamber (Sangely et al., 2005; Rollion-Bard et al., 2007a,b), and the presputtering time (time spent to ablate the surface sample without data acquisition) (Rollion-Bard et al., 2007a) that could influence the sputter-ionization process due to the implantation of primary ions into the sample (Benninghoven et al., 1987; Janssens et al., 2003).

IMF depends also on the isotopic system analyzed, and in general, the more lighter the element is, greater is the IMF. As shown in Figure 16.5, the magnitude of the IMF is proportional to the difference between the isotopes of the same element (Slodzian et al., 1980).

The matrix effect on the IMF is the fractionation caused by the sample chemistry because of the bond energy difference and

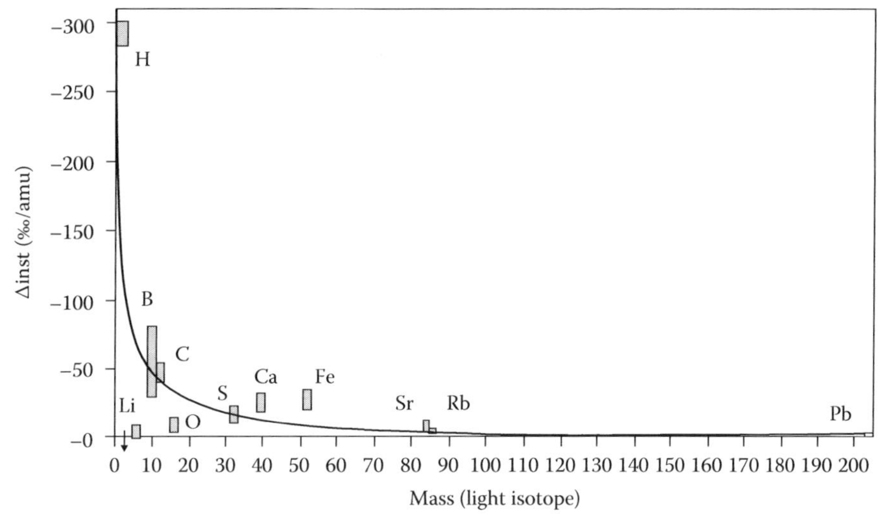

Figure 16.5 IMF (Δinst) as a function of the mass of the lighter isotope for different isotopic systems determined using the Cameca IMS 1270 ion probe in Nancy (France). Note that there is a global trend of decreasing Δinst from H to Pb isotopes. (Data from Chaussidon, M. et al., *Geochim. Cosmochim. Acta*, 70, 224, 2006; Deloule, E. et al., *Geochim. Cosmochim. Acta*, 62, 3367, 1998; Deloule, E. et al., *Geochim. Cosmochim. Acta*, 65, 1833, 2001; Decitre, S. et al., *Geochim. Geophys. Geosyst.*, 2002, 3:10.1029/2001GC000178; Giuliani, G. et al., *Science*, 287, 631, 2000; Marin-Carbonne, J. et al., *Chem. Geol.*, 285, 50, 2011; Rollion-Bard, C. et al., *Earth Planet. Sci. Lett.*, 215, 275, 2003a; Rollion-Bard, C. et al., *Coral Reefs*, 22, 405, 2003b; Rollion-Bard, C. et al., *Chem. Geol.*, 244, 679, 2007a; Rollion-Bard, C. et al., *Geost. Geoanal. Res.*, 31, 39, 2007b; Sangely, L. et al., *Chem. Geol.*, 223, 179, 2005; Thomassot, E. et al., *Earth Planet. Sci. Lett.*, 282, 79, 2009; Vielzeuf, D. et al., *Chem. Geol.*, 223, 208, 2005; Vigier, N. et al., *Geochim. Geophys. Geosyst.*, 8, 1, 2007.)

the crystallography effects on the sputtering. It has, for example, been shown in carbonates that the IMF for oxygen isotope measurements is strongly dependent on the Mg content of the RMs (Rollion-Bard and Marin-Carbonne, 2011) (Figure 16.6). It implies that for the accurate measurement of $\delta^{18}O$ of carbonates, the Mg content has to be taken into account for the correction of the IMF. It would be crucial in the measurement of high-Mg calcite organisms, such as gorgonian deep-sea corals (DSCs). It could also be noticed that the IMF for carbon and oxygen isotope measurements is not the same between aragonite and calcite (Rollion-Bard et al., 2007b), despite their similar chemical formula ($CaCO_3$). In that case, this IMF difference is likely due to the difference in the strength of the chemical bond of the analyzed ion and its matrix, because of the dissimilarity between the two crystal systems, trigonal for calcite and orthorhombic for aragonite.

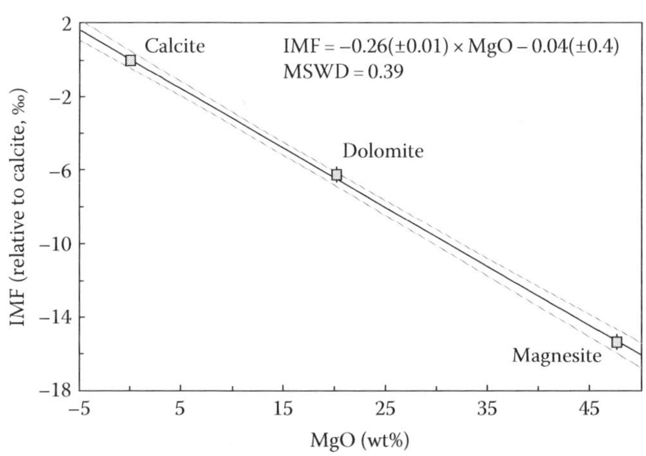

Figure 16.6 IMF of oxygen isotopes relative to the IMF determined of calcite RM as a function of MgO content (wt%). (After Rollion-Bard, C. and Marin-Carbonne, J., *J. Anal. At. Spectr.*, 26, 1285–1289, 2011.)

In summary, the main points that have to be taken into account to perform accurate and precise measurements using SIMS are (1) the polishing of the sample, (2) the choice of the primary beam source, (3) the determination of the mass resolution to avoid any isobaric interferences, and (4) the choice of the RMs that must have, as far as possible, the closest chemical formula and crystallographic system to the sample.

16.3 EXAMPLE OF APPLICATION OF SIMS IN BIOMINERALIZATION: DEEP-SEA CORALS

Many proxies can be measured in biocarbonates. Some of them have been explored using SIMS method and are listed in Table 16.1. In the following sections, we only focus on the use of oxygen and boron isotope compositions in two DSCs, *Lophelia pertusa* and *Madrepora oculata*.

16.3.1 INTRODUCTION TO DEEP-SEA CORALS

DSCs are known to paleontologists and marine biologists from the sixteenth century, but have been of growing interest since about two decades through their potentials to be used as proxies in paleoceanography. Two main issues have been addressed, thanks to the study of DSCs. The first focuses on the determination of "^{14}C reservoir age." This age corresponds to the time required to integrate the atmospheric carbon to marine waters. To determine the reservoir age, one must access the difference between the measured ^{14}C apparent age and the true age of the sample. The true age is obtained with the U/Th dating method from which absolute age can be derived. The reservoir age of the modern ocean is about 400 years, but it has varied over time and during major climatic transitions.

Table 16.1 Proxies used in paleoenvironmental studies, the environmental parameter(s) that they record, the precision reached by SIMS, and selected reference(s) for each proxy measured by SIMS

PROXY	PARAMETER	SIMS PRECISION (1σ)	REFERENCE(S)
$\delta^{18}O$	Temperature Salinity pH	±0.2‰	Rollion-Bard et al. (2003a) Allison et al. (2010a)
$\delta^{13}C$	$\delta^{13}C$ of seawater DIC[a] Nutrient availability Oceanic currents	±0.6‰	Rollion-Bard et al. (2003b)
$\delta^{11}B$	pH	±0.9‰	Rollion-Bard et al. (2003a) Blamart et al. (2007) Rollion-Bard and Erez (2010)
$\delta^{7}Li$	Erosion rate	±1‰	Rollion-Bard et al. (2009)
$\delta^{44}Ca$	Temperature?	±0.25‰	Rollion-Bard et al. (2007a)
Mg/Ca	Temperature Salinity	≈3%	Allison and Finch (2007)
Sr/Ca	Temperature	≈1%	Hart and Cohen (1996) Allison and Austin (2003)
Ba/Ca	Alkalinity Oceanic upwelling Local rainfall and freshwater runoff	≈7%	Hart and Cohen (1996) Allison and Finch (2007)
B/Ca	$[CO_3^{2-}]$	2%–3%	Hart and Cohen (1996)

[a] DIC, dissolved inorganic carbon.

The second subject was linked to the role played by the intermediate waters (750–1500 m water depth) in ocean circulation and its impact on the climate variability (Lutringer, 2005). The main purpose was to document the poorly known intermediate waters and to understand the state of the water column during the transmission of temperature–salinity signals from the surface to the depth during the last 10,000 years. Paleotemperature reconstructions using O-isotopes have been performed on DSC collected from different sedimentary cores coming from North Atlantic, a key area in the global ocean circulation. Lutringer et al. (2005) demonstrated that the sampling strategy was very important in the determination of accurate and precise paleotemperatures from O-isotopes. The study revealed that some parts of the coral skeleton were not appropriate to an accurate determination of paleotemperature. Thus, we conducted several SIMS studies on two different DSCs (*Lophelia pertusa* and *Madrepora oculata*) to understand the behavior of O and later B isotopes.

From a classification based on skeletal morphology (Cairns, 1994), the two corals in common belong to the phylum Cnidaria (Hatschek, 1888), the class Anthozoa (Ehrenberg, 1834), the subclass Hexacorals, and order scleractinian (Bourne, 1900). *Lophelia pertusa* belongs to the family Caryophylliidae (Dana, 1846), genus Lophelia (Milne-Edwards and Haime, 1849), and species *Lophelia pertusa* (Linnaeus, 1758). *Madrepora oculata* is from the family of Oculinidae (Gray, 1847), genus Madrepora (Linnaeus, 1758), and species *Madrepora oculata* (Linnaeus, 1758). More recently, using molecular studies (sequences of the mitochondrial 16S rDNA), Le Goff Vitry et al. (2005) investigated the phylogenetic relationships of *Lophelia* and *Madrepora*, confirming the existing classification (de Portales, 1871; Cairns, 1994).

Lophelia pertusa

Kingdom: Animalia
Phylum: Cnidaria (Hatschek, 1888)—cnidarians, coelenterates, anemone, coral
Class: Anthozoa (Ehrenberg, 1834)—corals, flower animals, sea anemones
Subclass: Hexacorallia
Order: Scleractinia (Bourne, 1900)—stony corals, madrepora
Suborder: Caryophylliina (Vaughan and Wells, 1943)
Family: Caryophylliidae (Dana, 1846)
Genus: Lophelia (Milne-Edwards and Haime, 1849)
Species: *Lophelia pertusa* (Linnaeus, 1758)

Madrepora oculata

Kingdom: Animalia
Phylum: Cnidaria (Hatschek, 1888)—cnidarians, coelenterates, anemone, coral
Class: Anthozoa (Ehrenberg, 1834)—corals, flower animals, sea anemones
Subclass: Hexacorallia
Order: Scleractinia (Bourne, 1900)—stony corals, madrepora
Suborder: Faviina (Vaughan and Wells, 1943)
Family: Oculinidae (Gray, 1847)
Genus: Madrepora (Linnaeus, 1758)
Species: *Madrepora oculata* (Linnaeus, 1758)

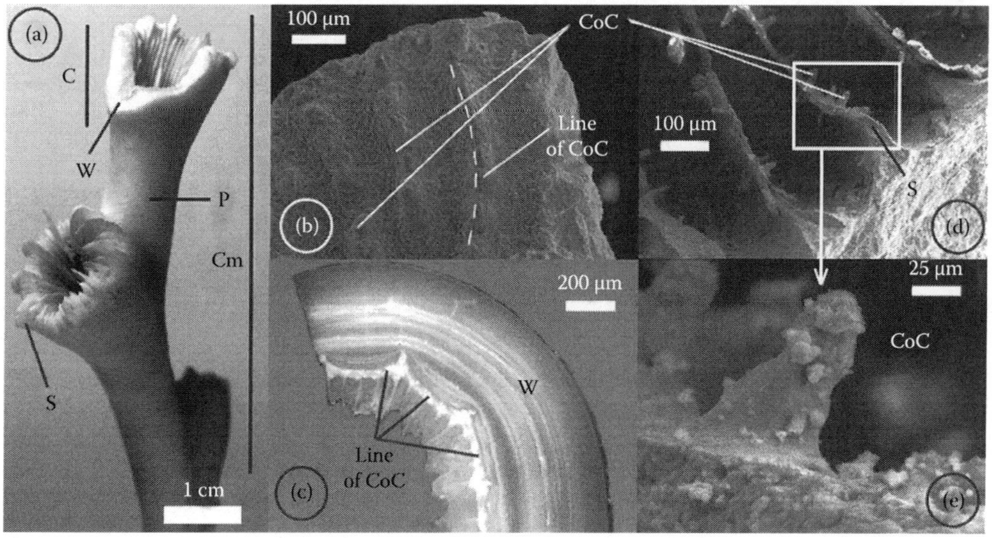

Figure 16.7 Different photographs of deep-sea coral *Lophelia pertusa* showing different aspects of the macro- and microstructures of the aragonitic skeleton: (a) Natural light photograph of the coral skeleton, also called coralum (Cm), composed of the pedicel (P), a calyx (C), with walls (W) and septa (S). (b–e) Detailed views with the microstructure of the skeleton with (b) scanning electron microscope (SEM) image of a septa where the CoCs are underlined by dashed line, (c) natural light view in cross section of the skeleton showing the clear arrangement of the lines of CoCs within the calyx, and (d and e) sagittal SEM details from the area shown in (b), where CoCs produce the line of CoC.

16.3.2 STRUCTURE AND MICROSTRUCTURE OF DSC

Deep-water corals, like all biocarbonates, present a skeleton organized with, in our case, well-expressed structures and microstructures (Figure 16.7) that develop during skeleton growth. The coral skeleton is called the corallum with a base attached to the substrate. This part is topped by a pedicel, which supports the calyx with a cylindrical or trochoidal shape. The calyx is formed by walls, which give the main form of the skeleton; laterally, the walls form the theca. Within the calyx are vertical centripetal strips called septa. Sometimes, as in the case of *Lophelia pertusa*, the septa extend into the theca and form vertical stripes (Figure 16.7) (Cairns, 1981; Lazier et al., 1999).

Basically, the development of the coral skeleton starts after the fixation of the larvae on a hard substrate of the sea-bed, followed by the formation of the base. The polyp secretes calcium carbonate fibers from the basal endodermal cells, also called calicoblastic cells. The development of the septa takes place in two steps. First, the polyp produces small-sized fibers agglomerated into the center of calcification (CoC) (Ogilvie, 1896), corresponding to the early biomineralization zones (EMZ). Then, the coral produces larger surrounding fibers (Figure 16.7). These two stages thus contribute to the construction of the septa and are controlled by crystallization mechanism leading to two crystallographic microstructures whose characteristics are clearly different. The organization with CoC or EMZ and the surrounding fibers suggests that calcification is initiated by calcification centers and contributes to the lateral and vertical growth of the coral. The initiation of the vertical growth through the formation of a floor is still poorly understood and probably happens when the vertical growth of the theca and septa would become incompatible with the stretching of the polyp.

16.3.3 PROCESSES OF BIOMINERALIZATION IN DSC: WHAT CAN WE LEARN FROM O AND B ISOTOPE COMPOSITIONS?

A decade ago, studies have shown that CoC and surrounding fibers have different optical chemical and crystallographic properties. It is therefore natural to ask if these two microstructures are fundamentally different also in their isotopic compositions. Here, we have deliberately chosen two examples, which are both very important in the use of proxies, not only for understanding the evolution of marine paleoenvironments, but also to understand the fundamental mechanisms involved in the biomineralization process. These two examples are first the use of boron isotopes as an indicator of paleoocean pH and the use of oxygen isotopes as paleotemperature indicator of the ocean on the other side; the mechanisms of biomineralization will be addressed by the joint study of boron and oxygen isotopes. The main interest for the use of SIMS is that we have the possibility to focus the measurements in each microstructure and then to decipher the role of biomineralization processes from the impact of changes in environmental parameters.

16.3.3.1 B isotope composition in carbonates as pH proxy

Boron isotope composition of carbonates can be used as pH proxy of the solution from which the carbonate precipitates. This is due to the fact that the distribution of the two dissolved boron species, that is, borate ion $B(OH)_4^-$ and boric acid $B(OH)_3$, is pH-dependent and that an isotopic fractionation of 27.2‰ (Klochko et al., 2006) exists between these two species, $B(OH)_3$ being enriched in ^{11}B relative to $B(OH)_4^-$. As it is assumed that only $B(OH)_4^-$ is incorporated into the carbonate (Hemming and

Hanson, 1992), the carbonate $\delta^{11}B$ reflects borate ion $\delta^{11}B$, and thus the pH of the solution of precipitation according to

$$pH = pK_B - \log\left(\frac{\delta^{11}B_{sw} - \delta^{11}B_{carb}}{\alpha_{3/4}^{-1} \times \delta^{11}B_{carb} - \delta^{11}B_{sw} + 1000 \times \left(\alpha_{3/4}^{-1} - 1\right)}\right),$$

where

$\delta^{11}B_{carb}$ is the boron isotope composition of the carbonate
$\delta^{11}B_{sw}$ is the boron isotopic composition of seawater, which is measured at 39.5‰
pK_B is the dissociation constant of boric acid and is dependent on the salinity and the temperature of the solution (Dickson, 1990)
$\alpha_{3/4}$ is the isotopic fractionation factor between $B(OH)_3$ and $B(OH)_4^-$

16.3.3.2 B and O isotopic compositions in deep-sea corals and their link to the microstructure

We measured boron isotopic composition by SIMS in two DSCs, *Lophelia pertusa* and *Madrepora oculata*, and oxygen isotopic compositions in *Lophelia pertusa* only. The results with *Lophelia pertusa* were already published by Blamart et al. (2007), and Rollion-Bard et al. (2010). We will here highlight the main points of these studies and compare them to the results with *Madrepora oculata*:

1. The overall variability in $\delta^{11}B$ in *Lophelia pertusa* is ≈10‰ (from 28‰ to 38‰), and about 4‰ in *Madrepora oculata* (from 28.7‰ to 32.8‰) (Figure 16.8). In both cases, these variations cannot be related to changes in pH of the surrounding seawater, as this parameter is almost constant in the deep sea during the lifetime of the coral specimens.

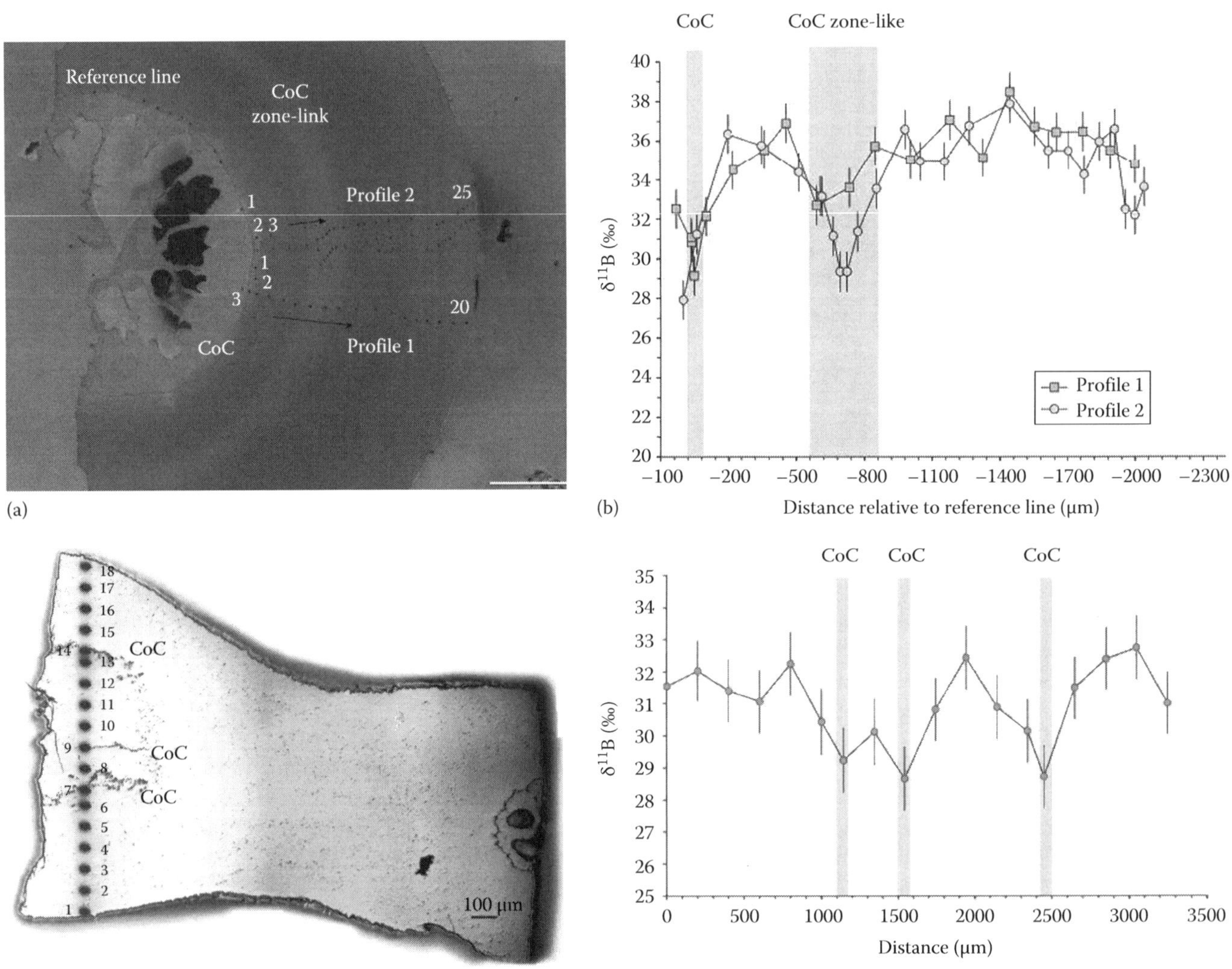

Figure 16.8 $\delta^{11}B$ measurements and their relation to the microstructure of deep-sea coral skeleton. (a) Picture in fluorescence of *Lophelia pertusa*. (Courtesy of Jarek Stolarski, Institute of Paleobiology, Poland.) The zone of CoC and the CoC zone-like appeared as white zones. The CoC zone-like corresponds to a zone that is geochemically similar to the CoCs, but that does not show any morphological aspect of CoCs. Dark ellipses correspond to ion-probe spots. They are organized in two profiles (Profile 1 and Profile 2) that go from the CoC zone toward the outermost part of the calyx. (b) Measured $\delta^{11}B$ values relative to their position from the reference line indicated in (a). Note that the $\delta^{11}B$ values drop in the CoC and the CoC-like zones. (Data from Blamart, D. et al., *Geochem. Geophys. Geosys.*, 8, Q12001, 2007.) (c) Picture of the *Madrepora oculata* sample. The black ellipses represent the spot made by the ion probe. CoCs are indicated and are represented in the picture by black zones of grain clusters. (d) Measured $\delta^{11}B$ values in function of distance in μm. The presence of lines of CoCs are indicated in light gray.

2. Oxygen isotopic composition in *Lophelia pertusa* also recorded a large variability, up to 6‰ (from −2.6‰ to 3.4‰) (Rollion-Bard et al., 2010), which cannot be related to changes in the surrounding temperature, that is, it would correspond to a change of 30°C, which is impossible.
3. $\delta^{11}B$ and $\delta^{18}O$ show a drop in their values corresponding to the localization of EMZ in both the coral species (Figure 16.8).

16.3.3.3 Implications for the processes of biomineralization

Here, we only focused on $\delta^{18}O$ and $\delta^{11}B$, but all geochemical systems measured so far (except $\delta^{7}Li$ [Rollion-Bard et al., 2009]) show a significant difference between EMZ and the surrounding fibers. Three main geochemical models of biomineralization processes are used to explain this difference between EMZ and fibers: (1) Rayleigh fractionation (Cohen et al., 2006), (2) variation of pH of the calcifying fluid, also called the "carbonate" model (Adkins et al., 2003), and (3) a kinetic model (McConnaughey, 2003):

1. Based on the relationships between elemental ratios (Mg/Ca and Sr/Ca) measurements, a Rayleigh fractionation process model was proposed. In this model, the main assumption is that the fluid of calcification forms a closed or semiclosed system during the precipitation of the skeleton. During carbonate precipitation, there is an evolution of the fluid in Sr/Ca and Mg/Ca ratios, and thus an evolution of these ratios in the carbonate precipitated. This model was shown to not describe the processes of biomineralization in numerous studies (Rollion-Bard et al., 2009, 2010; Allison et al., 2010b, 2011; Brahmi et al., 2012).
2. In the "carbonate" model, the different geochemical behaviors of EMZ relative to the fibers are explained by an evolution of the pH of the calcifying fluid, with an increase of the pH at the sites of calcification of the EMZ. pH increase should shift $\delta^{18}O$ toward lighter values (McCrea, 1950; Usdowski et al., 1991; Usdowski and Hoefs, 1993; Zeebe, 1999), and this model was developed to explain lighter $\delta^{18}O$ in EMZ relative to the fibers. As shown earlier (see Section 16.3.3.1), $\delta^{11}B$ values can be used as pH proxy and then give some information of the pH at the sites of calcification. If this model is right, we can predict that $\delta^{11}B$ values should be higher in the EMZ than in the fibers. But, this is the complete opposite to that we observed, with a systematic drop of $\delta^{11}B$ data in EMZ. This precludes the pH evolution of a common fluid of calcification between EMZ and fibers.
3. The kinetic model is based on the possibility that the carbonate species in solution (mainly HCO_3^- and CO_3^{2-}) could be incorporated into the coral skeleton before they are in isotopic equilibrium with H_2O (McConnaughey, 1989). Increasing precipitation rate results in lighter oxygen (and carbon) isotopic composition. Thus, to explain the lighter O composition of EMZ, it would be the result of a faster precipitation rate in EMZ compared to that in fibers. This would not have any effect on B isotopic compositions, as the equilibrium is almost instantaneous between $B(OH)_3$ and $B(OH)_4^-$ (Zeebe, 2005). The $\delta^{11}B$ results seem to show that the EMZ pH is lower than the pH of precipitation of fibers. So, if the kinetic model is applicable, EMZ precipitation would be quasi-instantaneous, taking into account kinetic calculations (Rollion-Bard et al., 2003a, 2011) and the impact of the presence of Ca–ATPase on the time of equilibration (Uchikawa and Zeebe, 2012). Nevertheless, this kinetic model cannot explain the behavior of some other elements like lithium (Rollion-Bard et al., 2009).

So, based on the *in situ* data (isotopic and elemental ratios), it is not possible to reconcile any of the different geochemical models with the different geochemical and morphological characteristics of EMZ and surrounding fibers (Figures 16.7 and 16.9). Then we proposed (Rollion-Bard et al., 2010) that EMZ calcification is not derived from the same fluid involved in the precipitation of fibers. EMZ and fibers precipitate from different "domains." These domains could be different cells or the same cells but applying different modes of calcification. This model needs to be validated by biological observations. Nevertheless, whatever be the processes of biomineralization, a careful sampling is necessary before any paleoenvironmental reconstruction. The mixing

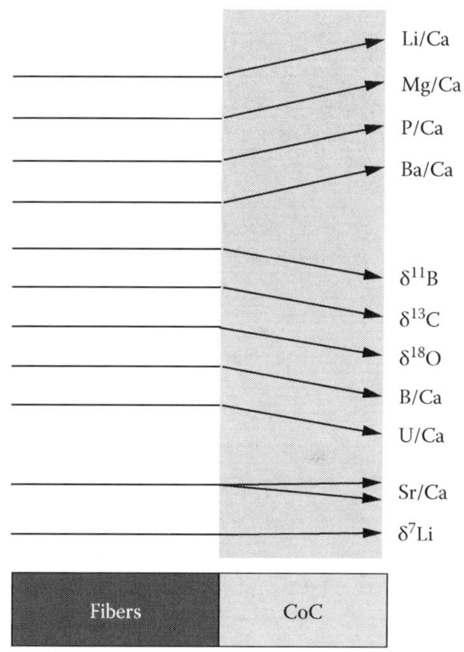

Figure 16.9 Schematic representation of different elemental and isotopic ratio signatures (not at scale) measured by *in situ* methods (micromill, SIMS, and LA-ICP-MS [laser inductively coupled plasma mass spectrometry]), compared between CoCs (left) and fibers (right, with trends shown by arrows) in DSCs. (Data from Adkins, J.F. et al., *Geochim. Cosmochim. Acta*, 67, 1129, 2003; Anagnostou, E. et al., *Geochim. Cosmochim. Acta*, 75, 2529, 2011; Blamart, D. et al., C and O isotopes in a deep-sea coral (*Lophelia pertusa*) related to skeletal microstructure, in *Cold-water Corals and Ecosystems*, Freiwald, A. and Roberts, J.M. (eds.), Springer-Verlag, Berlin, Germany, pp. 1005–1020, 2005; Blamart, D. et al., *Geophys. Geosys*., 8, Q12001, 2007; Case, D.H. et al., *Earth Planet. Sci. Lett*., 300, 215, 2010; Gagnon, A.C. et al., *Earth Planet. Sci. Lett*., 261, 280–295, 2007; Lopez Correa, M. et al., *Deep-Sea Res. II*, 57, 471, 2010; Montagna, P. et al., High-resolution trace and minor element in deep-water scleractinian corals (*Desmophyllum dianthus*) from the Mediterranean Sea and the Great Australian Bight, in *Cold-Water Corals and Ecosystems*, Freiwald, A. and Roberts, J.M. (eds.), Springer-Verlag, Berlin, Germany, pp. 1109–1126, 2005; Rollion-Bard, C. et al., *Coral Reefs*, 22, 405, 2003b; Rollion-Bard, C. et al., *Earth Planet. Sci. Lett*., 286, 63, 2009; Rollion-Bard, C. et al., *Geochim. Cosmochim. Acta*, 74, 1338, 2010; Shirai, K. et al., *Chem. Geol*., 224, 212, 2005; Sinclair, D.J. et al., *Geophys. Res. Lett*., 33, L17707, 2006.)

in various proportions of signatures of EMZ and fibers can introduce some bias in the interpretations of the environmental parameter variations. We suggest to sample only the fibers that seem to be closer to equilibrium values. This operation is quite easy in DSCs, because the fibers zone is very large and easily identifiable, but is more challenging in tropical corals in which CoC (equivalent of EMZ) and fibers are tightly imbricated.

16.4 PERSPECTIVES

We have shown that SIMS (or more generally, *in situ*) measurements can give us some information about the processes of biomineralization, and that the bulk data represent an average of all the different mechanisms present at the interface calicoblastic cells and coral skeleton, that is, pH variation, kinetic effects, amorphous precursors, etc. Nevertheless, it is important to note that some geochemical signatures cannot be measured by SIMS, and this technique represents then also a bulk analysis. To go deeper into the biomineralization processes and the impact of these processes on geochemical proxies, a technique with a higher spatial resolution is needed. The NanoSIMS is the instrument dedicated to high spatial resolution. The basic principles are the same as the SIMS, but as the primary beam is perpendicular to the sample, the size of the beam is smaller, typically about 100–500 nm. With this type of instrument, it is possible, for example, to study the formation dynamics of different parts of the coral skeleton and to determine the growth rate of CoC and fibers (Brahmi et al., 2012), or to examine the distribution of Mg and Sr into the coral skeleton (Meibom et al., 2004) or in foraminifera. This study showed that the Mg distribution is strongly dependent on the microstructure of the coral and *corresponds to the layered organization of aragonite fibers surrounding the centers of calcification* (Meibom et al., 2004). From this study, it was proposed that Mg may be biologically used to control the growth of the coral. This kind of technique can be also adapted to the measurements of elements on cells (Pernice et al., 2012), and so on the calcifying interface of the coral (i.e., calicoblastic cells—coral skeleton—space between these two components).

This interface between cells and skeleton is the key space to understand, to have a better view of the different processes of biomineralization, and to reconcile the calcification model derived from biology and geochemistry communities. Two major points are still under discussion not only between biologists and geochemists, but also between geochemists themselves. The two questions are: what is the thickness of the space between the skeleton and the tissues, and what is the nature of the so-called extra calcifying fluid (ECF), if this fluid exists. At the moment, in almost all geochemical models, the ECF is thick and is constituted of seawater. Indeed, geochemical models are based on aqueous solution chemistry to precipitate carbonates in a large free space, applying, for example, Rayleigh distillation (Gagnon et al., 2008). This space between the newly formed skeleton and the calicoblastic cells, where ECF is found, is believed to be in the range of 1.5 cm thick if Rayleigh geochemical model is applied (Meibom et al., 2008). This unrealistic large space is at the opposite to the biologists' observations, where the ECF appears, by microscopy, thinner than the calicoblastic cells, which are ≈1 μm thick (Barnes, 1972), or even does not exist (Tambutté et al., 2007). Moreover, the nature of the medium at the interface is not really identified even if some parameters like pH can be measured (Al-Horani et al., 2003; de Nooijer et al., 2009). At that stage of understanding, it is of prime importance to characterize this medium: Is it seawater? A modified seawater? An organic hydrogel? Then, geochemists could work on a model that takes into account the different parameters of the interface. From *in situ* analyses, we have shown that it is not possible and that a specialization of the calicoblastic cells is needed to explain the different signatures of CoC and fibers. Is this specialization due to different cells present in the calicoblastic epithelium, or is it due to a specialization of the cells during their life? What is the importance of the organic matter in the formation of the skeleton? All these open questions need a close collaboration between geochemists and biologists to get answers. Then, when processes of biomineralization will be more constrained, it will be easier to evaluate their impact on environmental proxies.

Actually, corals and foraminifera are the two main proxies used in paleoenvironmental reconstructions. Many other organisms are able to bring precious information on past environment. However, they need to be carefully studied to be efficient tools. Among them, some have been already used with limited success like phosphates (bones, teeth) or silicates (diatoms). As with corals and foraminifera, the microstructure as well as the mineralogy of these proxies is complex. A clear characterization of the isotopic behavior of O-stable isotopes in relation with the mineralogy and the microstructure is the basis of pertinent studies on bones and teeth; both are generally used either to reconstruct past seawater or precipitation of O-isotopic composition.

One another important aspect that can be studied using SIMS is related to diagenesis, that is, all the chemical and/or physical processes that are involved after the death and the deposition of the organism. The mechanisms involved during the diagenesis of the samples are of prime importance in our studies. Diagenitized samples are in most of the cases "chemically" altered, and the initial chemical or isotopic composition of the proxies is affected. Bulk analysis can give us some information on the shift of modified values; however, it will never give us information on the mechanism involved during diagenesis.

Biomineralization will be in the next future a subject of intensive research, especially at the frontiers between medical solid materials sciences, where, for example, corals can be used in the reconstruction of broken facial bones.

ACKNOWLEDGMENT

The authors thank Jarek Stolarski (Institute of Paleobiology, Poland) for providing the fluorescence image of the *Lophelia pertusa* specimen. This is a CRPG contribution n°2276.

REFERENCES

Adkins, J.F., Boyle, E.A., Curry, W.B., and Lutringer, A. 2003. Stable isotopes in deep-sea corals and a new mechanism for "vital effects". *Geochim. Cosmochim. Acta* 67:1129–1143.

Adkins, J.F., Griffin, S., Kashagarian, H., Cheng, E.R.M., Druffel, E.A., Boyle, R.L., and Shen, C.-C. 2002. Radiocarbon dating of deep-sea corals. *Radiocarbon* 44:567–580.

Al-Horani, F.A., Al-Moghrabi, S.M., and de Beer, D. 2003. The mechanism of calcification and its relation to photosynthesis and respiration in the scleractinian coral *Galaxea fascicularis*. *Mar. Biol.* 142:419–426.

Allison, N. and Austin, W.E.N. 2003. The potential of ion microprobe analysis in detecting geochemical variations across individual foraminifera tests. *Geochem. Geophys. Geosyst.* 4:8403, doi:10.1029/2002GC000430.

Allison, N., Cohen, I., Finch, A.A., Erez, J., and EIMF. 2011. Controls on Sr/Ca and Mg/Ca in scleractinian corals: The effect of Ca-ATPase and transcellular Ca channels on skeletal chemistry. *Geochim. Cosmochim. Acta* 75:6350–6360.

Allison, N. and Finch, A.A. 2007. High temporal resolution Mg/Ca and Ba/Ca records in modern *Porites lobata* corals. *Geochem. Geophys. Geosyst.* 8:Q05001, doi:10.1029/2006GC001477.

Allison, N., Finch, A.A., and EIMF. 2010a. The potential origins and paleoenvironmental implications of high temporal resolution $\delta^{18}O$ heterogeneity in coral skeletons. *Geochim. Cosmochim. Acta* 74:5537–5548.

Allison, N., Finch, A.A., and EIMF. 2010b. $\delta^{11}B$, Sr, Mg and B in a modern *Porites* coral: The relationship between calcification site pH and skeletal chemistry. *Geochim. Cosmochim. Acta* 74:1790–1800.

Anagnostou, E., Sherrell, R.M., Gagnon, A., LaVigne, M., Field, P.M., and McDonough, W.F. 2011. Seawater nutrient and carbonate ion concentrations recorded as P/Ca, Ba/Ca, and U/Ca in the deep-sea coral *Desmophyllum dianthus*. *Geochim. Cosmochim. Acta* 75:2529–2543.

Andrews, A.H., Cordes, E., Mahoney, M.M., Munk, K., Coale, K.H., Cailliet, G.M., and Heifetz, J. 2002. Age and growth and radiometric age validation of a deep-sea, habitat-forming gorgonian (*Primnoa resedaeformis*) from the Gulf of Alaska. In *Biology of Cold Water Corals*, eds. Watling, L. and Risk, M., *Hydrobiologia* 471:101–110.

Bard, E. 1988. Correction of accelerator mass spectrometry ^{14}C ages measured in planktonic foraminifera: Paleoceanographic implications. *Paleoceanography* 3:635–645.

Barnes, D.J. 1972. The structure and formation of growth-ridges in scleractinian coral skeletons. *Proc. R. Soc. Lond. B.* 182:331–350.

Benninghoven, A., Rüdenauer, F.G., and Werner, H.W. 1987. *Secondary Ion Mass Spectrometry: Basic Concepts, Instrumental Aspects, Applications, and Trends*. Wiley, New York, 1227pp.

Blamart, D., Rollion-Bard, C., Cuif, J.-P., Juillet-Leclerc, A., Lutringer, A., van Weering, T.C.E., and Henriet, J.-P. 2005. C and O isotopes in a deep-sea coral (*Lophelia pertusa*) related to skeletal microstructure. In *Cold-water Corals and Ecosystems*, eds. Freiwald, A. and Roberts, J.M., Springer-Verlag, Berlin, Germany, pp. 1005–1020.

Blamart, D., Rollion-Bard, C., Meibom, A., Cuif, J.-P., Juillet-Leclerc, A., and Dauphin, Y. 2007. Correlation of boron isotopic composition with ultrastructure in the deep-sea coral *Lophelia pertusa*: Implications for biomineralization and paleo-pH. *Geochem. Geophys. Geosys.* 8:Q12001, doi:10.1029/2007GC001686.

Brahmi, C., Kopp, C., Domart-Coulon, I., and Meibom, A. 2012. Skeletal growth dynamics linked to trace-element composition in the scleractinian coral *Pocillopora damicornis*. *Geochim. Cosmochim. Acta* 99:146–158.

Cairns, S.D. 1981. *Marine Flora and Fauna of the Northeastern United States. Scleractinia.* NOAA technical report NMFS Circular 438, Silver Spring.

Cairns, S.D. 1994. Scleractinia of the temperate North Pacific. *Smiths. Contrib. Zool.* 557:150pp.

Case, D.H., Robinson, L.F., Auro, M.E., and Gagnon, A.C. 2010. Environmental and biological control on Mg and Li in deep-sea scleractinian corals. *Earth Planet. Sci. Lett.* 300:215–225.

Castaing, R., Bizouard, H., Clochiatti, R., and Havette, A. 1978. Quelques applications de la microsonde ionique et de l'analyseur ionique en minéralogie. *Bull. Miner.* 101:245–262.

Chaussidon, M., Robert, F., and McKegan, K.D. 2006. Li and B isotopic variations in an Allende CAI: Evidence for the in situ decay of short-lived ^{10}Be and for the possible presence of the short-lived nuclide ^{7}Be in the early solar system. *Geochim. Cosmochim. Acta* 70:224–245.

Cheng, H., Adkins, J.F., Edwards, R.L., and Boyle, E.A. 2000. U-Th dating of deep-sea corals. *Geochim. Cosmochim. Acta* 64:2401–2416.

Cohen, A.L., Gaetini, G.A., Lundälv, T., Corliss, B.H., George, R.Y. 2006. Compositional variability in a cold-water scleractinian *Lophelia pertusa*: New insights into "vital effects". *Geochim. Geophys. Geosyst.* 7:Q12004, doi:10.1029/2006GC001354.

Cohen, A.L. and McConnaughey, T.A. 2003. Geochemical perspectives on coral mineralization. *Rev. Min. Geochem.* 54:151–187.

Decitre, S., Deloule, E., Reisberg, L., James, R., and Mevel, C. 2002. Behavior of Li and its isotopes during serpentinisation of oceanic peridotites. *Geochim. Geophys. Geosyst.* 3:10.1029/2001GC000178.

Deloule, E., Chaussidon, M., Glass, B.P., and Koeberg, C. 2001. U-Pb isotopic study of relict zircon inclusions recovered from Muong Nong-type tektites. *Geochim. Cosmochim. Acta* 65:1833–1838.

Deloule, E., Robert, F., and Doukhan, J.C. 1998. Interstellar hydroxyl in meteoritic chondrules: Implications for the origin of water in the inner solar system. *Geochim. Cosmochim. Acta* 62:3367–3378.

de Nooijer, L.J., Toyofuku, T., and Kitazato, H. 2009. Foraminifera promote calcification by elevating their intracellular pH. *Proc. Natl. Acad. Sci. USA* 106:15374–15378.

Dodge, R.E. and Thomson, J. 1974. The natural radiochemical and growth records in contemporary hermatypic corals from the Atlantic and Caribbean. *Earth Planet. Sci. Lett.* 23:313–322.

Druffel, E.M. and Linick, T.W. 1978. Radiocarbon in annual coral rings of Florida. *Geophys. Res. Lett.* 5:913–916.

Emiliani, C. 1954. Temperature of Pacific bottom waters and polar superficial waters during the Tertiary. *Science* 119:853–855.

Erez, J. 2003. The source of ions for biomineralization in foraminifera and their implications for paleoceanographic proxies. *Rev. Min. Geochem.* 54:115–149.

Gagnon, A.C., Adkins, J.F., Fernandez, D.P., and Robinson, L.F. 2007. Sr/Ca and Mg/Ca vital effects correlated with skeletal architecture in a scleractenian deep-sea coral and the role of Rayleigh distillation. *Earth Planet. Sci. Lett.* 261:280–295.

Giuliani, G., Chaussidon, M., Schubnel, H.J., Piat, D.H., Rollion-Bard, C., France-Lanord, C., Giard, D., de Narvaez, D., and Rondeau, B. 2000. Oxygen isotopes and emerald trade routes since antiquity. *Science* 287:631–633.

Hart, S.R. and Cohen, A.L. 1996. An ion probe study of annual cycles of Sr/Ca and other trace elements in corals. *Geochim. Cosmochim. Acta* 60:3075–3084.

Hays, J.D., Imbrie, J., and Shackleton, N.J. 1976. Variations in the Earth's orbit: Pacemaker of the Ice Ages. *Science* 194:1121–1132.

Hinton, R.W. 1995. Ion microprobe analysis in geology. In *Microprobe Techniques in the Earth Sciences*, eds. Potts, P.J., Bowles, J.F.W., Reed, S.J.B., and Cave, M.R.. Chapman and Hall, London, U.K., pp. 235–285.

Ireland, T.R. 1995. Ion microprobe mass spectrometry: Techniques and applications in cosmochemistry, geochemistry, and geochronology. *Adv. Anal. Geochem.* 2:1–118.

Janssens, T., Huyghebaert, C., Vandervorst, W., Gildenpfenning, A., and Bronsgerma, H.H. 2003. On the correlation between Si^{+} yields and surface oxygen concentration using in-situ SIMS-LEIS. *Appl. Surf. Sci.* 203–204:30–34.

Kita, N.T., Ushikubo, T., Fu, B., and Valley, J.W. 2009. High precision SIMS oxygen isotope analysis and the effect of sample topography. *Chem. Geol.* 264:43–57.

Klochko, K., Kaufman, A.J., Yao, W., Byrne, R.H., and Tossell, J.A. 2006. Experimental measurement of boron isotope fractionation. *Earth Planet. Sci. Lett.* 248:276–285.
Lazier, A.V., Smith, J.E., Risk, M.J., and Schwarcz, P. 1999. The skeletal structure of *Desmophyllum cristagalli*: The use of deep-water corals in sclerochronology. *Lethaia* 32:119–130.
Long, J.V.P. 1995. Microanalysis from 1950 to the 1990s. In *Microprobe Techniques in the Earth Sciences*, eds. Potts, P.J., Bowles, J.F.W., Reed, S.J.B., and Cave, M.R., pp. 1–48. Chapman and Hall, London, U.K.
Lopez Correa, M., Montagna, P., Vendrell-Simon, B., McCulloch, M., and Taviani, M. 2010. Stable isotopes ($\delta^{18}O$ and $\delta^{13}C$), trace and minor element compositions of recent scleractinians and last glacial bivalves at the Santa Maria di Leuca deep-water coral province, Ionian Sea. *Deep-Sea Res. II* 57:471–486.
Lutringer, A., Blamart, D., Frank, N., and Labeyrie, L. 2005. Paleotemperatures from deep-sea corals: Scale effects. In *Cold-water Corals and Ecosystems*, eds. Freiwald, A. and Roberts, J.M., Springer-Verlag, Berlin, Germany, pp. 1081–1096.
Marin-Carbonne, J., Rollion-Bard, C., and Luais, B. 2011. In-situ measurements of iron isotopes by SIMS: MC-ICPMS intercalibration and application to a magnetite crystal from the Gunflint chert. *Chem. Geol.* 285:50–61.
McConnaughey, T. 1989. ^{13}C and ^{18}O isotopic disequilibrium in biological carbonates: II. In vitro simulations of kinetic isotopic effects. *Geochim. Cosmochim. Acta* 53:163–171.
McConnaughey, T. 2003. Sub-equilibrium oxygen-18 and carbon-13 levels in biological carbonates: Carbonate and kinetic isotope effects. *Coral Reefs* 22:316–327.
McCrea, J.M. 1950. On the isotopic chemistry of carbonates and a paleotemperature scale. *J. Chem. Phys.* 18:849–857.
Meibom, A., Cuif, J.-P., Hillion, F., Constantz, B.R., Juillet-Leclerc, A., Dauphin, Y., Watanabe, T., and Dunbar, R.B. 2004. Distribution of magnesium in coral skeleton. *Geophys. Res. Lett.* 31:L23306, doi:10.1029/2004GL021313.
Meibom, A., Cuif, J.-P., Houlbreque, F., Mostefaoui, S., Dauphin, Y., Meibom, K.L., and Dunbar, R. 2008. Compositional variations at ultra-structure length scales in coral skeleton. *Geochim. Cosmochim. Acta* 72:1555–1569.
Montagna, P., McCulloch, M., Taviani, M. Remia, A., and Rouse, G. 2005. High-resolution trace and minor element in deep-water scleractinian corals (*Desmophyllum dianthus*) from the Mediterranean Sea and the Great Australian Bight. In *Cold-water Corals and Ecosystems*, eds. Freiwald, A. and Roberts, J.M., Springer-Verlag, Berlin, Germany, pp. 1109–1126.
Moore, W.S. and Krishnaswami, S. 1972. Coral growth rates using Ra-228 and Pb-210. *Earth Planet. Sci. Lett.* 15:187–190.
Ogilvie, M. 1896. Microscopic and systematic study of madreporian types of corals. *R. Soc. London Phil. Trans.* 187(B):83–345.
Pernice, M., Meibom, A., Van Den Heuvel, A., Kopp, C., Domart-Coulon, I., Hoegh-Gulbert, O., and Dove, S. 2012. A single-cell view of ammonium assimilation in coral-dinoflagellate symbiosis. *ISME J.* 6:1314–1324.
Reed, S.J.B. 1984. Geological applications of SIMS. In *SIMS IV*, eds. Benninghoven, A., Okano, J., Shimizu, R., and Werner, H.W., Springer Series in Chemical Physics, Vol. 36. Springer-Verlag, Berlin, Germany, pp. 451–455.
Rollion-Bard, C., Blamart, D., and Cuif, J.-P. 2010. In situ measurements of oxygen isotopic composition in deep-sea coral, *Lophelia pertusa*: Re-examination of the current geochemical models of biomineralization. *Geochim. Cosmochim. Acta* 74:1338–1349.
Rollion-Bard, C., Blamart, D., Cuif, J.-P., Juillet-Leclerc, A. 2003b. Microanalysis of C and O isotopes of azooxanthellate and zooxanthellate corals by ion microprobe. *Coral Reefs* 22:405–415.
Rollion-Bard, C., Chaussidon, M., and France-Lanord, F. 2003a. pH control on oxygen isotopic composition of symbiotic corals. *Earth Planet. Sci. Lett.* 215:275–288.
Rollion-Bard, C., Chaussidon, M., and France-Lanord, C. 2011. Biological control of internal pH in scleractinian corals: Implications on paleo-pH and paleo-temperature reconstructions. *C. R. Geosci.* 343:397–405.
Rollion-Bard, C. and Erez, J. 2010. Intra-shell boron isotope ratios in the symbiont-bearing benthic foraminiferan *Amphistegina lobifera*: Implications for $\delta^{11}B$ vital effects and paleo-pH reconstructions. *Geochim. Cosmochim. Acta* 74:1530–1536.
Rollion-Bard, C., Mangin, D., and Champenois, M. 2007b. Development and applications of oxygen and carbon isotopic measurements of biogenic carbonates by ion microprobe. *Geost. Geoanal. Res.* 31:39–50.
Rollion-Bard, C. and Marin-Carbonne, J. 2011. Determination of SIMS matrix effects on oxygen isotopic compositions in carbonates. *J. Anal. At. Spectr.* 26:1285–1289.
Rollion-Bard, C., Vigier, N., Meibom, A., Blamart, D., Reynaud, S., Rodolfo-Metalpa, R., Martin, S., and Gattuso, J.-P. 2009. Effect of environmental conditions and skeletal ultrastructure on the Li isotopic composition of scleractinian corals. *Earth Planet. Sci. Lett.* 286:63–70.
Rollion-Bard, C., Vigier, N., and Spezzaferri, S. 2007a. In situ measurements of calcium isotopes by ion microprobe in carbonates and application to foraminifera. *Chem. Geol.* 244:679–690.
Sabatier, P., Reyss, J.-L., Hall-Spencer, J.-M., Colin, C., Frank, N., Tisnerat-Laborde, N., Bordier, L., and Douville, E. 2012. ^{210}Pb-^{226}Ra chronology reveals rapid growth rate of *Madrepora oculata* and *Lophelia pertusa* on world's largest cold-water coral reef. *Biogeosciences* 9:1253–1265.
Sangely, L., Chaussidon, M., Michels, R., and Huault, V. 2005. Microanalysis of carbon isotope composition in organic matter by secondary ion mass spectrometry. *Chem. Geol.* 223:179–195.
Shackleton, N.J. 1973. Oxygen isotope and paleomagnetic stratigraphy of Equatorial Pacific core V28–238: Oxygen isotope temperatures and ice volumes on a 10^5 year and 10^6 year scale. *Quat. Res.* 3:39–55.
Shimizu, N. and Hart, S.R. 1982. Isotope fractionation in secondary ion mass spectrometry. *J. Appl. Phys.* 53:1303–1311.
Shimizu, N., Semet, M.P., and Allègre, C.J. 1978. Geochemical applications of quantitative ion-microprobe analysis. *Geochim. Cosmochim. Acta* 42:1321–1334.
Shirai, K., Kasukabe, M., Nakai, S., Ishii, T., Watanabe, T., Hiyagon, H., and Sano, Y. 2005. Deep-sea coral geochemistry: Implication for the vital effect. *Chem. Geol.* 224:212–222.
Sinclair, D.J., Williams, B., and Risk, M. 2006. A biological origin for climate signals in coral—Trace element "vital effects" are ubiquitous in Scleractinian coral skeleton. *Geophys. Res. Lett.* 33:L17707, doi:10.1029/2006GL027183.
Slodzian, G., Lorin, J.C., and Havette, A. 1980. Effet isotopique sur les probabilities d'ionisation en émission secondaire. *C. R. Acad. Sci.* 291:121–124.
Storms, H.A., Brown, K.F., and Stein, J.D. 1977. Evaluation of a cesium positive ion source for secondary ion mass spectrometry. *Anal. Chem.* 49:2023–2030.
Tambutté, E., Allemand, D., Zoccola, D., Meibom, A., Lotto, S., Caminiti, N., and Tambutté, S. 2007. Observations of the tissue-skeleton interface in the scleractinian coral *Stylophora pistillata*. *Coral Reefs* 26:517–529.
Thomassot, E., Cartigny, P., Harris, J.W., Lorand, J.-P., Rollion-Bard, C., and Chaussidon, M. 2009. Metasomatic diamond growth: A multi-isotope study (^{13}C, ^{15}N, ^{33}S, ^{34}S) of sulphide inclusions and their host diamonds from Jwaneng (Botswana). *Earth Planet. Sci. Lett.* 282:79–90.

Uchikawa, J. and Zeebe, R.E. 2012. The effect of carbonic anhydrase on the kinetics and equilibrium of the oxygen isotope exchange in the CO_2-H_2O system: Implications for $\delta^{18}O$ vital effects in biogenic carbonates. *Geochim. Cosmochim. Acta* 95:15–34.

Usdowski, E. and Hoefs, J. 1993. Oxygen isotope exchange between carbonic acid, bicarbonate, carbonate, and water: A re-examination of the data of McCrea (1950) and an expression for the overall partitioning of oxygen isotopes between the carbonate species and water. *Geochim. Cosmochim. Acta* 57:3815–3818.

Usdowski, E., Michaelis, J., Böttcher, M.E., and Hoefs, J. 1991. Factors for the oxygen isotope equilibrium between aqueous and gaseous CO_2, carbonic acid, bicarbonate, carbonate, and water (19°C). *Zeit. Phys. Chem.* 170:237–249.

Vielzeuf, D., Champenois, M., Valley, J.W., and Brunet, F. 2005. SIMS analyses of oxygen isotopes: Matrix effects in Fe-Mg-Ca garnets. *Chem. Geol.* 223:208–226.

Vigier, N., Rollion-Bard, C., Spezzaferri, S., and Brunet, F. 2007. In-situ measurements of Li isotopes in foraminifera. *Geochim. Geophys. Geosyst.* 8:1–9.

Weiner, S. and Dove, P.M. 2003. An overview of the biomineralization processes and the problem of the vital effect. *Rev. Min. Geochem.* 54:1–29.

Williams, P. 1983. Secondary ion mass spectrometry. *App. Atom. Coll. Phys.* 4:327–377.

Zeebe, R.E. 1999. An explanation of seawater carbonate concentration on foraminiferal oxygen isotopes. *Geochim. Cosmochim. Acta* 63:2001–2007.

Zeebe, R.E. 2005. Stable boron isotope fractionation between dissolved $B(OH)_3$ and $B(OH)_4^-$. *Geochim. Cosmochim. Acta* 69:2753–2766.

Part IV

Properties of the composite: Energetics and forces in assembly

17 Molecular simulation of biomineral nucleation and crystal growth: Modern computational challenges and approaches

Yang Yang, Zhijun Xu, Qiang Cui, and Nita Sahai

Contents

17.1 INTRODUCTION

One of the ultimate goals of the study on biomineralization is to reveal the interaction mechanism *at the atomic level* between biological organic molecules, such as proteins, glycosaminoglycans, and citrate; inorganic minerals like hydroxyapatite (HAP), calcite, and silica; and their constituent ions in solutions such as Ca^{2+}, PO_4^{3-}, and CO_3^{2-} (Mann 2001). As Richard Feynman stated in his seminal 1963 *Lectures on Physics* (Feynman 1970), "everything that living things do can be understood in terms of the jiggling and wiggling of atoms," the biomineralization process can be explored through molecular simulations, where the atomic motions (jiggling and wiggling) are demonstrated explicitly (Harding et al. 2008).

Since the first atomistic theoretical study of protein dynamics by McCammon and Karplus in 1977 (McCammon et al. 1977), molecular simulations have been extensively and successfully applied to solve various biological problems, including protein folding (Das and Baker 2008; Freddolino et al. 2008, 2009, 2010; Freddolino and Schulten 2009; Shaw et al. 2010; Voelz et al. 2010; Bowman et al. 2011; Piana et al. 2012), enzymatic catalysis (Bruice 2002; Benkovic and Hammes-Schiffer 2003; Warshel 2003; Friesner and Guallar 2005; Gao et al. 2006; Riccardi et al. 2006; Senn and Thiel 2007; Hu and Yang 2008), biomedical drug design (Aqvist et al. 1994; Goodsell et al. 1996; Gilson et al. 1997; Wlodawer and Vondrasek 1998; Carlson and McCammon 2000; Taylor et al. 2002; Jorgensen 2004; Kitchen et al. 2004), protein–membrane interactions (Woolf and Roux 1994, 1996; Tobias et al. 1997; Tieleman et al. 1999; Im et al. 2003; Ash et al. 2004; Bond and Sansom 2006), and so on. With the rapid growth of computer hardware resources and development of efficient parallel calculation algorithms, more and more complicated biomolecular systems (over 10^9 atoms) can now be simulated (Sanbonmatsu et al. 2005; Freddolino et al. 2006; Dror et al. 2012). A large amount of critical information on structures, energetics, and interactions with atomic-level details that cannot be resolved easily by experiments has been obtained with such "computational microscopes" (Lee et al. 2009; Dror et al. 2012). Encouraged by these successes, molecular simulations have been employed to investigate biomineralization processes for the past decade and have started to shed insightful lights (Harding et al. 2008). However, molecular simulation methods cannot be directly adopted from other areas of application, and they must be carefully developed and validated before being applied to biomineralization (Latour 2008). Therefore, in this chapter, we start with a general introduction to the conventional molecular simulation approaches, which have been used routinely in biomineralization studies. We then discuss the scientific and technical challenges of applying these methods to investigate biomineralization processes where we focus specifically on nucleation and crystal growth. We describe some advanced molecular dynamics (MD) approaches as potential solutions to the challenges in modeling biomineralization and provide examples from our own research to highlight these points. In particular, the initial stages of nucleation and crystal growth are studied by examining (1) the effect of bone sialoprotein (BSP) conformation on HAP nucleation and (2) the interactions of small molecules or BSP with HAP surfaces in affecting crystal growth and morphology.

17.1.1 MOLECULAR DYNAMICS SIMULATION

To reveal the atomic "jiggling and wiggling" as a function of time *in silico*, MD simulation is one of the most direct and powerful techniques (Frenkel and Smit 2001; Leach 2001). In atomistic MD simulations, a sequence of configurations of the molecular system is produced by integrating Newton's laws of motion. The resulting trajectory specifies how the positions and velocities of the atoms in the system change with time. In practice, the obtained trajectory can be visualized with a computer program, such as Visual Molecular Dynamics (VMD) (Humphrey et al. 1996). A movie, thus, can be generated to illustrate the atomic motions. Important structural information, including molecular geometry, conformational transition, intermolecular interactions, water/ion distribution, particle mobility, and so on, can be observed directly in the movie and analyzed quantitatively. MD simulations are routinely used to study various biological problems as a valuable complement to experiments (Karplus and McCammon 2002; Karplus and Kuriyan 2005; Sotomayor and Schulten 2007; Lee et al. 2009; Dror et al. 2012).

17.1.1.1 Calculation algorithm

A set of assumptions is applied in classical atomistic MD simulations (Frenkel and Smit 2001; Leach 2001). First, the Born–Oppenheimer approximation is assumed, so that the potential energy of the system can be written as a function of the nuclear coordinates, and the electronic motions are ignored. Second, an atom is modeled as a single particle with an assigned radius (like a "ball"), and a molecule is modeled as a collection of these single particles connected with "springs" that represent the covalent chemical bonds. Third, the total potential energy of the system comes from the contributions of both intramolecular and intermolecular interactions. Specifically, the intramolecular contributions include the terms of bond stretching, bond angle bending, and bond torsion. The intermolecular interactions involve van der Waals and electrostatic interactions between the nonbonded parts of the system (see following sections for more discussions).

The simulation trajectory of the system is achieved with the flowchart shown in Figure 17.1. At the beginning of any MD simulation, the atomic coordinates of the entire system need to be provided along with the assignment of the initial velocity to atoms. The total potential energy of the system is determined. Consequently, the force acting on each atom can be calculated as the negative gradient of the potential energy, and thus, the acceleration can be obtained through Newton's second law of classical mechanics. For every time step, each atom's position and velocity need to be updated with a symplectic method, such as velocity Verlet algorithm based on the Taylor series expansion (Swope et al. 1982). By iterating such a calculation cycle, the atomic motions (position and velocity) as a function of time are obtained and recorded as "frames" in a movie. System properties are then analyzed with the collected simulation trajectory.

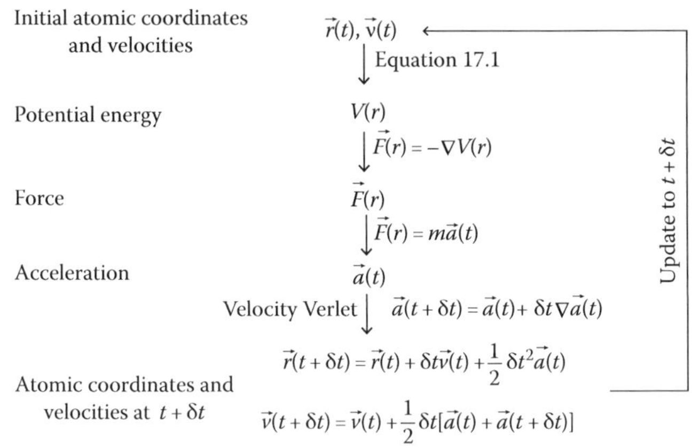

Figure 17.1 The logic flow for an MD simulation.

In practice, 1 or 2 fs is typically set as the integration time step. Because electronic motions are ignored and the calculation algorithm is simple (Figure 17.1), with recent advances in CPU speeds and computational algorithms, the classical all-atom MD simulation can now be used to investigate the behavior of systems with hundreds of thousands of atoms up to tens to hundreds of nanoseconds (Lee et al. 2009; Dror et al. 2012). One of the most outstanding advantages of the atomistic MD simulation over experiments is the capability to explore the solvent effect explicitly. The dynamic behavior of solvent as individual molecules can be examined directly, which cannot be studied conveniently in many experiments. For example, high-resolution x-ray crystallography can resolve the positions of the oxygen atoms of water molecules associated with proteins. However, it does not provide the dynamic information of these water molecules. For many biological systems, water molecules play an essential role in the organization and functions of the biomolecules, including the process of biomineralization (Mann 2001; Harding et al. 2008).

17.1.1.2 Molecular mechanics force field

From the previous discussion, it is evident that the quality of the potential energy calculation is one of the most fundamental factors to ensure accurate results from MD simulations. It is certainly necessary to have more discussions on the potential energy function. As shown in Equation 17.1, $V(r)$ is the potential energy of the system and depends only on the atomic positions (r). The potential energy function plus the associated parameter set is called the molecular mechanics (MM) force field (FF). Bond stretching

($V_{stretch}$) and bond angle bending ($V_{bending}$) contributions are modeled with harmonic potentials around the equilibrium bond length/angle of l_{eq}/θ_{eq}, respectively. $k_{stretch}$ and k_{bend} are the associated force constants. The torsion term models how the energy changes as a bond rotates around the solid angle ϕ. The cosine function captures the periodicity ($1/n$) of the energy variation along the bond rotation. The intermolecular interactions account for all pairwise interactions between atoms (i and j) that are in different molecules or in the same molecule, but separated by several bonds (e.g., at least three). Commonly, a Lennard-Jones potential with the associated ε_{ij} and σ_{ij} parameters and a Coulombic potential with the q_i parameter representing the atomic charge are used, respectively, to describe the van der Waals and electrostatic interactions.

$$V(r) = V_{intramolecular} + V_{intermolecular} \tag{17.1}$$

$$= \sum_{bonds} V_{stretch} + \sum_{angles} V_{bend} + \sum_{dihedrals} V_{torsion} + \sum_{pairs} V_{vdW} + \sum_{pairs} V_{electrostatic} \tag{17.2}$$

$$V_{stretch} = k_{stretch} \ (l - l_{eq})^2 \tag{17.3}$$

$$V_{bend} = k_{bend} \ (\theta - \theta_{eq})^2 \tag{17.4}$$

$$V_{torsion} = A[1 + \cos(n\phi)] \tag{17.5}$$

$$V_{vdW} = \varepsilon_{ij} \left[\left(\frac{\sigma_{ij}}{r_{ij}} \right)^{12} - \left(\frac{\sigma_{ij}}{r_{ij}} \right)^{6} \right] \tag{17.6}$$

$$V_{electrostatic} = \frac{q_i q_j}{4\pi\varepsilon_0 r_{ij}} \tag{17.7}$$

The values for the parameters in Equations 17.3 through 17.7, such as the constants l_{eq}, θ_{eq}, $k_{stretch}$, k_{bend}, ε_{ij}, σ_{ij}, and q_i need to be developed systematically in order to reproduce relevant experimental data or results from high-level quantum-mechanical (QM) calculations, where the electronic degree of freedom is taken into account by solving the Schrödinger equation. A number of important general features of FFs need to be noted:

1. Potential energy functions and a particular set of parameter values are tightly coupled. Different FFs may have an identical functional form, but with different parameter values.
2. The potential energy functions are completely empirical. Researchers are free to modify these mathematical expressions or add new terms to treat their specific systems. For example, the Born–Mayer–Huggins function may be used to represent van der Waals interactions instead of the Lennard-Jones form. In practice, the functional forms are chosen as a balance of computational accuracy and efficiency in most of the popular FFs, such as CHARMM (Brooks et al. 1983, 2009; MacKerell et al. 1998), AMBER (Weiner et al. 1984; Cornell et al. 1995), and OPLS (Jorgensen and Tirado-Rives 1988; Jorgensen et al. 1996).
3. Without careful validation, the parameters from one FF cannot be directly used in another due to the different functional forms or different parameterization strategies.
4. The different FFs have their specific applications. For example, MM2 FF (Allinger 1977) is particularly good at simulating hydrocarbons and other small organic molecules. CHARMM is designed for proteins and nucleic acids (Brooks et al. 1983, 2009; MacKerell et al. 1998). CLAYFF is excellent for modeling hydrated aluminosilicate minerals (Cygan et al. 2004), such as clays. It is important to note here that FFs originally designed to address molecular behavior in aqueous solutions are not guaranteed to show good performance when used to study systems involving an inorganic solid surface. Therefore, when treating a new molecular system, a FF benchmark with respect to available experimental data or calculation results with the other, more sophisticated methods, such as quantum calculations, is always needed to obtain accurate and meaningful data from the simulations (Latour 2008).
5. For most of the traditional FFs, such as CHARMM, AMBER, OPLS, GROMOS (Scott et al. 1999), the atomic charges (q), which dictate the electrostatic interactions, are fixed (i.e., constants) during the entire simulation. Electronic polarizations and charge transfers due to the interactions with the dynamic environment are ignored. For systems that feature highly polarizable chemical groups (such as the $-COO^-$ group of the glutamate side chain when approaching the HAP surface in biomineralization), simulation artifacts may be introduced with the traditional FFs. Polarizable FFs development thus has become one of the mainstream subjects in molecular simulations methodology (Anisimov et al. 2005; Xie and Gao 2007; Yu et al. 2010).

17.1.2 FREE ENERGY CALCULATION

The classical atomistic MD simulation provides a direct method to explore the structural information of a molecular system. However, in many cases, the kinetic and thermodynamic properties of a system are of key significance to the process of interest. For example, what is the relative stability of two protein conformations and how fast can such a structural transition occur? Or, what is the binding energy for peptide adsorption at a solid surface? Such information is critical for understanding the mechanisms of these dynamic processes at the atomic level.

17.1.2.1 Importance of sampling

To answer the previous questions, it is necessary to compute the relative free energies between the reactant state (e.g., system ensembles for the starting state), the product state (e.g., ensembles for the ending state), and the transition state (e.g., ensembles for the intermediate state with the highest free energy). Based on statistical mechanics (McQuarrie 1973), the Gibbs free energy difference between any two states (A and B) is related to the relative probability (P_{AB}) of a given energy state (P_B) being sampled with respect to another (P_A), according to Equation 17.8:

$$P_{AB} = \frac{P_B}{P_A} = e^{-\Delta G_{AB}/RT}, \tag{17.8}$$

where

ΔG_{AB} is the free energy difference between states A and B (i.e., $\Delta G_{AB} = G_B - G_A$)
R is the ideal gas constant
T is the absolute temperature

If a MD simulation is sufficiently long for the transition between the molecular states to occur (i.e., where both A and B states

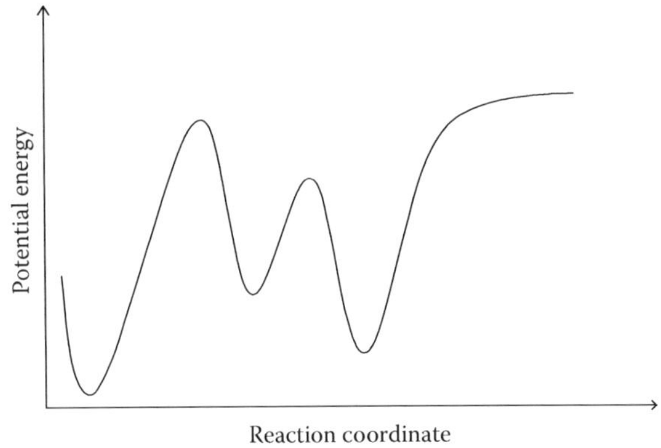

Figure 17.2 "Bumpy" potential energy surface.

are adequately sampled) and, thus, "chemical equilibrium is achieved," then ΔG_{AB} can be obtained in a straightforward manner with Equation 17.8.

Unfortunately, though, the intrinsic complexity of biomolecular systems results in a very bumpy potential energy surface (Figure 17.2). The potential energy surface represents the relationship between the potential energy as a function of the system coordinate. Many local energy minima are present on the surface, separated by relatively high energy barriers (Figure 17.2). Thus, a drawback of conventional MD simulations is that the system is trapped in one or a few low-energy states for most of the calculation time, without sufficient sampling of the other states. This leads to a poor representation of the correct ensemble average of the system; so, Equation 17.8 cannot be used to obtain accurate values of ΔG_{AB}.

17.1.2.2 Umbrella sampling

The umbrella sampling (Torrie and Valleau 1977) technique is one approach to overcome the problem of insufficient sampling with conventional MD simulations. In umbrella sampling simulations, a reaction coordinate is defined to represent progress along a reaction or physical change pathway. An artificial driving force is introduced along the defined reaction coordinate to allow the system to escape from the local minima, thus providing an enhanced sampling of low-energy states separated by high barriers. However, the application of the driving force results in a biased sampling as shown in Equation 17.9:

$$P_{AB} = \frac{\overline{P}_{AB}\exp\left(W_{AB}/RT\right)\left\langle\exp\left(W_{A}/RT\right)\right\rangle_{biased,\,A}}{\left\langle\exp\left(W_{B}/RT\right)\right\rangle_{biased,\,B}}, \quad (17.9)$$

where

$\overline{P}_{AB}$ is the biased relative probability between the states A and B (i.e., $\overline{P}_{AB} = \overline{P}_B/\overline{P}_A$)

$W_{AB} = W_B - W_A$, where $W_{A/B}$ is the biasing potential added to the regular unbiased potential energy function at state A or B

$\langle\ \rangle_{biased,\,A/B}$ refers to the ensemble average from the biased simulation at stage A or B

In a postsimulation process, ΔG_{AB} can be obtained from $\overline{P}_{AB}$ with proper removal of the effects from W_{AB} (e.g., $\Delta G_{AB} = -RT\ln \overline{P}_{AB} - W_{AB} - C$, where C is a correction accounting for the biased ensemble average terms $\langle\ \rangle_{biased,\,A/B}$ in Equation 17.9). In practice, a harmonic restraining potential energy is typically employed in an umbrella sampling simulation. The obtained biased probability distribution is corrected by the weighted histogram analysis method (WHAM) (Kumar et al. 1992) to obtain unbiased statistics. Thus, we obtain the potential of mean force (PMF) of the system, which is the free energy of the system along the projection of the defined reaction coordinate. A specific application of this technique to study the binding energy of an amino acid upon adsorption at a HAP surface is illustrated in Section 17.3.1.

More sophisticated methods such as metadynamics (Laio and Parrinello 2002; Laio and Gervasio 2008) have been developed to describe the free energy profile spanned by complex order parameters, which describe the process of a reaction or a physical change. The order parameter can be any explicit function of the atom coordinates, such as distance and angle, or potential energy. In metadynamics, a history-dependent potential, which is constructed as a sum of Gaussian functions along a defined reaction coordinate, iteratively and adaptively compensates the underlying free energy to help the system escape from the free energy minima. The free energy profile can finally be reconstructed through the Gaussian potentials (Ensing et al. 2006).

17.1.3 BONE SIALOPROTEIN: A MODEL SYSTEM

In this section, we use BSP (Ganss et al. 1999) as a model system to illustrate the scientific and technical challenges in applying molecular simulation approaches to study biomineralization. Theoretical solutions to these challenges are proposed and discussed based on our own experience. Because BSP shares various typical characteristics with many other acidic noncollagenous proteins (NCPs), understanding the functional mechanisms of BSP as obtained from our current research should shed light on biomineralization protein systems in general (George and Veis 2008).

Before proceeding to the technical details, we provide a brief background to bone structure at the molecular to microscale and to the target protein system BSP.

Bone structure is comprised of Type I collagen fibrils, which constitute a major part of the extracellular matrix, along with various NCPs, the mineral phase, and small molecules, such as citrate (Boskey 2000, 2003; Hu et al. 2010). The inorganic phase is a nonstoichiometric nanocrystalline, HAP-like ($Ca_5(PO_4)_3OH$) mineral in specific spatial registry with the collagen matrix (Knott and Bailey 1998). Several acidic NCPs, such as osteopontin (OPN), matrix extracellular phosphoglycoprotein, and BSP are also associated with the matrix and have been suggested to control mineral nucleation and growth by different mechanisms and also to play some role in intercellular signaling (George and Veis 2008). Malfunction of these proteins or calcium or vitamin D deficiencies may lead to severe skeleton-related diseases (Boskey 2000), such as rickets, osteogenesis imperfecta, and osteoporosis. Thus, it is of great fundamental biomedical importance to understand the mechanism of bone formation and growth *in vivo*, particularly, how these naturally evolved proteins interact with the inorganic ions and/or nanocrystals. The study of protein-controlled HAP mineralization is also interesting to understand from a physical chemistry point of view due to the nature of such protein–ion and protein–mineral surface interactions (Gray 2004).

Blood and the extracellular medium from which bone grows is supersaturated with respect to HAP (K_{sp} of HAP is 6.8 × 10^{-37}; in blood, [Ca^{2+}] = 2.2–2.6 mM, and [Phosphate] = 0.26–0.47 mM). However, the presence of inhibitors, such as fetuin, prevents nucleation, except at specific bone-forming sites. So, almost any factor such as a ligand, protein, or inorganic surface capable of lowering the nucleation barrier may induce nucleation. It is, therefore, important to identify which one of the several potential promoters is the most efficient. Alternatively, or in addition, mineralization may occur when the inhibitors are somehow excluded from the nucleation sites in the collagen fibrils, as proposed by Price et al. (2009). We consider the first mechanism where BSP may act as a nucleator and growth modifier of the mineral nanocrystals.

BSP was the first NCP identified from bone and is the most abundant (8%–12%) of the total NCP content (Ganss et al. 1999). Extensive experimental studies have been performed since its first isolation from bovine cortical bone in 1972 (Fisher et al. 1987, 1990, 2001; Oldberg et al. 1988; Somerman et al. 1988; Bianco et al. 1991; Chen et al. 1992a,b; Flores et al. 1992; Hunter and Goldberg 1993, 1994; Ross et al. 1993; Roach 1994; Tye et al. 2003; Yang et al. 2011). As one of the most intriguing discoveries, it was found that BSP transcripts are expressed at a high level in areas of de novo bone formation (Bianco et al. 1991, 1993; Chen et al. 1991; Chen et al. 1993, 1994; Hultenby et al. 1994) during the early phase of bone deposition (Kasugai et al. 1992; Riminucci et al. 1998). BSP has been suggested as a potential nucleator of HAP based on such spatial and temporal expression levels associated with bone mineralization. *In vitro*, this proposal has been verified by Hunter and Goldberg (1993), who showed that BSP is capable of promoting HAP nucleation in a steady-state agarose gel.

Some of the important structural features of BSP, which may dictate its role in HAP biomineralization, are highlighted in this chapter. First, BSP is highly acidic. The nascent mammalian BSP (i.e., without any posttranslational modifications) has ~327 amino acids, with a molecular weight of 33–34 kDa (Ganss et al. 1999; Tye et al. 2003; George and Veis 2008). Around 22% of the total amino acids are glutamic acid (Glu) residues, which is reflected in the protein's low isoelectric point of 3.9 (Stubbs et al. 1997). In addition, extensive posttranslational modifications, which involve serine/threonine/tyrosine (Ser/Thr/Tyr) phosphorylation, Tyr sulfation, and N/O-linked glycosylation, increase the apparent molecular weight to ~75 kDa for a mature BSP (Franzen and Heinegard 1985; Ecarotcharrier et al. 1989; Fisher et al. 1990; Sorensen et al. 1995; Salih et al. 1996; Zaia et al. 2001; Zhu et al. 2001; Salih 2003) and, more importantly, enhance the acidity even further. Second, unlike many other proteins, where stable three-dimensional (3D) structures exist to provide structural support for functions, based on the secondary structure and hydrophilicity analyses and 1D proton NMR study (Shapiro et al. 1993; Fisher et al. 2001), BSP is predicted to have an open, extended, and flexible structure with the potential to form regions of α-helix and β-sheet in solution. By using electron microscopy, Franzén and Heinegard (1985) observed that BSP appeared as an extended rod, having a core with an average length of 40 nm. With circular dichroism and small-angle x-ray scattering, Tye et al. studied the structure of rat BSP in solution, and observed an unfolded protein with no evident secondary structure (Tye et al. 2003). In contrast to BSP, many other proteins in the body have stable 3D structures, so that the structure–function relationships are much easier to determine. Third, with primary sequence alignment, BSP shows significant homology across the different mammalian species with 45% identity and a further 10%–23% conservative replacements of the protein sequence (Ganss et al. 1999). Two highly conserved Glu-rich domains were identified and demonstrated to have nucleating activity. In human BSP, for example, contiguous Glu sequences are found from residue 78 to 85 and from residue 158 to 185 (Fisher et al. 1990). An earlier site-directed mutagenesis study by Tye et al. indicated that a sequence of at least eight contiguous Glu residues is required for BSP to nucleate HAP *in vitro* (Tye et al. 2003).

It is worth mentioning that the aforementioned structural characters of BSP are shared by many other acidic NCPs involved in biomineralization of bone and teeth (George and Veis 2008). Aspartate and glutamate account for ~20% of the total amino acids in OPN. Random flexible tertiary structures are also observed for OPN, dentin matrix protein 1 (DMP1), and dentin phosphophoryn (DPP), although some domains may form ordered secondary structures. Therefore, BSP serves as an appropriate, representative model system. Intrinsically disordered proteins have also been identified to play a role in calcite biomineralization (Ndao et al.).

One of the major obstacles in understanding the precise role of BSP in HAP mineralization is that, with few exceptions, experimental structural information is lacking *at the atomic level* in solution, bound to the collagen matrices or bound to the HAP surface. The NCPs may be intrinsically disordered, which may explain the difficulty in obtaining the crystal structure from x-ray crystallography analyses. Several potential reasons may be responsible for the flexible protein structure. Under physiological conditions, the acidic groups are largely deprotonated resulting in strong electrostatic repulsions from adjacent charged side chains, which could hinder the formation of a stable 3D structure. Therefore, quenching the electrostatic repulsions should promote the formation of more ordered and stable protein conformations. Burke et al. (2000) found that the addition of Ca^{2+} (even at mM concentrations) to DPP in a dilute solution induces a structural change of DPP from a random coil to an ordered β-sheet-like conformation. Such charge-induced conformational changes are also observed when proteins bind to HAP surfaces. Statherin, a protein which inhibits nucleation and growth of HAP in the saliva, has a helical structure when bound to the HAP surface as determined by solid-state NMR (Long et al. 2001). Therefore, it is not unreasonable to speculate on a correlation between the acidic nature and the random structures of the protein in solution.

Biomineralization proteins may have evolved to be *intrinsically* flexible and disordered. The disordered structure has a larger solvent-accessible surface area, which may enhance their exposure to Ca^{2+} and phosphate ions in the local environment (Yang et al. 2010). Moreover, the functional sequences in biomineralization proteins are relatively short; thus, the local secondary structures may be more crucial for the potential functions of the proteins.

Different domains in proteins need to interact with different binding partners. BSP has an arginine–glycine–aspartic acid (Arg–Gly–Asp or RGD) sequence close to the C-terminus, which is the major cell attachment site (Ganss et al. 1999). The conserved poly-Glu (~10 amino acids) sequences are close to the N-terminus, which have been proposed as the HAP nucleating motifs (Goldberg and Hunter 1994; Goldberg et al. 1996). A flexible tertiary structure is thus ideal to optimize multiple functions at the same time, especially considering that a well-defined, 3D active site is not necessary to fulfill these functions. In fact, many proteins display functions requiring intrinsic disorder, which has been discussed in a study by Dunker et al. (2002).

In our research, we tried to reveal the mechanisms of BSP in bone mineralization, where we targeted the previously reported conserved nucleating motifs. Our goal was to understand how these functional motifs interact with ions (Ca^{2+} and phosphate) in solution to promote HAP nucleation, and further, how these motifs bind to the incipient HAP crystals to regulate growth. Because of the anionic nature of the polyglutamic acid sequence, it has been hypothesized in the literature that an appropriate spatial arrangement of γ-carboxylate groups may favor specific Ca^{2+} distribution and, therefore, promote nucleation or bind to specific HAP faces to modulate the crystal growth (Boskey 2003). However, whether such a hypothesis is valid, and if so, what the precise stereochemical arrangement of functional groups is, remain elusive (Ganss et al. 1999; Boskey 2003).

17.1.4 THEORETICAL STUDY OF BIOMINERALIZATION

Theoretical modeling affords us a means to assemble the available experimental information and provide insights into the nucleating mechanisms and other biomineralization processes with atomic details (Latour 1999, 2008; Cygan et al. 2004; Gray 2004; Harding and Duffy 2006; Greathouse and Cygan 2007; Di Tommaso and de Leeuw 2008, 2009; Harding et al. 2008). Unfortunately, both nucleation as well as protein–solid surface interactions are nontrivial to study *at the atomic level* with reliable accuracy using conventional MD approaches (Harding et al. 2008). We will address each of these challenges in the following two subsections.

17.1.4.1 Challenges and approaches in nucleation studies

For nucleation, several challenges deter the development and application of atomistic modeling approaches. As pointed out by Harding et al. (2008), the scientific understanding of heterogeneous nucleation mechanism is limited in general. Furthermore, the opinions on nucleation mechanisms in the biomineralization field are far from being mature and sometimes change over time. This makes it difficult to find an ideal experimental benchmarking system to gauge the accuracy of the theoretical methods. The concept of structural pattern matching (i.e., templating) for promoting nucleation was popular for many decades in the biomineralization field (Addadi et al. 1989). One of the "best" characterized systems is the nacre of the mollusk shell. Based on x-ray and electron diffraction studies, a strong correlation was observed between the spatial arrangement of carboxylate groups on the antiparallel β-pleated sheet and Ca^{2+} ion distributions in the (001) surface of aragonite (Weiner et al. 1983). However, recent experimental work suggests the formation of an amorphous $CaCO_3$ precursor, which transforms to aragonite in mollusk shell nacre (Gotliv et al. 2003) and to calcite in sea-urchin spine (Weiss et al. 2002; Politi et al. 2004, 2006, 2008; Killian et al. 2009). Similar debates also apply to vertebrate biomineralization as to whether the first solid phase deposit is crystalline or an amorphous precursor (Termine and Posner 1966a,b; Brown and Chow 1976; Glimcher 1984; Crane et al. 2006; Weiner 2006; Grynpas 2007). Results of many experimental studies have supported the notion that an amorphous calcium phosphate phase is first formed in the vertebrate bone formation. Using infrared spectroscopy, Lowenstam and coworkers demonstrated that the initial precipitate of a biologically formed calcium phosphate mineral is amorphous in chiton teeth (Lowenstam and Weiner 1985). Mahamid et al. (2008) found an abundant amorphous calcium phosphate phase in newly formed and continuously growing fin bones of the Tuebingen long-fin zebra fish, as a model for bone mineralization. Raman spectroscopy also showed evidence for amorphous calcium phosphate during intramembranous mineralization in matrix vesicles, although the observed mechanism might not be the same as that for endochondral bone formation in the extracellular environment (Crane et al. 2006). Very recently, Yang et al. (2010) showed that amelogenin, an extracellular matrix protein found in developing tooth enamel, forms stable interactions with amorphous calcium phosphate at the early stages of HAP formation. Thus, opinion in the literature has swung from a strong role for structural templating of proteins involved in mineralization of a wide range of tissues in both vertebrates and invertebrates, to the present widely cited opinion that amorphous precursors are formed first.

We believe that some of the earlier-described controversies may simply be a result of the improved high resolution of modern microscopy and spectroscopy techniques, in terms of both spatial as well as temporal resolution. It is obvious that for any solid precipitate to form, an initial cluster of atoms is necessary. The existence of such stable, multinuclear clusters, especially at high concentrations, is a well-recognized fact in the aqueous geochemistry literature. Aluminum oxyhydroxides are an excellent example, where multinuclear clusters are involved in both precipitation and dissolution of the solid phases. We believe that with improved instrumental resolution, these clusters are simply being renamed in the literature as amorphous precursor phases or "prenucleation clusters." Eventually, of course, these aqueous multinuclear clusters or amorphous prenucleation clusters transform to the most thermodynamically stable crystalline phase. Thermodynamically, the solubility product (K_{sp}) is defined for a bulk phase, and nucleation energy includes the additional interfacial energy term of a stable cluster or "nucleus." "Stable" does not mean that the cluster is static once formed. Ions may dynamically add to or leave from the cluster, but the cluster reaches a critical size (at which point it is a "nucleus") and then grows as time progresses. Thus, the apparent problem of conceptual models for nucleation mechanisms arises, in part, because the high spatial and temporal resolution spectroscopic and microscopic methods can now "see" these multinuclear, aqueous clusters, even though they may not be of the critical size (i.e., may not be a true nucleus). Thus, we are now running up against *what is the experimental definition of a critical cluster (nucleus) in terms of size and over what timescale should we observe it to claim that it is a stable nucleus?*

A second hurdle to using computational methods to study nucleation is that nucleation is a dynamic process, where the assembly of ions to form either crystalline or amorphous structures needs to be investigated with the inclusion of thermal fluctuations. Potential-energy-based static structure studies with the conventional quantum calculations using small model complexes are insufficient to provide direct illustration of the dynamic nucleation process. In a previous work, *ab initio* cluster calculations were used to study a fragment of the calcium orthophosphoserine crystal structure as a small molecule model of the purported nucleating motif of BSP. The fragment consisted of one phosphorylated serine and two glutamate residues. Interactions of this fragment with Ca^{2+} and phosphate ions were examined (Sahai 2005). An interesting structural matching pattern was noted between the arrangement of phosphate and carboxylate groups of the fragment and the phosphate ion distributions on the (001) face of the HAP. However, due to the CPU speed limitation at that time, solvation and thermal fluctuations were not included in that work; so, the studied conformation of the model system may not represent the dominant ensemble of the BSP nucleating motif in solution with the presence of water and nucleating ions. Therefore, the potential role of the BSP motif's conformation in nucleation could not be confirmed.

Furthermore, nucleation is sometimes a long process, taking up to hours in the laboratory and also for calcification on the extracellular matrix in bone biomineralization *in vivo*. Thus, nucleation is a rare event to sample in simulations. It is practically impossible for the conventional MD simulation to demonstrate all the stages involved in nucleation process from the formation of initial multinuclear clusters to the formation of a true nucleus, even with the present high-performance supercomputers.

Two computational strategies may be employed to resolve the aforementioned problems. One is to design the "*in silico* experiment" carefully, and to address those problems, which *can* be answered by conventional MD with the accessible computational resources (Duffy and Harding 2005; Yang et al. 2010). With parallel computations for an all-atom simulation consisting of $\sim 10^4$ atoms, it is not unreasonable to reach a timescale up to tens to hundreds of nanoseconds (Freddolino et al. 2010; Shaw et al. 2010). Such a system could perhaps help us understand the early stage of nucleation when the protein sequesters ions from solution and manipulates the distributions of ions and ion clusters around the protein surface (Yang et al. 2010). Within the rest of this chapter (Section 17.2), we demonstrate one of our studies using the first of the aforementioned strategies to explore the conformational effect of BSP in modulating the distributions of Ca^{2+} and phosphate ions in solution. Specifically, we explore whether a nucleating template for orientated HAP could be formed in different peptide conformations.

The second way to investigate the rare events leading to nucleation relies on the application of special simulation techniques, including umbrella sampling (Torrie and Valleau 1977), metadynamics (Laio and Parrinello 2002; Laio and Gervasio 2008), Monte Carlo (Landau and Binder 2002), transition path sampling methods (Bolhuis et al. 1998; Dellago et al. 1998a,b), etc. Nucleation could thus be investigated without imposing special constraints on simulation temperature and/or ion concentrations. One of the challenges for metadynamics is to define an "order parameter" to describe the nucleation process and potential phase transitions among different polymorphs. A qualified order parameter should be able to clearly distinguish various stages along the nucleation process. The potential energy, the Steinhardt order parameter (Steinhardt et al. 1983), and a tetrahedral order parameter are usually employed in nucleation studies. Advantages and disadvantages of each approach to study nucleation have been reviewed by Harding et al. (2008).

17.1.4.2 Challenges and approaches in studying protein–water–solid surface interactions

As discussed earlier, nucleation is only one of the challenges in applying modeling methods to biomineralization. An equally difficult task is to model accurately the interactions of proteins at the solid–water interface. Studies on protein–solid surface interactions are not limited to the biomineralization systems. The adsorption of proteins on synthetic materials is also of general interest in biomaterials development, where the protein does not necessarily share the common characteristics of biomineralization proteins as discussed earlier, and the solid surfaces could be ceramics, carbon nanostructures, polymers, metals, or ionic/molecular crystals (Latour 2005). Consequently, the technical concerns raised here for biomineralization may also apply to many other pertinent fields.

Modeling of protein–water–solid surface interactions involves several challenges. The "real" solid surface structure is complicated and often not known in detail at the atomic level (Mann 2001). For instance, the solid surface is usually not perfectly flat. Some atoms may lie slightly above others, even within an ideal crystallographic surface; in addition, surface defects such as steps, edges, terraces, etch pits, and growth spirals are commonly present. These different local structures affect interactions with proteins. Step directions and detailed atomic level structures are well-defined for calcite, but little is known for HAP. More importantly, the chemical modifications of the surface arising from interactions with water, such as degree and sites of nondissociative hydration as well as dissociative protonation and hydroxylation, are not well-characterized. By using *ab initio* density functional method, Corno et al. (2009) studied water adsorption at HAP (001) and (010) faces and noted that water dissociates on the (010) surface, giving rise to new surface terminations. It has been reported that the surface Ca^{2+} is hydroxylated and the PO_4^{3-} group is protonated to form –Ca-OH and $-HPO_4$ surface sites *in vitro* (Van Cappellen et al. 1993), and the PO_4^{3-} group of the earliest mineral crystals in bone is also protonated (Shimabayashi et al. 1981; Rey et al. 1990, 1991; Wu et al. 2003; Glimcher 2006; Jaeger et al. 2006; Bertinetti et al. 2009). The surface hydroxylation/protonation results in different charges from the original ionic sites. Most experimental results, however, have provided only average properties obtained on HAP powders, so that the quantitative information like density and distributions of these modified sites on specific surfaces is not available. This limits the ability to accurately represent the HAP faces in molecular simulations. In practice, for the sake of simplicity, most of the atomistic simulations assume a flat surface without hydroxylation/protonation, except in cases where more detailed information is available, such as specific step directions on the calcite (104) surface.

A second challenge in modeling interfacial interactions is that the 3D structure of the protein at the crystal surface is unknown at the atomic level for most biomineralization systems (George and Veis 2008). One of the few exceptions is the availability of high-resolution (2.0 Å) x-ray crystal structure of porcine osteocalcin coordinating five Ca^{2+} ions in a spatial orientation that is complementary to Ca^{2+} sites in the HAP (100) surface (Hoang et al. 2003). The unique successful structural determination of porcine osteocalcin is probably due to the relatively short sequence (49 amino acids), addition of $CaCl_2$ (10 mM), and the existence of three stable helices at the level of the secondary structure. In MD simulations for biomineralization systems, the protein structure at the solid surface is needed as part of the initial setup. Based on such initial configurations, where the protein is already in a thermodynamic local minimum, the protein dynamics and its interactions with solid surface and water environment can be investigated at the atomic level.

Due to the lack of relevant experimental information, several computational schemes are applied in practice to obtain the initial protein conformation. When studying the interactions between HAP surfaces and model peptides from the phosphoprotein, OPN, Azzopardi et al. (2010) used the fully extended conformations as the initial peptide structures. However, in situations where the protein structure is resolved in solution, that known structure is used as the initial guess for the protein adsorbed at the solid surface, even though the protein structure could be largely different in the two situations. For example, Zhou et al. (2007) investigated the adsorption mechanism of bone morphogenetic protein-7 (BMP-7) on the HAP (001) surface. The initial structure of BMP-7 was obtained from the crystal structure of the BMP-7–noggin complex, where no HAP was present (noggin is an antagonist of BMP-7). In recent years, led by Baker and Gray's groups (Kim et al. 2004; Rohl et al. 2004; Bradley et al. 2005; Schueler-Furman et al. 2005; Makrodimitris et al. 2007; Masica and Gray 2009; Masica et al. 2010), a bioinformatics-based technique is being used to predict protein structures upon adsorption on different solid surfaces (e.g., HAP and calcite) (Makrodimitris et al. 2007; Masica and Gray 2009; Masica et al. 2010). Coupled with experimental results, like solid-state NMR data, such a computational method starts to shed light on protein–surface interaction mechanisms at the atomic level. However, due to the simplified calculation algorithm and the associated scoring function, the solvent effect is not treated explicitly, but rather, is taken into account with implicit solvent model. Thus, the bioinformatics approach does not account for specific hydrogen bond networks among water and ions on the solid–solution interface, which are crucial for a correct description of the protein–surface interactions. In Section 17.3.2, we describe our application of such a bioinformatics method to BSP adsorption on the HAP (001) surface, where the obtained peptide–surface complexes were then used as the initial structures for long-time, conventional MD simulation with explicit water models.

In addition to the aforementioned bottlenecks on protein–water–surface simulations, another significant issue is the lack of reliable empirical all-atom MM FFs (Harding et al. 2008; Latour 2008). As discussed earlier, many routinely used MM FFs, such as AMBER, CHARMM, OPLS, and GROMOS, are typically applied to proteins, DNA, RNA, and other organic or biomolecular systems. These FFs do not provide a reasonable description for the bulk or surface properties of biominerals, most of which are ionic solids, such as HAP, calcite, calcium oxalate, and iron oxyhydroxides. Effective or partial ion charges in these solids could also be dramatically different from charges in bulk solution, and proper ion charges in the FFs are crucial for accurate calculations of solid properties. Special FFs to treat various ionic solids and surfaces have been developed, including Hauptmann's FF for HAP (Hauptmann et al. 2003), the several FFs for calcite (Freeman et al. 2007; Raiteri et al. 2010; Gale et al. 2011), and a more general FF like CLAYFF (Cygan et al. 2004) which is suitable for the simulation of hydrated, aluminosilicate minerals and their interfaces with aqueous solutions, and the INTERFACE FF (Heinz et al. 2013) which enables simulations of inorganic–organic and inorganic–biomolecular interfaces for silicates, aluminates, metals, oxides, sulfates, and apatites. Even with these FFs, some were developed solely for modeling solid bulk properties, and interactions with water at the surface were not considered. Furthermore, the compatibility between FFs describing different types of species (protein, water, and solid surface) in the system needs to be validated before applying to any realistic systems.

Although the importance of such compatibility was stressed heavily in Latour's work (Latour 2008), systematic benchmark calculations for the applied FFs in protein–water–solid systems are elusive in many studies. The normal treatments to combine the different FFs for the nonbonded interfacial interactions include: (1) the application of Lorentz–Berthelot mixing rule for the Lennard–Jones (LJ) potential parameters describing van der Waals interactions; (2) the adoption of the partial charges from different FFs for cross-Coulombic interactions (e.g., the charges of the peptide are from CHARMM, and the charges of the mineral surface ions are from CLAYFF). Without FF benchmarks, the simulation results could deviate from experimental measurements significantly. One example, as highlighted in Latour's work (Latour 2008), is the theoretical prediction of a strong adsorption of a $(Gly)_4$–Lys–$(Gly)_4$ (Gly = glycine; Lys = lysine) peptide on an oligo(ethylene-oxide)-self-assembled monolayer (SAM) surface, whereas the surface is observed experimentally to be more nonadsorbing for peptides and proteins (Prime and Whitesides 1993; Raut et al. 2005). Similar examples include recent theoretical studies on amino acid–HAP adsorption. The binding free energy for Glu on HAP (001) surface from calculations is about 400 $kJ{\cdot}mol^{-1}$, which is substantially larger than the experimental determinations (Pan et al. 2007). Therefore, to obtain reliable structural and energetic information for protein–water–surface interactions, the importance of carefully validating the applied FFs cannot be overemphasized. Along with high-level QM calculation results, quality experimental measurements can be used as the benchmark references as summarized in Table 17.1. We will describe in Section 17.3.1.2 our strategies

Table 17.1 Experimental measurements that can be used as benchmark references for the theoretical study of the protein–water–surface interactions

EXPERIMENT	BENCHMARKING INFORMATION
X-ray crystallography	Crystal lattice structure
Solid-state NMR	Crystal structure, sites of functional group interaction with crystal lattice sites, protonation of surface anionic groups on mineral, conformation, and secondary structure of adsorbed protein/peptide
Infrared spectroscopy (IR)	Sites of functional group interaction with crystal lattice sites, protonation of surface anionic groups on mineral
Circular dichroism (CD)	Secondary structure of peptides and proteins
Optical waveguide light mode spectroscopy (OWLS)	Rates of protein or peptide adsorption/desorption from surface, changes in conformation upon adsorption/desorption, thickness and mass of adsorbed layer (without mass of adsorbed water)
Quartz crystal microbalance with dissipation (QCM-D)	Same as for OWLS, except that the mass of the adsorbed layer includes the mass of adsorbed water
Total internal reflection fluorescence (TIRF) microscopy	Visualize and identify binding regions of peptide/protein to mineral surface
Atomic force microscopy (AFM), *in situ* cell	Conformation and size of adsorbed protein or peptide, identity of surface sites involved in binding to protein/peptide, force of adhesion/pull-off (adsorption/desorption)

to benchmark systematically the applied FFs with a model amino acid–HAP system.

An alternative route to avoid the limitations of the classical FF issue is to use QM potentials, where the electronic energy is obtained by solving the Schrödinger equation. *Ab initio* quantum dynamics, such as Car–Parrinello molecular dynamics (CPMD) (Car and Parrinello 1985), has been applied successfully to study solution dynamics and enzymatic catalysis mechanisms. Hug et al. studied the interactions between model amino acids and a calcium oxalate monohydrate (COM) surface with CPMD (Hug et al. 2010). Due to the very high computational expense of these calculations (120,000 CPU h), the simulation system size was limited: less than 100 water molecules were used to solvate the entire system; only the top three layers of the crystal surface were taken into account, and the simulation time ranged from 3.4 to 8.64 ps. The simulation time was much shorter than the timescale for the conformational relaxation of an amino acid upon adsorption at the crystal surface, especially considering that the water structure would also be rearranged and at a slower rate. Thus, although tremendous effort has been put along this direction, *ab initio* QM-based approaches are not viable at this stage for protein–water–surface investigations.

Semiempirical quantum potentials may provide a useful approach in studying protein–water–surface interactions, because of their reasonable balance between calculation accuracy and efficiency, afforded by various simplifications when solving the electronic Schrödinger equation (Leach 2001). Semiempirical QM methods have been extensively employed to study enzymatic systems (Riccardi et al. 2006; Senn and Thiel 2007). Popular methods include AM1 (Dewar et al. 1985), PM3 (Stewart 1989), self-consistent-charge density functional tight binding (SCC-DFTB) (Elstner et al. 1998) methods, and so on. Recently, Frauenheim and coworkers used the SCC-DFTB/MD to study the adsorption of zwitterionic glycine on a geminal hydroxylated silica surface in an explicit water environment (Zhao et al. 2011). The entire system (including 140 water molecules) was treated quantum mechanically, and tens of picoseconds were simulated.

One conceivable method to benefit the well-validated semiempirical method is to build a hybrid system within the QM/MM framework (Warshel and Levitt 1976; Field et al. 1990), where the adsorbate, the neighboring crystal surface atoms, and water molecules are described with semiempirical methods, and the rest of the system is treated with an empirical MM FF. Such a scheme is already well-established in the simulation of biological systems for decades (Riccardi et al. 2006; Senn and Thiel 2007); so, adopting this approach to explore protein–mineral interactions is conceptually straightforward (Harding et al. 2008). However, specific technical challenges are expected, such as a reliable treatment of interactions between QM and MM atoms at the mineral interface, which features strong electrostatic, polarization, and charge transfer effects. The performance of semiempirical QM methods for these systems also needs to be carefully validated before any serious applications (Riccardi et al. 2006).

The fourth challenge in studying protein–water–surface interactions is similar to the issue for modeling nucleation, namely, sufficient sampling. This problem is exacerbated when the surface is composed of ions with large charges, which is true for biominerals, such as HAP, calcite, and calcium oxalate. Due to the strong electrostatic interactions between the charged groups on the protein and the ions at the solid surface, once adsorbed, the protein takes a long time to sample other binding modes. To avoid such limitations, special techniques like replica exchange molecular dynamics (REMD) (Sugita and Okamoto 1999; Garcia and Sanbonmatsu 2001) need to be applied to enhance sampling of different binding modes separated by significant energy barriers. A recent effort by Yu and Schmidt (2011) along this direction was to study the binding free energy between methyl butyrate and Mg–Al hydrotalcites in various solvents in conjunction with umbrella sampling. We are adopting a similar scheme in an ongoing project, where the combination of REMD and metadynamics is being used to explore the adsorption path and binding sites of citrate on HAP surfaces. A series of structures along the adsorption path is obtained for the umbrella sampling to get a reliable free energy profile.

Within the rest of the chapter, we will use the BSP–water–HAP system to demonstrate our specific calculation strategies and techniques, including the benchmarking of our modified Hauptmann et al. FF (Hauptmann et al. 2003), referred to in Section 17.3. Some of these methods are not necessarily the most ideal solutions, but they serve as excellent examples to inspire further improvement in the application of computational methods to the biomineralization field.

17.2 POTENTIAL ROLE OF BONE SIALOPROTEIN IN HYDROXYAPATITE NUCLEATION BY CONVENTIONAL MOLECULAR DYNAMICS SIMULATIONS

Using conventional MD simulations for long timescales (20–40 ns per trajectory), we have attempted to model HAP nucleation in the presence of one of the two acidic functional motifs of BSP (Yang et al. 2010). In particular, we aimed to understand whether templated nucleation of HAP is promoted by specific peptide conformations.

We used several preliminary steps to make a nucleation study technically possible, while not compromising scientific significance, even though we were using only conventional MD simulations. To reduce the size of the system, a model peptide, $(SerP)_2(Glu)_8$ (where SerP = phosphoserine; Glu = Glutamate) was used to mimic the purported BSP nucleating motif instead of including the entire protein sequence. Second, we focused on specific peptide conformations, including α-helix and random-coil structures. Third, high Ca^{2+} and inorganic phosphate (Pi) concentration solutions (0.125 M and 0.084 M, respectively; note 0.1 M NaCl is also included) were used to reflect supersaturations with respect to HAP. The applied Ca^{2+} and Pi concentrations were ~100 times higher than those in blood, but the simulated concentrations near the peptide surface might not be too far from the physiological range at the early stage of nucleation (which is undetermined experimentally). Fourth, multiple simulation trajectories were collected with different initial ion distributions for both initial peptide configurations. Furthermore, the simulations were run for long times (20–40 ns per trajectory). Finally, reference solutions with only inorganic ions (Ca^{2+}, Pi, Na^+, and Cl^-) were investigated to elucidate the effects of the peptide on the distribution of Ca^{2+} and Pi ions and, thus, on nucleation.

MD simulations provided structural insights into ion distributions around the peptide. In our study, both −1 and −2 protonation states of phosphate groups (for both inorganic phosphate Pi and SerP) were considered. However, both cases demonstrated similar results. We discuss in the following paragraphs the results for the −2 protonation state, which is closest to the physiological condition. Several interesting structural features were obtained from the MD simulations. Independent of peptide conformations, Ca^{2+} ions formed direct and persistent electrostatic interactions with the acidic side chains from both Glu and SerP of the peptide. The Pi ions formed extensive networks with Ca^{2+} around the peptide in both α-helix and random-coil conformations, and SerP side chains may also participate in the Ca^{2+}–Pi networks (Figure 17.3). The background electrolyte ions, Na^+ and Cl^-, did not form stable interactions with the Ca–Pi networks. The largest Ca–Pi cluster around the peptide surface was slightly larger than those obtained in the reference solution. Further, the random-coil conformation was more efficient in promoting the Ca^{2+}–Pi networks compared to the α-helix structure, because the former conformation showed significant structural fluctuation and the ability to extend in space. Third, three Ca^{2+} ions formed an equilateral triangle by coordinating to the SerP2, Glu6, and Glu9 side chains of the peptide in the α-helical conformation, which matched a Ca^{2+} equilateral triangle ion distribution on the HAP (001) face (Figure 17.3). This triangular structure was stable within the entire 25 ns simulation. An additional Ca^{2+} equilateral triangle, coordinating to SerP1, Glu5, and Glu8, formed up to ~5 ns, but the triangle was not stable and broke apart as the simulation proceedings for periods > ~5 ns. Also, only one out of three

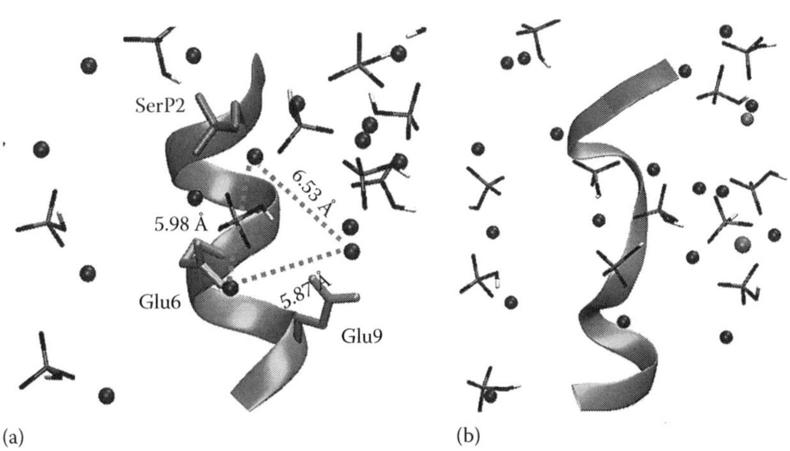

Figure 17.3 Snapshots that illustrate the interactions between Ca^{2+}, Pi, and the peptide in the MD simulations. The peptide backbone is rendered in ribbons. The Ca^{2+} and Pi ions are in the CPK and licorice representations, respectively. (a) The peptide with an α-helix conformation, where the side chain heavy atoms of SerP2, Glu6, and Glu9, which coordinate to three Ca^{2+} ions in an equilateral triangle (indicated with dotted lines, distances in Å), are highlighted in licorice representation. (b) The peptide with a random-coil structure.

additional simulations with different initial configurations gave a stable Ca^{2+} equilateral triangle throughout the entire 17 ns trajectory. Thus, it appeared that, although the α-helix peptide conformation may promote the template for the (001) face, template formation was not universal for most of the ion distributions in solution, nor was it stable over long time periods.

Based on our conventional simulations, we concluded that no apparent nucleation template for the HAP crystal lattice was observed. We based this conclusion on the result that template formation was neither general (not observed for all starting configurations) nor robust (not observed over long period simulations for all trajectories), and the templating effect with the helical conformation was not likely to be the dominant mechanism. Thus, BSP may be more likely to promote the nucleation of the amorphous calcium phosphate. On the other hand, it is possible that advanced MD methods that provide better sampling of the rare nucleation events and that can be run for much longer times may be able to capture a templating role for BSP. From the experimental perspective, it would be particularly interesting and important if the nucleation process of HAP promoted by the model peptide could be studied, especially with a careful characterization of the first-formed phase of calcium phosphate.

17.3 THEORETICAL STUDY OF INTERACTIONS BETWEEN AMINO ACID OR PEPTIDE WITH HYDROXYAPATITE USING ADVANCED MOLECULAR DYNAMICS APPROACHES

One of our specific research interests is to investigate the interaction mechanisms between the BSP protein and HAP surfaces with theoretical means, which includes careful characterizations of binding structures and energetics (Yang et al. 2011). As discussed earlier, establishing a reliable computational scenario is a prerequisite to achieve this goal. Therefore, our MM FF was first benchmarked using simple amino acid–water–HAP model systems (Xu et al. submitted) before it was applied to the BSP peptide–water–HAP system.

17.3.1 AMINO ACID–WATER–HYDROXYAPATITE INTERACTIONS

We chose the charge-neutral serine amino acid in zwitterionic form as the model amino acid and the HAP (100) face, because it has the largest surface area in HAP nanocrystals of bone.

One of the ultimate goals in biomineralization studies to reveal protein–crystal surface interaction mechanism is to characterize the binding free energies upon protein adsorption at different crystal faces. The adsorption free energy is important in understanding crystal growth modification and morphology control. If an adsorbate (e.g., small molecules, peptide, protein, etc.) binds preferentially to specific crystal faces, the molecule will inhibit crystal growth in the direction perpendicular to that face, while allowing growth parallel to that face. As a result, a unique crystal morphology is formed relative to the inorganically grown crystal.

17.3.1.1 Calculation setup

As emphasized earlier, reliable calculations of such binding free energies on specific crystal faces depend on a high-quality FF and a reliable free energy computation algorithm that provides sufficient sampling of the configurational space. To describe the HAP surface, in our study we used a modified version of the rigid HAP (the structures for ionic groups are fixed) FF developed by Hauptmann et al. (Hauptmann et al. 2003), because their FF accurately reproduces bulk crystal parameters obtained from experiment. In the original formulation of Hauptmann et al., the intermolecular interactions between the ionic groups in HAP were described as a summation of Coulombic and Born–Mayer–Huggins (BMH) potentials. In our calculations, we converted the BMH potential to the classical Lennard-Jones (LJ) potential. This modification allowed us to apply the Lorentz–Berthelot mixing rule to treat the cross-van der Waals interactions between nonbonded atoms. The CHARMM22 FF (Brooks et al. 2009) and TIP3P models (Jorgensen et al. 1983) were used to treat serine and water molecules, respectively, because of the successful application of these FFs in studying biological systems in the condensed phase. We used a number of systematic benchmark calculations, including crystal bulk properties and water–HAP interactions, to validate these FFs for our system as presented in Section 17.3.1.

Benchmarking of the peptide–water–HAP system was accomplished by calculating serine–HAP binding free energy, with the inclusion of explicit solvation and comparing these results to density functional theory–Poisson Boltzmann (DFT–PB) calculation results. In the MD simulation, the binding free energy was obtained through the PMF calculation using the umbrella sampling approach (Torrie and Valleau 1977) combined with WHAM (Kumar et al. 1992). The PMF is defined as the free energy of a chemical/physical process along a chosen reaction coordinate (McQuarrie 2000), and, thus, it includes both enthalpic and entropic contributions. In the DFT-PB calculations, the system energy was obtained through solving the Schrödinger equation, and the solvent effect was taken into account with the implicit model.

In our system, we defined the reaction coordinate as the separation between the center of mass of serine and the uppermost Ca^{2+} layer at the HAP crystal face. With statistical mechanics, PMF can be computed through the probability distribution of the defined reaction coordinate. The PMF profile can be used to determine the energetically most favorable binding position, orientation, and structure of the absorbed amino acid near the crystal surfaces (Xu et al. 2008). We benchmarked our PMF results with DFT-PB QM calculations described in Section 17.3.1.2.

17.3.1.2 Force-field benchmarking and potential of mean force

Systematic benchmark calculations were performed to ensure the quality of our calculation setup.

Bulk HAP crystal lattice parameters computed with our modified Hauptmann et al. FF (converted LJ parameters) were in good agreement with the experimental values. The average deviation was 0.28% for lattice parameters, *a*, *b*, and *c*, over a broad range of temperatures (73, 300, 600, and 1273 K), which was even smaller than that obtained by Hauptmann et al. with their original BMH formalism, where the deviation was 0.42%. The standard molar lattice enthalpy of HAP crystal at room temperature was computed to be –34,077.40 kJ · mol^{-1} comparable to the available experimental measurement of –34,183 ± 134 kJ · mol^{-1}, and the computed isobaric thermal expansivity coefficient of 1.535×10^{-5} K^{-1} was also consistent with the previous MD calculation results obtained by Cruz et al. (1.82×10^{-5} K^{-1}) (Cruz et al. 2005).

Water–HAP surface interactions were examined from the structural perspective. Over 10 ns MD simulations ensured that the water structure was fully equilibrated on HAP (001) and (100) faces. Consistent with our expectation, water molecules formed favorable interactions with both HAP faces because of strong ion–dipole interactions. We observed two stable water layers at both HAP surfaces, which was consistent with the previous computational and experimental studies (Park et al. 2002; Pareek et al. 2007). The interfacial structure of the fluorapatite–water system studied with high-resolution specular x-ray reflectivity showed two distinct water layers on the (100) face. This result was later confirmed by Pareek et al. (2007) using the grazing incidence x-ray diffraction (GIXRD) technique.

The PMF obtained for serine adsorption on HAP (100) surface is shown in Figure 17.4. For the binding state, the COO^- group of serine has a direct interaction with the Ca^{2+} on the HAP (100) surface. An adsorption free energy of about –11.0 kcal·mol^{-1} was obtained. In the DFT QM calculation, the long-range electrostatic solvation effect was taken into account by using a periodic continuum solvation model based on the modified PB equation. The energy obtained from the DFT–PB calculation of –10.1 kJ · mol^{-1} compared quite well with the PMF result.

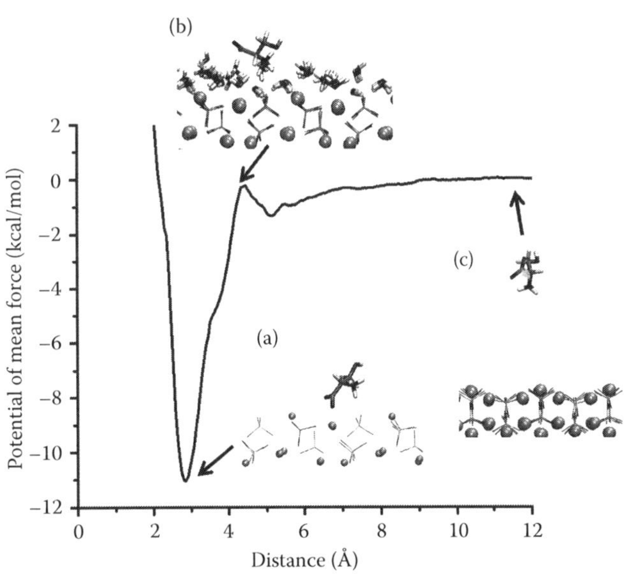

Figure 17.4 Free energy profile of serine adsorption at the HAP (100) face. The representative structures for (a) adsorbed state, (b) transition state, and (c) bulk state along the adsorption pathway are illustrated. The stable "sandwiched" water layers between serine and the HAP (100) surface are shown in (b); the other water molecules in that structure and water in structures (a) and (c) are not shown for clarity. Serine and water molecules are rendered in licorice representation. PO_4^{3-} and OH^- are in line representation. Ca^{2+} ions are in CPK representation.

In summary, careful benchmarking of classical FFs is imperative to study organic–water–mineral interfacial processes, such as biomineralization. Furthermore, we have attempted to demonstrate that reliable characterization of multiple closely spaced free energy minima upon adsorption on solid surfaces requires the use of computational methods such as umbrella sampling and WHAM for relatively long period simulations (~160–240 ns for one PMF calculation) to ensure sufficient sampling of various binding modes. This level of computational rigor has not been the norm in most previous computational studies of biomineralization.

17.3.2 BONE SIALOPROTEIN PEPTIDE–WATER–HYDROXYAPATITE INTERACTIONS

We could now use the benchmarked FF to explore, with confidence in the accuracy of our results, BSP peptide binding conformations at various HAP faces. The model peptide, $(SerP)_2(Glu)_8$, for the BSP nucleating motif was examined in our study. As discussed earlier, there is a lack of experimental information on peptide/protein structures *at the atomic level* when adsorbed at the surface; so, the preparation of the initial peptide structure was one of the major obstacles for our MD simulations.

17.3.2.1 Bioinformatics-based binding study

We tackled the challenge of lacking an initial structure for the peptide by using a bioinformatics-based method implemented in the RosettaSurface package (Makrodimitris et al. 2007; Masica and Gray 2009; Masica et al. 2010) to predict the model peptide structure adsorbed at the (001), (100), and (110) faces of HAP. The algorithm runs an initial, fast, low-resolution (centroid mode) calculation to generate diverse peptide folds, and subsequent high-resolution (full-atom mode) steps for improved structures. Each HAP face is then introduced into the peptide system with a random orientation to interact with the peptide for further structure refinement and energy minimization. The peptide's conformational space is thus sampled broadly with discrete fragment and side chain rotamer libraries, and the obtained structures are evaluated with a semiempirical energy function. We began with the peptide in a fully extended conformation. At the end of each execution of the RosettaSurface algorithm, an energy-minimized peptide–crystal structure was obtained (Masica and Gray 2009). In total, 80,000 energy-minimized structures were generated for each crystal face, from which we selected the 10 structures with the lowest energies for further structural analysis.

The 10 peptide structures with the lowest energies generated from the bioinformatics calculations were examined in detail (Table 17.1). A number of interesting observations were obtained (Figure 17.5). In most of these structures, regardless of the peptide conformation and the HAP face, the peptide lay elongated along the surface. An exception is the HAP (110) face, where the lowest energy peptide had a helical conformation, and the long axis of the helix tilted on the crystal surface with the peptide C-terminus oriented away from HAP. Second, independent of HAP faces, we observed both random-coil structure and 3_{10} helical conformations. Further, the SerP side chains showed a strong propensity to bind to ions (Ca^{2+} or the hydrogen of OH^-)

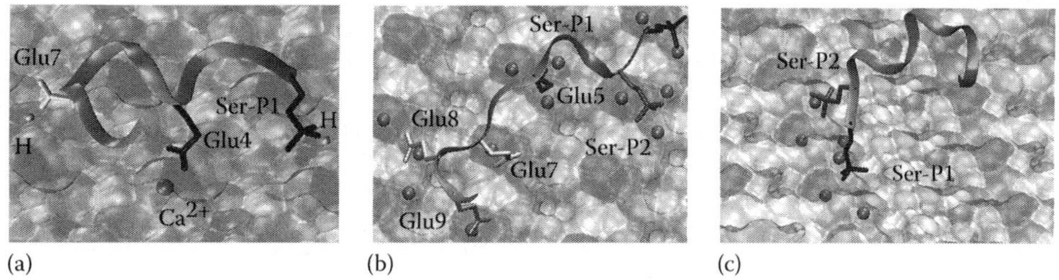

Figure 17.5 The lowest-energy peptide structure adsorbed on HAP crystal faces: (a) (001), (b) (100), and (c) (110), as predicted by RosettaSurface. The peptide backbone is rendered in ribbons. The side chains, which form direct interactions with HAP surface ions (represented in CPK), are rendered in licorice with their names labeled. The HAP surface is rendered in the surf mode. See structural details in Table 17.2.

Table 17.2 Peptide structures upon binding to different HAP faces

FACES	% OF THE 10 LOWEST ENERGY PEPTIDE STRUCTURES IN		IN THE STRUCTURE WITH THE LOWEST ENERGY			
			PEPTIDE		INTERACTING MOTIFS OF	
	3_{10} HELIX CONFORMATION	RANDOM-COIL CONFORMATION	CONFORMATION	ORIENTATION	PEPTIDE RESIDUES	HAP IONS
(001)	50	50	3_{10} Helix	Parallel	SerP1, Glu4, Glu7	OH^-, Ca^{2+}
(100)	50	50	Random coil	Parallel	SerP1, SerP2, Glu5, Glu7, Glu8, Glu9	Ca^{2+}
(110)	40	60	3_{10} Helix	Tilt	SerP1, SerP2	Ca^{2+}

on the (001), (100), and (110) faces, independent of peptide conformation. Significantly, no apparent geometrical matching pattern (template) was observed between the peptide residues and the HAP surface sites.

The bioinformatics-predicted binding complexes provided structural insight into peptide–HAP interactions. However, the present RosettaSurface approach did not include explicit solvents; therefore, specific H-bond interactions between water, the peptide, and the HAP surface were not directly addressed. Nonetheless, the bioinformatics approach provided a good starting point for further MD simulations, especially, given the dearth of experimental structural information *at the atomic level.*

17.3.2.2 Molecular dynamics simulation of peptide–water–hydroxyapatite interactions

Using the lowest energy peptide structure at the HAP (001) surface as predicted by the bioinformatics approach, we performed long-time atomistic MD simulations with the benchmarked FF (see the previous discussions). Specifically, the water molecules were treated explicitly with the TIP3P models. The peptide and HAP (001) slab were described with the CHARMM22 and the modified Hauptmann et al. FFs, respectively. A 50 ns conventional MD simulation was carried out.

A number of interesting observations were obtained from the simulation. First, one to two stable water layers were captured between the BSP peptide and the HAP (001) surface, which was consistent with the result from the amino acid–water–HAP simulations. Second, the peptide conformation experienced minor changes along the MD simulation relative to the starting structure predicted by the bioinformatics approach. Figure 17.6 shows that the peptide backbone structure from the MD

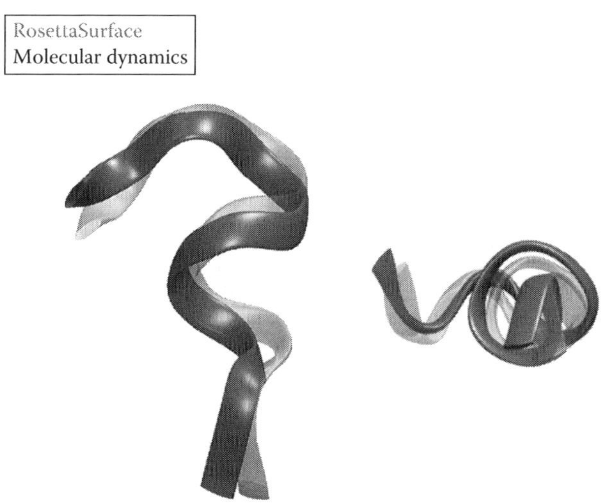

Figure 17.6 Superposition of the BSP peptide backbone structures from the MD simulation (dark gray) versus the RosettaSurface prediction (light gray).

simulation remains very similar to that from the bioinformatics prediction at the timescale of 50 ns. The root-mean-square deviation (RMSD) for the backbone atoms of the BSP peptide for the last 40 ns simulation showed an average value of 0.76 Å. Whether this similarity holds at longer timescales or with the use of enhanced sampling techniques remains to be clarified in the future. Third, the important interactions between the peptide and the HAP surface predicted by RosettaSurface involving SerP1 and Glu4 were maintained within the present simulation timescale (Figure 17.7). No apparent interaction matching template between the peptide side chain orientation and the ion distribution on the HAP (001) surface was observed from the current theoretical study.

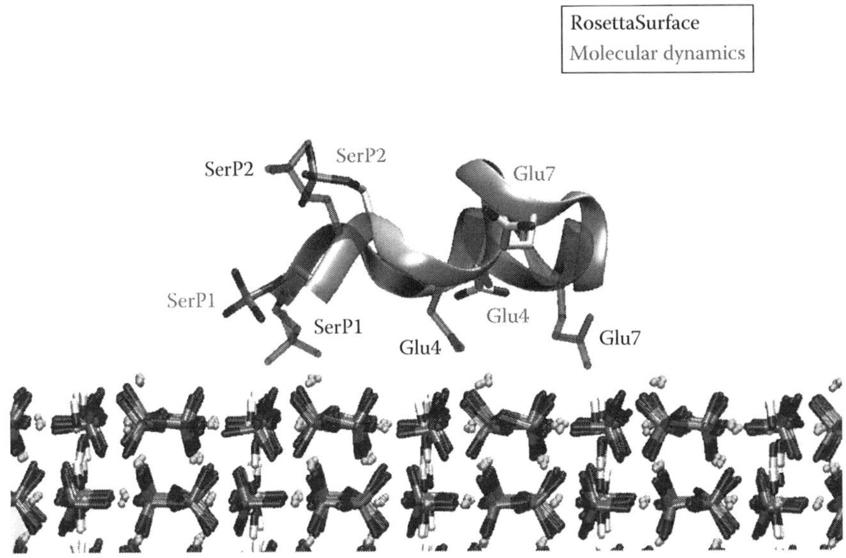

Figure 17.7 Superposition of the BSP peptide structures from the MD simulation (light gray) versus the RosettaSurface prediction (dark gray) upon adsorption on the HAP (001) surface. Water molecules are not showed for clarity. The side chains for the important residues (labeled on the figure) of the peptide and the HAP surface are rendered in licorice representation, with the calcium atoms shown in light gray, and phosphorus, oxygen (dark gray), hydrogen, and carbon depicted. The backbone structure is rendered in ribbon.

17.4 CONCLUDING DISCUSSIONS AND FUTURE PERSPECTIVES

We have attempted in this chapter to describe the challenges and opportunities, both computational and experimental, in studying biomineralization mechanisms at the atomic level. In particular, we have emphasized the critical importance of accurate benchmarking of FFs with both quantum calculations and experimental results, as well as the more efficient sampling of rare events (processes that have a low probability of occurring and therefore a long waiting time) such as formation of a nucleus and of crossing closely spaced energy barriers in a reaction pathway, by adapting advanced MD simulation methods to the specific challenges in modeling biomineralization. Experimentally, there is a dire need for atomic-level energetic and structural information specific to the dominant crystallographic faces of HAP and in the presence of aqueous solution, for proper benchmarking of existing FFs and guiding the developments of more sophisticated models (e.g., hybrid QM/MM) that include polarization and charge-transfer effects.

Even if we succeed in obtaining such experimental data *in vitro* and conduct detailed benchmarking of FFs, one of the most significant hurdles in biomineralization studies in general, and for bone biomineralization in particular, is to relate these results to *in vivo* mineralization pathways. For example, the mineral in bone is a nonstoichiometric and nonideal apatite phase, and the chemical composition becomes more ideal as the mineral matures. At present, it is impossible to model the nonstoichiometric phase, because the precise sites and extent of atomic substitutions are not known.

The role of NCPs in biomineralization is often based on results of *in vitro* studies and usually emphasizes the interactions with HAP. However, many of the NCPs interact with multiple components such as cell surface ligands, collagen, and the HAP mineral phase of bone. The location, sites of binding, and hence, functions of NCPs in biomineralization *in vivo* are hardly known at the atomic or nanometer-level, and only now are some studies starting to probe these questions. Modeling the extracellular matrix, including collagen and NCPs, and/or portions of the cell surface will require bridging multiple length scales from the atomic to the submicron, and this will require different computational approaches from the ones discussed in this chapter.

ACKNOWLEDGMENTS

This research was funded by a NSF CAREER Award (EAR 0346689), NSF DMR ARRA grant (DMR 0906817), ACS-PRF (41777-AC2) grant, and start-up funds from University of Akron to Nita Sahai, and NSF-CHEM0957285 to Qiang Cui. Yang Yang acknowledges the start-up funds and the Frances R. Lax research award fund from Rowan University. Yang Yang thanks Prof. Jeffrey J. Gray, Dr. David L. Masica, and Dr. Liang Ma for helpful technical discussions. Prof. Zhi-Pan Liu and Mr. Cheng Shang are acknowledged for their DFT-PB calculations. Dr. Donald Mkhonto is thanked for providing HAP surface structures. The authors are grateful to the National Center for Supercomputing Applications at the University of Illinois at Urbana-Champaign, and the Condor High Throughput Computing at the University of Wisconsin-Madison, and Ohio Supercomputer Center at The Ohio State University for the use of computational resources.

REFERENCES

Addadi, L., A. Berman, J.M. Oldak, and S. Weiner. 1989. Structural and stereochemical relations between acidic macromolecules of organic matrices and crystals. *Connective Tissue Research* 21(1–4):457–465.

Allinger, N.L. 1977. Conformational-analysis.130. Mm2—Hydrocarbon force-field utilizing V1 and V2 torsional terms. *Journal of the American Chemical Society* 99(25):8127–8134.

Anisimov, V.M., G. Lamoureux, I.V. Vorobyov, N. Huang, B. Roux, and A.D. MacKerell. 2005. Determination of electrostatic parameters for a polarizable force field based on the classical Drude oscillator. *Journal of Chemical Theory and Computation* 1(1):153–168.

Aqvist, J., C. Medina, and J.E. Samuelsson. 1994. New method for predicting binding-affinity in computer-aided drug design. *Protein Engineering* 7(3):385–391.

Ash, W.L., M.R. Zlomislic, E.O. Oloo, and D.P. Tieleman. 2004. Computer simulations of membrane proteins. *Biochimica et Biophysica Acta-Biomembranes* 1666(1–2):158–189.

Azzopardi, P.V., J. O'Young, G. Lajoie, M. Karttunen, H.A. Goldberg, and G.K. Hunter. 2010. Roles of electrostatics and conformation in protein-crystal interactions. *Plos One* 5(2):e9330.

Benkovic, S.J. and S. Hammes-Schiffer. 2003. A perspective on enzyme catalysis. *Science* 301(5637):1196–1202.

Bertinetti, L., C. Drouet, C. Combes, C. Rey, A. Tampieri, S. Coluccia, and G. Martra. 2009. Surface characteristics of nanocrystalline apatites: Effect of Mg surface enrichment on morphology, surface hydration species, and cationic environments. *Langmuir* 25(10):5647–5654.

Bianco, P., L.W. Fisher, M.F. Young, J.D. Termine, and P.G. Robey. 1991. Expression of bone sialoprotein (BSP) in developing human tissues. *Calcified Tissue International* 49(6):421–426.

Bianco, P., M. Riminucci, G. Silvestrini, E. Bonucci, J.D. Termine, L.W. Fisher, and P.G. Robey. 1993. Localization of bone sialoprotein (BSP) to Golgi and post-Golgi secretory structures in osteoblasts and to discrete sites in early bone-matrix. *Journal of Histochemistry & Cytochemistry* 41(2):193–203.

Bolhuis, P.G., C. Dellago, and D. Chandler. 1998. Sampling ensembles of deterministic transition pathways. *Faraday Discussions* 110:421–436.

Bond, P.J. and M.S.P. Sansom. 2006. Insertion and assembly of membrane proteins via simulation. *Journal of the American Chemical Society* 128(8):2697–2704.

Boskey, A.L. 2000. Bone mineralization. In *Bone Mechanics Handbook*, 2nd ed., S.C. Cowin (ed.) (Boca Raton, FL: CRC Press), pp. 5.1–5.31.

Boskey, A.L. 2003. Biomineralization: An overview. *Connective Tissue Research* 44:5–9.

Bowman, G.R., V.A. Voelz, and V.S. Pande. 2011. Taming the complexity of protein folding. *Current Opinion in Structural Biology* 21(1):4–11.

Bradley, P., K.M.S. Misura, and D. Baker. 2005. Toward high-resolution de novo structure prediction for small proteins. *Science* 309(5742):1868–1871.

Brooks, B.R., C.L. Brooks, A.D. Mackerell, L. Nilsson, R.J. Petrella, B. Roux, Y. Won et al. 2009. CHARMM: The biomolecular simulation program. *Journal of Computational Chemistry* 30(10):1545–1614.

Brooks, B.R., R.E. Bruccoleri, B.D. Olafson, D.J. States, S. Swaminathan, and M. Karplus. 1983. CHARMMA program for macromolecular energy, minimization, and dynamics calculations. *Journal of Computational Chemistry* 4(2):187–217.

Brown, W.E. and L.C. Chow. 1976. Chemical properties of bone-mineral. *Annual Review of Materials Science* 6:213–236.

Bruice, T.C. 2002. A view at the millennium: The efficiency of enzymatic catalysis. *Accounts of Chemical Research* 35(3):139–148.

Burke, E.M., Y. Guo, L. Colon, M. Rahima, A. Veis, and G.H. Nancollas. 2000. Influence of polyaspartic acid and phosphophoryn on octacalcium phosphate growth kinetics. *Colloids and Surfaces B-Biointerfaces* 17(1):49–57.

Car, R. and M. Parrinello. 1985. Unified approach for molecular-dynamics and density-functional theory. *Physical Review Letters* 55(22):2471–2474.

Carlson, H.A. and J.A. McCammon. 2000. Accommodating protein flexibility in computational drug design. *Molecular Pharmacology* 57(2):213–218.

Chen, J., M.D. Mckee, A. Nanci, and J. Sodek. 1994. Bone sialoprotein messenger-RNA expression and ultrastructural-localization in fetal porcine calvarial bone—Comparisons with osteopontin. *Histochemical Journal* 26(1):67–78.

Chen, J.K., H.S. Shapiro, and J. Sodek. 1992a. Developmental expression of bone sialoprotein messenger-RNA in rat mineralized connective tissues. *Journal of Bone and Mineral Research* 7(8):987–997.

Chen, J.K., H.S. Shapiro, J.L. Wrana, S. Reimers, J.N.M. Heersche, and J. Sodek. 1991. Localization of bone sialoprotein (BSP) expression to sites of mineralized tissue formation in fetal-rat tissues by in situ hybridization. *Matrix* 11(2):133–143.

Chen, J.K., K. Singh, B.B. Mukherjee, and J. Sodek. 1993. Developmental expression of osteopontin (OPN) messenger-RNA in rat-tissues—Evidence for a role for OPN in bone-formation and resorption. *Matrix* 13(2):113–123.

Chen, Y., B.S. Bal, and J.P. Gorski. 1992b. Calcium and collagen binding-properties of osteopontin, bone sialoprotein, and bone acidic glycoprotein-75 from bone. *Journal of Biological Chemistry* 267(34):24871–24878.

Cornell, W.D., P. Cieplak, C.I. Bayly, I.R. Gould, K.M. Merz, D.M. Ferguson, D.C. Spellmeyer, T. Fox, J.W. Caldwell, and P.A. Kollman. 1995. A 2nd generation force-field for the simulation of proteins, nucleic-acids, and organic-molecules. *Journal of the American Chemical Society* 117(19):5179–5197.

Corno, M., C. Busco, V. Bolis, S. Tosoni, and P. Ugliengo. 2009. Water adsorption on the stoichiometric (001) and (010) surfaces of hydroxyapatite: A periodic B3LYP study. *Langmuir* 25(4):2188–2198.

Crane, N.J., V. Popescu, M.D. Morris, P. Steenhuis, and M.A. Ignelzi. 2006. Raman spectroscopic evidence for octacalcium phosphate and other transient mineral species deposited during intramembranous mineralization. *Bone* 39(3):434–442.

Cruz, F.J.A.L., J.N. Canongia Lopes, J.C.G. Calado, and M.E. Minas da Piedade. 2005. A molecular dynamics study of the thermodynamic properties of calcium apatites. 1. Hexagonal phases. *Journal of Physical Chemistry B* 109:24473–24479.

Cygan, R.T., J.J. Liang, and A.G. Kalinichev. 2004. Molecular models of hydroxide, oxyhydroxide, and clay phases and the development of a general force field. *Journal of Physical Chemistry B* 108(4):1255–1266.

Das, R. and D. Baker. 2008. Macromolecular modeling with Rosetta. In *Annual Review of Biochemistry*. Palo Alto, CA: Annual Reviews.

Dellago, C., P.G. Bolhuis, and D. Chandler. 1998a. Efficient transition path sampling: Application to Lennard-Jones cluster rearrangements. *Journal of Chemical Physics* 108(22):9236–9245.

Dellago, C., P.G. Bolhuis, F.S. Csajka, and D. Chandler. 1998b. Transition path sampling and the calculation of rate constants. *Journal of Chemical Physics* 108(5):1964–1977.

Dewar, M.J.S., E.G. Zoebisch, E.F. Healy, and J.J.P. Stewart. 1985. The development and use of quantum-mechanical molecular-models.76. Am1—A new general-purpose quantum-mechanical molecular-model. *Journal of the American Chemical Society* 107(13):3902–3909.

Di Tommaso, D. and N.H. de Leeuw. 2008. The onset of calcium carbonate nucleation: A density functional theory molecular dynamics and hybrid microsolvation/continuum study. *Journal of Physical Chemistry B* 112(23):6965–6975.

Di Tommaso, D. and N.H. de Leeuw. 2009. Theoretical study of the dimerization of calcium carbonate in aqueous solution under natural water conditions. *Geochimica et Cosmochimica Acta* 73(18):5394–5405.

Dror, R.O., R.M. Dirks, J.P. Grossman, H.F. Xu, and D.E. Shaw. 2012. Biomolecular simulation: A computational microscope for molecular biology. In *Annual Review of Biophysics,* Vol. 41. Palo Alto, CA: Annual Reviews.

Duffy, D.M. and J.H. Harding. 2005. Modeling the properties of self-assembled monolayers terminated by carboxylic acids. *Langmuir* 21(9):3850–3857.

Dunker, A.K., C.J. Brown, J.D. Lawson, L.M. Iakoucheva, and Z. Obradovic. 2002. Intrinsic disorder and protein function. *Biochemistry* 41(21):6573–6582.

Ecarotcharrier, B., F. Bouchard, and C. Delloye. 1989. Bone sialoprotein-Ii synthesized by cultured osteoblasts contains tyrosine sulfate. *Journal of Biological Chemistry* 264(33):20049–20053.

Elstner, M., D. Porezag, G. Jungnickel, J. Elsner, M. Haugk, T. Frauenheim, S. Suhai, and G. Seifert. 1998. Self-consistent-charge density-functional tight-binding method for simulations of complex materials properties. *Physical Review B* 58(11):7260–7268.

Ensing, B., M. De Vivo, Z.W. Liu, P. Moore, and M.L. Klein. 2006. Metadynamics as a tool for exploring free energy landscapes of chemical reactions. *Accounts of Chemical Research* 39(2):73–81.

Feynman, R.P. 1970. *The Feynman Lectures on Physics* (Boston, MA: Addison Wesley Longman).

Field, M.J., P.A. Bash, and M. Karplus. 1990. A combined quantum-mechanical and molecular mechanical potential for molecular-dynamics simulations. *Journal of Computational Chemistry* 11(6):700–733.

Fisher, L.W., G.R. Hawkins, N. Tuross, and J.D. Termine. 1987. Purification and partial characterization of small proteoglycan-I and proteoglycan-Ii, bone sialoprotein-I and sialoprotein-Ii, and osteonectin from the mineral compartment of developing human-bone. *Journal of Biological Chemistry* 262(20):9702–9708.

Fisher, L.W., O.W. Mcbride, J.D. Termine, and M.F. Young. 1990. Human-bone sialoprotein-deduced protein-sequence and chromosomal localization. *Journal of Biological Chemistry* 265(4):2347–2351.

Fisher, L.W., D.A. Torchia, B. Fohr, M.F. Young, and N.S. Fedarko. 2001. Flexible structures of SIBLING proteins, bone sialoprotein, and osteopontin. *Biochemical and Biophysical Research Communications* 280(2):460–465.

Flores, M.E., M. Norgard, D. Heinegard, F.P. Reinholt, and G. Andersson. 1992. RGD-directed attachment of isolated rat osteoclasts to osteopontin, bone sialoprotein, and fibronectin. *Experimental Cell Research* 201(2):526–530.

Franzen, A. and D. Heinegard. 1985. Isolation and characterization of 2 sialoproteins present only in bone calcified matrix. *Biochemical Journal* 232(3):715–724.

Freddolino, P.L., A.S. Arkhipov, S.B. Larson, A. McPherson, and K. Schulten. 2006. Molecular dynamics simulations of the complete satellite tobacco mosaic virus. *Structure* 14(3):437–449.

Freddolino, P.L., C.B. Harrison, Y.X. Liu, and K. Schulten. 2010. Challenges in protein-folding simulations. *Nature Physics* 6(10):751–758.

Freddolino, P.L., F. Liu, M. Gruebele, and K. Schulten. 2008. Ten-microsecond molecular dynamics simulation of a fast-folding WW domain. *Biophysical Journal* 94(10):L75–L77.

Freddolino, P.L., S. Park, B. Roux, and K. Schulten. 2009. Force field bias in protein folding simulations. *Biophysical Journal* 96(9):3772–3780.

Freddolino, P.L. and K. Schulten. 2009. Common structural transitions in explicit-solvent simulations of villin headpiece folding. *Biophysical Journal* 97(8):2338–2347.

Freeman, C.L., J.H. Harding, D.J. Cooke, J.A. Elliott, J.S. Lardge, and D.M. Duffy. 2007. New forcefields for modeling biomineralization processes. *Journal of Physical Chemistry C* 111(32):11943–11951.

Frenkel, D. and B. Smit. 2001. *Understanding Molecular Simulation: From Algorithms to Applications.* (San Diego, CA: Academic Press).

Friesner, R.A. and V. Guallar. 2005. Ab initio quantum chemical and mixed quantum mechanics/molecular mechanics (QM/MM) methods for studying enzymatic catalysis. In *Annual Review of Physical Chemistry*. Palo Alto, CA: Annual Reviews.

Gale, J.D., P. Raiteri, and A.C.T. van Duin. 2011. A reactive force field for aqueous-calcium carbonate systems. *Physical Chemistry Chemical Physics* 13(37):16666–16679.

Ganss, B., R.H. Kim, and J. Sodek. 1999. Bone sialoprotein. *Critical Reviews in Oral Biology & Medicine* 10(1):79–98.

Gao, J.L., S.H. Ma, D.T. Major, K. Nam, J.Z. Pu, and D.G. Truhlar. 2006. Mechanisms and free energies of enzymatic reactions. *Chemical Reviews* 106(8):3188–3209.

Garcia, A.E. and K.Y. Sanbonmatsu. 2001. Exploring the energy landscape of a beta hairpin in explicit solvent. *Proteins-Structure Function and Genetics* 42(3):345–354.

George, A. and A. Veis. 2008. Phosphorylated proteins and control over apatite nucleation, crystal growth, and inhibition. *Chemical Reviews* 108(11):4670–4693.

Gilson, M.K., J.A. Given, B.L. Bush, and J.A. McCammon. 1997. The statistical-thermodynamic basis for computation of binding affinities: A critical review. *Biophysical Journal* 72(3):1047–1069.

Glimcher, M.J. 1984. Recent studies of the mineral phase in bone and its possible linkage to the organic matrix by protein-bound phosphate bonds. *Philosophical Transactions of the Royal Society of London Series B-Biological Sciences* 304(1121):479–508.

Glimcher, M.J. 2006. Bone: Nature of the calcium phosphate crystals and cellular, structural, and physical chemical mechanisms of their formation. In *Medical Mineralogy and Geochemistry, Reviews in Mineralogy & Geochemistry*, Vol. 64, N. Sahai, M.A.A. Schoonen (eds), pp 223–282.

Goldberg, H.A. and G.K. Hunter. 1994. Nucleation of hydroxyapatite by bone sialoprotein: Role of glutamate-rich sequences. *Journal of Dental Research* 73:282–282.

Goldberg, H.A., K.J. Warner, M.J. Stillman, and G.K. Hunter. 1996. Determination of the hydroxyapatite-nucleating region of bone sialoprotein. *Connective Tissue Research* 35(1–4):385–392.

Goodsell, D.S., G.M. Morris, and A.J. Olson. 1996. Automated docking of flexible ligands: Applications of AutoDock. *Journal of Molecular Recognition* 9(1):1–5.

Gotliv, B.A., L. Addadi, and S. Weiner. 2003. Mollusk shell acidic proteins: In search of individual functions. *Chembiochem* 4(6):522–529.

Gray, J.J. 2004. The interaction of proteins with solid surfaces. *Current Opinion in Structural Biology* 14(1):110–115.

Greathouse, J.A. and R.T. Cygan. 2007. Computational and spectroscopic studies of dichlorofluoroethane hydrate structure and stability. *Journal of Physical Chemistry C* 111(45):16787–16795.

Grynpas, M.D. 2007. Transient precursor strategy or very small biological apatite crystals? *Bone* 41(2):162–164.

Harding, J.H. and D.M. Duffy. 2006. The challenge of biominerals to simulations. *Journal of Materials Chemistry* 16(12):1105–1112.

Harding, J.H., D.M. Duffy, M.L. Sushko, P.M. Rodger, D. Quigley, and J.A. Elliott. 2008. Computational techniques at the organic-inorganic interface in biomineralization. *Chemical Reviews* 108(11):4823–4854.

Hauptmann, S., H. Dufner, J. Brickmann, S.M. Kast, and R.S. Berry. 2003. Potential energy function for apatites. *Physical Chemistry Chemical Physics* 5(3):635–639.

Heinz, H., T.J. Lin, R.K. Mishra, and F.S. Emami. 2013. Thermodynamically consistent force fields for the assembly of inorganic, organic, and biological nanostructures: The INTERFACE force field. *Langmuir* 29(6):1754–1765.

Hoang, Q.Q., F. Sicheri, A.J. Howard, and D.S.C. Yang. 2003. Bone recognition mechanism of porcine osteocalcin from crystal structure. *Nature* 425(6961):977–980.

Hu, H. and W.T. Yang. 2008. Free energies of chemical reactions in solution and in enzymes with ab initio quantum mechanics/molecular mechanics methods. In *Annual Review of Physical Chemistry* 59:573–601.

Hu, Y.Y., A. Rawal, and K. Schmidt-Rohr. 2010. Strongly bound citrate stabilizes the apatite nanocrystals in bone. *Proceedings of the National Academy of Sciences of the United States of America* 107(52):22425–22429.

Hug, S., G.K. Hunter, H.A. Goldberg, and M. Karttunen. 2010. *Ab initio* simulations of peptide-mineral interactions. *Physics Procedia* 4:51–60.

Hultenby, K., F.P. Reinholt, M. Norgard, A. Oldberg, M. Wendel, and D. Heinegard. 1994. Distribution and synthesis of bone sialoprotein in metaphyseal bone of young-rats show a distinctly different pattern from that of osteopontin. *European Journal of Cell Biology* 63(2):230–239.

Humphrey, W., A. Dalke, and K. Schulten. 1996. VMD: Visual molecular dynamics. *Journal of Molecular Graphics & Modelling* 14(1):33–38.

Hunter, G.K. and H.A. Goldberg. 1993. Nucleation of hydroxyapatite by bone sialoprotein. *Proceedings of the National Academy of Sciences of the United States of America* 90(18):8562–8565.

Hunter, G.K. and H.A. Goldberg. 1994. Modulation of crystal-formation by bone phosphoproteins—Role of glutamic acid-rich sequences in the nucleation of hydroxyapatite by bone sialoprotein. *Biochemical Journal* 302:175–179.

Im, W., M. Feig, and C.L. Brooks. 2003. An implicit membrane generalized born theory for the study of structure, stability, and interactions of membrane proteins. *Biophysical Journal* 85(5):2900–2918.

Jaeger, C., S. Maltsev, and A. Karrasch. 2006. Progress of structural elucidation of amorphous calcium phosphate (ACP) and hydroxyapatite (HAp): Disorder and surfaces as seen by solid state NMR. *Bioceramics* 18, Pts 1 and 2(309–311):69–72.

Jorgensen, W.L. 2004. The many roles of computation in drug discovery. *Science* 303(5665):1813–1818.

Jorgensen, W.L., J. Chandrasekhar, J.D. Madura, R.W. Impey, and M.L. Klein. 1983. Comparison of simple potential functions for simulating liquid water. *Journal of Chemical Physics* 79:926–935.

Jorgensen, W.L., D.S. Maxwell, and J. Tirado-Rives. 1996. Development and testing of the OPLS all-atom force field on conformational energetics and properties of organic liquids. *Journal of the American Chemical Society* 118(45):11225–11236.

Jorgensen, W.L. and J. Tirado-Rives. 1988. The OPLS potential functions for proteins—Energy minimizations for crystals of cyclic-peptides and crambin. *Journal of the American Chemical Society* 110(6):1657–1666.

Karplus, M. and J. Kuriyan. 2005. Molecular dynamics and protein function. *Proceedings of the National Academy of Sciences of the United States of America* 102(19):6679–6685.

Karplus, M. and J.A. McCammon. 2002. Molecular dynamics simulations of biomolecules. *Nature Structural Biology* 9(9):646–652.

Kasugai, S., T. Nagata, and J. Sodek. 1992. Temporal studies on the tissue compartmentalization of bone sialoprotein (BSP), osteopontin (OPN), and SPARC protein during bone-formation in vitro. *Journal of Cellular Physiology* 152(3):467–477.

Killian, C.E., R.A. Metzler, Y.U.T. Gong, I.C. Olson, J. Aizenberg, Y. Politi, F.H. Wilt et al. 2009. Mechanism of calcite Co-orientation in the sea urchin tooth. *Journal of the American Chemical Society* 131(51):18404–18409.

Kim, D.E., D. Chivian, and D. Baker. 2004. Protein structure prediction and analysis using the Robetta server. *Nucleic Acids Research* 32:W526–W531.

Kitchen, D.B., H. Decornez, J.R. Furr, and J. Bajorath. 2004. Docking and scoring in virtual screening for drug discovery: Methods and applications. *Nature Reviews Drug Discovery* 3(11):935–949.

Knott, L. and A.J. Bailey. 1998. Collagen cross-links in mineralizing tissues: A review of their chemistry, function, and clinical relevance. *Bone* 22(3):181–187.

Kumar, S., J.M. Rosenberg, D. Bouzida, R.H. Swendsen, and P.A. Kollman. 1992. The weighted histogram analysis method for free-energy calculations on biomolecules. I. The method. *Journal of Computational Chemistry* 13(8):1011–1021.

Laio, A. and F.L. Gervasio. 2008. Metadynamics: A method to simulate rare events and reconstruct the free energy in biophysics, chemistry and material science. *Reports on Progress in Physics* 71(12).

Laio, A. and M. Parrinello. 2002. Escaping free-energy minima. *Proceedings of the National Academy of Sciences of the United States of America* 99(20):12562–12566.

Landau, D.P. and K. Binder. 2002. *A Guide to Monte Carlo Simulations in Statistical Physics.* (Cambridge, U.K.: Cambridge University Press).

Latour, R.A. 1999. Molecular modeling of biomaterial surfaces. *Current Opinion in Solid State & Materials Science* 4(4):413–417.

Latour, R.A. 2005. *The Encyclopedia of Biomaterials and Bioengineering.* (New York: Taylor & Francis), pp. 1–15.

Latour, R.A. 2008. Molecular simulation of protein-surface interactions: Benefits, problems, solutions, and future directions. *Biointerphases* 3(3):Fc2–Fc12.

Leach, A.R. 2001. *Molecular Modelling: Principles and Applications.* (Harlow, U.K./New York, NY: Prentice Hall).

Lee, E.H., J. Hsin, M. Sotomayor, G. Comellas, and K. Schulten. 2009. Discovery through the computational microscope. *Structure* 17(10):1295–1306.

Long, J.R., W.J. Shaw, P.S. Stayton, and G.P. Drobny. 2001. Structure and dynamics of hydrated statherin on hydroxyapatite as determined by solid-state NMR. *Biochemistry* 40(51):15451–15455.

Lowenstam, H.A. and S. Weiner. 1985. Transformation of amorphous calcium-phosphate to crystalline dahllite in the radular teeth of chitons. *Science* 227(4682):51–53.

MacKerell, A.D., D. Bashford, M. Bellott, R.L. Dunbrack, J.D. Evanseck, M.J. Field, S. Fischer et al. 1998. All-atom empirical potential for molecular modeling and dynamics studies of proteins. *Journal of Physical Chemistry B* 102(18):3586–3616.

Mahamid, J., A. Sharir, L. Addadi, and S. Weiner. 2008. Amorphous calcium phosphate is a major component of the forming fin bones of zebrafish: Indications for an amorphous precursor phase. *Proceedings of the National Academy of Sciences of the United States of America* 105(35):12748–12753.

Makrodimitris, K., D.L. Masica, E.T. Kim, and J.J. Gray. 2007. Structure prediction of protein-solid surface interactions reveals a molecular recognition motif of statherin for hydroxyapatite. *Journal of the American Chemical Society* 129(44):13713–13722.

Mann, S. 2001. *Biomineralization Principles and Concepts in Bioinorganic Materials Chemistry.* (New York: Oxford University Press).

Masica, D.L. and J.J. Gray. 2009. Solution- and adsorbed-state structural ensembles predicted for the statherin-hydroxyapatite system. *Biophysical Journal* 96(8):3082–3091.

Masica, D.L., S.B. Schrier, E.A. Specht, and J.J. Gray. 2010. De novo design of peptide-calcite biomineralization systems. *Journal of the American Chemical Society* 132(35):12252–12262.

McCammon, J.A., B.R. Gelin, and M. Karplus. 1977. Dynamics of folded proteins. *Nature* 267(5612):585–590.

McQuarrie, D.A. 1973. *Statistical Thermodynamics.* (Mill Valley, CA: University Science Books).

McQuarrie, D.A. 2000. *Statistical Mechanics.* (Sausolito, CA: University Science Books).

Ndao, M., E. Keene, F.F. Amos, G. Rewari, C.B. Ponce, L. Estroff, and J.S. Evans. 2010. Intrinsically disordered mollusk shell prismatic protein that modulates calcium carbonate crystal growth. *Biomacromolecules* 11(10):2539–2544.

Oldberg, A., A. Franzen, and D. Heinegard. 1988. The primary structure of a cell-binding bone sialoprotein. *Journal of Biological Chemistry* 263(36):19430–19432.

Pan, H.H., J.H. Tao, X.R. Xu, and R.K. Tang. 2007. Adsorption processes of Gly and Glu amino acids on hydroxyapatite surfaces at the atomic level. *Langmuir* 23(17):8972–8981.

Pareek, A., X. Torrelles, J. Rius, U. Magdans, and H. Gies. 2007. Role of water in the surface relaxation of the fluorapatite (100) surface by grazing incidence x-ray diffraction. *Physical Review B* 75(3):035418.

Park, C., P. Fenter, Z. Zhang, L. Cheng, and N.C. Sturchio. 2002. Structure of the fluorapatite (100)—Water interface by high resolution x-ray reflectivity. *American Geophysical Union, Fall Meeting Abstracts* 89:1647–1654.

Piana, S., K. Lindorff-Larsen, and D.E. Shaw. 2012. Protein folding kinetics and thermodynamics from atomistic simulation. *Proceedings of the National Academy of Sciences of the United States of America* 109(44):17845–17850.

Politi, Y., T. Arad, E. Klein, S. Weiner, and L. Addadi. 2004. Sea urchin spine calcite forms via a transient amorphous calcium carbonate phase. *Science* 306(5699):1161–1164.

Politi, Y., Y. Levi-Kalisman, S. Raz, F. Wilt, L. Addadi, S. Weiner, and I. Sagi. 2006. Structural characterization of the transient amorphous calcium carbonate precursor phase in sea urchin embryos. *Advanced Functional Materials* 16(10):1289–1298.

Politi, Y., R.A. Metzler, M. Abrecht, B. Gilbert, F.H. Wilt, I. Sagi, L. Addadi, S. Weiner, and P. Gilbert. 2008. Transformation mechanism of amorphous calcium carbonate into calcite in the sea urchin larval spicule. *Proceedings of the National Academy of Sciences of the United States of America* 105(45):17362–17366.

Price, P.A., D. Toroian, and J.E. Lim. 2009. Mineralization by inhibitor exclusion the calcification of collagen with fetuin. *Journal of Biological Chemistry* 284(25):17092–17101.

Prime, K.L. and G.M. Whitesides. 1993. Adsorption of proteins onto surfaces containing end-attached oligo(ethylene oxide)—A model system using self-assembled monolayers. *Journal of the American Chemical Society* 115(23):10714–10721.

Raiteri, P., J.D. Gale, D. Quigley, and P.M. Rodger. 2010. Derivation of an accurate force-field for simulating the growth of calcium carbonate from aqueous solution: A new model for the calcite-water interface. *Journal of Physical Chemistry C* 114(13):5997–6010.

Raut, V.P., M.A. Agashe, S.J. Stuart, and R.A. Latour. 2005. Molecular dynamics simulations of peptide-surface interactions. *Langmuir* 21(4):1629–1639.

Rey, C., M. Shimizu, B. Collins, and M.J. Glimcher. 1990. Resolution-enhanced Fourier-transform infrared-spectroscopy study of the environment of phosphate ions in the early deposits of a solid-phase of calcium-phosphate in bone and enamel, and their evolution with age.1. Investigations in the V4 Po4 domain. *Calcified Tissue International* 46(6):384–394.

Rey, C., M. Shimizu, B. Collins, and M.J. Glimcher. 1991. Resolution-enhanced Fourier-transform infrared-spectroscopy study of the environment of phosphate ion in the early deposits of a solid-phase of calcium-phosphate in bone and enamel and their evolution with age. 2. Investigations in the Nu-3 Po4 domain. *Calcified Tissue International* 49(6):383–388.

Riccardi, D., P. Schaefer, Y. Yang, H.B. Yu, N. Ghosh, X. Prat-Resina, P. Konig, G.H. Li, D.G. Xu, H. Guo, M. Elstner, and Q. Cui. 2006. Development of effective quantum mechanical/molecular mechanical (QM/MM) methods for complex biological processes. *Journal of Physical Chemistry B* 110(13):6458–6469.

Riminucci, M., J.N. Bradbeer, A. Corsi, C. Gentili, F. Descalzi, R. Cancedda, and P. Bianco. 1998. Vis-a-vis cells and the priming of bone formation. *Journal of Bone and Mineral Research* 13(12):1852–1861.

Roach, H.I. 1994. Why does bone-matrix contain noncollagenous proteins—The possible roles of osteocalcin, osteonectin, osteopontin and bone sialoprotein in bone mineralization and resorption. *Cell Biology International* 18(6):617–628.

Rohl, C.A., C.E.M. Strauss, K.M.S. Misura, and D. Baker. 2004. Protein structure prediction using Rosetta. *Numerical Computer Methods,* Pt D 383:66–93.

Ross, F.P., J. Chappel, J.I. Alvarez, D. Sander, W.T. Butler, M.C. Farachcarson, K.A. Mintz, P.G. Robey, S.L. Teitelbaum, and D.A. Cheresh. 1993. Interactions between the bone-matrix proteins osteopontin and bone sialoprotein and the osteoclast integrin alpha-V-beta-3 potentiate bone-resorption. *Journal of Biological Chemistry* 268(13):9901–9907.

Sahai, N. 2005. Modeling apatite nucleation in the human body and in the geochemical environment. *American Journal of Science* 305(6–8):661–672.

Salih, E. 2003. In vivo and in vitro phosphorylation regions of bone sialoprotein. *Connective Tissue Research* 44:223–229.

Salih, E., H.Y. Zhou, and M.J. Glimcher. 1996. Phosphorylation of purified bovine bone sialoprotein and osteopontin by protein kinases. *Journal of Biological Chemistry* 271(28):16897–16905.

Sanbonmatsu, K.Y., S. Joseph, and C.S. Tung. 2005. Simulating movement of tRNA into the ribosome during decoding. *Proceedings of the National Academy of Sciences of the United States of America* 102(44):15854–15859.

Schueler-Furman, O., C. Wang, P. Bradley, K. Misura, and D. Baker. 2005. Progress on modeling of protein structures and interactions. *Science* 310(5748):638–642.

Scott, W.R.P., P.H. Hunenberger, I.G. Tironi, A.E. Mark, Billeter, Sr., J. Fennen, A.E. Torda, T. Huber, P. Kruger, and W.F. van Gunsteren. 1999. The GROMOS biomolecular simulation program package. *Journal of Physical Chemistry A* 103(19):3596–3607.

Senn, H.M. and W. Thiel. 2007. QM/MM methods for biological systems. In *Atomistic Approaches in Modern Biology: From Quantum Chemistry to Molecular Simulations*, pp. 173–290.

Shapiro, H.S., J.K. Chen, J.L. Wrana, Q. Zhang, M. Blum, and J. Sodek. 1993. Characterization of porcine bone sialoprotein—Primary structure and cellular expression. *Matrix* 13(6):431–440.

Shaw, D.E., P. Maragakis, K. Lindorff-Larsen, S. Piana, R.O. Dror, M.P. Eastwood, J.A. Bank, J.M. Jumper, J.K. Salmon, Y.B. Shan, and W. Wriggers. 2010. Atomic-level characterization of the structural dynamics of proteins. *Science* 330(6002):341–346.

Shimabayashi, S., C. Tamura, and M. Nakagaki. 1981. Adsorption of hydroxyl ion on hydroxyapatite. *Chemical & Pharmaceutical Bulletin* 29(11):3090–3098.

Somerman, M.J., L.W. Fisher, R.A. Foster, and J.J. Sauk. 1988. Human-bone sialoprotein-I and sialoprotein-Ii enhance fibroblast attachment in vitro. *Calcified Tissue International* 43(1):50–53.

Sorensen, E.S., P. Hojrup, and T.E. Petersen. 1995. Posttranslational modifications of bovine osteopontin—Identification of 28 phosphorylation and 3 O-glycosylation sites. *Protein Science* 4(10):2040–2049.

Sotomayor, M. and K. Schulten. 2007. Single-molecule experiments in vitro and in silico. *Science* 316(5828):1144–1148.

Steinhardt, P.J., D.R. Nelson, and M. Ronchetti. 1983. Bond-orientational order in liquids and glasses. *Physical Review B* 28(2):784–805.

Stewart, J.J.P. 1989. Optimization of parameters for semiempirical methods.1. Method. *Journal of Computational Chemistry* 10(2):209–220.

Stubbs, J.T., K.P. Mintz, E.D. Eanes, D.A. Torchia, and L.W. Fisher. 1997. Characterization of native and recombinant bone sialoprotein: Delineation of the mineral-binding and cell adhesion domains and structural analysis of the RGD domain. *Journal of Bone and Mineral Research* 12(8):1210–1222.

Sugita, Y. and Y. Okamoto. 1999. Replica-exchange molecular dynamics method for protein folding. *Chemical Physics Letters* 314(1–2):141–151.

Swope, W.C., H.C. Andersen, P.H. Berens, and K.R. Wilson. 1982. A computer-simulation method for the calculation of equilibrium-constants for the formation of physical clusters of molecules—Application to small water clusters. *Journal of Chemical Physics* 76(1):637–649.

Taylor, R.D., P.J. Jewsbury, and J.W. Essex. 2002. A review of protein-small molecule docking methods. *Journal of Computer-Aided Molecular Design* 16(3):151–166.

Termine, J.D. and A.S. Posner. 1966a. Infra-red determination of percentage of crystallinity in apatitic calcium phosphates. *Nature* 211(5046):268–270.

Termine, J.D. and A.S. Posner. 1966b. Infrared analysis of rat bone—Age dependency of amorphous and crystalline mineral fractions. *Science* 153(3743):1523–1525.

Tieleman, D.P., M.S.P. Sansom, and H.J.C. Berendsen. 1999. Alamethicin helices in a bilayer and in solution: Molecular dynamics simulations. *Biophysical Journal* 76(1):40–49.

Tobias, D.J., K.C. Tu, and M.L. Klein. 1997. Atomic-scale molecular dynamics simulations of lipid membranes. *Current Opinion in Colloid & Interface Science* 2(1):15–26.

Torrie, G.M. and J.P. Valleau. 1977a. Non-physical sampling distributions in Monte-Carlo free-energy estimation—Umbrella sampling. *Journal of Computational Physics* 23(2):187–199.

Torrie, G.M. and J.P. Valleau. 1977b. Nonphysical sampling distributions in Monte Carlo free-energy estimation: Umbrella sampling. *Journal of Computational Physics* 23(2):187–199.

Tye, C.E., K.R. Rattray, K.J. Warner, J.A.R. Gordon, J. Sodek, G.K. Hunter, and H.A. Goldberg. 2003. Delineation of the hydroxyapatite-nucleating domains of bone sialoprotein. *Journal of Biological Chemistry* 278(10):7949–7955.

Van Cappellen, P., L. Charlet, W. Stumm, and P. Wersin. 1993. A surface complexation model of the carbonate mineral-aqueous solution interface. *Geochimica et Cosmochimica Acta* 57(15):3505–3518.

Voelz, V.A., G.R. Bowman, K. Beauchamp, and V.S. Pande. 2010. Molecular simulation of ab initio protein folding for a millisecond folder NTL9(1–39). *Journal of the American Chemical Society* 132(5):1526–1528.

Warshel, A. 2003. Computer simulations of enzyme catalysis: Methods, progress, and insights. *Annual Review of Biophysics and Biomolecular Structure* 32:425–443.

Warshel, A. and M. Levitt. 1976. Theoretical studies of enzymic reactions—Dielectric, electrostatic and steric stabilization of carbonium-ion in reaction of lysozyme. *Journal of Molecular Biology* 103(2):227–249.

Weiner, S. 2006. Transient precursor strategy in mineral formation of bone. *Bone* 39(3):431–433.

Weiner, S., Y. Talmon, and W. Traub. 1983. Electron-diffraction of mollusk shell organic matrices and their relationship to the mineral phase. *International Journal of Biological Macromolecules* 5(6):325–328.

Weiner, S.J., P.A. Kollman, D.A. Case, U.C. Singh, C. Ghio, G. Alagona, S. Profeta, and P. Weiner. 1984. A new force-field for molecular mechanical simulation of nucleic-acids and proteins. *Journal of the American Chemical Society* 106(3):765–784.

Weiss, I.M., N. Tuross, L. Addadi, and S. Weiner. 2002. Mollusc larval shell formation: Amorphous calcium carbonate is a precursor phase for aragonite. *Journal of Experimental Zoology* 293(5):478–491.

Wlodawer, A. and J. Vondrasek. 1998. Inhibitors of HIV-1 protease: A major success of structure-assisted drug design. *Annual Review of Biophysics and Biomolecular Structure* 27:249–284.

Woolf, T.B. and B. Roux. 1994. Molecular-dynamics simulation of the gramicidin channel in a phospholipid-bilayer. *Proceedings of the National Academy of Sciences of the United States of America* 91(24):11631–11635.

Woolf, T.B. and B. Roux. 1996. Structure, energetics, and dynamics of lipid-protein interactions: A molecular dynamics study of the gramicidin A channel in a DMPC bilayer. *Proteins-Structure Function and Genetics* 24(1):92–114.

Wu, Y., J.L. Ackerman, E.S. Strawich, C. Rey, H.M. Kim, and M.J. Glimcher. 2003. Phosphate ions in bone: Identification of a calcium-organic phosphate complex by P-31 solid-state NMR spectroscopy at early stages of mineralization. *Calcified Tissue International* 72(5):610–626.

Xie, W.S. and J.L. Gao. 2007. Design of a next generation force field: The X-POL potential. *Journal of Chemical Theory and Computation* 3(6):1890–1900.

Xu, Z., Y. Yang, D. Mkhonto, C. Shang, Z.P. Liu, Q. Cui, and N. Sahai. Small molecule-mediated control of hydroxyapatite growth: Free energy calculations benchmarked to density functional theory. *Journal of Physical Chemistry C*. Accepted for publication.

Xu, Z.J., X.N. Yang, and Z. Yang. 2008. On the mechanism of surfactant adsorption on solid surfaces: Free-energy investigations. *Journal of Physical Chemistry B* 112(44):13802–13811.

Yang, X.D., L.J. Wang, Y.L. Qin, Z. Sun, Z.J. Henneman, J. Moradian-Oldak, and G.H. Nancollas. 2010. How amelogenin orchestrates the organization of hierarchical elongated microstructures of apatite. *Journal of Physical Chemistry B* 114(6):2293–2300.

Yang, Y., Q.A. Cui, and N. Sahai. 2010. How does bone sialoprotein promote the nucleation of hydroxyapatite? A molecular dynamics study using model peptides of different conformations. *Langmuir* 26(12):9848–9859.

Yang, Y., D. Mkhonto, Q. Cui, and N. Sahai. 2011. Theoretical study of bone sialoprotein in bone biomineralization. *Cells Tissues Organs* 194(2–4):182–187.

Yu, H.B., T.W. Whitfield, E. Harder, G. Lamoureux, I. Vorobyov, V.M. Anisimov, A.D. MacKerell, and B. Roux. 2010. Simulating monovalent and divalent ions in aqueous solution using a drude polarizable force field. *Journal of Chemical Theory and Computation* 6(3):774–786.

Yu, K.A. and J.R. Schmidt. 2011. Elucidating the crystal face- and hydration-dependent catalytic activity of hydrotalcites in biodiesel production. *Journal of Physical Chemistry C* 115(5):1887–1898.

Zaia, J., R. Boynton, D. Heinegard, and F. Barry. 2001. Posttranslational modifications to human bone sialoprotein determined by mass spectrometry. *Biochemistry* 40(43):12983–12991.

Zhao, Y.L., S. Koppen, and T. Frauenheim. 2011. An SCC-DFTB/MD study of the adsorption of zwitterionic glycine on a geminal hydroxylated silica surface in an explicit water environment. *Journal of Physical Chemistry C* 115(19):9615–9621.

Zhou, H., T. Wu, X. Dong, Q. Wang, and J.W. Shen. 2007. Adsorption mechanism of BMP-7 on hydroxyapatite (001) surfaces. *Biochemical and Biophysical Research Communications* 361(1):91–96.

Zhu, X.L., B. Ganss, H.A. Goldberg, and J. Sodek. 2001. Synthesis and processing of bone sialoproteins during de novo bone formation in vitro. *Biochemistry and Cell Biology—Biochimie et Biologie Cellulaire* 79(6):737–746.

18 Application of enhanced sampling approaches to the early stages of mineralization

Adam F. Wallace

Contents

18.1 INTRODUCTION

18.1.1 GLOBAL SIGNIFICANCE OF $CaCO_3$ BIOMINERALIZATION

The intricate skeletal structures that are formed as the products of controlled biomineralization have long-fascinated researchers and inspired scientific inquiries. Due to the technological limitations of the age, the earliest studies of biominerals aspired only to describe and classify specimens based on morphological considerations (i.e., Haeckel, 1887; Schmidt, 1924); however, with the advent of high-resolution microanalytical techniques and significant advances in the ability to identify and characterize the structure, function, and genetic expression of cellular constituents involved in all aspects of skeletal formation, the focus of contemporary research has begun to shift toward understanding the physical and chemical basis of biomineralization at the microscale.

Of particular interest, perhaps most obviously to those within the materials community who seek biomimetic approaches to construct novel materials for a variety of potential applications, are the molecular-, nano-, and mesoscale processes that underlie the nucleation, growth, and assembly of biomineral phases. However, it is also notable that biominerals, as preserved in the sedimentary rock record, also tell the story of the history of life and environmental change on Earth. Among the most abundant biomineral phases, calcium carbonate holds a special significance in this regard as the trace element and isotope compositions of these phases are commonly used to reconstruct past environmental conditions and comprise one of the primary sources of paleoclimate information in marine and terrestrial settings (Finch et al., 2001; Klein et al., 1996). Such correlations are complicated by well-known *vital effects* that arise as a consequence of differing conditions between the internal mineral deposition sites controlled by the organism and the ambient extracellular environment (Weiner and Dove, 2003). Further, with recognition that many organisms such as the sea urchin utilize transient amorphous phases in the deposition of their skeletal elements (spicules) (Beniash et al., 1997; Gong et al., 2012; Radha et al., 2010) comes the realization that the crystallization pathway itself likely influences the composition of biogenic and authigenic carbonates alike. Indeed, recent evidence suggests that the magnesium content of calcite may be greatly enhanced over traditional mechanisms of crystal growth when crystallization proceeds via an amorphous calcium carbonate (ACC) precursor (Radha et al., 2012; Raz et al., 2003; Wang et al., 2009a; Weiner et al., 2003). Additionally, because the carbonate system also acts as the primary buffer system in natural waters, a robust understanding of the calcium carbonate mineralization pathway is critical not only to aid in the development of novel materials or the interpretation of biomineral-based climate proxy data, but to make informed predictions about the response of the surface ocean to rising atmospheric CO_2 levels (Ridgwell and Zeebe, 2005; Ridgwell et al., 2003; Schuster and Watson, 2007) and the extent to which geological formations can act as carbon storage reservoirs (De Silva and Ranjith, 2012).

18.1.2 SPECIATION IN THE $CaCO_3$ SYSTEM

For the reasons outlined above, the formation and dissolution of calcium carbonate are among the most important chemical processes occurring in natural and engineered environments. Consequently, much work has been invested over the past century into quantifying the phase equilibria of hydrous and anhydrous calcium carbonates (calcite, aragonite, vaterite, monohydrocalcite, ikaite, ACC) and their interactions with aqueous solutions and nonaqueous solvents over a range of temperature, pressure, and atmospheric conditions. In solution, the solubility of calcium carbonate is equal to the sum over the activities (A_i) of all calcium-bearing species:

$$S = \sum A_i = \sum \gamma_i C_i \tag{18.1}$$

where

γ_i is the activity coefficient
C_i is the concentration of the ith species

Although evermore sophisticated activity coefficient models must be applied to account for nonideal interactions between species at higher concentrations (Davies, 1962; Pitzer, 1973, 1991), ion-activity models have proven to be effective tools for quantifying the behavior of carbonate buffer solutions. One of the simplest possible representations of the $CaCO_3$–water system arises from the analytic solution of a linear set of equations that includes Henry's law relationship for $CO_2(g)$ solubility in water,

$$CO_2(g) + H_2O \rightleftarrows H_2CO_3^{\circ} \quad K_H \tag{18.2}$$

the solubility product of $CaCO_3$,

$$CaCO_3(S) \rightleftarrows Ca^{2+} + CO_3^{=} \quad K_{sp} \tag{18.3}$$

the stepwise dissociation of carbonic acid,

$$H_2CO_3^{\circ} \rightleftarrows HCO_3^{-} + H^{+} \quad K_1 \tag{18.4}$$

$$HCO_3^{-} \rightleftarrows CO_3^{=} + H^{+} \quad K_2 \tag{18.5}$$

and the complexation of calcium by bicarbonate and carbonate ions,

$$Ca^{2+} + HCO_3^{-} \rightleftarrows CaHCO_3^{-} \quad K_3 \tag{18.6}$$

$$Ca^{2+} + CO_3^{=} \rightleftarrows CaCO_3^{\circ} \quad K_4 \tag{18.7}$$

Given these basis reactions, the total solubility of calcium carbonate in water is defined by the sum of the Ca^{2+}, $CaHCO_3^{+}$, and $CaCO_3^{\circ}$ species activities and their respective dependencies on the solution pH (as the activity of the hydrogen ion) and carbon dioxide fugacity f_{CO_2} (Figure 18.1).

$$S = A_{Ca^{2+}} + A_{CaHCO_3^{+}} + A_{CaCO_3^{\circ}} = \frac{K_{sp} A_{H^+}^2}{f_{CO_2} K_H K_1 K_2} + \frac{K_{sp} K_3 A_{H^+}}{K_2} + K_{sp} K_4 \tag{18.8}$$

Recent experimental work is challenging the standard treatment of electrolyte solutions as presented here. The detailed titration work of Gebauer et al. (2008) shows that the calcium-ion activity of a solution deviates from the dosed amount of calcium. On its own, this result is completely consistent with traditional speciation models that predict nonideal solution behavior. However, analytical ultracentrifugation (Gebauer et al., 2008; Pouget et al., 2009) and cryo-TEM (Pouget et al., 2009) also suggest that ions are concentrated in clusters that are not

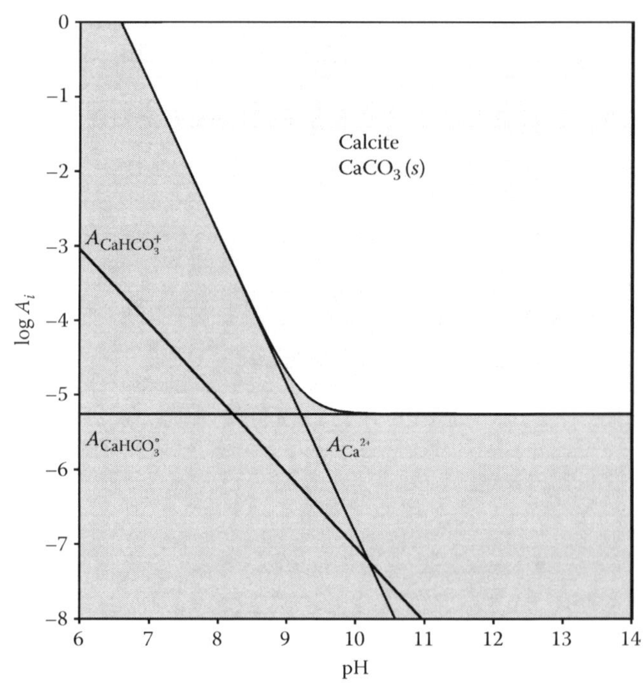

Figure 18.1 Thermodynamic speciation model of the $CaCO_3$–water system showing the activities of ions and ion pairs in equilibrium with calcite at 25°C and $P_{CO_2} = 10^{-3.5}$ atm as a function of the solution pH (see text for additional details). This particular model is as simple as possible but depicts the basic behavior of the system. More sophisticated models account for the presence of additional ion and ion pair species; however, larger ion species and clusters are typically not required. Rather, activity coefficient models are used to account for increasingly nonideal solution behavior with increasing concentration.

predicted by ion-activity models. While the presence of clusters is expected in supersaturated solutions such as those employed in these experiments, the narrow range of sedimentation coefficients has been interpreted as evidence of a long-lived, narrowly distributed, and therefore thermodynamically stable population of *prenucleation* clusters (PNCs). If indeed ions more favorably reside in cluster species than as ion pairs and free ions, perhaps even at equilibrium (Demichelis et al., 2011; Gebauer et al., 2008), then presumably, the formation of calcium carbonate from solution follows a nonclassical nucleation pathway dominated by cluster–cluster aggregation (PNC theory) rather than ion-by-ion addition as assumed by classical theory. However, these data have more recently been interpreted (Discussions, 2012) to represent an average over an unknown distribution of clusters detected over the course of many hours. Moreover, the ultracentrifugation data display sample-to-sample variations larger than the apparent cluster size distributions themselves (Pouget et al., 2009). Further, Pouget et al.'s cryo-TEM results reveal both small (~0.7–1.1 nm) and large clusters (30–250 nm) and demonstrate cluster coalescence. Consequently, while existing data conclusively show that clusters are present in solution, neither the actual cluster size distribution nor the stability of the clusters has been definitively determined.

18.1.3 CLASSICAL DESCRIPTION OF THE MINERALIZATION PATHWAY

Nucleation is a process by which the thermodynamic potential or free energy of a metastable parent medium is reduced by the appearance of a more stable phase. Physically, nucleation occurs as thermally driven fluctuations induce the formation of unstable clusters, which grow ion-by-ion against a free energy barrier until a critical size threshold is surpassed, beyond which the clusters are stable with respect to dissolution and unstable with respect to continued growth of the new phase. The unstable precritical clusters are always present in solution and their size distribution is stable; however, the existence of an individual cluster may be quite brief. The classical nucleation theory (CNT) provides a means of describing the free energy landscape upon which the nascent clusters grow into the bulk phase (Markov, 2004). In the most general sense, the evolution of the free energy along the nucleation pathway follows the sum of two quantities:

$$\Delta G = \Delta G_{bulk} + \Delta G_{interface} \tag{18.9}$$

The first, ΔG_{bulk}, is the free energy change due to the formation of the bulk phase. This term is always negative in a supersaturated system and dominates the behavior of the total free energy function as clusters tend toward larger sizes. The second, $\Delta G_{interface}$, is the free energy associated with the formation of a new interface; it is positive under all circumstances and works opposite ΔG_{bulk} to establish a free energy barrier opposing nucleation at the low end of the cluster size spectrum (Figure 18.2a). The peak height of the nucleation barrier ΔG^* scales with the supersaturation, σ, and the energy per unit area of the newly formed interface, α:

$$\Delta G^* \propto \frac{\alpha^3}{\sigma^2} \tag{18.10}$$

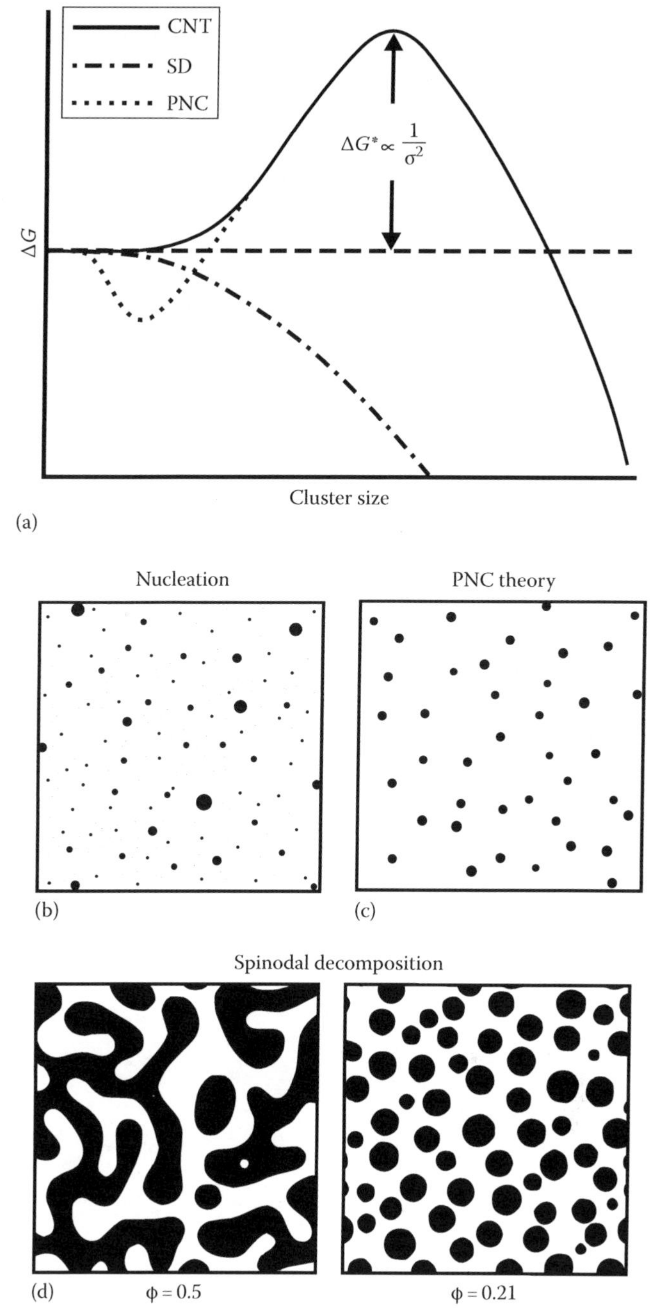

Figure 18.2 Classical and nonclassical mechanisms of phase separation. (a) The free energy as a function of cluster size as described by CNT, SD, and PNC theory. CNT predicts that the height of the free energy barrier opposing nucleation scales with the supersaturation (σ). As σ increases, the barrier decreases and ultimately vanishes. When the barrier goes to zero, the mechanism of phase separation transitions from nucleation to SD. PNC theory supposes rather that the clusters remain trapped in a metastable state prior to aggregation. The other figure panels schematically represent the distribution of cluster sizes for (b) nucleation (exponential size distribution), (c) PNC theory (stationary distribution focused about a mean cluster size), and (d) SD. At high volume fractions (ϕ = 0.5), a percolation texture characteristic of SD develops. However, at low volume fractions (ϕ = 0.21), a relatively monodisperse size distribution develops whose mean size evolves in time as $t^{1/3}$ in the absence of hydrodynamic effects. The observation that clusters exist in supersaturated calcium bicarbonate solutions is consistent with nucleation, SD at low volume fractions, and PNC theory. (Panel D is redrawn after Binder, K. and Fratzl, P.: *Phase Transformations in Materials*. 409–480. 2005. Copyright Wiley-VCH Verlag GmbH & Co. KGaA. Reproduced with permission.)

The CNT assumes that α is constant, which corresponds to a condition where the structure of the developing nucleus is identical to that of the bulk phase. The likely inaccuracy of this simplification is well known and has been recognized since the inception of the theory. More accurate descriptions of the interfacial tension as a function of increasing size are found in the literature (Laaksonen and Mcgraw, 1996; Lu and Jiang, 2004; Magomedov, 2004; Ouyang et al., 2006; Treivus, 2002; Xiong et al., 2011; Zhang et al., 2009); however, regardless of whether α is described with high fidelity over the full range of cluster sizes involved in nucleation is an aside to the most salient aspect of Equation 18.10, which is as follows: as the supersaturation increases, ΔG^* tends toward zero. The supersaturation threshold where the free energy barrier opposing nucleation vanishes is critical because it denotes the boundary, or spinodal line, where the chemical stability limit of the solution is reached and the mechanism of the phase-separation transitions from nucleation to spinodal decomposition (SD). At the spinodal line, small fluctuations that during nucleation generate clusters that are unstable with respect to dissolution, instead produce clusters that are unstable with respect to growth. Furthermore, once the system enters the spinodal regime, the assumption of ion-by-ion growth that is implicit in the classical kinetic treatment of nucleation no longer applies in a strict sense and phase separation may proceed by cluster coalescence as well as single ion addition (Bhimalapuram et al., 2007; Gould and Klein, 1993; Gunton, 1999; Monette and Klein, 1992; Trudu et al., 2006; Yang et al., 2006). At high volume fractions, a case of SD is often diagnosed by the presence of a characteristic percolation texture (Figure 18.2d) (Binder and Fratzl, 2005). However, at small volume fractions, nucleation and SD have very similar topological expressions (Figure 18.2b and d) despite the vast differences between their respective free energy landscapes. Likewise, because the expectation of both PNC theory (Figure 18.2c) and SD is that phase separation will proceed largely by cluster coalescence, these processes may appear similar from an experimentalist's perspective, even though one occurs in the presence of a free energy barrier and the other does not.

18.1.4 ROLE OF THEORY AND SIMULATION

While there are many instances where the mechanism of phase separation is readily apparent, the process at work in $CaCO_3$–water system is not so easily discerned because the most fundamental observation of the system's behavior (i.e., that clusters are present in supersaturated solutions) is not sufficient enough to implicate any particular mechanism. Therefore, knowledge of the free energy landscape is required to determine whether the phase-separation pathway is classical (nucleation or SD) or nonclassical (i.e., PNC theory). Experimental approaches face enormous challenges against directly quantifying the energetics of transient nanoscale cluster species that form during the nucleation process; however, simulation techniques can in principle provide such information.

Nonetheless, the study of rare events is also challenging for theorists. More exact approaches (i.e., electronic structure methods) are too computationally intensive to be tractable for low-frequency events that involve the collective rearrangement of many particles; such processes must be treated with methods that use simplified representations of the particle–particle interactions to reduce computational overhead and extend the simulation accessible timescale. Even classical molecular dynamics (MD), which is the method of choice for studying the chemical evolution of complex systems, probes relatively short periods of time (typically up to ~100 ns in explicitly solvated systems), such that rare events like nucleation and protein folding generally remain significant computational challenges. Longer trajectories can be attained by further simplifying the description of the particle–particle interactions through development of coarse-grained models. While this is achieved at the expense of atomistic detail and chemical specificity in the models, such methods are useful for exploring the general consequences of physical processes.

In practice, there are two commonly used means of trying to overcome the time- and length-scale limitations that are inherent to MD simulations. The brute-force approach, in which as large a simulation as possible is run for as long as possible, is sufficient enough to characterize the nucleation pathway in some instances (i.e., Trudu et al., 2006). However, high concentrations are utilized to enable cluster formation during these simulations, and in the case of low solubility materials like $CaCO_3$ and other biomineral phases, very high supersaturations result that are not comparable with experimental conditions. This is important because as discussed earlier, the mechanism of phase separation depends upon the chemical driving force (σ). Methods that more thoroughly explore the energy landscape than traditional MD can be used as an alternative to the brute-force approach. Collectively, these enhanced sampling protocols comprise a varied set of approaches that are designed to increase the occurrence of rare events during molecular simulations (Dellago and Bolhuis, 2008). The remainder of this chapter is dedicated to these methods but is by no means a comprehensive review of them. Instead, focus is given to those approaches that are most widely used in simulations of biominerals and biomolecules.

18.2 MOLECULAR DYNAMICS

MD simulations rely upon the rapid solution of the classical equations of motion for a many-body system consisting of N particles with coordinates, r, interacting via a potential $U(r^N)$:

$$U\left(r^N\right) = U(r_1, r_2, \ldots, r_N) \tag{18.11}$$

$U(r^N)$ is itself comprised of functions describing the effective interactions among the particles:

$$U\left(r^N\right) = \sum_i \sum_{j>i} u_2(r_i, r_j) + \sum_i \sum_{j>i} \sum_{k>j>i} u_3(r_i, r_j, r_k) + \cdots \tag{18.12}$$

where the individual u_n terms describe the constituent 2, 3, 4,..., n-body interaction potentials. Contributions from a background field may also be included but are omitted here. The fitting and parameterization of intra- and intermolecular potentials is nontrivial and beyond the scope of this chapter. Interested readers are referred to several comprehensive treatments of this topic (Allen and Tildesley, 1987; Frenkel and Smit, 2002; Haile, 1997). The equations of motion can be described in several ways (i.e., Newtonian, Hamiltonian, or Lagrangian formulations of classical mechanics). For a thorough treatment of these, see Allen and Tildesley (1987). Using the formalism of Hamiltonian dynamics, the total Hamiltonian, H, which is a function of both the positions, r^N, and corresponding momenta, P^N, of the particles in the system, is defined for an isolated system (i.e., no exchange of energy with the surroundings) as the sum of the system kinetic and potential energies, $U(P^N)$ and $U(r^N)$:

$$H(r^N, p^N) = U(p^N) + U(r^N) \tag{18.13}$$

The Hamiltonian is a time-invariant quantity, such that Equation 18.13 is always equal to a constant. For an isolated system, the Hamiltonian is equal to the total system energy, E, and it can be shown that the velocity, v_i, and corresponding force, f_i, acting on each particle are (m_i = mass):

$$v_i = \frac{\partial H}{\partial p_i} = \frac{p_i}{m_i} = \dot{r}_i \tag{18.14}$$

$$f_i = -\frac{\partial H}{\partial r_i} = \dot{p}_i \tag{18.15}$$

Taken together, (18.14) and (18.15) are Hamilton's equations of motion. For each particle, N, there are $6N$ such first-order differential equations. Alternatively, there are $3N$ equivalent second-order equations that result from Newton's second law:

$$f_i = m_i\ddot{r}_i \tag{18.16}$$

In an MD simulation, the classical equations of motion are integrated to obtain a phase space trajectory (i.e., positions and velocities/momenta) that can be used to calculate the system's properties within a given statistical mechanical ensemble (usually, the microcanonical (NVE), canonical (NVT), or isothermal–isobaric (NPT) is chosen for MD (V = volume, T = temperature, P = pressure)). In practice, it is most common to integrate Equation 18.16 directly using a finite-differences numerical procedure that preserves its time-reversible character. Provided that a sufficiently small time step is chosen for the integration, standard algorithms (i.e., Verlet) maintain conserved quantities (such as the Hamiltonian or total energy in the earlier discussion) to a high degree of fidelity. Properties are obtained from the simulation by assuming that the ergodic hypothesis applies. That is, given a trajectory of sufficient length, the system will visit every phase point on the constant Hamiltonian surface (Haile, 1997). This original definition of ergodicity has been relaxed considerably in contemporary literature to mean simply that the ensemble average of a given property (i.e., as obtained from a Monte Carlo simulation), $\langle A\rangle_{ens}$, is equal to the time average of that property over the length of the trajectory (Allen and Tildesley, 1987; Frenkel and Smit, 2002; Haile, 1997). Generally, we assume the value of an observable property, A_{obs}, is given by the time average, as compiled over a sufficiently large number of independent observations, τ_{obs}, sampled at discrete points along the phase space trajectory:

$$A_{obs} = \langle A\rangle_{ens} = \frac{1}{\tau_{obs}}\sum_{\tau=1}^{\tau_{obs}} A(\tau) \tag{18.17}$$

18.3 ENHANCED SAMPLING

Although simulation continues to assume evermore prominent roles in understanding the nature of chemical phenomena at small length scales, there are significant limitations that inhibit treatment of many problems. Complications typically occur when the system is restricted to a subregion of the energy landscape over the length of the simulation rather than being free to explore the full extent of the landscape. Take for example Figure 18.3, which displays the energy landscape for the interconversion of an arbitrary system between two states, A and B, that can be distinguished by an order parameter, Q ($Q_A = 0; Q_B = 1$). Because A and B both reside in energy minima, they are long-lived states compared to the intermediate states that lie between them. Therefore, prior to the onset of any modeling effort, pertinent structure information for states A and B is most likely to be available. Then, it is also probable that simulated trajectories will originate from initial conditions that reside within the corresponding energy wells for either A or B. Provided that the barrier between A and B is low, a standard MD simulation may sample all states frequently, and a time average of over the computed trajectories may accurately represent the properties of the system. However, as the height of the energy barrier becomes large compared to the ambient thermal energy $\beta = (k_BT)^{-1}$, it becomes less likely that trajectories will surmount the barrier, and the system can become confined to one side or the other for the full length of the simulation. Moreover, the higher energy states lying between the wells also become underrepresented in the trajectories such that the time averages computed from them only reflect the properties of the system when it is in either A or B. This is not concerning at all if the objective is to characterize A or B alone; however, it is most certainly problematic if the objective is to observe the evolution of the system from A to B and vice versa.

There are currently a large number of methods that enable a more thorough exploration of the energy landscape than standard MD. Some approaches simply seek to accelerate MD through the use of multiple time step algorithms (Marchi and Procacci, 1998; Schlick, 2001) or by improving the performance of other aspects of the general simulation protocol (i.e., treatment of long-range forces [Schlick et al., 1999]). However, enhanced sampling methods take the alternative approach of accelerating the rate at which the system traverses the energy landscape itself. In the remainder of this section, three methods are discussed

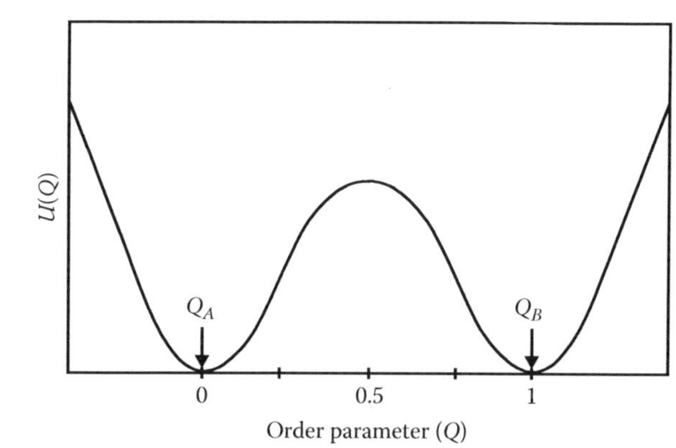

Figure 18.3 Fictitious energy landscape between two states, A and B, that can be distinguished by an order parameter Q.

that have each provided insight into the early stages of calcium carbonate formation and that will provide the reader with a firm introduction to enhanced sampling techniques.

18.3.1 UMBRELLA SAMPLING

Umbrella sampling determines the potential of mean force (PMF) or equivalently the free energy (Helmholtz) along a specified reaction coordinate, ξ. The full energy landscape along ξ is sampled by segmenting the reaction coordinate into discrete windows that are simulated independently (Torrie and Valleau, 1974). In each window, a restraining potential, $\omega_i(\xi)$ is applied so that the simulation does not drift away from the small segment of the reaction coordinate it is intended to sample. The restraining potential may take several forms; however, the harmonic variety is usually employed:

$$\omega_i(\xi) = \frac{1}{2}K(\xi - \xi_i)^2 \tag{18.18}$$

where

K is a force constant

ξ_i is the value of the reaction coordinate that defines the center of the umbrella sampling window

However, while the use of the restraining potential enables the entire landscape to be sampled, it also augments the description of the system's potential energy, $U(r^N)$:

$$U^b(r^N) = U(r^N) + \omega_i(\xi) \tag{18.19}$$

Since each window samples this biased potential function, $U^b(r^N)$, rather than the true potential, the resulting probability distributions, $P_i^b(\xi)$, also contain a bias that must be removed in order to obtain the natural unbiased free energy landscape. For a given window, the unbiased distribution, $P_i(\xi)$, and hence the free energy, $A_i(\xi)$, is given by the following:

$$A_i(\xi) = -\beta^{-1}\ln[P_i(\xi)] = -\beta^{-1}\ln[P_i^b(\xi)] - \omega_i(\xi) + F_i \tag{18.20}$$

where F_i is a constant that defines the position of the free energy segment on the energy scale relative to all the $A_i(\xi)$ obtained from other windows. Unfortunately, the F_i values cannot be obtained directly from the simulation results; however, it is possible to estimate them. To that end, the weighted histogram analysis method (WHAM) is perhaps the most widely utilized approach in the literature (Rosenbergl, 1992; Souaille and Roux, 2001). The WHAM equations (shown here for a 1D reaction coordinate although extension to multidimensional coordinates is straightforward) operate to minimize the total statistical error in the total unbiased probability distribution, $P(\xi)$, which is described as the weighted sum of the individual unbiased distributions from N_{win} umbrella sampling windows:

$$P(\xi) = \sum_{i=1}^{N_{win}} n_i P(\xi)_i \times \left[\sum_{j=1}^{N_{win}} n_j e^{-\beta(\omega_j(\xi) - F_i)}\right]^{-1} \tag{18.21}$$

where n_i is the total number of data points used to compile the distributions. The free energy constants may then be determined from the best estimate of the unbiased probability distribution:

$$e^{-\beta F_i} = \int P(\xi) e^{-\beta(\omega_i(\xi))} d\xi \tag{18.22}$$

To obtain a good estimate for the total probability distribution, it is critical that there be adequate overlap between the distributions of neighboring windows. Since $P(\xi)$ and F_i appear in both (18.21) and (18.22), the WHAM equations must be solved iteratively. However, with all the F_i in hand, the free energy along the full reaction coordinate, $A(\xi)$, may be obtained by stitching together the previously obtained free energy segments from all of the individual windows.

18.3.1.1 Comparison of umbrella sampling and brute force

As a demonstration of how umbrella sampling works in practice, consider the process of ligand exchange about a single calcium ion in solution. The distribution of water molecules about the ion is described by the radial pair distribution function (PDF), $g(r_{Ca-O_w})$, computed between the ion and the oxygen atom, O_w, of each surrounding solvent molecule. $g(r_{Ca-O_w})$ is easily obtained from a standard MD simulation and shows two prominent peaks that denote the location of concentric hydration shells about the ion (Figure 18.4a). As it turns out, the rate of water exchange within the first hydration shell of the calcium ion is sufficiently fast (an exchange per ~100 ps) that many exchange events can be observed during a standard simulation run. If enough exchanges occur over the length the computed trajectory, then it is even possible to construct the full energy landscape for the solvent exchange without using umbrella sampling at all. Since $g(r_{Ca-O_w})$ is fundamentally a probability distribution over all the possible states of the system, it can be directly related to the free energy:

$$A(r_{Ca-O_w}) = -\beta^{-1}\ln g(r_{Ca-O_w}) \tag{18.23}$$

where r is both the Ca–O_w distance and the reaction coordinate. However, even a problem such as this that can be treated by standard MD may require computation of long trajectories in order to acquire enough observations of the system in the low-density region between the first and second solvent shells to construct a reliable depiction of the energy landscape. Figure 18.4b shows the free energy as determined from $g(r_{Ca-O_w})$ compiled at several different simulation times. As shown on the plot, after several nanoseconds of simulation, the high energy portion of the landscape remains rough due to the low number of observations of the system in those states.

While it is not necessary in this case, the energy landscape can also be determined by umbrella sampling. Here, the reaction coordinate, r_{Ca-O_w}, is divided into several windows that are centered 0.25 Å apart over the interval 2.0 Å $\le r_{Ca-O_w} \le$ 6.0 Å. Prior to data collection, each window is equilibrated in the presence of a harmonic restraining potential, $\omega_i(r_{Ca-O_w})$. After equilibration, the biased system is sampled for 0.5 ns in each window. From the computed trajectories, the unbiased probability distributions, $P_i(r_{Ca-O_w})$, are obtained by subtracting

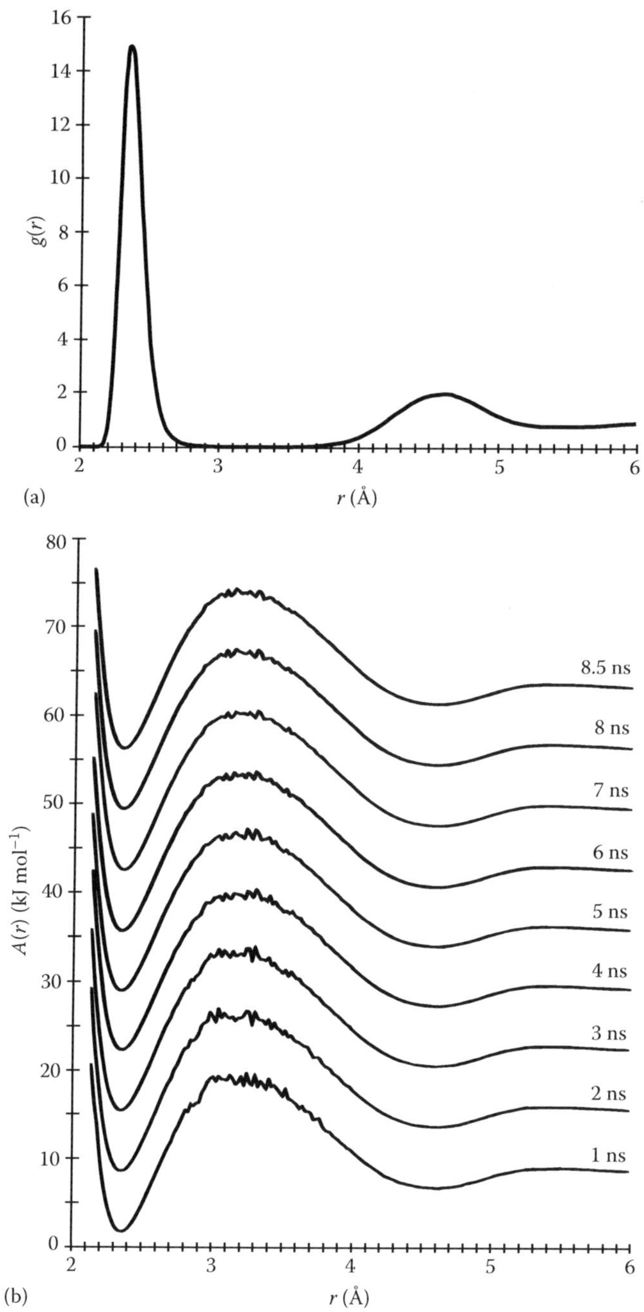

Figure 18.4 (a) The radial PDF computed between a calcium ion and the oxygen atoms belonging to surrounding water molecules in solution. (b) The free energy landscape for the exchange of a water molecule from within the first hydration shell as determined directly from the pair distribution (Equation 18.23) function as compiled over various simulation lengths. The region of the landscape between 3 and 4 Å remains rough even after 8.5 ns of simulation time because it is rare to observe the system occupying higher energy states.

the restraining potential from the biased distributions, $P_i^b(r_{Ca-O_w})$. For each of the unbiased distributions, a small segment of the free energy landscape, $A_i(r_{Ca-O_w})$, is acquired as in Equation 18.20. These free energy curves are displayed in the top portion of Figure 18.5, along with the corresponding restraining potentials and biased/unbiased probability distributions. In order to combine the small free energy segments into a single energy landscape, $A(r_{Ca-O_w})$, the free energy associated with introducing the restraining potential, F_i, must be computed for each window. These constants are estimated in a straightforward fashion by iteratively solving the WHAM equations as described earlier. As is shown in the lower portion of Figure 18.5, the full energy landscape is constructed by summing each of the F_i into the $A_i(r_{Ca-O_w})$ from their respective windows and discarding the overlapping portions of the free energy segments. A correction (the Jacobian [Henin and Chipot, 2004]) is also typically applied to the PMF between species in solution to ensure that the PMF tends towards zero at large r. This is readily done by subtracting $2k_bT\ln(r/r_{eq})$ from the computed PMF, however, for the sake of presentation this correction has been omitted from the PMFs presented in this chapter. Note that in this particular example, not much is gained by umbrella sampling because the brute-force approach is adequate; however, had the rate of solvent exchange been slower as it is about iron or magnesium (exchange rate on the order of microseconds), the free energy landscape could not have been obtained without using an enhanced sampling approach like umbrella sampling.

18.3.1.2 Application to complex processes

Although the example presented earlier focuses on a relatively simple process, umbrella sampling can in principle be used to explore complex reactions. In the case of solvent exchange about an ion, the use of the ion–water separation distance is a fairly obvious choice of reaction coordinate, although it may not be the most ideal option possible. Indeed, the validity of the result depends largely on whether the chosen reaction coordinate is a reasonably accurate approximation of the actual reaction coordinate. For a process such as nucleation, the reaction coordinate is not so easily surmisable because many degrees of freedom are potentially important. One solution that has emerged in the literature is based on the assumption that during nucleation, large clusters are rare (Allen et al., 2006; Auer and Frenkel, 2004; Filion et al., 2010; Leyssale et al., 2007). Then, the restraining potential can be defined so that each umbrella sampling window is centered on a certain cluster size. Defining the windows in this way enables the free energy to be determined as a function of cluster size so that the simulation results can be directly compared to the expectations of CNT. Alternatively, a local order parameter may be used to drive the transition between two or more states (see Section 18.3.2.1); however, in this case, the choice of order parameter limits which states the system can visit, and that may be undesirable if the system is host to polymorphic phases (Quigley and Rodger, 2009). This approach is popular for understanding nucleation in systems composed of hard spheres (Auer and Frenkel, 2004; Filion et al., 2011; Wang et al., 2009b) and Lennard–Jones particles (Ten Wolde et al., 1996). However, even in these simple systems, very long equilibration times are required in each window, a condition that has so far inhibited the application of umbrella sampling to the problem of calcium carbonate nucleation.

18.3.1.3 Calculation of stability constants for aqueous complexes in the $CaCO_3$–H_2O system

Umbrella sampling is commonly used to obtain the PMF between calcium and carbonate/bicarbonate ions in solution and to characterize the free energy landscape for the attachment of these ions to calcite surfaces and disordered carbonate clusters

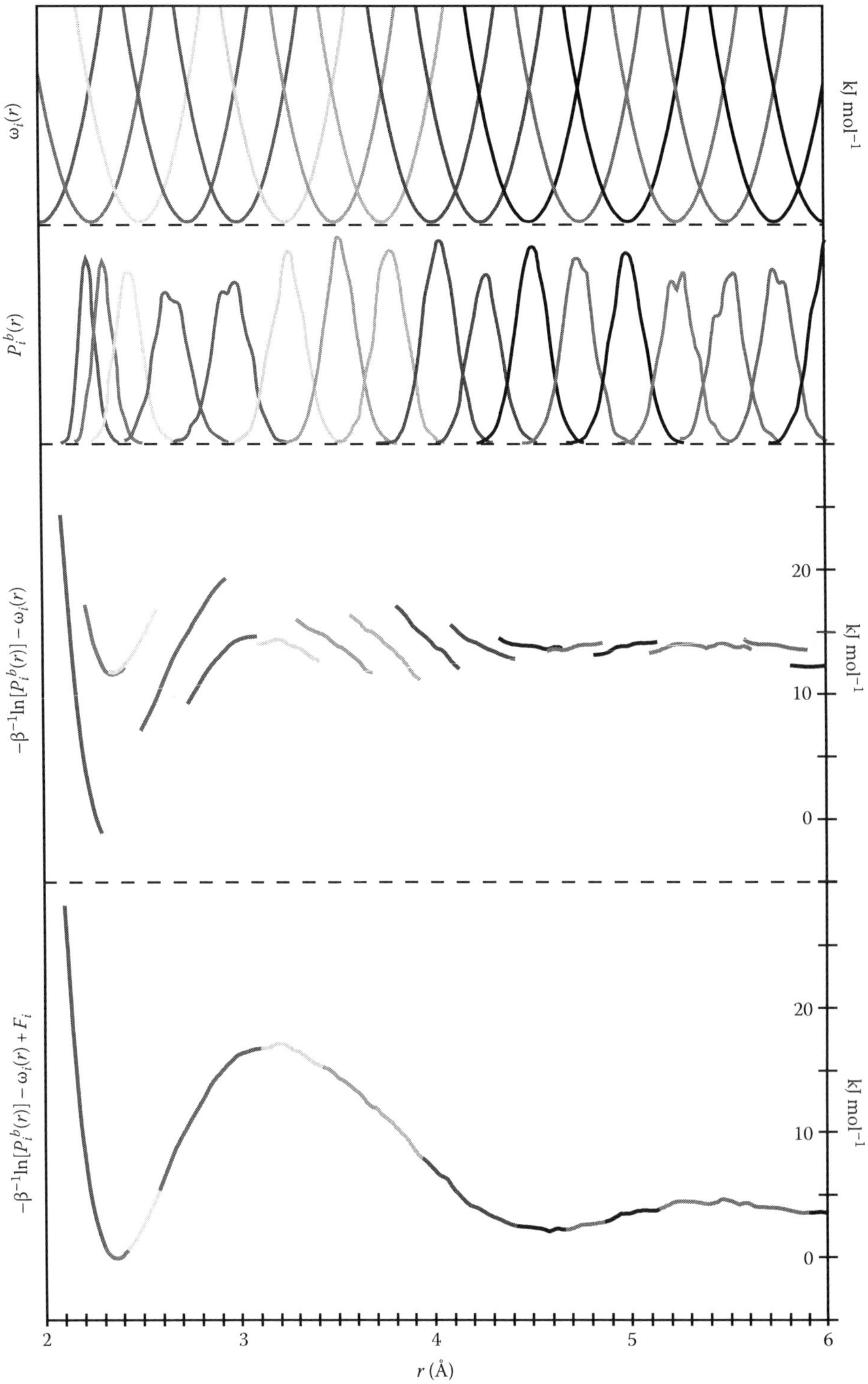

Figure 18.5 (See color insert.) Construction of the free energy landscape for solvent exchange about calcium as determined by umbrella sampling without the Jacobian correction. The topmost portion shows the individual restraining potentials $\omega_i(r)$ and the biased probability distributions $P_i^b(r)$ that result from running an 0.5 ns MD simulation in each of the windows. Respectively, the bottom two panels show the individual free energy segments that are computed for each window and the assembly of those segments into the full landscape using the WHAM.

(Raiteri and Gale, 2010; Tribello et al., 2009a). The free energy of association between the ions in solution is of particular importance because it can be used to determine whether the simulated ion–ion interactions compare well with standard thermodynamic speciation models (see Section 18.1.2). This is accomplished by integrating either the radial PDF or the PMF, $W(r)$, to obtain equilibrium constants for reactions such as (18.6) and (18.7) earlier (Matthews and Naidoo, 2010):

$$K_{eq} = 4\pi R \int_0^{r_{max}} g(r) r^2 dr = 4\pi R \int_{r_{min}}^{r_{max}} e^{\left[\frac{-W(r)+W(r_{max})}{k_b T}\right]} r^2 dr \quad (18.24)$$

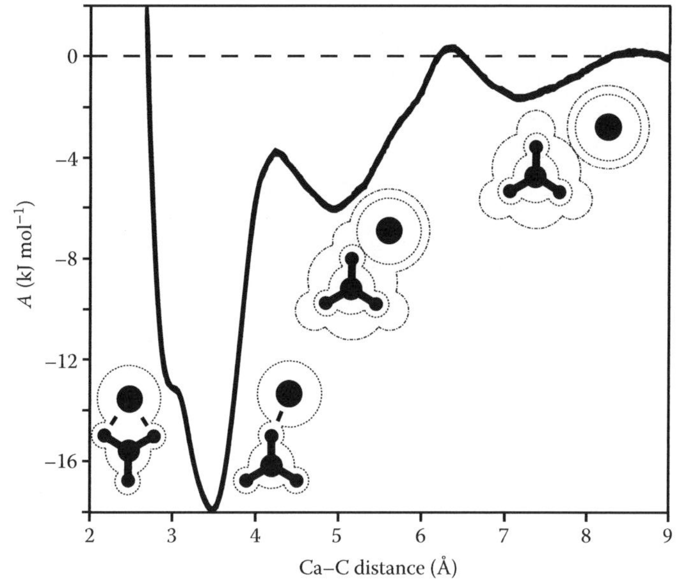

Figure 18.6 Free energy landscape between calcium and carbonate ions in solution as determined by umbrella sampling along the Ca–C separation distance without the Jacobian correction. While none of the features are significant compared to k_bT, there are several local features corresponding to various ion pair species. The CIPs are the most stable, followed by the solvent-shared and solvent-separated species at ~5 and 7 Å, respectively. In the inset cartoon representations, the dashed lines represent solvation shells surrounding the ions. Integration of the landscape provides the equilibrium constant for the $CaCO_3^\circ$ ion pair (as described in the text), which can be used to benchmark atomistic models against thermodynamic speciation models as in Figure 18.2.

where $R = 6.0221 \times 10^{-4}$ L mol^{-1} Å^{-3} mol^{-1} is the conversion factor from cubic angstroms per molecule to liters per mole, r_{min} is the minimum separation distance on the free energy landscape, and $W(r_{max})$ is the ion–ion coulomb interaction energy at the maximum separation distance in the computed free energy profile. The energy scale is also set to zero at r_{max} prior to integration. Figure 18.6 shows the PMF for the formation of a $CaCO_3^\circ$ ion pair in solution. The landscape itself is quite interesting because it shows that the ions are loosely associated at separation distances upward of ~7 Å. The global energy minimum corresponds to the Ca–C distance where the carbonate ion complexes calcium as a monodentate contact ion pair (CIP). The local minimum at a somewhat shorter separation distance represents the bidentate CIP. At larger separations, there are two additional local energy minima corresponding to the presence of solvent-shared ion pair (SSHIP) and solvent-separated ion pair (SSIP) species. It is also notable that the free energy barriers opposing transitions between the loosely associated states are all on the order of k_bT such that there is no significant thermodynamic force opposing the formation of CIPs. Likewise, umbrella sampling simulations of ion addition to disordered carbonate clusters show that the PMF for association of an ion with a cluster is similar, which suggests that the growth of a phase such as ACC is transport limited. Comparable simulations of ion addition to the calcite (104) surface show conversely that calcite growth is reaction limited due to the presence of a sizable free energy barrier opposing ion attachment (Tribello et al., 2009a); this barrier is apparently associated with the unfavorable perturbation of well-ordered water layers at the crystal surface that do not form in the association with amorphous carbonate clusters (Demichelis et al., 2011; Raiteri and Gale, 2010).

18.3.2 METADYNAMICS

Like umbrella sampling, metadynamics provides information about the free energy landscape by applying a bias to the system's potential energy. However, instead of augmenting the potential function to restrain the system to a given region of the energy landscape, metadynamics employs a repulsive bias that discourages the system from revisiting previously occupied states (Ensing et al., 2006; Iannuzzi et al., 2003; Laio and Gervasio, 2008; Laio and Parrinello, 2002; Laio et al., 2005; Quigley and Rodger, 2009). The bias potential, $V(s(r^N), t)$, is a function of time and one or more *collective variables*, $s_i(r)$, that can distinguish between the reactant and product states. $V(s(r^N), t)$ is built up during the otherwise normal MD simulation by periodically adding a Gaussian potential or *hill* centered about the current value(s) of the collective variable(s). The total value of the potential at a given time is taken as the sum over all the individual Gaussian hills:

$$V(s,t) = h_G \sum_{k=1}^{N_G} e^{\left[\frac{-|s(k\tau_G)-s(t)|^2}{2\delta w_G^2}\right]} \tag{18.25}$$

where

τ_G is the frequency of Gaussian addition

h_G and δw_G are the respective height and width of the Gaussian functions

Over time, the hills are deposited in the lowest lying accessible regions of the energy landscape; as the basins reach capacity, the bias potential drives the system into new regions of configuration space that are poorly sampled by brute-force approaches. Moreover, if the bias potential is built up over a long period of time, it accurately compensates the underlying energy landscape by filling in the potential wells and provides an estimate of the free energy, $F(s)$:

$$\lim_{t\to\infty} V(s,t) \sim F(s) \quad (18.26)$$

Figure 18.7 demonstrates how the metadynamics bias potential both accelerates the rate at which the underlying potential is traversed and provides a high-fidelity representation of the energy landscape.

18.3.2.1 Extended system and direct metadynamics

In the original implementation of the continuous metadynamics algorithm (Iannuzzi et al., 2003; Laio and Parrinello, 2002), the bias is applied to a set of additional auxiliary degrees of freedom, $\tilde{s}_i$, that provide an adequate coarse-grained representation of the system. For every $\tilde{s}_i$, there is also a collective variable $s_i(r)$, which is an explicit function of the particle positions in the system. The time evolution of all the degrees of freedom is described by an extended system Hamiltonian:

$$H = H_0 + \sum_i \frac{1}{2} M_i \dot{\tilde{s}}_i^2 + \sum_i \frac{1}{2} k\left(\tilde{s}_i - s_i(r)\right)^2 + V(\tilde{s}(r^N),t) \quad (18.27)$$

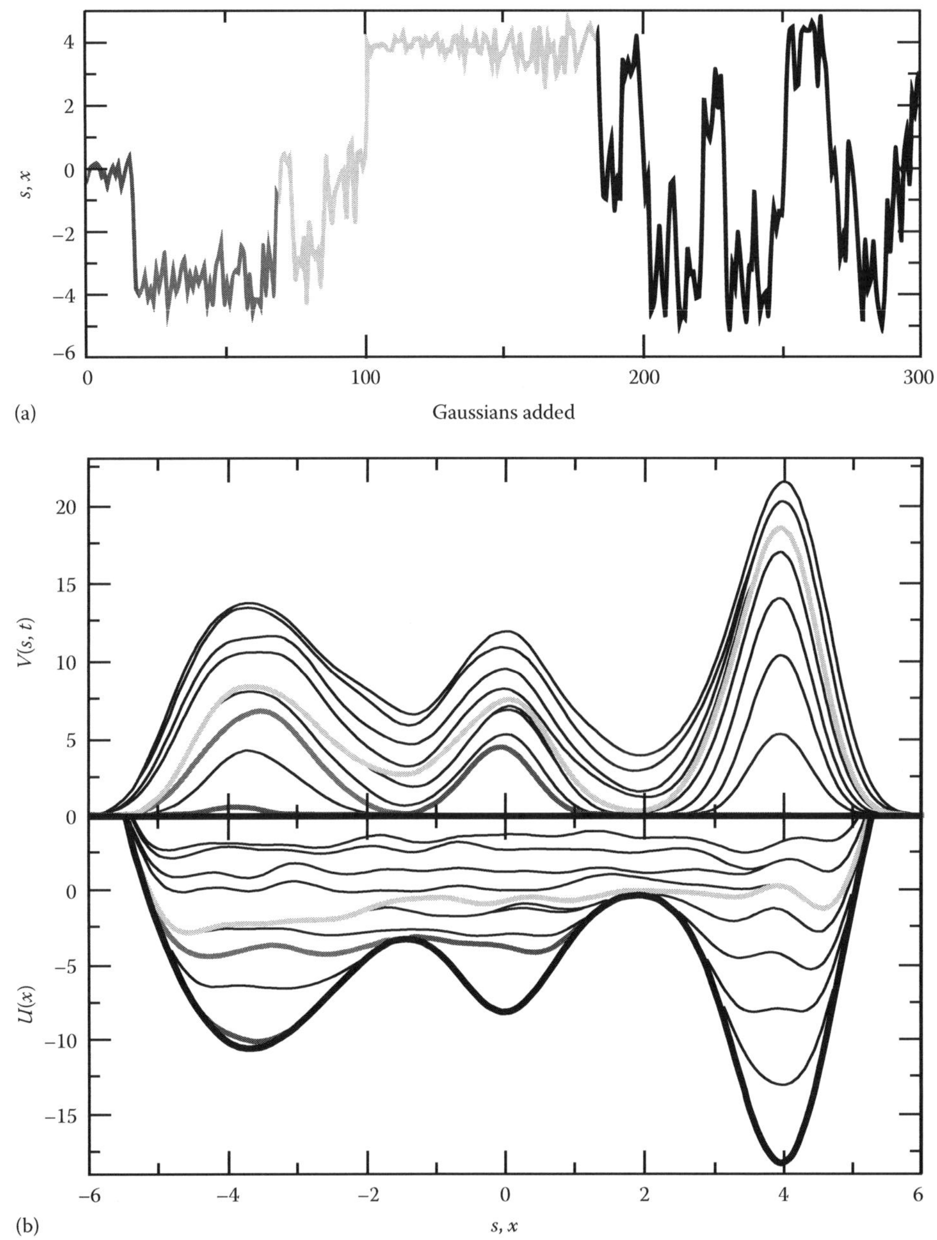

Figure 18.7 (See color insert.) Compensation of the energy landscape, $U(x)$, by the history-dependent metadynamics bias potential $V(s, t)$. In the direct variant of metadynamics, $s = x$. (a) Evolution of the collective variable, s, throughout the simulation. (b) The upper panel shows the bias potential at several intermediate states during the simulation. The lower panel shows the current estimate of the energy landscape at each intermediate time. (Reproduced from Laio, A. and Gervasio, F.L., *Rep. Prog. Phys.*, 71(12), 126601, doi:10.1088/0034-4885/71/12/126601, 2008. With permission from IOP.)

where the first term on the right is the standard system Hamiltonian as in Equation 18.13, the second is the fictitious kinetic energy ascribed to the auxiliary degrees of freedom, and the third is a harmonic potential that couples each $s_i(r)$ to its respective $\tilde{s}_i$. Provided that the mass, M_i, attributed to each auxiliary variable is sufficiently large, the dynamics of the $\tilde{s}_i$ remain adiabatically separated from the other degrees of freedom in the system. In more straightforward terms, all the coordinates relax to their equilibrium distributions before the $\tilde{s}_i$ evolve significantly.

It is now known that the success of the metadynamics approach does not strictly require the use of auxiliary variables (Laio and Gervasio, 2008; Quigley and Rodger, 2009). In the limit that the spring constants harmonically coupling the $\tilde{s}_i$ with their respective $s_i(r)$ are infinitely large, the auxiliary variables become equivalent to their collective coordinate counterparts (i.e., $\tilde{s}_i = s_i(r)$), and the system Hamiltonian collapses to

$$H = H_0 + V(s(r^N), t) \tag{18.28}$$

such that the force on each particle is directly influenced by the bias potential (Quigley and Rodger, 2009; Quigley et al., 2011):

$$f_i = -\nabla_{r_i} U(r^N) - \sum_{j=1}^{M} \frac{\delta V}{\delta s_j} \nabla_{r_i} s_j(r) \tag{18.29}$$

As it turns out, it is most critical that the bias potential is built up slowly enough that the collective variables can diffuse to the current energy minimum before the potential is modified. This condition can be achieved by ensuring that the rate of Gaussian addition to the bias potential is slow. This so-called *direct* form of the continuous metadynamics algorithm is now widely applied in models that rely upon the classical description of particle dynamics.

18.3.2.2 Onset of order in $CaCO_3$ clusters and nanoparticles

Perhaps one of the most critical aspects of metadynamics simulations is the choice of collective variables. There are many options (distances, angles, dihedral angles, coordination numbers, and more), and metadynamics is readily applied to multiple collective variables at once. However, the efficiency of the algorithm decreases exponentially as the number of collective variables is increased so it is most practical to use as few variables as are necessary to uniquely define all the states in the system. In metadynamics simulations of nanoscopic $CaCO_3$, the set of collective variables typically includes one or more smoothly varying bond-order parameters of the Steinhardt variety (Quigley and Rodger, 2008; Quigley et al., 2011):

$$Q_l^{\alpha\beta} = \left[\frac{4\pi}{2l+l} \sum_{m=-l}^{l} \left| \frac{1}{N_c N_\alpha} \bar{Q}_{lm}^{\alpha\beta} \right|^2 \right]^{1/2} \tag{18.30}$$

where

$$\bar{Q}_{lm}^{\alpha\beta} = \sum_{b=1}^{N_b} f(r_b) Y_{lm}(\theta_b, \phi_b) \tag{18.31}$$

In this description, the spherical harmonic functions $Y_{lm}(\theta_b, \phi_b)$ are summed over all vector distances, r_b, between atom types α and β (i.e., Ca–Ca, Ca–C, Ca–O, C–C, C–O), with the polar angles of each vector being determined with respect to an arbitrarily chosen reference axis. Furthermore, the contribution of a given spherical harmonic is forced to zero over a certain distance by a tapering function, $f(r_b)$, so that the order parameters describe the local rather than the global state of the system (for a detailed description of the usual tapering function and the effect of its parameterization on the simulation results, refer to Quigley et al. [2011]). In addition, the local potential energy (i.e., bond, angle, van der Waals, and short-range electrostatic energies) of all atoms involved in these local bond-order parameters is used as a collective variable.

To date, there are several metadynamics studies of calcium carbonate nanoparticles that share the general methodology outlined earlier (Darkins et al., 2013; Freeman et al., 2010; Quigley and Rodger, 2008; Quigley et al., 2009, 2011). The common objective of almost all of these studies is to determine which phase of $CaCO_3$ is thermodynamically preferred at a given nanoparticle size. The initial findings of Quigley and Rodger, performed at constant density (Quigley and Rodger, 2008), display two free energy minima for a ~2 nm diameter particle (75 $CaCO_3$ units) that correspond to separate disordered and partially ordered states, with the disordered state being overwhelmingly preferred. Follow-up work shows that for larger particles (192 and 300 formula units), the landscape is possibly reversed, with the amorphous state being metastable with respect to both vaterite- and calcite-like states (Freeman et al., 2010). Additionally, the thermodynamic barriers associated with transforming between the disordered and semiordered states are on the order of several hundred times k_bT, suggesting that the reordering of the particles in the solid state is unlikely. However, in the most recent work (Quigley et al., 2011), these authors show that at constant pressure rather than constant density (i.e., *NPT* versus *NVT*), this trend also reverses for the 75 formula unit cluster and the disordered state is predicted as metastable with respect to a partially ordered state where more than 25% of the ions occupy calcite-like environments (Figure 18.8). Additionally, in the NPT simulations, the barrier between the disordered and partially ordered states is substantially reduced. It is suggested that the enhanced stability of the amorphous phase at constant density arises from spatial confinement of the particle and solvent, such as is observed experimentally (Loste and Meldrum, 2001; Loste et al., 2004; Stephens et al., 2010; Yashina et al., 2012). However, it is well known that synthetic and biogenic ACC contains a certain amount of water that may likely influence the stability of these phases; to date, only anhydrous $CaCO_3$ particles have been investigated with metadynamics. It is also important to consider whether the underlying potential model predicts the proper relative thermodynamic relationships between the bulk phases of calcium carbonate and accurate ion-hydration properties (Raiteri et al., 2010). The recent potentials developed by Raiteri and Gale (Demichelis et al., 2011; Raiteri and Gale, 2010; Raiteri et al., 2010) address these issues, and results utilizing these models suggest that disordered states are thermodynamically preferred up to ~4 nm in diameter (especially when the effects of water incorporation into the bulk structure are accounted for). Nonetheless, the metadynamics

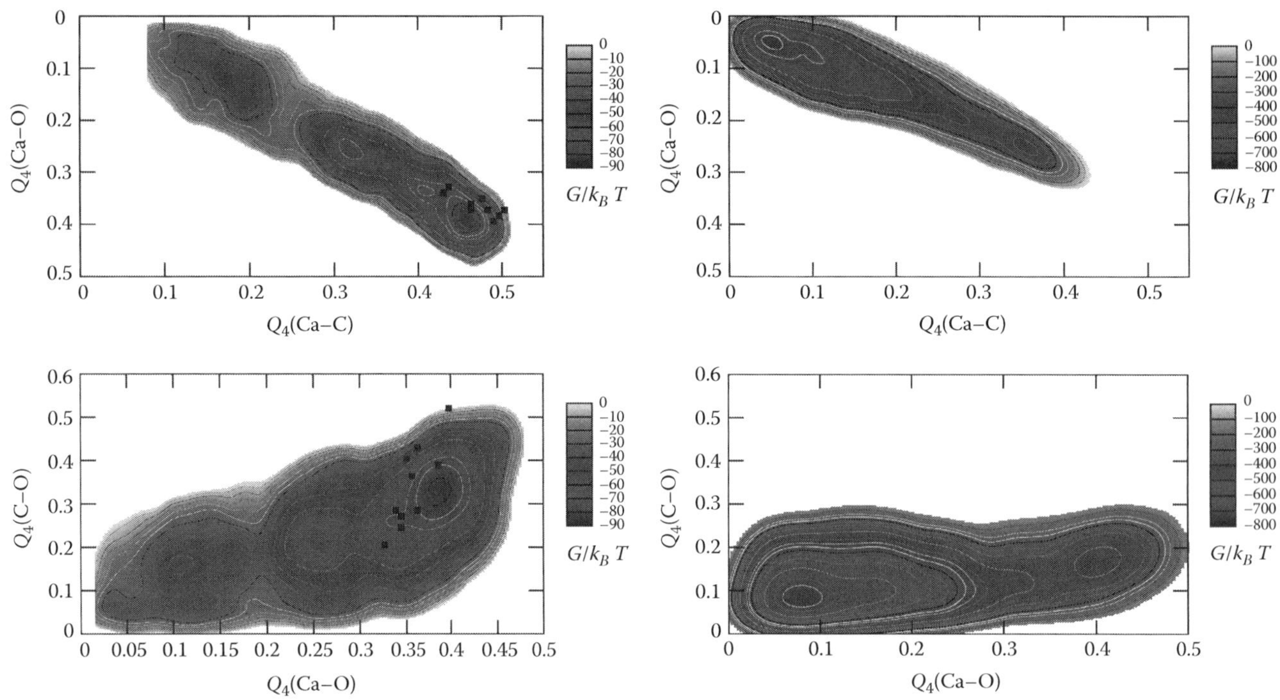

Figure 18.8 The metadynamics-derived free energy landscape of a 75 formula unit $CaCO_3$ nanoparticle projected with respect to the Steinhardt-type order parameters (Q_4(Ca–O), Q_4(Ca–C) and Q_4(C–O), Q_4(Ca–O)). The left side of the figure displays results obtained at constant pressure (*NPT*), while the right side shows the constant density result. The black dots on the landscape represent points where >25% (but not more than 35%) of the calcium ions in the particle occupy calcite-like environments. (Reprinted with permission from Quigley, D., Freeman, C.L., Harding, J.H., and Rodger, P.M., Sampling the structure of calcium carbonate nanoparticles with metadynamics, *J. Chem. Phys.*, 134, 044703, doi:10.1063/1.3530288, 2011. Copyright 2011, American Institute of Physics.)

approach is a powerful methodology that is also shedding light on the influence of organic interfaces on the crystallization of calcium carbonate. Notably, Freeman et al. (Quigley et al., 2009) show that metadynamics can predict the experimentally observed orientation of calcite crystals on carboxyl-terminated self-assembled monolayers and that the crystallization mechanism is not strictly a matter of lattice epitaxy but rather involves the concomitant restructuring of both the monolayer and amorphous phase throughout the phase transition. Freeman et al. are also applying metadynamics to probe the interactions between the eggshell protein OC-17 and anhydrous ACC (Freeman et al., 2010, 2011, 2012) and have shown in this particular instance that binding to the protein likely destabilizes the amorphous phase and accelerates the formation of calcite (Freeman et al., 2010).

18.3.3 PARALLEL TEMPERING AND REPLICA-EXCHANGE MOLECULAR DYNAMICS

In the previous sections, two methods are presented (umbrella sampling and metadynamics) by which passage of the system through bottlenecks in the free energy landscape is facilitated by the application of a suitable bias. Replica-exchange molecular dynamics (REMD) (Sugita and Okamoto, 1999), which is based on the parallel tempering Monte Carlo sampling scheme (Earl and Deem, 2005; Freeman, 2000; Frenkel and Smit, 2002), is a fundamentally different approach whereby exploration of the energy landscape is enhanced, in principle, without introducing a bias to the system. In standard parallel tempering, several noninteracting copies (or replicas) of the system are all initiated at once and run in parallel, with each replica in the series equilibrated at progressively higher-temperature conditions or with different Hamiltonians, such that there is an overlap in the potential energy distributions of the neighboring replicas. Periodically, during the simulation, the potential energies of the various replicas are compared; when a given replica finds a state that is lower in energy than its neighbor, the two systems are allowed to swap states according to a conditional probability rule that preserves proper sampling of the Boltzmann distribution and upholds the principle of detailed balance (microscopic reversibility). In the canonical ensemble (*NVT*), the probability of swapping between two replicas *i* and *j* is most generally given by the following Metropolis criterion:

$$P(i \to j) = \min \begin{cases} 1, & \text{for } \Delta \le 0 \\ e^{-\Delta} & \text{for } \Delta > 0 \end{cases} \tag{18.32}$$

where

$$\Delta = \beta_i[U_i(x_j) - U_i(x_i)] + \beta_j[U_j(x_i) - U_j(x_j)] \tag{18.33}$$

where

β_i represents the thermodynamic temperature of replica i
$U_i(x_j)$ is the value of the potential energy function of replica i operating on the current configuration, x_j, of replica j

For variants of parallel tempering that enhance the sampling of the system by differentiating between replicas on the basis of their respective Hamiltonians rather than their temperatures, it is typical for all replicas to be thermostated at the same value (i.e., $\beta_i = \beta_j$) and Δ becomes

$$\Delta = \beta[U_i(x_j) - U_i(x_i) + U_j(x_i) - U_j(x_j)] \tag{18.34}$$

However, for conventional parallel tempering, the potential energies of all the replicas are evaluated with the same Hamiltonian (i.e., $U_i = U_j$), and Δ takes the following form:

$$\Delta = (\beta_i - \beta_j)[U(x_j) - U(x_i)] \tag{18.35}$$

With the convention that $i < j$, if the potential energy of configuration x_j is less than or equal to that of configuration x_i, the swap is always accepted. However, in the event that $U(x_j) > U(x_i)$, the swap may still be accepted provided that $e^{-\Delta}$ is greater than a random number generated on the interval [0, 1]. This procedure ensures that the simulation does not get trapped in a local energy minimum and that the lowest energy configurations are continually promoted toward lower temperature replicas as they are encountered. In practice, swaps are usually only attempted between neighboring replicas because the probability of accepting a swap between nonadjacent pairs diminishes rapidly.

For a simple example of parallel tempering and how it enhances the exploration of energy landscapes, refer to Figure 18.9. In Figure 18.9a, a 1D energy surface, $U(x)$, is presented that displays several degenerate local energy minima, which are separated by barriers that increase in magnitude as x tends toward higher values. First, consider the behavior of a particle operating on $U(x)$ at several increasing temperatures from T_1 to T_5. Also let the initial position of the particle at each temperature be the minimum centered at $x = -1.25$. If an ordinary Monte Carlo simulation is performed in the canonical ensemble with $T = T_1$, the particle is unable to overcome many of the barriers on the landscape and according to the probability distribution, $P(x)$, shown in Figure 18.9b, resides only in the two leftmost wells throughout the run. However, if $T = T_5$, the particle easily traverses the full potential. The parallel tempering method harnesses the ability of the higher-temperature simulations to easily traverse barriers to overcome the sampling deficiency that exists at lower temperatures. Figure 18.9c shows the corresponding probability distributions that are obtained at each temperature from parallel tempering. In contrast to those obtained from standard sampling, the distributions show that all the potential wells are sampled at every temperature. Moreover, the probability of finding the particle in any one of the energy wells is approximately the same because all the wells are at the same energy level, and since parallel tempering ensures proper sampling of the Boltzmann distribution, the system properties determined from ensemble averages taken at each temperature are also valid. If the sampling is thorough enough, it is even possible to obtain the energy landscape directly from any one of the finite temperature probability distributions (i.e., $A(x) = -\beta^{-1} \ln P(x)$), particularly the high-temperature ones; however, at lower temperatures $P(x)$ is more sparse in the high energy

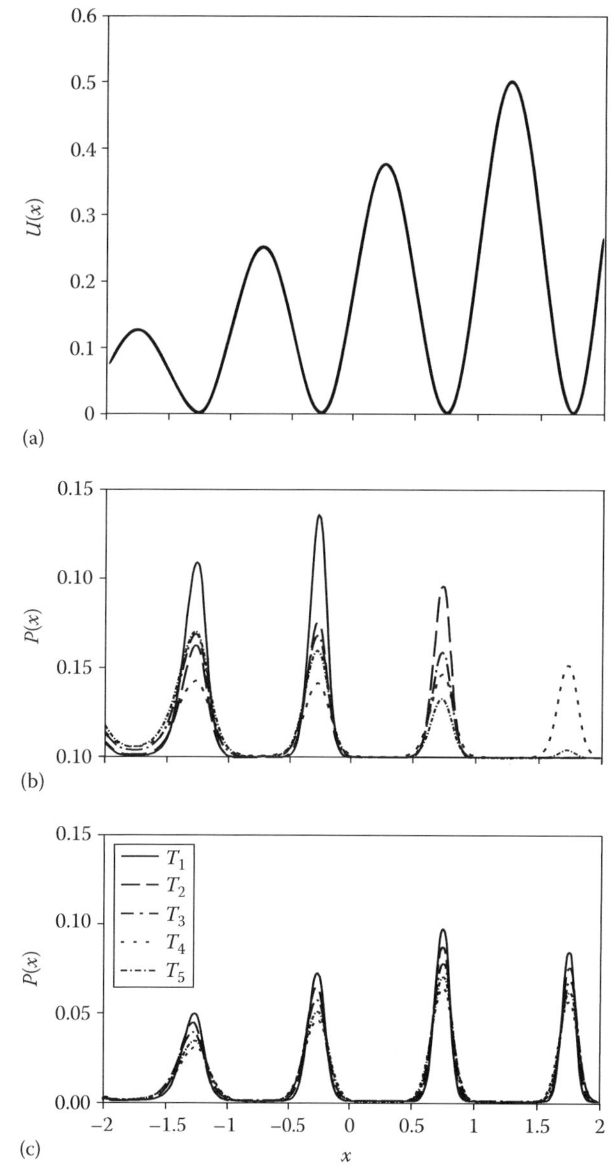

Figure 18.9 Parallel tempering/replica-exchange sampling of a single particle on a 1D energy landscape, $U(x)$, as depicted in (a). The value of the potential goes to infinity at the plot boundaries so that the system is confined to the region shown. (b) Probability distributions compiled for five standard Monte Carlo sampling runs with temperatures (T_1 through T_5) that all originate in the potential well centered at $x = -1.25$. (c) Probability distributions compiled from analogous Monte Carlo runs with replica-exchange. In contrast to the standard runs, all the trajectories sample each of the potential wells.

regions of the potential because the particle avoids the barriers entirely by hopping between the local minima.

18.3.3.1 REMD: Parallel tempering for MD

The problem in adapting the Monte Carlo–based parallel tempering scheme into an appropriate MD-based implementation is that in MD, a particle has both kinetic and potential energy, whereas in Monte Carlo, a particle has only potential energy. Sugita and Okamoto (1999) show in their derivation of the conditional exchange probability for REMD (Equation 18.32) that the kinetic energy components of the replica Hamiltonians cancel; however, in order to ensure that the average kinetic energy, $\langle U(P^N)\rangle_T$, of each temperature replica remains constant,

$$\left\langle U(P^N)\right\rangle_T = \frac{3}{2}\beta^{-1}N \tag{18.36}$$

it is still necessary to rescale the particle momenta when a configuration swap is accepted. Conventionally, the individual particle momenta are rescaled as

$$p^{[i]'} = p^{[i]}\sqrt{\frac{T_n}{T_m}}, \qquad p^{[j]'} = p^{[j]}\sqrt{\frac{T_m}{T_n}} \tag{18.37}$$

where

$p^{[i]}$ is the momentum of a particle in replica i at temperature m

$p^{[i]'}$ is the scaled momentum of that particle after the configuration is swapped with replica j at temperature n

Alternatively, all the particle momenta in a given configuration can also be randomly reassigned at the new temperature; however, there is some indication that the sampling efficiency may be negatively impacted by such a procedure (Cooke and Schmidler, 2008; Rosta and Hummer, 2009; Sindhikara et al., 2010).

18.3.3.2 Overcoming system size limitations

In order for REMD to be effective, replicas must swap frequently enough that the individual trajectories can diffuse over the entire temperature distribution. The acceptance probability associated with exchanging replicas depends on the temperature interval between the adjacent replicas as well as the width of the potential energy distributions at each temperature. If the replicas are separated by too large a temperature interval such that the potential energy distributions do not overlap, the acceptance probability drops to zero and there is no advantage in running parallel simulations. Conversely, if the replica temperatures are spaced too closely such that there is substantial overlap in the energy distributions of the neighboring replicas, swaps are accepted at a high rate, but the efficiency of the REMD approach decreases as resources must be allocated to simulate more replicas than necessary. Ideally, as few replicas are chosen as is necessary to span the temperature distribution. Typically, temperatures are chosen from an exponential distribution, $T_i = T_0e^{ki}$ (with k and T_0 as adjustable parameters), which helps ensure that the exchange probability is roughly consistent between all replica pairs. Unfortunately, the total value of the Hamiltonian increases with the number of degrees of freedom in the system such that obtaining reasonable exchange probabilities requires a large number of closely spaced replicas. Naturally, this places a practical limit on the size of systems where the application of REMD is a tractable solution. This is especially problematic for solvated systems because the solvent–solvent interactions dominate the potential energy. Therefore, the REMD literature is primarily focused on thoroughly exploring the conformational dynamics of coarse-grained systems with reduced degrees of freedom and small organic molecules, peptides, and proteins in the presence of implicit solvation fields and/or a small number of explicit solvent molecules (Chaudhury et al., 2012; Chebaro et al., 2009; Jiang and Roux, 2010; Nymeyer et al., 2004; Okur et al., 2006; Sabri Dashti et al., 2012; Sugita and Okamoto, 1999; Zhang et al., 2005).

However, there are also recent efforts aimed at extending the applicability of replica-exchange methods to larger explicitly solvated systems. One such approach is replica-exchange with solute tempering (REST) (Liu et al., 2005; Terakawa et al., 2010). In REST, temperature exchanges are performed using a temperature-scaled potential energy function that is designed to nullify the contribution of the solvent–solvent interactions in the evaluation of the exchange probability. The potential energy of a given configuration is decomposed into contributions from the solute (S), solute–water (SW), and water–water (WW) interactions:

$$U_0(x) = U_S(x) + U_{SW}(x) + U_{WW}(x) \tag{18.38}$$

As the temperature of the ith replica increases away from the target temperature (β_0), the potential energy is scaled as follows:

$$U_i(x) = U_S(x) + \left[\frac{\beta_0 + \beta_i}{2\beta_i}\right]U_{SW}(x) + \left[\frac{\beta_0}{\beta_i}\right]U_{WW}(x) \tag{18.39}$$

such that the solvent–solvent interaction term cancels out of the resulting expression for the exchange probability (obtained by substituting Equation 18.39 into Equation 18.33):

$$\Delta = (\beta_i - \beta_j)\left[\left(U_S(x_j) + \frac{1}{2}U_{SW}(x_j)\right) - \left(U_S(x_i) + \frac{1}{2}U_{SW}(x_i)\right)\right] \tag{18.40}$$

The consequence of eliminating $U_{WW}(x)$ from the aforementioned expression is that solvent–solvent interactions that traditionally cause poor scaling with respect to system size no longer contribute to declining acceptance probabilities in explicitly solvated systems. For a given system size, REST acceptance probabilities are significantly larger than traditional REMD and far fewer replicas are needed to span the temperature distribution (Liu et al., 2005) (Figure 18.10).

Although it is a more approximate approach, a similar effect can be achieved by utilizing an implicit or hybrid implicit/explicit treatment of the solvent throughout the simulation. However, while such simulations of the alanine dipeptide and other model systems show that the results generally bear a qualitative similarity to explicit solvent simulations, there can be significant quantitative differences in some instances (Chaudhury et al., 2012). The *semi-hybrid* approach substantially reduces quantitative errors due to sampling under solvent-deficient conditions by maintaining each replica in explicit solvent at all times while employing a hybrid implicit/explicit solvation model only during the evaluation of the exchange probability; if one to two layers of explicit solvent molecules are maintained about the solute as the energy is calculated in the implicit solvent field, comparable results to explicitly solvated REMD simulations can be obtained using fewer replicas and at a significantly reduced computational expense (Chaudhury et al., 2012; Okur et al., 2006).

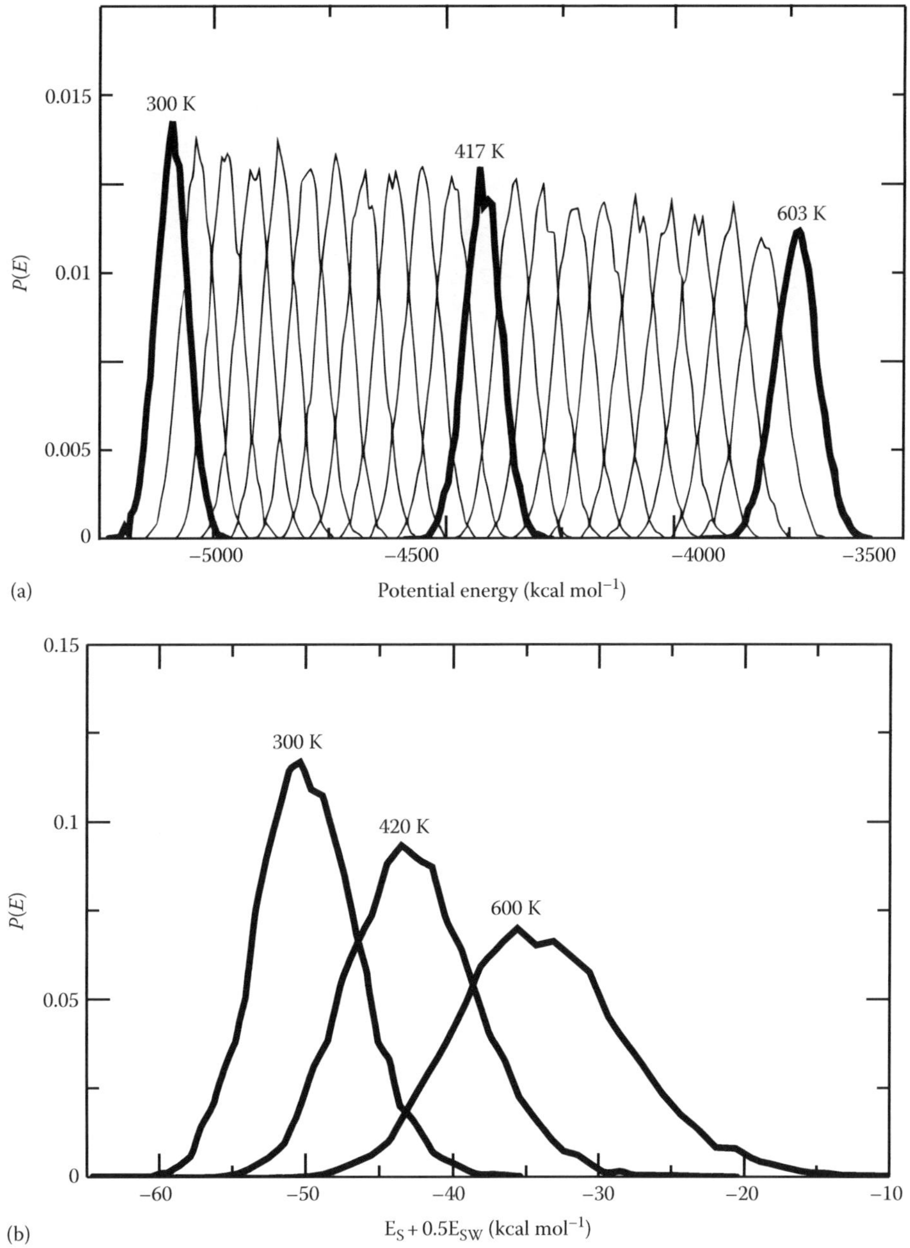

Figure 18.10 Potential energy distributions obtained from REMD simulations of the alanine dipeptide: (a) Distributions obtained from standard REMD. (b) Distributions obtained from REST. This approach extends the applicability of replica-exchange methods to larger explicitly solvated systems by eliminating the solvent–solvent interactions from the evaluation of the exchange probability (see main text). Using the effective potential energy (solute potential energy + 1/2(solute–water potential energy)) enables the full temperature range to be spanned by 3 replicas, rather than 22 as needed by standard REMD. (Reproduced from Liu, P., Kim, B., Friesner, R.A., and Berne, B.J., Replica exchange with solute tempering: A method for sampling biological systems in explicit water, *Proc. Natl. Acad. Sci. USA*, 102(39), 13749–13754, doi:10.1073/pnas.0506346102, 2005. Copyright 2005, National Academy of Sciences, U.S.A.)

18.3.3.3 Early stage growth of $CaCO_3(nH_2O)$

In Wallace et al. (2013), a semi-hybrid temperature-based REMD approach is used to probe the growth and conformational dynamics of hydrated calcium carbonate clusters up to ~2 nm in diameter. This size range encompasses the full size distribution of the *prenucleation* clusters as determined from cryo-TEM (Gebauer and Coelfen, 2011; Gebauer et al., 2008) and analytical ultracentrifugation (Gebauer et al., 2008; Pouget et al., 2009). It also overlaps with the size of anhydrous mineral clusters sampled by metadynamics (Quigley and Rodger, 2008; Quigley et al., 2011); however, the REMD approach is distinguished by its ability to explore the phase space available to the clusters while simultaneously allowing water to be incorporated during growth. Structural analysis of the clusters as a function of size shows that the average Ca–C coordination number increases over the range of cluster sizes investigated and that there are no abrupt structural transitions. Likewise, calcium ion diffusivities calculated over the same size range are also smoothly trending; they decrease with increasing cluster size and slowly approach a constant value that is significantly more diffusive than calculated for solid phases of $CaCO_3$ (ACC and calcite). Based on this observation, and the similarity of the ion diffusivities with the self-diffusivities of

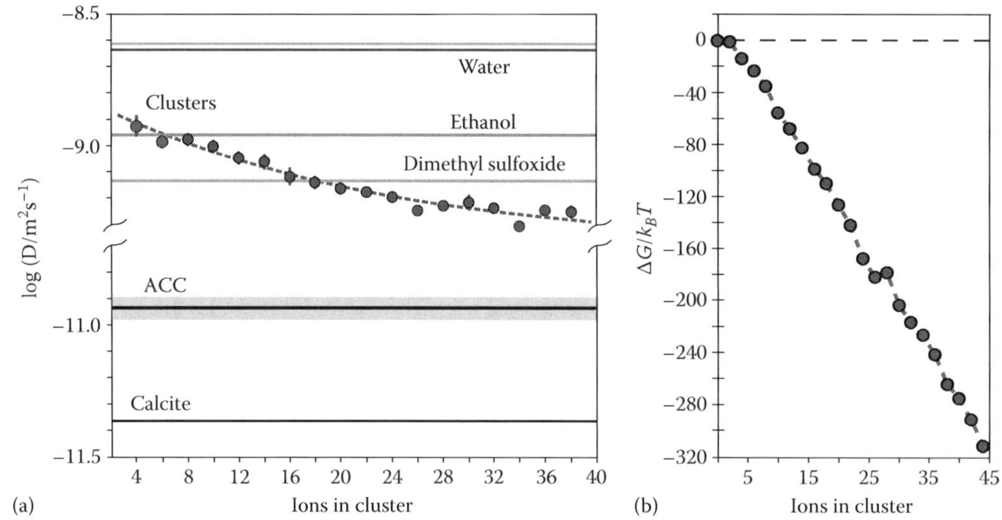

Figure 18.11 Results of REMD simulations performed on hydrated calcium carbonate clusters: (a) The diffusivity of the calcium ions is plotted as a function of size and compared against calculated values for calcite and hydrous ACC. The significantly greater diffusivity of the ions in the cluster phase along with the similarity with the self-diffusivities of several common solvents suggests that the clusters are a dense liquid phase. (b) The free energy landscape underlying phase separation and $[Ca^{2+}] = [CO_3^{2-}] = 15$ mmol L^{-1}. The slight feature at ~26 ions is within the statistical uncertainties and does not suggest clusters of that size have any special thermodynamic significance, as is the hallmark of PNC theory. Rather, the energy landscape is entirely consistent with SD. (Reproduced from Wallace, A.F. et al., Microscopic evidence for liquid-liquid separation at the onset of calcium carbonate mineralization, In Review.)

several common solvents, Wallace et al. argue, as did Demachialis et al. (2011), that the clusters occupy a liquid-like if not bona fide liquid state (Figure 18.11a). Additionally, calculation of the free energy as a function of cluster size using an absolute two-phase thermodynamic model (Huang et al., 2011; Lin et al., 2003, 2010) reveals that under the concentration conditions of the simulations (which are comparable to those where PNCs are observed experimentally), there is no free energy barrier opposing cluster formation nor is there any feature on the landscape that suggests any particular cluster size has any special thermodynamic significance (Figure 18.11b). Based on these results, Wallace et al. suggest that the observation of nonclassical nucleation of $CaCO_3$ through the aggregation of stable PNCs may not be nucleation at all, but rather an undiagnosed case of liquid–liquid phase separation by SD. To support this argument, they also present the results of lattice-gas simulations using the 2D and 3D Ising models, which explores the general consequences of phase separation without imposing any chemical specificity. These results show that when phase separation proceeds via SD at low volume fractions, a cluster population is established that grows primarily through cluster coalescence. Moreover, the cluster population consists of two distinct distributions corresponding to precritical fluctuations and the larger products of SD, respectively. These results support the conclusions of Faatz et al. (2004) who argued based on the temperature dependence of size distributions obtained by dynamic light scattering that liquid–liquid separation and SD occurs in the $CaCO_3$–water system and that the critical temperature is ~10°C. Although the initial microscopic phase observed by Faatz et al. was not clearly identifiable as either liquid or solid, the evolution of the particle size distribution is consistent with the rate of coarsening that is expected for SD, and the apparent proximity of the critical temperature to ambient conditions provides an additional means of producing nanoscale clusters by entirely classical processes. Moreover, a recent study using NMR to quantify the diffusion of ions in solution following the introduction of calcium to a bicarbonate buffer solution (Bewernitz et al., 2012) concluded that a bicarbonate-rich liquid phase may exist at conditions similar to those where PNCs are reported.

18.3.3.4 Structure of amorphous calcium carbonate

Among the many varied experimental observations concerning the early stages of calcium carbonate crystallization, the structure of ACC is by far the most consistent. Although there is some suggestion in the literature that ACC exhibits polyamorphism (Gebauer et al., 2010), at ambient conditions, the PDF obtained from total x-ray scattering consistently displays the same reproducible features (Goodwin et al., 2010; Michel et al., 2008; Reeder et al., 2013). Moreover, a review of the available solid-state NMR data (Gebauer et al., 2010; Michel et al., 2008; Singer et al., 2012) suggests that the broadening of the line widths about the reference spectra for calcite and vaterite observed in one instance (Gebauer et al., 2010) likely results from the microcrystalline character of the samples rather than variations in the short-range order between two disparate protocrystalline amorphous phases, as observed in the calcium phosphate system (Jäger et al., 2006). However, at elevated pressures, there is now convincing evidence that a partially reversible phase transition takes place as ACC adopts a somewhat denser configuration (Fernandez-Martinez et al., 2013). To a certain extent, total x-ray scattering characterizes the local order within ACC. However, because the 3D structure that exists in the sample is spherically averaged and reduced to a 1D representation (i.e., the PDF), the actual structure cannot be determined directly from the PDF. Rather, candidate structures are identified through the application of a Reverse Monte Carlo (RMC) procedure (Goodwin et al., 2010; McGreevy and Howe, 1992), whereby atom positions are adjusted at random until the calculated PDF is in agreement with the experimental PDF. RMC modeling generally produces well-converged PDFs; however, the

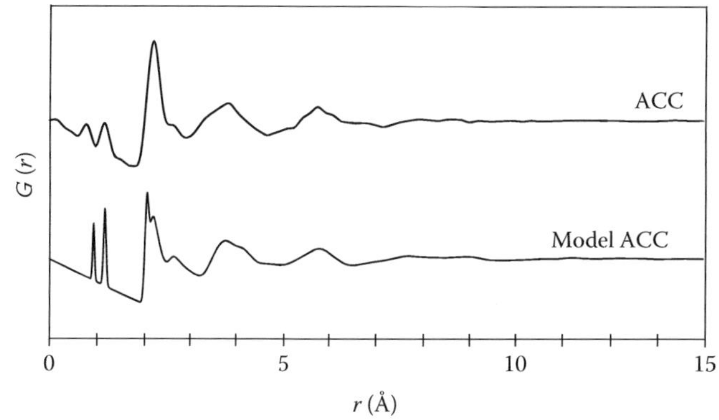

Figure 18.12 The total PDF of hydrated ACC as measured by total x-ray scattering. The model ACC structure derived by Wallace et al. (2013) through the process discussed in the text is shown for comparison.

results are not guaranteed to be unique or physically meaningful. Indeed, the results of MD simulations by Singer et al. (2012) show that the RMC-derived structure of ACC (Goodwin et al., 2010) is unstable.

Recognizing that the coherent x-ray scattering length in ACC is comparable to the diameter of the larger hydrated calcium carbonate clusters grown in their REMD simulations (1.5–2 nm), Wallace et al. (n.d.) constructed a model aggregate structure for the purpose of simulating the formation of ACC by cluster coalescence. The model ACC structure they obtain is stable and faithfully reproduces all the salient aspects of the experimental PDF (Figure 18.12). In their approach, the initial aggregate structure is obtained by randomly packing copies of a single $CaCO_3(nH_2O)$ cluster into a solvent-filled cell. Subsequently, the calcium-to-water ratio is slowly adjusted to match that of synthetic ACC ($Ca/H_2O \sim 1.0$) over a series of constant pressure MD simulations. At each iteration, a few solvent molecules are preferentially removed from the solvent-rich volumes between the clusters. The net effect of this procedure is that the clusters are driven to coalesce without disturbing the structural water molecules that are intrinsic to their character. Once this process is complete, computing the individual PDFs, $g_{ij}(r)$, between all atom pairs in the system is straightforward:

$$g_{ij}(r) = \frac{n_{ij}(r)}{4\pi r^2 \rho_i dr} \tag{18.41}$$

where

- $n_{ij}(r)$ is the number of atoms of type j located within the radial distances r
- $r + dr$ is the atom of type i
- ρ_i is the total number density of atom type i

However, in order to facilitate the comparison of the theoretical results with the experimental PDF, the individual $g_{ij}(r)$ functions for all m atom pairs must be summed up and properly weighted (Egami and Billinge, 2003; Fernandez-Martinez et al., 2010) (ρ_0 is the total atom number density):

$$G(r) = 4\pi\rho_0 \sum_{i,j}^{m} w_{ij}[g_{ij}(r) - 1] \tag{18.42}$$

The weighting factors, w_{ij}, are defined as follows, where c_i and c_j are the concentrations of atom types i and j, respectively, and f_i and f_j are their corresponding atomic form factors evaluated when the scattering vector, $q = 0$ (under this condition, the atomic form factors are equal to the atomic number):

$$w_{ij} = \frac{c_i c_j f_i f_j}{\left(\sum_i^m c_i f_i\right)^2} \tag{18.43}$$

18.4 SUMMARY AND FUTURE CHALLENGES

This chapter opens with a discussion of thermodynamic speciation models and the classical theories of nucleation and SD in contrast to the recently asserted PNC theory that is challenging them. The arguments presented therein maintain that the experimental observations in the $CaCO_3$ system are insufficient to classify the phase-separation pathway as either classical or nonclassical, while atomistic simulations are uniquely situated to do so because in principle, they can quantify the underlying free energy landscape. However, because of the time and length-scale limitations inherent to such methods, advanced simulation techniques are often required to explore even the earliest stages of the crystallization process. An introduction to enhanced sampling methods is provided by focusing on those approaches that are utilized in the $CaCO_3$ literature (umbrella sampling, metadynamics, and replica-exchange techniques); following the presentation of each technique, the most salient insights each brings to the table regarding the $CaCO_3$ mineralization pathway are discussed.

With regard to particles that form from homogeneous solution, there remains some minor disagreement as to exactly what size crystalline states become energetically favored over disordered states; however, modeling results to date concur that the threshold is in the neighborhood of a few nanometers. Metadynamics simulations show that the free energy barriers separating amorphous and crystalline states can be several hundred times k_bT (at least for anhydrous particles), which indicates that the disorder to order transition is unlikely to occur as a solid-state process, but rather by dissolution/reprecipitation or secondary nucleation in solution or at ACC surfaces. This result is in line with the common experimental observation that ACC often persists to the micron scale where it is undoubtedly metastable with respect to crystalline carbonates. Recent studies also suggest that the initially formed clusters are dense liquid or liquid-like phases that persist into the nanometer size regime where *prenucleation* clusters are observed. This phase is distinct from the well-known polymer-induced liquid-precursor (PILP) phase (Gower and Odom, 2000), which is observed only in the presence of organics; however, it is likely that PILP originates from the same dense liquid phase, which is able to occupy larger volumes due to the stabilizing activity of the organics. Wallace et al. (2013) demonstrated that the coalescence and solidification of these dense liquid clusters results in the formation an amorphous phase whose structure is consistent with hydrated ACC. Their REMD-based simulations show also that the structural and dynamical properties of the clusters vary continuously as a function of size and that the energy landscape upon which the clusters form is smooth and monotonically decreasing up to at least 2 nm in diameter. These results are consistent with SD and supports the experimental observations of Faatz et al. (2004) who suggested that the critical temperature of the liquid–liquid binodal resides near ambient conditions. If so, clusters may be produced by large but unstable precritical fluctuations (arising from the close proximity of the critical temperature to room temperature) as well as liquid–liquid separation, as has recently been suggested by Bewernitz et al. (2012).

Certainly, much progress is being made toward characterizing the early stages of carbonate mineralization with simulations; however, there are still many significant challenges ahead. Chief among these perhaps is the need to divine approaches capable of probing mineralization processes at both high and low driving force. As yet, there are no methods that enable the mineralization pathway to be simulated in its entirety; progress on this front is necessary to ultimately map out changes in system behavior across the range of conditions that occur in biological and laboratory environments. While achieving this goal may require some sacrifice of atomistic detail through development of coarse-grained models, there are currently efforts underway to circumvent this problem through development of approaches that can reliably map atomistic detail back onto coarse-grained system representations (Chng and Yang, 2008; Christen and van Gunsteren, 2006; Doruker and Mattice, 1999; Ghanbari et al., 2011; Romano et al., 2013; Spyriouni et al., 2007). Additional challenges center on furthering understanding the influence of impurities, organics, and interfaces (both hard and soft) that induce many varied responses in the $CaCO_3$ system. To date, there are several insightful studies that target specific mineral–organic interactions (Duffy and Harding, 2002; Freeman et al., 2008, 2009, 2010, 2012; Hamm et al., 2010; Harding et al., 2008; Piana et al., 2007; Quigley et al., 2009; Raiteri et al., 2012; Tribello et al., 2009b), but we are far from being able to predict the influence of an organic on the mineralization pathway based on its molecular properties.

REFERENCES

Allen, M.P. and Tildesley, D.J. 1987. *Computer Simulations of Liquids* (p. 385). New York: Oxford University Press.

Allen, R.J., Frenkel, D., and Ten Wolde, P.R. 2006. Simulating rare events in equilibrium or nonequilibrium stochastic systems. *J. Chem. Phys.*, 124(2), 024102. doi:10.1063/1.2140273.

Auer, S. and Frenkel, D. 2004. Quantitative prediction of crystal-nucleation rates for spherical colloids: A computational approach. *Annu. Rev. Phys. Chem.*, 55, 333–361. doi:10.1146/annurev.physchem.55.091602.094402.

Beniash, E., Aizenberg, J., Addadi, L., and Weiner, S. 1997. Amorphous calcium carbonate transforms into calcite during sea urchin larval spicule growth. *Proc. Royal Soc. B*, 264, 461–465.

Bewernitz, M.A., Gebauer, D., Long, J., Coelfen, H., and Gower, L. 2012. A meta-stable liquid precursor phase of calcium carbonate and its interactions with polyaspartate. *Faraday Discuss.* doi:10.1039/c2fd20080e

Bhimalapuram, P., Chakrabarty, S., and Bagchi, B. 2007. Elucidating the mechanism of nucleation near the gas-liquid spinodal. *Phys. Rev. Lett.*, 98(20), 206104. doi:10.1103/PhysRevLett.98.206104.

Binder, K. and Fratzl, P. 2005. Spinodal decomposition. In G. Kostorz (ed.), *Phase Transformations in Materials* (pp. 409–480). Weinheim, Germany: Wiley-VCH Verlag GmbH & Co. KGaA. doi:10.1002/352760264X.ch6.

Chaudhury, S., Olson, M.A., Tawa, G., Wallqvist, A., and Lee, M.S. 2012. Efficient conformational sampling in explicit solvent using a hybrid replica exchange molecular dynamics method. *J. Chem. Theory Comput.*, 8(2), 677–687. doi:10.1021/ct200529b.

Chebaro, Y., Dong, X., Laghaei, R., Derreumaux, P., and Mousseau, N. 2009. Replica exchange molecular dynamics simulations of coarse-grained proteins in implicit solvent. *J. Phys. Chem. B*, 113(1), 267–274. doi:10.1021/jp805309e.

Chng, C.-P. and Yang, L.-W. 2008. Coarse-grained models reveal functional dynamics-II. Molecular dynamics simulation at the coarse-grained level—Theories and biological applications. *Bioinform. Biol. Insights*, 2, 171–185.

Christen, M. and Van Gunsteren, W. F. 2006. Multigraining: An algorithm for simultaneous fine-grained and coarse-grained simulation of molecular systems. *J. Chem. Phys.*, 124(15), 154106. doi:10.1063/1.2187488.

Cooke, B. and Schmidler, S.C. 2008. Preserving the Boltzmann ensemble in replica-exchange molecular dynamics. *J. Chem. Phys.*, 129(16), 164112. doi:10.1063/1.2989802.

Darkins, R., Côté, A.S., Freeman, C.L., and Duffy, D.M. 2013. Crystallisation rates of calcite from an amorphous precursor in confinement. *J. Cryst. Growth*, 367, 110–114. doi:10.1016/j.jcrysgro.2012.12.027.

Davies, C.W. 1962. *Ion Association*. London, U.K.: Butterworths.

De Silva, P.N.K. and Ranjith, P.G. 2012. A study of methodologies for CO_2 storage capacity estimation of saline aquifers. *Fuel*, 93, 13–27. doi:10.1016/j.fuel.2011.07.004.

Dellago, C. and Bolhuis, P.G. 2008. Transition path sampling and other advanced simulation techniques for rare events. In C. Holm and K. Kremer (eds.), *Advanced Computer Simulation Approaches for Soft Matter Sciences III, Advances in Polymer Science*, Vol. 221 (pp. 167–233). Berlin, Germany: Springer-Verlag. doi:10.1007/12.

Demichelis, R., Raiteri, P., Gale, J.D., Quigley, D., and Gebauer, D. 2011. Stable prenucleation mineral clusters are liquid-like ionic polymers. *Nat. Commun.*, 2(590), doi:10.1038/ncomms1604.

Discussions, F. 2012. General discussion. *Faraday Discuss.*, 155, 103. doi:10.1039/c1fd90054d.

Doruker, P. and Mattice, W.L. 1999. Feature article A second generation of mapping/reverse mapping of coarse-grained and fully atomistic models of polymer melts. *Macromol. Theory Simul.*, 8(5), 463–478.

Duffy, D.M. and Harding, J.H. 2002. Modelling the interfaces between calcite crystals and Langmuir monolayers. *J. Mater. Chem.*, 12(12), 3419–3425. doi:10.1039/b205657g.

Earl, D.J. and Deem, M.W. 2005. Parallel tempering: Theory, applications, and new perspectives. *Phys. Chem. Chem. Phys.*, 7(23), 3910–3916.

Egami, T. and Billinge, S.J.L. 2003. *Underneath the Bragg Peaks: Structural Analysis of Complex Materials*. Amsterdam, the Netherlands: Elsevier.

Ensing, B., De Vivo, M., Liu, Z., Moore, P., and Klein, M.L. 2006. Metadynamics as a tool for exploring free energy landscapes of chemical reactions. *Acc. Chem. Res.*, 39(2), 73–81. doi:10.1021/ar040198i.

Faatz, M., Gröhn, F., and Wegner, G. 2004. Amorphous calcium carbonate: Synthesis and potential intermediate in biomineralization. *Adv. Mater.*, 16(12), 996–1000. doi:10.1002/adma.200306565.

Fernandez-Martinez, A., Kalkan, B., Clark, S.M., and Waychunas, G.A. 2013. Pressure-induced polyamorphism and formation of "aragonitic" amorphous calcium carbonate. *Angew. Chem. Int. Ed.* 52(32), 8354–8357.

Fernandez-Martinez, A., Timon, V., Roman-Ross, G., Cuello, G.J., Daniels, J.E., and Ayora, C. 2010. The structure of schwertmannite, a nanocrystalline iron oxyhydroxysulfate. *Am. Mineral.*, 95(8–9), 1312–1322. doi:10.2138/am.2010.3446.

Filion, L., Hermes, M., Ni, R., and Dijkstra, M. 2010. Crystal nucleation of hard spheres using molecular dynamics, umbrella sampling, and forward flux sampling: A comparison of simulation techniques. *J. Chem. Phys.*, 133(24), 244115. doi:10.1063/1.3506838.

Filion, L., Ni, R., Frenkel, D., and Dijkstra, M. 2011. Simulation of nucleation in almost hard-sphere colloids: The discrepancy between experiment and simulation persists. *J. Chem. Phys.*, 134(13), 134901. doi:10.1063/1.3572059.

Finch, A.A., Shaw, P.A., Weedon, G.P., and Holmgren, K. 2001. Trace element variation in speleothem aragonite: Potential for paleoenvironmental reconstruction. *Earth Planet. Sci. Lett.*, 186(2), 255–267. doi:10.1016/S0012-821X(01)00253-9.

Freeman, C.L., Asteriadis, I., Yang, M., and Harding, J.H. 2009. Interactions of organic molecules with calcite and magnesite surfaces. *J. Phys. Chem. C*, 113(9), 3666–3673. doi:10.1021/jp807051u.

Freeman, C.L., Harding, J.H., and Duffy, D.M. 2008. Simulations of calcite crystallization on self-assembled monolayers. *Langmuir*, 24(17), 9607–9615. doi:10.1021/la800389g.

Freeman, C.L., Harding, J.H., Quigley, D., and Rodger, P.M. 2010. Structural control of crystal nuclei by an eggshell protein. *Angew. Chem. Int. Ed.*, 49(30), 5135–5137. doi:10.1002/anie.201000679.

Freeman, C.L., Harding, J.H., Quigley, D., and Rodger, P.M. 2011. Simulations of Ovocleidin-17 binding to calcite surfaces and its implications for eggshell formation. *J. Phys. Chem. C*, 115(16), 8175–8183. doi:10.1021/jp200145m.

Freeman, C.L., Harding, J.H., Quigley, D., and Rodger, P.M. 2012. Protein binding on stepped calcite surfaces: Simulations of ovocleidin-17 on calcite {31.16} and {31.8}. *Phys. Chem. Chem. Phys.*, 14(20), 7287–7295. doi:10.1039/c2cp23987f.

Freeman, D.L. 2000. Phase changes in 38-atom Lennard-Jones clusters. I. A parallel tempering study in the canonical ensemble. *Chem. Phys.*, 112(23), 10340–10349.

Frenkel, D. and Smit, B. 2002. *Understanding Molecular Simulation: From Algorithms to Applications* (p. 638). London, U.K.: Academic Press.

Gebauer, D. and Coelfen, H. 2011. Prenucleation clusters and non-classical nucleation. *Nano Today*, 6(6), 564–584. doi:10.1016/j.nantod.2011.10.005.

Gebauer, D., Gunawidjaja, P.N., Ko, J.Y.P., Bacsik, Z., Aziz, B., Liu, L. et al. 2010. Proto-calcite and proto-vaterite in amorphous calcium carbonates. *Angew. Chem. Int. Ed.*, 49(47), 8889–8891.

Gebauer, D., Voelkel, A., and Coelfen, H. 2008. Stable prenucleation calcium carbonate clusters. *Science*, 322(5909), 1819–1822.

Ghanbari, A., Böhm, M.C., and Müller-Plathe, F. 2011. A simple reverse mapping procedure for coarse-grained polymer models with rigid side groups. *Macromolecules*, 44(13), 5520–5526. doi:10.1021/ma2005958.

Gong, Y.U.T., Killian, C.E., Olson, I.C., Appathurai, N.P., Amasino, A.L., Martin, M.C. et al. 2012. Phase transitions in biogenic amorphous calcium carbonate. *Proc. Natl. Acad. Sci. USA*, 109(16), 6088–6093. doi:10.1073/pnas.1118085109.

Goodwin, A.L., Michel, F.M., Phillips, B.L., Keen, D.A., Dove, M.T., and Reeder, R.J. 2010. Nanoporous structure and medium-range order in synthetic amorphous calcium carbonate. *Chem. Mater.*, 22(10), 3197–3205.

Gould, H. and Klein, W. 1993. Spinodal nucleation effects in systems with long-range interactions. *Physica D*, 66, 61–70.

Gower, L.B. and Odom, D.J. 2000. Deposition of calcium carbonate films by a polymer-induced liquid-precursor (PILP) process. *J. Cryst. Growth*, 210(4), 719–734. doi:10.1016/S0022-0248(99)00749–6.

Gunton, J.D. 1999. Homogeneous nucleation. *J. Stat. Phys.*, 95(5/6), 903–923.

Haeckel, E. 1887. Report on the Radiolaria collected by the H.M.S. Challenger during the Years 1873–1876. Report on the Scientific Results of the Voyage of the H.M.S. Challenger, Zoology, Volume XVIII. Thompson and J. Murray (eds.), (pp. 1–1760). London, U.K.: Her Majesty's Stationery Office.

Haile, J.M. 1997. *Molecular Dynamics Simulation: Elementary Methods* (p. 489). New York: John Wiley & Sons.

Hamm, L.M., Wallace, A.F., and Dove, P.M. 2010. Molecular dynamics of ion hydration in the presence of small carboxylated molecules and implications for calcification. *J. Phys. Chem. B*, 114(32), 10488–10495.

Harding, J.H., Duffy, D.M., Sushko, M.L., Rodger, P.M., Quigley, D., and Elliott, J.A. 2008. Computational techniques at the organic-inorganic interface in biomineralization. *Chem. Rev.*, 108(11), 4823–4854. doi:10.1021/cr078278y.

Huang, S.-N., Pascal, T.A., Goddard, W.A., Maiti, P.K., and Lin, S.-T. 2011. Absolute entropy and energy of carbon dioxide using the two-phase thermodynamic model. *J. Chem. Theory Comput.*, 7(6), 1893–1901. doi:10.1021/ct200211b.

Iannuzzi, M., Laio, A., and Parrinello, M. 2003. Efficient exploration of reactive potential energy surfaces using car-parrinello molecular dynamics. *Phys. Rev. Lett.*, 90(23), 238302. doi:10.1103/PhysRevLett.90.238302.

Jäger, C., Welzel, T., Meyer-Zaika, W., and Epple, M. 2006. A solid-state NMR investigation of the structure of nanocrystalline hydroxyapatite. *Magn. Reson. Chem.*, 44(6), 573–580. doi:10.1002/mrc.1774.

Jiang, W. and Roux, B. 2010. Free energy perturbation Hamiltonian replica-exchange molecular dynamics (FEP/H-REMD) for absolute ligand binding free energy calculations. *J. Chem. Theory Comput.*, 6(9), 2559–2565. doi:10.1021/ct1001768.

Klein, R., Lohmann, K., and Thayer, C. 1996. Bivalve skeletons record sea-surface temperature and delta O-18 via Mg/Ca and O-18/O-16 ratios. *Geology*, 24(5), 415–418.

Laaksonen, A. and Mcgraw, R. 1996. Size-dependent surface tension. *Europhys. Lett.*, 35(5), 367–372.

Laio, A. and Gervasio, F.L. 2008. Metadynamics: A method to simulate rare events and reconstruct the free energy in biophysics, chemistry and material science. *Rep. Prog. Phys.*, 71(12), 126601. doi:10.1088/0034-4885/71/12/126601.

Laio, A. and Parrinello, M. 2002. Escaping free-energy minima. *Proc. Natl. Acad. Sci. USA.*, 99(20), 12562–12566. doi:10.1073/pnas.202427399.

Laio, A., Rodriguez-Fortea, A., Gervasio, F.L., Ceccarelli, M., and Parrinello, M. 2005. Assessing the accuracy of metadynamics. *J. Phys. Chem. B*, 109(14), 6714–6721. doi:10.1021/jp045424k.

Leyssale, J.-M., Delhommelle, J., and Millot, C. 2007. Hit and miss of classical nucleation theory as revealed by a molecular simulation study of crystal nucleation in supercooled sulfur hexafluoride. *J. Chem. Phys.*, 127(4), 044504. doi:10.1063/1.2753147.

Lin, S.-T., Blanco, M., and Goddard, W.A. 2003. The two-phase model for calculating thermodynamic properties of liquids from molecular dynamics: Validation for the phase diagram of Lennard-Jones fluids. *J. Chem. Phys.*, 119(22), 11792. doi:10.1063/1.1624057.

Lin, S.-T., Maiti, P.K., and Goddard, W.A. 2010. Two-phase thermodynamic model for efficient and accurate absolute entropy of water from molecular dynamics simulations. *J. Phys. Chem. B*, 114(24), 8191–8198. doi:10.1021/jp103120q.

Liu, P., Kim, B., Friesner, R.A., and Berne, B.J. 2005. Replica exchange with solute tempering: A method for sampling biological systems in explicit water. *Proc. Natl. Acad. Sci. USA*, 102(39), 13749–13754. doi:10.1073/pnas.0506346102.

Loste, E. and Meldrum, F.C. 2001. Control of calcium carbonate morphology by transformation of an amorphous precursor in a constrained volume. *Chem. Commun.*, 1(10), 901–902. doi:10.1039/b101563j.

Loste, E., Park, R.J., Warren, J., and Meldrum, F.C. 2004. Precipitation of calcium carbonate in confinement. *Adv. Funct. Mater.*, 14(12), 1211–1220. doi:10.1002/adfm.200400268.

Lu, H.M. and Jiang, Q. 2004. Size-dependent surface energies of nanocrystals. *J. Phys. Chem. B*, 108(18), 5617–5619. doi:10.1021/jp0366264.

Magomedov, M.N. 2004. Dependence of the surface energy on the size and shape of a nanocrystal. *Phys. Solid State*, 46(5), 954–968. doi:10.1134/1.1744976.

Marchi, M. and Procacci, P. 1998. Coordinates scaling and multiple time step algorithms for simulation of solvated proteins in the NPT ensemble. *J. Chem. Phys.*, 109(13), 5194. doi:10.1063/1.477136.

Markov, I.V. 2004. *Crystal Growth for Beginners: Fundamentals of Nucleation, Crystal Growth and Epitaxy*, 2nd ed. (p. 546). Singapore: World Scientific Publishing Company.

Matthews, R.P. and Naidoo, K.J. 2010. Experimentally consistent ion association predicted for metal solutions from free energy simulations. *J. Phys. Chem. B*, 114(21), 7286–7293. doi:10.1021/jp911823x.

McGreevy, R.L. and Howe, M.A. 1992. RMC: Modeling disordered structures. *Annu. Rev. Mater. Sci.*, 22, 217–242.

Michel, F.M., MacDonald, J., Feng, J., Phillips, B.L., Ehm, L., Tarabrella, C. et al. 2008. Structural characteristics of synthetic amorphous calcium carbonate. *Chem. Mater.*, 20(14), 4720–4728. doi:10.1021/cm800324v.

Monette, L. and Klein, W. 1992. Spinodal nucleation as a coalescence process. *Phys. Rev. Lett.*, 68(15), 13–16.

Nymeyer, H., Gnanakaran, S., and García, A.E. 2004. Atomic simulations of protein folding, using the replica exchange algorithm. *Methods Enzymol.*, 383(2000), 119–149. doi:10.1016/S0076-6879(04)83006-4.

Okur, A., Wickstrom, L., Layten, M., Geney, R., Song, K., Hornak, V. et al. 2006. Improved efficiency of replica exchange simulations through use of a hybrid explicit/implicit solvation model. *J. Chem. Theory Comput.*, 2(2), 420–433. doi:10.1021/ct050196z.

Ouyang, G., Liang, L.H., Wang, C.X., and Yang, G.W. 2006. Size-dependent interface energy. *Appl. Phys. Lett.*, 88(9), 091914. doi:10.1063/1.2172396.

Piana, S., Jones, F., and Gale, J.D. 2007. Aspartic acid as a crystal growth catalyst. *Cryst. Eng. Commun.*, 9(12), 1187–1191.

Pitzer, K.S. 1973. Thermodynamics of electrolytes. I. Theoretical basis and general equations. *J. Phys. Chem.*, 77(2), 268–277. doi:10.1021/j100621a026.

Pitzer, K.S. (ed.). 1991. *Activity Coefficients in Electrolyte Solutions*, 2nd ed. (p. 542). Boca Raton, FL: CRC Press.

Pouget, E.M., Bomans, P.H.H., Goos, J.A.C.M., Frederik, P.M., De With, G., and Sommerdijk, N.A.J.M. 2009. The initial stages of template-controlled $CaCO_3$ formation revealed by cryo-TEM. *Science*, 323(5920), 1455–1458.

Quigley, D., Freeman, C.L., Harding, J.H., and Rodger, P.M. 2011. Sampling the structure of calcium carbonate nanoparticles with metadynamics. *J. Chem. Phys.*, 134, 044703. doi:10.1063/1.3530288.

Quigley, D. and Rodger, P.M. 2008. Free energy and structure of calcium carbonate nanoparticles during early stages of crystallization. *J. Chem. Phys.*, 128(22), 221101. doi:10.1063/1.2940322.

Quigley, D. and Rodger, P.M. 2009. A metadynamics-based approach to sampling crystallisation events. *Mol. Simul.*, 35(7), 613–623. doi:10.1080/08927020802647280.

Quigley, D., Rodger, P.M., Freeman, C.L., Harding, J.H., and Duffy, D.M. 2009. Metadynamics simulations of calcite crystallization on self-assembled monolayers. *J. Chem. Phys.*, 131(9), 094703. doi:10.1063/1.3212092.

Radha, A.V., Fernandez-Martinez, A., Hu, Y., Jun, Y.-S., Waychunas, G.A., and Navrotsky, A. 2012. Energetic and structural studies of amorphous Ca1–xMgxCO_3·nH_2O. *Geochim. Cosmochim. Acta*, 90, 83–95. doi:10.1016/j.gca.2012.04.056.

Radha, A.V., Forbes, T.Z., Killian, C.E., Gilbert, P.U.P.A., and Navrotsky, A. 2010. Transformation and crystallization energetics of synthetic and biogenic amorphous calcium carbonate. *Proc. Natl. Acad. Sci. USA.*, 107(38), 16438–16443. doi:10.1073/pnas.1009959107.

Raiteri, P., Demichelis, R., Gale, J.D., Kellermeier, M., Gebauer, D., Quigley, D. et al. 2012. Exploring the influence of organic species on pre- and post-nucleation calcium carbonate. *Faraday Discuss.*, 159, 61–85. doi:10.1039/c2fd20052j.

Raiteri, P. and Gale, J.D. 2010. Water is the key to nonclassical nucleation of amorphous calcium carbonate. *J. Am. Chem. Soc.*, 132(49), 17623–17634.

Raiteri, P., Gale, J.D., Quigley, D., and Rodger, P.M. 2010. Derivation of an accurate force-field for simulating the growth of calcium carbonate from aqueous solution: A new model for the calcite–water interface. *J. Phys. Chem. C*, 114(13), 5997–6010. doi:10.1021/jp910977a.

Raz, S., Hamilton, P.C., Wilt, F.H., Weiner, S., and Addadi, L. 2003. The transient phase of amorphous calcium carbonate in sea urchin larval spicules: The involvement of proteins and magnesium ions in its formation and stabilization. *Adv. Funct. Mater.*, 13(6), 480–486. doi:10.1002/adfm.200304285.

Reeder, R., Tang, Y., Schmidt, M.P., Kubista, L.M., Cowan, D.M., and Phillips, B.L. 2013. Characterization of structure in biogenic amorphous calcium carbonate: Pair distribution function and nuclear magnetic resonance studies of Lobster Gastrolith. *Cryst. Growth Des.*, 13(5), 1905–1914, 130401171732005. doi:10.1021/cg301653s.

Ridgwell, A. and Zeebe, R. 2005. The role of the global carbonate cycle in the regulation and evolution of the Earth system. *Earth Planet. Sci. Lett.*, 234(3–4), 299–315. doi:10.1016/j.epsl.2005.03.006.

Ridgwell, A.J., Kennedy, M.J., and Caldeira, K. 2003. Carbonate deposition, climate stability, and Neoproterozoic Ice Ages. *Science*, 302(5646), 859–862. doi:10.1126/science.1088342.

Romano, F., Chakraborty, D., Doye, J.P.K., Ouldridge, T.E., and Louis, A.A. 2013. Coarse-grained simulations of DNA overstretching. *J. Chem. Phys.*, 138(8), 085101. doi:10.1063/1.4792252.

Rosenbergl, J.M. 1992. The weighted histogram analysis method for free-energy calculations on biomolecules. I. The method, *J. Comput. Chem.*, 13(8), 1011–1021.

Rosta, E. and Hummer, G. 2009. Error and efficiency of replica exchange molecular dynamics simulations. *J. Chem. Phys.*, 131(16), 165102. doi:10.1063/1.3249608.

Sabri Dashti, D., Meng, Y., and Roitberg, A.E. 2012. pH-replica exchange molecular dynamics in proteins using a discrete protonation method. *J. Phys. Chem. B*, 116(30), 8805–8811. doi:10.1021/jp303385x.

Schlick, T. 2001. Time-trimming tricks for dynamic simulations: Splitting force updates to reduce computational work ways & means. *Structure*, 9(01), 45–53.

Schlick, T., Skeel, R.D., Brunger, A.T., Kal, L.V., Board, J.A., Hermans, J., and Schulten, K. 1999. Algorithmic challenges in computational molecular biophysics. *J. Comput. Phys.*, 48(1), 9–48.

Schmidt, W.J. 1924. *Die Bausteine des Tierkorpers in Polarisiertem Lichte*. Bonn, Germany: Verlag von Friedrich Cohen.

Schuster, U. and Watson, A.J. 2007. A variable and decreasing sink for atmospheric CO_2 in the North Atlantic. *J. Geophys. Res.*, 112(C11). doi:10.1029/2006JC003941.

Sindhikara, D.J., Emerson, D.J., and Roitberg, A.E. 2010. Exchange often and properly in replica exchange molecular dynamics. *J. Chem. Theory Comput.*, 6(9), 2804–2808. doi:10.1021/ct100281c.

Singer, J.W., Yazaydin, A.O., Kirkpatrick, R.J., and Bowers, G.M. 2012. Structure and transformation of amorphous calcium carbonate: A solid-state 43 Ca NMR and computational molecular dynamics investigation. *Chem. Mater.*, 24(10), 1828–1836.

Souaille, M. and Roux, B. 2001. Extension to the weighted histogram analysis method: Combining umbrella sampling with free energy calculations. *Comput. Phys. Commun.*, 135(1), 40–57. doi:10.1016/S0010-4655(00)00215-0.

Spyriouni, T., Tzoumanekas, C., Theodorou, D., Müller-Plathe, F., and Milano, G. 2007. Coarse-grained and reverse-mapped united-atom simulations of long-chain atactic polystyrene melts: Structure, thermodynamic properties, chain conformation, and entanglements. *Macromolecules*, 40(10), 3876–3885. doi:10.1021/ma0700983.

Stephens, C.J., Ladden, S.F., Meldrum, F.C., and Christenson, H.K. 2010. Amorphous calcium carbonate is stabilized in confinement. *Adv. Funct. Mater.*, 20(13), 2108–2115. doi:10.1002/adfm.201000248.

Sugita, Y. and Okamoto, Y. 1999. Replica-exchange molecular dynamics method for protein folding. *Chem. Phys. Lett.*, 314(1–2), 141–151. doi:10.1016/S0009-2614(99)01123-9.

Ten Wolde, P., Ruiz-Montero, M.J., and Frenkel, D. 1996. Numerical calculation of the rate of crystal nucleation in a Lennard-Jones system at moderate undercooling. *J. Chem. Phys.*, 104(24), 9932. doi:10.1063/1.471721.

Terakawa, T., Kameda, T., and Takada, S. 2010. On easy implementation of a variant of the replica exchange with solute tempering in GROMACS. *J. Comput. Chem.*, 32(7), 1228–1234. doi:10.1002/jcc.21703.

Torrie, G.M. and Valleau, J.P. 1974. Monte Carlo free energy estimates using non-Boltzmann sampling: Application to the sub-critical Lennard-Jones fluid. *Chem. Phys. Lett.*, 28(4), 578–581.

Treivus, E.B. 2002. On the thermodynamics of homogeneous crystal nucleation. *Crystallogr. Rep.*, 47(6), 1072–1075.

Tribello, G.A., Bruneval, F., Liew, C., and Parrinello, M. 2009a. A molecular dynamics study of the early stages of calcium carbonate growth. *J. Phys. Chem. B*, 113(34), 11680–11687. doi:10.1021/jp902606x.

Tribello, G.A., Liew, C., and Parrinello, M. 2009b. Binding of calcium and carbonate to polyacrylates. *J. Phys. Chem. B*, 113(20), 7081–7085. doi:10.1021/jp900283d.

Trudu, F., Donadio, D., and Parrinello, M. 2006. Freezing of a Lennard-Jones fluid: From nucleation to spinodal regime. *Phys. Rev. Lett.*, 97(10), 105701. doi:10.1103/PhysRevLett.97.105701.

Wallace, A.F., Hedges, L.O., Fernandez-Martinez, A., Raiteri, P., Waychunas, G.A., Gale, J.D. et al. (2013). Microscopic evidence for liquid-liquid separation in supersaturated $CaCO_3$ solutions. *Science*, 341(6148), 885–889.

Wang, D., Wallace, A.F., De Yoreo, J.J., and Dove, P.M. 2009a. Carboxylated molecules regulate magnesium content of amorphous calcium carbonates during calcification. *Proc. Natl. Acad. Sci. USA*, 106(51), 21511–6. doi:10.1073/pnas.0906741106.

Wang, Z.-J., Valeriani, C., and Frenkel, D. 2009b. Homogeneous bubble nucleation driven by local hot spots: A molecular dynamics study. *J. Phys. Chem. B*, 113(12), 3776–84. doi:10.1021/jp807727p.

Weiner, S. and Dove, P.M. 2003. An overview of biomineralization processes and the problem of the vital effect. In P.M. Dove, J.J. De Yoreo, and S. Weiner (eds.), *Biomineralization, Reviews in Mineralogy and Geochemistry*, Vol. 54 (pp. 1–29). Washington, DC: Mineralogical Society of America.

Weiner, S., Levi-Kalisman, Y., Raz, S., and Addadi, L. 2003. Biologically formed amorphous calcium carbonate. *Connect. Tissue Res.*, 44(1), 214–218. doi:10.1080/713713619.

Xiong, S., Qi, W., Huang, B., Wang, M., Cheng, Y., and Li, Y. 2011. Size and shape dependent surface free energy of metallic nanoparticles. *J. Comput. Theor. Nanosci.*, 8(12), 2477–2481. doi:10.1166/jctn.2011.1982.

Yang, J., McCoy, B.J., and Madras, G. 2006. Cluster kinetics and dynamics during spinodal decomposition. *J. Chem. Phys.*, 124(2), 024713. doi:10.1063/1.2151900.

Yashina, A., Meldrum, F., and Demello, A. 2012. Calcium carbonate polymorph control using droplet-based microfluidics. *Biomicrofluidics*, 6(2), 22001–2200110. doi:10.1063/1.3683162.

Zhang, H., Chen, B., and Banfield, J.F. 2009. The size dependence of the surface free energy of titania nanocrystals. *Phys. Chem. Chem. Phys.*, 11(14), 2553–2558. doi:10.1039/b819623k.

Zhang, W., Wu, C., and Duan, Y. 2005. Convergence of replica exchange molecular dynamics. *J. Chem. Phys.*, 123(15), 154105. doi:10.1063/1.2056540.

19 Direct measurement of interaction forces and energies with proximal probes

Raymond W. Friddle

Contents

19.1 WHY MEASURE BOND RUPTURE OF SINGLE BONDS?

The thought of capturing and rupturing molecular bonds may seem like a task too complicated for its worth. The size of a typical protein molecule is about 2 nm in diameter, or about 10,000 times smaller than the diameter of a human cell. And on a practical level, how can the strength of the bond connecting such a minute object be significant compared to the scale of objects that we are familiar with? But on the contrary, in situations that are vital to our lives, the breakage of a microscopic individual bond is the first step to a much larger event that we ultimately observe by eye.

Materials rarely, if ever, are known to fail by a one-step mechanism. Instead, failure usually occurs by way of cracks at interfaces or within the bulk of the object. This can be thought of as the "path of least resistance" to breaking something in two. Crack propagation occurs through the successive rupture of individual bonds at the crack tip. In order for the crack to propagate by a small amount, only a single bond need be broken. Hence, although the load is distributed across all bonds, the process leading to failure is dictated by the breakage of individual bonds—one at a time.

The fundamental nature of studying single-bond rupture cannot be overstated. When reviewing advanced textbooks on the kinetics and thermodynamics of reactions, the starting principle of the calculated quantities is a single-pair interaction. For example, the partition function, Z, of a molecule in its surface-bound state is calculated based on the energetic parameters of a single molecule. The scaling to an ensemble of molecules, which is what we are used to dealing with, is done through generic probability theory. Quantities like the lifetime of a bond, or free energy of a state, can all be calculated from the partition function, and hence are rooted in the fundamental single-molecule interaction they describe. Furthermore, details will invariably exist within the single-molecule interactions themselves—such as transient intermediate states—which are impossible to observe when looking at the averages of an ensemble.

Many of the theories used to describe modern-day bond rupture were derived during the 1930s to study the failure of nonbiological materials such as yarns and metals (Tobolsky and Eyring 1943; Zhurkov 1965). However, direct application of these theories to single-molecule measurements was ultimately realized by biophysicists to tweeze out the physics of life's machinery. The field of biomineralization is an exciting venue to bridge these two disciplines by bringing together the dynamics of inorganic mineral growth with the functional prowess of biological molecules.

19.2 ATOMIC FORCE MICROSCOPY

Prior to the 1980s, imaging of surfaces with resolution beyond the diffraction limit of light was limited to the various flavors of electron microscopes, which required special conditions such as a vacuum environment and samples bearing good electron-scattering elements. A major breakthrough in surface science came with the invention of the scanning tunneling microscope (STM) by Binnig and Rohrer in 1981, for which they received the Nobel Prize in Physics in 1986. Spectacular views of surface atoms became possible, such as the well-sought after structure of the Si(111)-(7 × 7) surface (Binnig et al. 1983). But the feedback mechanism in STM relies on the current passing through a small gap between the STM tip and the surface atoms. Therefore, STM is limited to conducting samples. STM experiments themselves revealed yet another possible route to investigating surfaces. When the STM tip was brought close to a surface, significant forces were found to act between the tip and the sample (Giessibl 2003). Shortly thereafter, Binnig, Gerber, and Quate developed the first prototype atomic force microscope (AFM) as a means to scan the topography of a surface simply through sensing the interatomic force between the sharp tip and the sample. The sheer simplicity of the technique opened scanned probe microscopies (SPM) to a much broader range of samples and environments.

The prototype AFM used gold foil as a flexible cantilever with a diamond tip attached to the end. An STM tip was placed above the foil to sense tiny deflections as the tip scanned over a surface. Later, Meyer and Amer developed the optical beam deflection method, an incredibly simple and effective approach to detecting miniscule deflections of a flexible cantilever (Meyer and Amer 1990). This approach focuses a laser beam on the back of a small cantilever, which reflects on to a split or quadrant photodiode. The difference between the signals at the photodiode segments is directly proportional to the change in the deflection of the cantilever. This approach to sensing AFM cantilever deflections is now standard in commercial and most custom AFMs today.

The primary components of an AFM are illustrated in Figure 19.1. The cantilever is mounted between the optical components and the sample stage. Lateral movement of the stage in x and y, as well as relative vertical displacement between the tip and sample in z is controlled by the voltage applied to the piezoelectric elements. Figure 19.1 shows the cantilever fixed while the piezo scans the sample; however, some AFMs mount the cantilever to the piezo from above. A laser is bounced off the back of a micromachined cantilever, typically made of Si or Si_3N_4, and then directed by a mirror onto a quadrant photodiode detector. The top–bottom and left–right halves of the detector are independently accessed such that the displacements of the laser

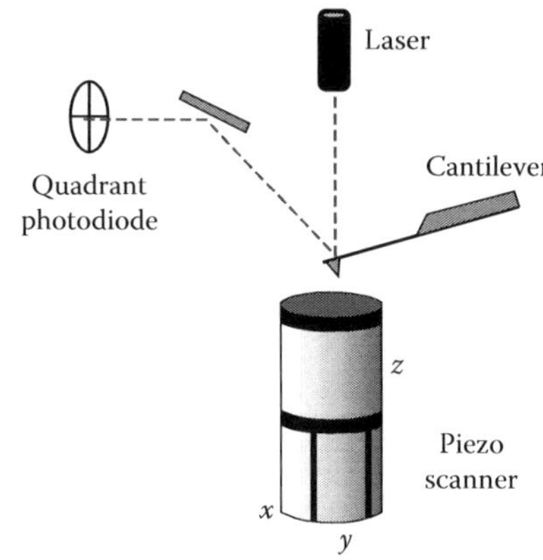

Figure 19.1 The principal components of an AFM. The optical lever detection system consists of a red laser focused on the back of the cantilever. The reflected laser beam is directed by a mirror on to a quadrant photodiode detector, which senses vertical and horizontal deflections of the cantilever. The segmented piezoelectric scanner allows displacements along the x, y, and z axes when bipolar high voltages are applied to the opposing sides of the segments. Shown here is a "sample-scanning" variety of AFM, whereas other systems are "tip-scanning," and some are a hybrid of sample (x–y) and tip (z) scanning.

spot produce a voltage difference, which is proportional to the vertical deflection or lateral torsion. This voltage signal is sent to the controller, which digitizes the signal for further processing. The typical imaging mode used in surface studies is to bring the cantilever tip in constant contact with the sample surface (contact mode). The user enters a set point for operation, which defines the vertical deflection of the cantilever to be maintained during scanning. The x–y piezo stage rasters the sample under the tip of the cantilever, while the z-piezo acts to null any deflections, due to the changes in the tip–sample force, of the cantilever away from the deflection set point. When in solid contact, changes in force are due to changes in sample topography, such as bumps or valleys, and therefore, the generated image represents the topography of the sample.

19.3 BASIC MECHANICS OF AFM CANTILEVERS

19.3.1 CANTILEVERS AS SPRINGS

The AFM is aptly named a *force* microscope due to the fact that an image is created through tip–sample forces transferred to the bending of the cantilever. A cantilever beam, conveniently, can be modeled as a spring (Figure 19.2), as long as the bending of the lever away from its force-free equilibrium is small:

$$V(x,t) = \frac{1}{2}k(x - vt)^2 \tag{19.1}$$

is the Hookean potential of the cantilever with stiffness k, which is time-dependent, and moving with velocity v. The force $f = -dV/dx$ exerted by the cantilever on an object located at, say $x = a$, is given by

$$f(x,t) = k(vt - a). \tag{19.2}$$

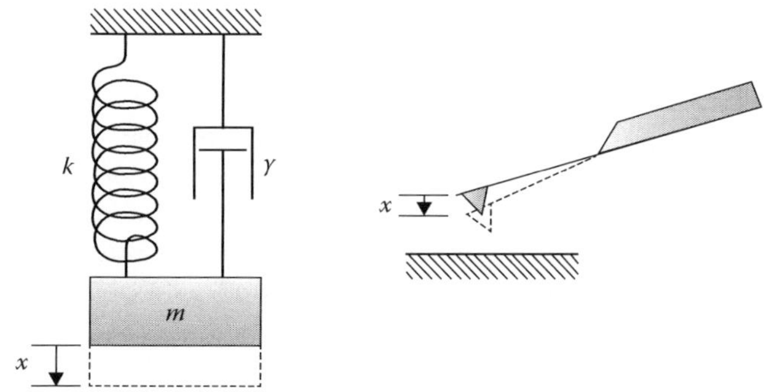

Figure 19.2 Basic mechanical model of an AFM cantilever. The restoring force is linear for small deflections of the cantilever beam, and therefore the cantilever sketched at the right is modeled as the spring on the left of stiffness *k*. The effective mass *m* relates to the cantilever tip and a large portion of the beam itself, as most of the beam bending occurs at the connection to the bulky chip. The model also includes a dashpot that adds damping of strength γ, which corresponds to the air or fluid surrounding the micron-scale cantilever.

Clearly, only the difference $\Delta x = vt - a$ is important in determining the applied force, and therefore, we have a linear relation between the force and deflection of the cantilever away from equilibrium:

$$\frac{f}{k} = \Delta x. \tag{19.3}$$

Hence, if the free tip-end of the cantilever is fixed, such as at the surface, while the clamped end of the cantilever is translated away from the surface at a constant velocity *v*, then the force applied to the surface is constantly ramped with time:

$$r = \frac{df}{dt} = kv. \tag{19.4}$$

19.3.2 FREQUENCY RESPONSE OF THE CANTILEVER AND THE THERMAL METHOD OF SPRING CONSTANT CALIBRATION

Two forces generally act on an AFM cantilever: forces due to interaction with the substrate, and random forces due to thermal motion of the surrounding molecules of the medium (i.e., air or fluid). The equation of motion describing the free end of the cantilever is modeled by the Langevin equation,

$$m\frac{d^2x}{dt^2} + \gamma\frac{dx}{dt} - \frac{dV(x,t)}{dx} = \xi(t), \tag{19.5}$$

where

- γ is the damping coefficient of the cantilever (Figure 19.2)
- $\xi(t)$ is a random force with zero mean
- $-dV/dx = kx$ is the restoring force of the cantilever when deflected away by a distance *x*

Rewriting, we have

$$\frac{d^2x}{dt^2} + \frac{\gamma}{m}\frac{dx}{dt} + \omega_0^2 x = \frac{\xi(t)}{m}, \tag{19.6}$$

where $\omega_0 = \sqrt{k/m}$ is the angular resonance frequency of the cantilever. Note that the usual frequency, *v*, used in practice is related by $\omega = 2\pi v$. The frequency response of this system, $S(\omega)$, known as the power spectral density (PSD), is found as the squared magnitude of the Fourier transform of the deflection with time $x(t)$. Making use of the Nyquist fluctuation–dissipation theorem (Gittes and Schmidt 1998) that states that the squared magnitude of the thermal white noise is given by $|\xi(\omega)|^2 = 4k_BTk/Q\omega_0$, the PSD is given by

$$S(\omega) = \frac{4k_BT}{\omega_0 Qk} \cdot \frac{1}{\left(1 - \frac{\omega^2}{\omega_0^2}\right)^2 + \frac{1}{Q^2}\frac{\omega^2}{\omega_0^2}}, \tag{19.7}$$

which is expressed in units of m^2/Hz. The PSD is the squared deflection of the cantilever per hertz—the density of noise at a given frequency. Equation 19.7 is used in the "thermal method" of estimating the spring constant of the cantilever, *k*. This is done by taking the squared magnitude of the Fourier transform of a time series of the cantilever deflection. It is crucial that the cantilever sensitivity (nm deflection/V deflection) is determined prior to performing the thermal method, since the units of Equation 19.7 are m^2/Hz, and not V^2/Hz. The leading factor in Equation 19.7 is known as the "thermomechanical noise limit" of the oscillator. It is the DC component of the spectrum ($S(\omega = 0)$) and is normally lumped into a fit parameter in practice. Various methods of finding the spring constant are available; however, the thermal method is the easiest to implement because it does not require an additional experimental setup—the spectral data is taken directly from the experiment at hand. However, a drawback of this technique is that measurements of the spring constant of a single cantilever can vary from 10% to 20% between measurements, and depending on whether it is in air or fluid (Kennedy et al. 2011). Figure 19.3 shows two thermal spectra taken on the same cantilever (microlever A, Bruker) in both air and toluene. Fitting Equation 19.7 to the spectra yields spring constants that differ by 7%, and clearly the fitting function

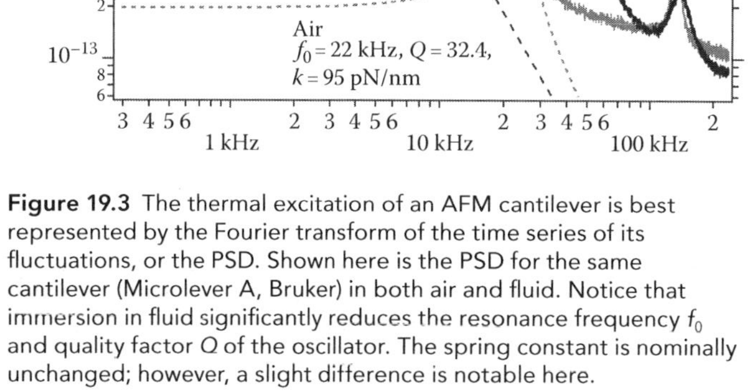

Figure 19.3 The thermal excitation of an AFM cantilever is best represented by the Fourier transform of the time series of its fluctuations, or the PSD. Shown here is the PSD for the same cantilever (Microlever A, Bruker) in both air and fluid. Notice that immersion in fluid significantly reduces the resonance frequency f_0 and quality factor *Q* of the oscillator. The spring constant is nominally unchanged; however, a slight difference is notable here.

is a poor representation of the actual spectrum. Techniques that include white noise and other modifications that can improve the results are currently under development (Pirzer and Hugel 2009; Kennedy et al. 2011; Dey and Szoszkiewicz 2012).

19.3.3 THERMAL AND EXTERNAL NOISE

The noise introduced into AFM bond rupture data arises from both internal and external sources. Internal sources of noise are those which are inherent to the technique itself. This is primarily due to the spectrum of thermal vibrations carried by the AFM cantilever when in contact with air or liquid (see preceding section). The amplitudes of these vibrations place a noise floor below which it is difficult to resolve rupture events. Additionally, external sources of noise arise from building and acoustic vibrations. These vibrations are coupled into the AFM housing and cantilever and result in differential oscillations, which are most notable when the tip is brought in contact. Both sources of noise limit the force resolution; however, internal sources are considered integral to the system being measured and nominally do not affect the quality of the data, whereas external sources of vibration do.

Minimizing the external sources of noise should be the first and foremost step to any force spectroscopy experiment. This is necessary to remove the oscillations the tip will experience when it is brought in contact with the surface—these oscillations will alter the bond rupture dynamics! Acoustic noise is remedied by an appropriate enclosure. It is important to consider the temperature and humidity variations, which can take place when the AFM is completely enclosed for extended periods of time. Vibration isolation can be accomplished by active or passive devices. Active vibration isolation stages can offer exceptional isolation; however, they can be expensive. Some passive examples that are tried and true are bungee cords suspended at least a few feet above the AFM, or negative-stiffness devices that use nonlinear springs designed to counteract the weight of the AFM in such a way that it rests on an effective spring of near-zero stiffness, and hence near-zero resonance.

It is well-known that the thermal noise of the cantilever itself is greater than the noise originating from the other parts of the AFM detection system (Smith 1995). As stated earlier, thermal noise does not manifest itself at a single frequency, but instead is distributed over a spectrum, as in Equation 19.7. Typical force curves are recorded at a sampling rate, B, well below the resonance frequency of the lever. In this case, it can be shown that the integral up to a bandwidth B of Equation 19.7 is approximately given by

$$\left\langle \Delta x^2 \right\rangle^{1/2} = \frac{\sqrt{4k_B T\gamma B}}{k}, \tag{19.8}$$

where γ is the damping coefficient of the cantilever, and here the relationship $\omega_0 Q = k/\gamma$ has been used. It follows directly that multiplying through by the spring constant, k, yields the thermal force noise,

$$\left\langle \Delta f^2 \right\rangle^{1/2} = \sqrt{4k_B T\gamma B}. \tag{19.9}$$

An important distinction here is that the root-mean-squared (rms) displacement of the cantilever is inversely related to the cantilever stiffness. Hence, the intuitive notion rings true that the softer the cantilever, the greater the fluctuations in the deflection signal. However, the relation between the spring constant and rms *force* fluctuations is lost. Making the cantilever softer or stiffer will not alter the force noise floor. The key to improving the force resolution at fixed bandwidth is through minimizing the damping coefficient, which is a function of the cantilever dimensions. For force measurements, the smaller the cantilever, the better the signal-to-noise resolution (Viani et al. 1999).

19.3.4 WORK OF ADHESION

For some applications, it is useful to not only measure rupture forces, but also the work done on pulling an adhesive bond apart. The thermodynamic work, W, performed on a system follows as the integral over the time rate of change of the Hamiltonian covering an observation time from 0 to t_f,

$$W = \int_0^{t_f} \frac{\partial H(x,t)}{\partial t} dt, \tag{19.10}$$

where in the case of bond rupture, the Hamiltonian is the sum of the time-invariant intermolecular potential of the bond, $U(x)$, and the time-dependent pulling potential applied to the bond, $V(x, t)$:

$$H(x,t) = U(x) + V(x,t). \tag{19.11}$$

Here, we consider a bimolecular bond with a single minimum in $U(x)$ located at $x = 0$. The total Hamiltonian evolves into a bistable system when the minimum of the pulling potential $V(x,t) = 1/2k(x - vt)^2$ is pulled sufficiently far past the barrier to rupture (see Figure 19.5). The minimum of the pulling potential defines the second state, and is located at vt. Therefore, assuming negligible displacement of the bound state minimum, the system will switch from $x = 0$ to $x = vt$ upon unbinding. Let unbinding occur at an arbitrary time t_s (with corresponding unbinding force $f_s = kvt_s$). The thermodynamic work is then

$$\begin{aligned} W &= \int_0^{t_s} kv^2 t dt + \int_{t_s}^{t_f} kv(vt - vt) dt \\ &= \frac{1}{2} kv^2 t_s^2 \\ &= \frac{f_s^2}{2k}. \end{aligned} \tag{19.12}$$

Thus, the thermodynamic work done on a bimolecular system is independent of the final observation time t_f, because no work is done on the molecule when it breaks away and resides around the minimum of the pulling potential. This is clear in the trivial second integral of Equation 19.12, where the location of the molecule at the tip moves along with the minimum of the cantilever at vt. While this is a sufficient approximation, it is not necessarily true in general, as hydrodynamic effects can contribute further dissipation as the probe is dragged through the fluid. Therefore, Equation 19.12 is valid at slow to moderate pulling speeds when viscous drag on the cantilever is negligible.

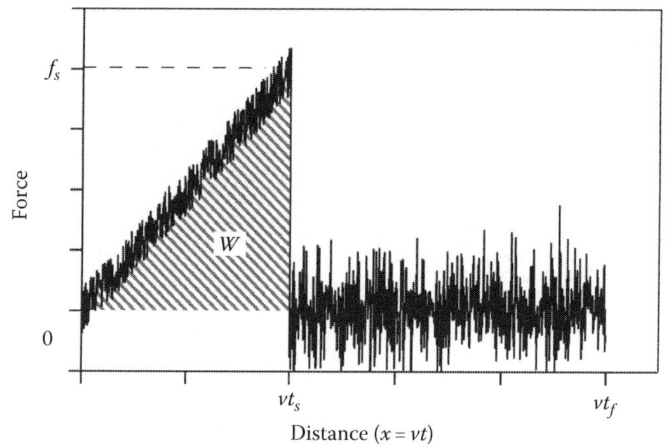

Figure 19.4 A typical force curve when pulling on a rigid substrate. The force is depicted as positive increasing here, whereas in most AFM systems it is presented as negative. The proportionality between the force, f, and displacement of the cantilever minimum, vt, means that one-half the area of the rectangle with sides f_s and vt_s defines the work done on the system (Equation 19.12). The net work done on rupturing the bond is independent of the final pulling distance, vt_f.

As summarized in Figure 19.4, the work done on bond rupture (dissociation of a bimolecular complex, or desorption from a surface) can be derived from the integral over the force-extension trajectory, or, in the case of a bond loaded by a spring of stiffness k, the work is simply one-half the square of the rupture force over the spring constant.

19.4 THEORETICAL MODELS OF DYNAMIC FORCE SPECTROSCOPY

19.4.1 POTENTIAL ENERGY MODEL AND ITS ASSUMPTIONS

The energy landscape of an archetypal constant loading-rate bond rupture experiment is depicted in Figure 19.5, where a quadratic pulling potential, the cantilever spring, is swept from left to right over the intermolecular potential of the bond. The sum of the two potentials at each point in time forms the resulting potential for which a theoretical particle must escape from the left well to the right. For the theory that follows, it is of utmost importance to consider the stiffness of the combined cantilever/linker relative to the typical stiffness of the intermolecular bonds. When the effective stiffness of the probe linkage is comparable to that of the bond, the bound state of the bond can be significantly perturbed by the steeply rising potential of the probe. When the effective stiffness of the probe is much less than that of the bond, the situation is much different. In this case, the bound state is negligibly moved by the probing potential, while the barrier to rupture is lowered. It is this second case which we consider here. From the perspective of the observer viewing the force trajectory, only two primary states are explored in this scenario: the bound state of interest and the unbound state. The measured trajectory, or force–distance curve, will nominally abruptly transition from one state to the other. Typically, theoretical modeling of driven bond rupture processes begins by recognizing these two primary states of the system, yet assuming that after the initial bond rupture, reverse transitions back into the bound state are negligible at experimental loading rates. Here, we will briefly derive the model based on the complete two-state system and show that the irreversible case emerges as a limit.

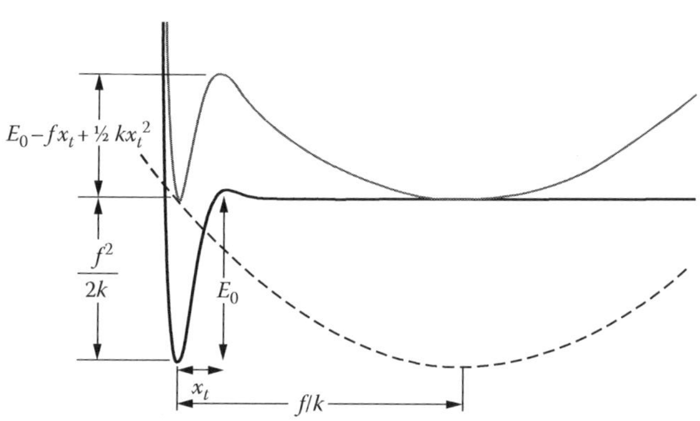

Figure 19.5 The potential energy of the bond (solid black), parabolic cantilever (dashed), and combined potentials (solid gray) used to model the forced escape process. The energy of the bound state (well at the left) is raised by $f^2/2k$, which is the energy of a spring extended by a distance f/k. The barrier height relative to the raised minimum is reduced in proportion to f, with a small correction of $kx_t^2/2$ accounting for the nonlinear shape of the forcing potential.

19.4.2 TRANSITION RATES AND THE TWO-STATE MODEL OF A SINGLE BOND

The equation of motion of the probabilities for residing in the bound p and unbound q states is given by the two-state master equation (Evans and Williams 2002; Friddle et al. 2008),

$$r\frac{dp(f)}{df} = -k_{off}(f)p(f) + k_{on}(f)q(f), \qquad (19.13)$$

where we have assumed that the loading rate is constant, and thus the time is linearly transformed to force, $r = df/dt$. For a minimum located at the origin and a barrier located at $x = x_t$, the effect of the pulling potential on an intrinsic energy barrier of E_0 is given to first order by $E(f) = E_0 - fx_t + kx_t^2/2$, where we include the finite effect of the pulling spring of stiffness k on the overall energy (Tshiprut et al. 2008; Walton et al. 2008). Assuming an Arrhenius form of thermally activated barrier crossing, the instantaneous kinetic unbinding rate follows as (Bell 1978; Evans 1998; Friddle et al. 2012)

$$k_{off}(f) = k_{off}^0 \exp\left[\frac{\left(fx_t - kx_t^2/2\right)}{k_BT}\right], \qquad (19.14)$$

where k_{off}^0 is the intrinsic unbinding rate of the bond. The rebinding rate follows as approximately the natural frequency of the cantilever, k_{on}^0, scaled by the Boltzmann-weighted energy of a spring extended between the barrier location x_t and the relative displacement of the spring minimum, f/k,

$$k_{on}(f) = k_{on}^0 \exp\left[-\frac{k}{2}\left(\frac{f}{k} - x_t\right)^2 \Big/ k_BT\right]$$
$$= k_{off}(f)\exp\left[\left(\Delta G_1 - \frac{f^2}{2k}\right)\Big/ k_BT\right], \qquad (19.15)$$

where $\Delta G_1 = kT \ln k_{on}^0 / k_{off}^0$ is the free energy difference between the single-molecule bound state and the potential of the cantilever. The second relation in Equation 19.15 shows immediately that the unbinding and rebinding rates are equal at a unique force given by (Evans 1998; Friddle et al. 2008)

$$f_{eq} = \sqrt{2k\Delta G_1}, \qquad (19.16)$$

which defines the equilibrium force for the combined cantilever/bond system.

19.4.3 FORCE-RAMP RUPTURE OF SINGLE BONDS: DYNAMIC FORCE SPECTROSCOPY

It is not possible to analytically solve the two-state master equation in Equation 19.13 due to the time dependence of the rates. It is, however, possible to derive a very accurate interpolative solution based on some straightforward observations of the behavior of the rates. It is clear from the second relation in Equation 19.15 that the ratio of rebinding to unbinding rates falls precipitously with increasing force, as $\exp - f^2/2k$, after the force exceeds f_{eq}. This suggests that the dominant contribution to the measured rupture force from the rebinding rate takes place for forces less than f_{eq}, whereas beyond f_{eq}, the unbinding rate dominates the dynamics. Indeed, we have recently shown (Friddle et al. 2012) that appropriately accounting for the contribution of the rebinding rate to the equilibrium force, and beginning the dynamics at f_{eq}, permits ignoring the rebinding rate for the remainder of the process and solving the first-order problem given by

$$\int_1^P \frac{dp'}{p'} \cong -\frac{1}{r} \int_{f_{eq}}^{f} k_{off}(f')\,df'. \qquad (19.17)$$

The mean force follows as $\langle f \rangle \cong f_{eq} + \int_{f_{eq}}^{\infty} p(f)\,df$, and we have (Friddle et al. 2012)

$$\langle f \rangle \cong f_{eq} + f_\beta e^{1/R(f_{eq})} E_1\left(\frac{1}{R(f_{eq})}\right),$$

$$R(f_{eq}) = \frac{r}{k_{off}(f_{eq}) f_\beta}, \qquad (19.18)$$

where

$f_\beta = k_B T/x_t$ is the thermal force scale

$E_1(z) = \int_z^{\infty} ds\, e^{-s}/s$ is the exponential integral

Some data analysis software such as Igor Pro (Wavemetrics) come standard with numerical routines for the exponential integral, and hence Equation 19.18 can be fit to data directly. However, in the event that such routines are not available, an analytical form

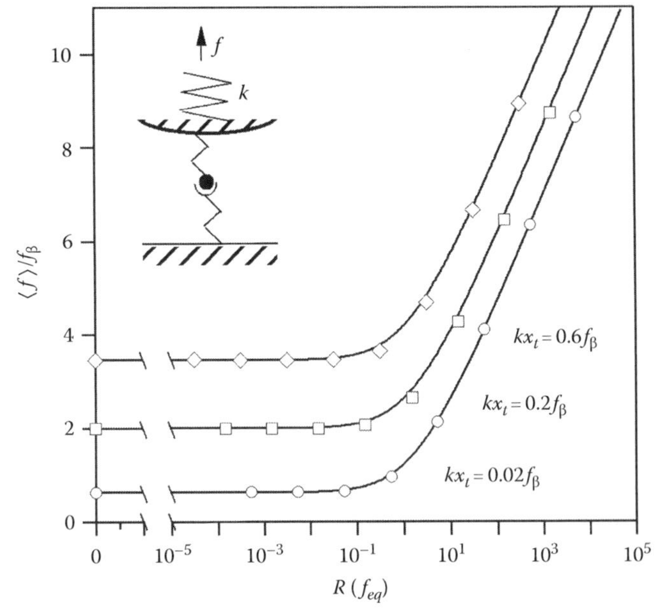

Figure 19.6 The differential equation defining the two-state master equation, Equation 19.13, is solved by numerical integration (symbols) with the Bulirsch–Stoer method using Richardson extrapolation (Igor Pro, Wavemetrics). Results are shown for $\Delta G_1 = 10k_BT$, and three different spring constants, k, given in units of the thermal force scale k_BT/x_t. The single-bond model in Equation 19.18 is calculated with identical parameters for comparison (solid curves). Note that the increased spring constant has two effects on the spectrum: the level of the equilibrium plateau is increased, and the entire spectrum is shifted to the left due to the decreased effective kinetic unbinding rate $k_{off}^0 \exp\left[-kx_t^2/2/k_BT\right]$.

for Equation 19.18 can be found by the following approximation (Friddle 2008; Friddle et al. 2012):

$$\langle f \rangle \cong f_{eq} + f_\beta \ln\left(1 + e^{-\gamma} R(f_{eq})\right), \qquad (19.19)$$

where $\gamma = 0.577\ldots$ is Euler's constant (not to be confused with the damping coefficient).

As shown in Figure 19.6, Equation 19.18 produces the two trends that are often observed in force spectroscopy data. In the limit of vanishing loading rate, the spectrum tends to the equilibrium force. In the fast-loading-rate limit, the spectrum reduces to the familiar $f \sim \ln r$ form for irreversible unbinding given by

$$\langle f \rangle_{r\to\infty} \cong f_\beta \ln[e^{-\gamma} R(0)], \qquad (19.20)$$

where again $\gamma = 0.577\ldots$ is Euler's constant.

19.4.3.1 Interpreting the single bond free energy

What is the meaning of the *single-molecule free energy* ΔG_1 that is presented in Equation 19.16? While it may appear to be a fundamental quantity, it is *not* a universal measure of free energy. As illustrated in Figure 19.7, the free end of the cantilever (bound by one of the interacting molecules) is converted from one state to another: the first being the bound state where the end of the cantilever is held to the surface through the adhesive bond at the tip, and the second being the unbound state where the end of the cantilever

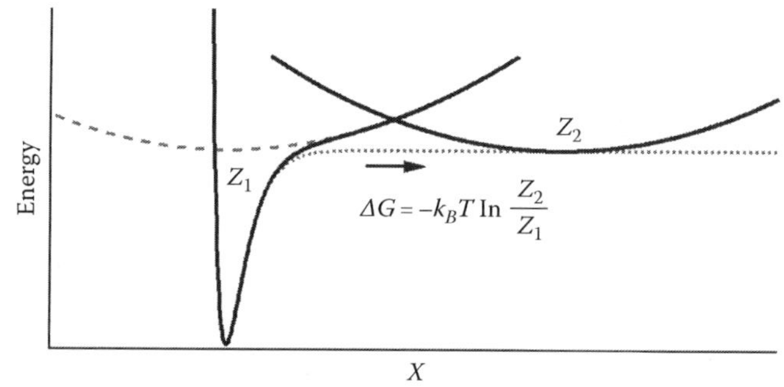

Figure 19.7 Statistical mechanics of the single-molecule free energy ΔG_1. Bond rupture proceeds by transferring the system from the bound state defined by the deep sharp well at left to the shallow wide well at the right. The free energy difference for this transformation is given by the logarithm of the ratio of partition functions of each state. The bound state partition function, Z_1, is defined by a high-frequency (low entropy) low-lying energy well. On the other hand, the unbound partition function Z_2 is given by a low frequency (high entropy) and higher energy. Thus, the balance between the depth of the bound state and its confinement against the dispersed unbound state dictates the value of the free energy change. When the probing potential is stiffer, the unbound state is more confined, and hence the change in entropy will be smaller—the reverse being true for softer probe potentials. Hence, in order to make proper use of the free energy measured by force spectroscopy, one must properly account for the contribution to the transformation by the entropy of the unbound state, defined by its stiffness, *k*.

fluctuates due to thermal motion but is not attached to the surface. Both of these two states have a defined absolute free energy, and ΔG_1 represents their difference. The driver for the conversion is due to the first state's free energy being raised by an amount $f^2/2k$ with translation of the cantilever a distance f/k away from the surface. Conveniently, however, the second state's free energy is unchanged with translation; it is simply a parabolic well defined by the stiffness of the cantilever, that is, after the cantilever potential is translated a sufficient distance to not overlap significantly with the bond potential. Because the free energy is related to the number of states—by way of the partition function, Z—a stiffer probe will create a less favorable, higher free energy state, whereas a softer probe produces a more favorable, lower free energy state. Hence, the following statement is crucial to our interpretation of energy in this context:

> The energy of the free cantilever end, defined solely by its stiffness *k*, is the reference state for the measurement of individual intermolecular bond free energy by force spectroscopy.

Just as binding free energy assays in bulk solution require a standard reference state to be meaningful (e.g., 1 M concentration), the single-molecule free energy measured by DFS requires a clear definition of its reference state to be comparable to the other measures of free energy.

19.4.4 FORCE-RAMP RUPTURE OF MULTIPLE BONDS

The most common scenario one encounters in practice is the measurement of multiple connections between the probe and the substrate. Evans and Williams discussed the difficulties in measuring a true single bond when blindly allowing molecules to connect between the probe and the substrate (Evans and Williams 2002). Their main conclusion was that, unless additional knowledge is used to determine bond valency (Sulchek et al. 2005), the actual experiment is more likely to measure multiple bonds than single bonds. It is therefore fitting that we consider the effects that multiple bonds have on the resulting rupture dynamics and how to interpret such data. In a continuum approach, the number of formed parallel bonds N_b, of which N_t total bonds are allowed to independently unbind and rebind, is represented by the differential equation (Bell 1978; Seifert 2000; Erdmann and Schwarz 2004; Liang and Chen 2011)

$$r\frac{dN_b}{df} = -N_b k_{off}\left(\frac{f}{N_b}\right) + (N_t - N_b)k_a, \qquad (19.21)$$

where

$k_{off}(f)$ is given by Equation 19.14

k_a is the constant rebinding rate of an individual molecule acting over a short distance from its retracted position, while the overall cluster remains bound (see inset to Figure 19.8)

Again, just as with the single-bond case, two primary regimes emerge. At slow loading rates, fluctuations in the number of closed bonds are relatively fast, and the system establishes a metastable number of bonds, $N(f) = N_t/(1 + k_{off}(f/N)/k_a)$. From here, we define N as the equilibrated number of formed bonds at zero force, $N = N_t/\left(1 + k_{off}^0/k_a\right)$. Stability of the cluster is

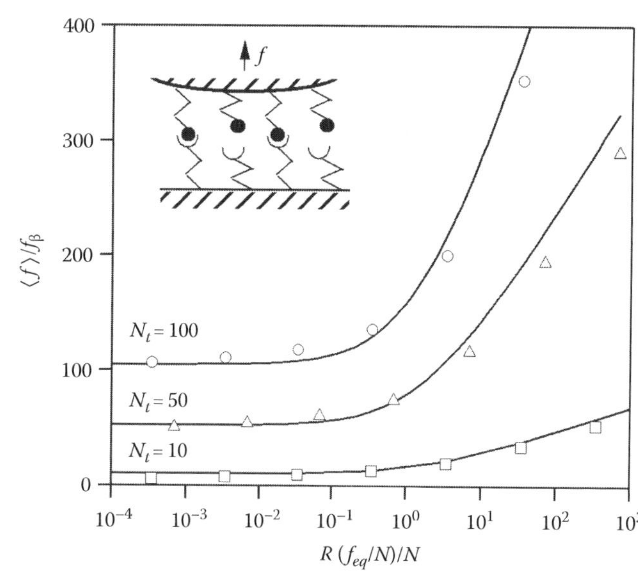

Figure 19.8 The multibond theory and comparison to Monte Carlo simulation of *N*-bonded systems. The mean rupture force for the indicated number of N_t bonds, as described by Equation 19.21, is solved by the Gillespie algorithm for $k_a/k_{off}^0 = 10$. Rupture forces for a given trajectory are determined when $N_b = 0$, and hundreds of trajectories are used to produce the mean rupture force (symbols) plotted here against the dimensionless loading rate. For comparison, the multibond model in Equation 19.23 is calculated (solid curves) for the same ratio of transition rates, and with $N = N_t/(1 + k_{off}^0/k_a)$. (Reproduced from Friddle, R.W., Noy, A., and De Yoreo, J.J., Interpreting the widespread nonlinear force spectra of intermolecular bonds, *Proc. Natl. Acad. Sci. USA*, 109 (34): 13573–13578, 2012. With permission from National Academy of Sciences, USA.)

completely lost when the force exceeds the equilibrium force (Bell 1978; Erdmann and Schwarz 2004; Lin et al. 2010),

$$f_{eq} = Nf_\beta W\left(e^{\beta \Delta G_N - 1}\right), \qquad (19.22)$$

where

$W(x)$ is the Lambert W-function, which is defined as the solution for W to the equation $We^W = x$

$\Delta G_N = k_B T \ln k_a / k_{off}^0$ defines the equilibrium-free energy of the closed bond relative to an open bond within the bound cluster

The second primary regime appears at fast loading rates where the rebinding rate k_a, and hence fluctuations in N_b, become less important. Here, the force reaches higher loads, and the bonds rupture rapidly after the first bond fails. An interpolation between these two regimes was found previously (Friddle et al. 2012):

$$\langle f \rangle \cong f_{eq} + Nf_\beta e^{N/R(f_{eq}/N)} E_1\left(\frac{N}{R(f_{eq}/N)}\right),$$

$$R\left(\frac{f_{eq}}{N}\right) = \frac{r}{k_{off}(f_{eq}/N) f_\beta}. \qquad (19.23)$$

A comparison of the spectrum predicted by Equation 19.23 to Monte Carlo simulation of the complete model in Equation 19.21 over seven decades in loading rate (Figure 19.8) shows that this model accurately captures the two primary trends.

19.4.4.1 Interpretation of multiple bond free energy

As we discuss below, in the case of an unknown number of N bonds, the true transition state x_t is not determined. However, the escape rate k_{off}^0 and the free energy can be determined from the fit of Equation 19.23 to a full data set plotted as mean rupture force versus loading rate. In this case, the free energy ΔG_N must be carefully understood to be the free energy for a single molecule between two states: (1) the bound state, bridging the probe to the surface, and (2) the unbound state, connected only to the probe. However, in the unbound state, the single molecule is still held closely to the surface, because we assume on average some fraction of the other $N - 1$ molecules in the cluster is still bridging the probe to the surface. Hence, the space available to an unbound molecule in the cluster can be highly constrained and biases the molecule to quickly rebind. Defining the energy of this unbound state is less trivial than the true single-bond case where it is simply defined by the cantilever potential. The free energy measured in the multibond case is, however, comparable to similar measurements in which the unbound state is equivalent (such as the same self-assembled monolayer on a probe against different surfaces).

19.4.5 DISCERNING SINGLE-BOND FROM MULTIBOND DATA

The multibond model of Equation 19.23 shows that if we combine the number of bonds N into an apparent force scale,

$$f_\beta^{app} = Nf_\beta = \frac{k_B T}{x_t / N}, \qquad (19.24)$$

along with a generic definition of an equilibrium force f_{eq}, then the multibond force spectrum takes the same mathematical form as the single-bond model in Equation 19.18. Equation 19.24 shows that the multiple bond model modifies the single-bond model simply through the parameter N, which factors inversely to the transition state x_t everywhere in the function. In some cases, unreasonably small transition state distances are reported in force spectroscopy experiments. We see here that if an N-bonded system is analyzed in the kinetic regime using the common practice single-bond model, the fitted apparent transition state, $x_t^{app} = x_t / N$, will be N times smaller than the true distance x_t. Potential of mean force calculations generally finds the distance from the minimum to the primary barrier to be on the order of 1 Å (Pettitt and Karplus 1985). Therefore, as a rule of thumb, if the bond rupture data is fitted to the single-bond force spectrum of Equation 19.18, and if one finds the best fit transition state falls several orders of magnitude less than 1 Å, it is most likely that multiple connections exist in the measurement. In such situations, it is difficult to determine the true value N. Note however, although the true distance x_t cannot be determined, all is not lost. Even without the knowledge of N, the intrinsic unbinding rate of a single bond k_{off}^0 is the fit parameter in the multibond model of Equation 19.23, as well as the single-bond free energy between the bound and closely unbound states ΔG_N (not to be confused with ΔG_1, the free energy between a single bond and the cantilever).

19.4.6 ANATOMY OF A FORCE SPECTRUM

To summarize this section on the theory of force-ramp bond rupture, Figure 19.9 shows an idealized spectrum and parameters

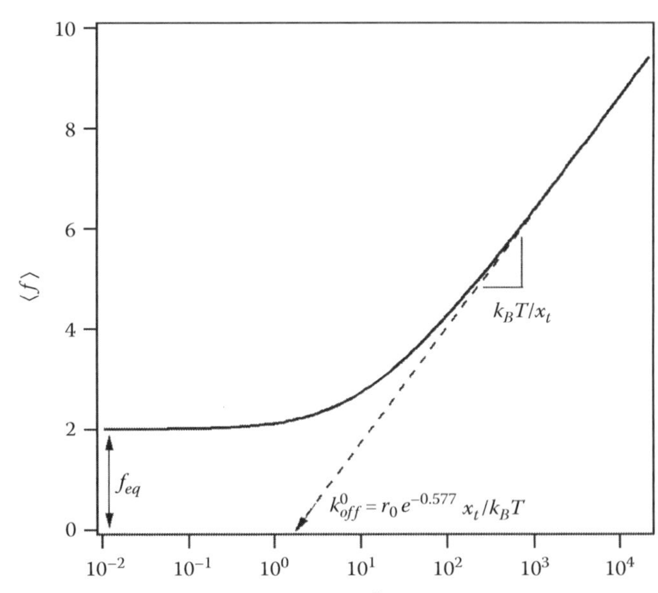

Figure 19.9 The three primary parameters of the system that define the force spectrum. In the high-force leg of the spectrum, the slope of the rupture force with log-loading rate is dictated inversely by the distance to the transition state x_t. Extending this trend downward (dashed line), the loading rate where the trend crosses zero force, r_0, is defined by the kinetic unbinding rate of the system. Hence, changes to k_{off}^0 shift the entire spectrum left and right. Finally, the height of the asymptotically flat lower leg is dictated by the equilibrium force, which encodes the free energy of the system.

which define its shape when it is cast into a spectrum of mean rupture force versus loading rate. This spectrum can be generated by either of the single-bond or multibond force spectra models when the transition state is assumed to be apparent, $x_t^{app} = x_t/N$, and the equilibrium force, f_{eq}, is generic to represent either the single-bond or multibond equilibrium force.

19.5 DATA ACQUISITION, ANALYSIS, AND INTERPRETATION

19.5.1 HOW MANY DATA POINTS?

A question that is often faced when taking force measurements is: "how many force curves should I record?" Obviously, higher number of measurements will improve the estimate of any parameter of interest. However, data storage, hours spent in the laboratory, and data analysis typically put a limit on the number of individual curves recorded per experiment. Unfortunately, this question does not have a straightforward answer for the following reasons:

1. Not all individual measurements are "hits." A number of tip–surface contacts will not result in forming the particular bond of interest. If a fixed number of samples is decided upon at the onset of experimentation, trouble may arise later if a significant fraction of those measurements are found to be "misses."
2. If the chance of multiple bonds is apparent, the mean for this distribution may be acquired with a modest number of samples, but finer details in the histogram, such as peaks that correspond to each bond valency, require smaller bin sizes, and hence more samples.
3. The desired information may require greater or fewer measurements. If the objective is to find the mean, this may be found with relatively fewer samples than those required to accurately fit the shape of a model distribution to a histogram of the samples.

We can however borrow from the statistics of normally distributed data to get a quantitative estimate of the required statistics. The number of samples that should be acquired, n, such that the true mean lies within a margin of error E, with a confidence defined by the z-score z^*, is given by

$$n = \left(\frac{z^*\sigma}{E}\right)^2,$$

where σ is the standard deviation of the population. The z-scores for commonly used confidence intervals are 90%: $z^* = 1.645$; 95%: $z^* = 1.96$; and 99%: $z^* = 2.576$. This relation shows rather intuitively that narrower distributions will require fewer data points to accurately estimate the mean. Note that for large loading rates, the distribution of rupture forces tends toward a standard deviation of $\sigma = f_\beta = k_B T/x_t$ (Evans 2001).

To get a sense for the number of measurements needed to estimate the *mean* rupture force, let us assume an ideal single bond with transition state $x_t = 2$ Å is repeatedly pulled at a single loading rate. Let us further require that the true mean be within ±10 pN of the measured value with 99% confidence. The value of $E = 10$ pN is a common noise threshold for commercial AFM/cantilever systems. Remarkably, these requirements only recommend a minimum sample size of about $n = 6$ data points per loading rate! Imposing a more stringent margin of error of ±1 pN raises the number of samples to $n = 53$, which is still reasonable considering most experimenters easily record 100s of force curves per loading rate. Such a relaxed requirement is encouraging after all, since carefully executed single-molecule experiments will nominally result in many "misses," where the low densities of molecules at the tip and substrate do not bind.

The "number of samples" question is less straightforward to answer in more complicated systems. When there are many bonds involved, the standard deviation can become quite large, thus requiring more measurements to accurately determine the mean. Furthermore, as mentioned earlier, if one wishes to further scrutinize the data beyond simply the mean, such as examining sub-peaks of the binned data, then a larger number of samples will be needed to resolve these details.

19.5.2 CURVE FITTING AND TRUSTING PARAMETERS

Fitting Equation 19.18 or 19.23 to experimental force spectra is a three-parameter fit: k_{off}^0, x_t/N, and f_{eq}, the kinetic unbinding rate, apparent distance to transition state, and equilibrium force, respectively. The quality of the fit will depend on the number of loading-rate points measured, and on how many decades in loading-rate those points cover. If too few decades are covered, usually only one of the two loading regimes will be amply defined by the data, and as such one or more fitting parameters will be ambiguous. Before attempting to fit Equation 19.18 or 19.23 to the data, be sure to display the mean rupture force against loading rate on a lin-log scale, and inspect the spectrum to ensure two clear regimes exist.

19.6 FORCE SPECTROSCOPY APPLIED: AN EXAMPLE AND PROCEDURAL TIPS

19.6.1 FACE-SPECIFIC PEPTIDE/MINERAL ADSORPTION FREE ENERGY: AMELOGENIN ADSORPTION TO TOOTH ENAMEL

During mineralization, organized proteins direct the formation of mineral components. As with all assembly processes, the free energy change provides the underlying thermodynamic driver, in this case, reflecting protein interactions with the emerging mineral. While a number of techniques exist to determine binding free energy, such as calorimetry, it is difficult to acquire the face-specific free energies of mineral binding by bulk methods. Furthermore, computational approaches struggle with the complexities of proteins, the inadequacies of water potential models, and the effects of background electrolytes. This is precisely the regime of experimentation where force spectroscopy excels. In the following, we briefly highlight an example of using the techniques discussed in this chapter to determine the free energy of adsorption of the Amelogenin protein C-terminal fragment bound to the (100) face of hydroxyapatite (HAp)—the mineral phase in tooth enamel (Margolis et al. 2006).

Box 19.1 Force Spectroscopy Step-by-Step

1. *Define the interaction.* Are single or many bonds the objective? Are flexible linkers needed? Does a specific location need to be targeted?
2. *Choose a cantilever.* Small cantilevers offer better signal to noise. Soft levers are appropriate for the models. Aim for less than 1 N/m.
3. *Clean the cantilever.* Even new cantilevers are coated in oils and polymers from the packaging. Clean by RF plasma, UV/ozone, or solvents.
4. *Coat and functionalize the parts.* Deposit Au for thiols, or activate by brief piranha etch for silanes. Add any linkers then molecules of interest. Sacrifice some tips to *perform control force measurements* at each chemical step to verify each reaction.
5. *Set up and wait.* Allow the cantilever to reach thermal equilibrium in solution with the laser aligned.
6. *Take some test shots.* Make sure the force curve looks good. Sometimes, the first few dislodge loose material.
7. *Calibrate the spring constant.* Measure the lever sensitivity (nm deflection/V deflection), then back away from the surface and take a thermal spectrum.
8. *Acquire the data thoughtfully.* Keeping the force spectrum in mind, consider how many force curves you want to punish the tip with at one velocity before moving to the next.
9. *Calibrate the spring constant (again).* It might have changed during the course of the experiment.
10. *Perform controls.* If the bond is specific, prove it by blocking it and repeating the experiment, preferably *with the same tip*.
11. *Analyze the data.* Use automated routines for hundreds of force curves, or by eye if looking for something other than simple rupture.

Box 19.1 provides the basic steps one should keep in mind when measuring a dynamic force spectrum. The experiment described in the following paragraphs was essentially performed following these steps. Figure 19.10 presents some of the major components: setup, data and controls, and analysis. Functionalization of the silicon nitride tips was carefully carried out by cleaning followed by a series of chemical steps. The tips were first cleaned under UV/ozone for 10 min, then immersed in acetone followed by ethanol, and dried under a nitrogen stream. Cleaned tips were then coated with 4 nm Cr followed by 40 nm Au by thermal evaporation. Gold-coated probes were then immersed for 30 min in a DMF solution containing 0.2 mM of the heterobifunctional cross-linker LC-SPDP consisting of a pyridyl disulfide, which adsorbs to Au, and an N-hydroxysuccinimide (NHS) ester that reacts with the N-terminal amine or lysine residues of the peptide to form an amide bond. After rinsing in DMF, followed by ethanol, the tips were immersed in 0.2 mM of the C-terminal peptide in phosphate-buffered saline (PBS) overnight. Functionalized tips (Figure 19.10a) were rinsed in pure water to remove any loosely bound peptide prior to use.

As a control, each functionalization step of the previously mentioned recipe was stopped and the tip used to acquire force spectroscopy data on the HAp crystal (Figure 19.10b). That is, a cleaned Au-coated tip and a Au/LC-SPDP tip were used, in addition to the complete Au/LC-SPDP/peptide tip. This is important to convince oneself that the functionalization reaction has been completed successfully. Fortunately, the strength of the pure Au and LC-SPDP functionalizations was much smaller than that of the peptide. This was determined from several independent tests of different cantilevers prepared with the same procedure, and provided a clear indicator of a properly functionalized tip.

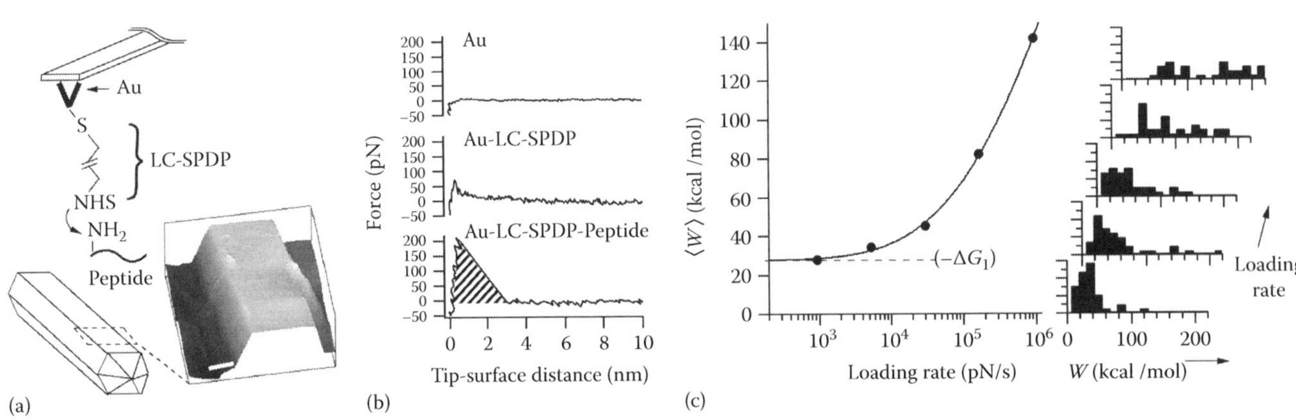

Figure 19.10 Measuring the work and free energy of binding the C-terminal sequence of Amelogenin to the (100) face of HAp. (a) The N-terminus of the peptide is linked to a gold-coated AFM cantilever by way of a bifunctional linker. Inset: AFM image of a HAp crystal used to characterize the surface and locate the target face before force measurements (scale bar 4 μm). (b) Force–distance curves for each level of functionalization (Au coating, LC-SPDP linker, and peptide) at approximately equivalent loading rates. The hatched region indicates the work, *W*, of adhesion. (c) Means (solid circles) and corresponding histograms (solid bars) of work measured from repeated force–distance trajectories for a peptide-functionalized tip (spring constant, 93 pN/nm). Solid curve is a fit to Equation 19.18 expressed as the work through $W = f^2/2k$. The mean work tends asymptotically to a finite value given by the free energy difference (dashed line) of $\Delta G_1 = -27.6$ kcal/mol. The histograms for increasing loading rate (0.93, 5.25, 29.5, 165.9, and 933 nN/s) are offset for clarity. (Data from Friddle, R.W. et al., *Angew. Chem. Int.*, 50(33), 7541, 2011.)

As shown in Figure 19.10a, the tip is placed directly on the (100) face of individual HAp crystals. The crystal diameters were on the order of 10 μm wide, making them visible under bright-field optics, and the quality of the target faces could be characterized by AFM imaging the crystals with bare tips. Force measurements between modified tips and HAp crystals were performed in a calcium phosphate solution at an approximately equilibrium saturation with the HAp crystal. Measurements were made with the MFP3D AFM (Asylum Research, Santa Barbara, CA). To account for any surface heterogeneity, a custom routine was used to randomly sample different locations on the surface to give a representative average. A constant approach velocity of 200 nm/s was used for every pulling velocity studied. A 2–3 nm deflection trigger was used to contact the surface and dwell for 1 s before pulling away.

The force–distance curves were converted directly into work, W, associated with the breaking of the Amel–HAp bond (see Equation 19.12). Because the bond rupture process is time dependent, this work is greater than the equilibrium free energy of binding by an amount of dissipated heat, which decreases with decreasing loading rate $r = df/dt$. Figure 19.10c shows the mean work against loading rate and the corresponding histograms for each loading rate. As expected, the mean work falls with decreasing loading rate and extrapolates to a finite value as r tends to zero. The fit of Equation 19.18 converted to work is excellent, and the asymptote gives $\Delta G_1 = -27.6$ kcal/mol.

Complementary MD simulations were carried out as a means to qualify the experimental free energy results against those found by computation. The details of these simulations are found in the study of Friddle et al. (2011) and in the detailed supplement to their publication. In brief, two simulation scenarios were devised: the first including water molecules explicitly, and the second including water implicitly through a dielectric environment of $\varepsilon = 80$. The range of computed binding energies for the (100) surfaces in explicit water was −83 to −104 kcal/mol, while estimates of −6 to −15 kcal/mol were found for implicit water. We see that the MD results bracket the experimental result of −28 kcal/mol found by force spectroscopy. This may highlight the limitation of MD methods for determining true energy magnitudes. Interaction energies of charged organic species on the surfaces of ionic crystals are overestimated in explicit water simulations due to a mismatch of force-field charge assignments between the organic and mineral components. On the other hand, implicit water quenches the normally solvent-inaccessible short-range interactions which dominate the binding energy.

Because force spectroscopy is the only direct experimental analog to MD simulations of molecular binding, this work established a unique link between experiment and computation that is otherwise inaccessible and should facilitate future experiments that elucidate how face-specific thermodynamics guide protein-directed growth of minerals (Friddle et al. 2011).

19.7 SUMMARY

What separates AFM-based measurements of kinetic and thermodynamic parameters from other bulk methods is the ability to (1) observe the complete trajectory of unbinding, offering the possibility of detecting intermediate events, (2) determine thermodynamic quantities with unparalleled range, whereas bulk techniques have limitations at extremely high or low concentrations, and (3) specifically target a microscopic location to measure these parameters locally, and eliminate averaging over multifaceted surfaces. These properties place dynamic force spectroscopy in a unique position for studying the physics of biomineralization through selecting the face-specific kinetics and free energies of biomolecule adsorption to inorganic minerals. Furthermore, the direct nature of the technique makes it the closest real-world embodiment of MD simulations of adsorption/desorption processes. Thus, the opportunity to bridge experiment to computation offers the benefits of measuring the true magnitude of energies and kinetics experimentally, while gaining the corresponding insight into the atomistic dynamics of the process.

Looking forward, combining computational and force spectroscopy techniques may lead to improved methods to account for water and other phenomena that render MD-determined energies quantitatively inadequate as of yet. Further work must be done in relating single-bond and multibond free energies to those measured by other techniques, and hence making them universally acceptable. Finally, improving procedures for linking only individual molecules to cantilever tips will greatly advance this technique from a specialized tool to a robust method for determining the fundamental parameters of intermolecular bonds.

REFERENCES

Bell, G.I. 1978. Models for the specific adhesion of cells to cells. *Science* 200 (4342) (May 12): 618–627. doi:10.1126/science.347575.

Binnig, G., H. Rohrer, Ch. Gerber, and E. Weibel. 1983. 7 × 7 reconstruction on Si(111) resolved in real space. *Physical Review Letters* 50 (2) (January 10): 120–123. doi:10.1103/PhysRevLett.50.120.

Dey, A. and R. Szoszkiewicz. 2012. Complete noise analysis of a simple force spectroscopy AFM setup and its applications to study nanomechanics of mammalian Notch 1 protein. *Nanotechnology* 23 (17) (May 4): 175101. doi:10.1088/0957-4484/23/17/175101.

Erdmann, T. and U.S. Schwarz. 2004. Adhesion clusters under shared linear loading: A stochastic analysis. *Europhysics Letters (EPL)* 66 (4) (May 2): 603–609. doi:10.1209/epl/i2003-10239-3.

Evans, E. 1998. Energy landscapes of biomolecular adhesion and receptor anchoring at interfaces explored with dynamic force spectroscopy. *Faraday Discussions* (111) (January): 1–16.

Evans, E. 2001. Probing the relation between force—lifetime—and chemistry in single molecular bonds. *Annual Review of Biophysics and Biomolecular Structure* 30 (1): 105–128. doi:10.1146/annurev.biophys.30.1.105.

Evans, E. and P. Williams. 2002. Dynamic force spectroscopy. In *Physics of Bio-molecules and Cells (Physique des biomolécules et des cellules)*, eds. F. Flyvbjerg, F. Jülicher, P. Ormos, and F. David, Vol. 75. Berlin, Heidelberg, Germany: Springer, November 6. doi:10.1007/3-540-45701-1.

Friddle, R.W. 2008. Unified model of dynamic forced barrier crossing in single molecules. *Physical Review Letters* 100 (13): 138302.

Friddle, R.W, K. Battle, V. Trubetskoy, J. Tao, E. Salter, J. Moradian-Oldak, J.J. De Yoreo, and A. Wierzbicki. 2011. Single-molecule determination of the face-specific adsorption of Amelogenin's C-terminus on hydroxyapatite. *Angewandte Chemie (International ed. in English)* 50 (33) (August 8): 7541–7545. doi:10.1002/anie.201100181.

Friddle, R.W., A. Noy, and J.J. De Yoreo. 2012. Interpreting the widespread nonlinear force spectra of intermolecular bonds. *Proceedings of the National Academy of Sciences of the United States of America* 109 (34) (August 6): 13573–13578.

Friddle, R.W., P. Podsiadlo, A.B. Artyukhin, and A. Noy. 2008. Near-equilibrium chemical force microscopy. *Journal of Physical Chemistry C* 112 (13) (April 3): 4986–4990. doi:10.1021/jp7095967.

Giessibl, F.J. 2003. Advances in atomic force microscopy. *Reviews of Modern Physics* 75 (3) (July 29): 949–983. doi:10.1103/RevModPhys.75.949.

Gittes, F. and C.F. Schmidt. 1998. Thermal noise limitations on micromechanical experiments. *European Biophysics Journal EBJ* 27 (1): 75–81.

Kennedy, S.J., D.G. Cole, and R.L. Clark. 2011. Note: Curve fit models for atomic force microscopy cantilever calibration in water. *The Review of Scientific Instruments* 82 (11) (November 17): 116107. doi:10.1063/1.3661130.

Liang, H.-H. and H.-Y. Chen. 2011. Strength of adhesion clusters under shared linear loading. *Physical Review E* 83 (6) (June). doi:10.1103/PhysRevE.83.061914.

Lin, H.-J., H.-Y. Chen, Y.-J. Sheng, and H.-K. Tsao. 2010. Free energy and critical force for adhesion clusters. *Physical Review E* 81 (6) (June). doi:10.1103/PhysRevE.81.061908.

Margolis, H.C., E. Beniash, and C.E. Fowler. 2006. Role of macromolecular assembly of enamel matrix proteins in enamel formation. *Journal of Dental Research* 85 (9) (September 1): 775–793. doi:10.1177/154405910608500902.

Meyer, G. and N.M. Amer. 1990. Optical-beam-deflection atomic force microscopy: The NaCl(001) surface. *Applied Physics Letters* 56 (21) (May 21): 2100. doi:10.1063/1.102985.

Pettitt, B.M. and M. Karplus. 1985. The potential of mean force between polyatomic molecules in polar molecular solvents. *The Journal of Chemical Physics* 83 (2) (July 15): 781. doi:10.1063/1.449493.

Pirzer, T. and T. Hugel. 2009. Atomic force microscopy spring constant determination in viscous liquids. *The Review of Scientific Instruments* 80 (3) (March 31): 035110. doi:10.1063/1.3100258.

Seifert, U. 2000. Rupture of multiple parallel molecular bonds under dynamic loading. *Physical Review Letters* 84 (12) (March): 2750–2753.

Smith, D.P.E. 1995. Limits of force microscopy. *Review of Scientific Instruments* 66 (5) (May 1): 3191. doi:10.1063/1.1145550.

Sulchek, T.A., R.W. Friddle, K. Langry, E.Y. Lau, H. Albrecht, T.V. Ratto, S.J. DeNardo, M.E. Colvin, and A. Noy. 2005. Dynamic force spectroscopy of parallel individual Mucin1-antibody bonds. *Proceedings of the National Academy of Sciences of the United States of America* 102 (46) (November 15): 16638–16643. doi:10.1073/pnas.0505208102.

Tobolsky, A. and H. Eyring. 1943. Mechanical properties of polymeric materials. *The Journal of Chemical Physics* 11 (3) (March 1): 125. doi:10.1063/1.1723812.

Tshiprut, Z., J. Klafter, and M. Urbakh. 2008. Single-molecule pulling experiments: When the stiffness of the pulling device matters. *Biophysical Journal* 95 (6) (September 15): L42–L44. doi:10.1529/biophysj.108.141580.

Viani, M.B., T.E. Schäffer, A. Chand, M. Rief, H.E. Gaub, and P.K. Hansma. 1999. Small cantilevers for force spectroscopy of single molecules. *Journal of Applied Physics* 86 (4): 2258. doi:10.1063/1.371039.

Walton, E.B., S. Lee, and K.J.V. Vliet. 2008. Extending Bell's model: How force transducer stiffness alters measured unbinding forces and kinetics of molecular complexes. *Biophysical Journal* 94 (7) (April 1): 2621–2630. doi:10.1529/biophysj.107.114454.

Zhurkov, S.N. 1965. Kinetic concept of the strength solids. *International Journal of Fracture Mechanics* 1 (4): 311–322.

Part V

Properties of the composite: Materials approaches to tissues and whole organs

20 Measuring forces between structural elements in composites: From macromolecules to bone

Philipp J. Thurner and Orestis L. Katsamenis

Contents

20.1 BONE: A HIERARCHICALLY STRUCTURED BIOCOMPOSITE MATERIAL

Bone is a highly complex hierarchical material (Rho et al. 1998, Fratzl and Weinkamer 2007) and an important component of the mammalian skeletal system. It provides mechanical stability, protection of vital organs, and a base structure enabling locomotion, and plays a significant role in the metabolic processes (Standring 2008). The main constituents of bone are inorganic mineral crystals, organic matrix, cells, and water. Due to the fact that bone hierarchy spans from the molecular scale upward, knowledge of nanoscale bone structure, composition, and mechanics is required to understand bone as a biomaterial and to understand its amazing properties. This chapter emphasizes the relevance of nanoscale interfaces in bone populated by the so-called noncollagenous proteins (NCPs) as well as experimental approaches for their mechanical characterization. In the first step, we will introduce the main components of the bone and its hierarchical structure. In the second step, we will outline how to conduct mechanical characterization of protein networks using of atomic force microscopy.

20.1.1 COMPOSITION OF BONE

20.1.1.1 Mineral

Bone consists of impure nonstoichiometric hydroxyapatite ($Ca_{10}(PO_4)_6(OH)_2$) crystals, 65% w/w, in the form of platelets, needles, and rods (LeGeros 1981, Eppell et al. 2001, Fratzl et al. 2004). Impurities include carbon (carbonate apatite, dahlite, or pseudoapatite—$Ca_5(PO_4,CO_3)_3F$), magnesium, fluorine (fluoroapatite—$Ca_5(PO_4)_3F$), and others of lesser abundance. These ions can substitute the hydroxide diatom of the hydroxyapatite (HAP) crystals (hydroxyl-deficient HAP), like fluoride (F^-) and carbonates (CO_3^{2-}), or they can be absorbed

Table 20.1 Chemical composition of bone

	BOVINE CORTICAL BONE (wt% OF WHOLE BONE)	RANGE IN LITERATURE FOR HEALTHY ADULT WHOLE BONE (wt% OF ASH)
Calcium	16.7	32.6–39.5
Magnesium	0.436	0.32–0.78
Sodium	0.731	0.26–0.82
Potassium	0.055	—
Strontium	0.035	—
Phosphorus	12.47	13.1–18.0
Carbonate	3.48	3.2–13.0
Citrate	0.863	0.04–2.67
Chloride	0.077	—
Fluorine	0.072	0.02–0.207

Source: Cowin, S.C. and Telega, J.J., *Appl. Mech. Rev.*, 56, B61, 2003.

into the crystal's surface (Cowin and Telega 2003). Table 20.1 summarizes the chemical composition of bone as obtained by numerous analytical techniques.

20.1.1.2 Collagen

Organic matrix and water comprise the other 35% w/w of bone. Ninety percent of the organic matrix is constituted by collagen, with the main component being type I collagen, while types II, V, and XII are also present. The collagen molecules are cross-linked in the extracellular matrix, and the resulting structure serves as a scaffold for the deposition of mineral crystals. The remaining 10% of the organic matrix consists of NCPs. Although the role of these proteins is not entirely clear yet (Cowin and Telega 2003), they appear to contribute to the organization and mineralization of bone (Hollinger 2005). The following section makes an effort to summarize the most important proteins from each family.

20.1.1.3 Noncollagenous proteins

Small integrin binding ligand with *N*-glycosylation (SIBLING) proteins, as named by Fisher et al. (Fisher et al. 2001, Hollinger 2005), are proteins that carry *N*-glycosylated residues—the term glycosylation refers to the enzymatic process that links saccharine to produce polysaccharides or oligosaccharides, attached to organic molecules such as proteins. *N*-glycosylated means that the produced glycan is attached to a nitrogen atom of the protein molecule. Members of this family are the osteopontin (OPN), bone sialoprotein (BSP), the dentin matrix protein-1 (DMP-1), the dentin sialoprotein, and matrix extracellular glycophosphoprotein (MEPE). Apart from the SIBLING proteins, there are also non-SIBLING proteins, such as the bone acidic glycoprotein-75 (BAG-75), which inhibits osteoclast activity, and osteonectin, which has the ability to bind (or, domain for binding) calcium and is one of the most abundant NCPs. In addition, a further group of NCPs are constituted by Gla-proteins (containing γ-carboxyglutamic acid—gla). In fact, osteocalcin or bone Gla-protein is the most abundant of all NCPs. It is secreted by osteoblasts, and studies show that it inhibits the mineralization process in bone (Hollinger 2005).

The exact role of many individual NCPs is still a matter of ongoing investigation; importantly, it has been shown to date that they are playing an essential role in controlling the nucleation and the growth of the mineral phase (Hunter and Goldberg 1993, Baht et al. 2008), and they also facilitate attachment between collagens and mineral crystals (Tye et al. 2005, Baht et al. 2008). Finally, cells and other biomolecules like proteoglycans and lipids are also present within the bone tissue.

20.1.2 HIERARCHICAL ORGANIZATION OF BONE: FROM THE NANO TO THE TISSUE LEVEL

The three main basic constituents of bone, that is, collagen fibrils, mineral crystals, and NCPs, assemble into the main building block of bone, the mineralized collagen fibril, essentially comprising bone's fundamental structural unit. It is from the organization and the arrangement of this unit that all of the higher hierarchical levels are derived. Figure 20.1 illustrates a bottom–up schematic representation of the hierarchical structure of a mature bone. It is important to note that in this illustration the mineralized collagen fibril (level 2) is presented according to the "gap-nucleation" model (Landis et al. 1993, 1996), as at present it is the most widely accepted one.

20.1.2.1 Levels 1, 2, and 3

The hierarchical level 1 refers to the total of the main constituents of bone as described in the previous sections, while level 2 refers to the first hierarchically structured entity, which is the mineralized collagen fibril. Level 3 corresponds to the bundling of these fibrils along their long axis to form fibers (Weiner and Wagner 1998). These fibers are assembled into lamellae which build up the level 4 structure. A lamellar unit is composed of sublayers. Each sublayer is an array of aligned mineralized collagen fibrils. The orientations of these arrays differ in each sublayer with respect to both collagen fibril axes and crystal layers, such that a complex rotated plywood-like structure is formed (Giraud-Guille 1988, Weiner et al. 1999).

20.1.2.2 Level 4: Subosteonal (or submicro) structure of bone

The term "subosteonal" structure or "submicrostructure" refers to the organization of the mineralized collagen fibers onto "sheets" of ~3–7 μm wide called "lamellae," which are essentially the

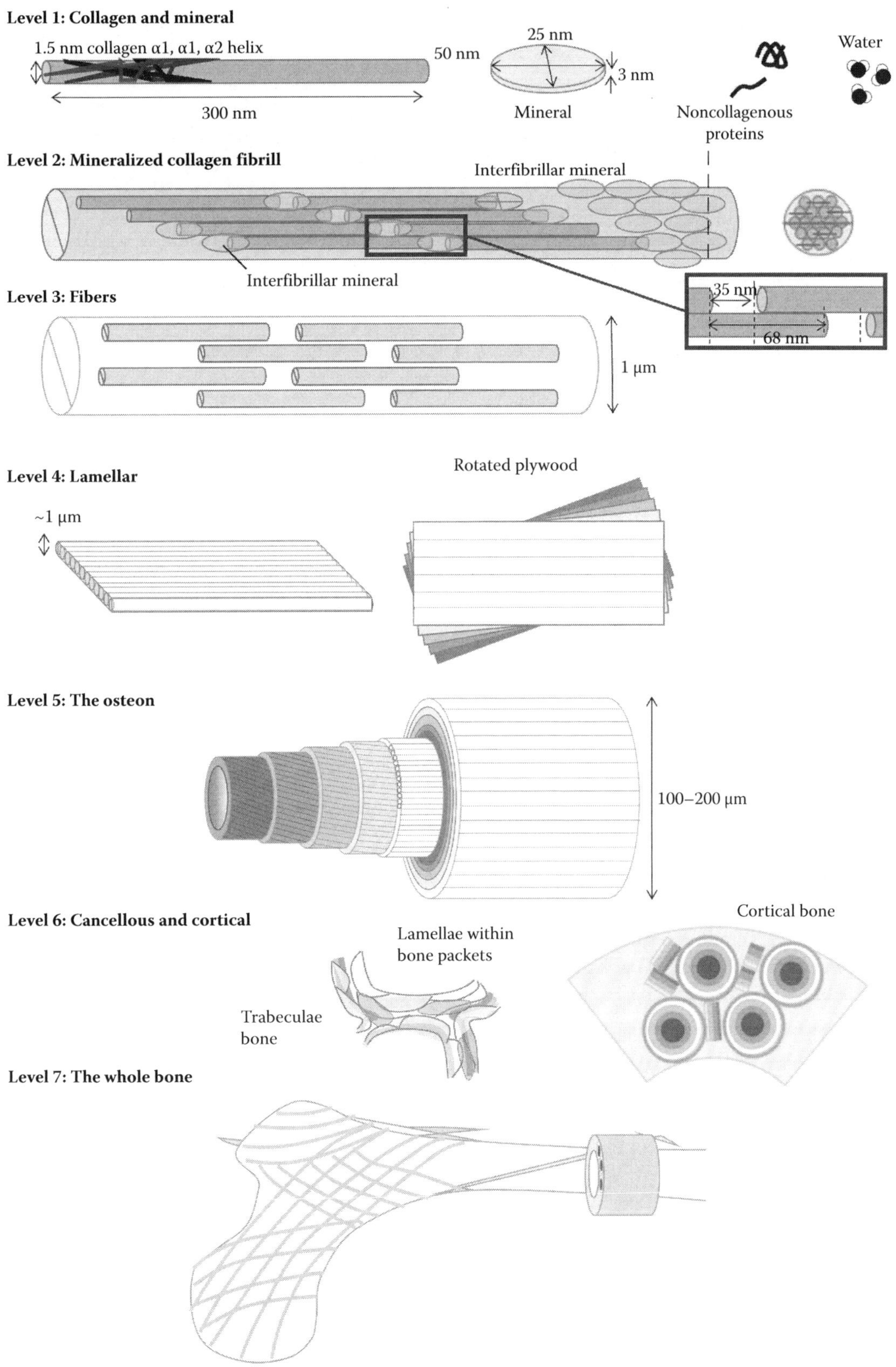

Figure 20.1 Hierarchical organization of bone. (Image courtesy of T. Jenkins.)

building blocks of almost all bone micro- and macrostructures (except woven bone) (Marotti 1993, Rho et al. 1998). The common perception of these structures is that of "twisted plywood" as described in the historic paper of Giraud-Guille. That is, each lamella consists of parallel-arranged mineralized fibers organized into five sublayers, each one of which gradually rotates "by a small and constant angle" with respect to the previous (Giraud-Guille 1988, Weiner et al. 1999). It is worth noting, though, that another less popular model argues that there is no preferential orientation of the collagen fibers within single lamellae (Marotti 1993). Currently, the actual structure of the single lamella is a matter of ongoing debate (Rho et al. 1998).

20.1.2.3 Level 5: Osteonal (or micro) structure of bone

The bone lamellae of level 4 further arrange in a concentric fashion around blood vessels to form a structure called osteon or Haversian system. They comprise the most complex structure of bone tissue and are the result of the remodeling process of the bone. During this process, bone is removed by specialized cells (osteoclasts) and redeposited (by osteoblasts) in concentric layers. Osteons are roughly 200–300 μm wide, and their long axis is oriented parallel to the long axis of the bone (Rho et al. 1998, Weiner and Wagner 1998).

20.1.2.4 Levels 6 and 7

In level 6, we find two major types of bone tissue: the cortical (or compact) and the trabecular (or cancellous) bone. As their names imply, the two types differ in density as a result of the different structure. Cortical bone is a highly dense bone tissue, consisting of osteons and some areas of lamellar bone (lamellae lay parallel instead of in concentric fashion). Trabecular bone is highly porous, and in contrast to the cortical bone, it lacks microstructural features like osteons. Instead, it consists of parallelly developed lamellae aligned with the orientation of the trabeculae (Rho et al. 1998, Weiner and Wagner 1998). Finally, level 7 refers to the "whole bones" of the skeleton, for example, a femur, a scapula, or a carpus.

20.1.3 MECHANICAL ROLE OF NCPs WITHIN THE HIERARCHICAL ORGANIZATION OF BONE

Throughout the hierarchical organization of bone, NCPs play a very important role, not only from the biological and biochemical points of view, but also from the mechanical one. From level 2 and all the way up to levels 5 and 6, NCPs facilitate attachment between various building blocks, stabilizing interfaces, insure optimal load transfer, and repeatedly dissipate significant amounts of energy, protecting bone against failure. Finally, NCP-rich interfaces are involved in the fracture behavior of the tissue, demonstrating the importance of NCPs to the ultimate mechanical properties of the material (Thurner et al. 2010).

20.1.3.1 Level 2: Mineral–collagen attachment

NCPs are believed to play an important role during the mineralization process by controlling nucleation and crystal growth (Roach 1994). On the other hand, recent studies report that the mineralization of collagen is possible even in the absence of "epitaxial nucleation of NCPs," showing that even collagen alone could act as an inhibitor of mineralization and a template of the mineral phase (Olszta et al. 2007, Thula et al. 2011, Wang et al. 2012).

Regardless of the role of NCPs within the mineralization process, these proteins are known to bind to both collagen and mineral crystals. Specifically, on the crystal side, the negatively charged NCPs bind to the positively charged domains of the crystals (Fujisawa and Kuboki 1991, Denhardt and Guo 1993, Kirkham et al. 2000b, Goldberg et al. 2001, Wallwork et al. 2002), whereas the collagen-binding domains attach to the collagen molecules (Tye et al. 2005). Hence, it seems very likely that these proteins act as adhesives between the two phases forming the "lower-level" interface within the bone structure.

20.1.3.2 Level 3: Building up the mineralized collagen fiber

Moving on to the next hierarchical structure, which is the mineralized fiber of level 3, another interface appears. This time, it is between the mineralized fibrils, which constitute the fiber. This interfibrillar space is filled with mineral crystals (also referred to as "extrafibrillar minerals") (Bonar et al. 1985, Pidaparti et al. 1996), NCPs (McKee et al. 1989, 1993), and probably other noncollagenous moieties like GAGs and/or lipids.

Fantner et al. (2005b) propose that this nonfibrillar organic component acts as a "glue" bonding the mineralized fibrils together (Figure 20.2). During loading, this glue is stressed resisting fibril separation and, at the same time, the rupture of ion-mediated sacrificial bonds within the glue dissipates significant amount of energy, adding to the toughness of the material. Importantly, when the stress is released and the surfaces put back together, these bonds can reform, and the glue self-heals. It has been estimated that less than 1% per weight of this matrix (glue) is enough to provide the bone with its known yield strength (~150 MPa) (Fantner et al. 2005b). Further to this point, shear transfer in this matrix redistributes the load between the fibrils, preventing stress concentration and consequently crack formation (Fantner et al. 2005b). In addition, a similar glue layer consisting of osteocalcin and OPN has recently been shown to exist between two neighboring extrafibrillar crystals on one collagen fibril, again resisting fracture, increasing fracture toughness, and being able to self-heal (Poundarik et al. 2012).

20.1.3.3 Levels 4 and 5: Bone fracture and the role of NCP-rich interfaces

In these levels, the role of NCPs in terms of mechanics is to a large extent still a matter of ongoing research. While both the mechanical and the fracture behavior of bone tissue have been studied at these scales, only a few studies attempt to correlate measured properties with composition and in particular NCPs. Level 4 consists of individual (thick) lamellae, interlamellar areas (or thin lamellae), and cement lines (the outer layer of an osteon), and all of these are known to have different mechanical properties. For example, single lamellae exhibit higher modulus of elasticity than interlamellar areas and cement lines (Donnelly et al. 2006, Gupta et al. 2006b, Katsamenis et al. 2013). The source of this dissimilarity is not entirely clear yet. It most likely is a combination of the inner architecture, that is, arrangement of the fibrils (Gupta et al. 2006b, Katsamenis et al. 2013), composition (Derkx 1998, Nanci 1999, Katsamenis et al. 2013), and degree of mineralization (Donnelly et al. 2006).

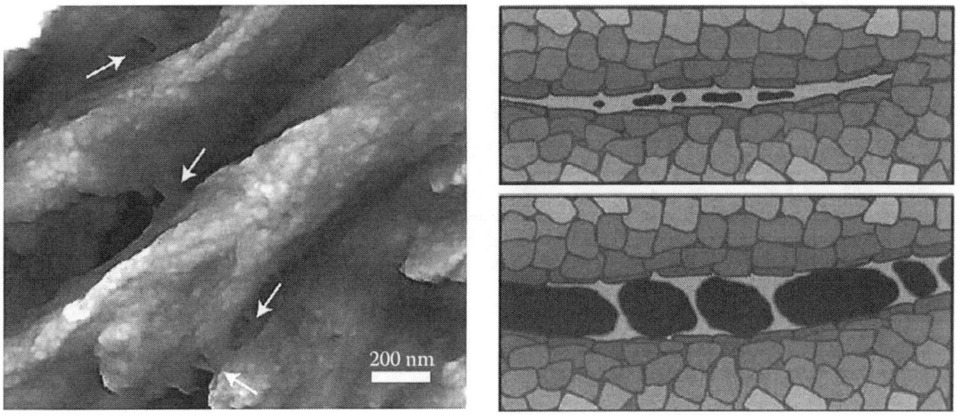

(a) (b)

Figure 20.2 (a) AFM image of a fractured bone surface showing the glue-like substance (arrows) between neighboring mineralized fibrils and (b) schematic representation of the "glue resistance" during fibril separation. (Reprinted by permission from Macmillan Publishers Ltd. *Nat. Mater.*, Fantner, G.E., Hassenkam, T., Kindt, J.H., Weaver, J.C., Birkedal, H., Pechenik, L., Cutroni, J.A. et al., Sacrificial bonds and hidden length dissipate energy as mineralized fibrils separate during bone fracture, 4(8), 612–616, 2005b, Copyright 2005.)

These features are known to affect bone fracture behavior. During failure, crack propagation preferably occurs through the interlamellar areas and cement lines (Peterlik et al. 2005, Katsamenis et al. 2013), while single lamellae act as crack barriers, deflecting, twisting, or even stopping the cracks (Gupta et al. 2006b, Koester et al. 2008, Launey et al. 2010). Likewise, at the osteonal level (level 5), cracks tend to deflect on the osteons and preferably propagate through the interstitial bone (Nalla et al. 2005). Interestingly, both cement lines and interlamellar areas, which facilitate crack propagation, are enriched with NCPs (Derkx et al. 1998, Nanci 1999, Katsamenis et al. 2013), suggesting their importance in the macroscopic fracture behavior of the tissue. In addition, recent results also suggest interlamellar areas to be elastoplastic-bearing pads in bone microstructure, allowing for increased deformation and repeated energy dissipation due to their increased NCP content, which could be beneficial for fatigue properties (Katsamenis et al. 2013).

20.1.3.4 Level 7: OPN deficiency and fracture toughness

The fracture of bone is a highly complex process that involves all the hierarchy levels of the structure. For fracture to occur at the higher level (level 7), failure also has to be taking place at the lower levels. This "lower level" failure generally occurs at the weak interfaces found in each level, which, as described earlier, are either populated by NCPs or are NCP rich. Therefore, it can be argued that variability, such as compositional variations, at these interfaces may influence fracture behavior and hence fracture toughness.

The influence of NCPs on elastic as well as fracture mechanics of bone has been demonstrated by several studies. Works presented by Kavukcuoglu et al. (2007), Thurner et al. (2010), and Poundarik et al. (2012) show that the deletion of OPN within the mouse bone or deletion of OPN and OC (Poundarik et al. 2012) leads to lowered fracture toughness, indentation modulus, strength, and work to fracture.

Due to all the implication of NCPs given previously, it is clear that nanoscale interface mechanics in bone and in particular the adhesion of proteins to mineral as well as protein–protein interactions are of utmost interest. The ultimate experiments would perhaps be the mechanical dissociation of two neighboring mineralized collagen fibrils as well as a tensile test of an individual mineralized collagen fibril. While such experiments would inevitably deliver new insights and allow specifically testing of NCP-rich interfaces, the reality unfortunately is that such experiments are currently very difficult to realize, if not impossible. Nevertheless, there is opportunity to classify interactions between protein and minerals as well as proteins themselves, as laid out in the following section.

20.2 CHARACTERIZATION OF NANOSCALE INTERFACE MECHANICS OF BONE

At the nanoscale, tools able to characterize interface mechanics are emerging. Various experimental approaches have been used in the past to characterize bone-related nanoscale interfaces by means of mechanical testing including force spectroscopy (Zappone et al. 2008b), high brilliance synchrotron small-angle x-ray scattering (SAXS) and wide angle x-ray diffraction (WAXD) (Gupta et al. 2006a), and the surface-forces apparatus (SFA) (Israelachvili and Adams 1978, Zappone et al. 2008b). In most cases, these approaches provide more qualitative than quantitative insights, but for the time being, this is a very important step to further our understanding of the mechanical behavior of the interfacial material of bone. Here, we focus on the usage of AFM force spectroscopy as a method to characterize the mechanical properties of thin NCP films and the effect of different chemical environments on them. For this purpose, a short description of the AFM apparatus is given, followed by a presentation of the main principles of force spectroscopy for bone-related materials. Finally, the main protocols and instrumentation for conducting force spectroscopy measurements on NCP layers are described.

20.2.1 MEASURING FORCES AT THE NANOSCALE BY MEANS OF AFM FORCE SPECTROSCOPY

20.2.1.1 Basic principles of atomic force microscopy

Atomic force microscopy (AFM) is a technique capable of obtaining nanoscale information, introduced in 1986 by Binning

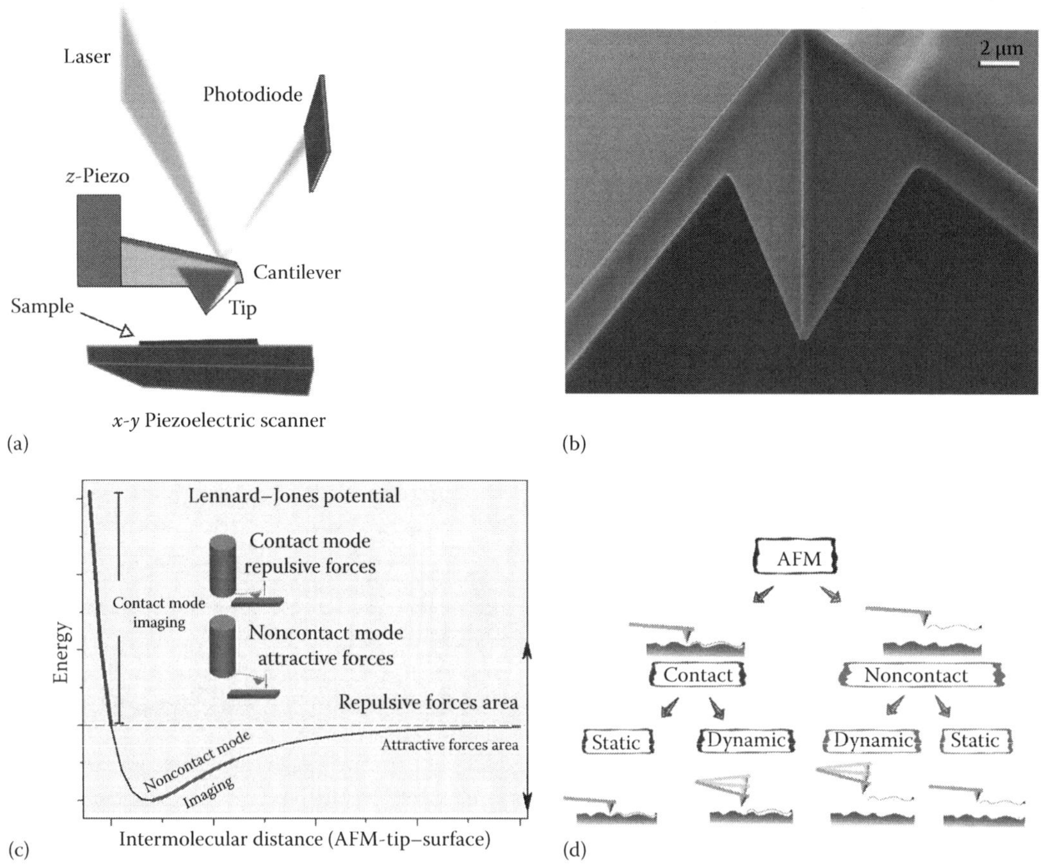

Figure 20.3 (a) Schematic representation of an AFM. The deflection of the cantilever is measured by a segmented photodiode. The actual deflection signal is calculated from the voltage that is produced in each segment of the photodetector. The controller approaches or withdraws the cantilever when the deflection is below or above the predefined set point by shrinking or expanding the z-piezo. (b) SEM image of a (used) AFM tip. (Olympus AC200TS; image courtesy of O. Andriotis.) (c) The Lennard–Jones potential, which describes the interactions between the tip and the surface. (d) Possible AFM operation modes.

et al. (1986). The type of information can be topographical (imaging), mechanical (stiffness, Young's modulus, hardness), tribological (scratch-resistance, friction), or chemical (composition—chemical force microscopy [CFM]) (Kirkham et al. 2000a, Jena and Hörber 2006, Thurner 2009).

In principle, AFM measures the relevant variation of the deflection (static mode) or oscillation amplitude (dynamic mode) of a small cantilever with a sharp tip at its end that is scanning across the sample's surface (Figure 20.3a). A feedback signal is then produced and, when needed, a controller adjusts the distance of the cantilever with respect to the sample surface in order to keep the deflection (or the oscillation amplitude) at a constant, predefined set point. Finally, depending on the nature of the tip–sample interaction, this information is translated into the equivalent quantity, most commonly topography (other quantities comprise, e.g., surface charge, chemical potential, stiffness, etc).

Scanning as well as cantilever positioning and oscillation are achieved by means of one to three piezoelectric transducers (depending on the scanner design). For simplicity, we will here assume three separate actuators: x-, y-, and z-piezo. By shrinking or expanding the z-piezo where the cantilever is mounted, the controller keeps the deflection or amplitude at the desired set point, whereas the x- and y-piezo generate the scanning motion. Finally, these adjustments are linked with the lateral x, y positions of the cantilever, converted from Volts (V) to nanometers (nm) based on the piezo calibration, and plotted (Figure 20.3a) (Meyer 1992).

AFM imaging can be accomplished in many different ways depending on the information which is being recorded. For surface topography, the most common modalities are the "contact mode" and the "noncontact mode" (Figure 20.3c and d). In each one of these, a static (the cantilever bends in response to the force which acts on the probing tip) or a dynamic (oscillation of the cantilever close to its resonance frequency) operation of the cantilever can be selected. In the contact mode, the AFM cantilever is in contact with the sample's surface (static, contact mode), or it contacts the surface once every oscillation (dynamic, tapping or AC mode). This way, the cantilever response is mainly due to the strong ionic repulsion forces, and high-resolution imaging can be achieved. In the noncontact mode, the cantilever scans or oscillates in a small distance above the sample's surface (typically 10–100 nm) without touching it. In this mode, the response of the cantilever is due to longer range attractive forces like van der Waals, electrostatic, magnetic, or capillary forces. Figure 20.3 depicts a schematic diagram of the different AFM imaging techniques as well as the Lennard–Jones potential which can be used to describe the interactions that take place at various modes (Meyer 1992).

A tip radius that can go down to ~3 nm allows the cantilever to feel topographical differences of several nanometers, while the low spring constant (0.006–48 N/m) ensures that the cantilever will react even in the presence of a tiny potential. In fact, the ultrahigh sensitivity of the cantilever in measuring accurately these potentials has been used to measure the forces involved in the antibody–antigen interactions or the unfolding of proteins with globular domains (Thurner 2009). For a more detailed description and technical information about AFM, the reader should refer to relevant textbooks (Birdi 2003, Braga and Ricci 2004).

20.2.1.2 Force spectroscopy

Force spectroscopy is an application of AFM which provides information about the magnitude of forces involved in the tip–substrate interaction. By using the cantilever as a force sensor, forces down to the scale of pN or even fN can be measured.

In the past, studies have implemented force spectroscopy to measure the interaction strength (bond energy) between molecules (intermolecular force spectroscopy) or between different domains of the same macromolecule (intramolecular force spectroscopy or single molecule force spectroscopy, SMFS) (Hugel and Seitz 2001). SMFS has been the more predominant method used. Due to the ability to chemically bind organic molecules onto gold-coated cantilever tips using thiol chemistry, studies have been presented utilizing such approaches to essentially graft one single molecule onto a cantilever tip. This molecule is then either investigated mechanically to study the unfolding of globular domains (Oberhauser et al. 2001), adhesion properties (Lee et al. 2006), or even antigen–antibody interaction on a specific substrate (Hinterdorfer et al. 1996, Hinterdorfer and Dufrêne 2006). Such investigations lead to direct measurements of protein morphology or forces required for dissociation of individual adhesion or antibody–antigen bonds, rendering SMFS a powerful technique. More recently, force spectroscopy has also been applied to investigate whole material systems like collagen or other protein films deposited on a substrate (Gutsmann et al. 2004, Fantner et al. 2007, Zappone et al. 2008b).

In principle, force spectroscopy assesses the tip–surface, molecule–substrate or molecule–molecule interaction by measuring cantilever deflection during vertical approach to or vertical retraction from the surface. This is achieved by subtracting the cantilever deflection from the *z*-piezo expansion or contraction. The force curve is comprised of approach, contact, and retract regions, all three of which can be governed by forces of different nature (Heinz and Hoh 1999). In more detail, for nonspecific protein film characterization, the tip approaches and touches the surface and then retracts pulling some of the molecules, which have been attached to it, away from the surface. The separation requires the rupture of intramolecular and intermolecular bonds forming the network as well as the ones which have been formed between the protein molecules and the surface or the tip (Zappone et al. 2008b). This results in a negative deflection of the cantilever, which can then be converted into force according to Hooke's law by multiplying it with the spring constant of the cantilever:

$$F = -kx, \tag{20.1}$$

where

k is the cantilever's spring constant
x is the deflection

The whole pulling process is schematically presented in Figures 20.4 and 20.5.

It is worth noting that during the approach of the cantilever and just before the contact, a jump-to-contact event can occur. The occurrence of such a jump depends on the charge of the substrate, the experimental conditions, for example, in-liquid/in-air. This is attributed to the attractive forces (van der Waals, electrostatic, etc.) between the tip and the sample (Heinz and Hoh 1999).

The area under the recorded force–retraction curve corresponds to the tip–surface separation energy and is attributed to the weak (~1 eV) intramolecular and intermolecular interactions (e.g., rapture of intermolecular and intramolecular bonds, unfolding domains, etc.) (Hugel and Seitz 2001, Fantner et al. 2007, Zappone et al. 2008a).

In previous studies, AFM force spectroscopy was utilized to evaluate the effect of cations such as Na^{1+}, Ca^{2+}, and La^{3+} on the energy dissipation during the separation of mineralized fibrils as well as on the adhesion and mechanical properties of adsorbed layers of human OPN (Fantner et al. 2005a, 2007; Adams et al. 2008; Zappone et al. 2008a). These studies proposed the presence of sacrificial cation-mediated bonds that break and reform during loading and unloading dissipating energy, thus contributing to the toughness of the material. Fantner et al. (2006) took an extra step forward by proposing a few simple general types of such sacrificial bonds and describing the morphology of the corresponding pulling curves.

20.2.2 CONDUCTING FORCE SPECTROSCOPY OF NONCOLLAGENOUS PROTEIN NETWORKS

The experimental approach for measuring these intermolecular and intramolecular interactions of protein networks such as the ones described earlier is given in the following sections in more detail. We discuss the preparation of the sample, the *ex situ* calibration of the AFM cantilever system, and the assembly of the fluid cell, prior to being able to perform the experiments. It should be noted that the AFM used for the experiments was an MFP3D (Asylum Research, Santa Barbara, CA), and some of the steps and part of the setup are instrument-specific and may vary from other AFM models.

20.2.2.1 Experimental setup

20.2.2.1.1 Sample preparation

The first step requires the deposition of the protein onto a freshly cleaned surface, which can either be freshly cleaved mica or another substrate of interest. For cleaving mica disks, only a bit of sticky tape is needed. Part of the tape is applied to the mica disk (usually glued to an AFM sample holder, i.e., a steel disk) and then ripped off in an upward motion while firmly gripping the steel/mica disk assembly. This generally creates smooth mica surfaces, which are essentially atomically flat. Once the substrate has been prepared, protein film deposition is the next step. This can be done via either the "drying droplet" technique or the "adsorption" method (Figure 20.6). For the first method, a solution of some mg/mL of the purified protein of interest is prepared, and a droplet of this solution ~5 μL is deposited on the surface. It is then left to dry resulting in the formation of a self-assembled protein layer. It should be noted here that higher concentrations generally lead to "thicker" networks and longer pulls with the AFM tip. Typical concentrations for experiments

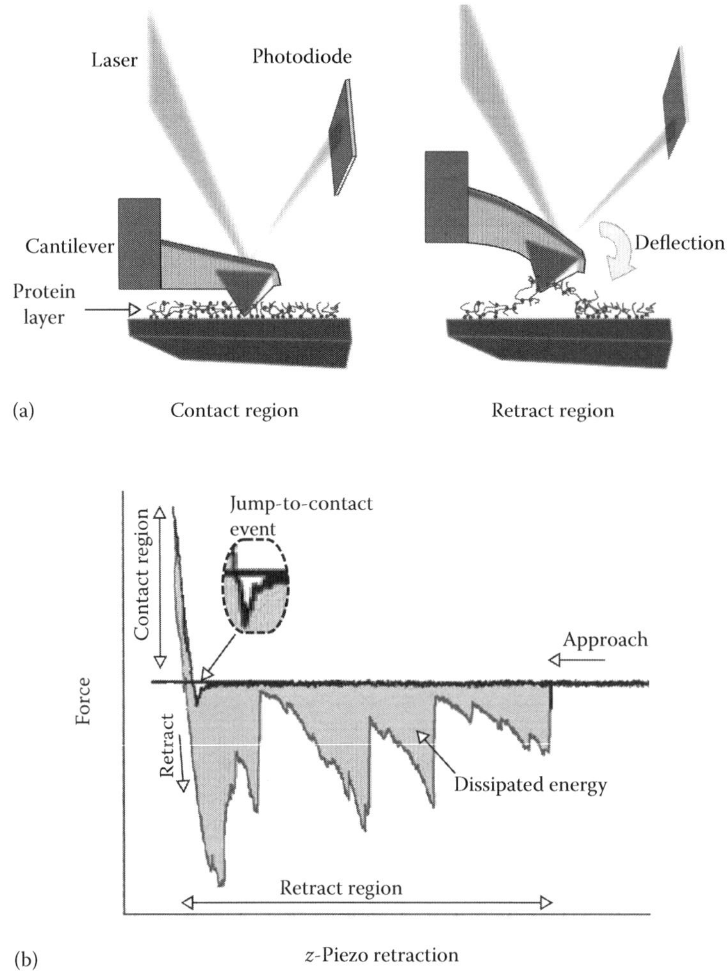

Figure 20.4 (a) Schematic representation of pulling experiment. Tip–surface separation requires energy to break the ion-mediated bonds (gray dots), which have been formed between the protein molecules as well as between the proteins and the surfaces. This results in the deflection of the cantilever, which can then be translated into the pulling force according to Hooke's law (Equation 20.1). (b) Example force curve obtained during the approach and retraction circle. (From *Biophys. J.*, 95(6), Zappone, B., Thurner, P.J., Adams, J., Fantner, G.E., and Hansma, P.K., Effect of Ca^{2+} ions on the adhesion and mechanical properties of adsorbed layers of human osteopontin, 2939–2950, Copyright 2008 with permission from Elsevier.)

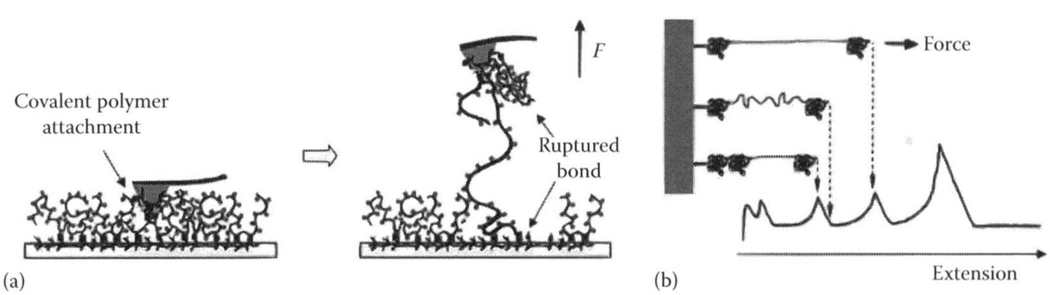

Figure 20.5 (a) Schematic representation of measurement of intermolecular forces. AFM's tip is sunk into the NCP layer and then retracted, pulling with it the attached proteins. The measurement is terminated after the rupture of a covalent bond along the last remaining stretched protein. (b) Force–displacement curve behavior (sawtooth pattern) produced during the unfolding of a macromolecule. The measured pattern represents the sequential unfolding of individual aggregates until the entire polymer chain is fully stretched and detached. (From Hugel, T. and Seitz, M.: The study of molecular interactions by AFM force spectroscopy. *Macromol. Rapid Commun.* 2001. 22(13). 989–1016. Copyright Wiley-VCH Verlag GmbH & Co. KGaA. Reprinted with permission.)

on OPN and DMP-1 were in the range of 0.2–2 mg/mL (Fantner et al. 2007, Adams et al. 2008, Zappone et al. 2008b), and were also higher on occasion (Fantner et al. 2007). The whole process ideally is taking place in a clean environment (e.g., a laminar flow hood) to avoid dust contamination. For the "adsorption" method, a protein solution droplet of some μL is inserted by capillarity between two freshly cleaved mica surfaces, separated, for example, by small spacers and left undisturbed for a certain amount of time (typically 1–2 h), depending on the surface and molecule charge, molecule size, temperature, etc., until adsorption of protein molecules onto the surface occurs (Zappone et al. 2008b). The advantage of the latter method is perhaps a

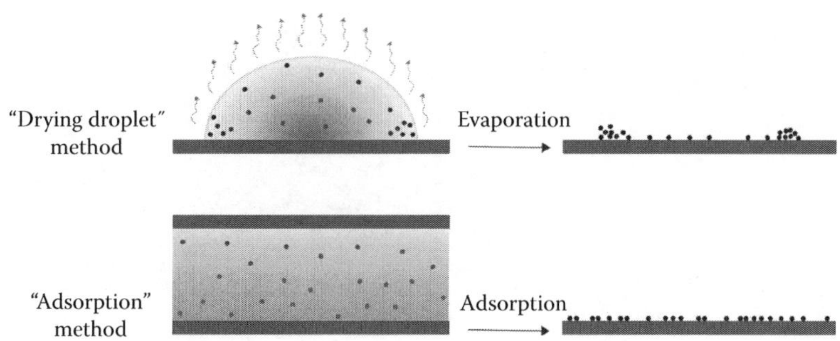

Figure 20.6 "Drying droplet" versus "adsorption" deposition method.

more homogenous coating in contrast to the "drying droplet" technique, where, because of the "coffee stain" effect (Deegan et al. 1997), higher concentrations of protein molecules locate at the outer perimeter. On the other hand, usually only small amounts of proteins are available, making the "drying droplet" technique the preferred one. In this case, the "coffee stain" effect becomes an advantage.

20.2.2.1.2 AFM calibration

Force spectroscopy experiments of protein networks require an aqueous environment. This allows for physiological functionality of the molecules, while at the same time preventing the potential impact of the capillary forces, acting between the tip and the sample in air, on the measurements (Hugel and Seitz 2001). Thermal stabilization of the system is also important to avoid inconsistency in cantilever behavior. Change in temperature usually leads to a drift of the deflection signal, which is converted directly into measured force. Hence, it is prudent to wait for the drift to subside before starting any measurements.

Appropriate selection of the cantilever as well as accurate calibration of the AFM in solution is essential for maximizing the accuracy and sensitivity of the technique. As intermolecular and intramolecular forces under consideration vary from some hundreds of pN up to only a few nN, the cantilevers used have usually a low typical spring constant (k) in the order of 0.005–0.05 N/m (Fantner et al. 2007, Zappone et al. 2008b).

As mentioned earlier, AFM force spectroscopy measurements are generated on the basis of Hooke's law (Equation 20.1) (Meyer and Amer 1988). The spring constant k is an intrinsic property of the cantilever, depending on the geometrical characteristics and cantilever material; the manufacturer usually provides an estimated value. However, this value can differ significantly from the actual one, and thus an accurate determination of the cantilever spring constant should ideally be carried out. For this, first the sensitivity of the cantilever (in nm/V) is determined.

This can easily be obtained by pressing the cantilever into a hard surface such as glass or mica. Assuming that the substrate does not deform, the sensitivity is directly determined through the slope of the force curve. The resulting value is the optical lever sensitivity (OLS) and expresses the deflection of the lever (in V) for a given expansion (in nm) of the z-piezo. The OLS is usually expressed in nm/V.

The spring constant is calculated from the thermal noise spectrum by means of equipartition theorem. This step requires the analysis of the thermal vibration (noise) spectrum of the free vibrating cantilever. Most of the commercially available AFM/SPM instruments have a built-in routine that uses this spectrum to work out the spring constant, k, by means of the equipartition theorem (Butt and Jaschke 1995). The main principle of this method relies on the fact that for a certain vibrational mode, the thermal fluctuation of a cantilever lever supported only at one of its ends (free-standing lever) is a function of its spring constant k. In the thermal spectrum (Figure 20.7), the higher oscillation amplitude corresponds to the first vibration mode of the cantilever, and from this the spring constant can be calculated. A detailed presentation of this calculation is beyond the scope of this chapter, but interested readers may consult the key publications detailing the thermal noise approach (Hutter and Bechhoefer 1993, Levy and Maaloum 2002). If for some reason,

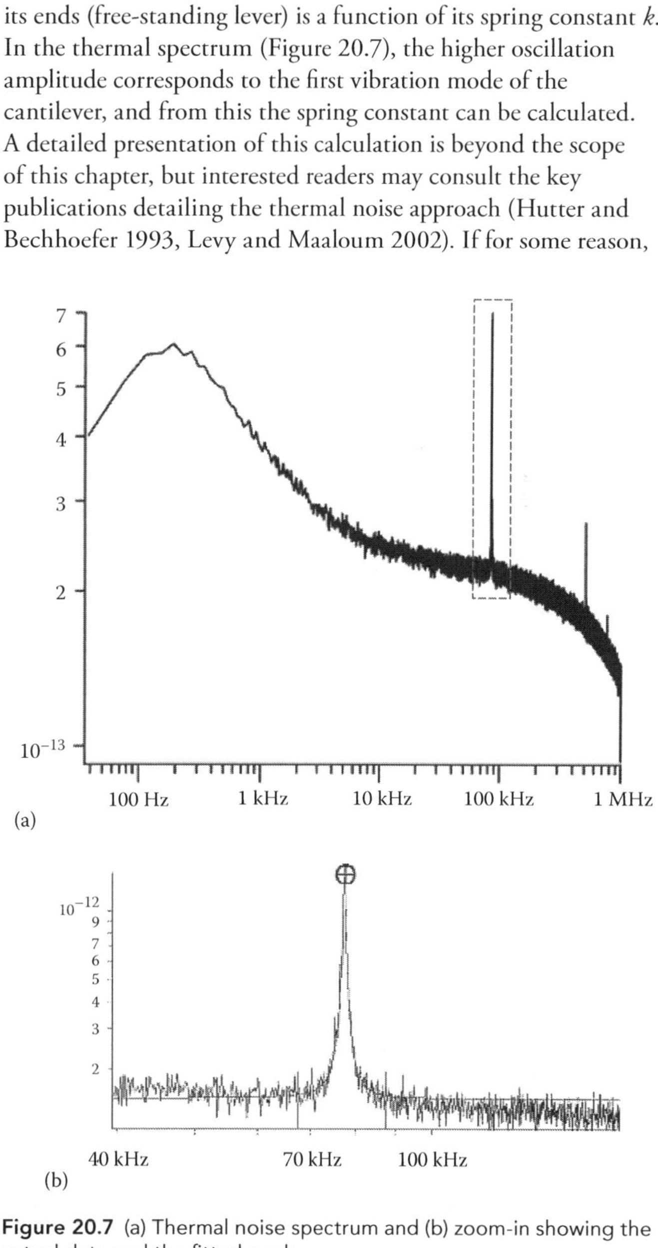

Figure 20.7 (a) Thermal noise spectrum and (b) zoom-in showing the actual data and the fitted peak.

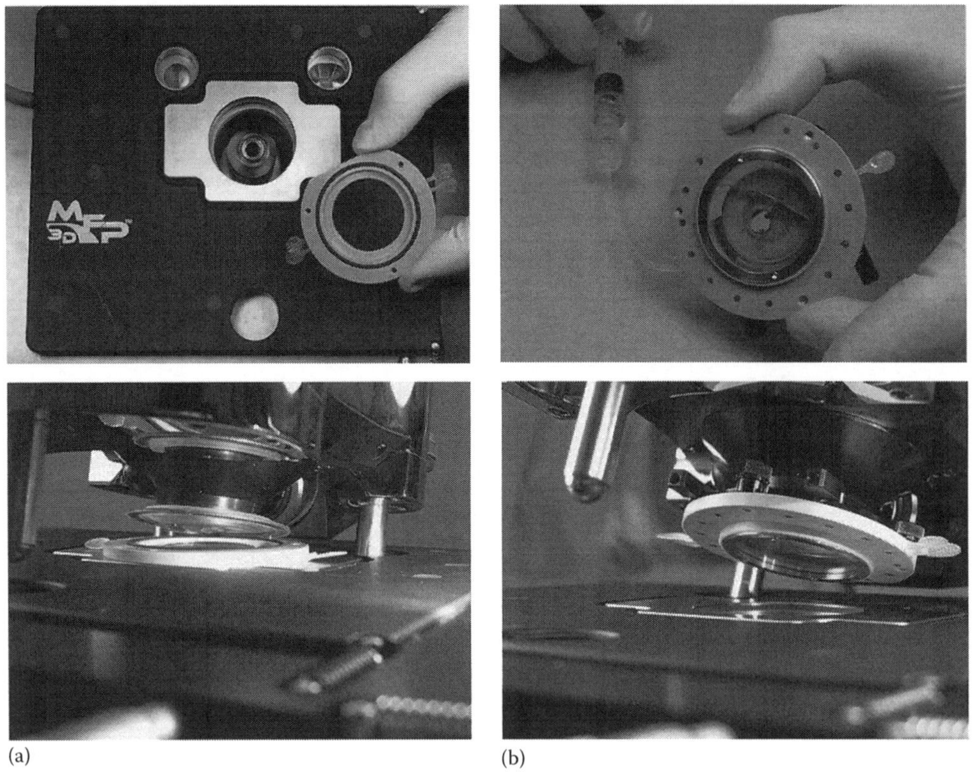

Figure 20.8 (a) Open fluid cell and MFP-3D head positioning; (b) closed fluid cell and head positioning. (Adapted from Asylum Research, *MFP-3D User Manual*, Santa Barbara, CA. With permission.)

a built-in function for the thermal noise method is not available, the experimenter may choose to use the spring constant provided by the manufacturer or simply use the cantilever geometry and material to calculate it (Lamprou et al. 2010). One should keep in mind though, that the actual spring constant of the cantilever may be different to the one stated by the manufacturer. However, it is reasonable to assume that the spring constants of cantilevers derived from the same silicon wafer do not vary significantly; hence, measurements done using different levers from the same batch would still be comparable when a single spring constant values is used.

20.2.2.1.3 Open versus closed fluid cell setup

The final step before starting to record force curves is to set up the fluid cell in which the experiments will be conducted. The two commonly used fluid cell setups are presented in Figure 20.8 (note that this is instrument-specific and shown for the MFP3D AFM). The open cell allows easier access to the sample and the head, as well as fast fluid exchange, but this cannot be done when the lever is engaged on the sample. Usually, the head of the AFM has to be lifted for the fluid exchange and then placed back, meaning that measuring the exact same spot under different environments is extremely difficult. The closed cell on the other hand allows fluid exchange "on the fly" at the cost of difficult access to the sample and the head.

20.2.2.2 Measured parameters in force spectroscopy experiments

Force spectroscopy allows for the evaluation of different parameters, each of which provides a different insight into the mechanical behavior of these films. Figure 20.9 illustrates a typical force–retract curve of an OPN film along with the parameters, which are usually extracted from such experiments (for a more complete example of such a research study, please refer to Section 20.2.3).

20.2.2.2.1 Dissipation energy

Perhaps, the most commonly used parameter is the dissipation energy, that is, the required separation/detachment energy between the tip and the surface, which is expressed by the area under the force–retraction curve. This energy depends on the nature and magnitude of the interactions between the molecules as well as on the presence or absence of hidden length (to be explained later). Specifically, it has been shown that the molecules of the unstructured NCPs of many biogenic materials interact via nonspecific ion-mediated bonds which act as cross-linkers between different molecules (intermolecular cross-links) or the domains of the same molecule (intramolecular cross-links) (Smith et al. 1999, Fantner et al. 2007). The intermolecular cross-links are responsible for the creation of a protein network by linking the molecules together, while the intramolecular ones are responsible for the formation of unstructured folded domains within a single molecule. As the retracting cantilever pulls the attached molecules away from the surface, the applied force is also transferred to the rest of the molecules via the intermolecular cross-links, forcing them to align with the loading axis. This process requires a significant amount of energy which is put into the system as work against entropic elasticity (Adams et al. 2008, Zappone et al. 2008b).

Further stretching results in the rupture of the weak ion-mediated cross-links, also called sacrificial bonds (Smith et al.

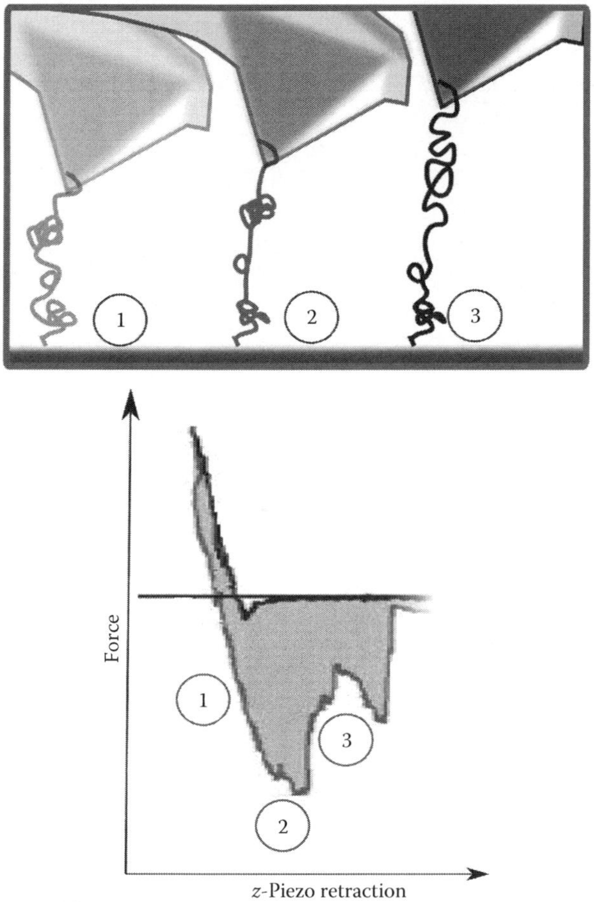

Figure 20.9 Schematic representation of the initial part of a force–retraction curve. During the retraction of the cantilever, stretching of the proteins causing the cantilever to bend and a negative (attractive) force is registered (region 1). While retraction continues, sacrificial bonds start rupturing, "hidden" length is exposed, and the attractive force steps down (region 2). The extra length is then stretched again giving rise to the next peak (region 3). This process continues until eventually the protein ligament that connects the tip with the surface ruptures. (From Zappone, B. et al., *Biophys. J.*, 95(6), 2939, 2008b.)

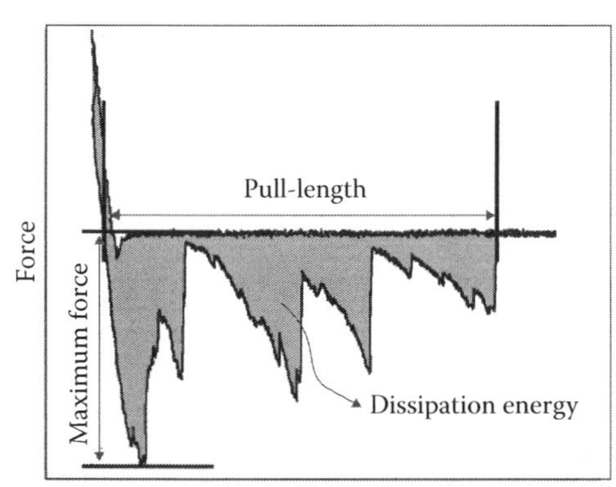

Figure 20.10 Three commonly used features extracted from force–retraction curves from AFM force spectroscopy measurements. (From Zappone, B. et al., *Biophys. J.*, 95(6), 2939, 2008b.)

1999), exposing additional "hidden length." This is registered as a drop on the force–retraction curve, which is then followed by an increase, as further work to stretch this "extra" length has to be done (Figure 20.10).

20.2.2.2.2 Maximum force

The maximum force is defined as the point of greatest difference between the retraction curve and the zero-deflection force level (Figure 20.10). This experiment constitutes a qualitative measure of the strength of the intermolecular and intramolecular sacrificial bonds. Interpretation of maximum force results has to be done with extra care, as changes in several noncontrollable parameters, like for example, the amount of protein molecules absorbed between the tip and the surface can affect these results (Adams et al. 2008).

20.2.2.2.3 Pulling length

The pulling length is the length from the surface contact to the maximum distance at which there is a nonzero force (Adams et al. 2008, Zappone et al. 2008b). It is a measure of the integrity of the protein network as well as the degree of the adhesion between the two surfaces (Zappone et al. 2008b). Low adhesion on the tip would cause easier detachment of the proteins from the tip-side, resulting in a lower pull length. Accordingly, low network integrity, for example, low degree of cross-links or low intermolecular bond strength, would cause easier rupture of the tip–network protein bridge again, resulting in a lower pulling length.

20.2.3 FORCE SPECTROSCOPY EXPERIMENTS ON BONE OPN AND DMP FILMS

Force spectroscopy has been utilized for the characterization of the mechanical behavior of bone NCPs. As mentioned before, these proteins are located at the interface between the mineralized fibrils facilitating adhesion. It is therefore believed that their mechanical behavior and their ability to dissipate energy during loading have a direct impact on the ultimate mechanical properties of the bone (Smith et al. 1999, Fantner et al. 2005b, Poundarik et al. 2012).

From this perspective, experiments have been carried out focusing on the effect of different ionic environments on the adhesion and cohesion properties of some SIBLING proteins, which are considered as being important for bone toughness. Fantner et al. (2007) and Zappone et al. (2008b) investigated the effect of solutions containing only Na^{1+} as well as ones containing both Na^{1+} and Ca^{2+} ions on the mechanical properties of OPN films adsorbed on freshly cleaved mica disks. Adams et al. (2008) used a similar approach to study the behavior of another SIBLING protein, the DMP-1. Looking at the study by Fantner et al. (2007) in more detail, samples were deposited using the drying droplet method (cf. Figure 20.6). As mentioned earlier, the substrate used was mica, whereas the AFM probe used was a soft (k = 0.02 N/m) cantilever with gold coating. The experiment is possible as the protein adheres to both the cantilever and the substrate, and yet while these materials do not constitute the actual substrate of interest, that is, carbonated apatite, this model system can still be used to interrogate the network mechanics of OPN films and their dependence on the chemical environment, namely the concentration and charge of cations in the solution.

Using the samples prepared, Fantner and coworkers went on to conduct force spectroscopy experiments in "sticky areas" of their samples. Such areas are localized by trial and error, that is, probing various locations and awareness of the coffee-stain effect (cf. Figure 20.6), for the drying droplet method will also help to identify good areas for an experiment. The study by Fantner et al. essentially provided three important insights.

First, they demonstrate that the "stickiness," that is, the energy dissipation ability, of the protein network depends on the valence of the ions, which are present in the surrounding solution. In the case of OPN, the dissipated energy in the presence of Ca^{2+} ions was significantly higher from that measured in the presence of Na^{1+} (Figure 20.11). Zappone et al. (2008b) and Adams et al. (2008) reported similar findings investigating OPN and DMP-1, respectively. It is worth noting that despite the fact of how coating of the OPN protein onto the mica surfaces was accomplished—adsorption (Zappone et al. 2008b) and dry droplet (Fantner et al. 2007, Adams et al. 2008, Zappone et al. 2008b)—the effects of ion-mediated cross-linking on the formation of a mechanical network with glue-like properties can be demonstrated by means of force spectroscopy. The main difference is that the values obtained from drying droplet generally yield thicker films and hence higher energy dissipation until disruption (Zappone et al. 2008b). Importantly, the underlying mechanism is a simple one:

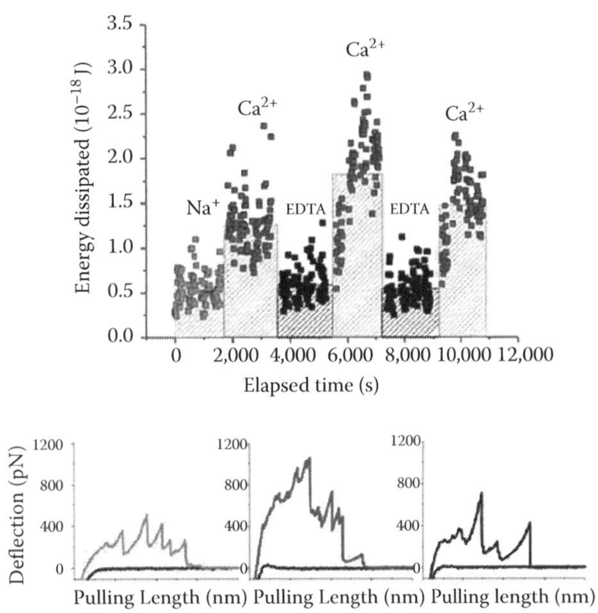

Figure 20.11 AFM force spectroscopy characterization of the network strength of an OPN layer deposited on freshly cleaned mica. The diagram at the top presents the dissipated energy per pull (scatter plot) and the mean energy dissipation (collared bars) for consecutive exposures of the OPN to different ionic environments, that is, $Na^{1+} \rightarrow Ca^{2+} \rightarrow$ EDTA $\rightarrow Ca^{2+} \rightarrow$ EDTA $\rightarrow Ca^{2+}$. At the bottom, representative pull curves in the presence of Na^{1+} and Ca^{2+} buffer as well as in EDTA are presented. In the presence of Ca^{2+} ions, the energy dissipation is increased due to the stronger electrostatic interaction between the positively charged Ca^{2+} ions and the acidic and negatively charged domains of the OPN. After chelation of the Ca^{2+} ions with EDTA, the energy dissipation decreases to its "original" levels. (Reprinted with permission from Fantner, G.E., Adams, J., Turner, P., Thurner, P.J., Fisher, L.W., and Hansma. P.K., Nanoscale ion mediated networks in bone: Osteopontin can repeatedly dissipate large amounts of energy, *Nano Lett.*, 7(8), 2491–2498, Copyright 2007 American Chemical Society.)

the presence of the Ca^{2+} ions in the environment overcomes the repulsive electrostatic interaction between the OPN molecules, and at the same time, the Ca^{2+} ions act as cross-linkers forming intermolecular and intramolecular bonds. During the pulling experiments, a more effective stress transfer occurs among the protein molecules because of the higher density of Ca^{2+}-mediated cross-links, which results in a higher maximum force, pulling length, and eventually dissipation energy to complete tip–surface separation, due to the sacrificial bond and hidden length mechanism previously discussed in this chapter. The fact that more than one protein molecule is involved in this process is demonstrated by pulling lengths of up to 3 μm on highly concentrated OPN films—for comparison an individual OPN molecule has an estimated contour length of ~100 nm (Fantner et al. 2007).

Second, Fanter et al. demonstrate that upon relaxation and waiting, the aforementioned weak ionic bonds can reform, such that energy can repeatedly be dissipated during further pulling events. Importantly, the amount of energy that can repeatedly be dissipated depends on the waiting, dwell, or surface retention time, that is, the time allowed for the network to reheal itself. The longer the time in between two pulling events, the higher is the energy dissipation, which increases in a nonlinear logarithmic-like fashion. This can be explained by the fact that most likely thermal energy is responsible for the movement of ions and proteins, and the longer these components are left relaxed, the higher is the likelihood of bond reformation.

Third, Fantner et al. demonstrate that the presence of Ca^{2+} ions is critical for energy dissipation and that when these ions are sequestered and washed away with the use of EDTA solution (acting as a chelator), the load-bearing and energy dissipation abilities of the OPN network are diminished, verifying the importance of ion-mediated bridges. In other words, manipulating the surrounding chemical environment, one can control the energy dissipation of an OPN film. This is shown in Figure 20.11, where dissipation energies are depicted versus elapsed time, measured on an OPN sample (Fantner et al. 2007). The system starts out with a solution containing only Na^{1+} ions, constituting a baseline; then after a given time, the liquid cell of the AFM is flushed with a solution containing both Na^{1+} and Ca^{2+} ions. This leads to an almost instant rise in energy dissipation of more than 100%. That this effect truly depends exclusively on the cations with more than one charge is evidenced by the next solution applied to the system; upon introduction of EDTA solution, the energy dissipation reduces back to the Na^{1+} baseline value as the EDTA chelates the Ca^{2+} ions from solution, and this effect can be reversed with reintroducing Ca^{2+} ions. It is interesting to note that the energy dissipation only gradually rises with time after reintroducing Ca^{2+} (cf. Figure 20.11); this is due to the much higher concentration of EDTA (250 mM) compared to Ca^{2+} ions (40 mM) and the fact that the solution change did happen in a series of flushes and not all at once.

As mentioned earlier, bone OPN and DMP-1 are both members of the SIBLING family (George et al. 1993, Fisher et al. 2001) and are known to bind to hydroxyapatite (Fisher et al. 2001). In mammalian skeletal tissue, the mineral crystal deposition occurs within the collagen fibrils, in the "gap" region between the tropocollagen molecules (Landis et al. 1993), and also at the fibril surface (Landis et al. 1996, Olszta et al. 2007). In the latter case, the crystal attachment to the collagen surface is likely stabilized via NCP bridges and hydrogen-bonding networks, which facilitate attachment (1) between the collagen and the mineral crystals (Girija et al. 2004, Wilson et al. 2006) and (2) between neighboring crystals (Boskey et al. 1993, Goldberg et al. 2001). During loading, this NCP matrix is subjected to shear, transferring the applied load between the mineralized fibrils (Gupta et al. 2006a). In addition, it has recently been pointed out that NCP aggregates also stabilize neighboring crystals on mineralized collagen fibrils (Poundarik et al. 2012) and hence also increase the toughness of these fibrils. When bone fails, fracture occurs at the very interfaces mentioned (Thurner et al. 2009, Poundarik et al. 2012), which clearly suggests that the mechanical behavior, integrity, and attachment strength of these proteins are of high importance for bone tissue material properties and fracture behavior. Hence, it is not surprising that OPN-deficient mice bones exhibit diminished mechanical properties (Kavukcuoglu et al. 2007, Thurner et al. 2010, Poundarik et al. 2012).

The approach for measuring the mechanical behavior of these proteins that is presented in this chapter is for the time being the only available experimental way to characterize the properties of the fibrillar or crystal–crystal interfaces. However, because of the nonspecific tip–protein and protein–surface interactions, the interpretation of the exact shape of the pulling curve is challenging. It is, for example, not possible to know the exact amount of protein molecules that are attached to the tip, and it is reasonable to believe that this changes on every pull. This also explains the fluctuation of the measured values for each feature, that is, maximum force, dissipation energy, etc., around an average value. Furthermore, measured pulling curves are the result of many different events (protein–protein, protein–substrate, and protein–tip interactions), which cannot be distinguished. Nevertheless, for a given system, this technique is able to accurately characterize the relative changes of these features under different experimental conditions.

20.3 OUTLOOK: FUTURE WORK

Although mica could serve as an adequate model system for studying the interaction between biological apatite and NCPs (Adams et al. 2008), mainly because of the similarities of the surface charge with that of apatite, further nonmodel-based studies would allow the analysis of additional parameters of biological importance. The effect of ion substitutions, crystal size, or surface roughness of the hydroxyapatite surface could be such parameters. Gluing hydroxyapatite in the form of single crystals or polycrystalline particles onto the cantilever would allow for similar experiments against calcium phosphate surfaces, that is, HAP–NCP–HAP interaction instead of mica–NCP–cantilever material. In addition, chemical fixation of a single protein or fragments of the same could give further insight into the interaction of a single protein with the whole protein film or substrate giving information on adhesion and cohesion properties in different chemical environments. Further, the adhesion strength of certain amino acid sequences can be tested for a given substrate (Lee et al. 2006), which might be of interest.

Overall, force spectroscopy allows many interesting experiments, and with the absence of better characterization tools, it is still one of the few methods that can produce nanomechanical data for protein assemblies.

REFERENCES

Adams, J., G.E. Fantner, L.W. Fisher, and P.K. Hansma. 2008. Molecular energy dissipation in nanoscale networks of dentin matrix protein 1 is strongly dependent on ion valence. *Nanotechnology* 19:384008.

Baht, G.S., G.K. Hunter, and H.A. Goldberg. 2008. Bone sialoprotein-collagen interaction promotes hydroxyapatite nucleation. *Matrix Biol* 27(7):600–608.

Binnig, G., C.F. Quate, and C. Gerber. 1986. Atomic force microscope. *Phys Rev Lett* 56(9):930–933.

Birdi, K.S. 2003. *Scanning Probe Microscopes: Applications in Science and Technology*. CRC, Boca Raton, FL.

Bonar, L.C., S. Lees, and H.A. Mook. 1985. Neutron diffraction studies of collagen in fully mineralized bone. *J Mol Biol* 181(2):265–270.

Boskey, A.L., M. Maresca, W. Ullrich, S.B. Doty, W.T. Butler, and C.W. Prince. 1993. Osteopontin-hydroxyapatite interactions in vitro: Inhibition of hydroxyapatite formation and growth in a gelatin-gel. *Bone Mineral* 22(2):147–159.

Braga, P.C. and D. Ricci. 2004. *Atomic Force Microscopy: Biomedical Methods and Applications*. Vol. 242. Humana Pr Inc., Totowa, NJ.

Butt, H.J. and M. Jaschke. 1995. Calculation of thermal noise in atomic force microscopy. *Nanotechnology* 6:1.

Cowin, S.C. and J.J. Telega. 2003. Bone mechanics handbook. *Appl Mech Rev* 56:61.

Deegan, R.D., O. Bakajin, T.F. Dupont, G. Huber, S.R. Nagel, and T.A. Witten. 1997. Capillary flow as the cause of ring stains from dried liquid drops. *Nature* 389(6653):827–828.

Denhardt, D.T. and X. Guo. 1993. Osteopontin: A protein with diverse functions. *FASEB J* 7(15):1475–1482.

Derkx, P., A.L. Nigg, F.T. Bosman, D.H. Birkenhäger-Frenkel, A.B. Houtsmuller, H.A.P. Pols, and J. Van Leeuwen. 1998. Immunolocalization and quantification of noncollagenous bone matrix proteins in methylmethacrylate-embedded adult human bone in combination with histomorphometry. *Bone* 22(4):367–373.

Donnelly, E., S.P. Baker, A.L. Boskey, and M.C.H. van der Meulen. 2006. Effects of surface roughness and maximum load on the mechanical properties of cancellous bone measured by nanoindentation. *J Biomed Mater Res Part A* 77(2):426–435.

Eppell, S.J., W. Tong, J.L. Katz, L. Kuhn, and M.J. Glimcher. 2001. Shape and size of isolated bone mineralites measured using atomic force microscopy. *J Ortho Res* 19(6):1027–1034.

Fantner, G.E., J. Adams, P. Turner, P.J. Thurner, L.W. Fisher, and P.K. Hansma. 2007. Nanoscale ion mediated networks in bone: Osteopontin can repeatedly dissipate large amounts of energy. *Nano Lett* 7(8):2491–2498.

Fantner, G.E., T. Hassenkam, J.H. Kindt, J.C. Weaver, H. Birkedal, L. Pechenik, J.A. Cutroni et al. 2005a. Sacrificial bonds and hidden length dissipate energy as mineralized fibrils separate during bone fracture. *Nat Mater* 4(8):612–616.

Fantner, G.E., T. Hassenkam, J.H. Kindt, J.C. Weaver, H. Birkedal, L. Pechenik, J.A. Cutroni et al. 2005b. Sacrificial bonds and hidden length dissipate energy as mineralized fibrils separate during bone fracture. *Nat Mater* 4(8):612–616.

Fantner, G.E., E. Oroudjev, G. Schitter, L.S. Golde, P. Thurner, M.M. Finch, P. Turner, T. Gutsmann, D.E. Morse, H. Hansma, and P.K. Hansma. 2006. Sacrificial bonds and hidden length: Unraveling molecular mesostructures in tough materials. *Biophys J* 90(4):1411–1418.

Fisher, L.W., D.A. Torchia, B. Fohr, M.F. Young, and N.S. Fedarko. 2001. Flexible structures of SIBLING proteins, bone sialoprotein, and osteopontin. *Biochem Biophys Res Commun* 280(2):460–465.

Fratzl, P., H.S. Gupta, E.P. Paschalis, and P. Roschger. 2004. Structure and mechanical quality of the collagen–mineral nano-composite in bone. *J Mater Chem* 14(14):2115–2123.

Fratzl, P. and R. Weinkamer. 2007. Nature's hierarchical materials. *Progr Mater Sci* 52(8):1263–1334.

Fujisawa, R. and Y. Kuboki. 1991. Preferential adsorption of dentin and bone acidic proteins on the (100) face of hydroxyapatite crystals. *Biochim Biophys Acta (BBA)—Gen Sub* 1075(1):56–60.

George, A., B. Sabsay, P.A. Simonian, and A. Veis. 1993. Characterization of a novel dentin matrix acidic phosphoprotein. Implications for induction of biomineralization. *J Biol Chem* 268(17):12624–12630.

Giraud-Guille, M.M. 1988. Twisted plywood architecture of collagen fibrils in human compact bone osteons. *Calcified Tissue Int* 42(3):167–180.

Girija, E.K., Y. Yokogawa, and F. Nagata. 2004. Apatite formation on collagen fibrils in the presence of polyacrylic acid. *J Mater Sci: Mater Med* 15(5):593–599.

Goldberg, H.A., K.J. Warner, M.C. Li, and G.K. Hunter. 2001. Binding of bone sialoprotein, osteopontin and synthetic polypeptides to hydroxyapatite. *Conn Tissue Res* 42(1):25–37.

Gupta, H.S., J. Seto, W. Wagermaier, P. Zaslansky, P. Boesecke, and P. Fratzl. 2006a. Cooperative deformation of mineral and collagen in bone at the nanoscale. *Proc Natl Acad Sci* 103(47):17741.

Gupta, H.S., U. Stachewicz, W. Wagermaier, P. Roschger, H.D. Wagner, and P. Fratzi. 2006b. Mechanical modulation at the lamellar level in osteonal bone. *J Mater Res* 21(8):1913–1921.

Gutsmann, T., G.E. Fantner, J.H. Kindt, M. Venturoni, S. Danielsen, and P.K. Hansma. 2004. Force spectroscopy of collagen fibers to investigate their mechanical properties and structural organization. *Biophys J* 86(5):3186–3193.

Heinz, W.F. and J.H. Hoh. 1999. Spatially resolved force spectroscopy of biological surfaces using the atomic force microscope. *Trends Biotechnol* 17(4):143–150.

Hinterdorfer, P., W. Baumgartner, H.J. Gruber, K. Schilcher, and H. Schindler. 1996. Detection and localization of individual antibody-antigen recognition events by atomic force microscopy. *Proc Natl Acad Sci* 93(8):3477–3481.

Hinterdorfer, P. and Y.F. Dufrêne. 2006. Detection and localization of single molecular recognition events using atomic force microscopy. *Nat Methods* 3(5):347–355.

Hollinger, J.O. 2005. *Bone Tissue Engineering*. CRC, Boca Raton, FL.

Hugel, T. and M. Seitz. 2001. The study of molecular interactions by AFM force spectroscopy. *Macromol Rapid Commun* 22(13):989–1016.

Hunter, G.K. and H.A. Goldberg. 1993. Nucleation of hydroxyapatite by bone sialoprotein. *Proc Natl Acad Sci* 90(18):8562.

Hutter, J.L. and J. Bechhoefer. 1993. Calibration of atomic-force microscope tips. *Rev Sci Instrum* 64(7):1868–1873.

Israelachvili, J.N. and G.E. Adams. 1978. Measurement of forces between two mica surfaces in aqueous electrolyte solutions in the range 0–100 nm. *J Chem Soc Faraday Trans 1* 74:975–1001.

Jena, B.P. and J.K.H. Hörber. 2006. *Force Microscopy: Applications in Biology and Medicine*. John Wiley & Sons, New York.

Katsamenis, O.L., H.M.H. Chong, O.G. Andriotis, and P.J. Thurner. 2013. Load-bearing in cortical bone microstructure: Selective stiffening and heterogeneous strain distribution at the lamellar level. *J Mech Behav Biomed Mater* 17:152–165. doi: http://dx.doi.org//10.1016//j.jmbbm.2012.08.016.

Kavukcuoglu, N.B., D.T. Denhardt, N. Guzelsu, and A.B. Mann. 2007. Osteopontin deficiency and aging on nanomechanics of mouse bone. *J Biomed Mater Res Part A* 83(1):136–144.

Kirkham, J., J. Zhang, S.J. Brookes, R.C. Shore, S.R. Wood, D.A. Smith, M.L. Wallwork, O.H. Ryu, and C. Robinson. 2000a. Evidence for charge domains on developing enamel crystal surfaces. *J Dental Res* 79(12):1943.

Kirkham, J., J. Zhang, S.J. Brookes, R.C. Shore, S.R. Wood, D.A. Smith, M.L. Wallwork, O.H. Ryu, and C. Robinson. 2000b. Evidence for charge domains on developing enamel crystal surfaces. *J Dental Res* 79(12):1943–1947.

Koester, K.J., J.W. Ager, and R.O. Ritchie. 2008. The true toughness of human cortical bone measured with realistically short cracks. *Nat Mater* 7(8):672–677.

Lamprou, D.A., J.R. Smith, T.G. Nevell, E. Barbu, C.R. Willis, and J. Tsibouklis. 2010. Self-assembled structures of alkanethiols on gold-coated cantilever tips and substrates for atomic force microscopy: Molecular organisation and conditions for reproducible deposition. *Appl Surf Sci* 256(6):1961–1968.

Landis, W.J., K.J. HodgensMin Ja, J. Arena, S. Kiyonaga, and M. MarkoCameron. 1996. Mineralization of collagen may occur on fibril surfaces: Evidence from conventional and high-voltage electron microscopy and three-dimensional imaging. *J Struct Biol* 117(1):24–35.

Landis, W.J., M.J. Song, A. Leith, L. McEwen, and B.F. McEwen. 1993. Mineral and organic matrix interaction in normally calcifying tendon visualized in three dimensions by high-voltage electron microscopic tomography and graphic image reconstruction. *J Struct Biol* 110(1):39–54.

Launey, M.E., M.J. Buehler, and R.O. Ritchie. 2010. On the mechanistic origins of toughness in bone. *Ann Rev Mater Res* 40:25–53.

Lee, H., N.F. Scherer, and P.B. Messersmith. 2006. Single-molecule mechanics of mussel adhesion. *Proc Natl Acad Sci* 103(35):12999–13003.

LeGeros, Z. 1981. Apatites in biological systems. *Progr Cryst Growth Char* 4(1–2):1–45.

Levy, R. and M. Maaloum. 2002. Measuring the spring constant of atomic force microscope cantilevers: Thermal fluctuations and other methods. *Nanotechnology* 13:33.

Marotti, G. 1993. A new theory of bone lamellation. *Calcified Tissue Int* 53:47–56.

McKee, M.D., M.C. Farach-Carson, W.T. Butler, P.V. Hauschka, and A. Nanci. 1993. Ultrastructural immunolocalization of noncollagenous (osteopontin and osteocalcin) and plasma (albumin and α2HS-glycoprotein) proteins in rat bone. *J Bone Min Res* 8(4):485–496.

McKee, M.D., A. Nanci, W.J. Landis, L.C. Gerstenfeld, Y. Gotoh, and M.J. Glimcher. 1989. Ultrastructural immunolocalization of a major phosphoprotein in embryonic chick bone. *Connect Tissue Res* 21(1–4):21–29.

Meyer, E. 1992. Atomic force microscopy. *Progr Surf Sci* 41(1):3–49.

Meyer, G. and N.M. Amer. 1988. Novel optical approach to atomic force microscopy. *Appl Phys Lett* 53:1045.

Nalla, R.K., J.J. Kruzic, J.H. Kinney, and R.O. Ritchie. 2005. Mechanistic aspects of fracture and R-curve behavior in human cortical bone. *Biomaterials* 26(2):217–231.

Nanci, A. 1999. Content and distribution of noncollagenous matrix proteins in bone and cementum: Relationship to speed of formation and collagen packing density. *J Struct Biol* 126(3):256–269.

Oberhauser, A.F., P.K. Hansma, M. Carrion-Vazquez, and J.M. Fernandez. 2001. Stepwise unfolding of titin under force-clamp atomic force microscopy. *Proc Natl Acad Sci* 98(2):468–472.

Olszta, M.J., X. Cheng, S.S. Jee, R. Kumar, Y.Y. Kim, M.J. Kaufman, E.P. Douglas, and L.B. Gower. 2007. Bone structure and formation: A new perspective. *Mater Sci Eng: R: Rep* 58(3–5):77–116.

Peterlik, H., P. Roschger, K. Klaushofer, and P. Fratzl. 2005. From brittle to ductile fracture of bone. *Nat Mater* 5(1):52–55.

Pidaparti, R.M.V., A. Chandran, Y. Takano, and C.H. Turner. 1996. Bone mineral lies mainly outside collagen fibrils: Predictions of a composite model for osternal bone. *J Biomech* 29(7):909–916.

Poundarik, A.A., T. Diab, G.E. Sroga, A. Ural, A.L. Boskey, C.M. Gundberg, and D. Vashishth. 2012. Dilatational band formation in bone. *Proc Natl Acad Sci* 109(47):19178–19183.

Rho, J.Y., L. Kuhn-Spearing, and P. Zioupos. 1998. Mechanical properties and the hierarchical structure of bone. *Med Eng Phys* 20(2):92–102.

Roach, H.I. 1994. Why does bone matrix contain non-collagenous proteins? The possible roles of osteocalcin, osteonectin, osteopontin and bone sialoprotein in bone mineralisation and resorption. *Cell Biol Int* 18(6):617–628.

Smith, B.L., T.E. Schäffer, M. Viani, J.B. Thompson, N.A. Frederick, J. Kindt, A. Belcher, G.D. Stucky, D.E. Morse, and P.K. Hansma. 1999. Molecular mechanistic origin of the toughness of natural adhesives, fibres and composites. *Nature* 399(6738):761–763.

Standring, S. 2008. *Gray's Anatomy*. Churchill Livingstone, Elsevier, Philadelphia, PA.

Thula, T.T., D.E. Rodriguez, M.H. Lee, L. Pendi, J. Podschun, and L.B. Gower. 2011. *In vitro* mineralization of dense collagen substrates: A biomimetic approach toward the development of bone-graft materials. *Acta Biomater* 7(8):3158–3169.

Thurner, P.J. 2009. Atomic force microscopy and indentation force measurement of bone. *Wiley Interdisciplinary Rev: Nanomed Nanobiotechnol* 1(6):624–649.

Thurner, P.J., C.G. Chen, S. Ionova-Martin, L. Sun, A. Harman, A. Porter, J.W. Ager III, R.O. Ritchie, and T. Alliston. 2010. Osteopontin deficiency increases bone fragility but preserves bone mass. *Bone* 46(6):1564–1573.

Thurner, P.J., S. Lam, J.C. Weaver, D.E. Morse, and P.K. Hansma. 2009. Localization of phosphorylated serine, osteopontin, and bone sialoprotein on mineralized collagen fibrils in bone. *J Adhes* 85:526–545.

Tye, C.E., G.K. Hunter, and H.A. Goldberg. 2005. Identification of the type I collagen-binding domain of bone sialoprotein and characterization of the mechanism of interaction. *J Biol Chem* 280(14):13487.

Wallwork, M.L., J. Kirkham, H. Chen, S.X. Chang, C. Robinson, D.A. Smith, and B.H. Clarkson. 2002. Binding of dentin noncollagenous matrix proteins to biological mineral crystals: An atomic force microscopy study. *Calcified Tissue Int* 71(3):249–256.

Wang, Y., T. Azaïs, M. Robin, A. Vallée, C. Catania, P. Legriel, and N. Nassif. 2012. The predominant role of collagen in the nucleation, growth, structure and orientation of bone apatite. *Nat Mater* 11(8):724–733.

Weiner, S., W. Traub, and H.D. Wagner. 1999. Lamellar bone: Structure-function relations. *J Struct Biol* 126(3):241–255.

Weiner, S. and H.D. Wagner. 1998. The material bone: Structure-mechanical function relations. *Ann Rev Mater Sci* 28(1):271–298.

Wilson, E.E., A. Awonusi, M.D. Morris, D.H. Kohn, M.M.J. Tecklenburg, and L.W. Beck. 2006. Three structural roles for water in bone observed by solid-state NMR. *Biophys J* 90(10):3722–3731.

Zappone, B., P.J. Thurner, J. Adams, G.E. Fantner, and P.K. Hansma. 2008a. Effect of Ca2+ ions on the adhesion and mechanical properties of adsorbed layers of human osteopontin. *Biophys J* 95(6):2939–2950. doi: biophysj.108.135889 [pii]10.1529/biophysj.108.135889.

Zappone, B., P.J. Thurner, J. Adams, G.E. Fantner, and P.K. Hansma. 2008b. Effect of Ca2+ ions on the adhesion and mechanical properties of adsorbed layers of human osteopontin. *Biophys J* 95(6):2939–2950.

21 Mechanical and interface properties of biominerals: Atomistic to coarse-grained modeling

Arun K. Nair, Flavia Libonati, Zhao Qin, Leon S. Dimas, and Markus J. Buehler

Contents

21.1 INTRODUCTION

The combination of two distinct materials, with a hard component interspersed into a soft organic matrix, is believed to be essential to many load-bearing materials, such as bone, which provide structural support for many organisms. The superior mechanical properties result from the hierarchical organization of bone, formed in a complex array of structures that spans the nanoscale to the macroscale, as described in detail in Chapter 20. Each level of organization has a crucial role in determining the overall behavior of bone at the macroscopic scale (Launey et al. 2010). From a fundamental point of view, collagen protein and hydroxyapatite (HAP) crystals are considered the primary building blocks of bone. Therefore, the mechanical behavior of these constituents as well as the interaction of the interface between them is believed to affect the mechanical response of bone at the nanoscale. However, many open questions remain with respect to the effects of various mechanisms that act on and govern the various length scales in bone.

The interface between the organic and inorganic phases in bone plays a critical role to bind the two materials together, and greatly defines the mechanical function of this composite material (Fratzl and Weinkamer 2007). The two different materials feature two extreme mechanical properties: the HAP, as a representative part of the inorganic phase, is hard and brittle, while the collagen, as a representative part of the organic phase, is soft and tends to be ductile. While the mechanical properties of any of the single material, either collagen or HAP, have been studied (Snyders et al. 2007; Gautieri et al. 2009a), the structure and interaction of these materials at their interface pose many challenging questions. Moreover, how are the thin flakes of HAP, as shown in Figure 21.1a, capable of making bone become mechanically strong and robust? To answer these questions, a deeper understanding of the interfacial functions at the molecular level is necessary. Here, we primarily discuss how the mechanical properties of the collagen–HAP interface and the molecular morphology of the crystals become critical in understanding the mechanism of the bone's mechanical function at the most fundamental level. Understanding of this mechanism is important to gain the materials' interfacial properties and go further to assess the composites' behavior with more empirical models on larger scales (Qin et al. 2012).

HAP, with the chemical formula $Ca_{10}(PO_4)_6(OH)_2$, has the form of a hexagonal crystal. It not only makes up to ~65% wt. of the natural bone but has also found applications as an important coating material. Many modern implants, including joint replacements and dental implants, are coated with HAP (Allegrini et al. 2006). In addition to its low propensity to wear, its chemical composition may help to form a stronger bonding with bone. HAP is fully biocompatible, and its porous form is used for producing artificial bones by mimicking the mechanical properties of real bones, as well as for controlled drug-delivery systems (Sopyan et al. 2007; Kundu et al. 2010). Therefore, understanding the collagen–HAP interface is also important for designing the coating and structure of advanced materials for various bioengineering applications.

Molecular simulations provide a useful tool to investigate the atomistic-level chemomechanical properties of the

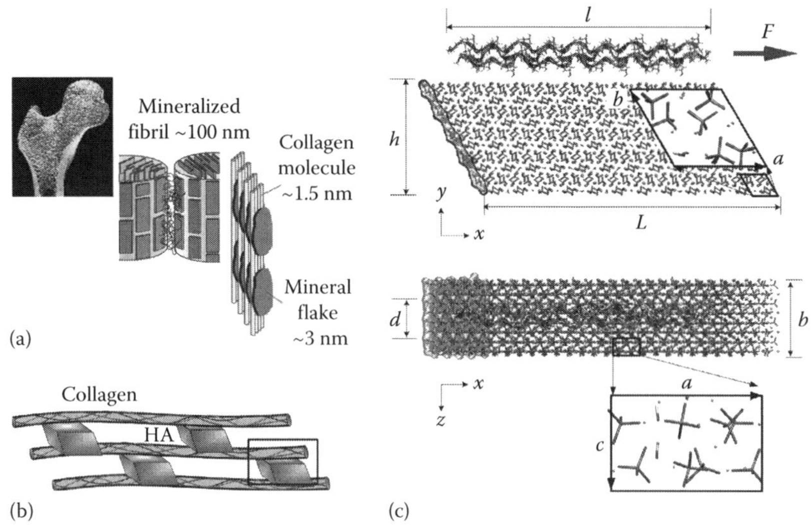

Figure 21.1 Schematic of the collagen–HAP molecular systems as reported in the study of Qin et al. (2012): (a) Microstructures of human femur bone at different scale levels (Adapted and reprinted by permission from Macmillan Publishers Ltd. *Nat. Mater.*, Fratzl, P., Bone fracture—When the cracks begin to show, 7(8), 610–612, 2008a; Fratzl, P., *Collagen: Structure and Mechanics*, Springer, New York, 2008b, Copyright 2012.) (b) Schematic arrangement of the assembly of tropocollagen molecule and HAP crystals in bone. We focus on the region selected by black rectangular as shown. (c) Lateral and top views of the systems with collagen molecule lying on the Ca surface (*y*+). A lateral total force *F* that is parallel to the surface in *x*-direction is applied to the three alpha carbon atoms at the right-end of the tropocollagen segment, while a single layer of atoms at the left-end of the HAP is fixed as shown by the bordering outlines. The OH surface is the bottom surface (*y*−) of the HAP crystal in this panel. We zoom in to show the projection of a unit cell of the HAP with its three lattice constants *a*, *b*, and *c*. (Panels b and c: Reprinted from Qin, Z. et al., *Langmuir*, 28(4), 1982, 2012.)

organic–inorganic interfaces and its correlation with the overall mechanical behavior of the composite materials. Studies to this level with detailed mechanisms are still difficult or impossible to achieve with experimental techniques alone. Atomistic simulations have also been recently used to gain insights into the mechanical properties of HAP (Snyders et al. 2007) and its interfacial property with short peptides. Previous works have used atomistic simulations to investigate the load-deformation behavior of tropocollagen molecules with HAP in proximity of their terminals (Bhowmik et al. 2009), the adsorption energy of tropocollagen molecules such as the common Gly–Pro–Hyp segment (Almora-Barrios and deLeeuw 2010b), and the description of model systems enamel (Zahn and Duchstein 2011). Those models are still quite far from the basic unit of the natural bone configuration.

In this chapter, we present models of bone-like composites, applied to shed light on the mechanisms by which HAP can provide stiffness to the collagen matrix. We first discuss the methodology that is utilized to describe the interactions between collagen molecules, HAP crystals, and the interactions between collagen and HAP. We then discuss the mechanical properties of a collagen–HAP composite, followed by a study of size effects in HAP crystals in resistance to fracture. Our basic model is set up by considering a unit cell of the nanostructures of tropocollagen and HAP crystal, as shown in Figure 21.1b, and their interface is parallel to the tropocollagen axis. By applying mechanical force to this model and studying its response, the model can be used to understand the tropocollagen–HAP interface properties under extreme mechanical conditions. We also introduce a microfibril model of bone and review the results for tensile tests. The latter study corresponds to nanoscale deformation mechanisms. Finally, a coarse grain model for bone is presented, which aims at uncovering macroscale level deformation mechanisms, and also leads a way to translate the structure of bone into synthetic materials that mimic some of the key mechanical traits.

21.2 METHODS

21.2.1 ATOMISTIC MODEL OF TROPOCOLLAGEN MOLECULE AND HYDROXYAPATITE

In atomistic modeling, the interactions between tropocollagen molecules and HAP at interfaces are decomposed into basic energy contributions, including van der Waals interactions, Coulombic interactions, and hydrogen bonds. The van der Waals interaction is computed pairwise for all the atom pairs in the tropocollagen molecule and HAP, while the Coulombic interactions mainly exist between charged side chains (e.g., NH^{3+} and COO^-) and surface ions of HAP (e.g., Ca^{2+}, PO_4^{3-}, and OH^-). Hydrogen bonds are formed between the uncharged polar side chains (e.g., OH in Hyp) and OH groups in HAP.

We build the atomistic structure of the tropocollagen molecule by using the software THeBuScr (triple helical collagen building script) (Rainey and Goh 2004). We focus on the simplest model of collagen, with only Gly–Pro–Hyp triplets on each of the three chains. The tropocollagen molecule model that we use, $[(Gly–Pro–Hyp)_{10}]_3$, is composed of only 30 amino acids per chain in order to reduce computational costs; such a length is comparable to the length used in previous studies (Buehler 2006; Gautieri et al. 2009a,b). Because the tropocollagen length (9 nm) is shorter than its persistence length (10–20 nm), entropic effects are not significant for our model. The computational model is developed with the aim of elucidating the generic behavior and deformation of a system containing the interaction between a single tropocollagen and a HAP nanocrystal, and as such is designed to be simple and not intended to be a direct representation of the actual bone nanostructure with a specific sequence. Our model enables us to perform a systematic study of bone nanomechanics from a fundamental point of view.

21.2.1.1 Tropocollagen parameterization

Collagen is the sole protein that features hydroxyproline (HYP), a nonstandard amino acid, resulting from hydroxylation of proline. Since it is rarely found, HYP is not parameterized in common biomolecular force fields like CHARMM (Brooks et al. 1983). A force field set for HYP has been developed by using quantum-mechanical simulations and subsequently deriving the atomistic parameters that best match the quantum-mechanics calculation (Park et al. 2005), with a particular focus on the correct modeling of the pucker of the HYP ring.

21.2.1.2 Hydroxyapatite force field parameterization

Most force fields for biomolecular simulations, like CHARMM (Brooks et al. 1983), do not include parameters for crystalline minerals such as HAP. Therefore, in order to model biomolecular systems including HAP, we extended the CHARMM force field. We use bond, angle, and dihedral parameters as reported earlier (Hauptmann et al. 2003), which are based on both quantum-mechanics calculations and empirical data. Therefore, for nonbonded terms, we use data from the study of Bhowmik et al. (2007a) in which the authors fitted the Born–Mayer–Huggins model (Hauptmann et al. 2003) with a simpler Lennard–Jones potential in CHARMM.

21.2.1.3 Crystal geometry

We generated the hexagonal HAP crystal unit cell by using Materials Studio 4.4 (Accelrys, Inc.), with the following lattice parameters: $a = 9.4214$ Å, $b = 9.4214$ Å, $c = 6.8814$ Å, $\alpha = 90°$, $\beta = 90°$, and $\gamma = 120°$. Based on this unit cell (44 atoms per unit cell), HAP crystals of varying thicknesses of $h = 0.7$, 1.4, 3.5, and 4.2 nm are generated for the purpose of this study. In our study, we focus on the interaction on the (010) plane, because this plane is dominant in the morphology of the biological material, due to the growth direction of the collagen matrix (Almora-Barrios and deLeeuw 2010a). Another reason to study this plane is that the surface charge for the two opposite surfaces of this plane is not neutral but negatively charged on the OH surface and positively charged on the Ca surface. Since the electrostatic interaction between the tropocollagen molecule and mineral part plays an important role to stabilize the interface, those two surfaces provide two extreme cases for our study. Moreover, since the (100) plane has similar geometric and chemical properties as the (010) plane and the surface charge of the (001) plane is neutral, by studying the two surfaces of the (010) plane, we may gain more representative results over the other two surfaces.

21.2.2 MOLECULAR DYNAMICS SIMULATIONS

Molecular dynamics calculations are performed using the LAMMPS code (Plimpton 1995) and the modified CHARMM force field. Lennard–Jones and Coulomb interactions are computed with a switching function that truncates the energy and force smoothly to 0 from 8 to 10 Å. This cutoff range is selected to include at least one complete lattice in the thickness direction. We have tested cutoff lengths longer than this value, and the tensile modulus and strength of the collagen–HAP model are not affected by this increment. The constructed collagen–HAP model is first geometrically optimized through energy minimization; then, an *NVT* (a canonical ensemble where moles, *N*, volume, *V*, and temperature, *T*, are conserved) equilibration is performed for 200 ps. The system temperature is set to constant 300 K with Langevin thermostat. After equilibration, we observe that the root-mean-square deviation of the crystal is stable and that no major changes occur in the crystal structure, confirming the reliability of the extended CHARMM force field. Visual molecular dynamics (VMD) (Humphrey et al. 1996) is a convenient tool for visualization, used here to show the snapshots of simulation and to compute hydrogen bonds. We count the hydrogen bonds within a cutoff distance of 3.5 Å and an angle range of 30°. We use MATLAB® (Mathworks, Inc.) to analyze the results of the mechanical testing.

21.3 CASE STUDIES USING MECHANICAL SIMULATIONS

21.3.1 MECHANICAL PROPERTIES OF COLLAGEN–HYDROXYAPATITE INTERFACE

Using the models of tropocollagen molecule and HAP as shown in Figure 21.1b and c, with the geometric parameters as summarized in Table 21.1, we study the effects of mineral surface, mineral thickness, and hydration state systematically, as reported by Qin et al. (2012). To obtain the force displacement relations, we use steered molecular dynamics (SMD) for loading the tropocollagen molecule and by fixing the left-end part of the HAP crystal and pulling the center of mass of the right-end alpha-carbon atoms of the tropocollagen molecule. The pulling velocity is set to 0.01 Å/ps (equivalent to 1 m/s), similar to that used in earlier simulation studies of the protein's tensile properties (Gautieri et al. 2009c; Qin et al. 2009). We keep a record of the applying force *F* as a function of the displacement of the SMD loading point *dx*. In the postprocessing stage, we calculate the applied stress to the end of the tropocollagen molecule via $\sigma = F/A$, where $A = \pi d^2/4$ is the cross-section area of the tropocollagen molecule, where *d* is the average diameter as shown in Figure 21.1c. Here, *A* only counts

Table 21.1 Basic geometric parameters of tropocollagen molecule and HAP for modeling, as shown in Figure 21.1

MATERIAL	CRYSTAL TYPE	GEOMETRIC PARAMETERS (nm)	BASIC UNIT
Tropocollagen molecule	Protein triple helix	$d = 1.5$ $l = 9.3$	GPO
HAP	Hexagonal crystal	$h = 0.7–4.2$ $b = 2.5$ $L = 10.5$	$Ca_{10}(PO_4)_6(OH)_2$

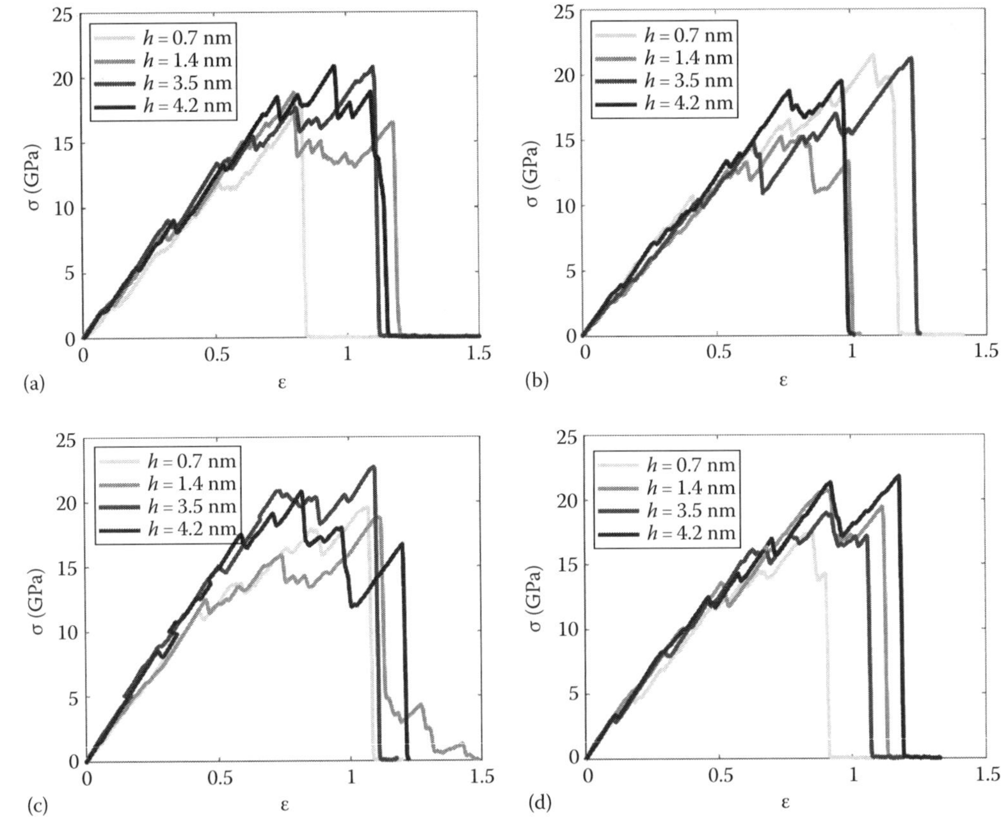

Figure 21.2 Stress–strain curves for collagen–HAP composites for HAP thicknesses *h* as indicated in the legends (Qin et al. 2012): (a) Stress–strain curves for Ca surface and (b) for the OH surface in dry conditions. (c) Stress–strain curves for Ca surface and (d) for the OH surface in hydrated conditions. (Reprinted from Qin, Z. et al., *Langmuir*, 28(4), 1982–1992, 2012.)

the cross-section of the tropocollagen molecule, because the force in our model is directly applied to the tropocollagen molecule end for the simulation set up. The tensile deformation is measured by the engineering strain calculated via $\varepsilon = dx/l$, where l is the length of the tropocollagen molecule.

First, we address the effect of HAP thickness h on the mechanical response of the model under tensile force. The force–extension curves in our simulations corresponding to various thicknesses and chemical surfaces of the HAP are summarized in Figure 21.2. All curves display a linear region from the beginning, followed by a "bumpy" region before failure of the material occurs. We observe that the deformation of the model in the first region originates from the extension of the atomic interactions, and rarely from the sliding between the tropocollagen molecule and HAP. For the second "bumpy" region, we observe that the collagen chains, after fully stretching, start to slide on top of the HAP surface. Notably, this sliding mechanism is consistent with the experimental observations (Gupta et al. 2013), where the intrafibrillar sliding between mineral and collagen leads to plastic strain in the fibril and the tissue level. Such a sliding mechanism occurs in a discontinuous way because of the atomically rough surface of the electrostatic interactions and hydrogen-bond forming patterns. The hydrogen bond analysis by Qin et al. (2012) shows that a difference in the number of hydrogen bonds for dry conditions and presence of water implies that the number of hydrogen bonds between the tropocollagen molecule and HAP contributes to the interaction force. We quantitatively analyze those strain–stress curves by focusing on the modulus and strength of this interface.

An analysis of tensile modulus E of the system for different surfaces and solvent conditions shows that the existence of HAP significantly increases with the stiffness of the system by comparing the tensile modulus to that of the pure collagen triple helix. We fit those results mathematically by using the exponential function

$$E = E_\infty - (E_\infty - E_{COL})\exp\left(\frac{-h}{h_{E0}}\right), \quad (21.1)$$

where

E_∞ is the converged value of the tensile modulus for larger h
E_{COL} (4.64 GPa, which agrees with the result of an earlier simulation study (Gautieri et al. 2009a)) is the tensile modulus of pure tropocollagen molecule
h_0 is a length constant as the characteristic thickness

It shows that for HAP crystals with a thickness beyond h_0, increasing the thickness can only increase the stiffness of the system up to 50%. We summarize E_∞ and h_0 for each of the two surfaces with and without water around the interface in Table 21.2. It can be seen that E_∞ is rather insensitive on the surface or hydration state, reaching a value of ≈31 GPa. On the other hand, the tensile modulus is shown to be dependent on the crystal thickness, in particular, for the Ca-rich surface in dry condition, where the tensile modulus in the case of the thinnest crystal (0.7 nm) is found to be about half the convergence value. Also, in the case of OH-rich surface (dry conditions), the thickness has little or no influence, possibly due to the higher

Table 21.2 Values of the parameters of the tensile modulus and strength of collagen–HAP interfaces as functions of the thickness of the HAP, defined in Equations 21.1 and 21.2

PARAMETER		TENSILE MODULUS		TENSILE STRENGTH	
		E_∞ (Gpa)	h_{E0} (nm)	σ_∞ (Gpa)	$h_{\sigma 0}$ (nm)
Ca surface	No water	30.16	0.86	20.79	0.84
	With water	31.72	0.51	21.41	0.68
OH surface	No water	31.72	0.02	19.37	0.02
	With water	31.87	0.36	20.62	0.60

number of collagen–HAP hydrogen bonds formed. In wet cases, the difference in the thickness effect (suggested by the differences by E_∞ and h_0) is less strong, due to the presence of water-mediated hydrogen bonds at the interface. This statement is supported by the evidence that water has a very strong attractive interaction with both collagen and HAP, especially for HAP (Katti et al. 2010), while at the same time, the existence of water can decrease the direct mechanical connection between the tropocollagen molecule and HAP with a lubricant effect (Gupta et al. 2006). Thus, by combining the analysis to our simulation results of E_∞ for each of the two surfaces, we can conclude that the existence of water increases the collagen–HAP interaction at the Ca-rich surface (indicated by the decreased h_0), but decreases the collagen–HAP interaction at the OH-rich surface (indicated by the increased h_0), leading to a more uniform E–h relationship for different surfaces as summarized in Table 21.2.

We fit the maximum stress, σ_C, reached by the strain–stress curve, as a function of the thickness h by using

$$\sigma_C = \sigma_\infty - (\sigma_\infty - \sigma_{COL})\exp\left(\frac{-h}{h_{\sigma 0}}\right), \tag{21.2}$$

where

σ_∞ is the converged maximum stress (Qin et al. 2012)
σ_{COL} (13 GPa from an earlier simulation study (Buehler 2006)) is the maximum stress of pure tropocollagen molecule
h_σ is the characteristic thickness

It shows that for HAP crystals with a thickness beyond h_σ, increasing the thickness can only increase the strength of the system up to 15%. We summarize σ_∞ and h_σ for each of the two surfaces with and without water around the interface in Table 21.2. It can be seen that σ_∞ is rather insensitive on the surface or hydration state, reaching a value of ≈20 GPa. Similarly, we find a significant dependence of σ_C on h (shown for large h_σ) for the Ca-rich surface in dry conditions. Also, in the case of OH-rich surface (dry conditions), the thickness has little or no influence. In the wet cases, the difference in the thickness effect (indicated by the differences in σ_∞ and h_σ) is less strong for these two different surfaces. This result agrees to the conclusion that the existence of water mediates the collagen–HAP interface and leads to a more uniform $\sigma_C - h$ relationship for different surfaces. The result also makes it clear that since the tensile strength of the pure tropocollagen molecule is weaker than this theoretical prediction of the collagen–HAP composite, the rupture event happens by breaking the tropocollagen molecule instead of by breaking the interface. The reason that our breaking force is larger than the tropocollagen strength is that the CHARMM force field we used here does not account for covalent bond breaking. Therefore, to make a more precise description of the rupture process of the interface (the competing process between collagen rupture and sliding apart at the interface), one may need a more fundamental reactive force field for the atomic interactions (Buehler 2006).

To summarize, this section discussed the effect of mineral surface, mineral thickness, and hydration state on the mechanical properties of collagen–mineral interface. We find that thin flakes of HAP crystals less than 1 nm can significantly improve the tensile modulus of the pure tropocollagen molecule. This observation is likely due to the fact that the tensile modulus of the HAP is much higher than that of the tropocollagen molecule, making the composite much stiffer than pure collagen. This result agrees with the experimental observations in which mineralized tendon shows a much higher tensile modulus (it can reach a factor of 10 times) than the unmineralized tendon (Weiner and Wagner 1998). On the other hand, the tensile modulus of the biomineral interface does not depend on the surface and hydration state, reaching a value of ≈31 GPa for increasing thicknesses. Specifically, the tensile modulus converges at a thickness of less than 2 nm, which agrees with the experimental observations that mineralized collagen structures in most forms of the natural morphology of bones are composed of collagen and thin HAP flakes of uniform thickness that varies between 1 and 4 nm for different bone types (Weiner and Wagner 1998; Fratzl 2008a; Hu et al. 2010). Such an agreement between simulations and experimental observations shows that the nanoscale structure of the collagen–HAP interface is important in defining the mechanical properties of bone. Our simulation results also show that the existence of water mediates the tensile modulus and strength of the collagen–HAP interface, leading to more homogeneous tensile modulus and tensile strength for various HAP surfaces with different chemical compositions. These results imply that the microscopic structure of bone evolved in nature largely according to its mechanical requirement to achieve its crucial function in supporting and protecting the body.

21.3.2 EFFECTS OF GEOMETRIC CONFINEMENT ON THE FRACTURE BEHAVIOR OF HYDROXYAPATITE PLATELETS: A CASE STUDY

The behavior of many biological materials is strongly affected by the geometric confinement at the nanometer length scale. As a result, the mechanical properties such as the strength and fracture toughness are strongly dependent on the characteristic size.

For instance, the human bone is well-known for its resistance to fracture, and it has been suggested that this mechanical property has its origin in a multitude of deformation and toughening mechanisms that occur at different scales, from the nanoscale to the macroscopic one (Ritchie et al. 2004, 2009; Dubey and Tomar 2008; Launey et al. 2010; Sen and Buehler 2011). The load transfer between the mineral and the protein phase has been described by the tension–shear model (Gao et al. 2003), where the mineral platelets carry most of the tensile load, and the protein matrix transfers the load between the platelets by shear (Ji and Gao 2004a,b; Gupta et al. 2005; Buehler et al. 2006; Gao 2006).

As was confirmed using a series of molecular dynamics simulations of HAP crystals in the study of Libonati et al. (2013), the geometric characteristics of the crystals have a strong influence on the mechanical properties. The elongated platelet shape leads to a high surface to volume ratio, ensuring a greater interaction with the collagen protein and enhanced mechanical properties. Indeed, using molecular models, we can directly address the effect of a systematic variation of the characteristic HAP crystal size on its mechanical behavior in the presence of an edge crack (Libonati et al. 2013). The effect of confining the size at the nanoscale on fracture response is studied by performing a series of *in silico* mechanical tests of tiny HAP samples of different sizes, with an edge crack. The dimensions of the samples (30.1 × 2.1 × 2*h* nm) are chosen so as to be similar to the HAP crystal platelets found in the bone. These platelets are generally tens of nanometers in width and length and 2–3 nm in thickness (Rho et al. 1998).

As presented in the study of Libonati et al. (2013), we consider a thin slab of HAP with an edge crack, which extends for half of the slab width, and we systematically vary the characteristic crystal size (i.e., the height of the crystal, 2*h*), keeping the sample length, the crack length, and the loading and boundary conditions constant for all the studied cases. An example of the tested sample is given in Figure 21.3a. The samples are created by replicating the unit cell (see Section 21.2.1 for the simulation cell dimensions) 32 times in the *x*-direction, 3 times in the *z*-direction, and *n* times in the *y*-direction, where *n* varies from 4 to 14. The HAP samples are generated by systematically increasing the height by two unit cells in the *y*-direction. The crack extends in the basal plane (001) with the main axes on the (010) plane. In all the cases, we consider a hexagonal close-packed (HCP) crystal, with $x = [100]$, $y = [010]$, and $z = [001]$. For simulations, we use an extended CHARMM force field discussed in Section 21.2.1.

By using a simple mechanical scheme, as shown in Figure 21.3b, each sample is clamped at the bottom and pulled at the top, in displacement control mode, applying a velocity of 1 Å/fs. We adopt a quasi-static loading approach, consisting of several loading steps followed by relaxation, through an energy minimization scheme, until failure occurred. The virial stress, calculated from the atomic positions and forces, and the engineering strain are used to obtain the stress–strain response. The mechanical properties (i.e., stiffness, maximum strength, strain at failure, etc.) are calculated from the stress–strain data. Stress maps are created by plotting the virial stresses (Tsai 1979; Zimmerman et al. 2004) on each atom, by means of the AtomEye tool (Li 2003). For analysis, the atomistic virial stress data computed in a thin strip ahead of the crack tip are averaged, and then fitted by using a power law of the form

$$y = (1-t)e^{-\beta_i x} + t, \tag{21.3}$$

where t represents the asymptotic value for each sample; β_i is defined in two ways:

$$\beta_1 = \frac{y_2 - y_1}{x_2 - x_1} \cdot (t-1)^{-1} \tag{21.4}$$

and

$$\beta_2 = \frac{1}{x_2} \cdot \ln\left(\frac{1-t}{y_2 - t}\right). \tag{21.5}$$

x_1 and x_2 represent the x-coordinates of the first two data, in each sample, and y_1 and y_2 represent the stress values of the first two data, in each sample.

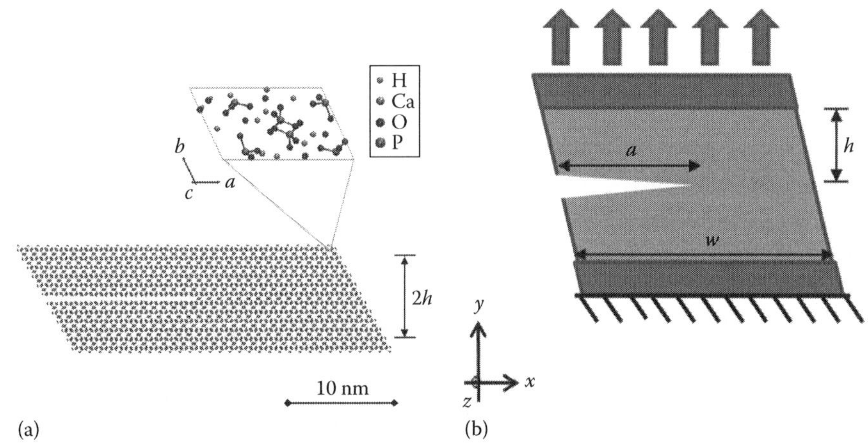

Figure 21.3 Geometry, dimensions, and loading and boundary conditions of a cracked HAP slab: (a) Snapshot of the sample in its initial geometry. The inset shows a zoomed view of the HAP unit cell, in a view parallel to the (001) plane (each atom is represented with a different size). (b) Model used for the atomistic study of fracture of HAP. Different samples are created by varying the sample aspect ratio (*h*/*w*). The width (*w*) and the crack length (*a*) are fixed for all specimens, and the crack extends over half of the platelet width. (Reprinted from Libonati, F. et al., *J. Mech. Behav. Biomed. Mater.*, 20, 184–191, 2013.)

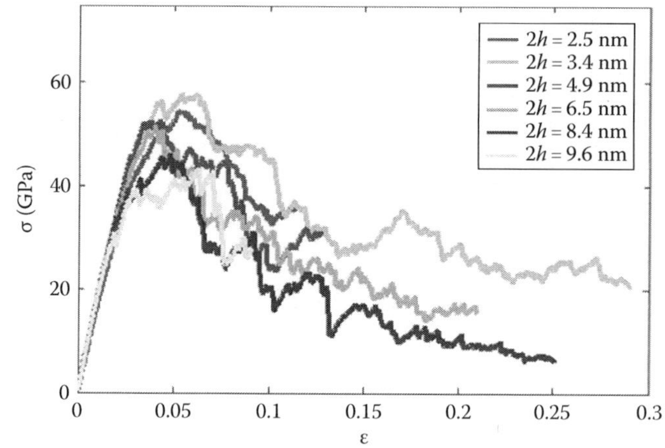

Figure 21.4 Stress–strain plot of the tested samples (Libonati et al. 2013). The stress plotted is σ_{yy}, and the strain represents the applied tensile strain in the y-direction. The initial slope of the stress–strain curve is not affected by the size of the sample, while the strength decreases as the sample height grows. (Reprinted from Libonati, N. et al., *J. Mech. Behav. Biomed. Mater.*, 20, 184–191, 2013.)

The results revealed that a decrease in the crystal size causes a change in the mechanical behavior of this material and the occurrence of new phenomena that are not present at larger scales:

1. A general increase in the mechanical properties (i.e., strength, toughness modulus)
2. A change in the failure mode, from a crack-driven failure in larger samples to a more spread failure mode in the smaller ones
3. A change in the stress field, from a heterogeneous stress field, characterized by a high stressed area near the crack tip, to a more homogeneous stress field

By varying the characteristic crystal size $2h$, we observe a variation in the mechanical response and properties (i.e., strength and toughness modulus), as shown in Figure 21.4, where the stress–strain curves of the tested samples are depicted. In all the cases, it is possible to recognize an initial linear region, where the material is slightly deformed in the elastic field. In this region, the material behavior is not dependent on the sample size; for instance, we observe that the slope of the stress–strain curves is similar for all cases; so, the stiffness of the sample is not affected by the sample size. This linear region is followed by a second region, characterized by a deviation from linearity, where the stress slowly increases until a maximum has been reached, and a subcritical propagation of the defect occurs. Indeed, initially we do not observe a clear crack propagation mechanism. Then, as the load increases, different phenomena besides crack opening occur. For small-sized cases, many defects emerge in different parts of the samples, growing until the maximum stress is reached. In this case, there is no sudden drop of the stress, like in larger samples, but it is "stepwise" decreasing. For larger samples, we find that failure is more localized around the crack tip region.

Beyond the maximum stress, failure occurs in different modes depending on the sample size. For small samples, it occurs more gradually, and it is not characterized by a clear crack path. Damage is widespread due to the formation and interaction of many small defects over the whole volume. The growth of these defects and their interaction lead to the global failure of the slab. By increasing the sample size, we observe a clearly distinguished crack path (Libonati et al. 2013). The difference in the failure mode is probably due to the different stress distributions reached in the various samples. Another effect of confining the size at the nanoscale is a change in the stress distribution. We observe that by decreasing the sample height, the stress concentration that characterizes the larger samples disappears. The more homogeneous stress distribution explains why small samples are characterized by a more spread failure damage and a slower failure mode. Figure 21.5 shows the stress–strain curves and stress fields for two samples of different sizes, confirming that there is a clear difference between the two cases. In the smaller sample (Figure 21.5c) with $2h$ = 2.5 nm, the stresses are homogenously distributed, whereas in the larger one (Figure 21.5d) with $2h$ = 8.4 nm, a concentration of higher stresses is clearly visible

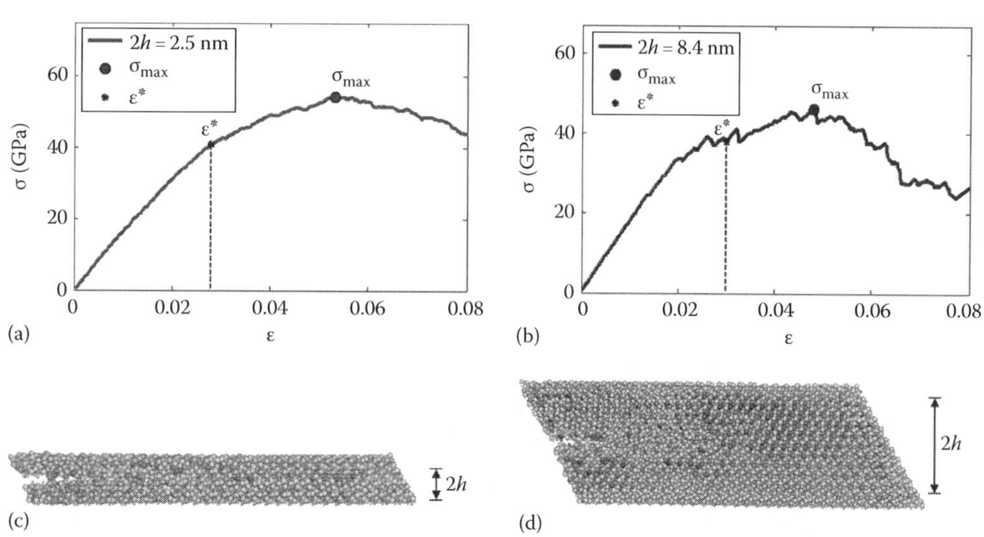

Figure 21.5 Stress–strain curve of two samples: (a) $2h$ = 2.5 nm and (b) $2h$ = 8.4 nm. The lower panels show maps of the stress distribution for two cases, (c) $2h$ = 2.5 nm and (d) $2h$ = 8.4 nm. The maps show the stress field (in the middle section of each sample) at the critical strain ε^*, which is highlighted in the earlier curves. The maximum stress (σ_{max}), reached before failure occurs, is highlighted in the two curves. (c) Depiction of a homogeneous stress distribution. (d) High stress concentration that forms ahead of the crack tip. (Reprinted from Libonati, N. et al., *J. Mech. Behav. Biomed. Mater.*, 20, 184–191, 2013.)

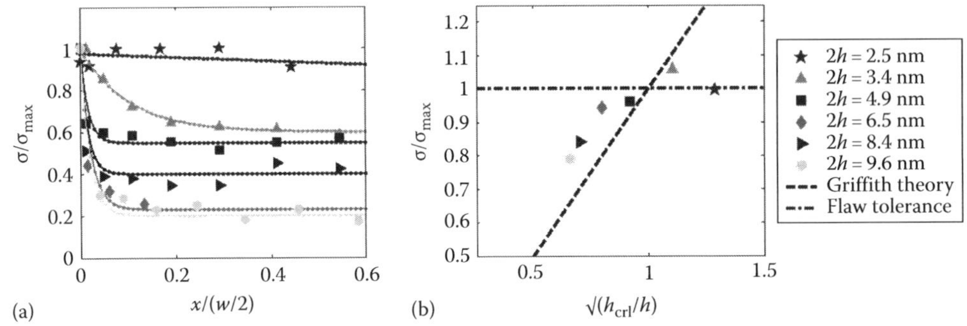

Figure 21.6 Stress fields for varying crystal dimensions (Libonati et al. 2013): (a) Stress field reached from the crack tip to the far field for different sample sizes. The longitudinal stress field, averaged over a thin strip in the middle and over the entire thickness, is plotted against the x-direction (the zero x-coordinate represents the crack tip). The values on both the axes are normalized by the maximum values, respectively. The continuous lines represent power-law fits to the atomistic data. (b) Normalized maximum strength over the normalized sample size. Comparisons with the Griffith theory prediction are given, showing a clear deviation at small length scales below h_{crl} = 4.15 nm. (Reprinted from Libonati, N. et al., *J. Mech. Behav. Biomed. Mater.*, 20, 184–191, 2013.)

ahead of the crack tip. This phenomenon is directly seen in Figure 21.6, where we plot the longitudinal stress distribution in a thin strip ahead of the crack tip just before failure occurs. The graph shown in Figure 21.6a confirms that the small sample is characterized by a quasi-homogeneous stress distribution, while the larger ones are characterized by a high-stressed region near the crack tip. The larger the size, the more localized is the area of high stress concentration. This analysis reveals that the stress field becomes increasingly homogeneous as the size is reduced.

From the stress–strain curves depicted in Figure 21.4, we calculate the strength, defined as the maximum stress reached during the test; we find that it is clearly affected by the size, being generally higher for smaller samples. As expected, the mechanical properties (i.e., the strength) also show a trend until a certain dimension is reached by the crystal. Below this characteristic size, the strength seems not to be affected by the variation of the size, hence approaching a constant value. Indeed, on the basis of the results presented earlier (e.g., failure mode, stress field, stress distribution from the crack tip to the far field), we observe that a distinct change in the mechanical behavior occurs at around $2h = 4.15$ nm. By considering the determined values of the strength (normalized by the maximum one) as a function of the inverse of the square root of the sample size, $\sqrt{h_{crl}/h}$, a bilinear trend is found (Figure 21.6b). Moreover, by comparing the data with the prediction of the Griffith's theory, we find that this theory is valid only for samples larger than the critical size, whereas for the smaller ones this theory does not hold, in agreement with the hypothesis put forth in the study by Gao et al. (2003). These findings corroborate the concept that, at the nanoscale, flaw tolerance occurs in HAP crystals below a critical size on the order of a few nanometers.

This study (Libonati et al. 2013) provided several important insights into the phenomena that occur at the nanoscale and on the different mechanisms governing the fracture at the nanoscale, where the concepts of stress concentration, crack-driven and brittle failure mode, and size-affected mechanical properties are replaced by quasi-homogeneous stress distribution, larger strain to failure, flaw tolerance, and constant mechanical properties, regardless of the presence of defect. This study may have large impact in understanding the high toughness of bone and bone-like materials, although they are mainly made of a brittle material.

21.3.3 ATOMISTIC MODELING OF A MINERALIZED COLLAGEN MICROFIBRIL

The model that was presented in Section 21.3.1 is limited to a single tropocollagen molecule (~nm scale). However, the assembly of several collagen molecules that form collagen fibrils has a diameter on the order of 100 nm (Ritchie et al. 2009). The knowledge about how collagen fibrils and HAP crystals interact at the molecular scale, and how they deform as an integrated system under external stress, are not well-understood. Several attempts have been made to develop a molecular model of bone resulting in a recently reported full-atomistic model of a mineralized collagen fibril that will be discussed here (Nair et al. 2013). The earlier studies fell short to providing a complete three-dimensional, chemically and structurally accurate model of the interactions of collagen with the mineral phase (Boskey 2003; Bhowmik et al. 2007b, 2009; Zhao et al. 2012). Coarse grain modeling (Buehler 2007) of nascent bone showed that the mineral crystals provide additional strength and also increases the Young's modulus and fracture strength. While these models provided some insight into the mechanics of bone, they failed to capture the atomic-scale mechanisms and did not allow for a direct comparison with the experimental work (e.g., *in situ* x-ray analysis of bone deformation [Weiner and Wagner 1998; Jager and Fratzl 2000; Fratzl et al. 2004; Gupta et al. 2006; Fratzl and Weinkamer 2007; Fratzl 2008b]).

The nucleation and growth of minerals in collagen fibril is a complex phenomenon (Colfen 2010) and involves timescales on the order of microseconds, and hence cannot be simulated directly by molecular dynamics with the current computational capabilities. Hence, in this model we apply an *in silico* mineralization scheme (Nair et al. 2013) that allows us to fill an initially unmineralized collagen fibril structure, up to a maximum mineral density of 40%. The geometry and composition of the model are illustrated in Figure 21.7, indicating varied levels of mineral content. The 0% case corresponds to the nonmineralized collagen microfibril and also shows the gap and the overlap regions (Figure 21.7), which emerge from the geometrical arrangement of collagen protein molecules. The *D*-period corresponds to the periodicity typically observed in collagen fibrils. The 20% and 40%

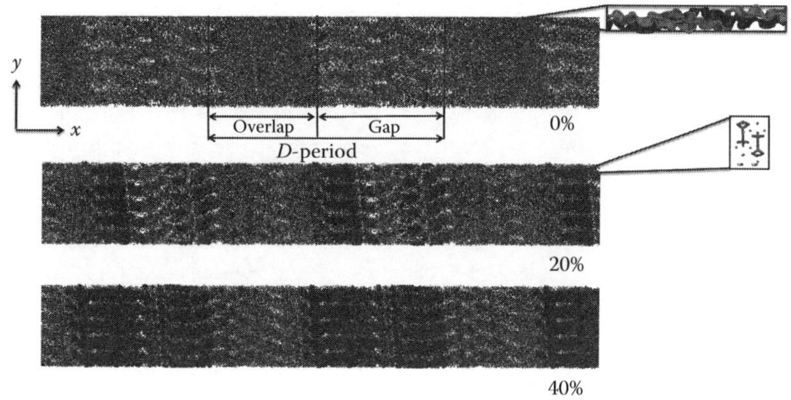

Figure 21.7 Full atomistic model of mineralized collagen fibrils as reported in the study of Nair et al. (2013). Collagen microfibril model with 0% mineralization (inset shows collagen triple helix structure), 20% mineral content (inset shows a HAP unit cell), and 40% mineral content. For mineralized cases, the dark region shows the presence of minerals in the gap and overlap regions. Inset shows that the HAP crystals are arranged such that the *c*-axis of the crystal aligns with the fibril axis. The HAP crystal comprises of C_α atoms, OH group and tetrahedron structure shows the PO_4 group. (Reprinted from Nair, A.K. et al., *Nat. Commun.*, 4, 1724, 2013.)

mineral density cases correspond to the two different mineral concentrations in the collagen microfibril.

As shown in Figure 21.8a, the mineral is deposited predominantly in the gap region along the fibril axis, with sparse deposition in the overlap region. Figure 21.8b shows a detailed analysis of the distribution of mineral crystals for 20% and 40% mineral densities. We perform a direct comparison with the experimental results that reported the distribution of mineral density as a function of the fibril axis (Nudelman et al. 2010), and plot the mineral distribution of 40% case normalized to its maximum value along with the experimental results along the fibril axis (Figure 21.8c). In good agreement with a rich set of experimental data (Weiner and Wagner 1998; Jager and Fratzl 2000; Fratzl et al. 2004; Gupta et al. 2006; Fratzl and Weinkamer 2007; Fratzl 2008b; Nudelman et al. 2010; Alexander et al. 2012), in the mineralized microfibril models with varying densities (10%, 20%, and 40%), the mineral deposition occurs primarily in the gap region. This observation is also consistent with the experimental finding (Nudelman et al. 2010) that the mineral nucleation point is close to the C-terminus (in the first section of the gap region, immediately after the gap/overlap transition). In their work, the authors suggest that the reason is the high concentration of positive charge, which attracts negatively charged peptides, in turn responsible for the onset of mineralization. Our model suggests that a concurrent mechanism is related to the larger voids found in this region of the fibrillar structure, which facilitates the onset of mineralization. We find that the 40% mineral density case shows more mineral deposition in the gap region compared to the experimental data (Nudelman et al. 2010). This could be due to the fact that in the experimental work the mineral density profile is measured for collagen that is mineralized for 24 h. Our model shows that the size of the mineral platelets in the gap region is $\approx 15 \times 3 \times 1.6$ nm^3 (for 40% mineral density). This is in good agreement with the experimental work that has shown that the size of mineral platelets is $15\text{–}55 \times 5\text{–}25 \times 2\text{–}3$ nm^3 (Fratzl et al. 1991; Nudelman et al. 2010; Alexander et al. 2012). Recent experiments (Alexander et al. 2012) show that the major fraction of the mineral in bone is outside the fibrils; this is substantiated by the presence of extrafibrillar bioapatite in nonmineralized fibrils. The inclusion

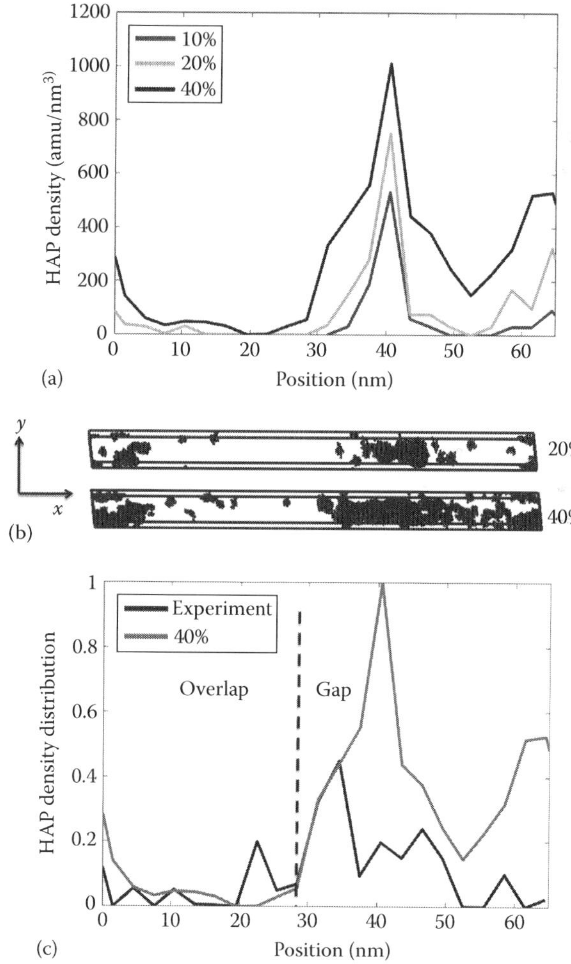

Figure 21.8 Mineral distribution in the collagen microfibril at different mineralization stages (Nair et al. 2013): (a) Distribution of HAP along the collagen fibril axis. The data shows that the maximum amount of HAP is deposited in the gap region (between 30 and 50 nm). (b) Spatial distribution of HAP in the unit cell for 20% and 40% mineral densities. (c) HAP density distribution along the fibril axis for the 40% case normalized (same data as depicted in panel (a)) compared to experimental data. (From Nudelman, F. et al., *Nat. Mater.*, 9(12), 1004, 2010.) The comparison confirms that maximum deposition is found in the gap region. (Reprinted from Nair, A.K. et al., *Nat. Commun.*, 4, 1724, 2013.)

of the extrafibrillar mineral content into the three-dimensional atomistic model is a challenging problem, considering the periodicity of the collagen microfibril model as well as the computational cost associated to perform the mechanical testing.

An *NVT* ensemble is used to first equilibrate the samples (0%, 20%, and 40%). In order to assess the mechanical properties of mineralized collagen fibrils, we perform stress-controlled (*NPT* or isothermal–isobaric, a canonical ensemble where moles, *N*, pressure, *P*, and temperature, *T*, are conserved) molecular dynamics simulations with increasing tensile stress applied along the *x*-axis of the unit cell as depicted in Figure 21.9a (Nair et al. 2013). The unit cell is under constant atmospheric pressure along the other two axes (*y* and *z*). We use an *NPT* ensemble (Plimpton 1995) for loading the samples with different stress states

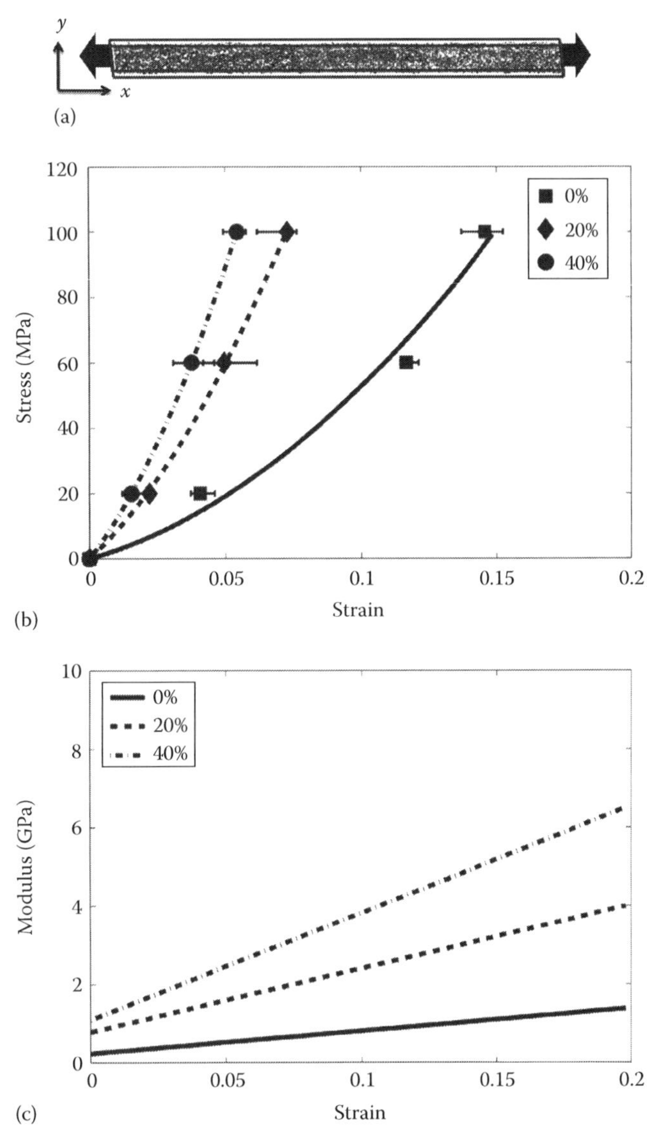

Figure 21.9 Mechanical properties of collagen fibrils at different mineralization stages as presented by Nair et al. (2013): (a) Fibril unit cell with mineral content used to perform tensile test by measuring stress versus strain. (b) Stress–strain plots for nonmineralized collagen fibril (0%), 20% mineral density and 40% mineral density cases. (c) Modulus versus strain for 0%, 20%, and 40% mineral densities showing an increase in modulus as the mineral content increases. (Reprinted from Nair, A.K. et al., *Nat. Commun.*, 4, 1724, 2013.)

($\sigma = -P$) ranging from (1) atmospheric pressure, (2) 20 MPa, (3) 60 MPa, and (4) 100 MPa. We observe that the samples reached equilibration under applied load at approximately 6 ns. The strain is computed as $\varepsilon = (L - L_o)/L_o$, where L_o is the equilibrium length identified at atmospheric pressure. To ensure that equilibrium is achieved, we monitor the pressure at equilibrium, the RMSD (root mean square deviation), and confirm that the size of the simulation cell reaches a steady-state value. Using the fibril strain ε associated with each applied stress σ, we obtain the stress–strain behavior for each case by plotting σ over ε. The modulus is computed from the first derivative of a polynomial function that is fitted to the stress–strain data. The computational time requirement for equilibration is 0.2 ns/week, using 24 processors for the 40% mineral density case.

As observed from Figure 21.9b, as the mineral content increases, the stress–strain behavior of the mineralized collagen microfibril also changes compared to pure collagen fibrils, with the mineralized cases showing an increasingly higher modulus as higher mineral densities are reached. In order to quantify the variation of the modulus for different strain levels, we plot the modulus as a function of strain, as shown in Figure 21.9c. The 0% case has an initial modulus of 0.5 GPa at a load less than 20 MPa, and increases to 1.1 GPa as the stress increases to 100 MPa. For larger deformation, the modulus approaches 2 GPa. These moduli are well within the range of values reported for collagen fibrils under tensile loading using both experiments and simulations (Van Der Rijt et al. 2006; Gautieri et al. 2011). For the 20% mineral density case, the initial modulus is 1.3 GPa (stress less than 20 MPa), and increases to 2.7 GPa at 100 MPa. This shows that even a relatively small mineral content greatly alters the stress–strain behavior of the collagen microfibril model and increases its modulus by ≈150%. Similarly, the 40% mineral density case has an initial modulus of 1.5 GPa, approaching 2.8 GPa. For stresses at 100 MPa, the 20% and 40% mineral density cases have very similar tangent moduli. However, as shown in Figure 21.9c, as the strain increases beyond 10%, the modulus for the 40% mineral density case also increases, indicating that at higher strains, the mineral content provides additional stiffness to the collagen–HAP composite. The moduli identified here for the cases with 20% and 40% mineral content are consistent with a recent experimental study (Hang and Barber 2011), which showed that mineralized collagen fibrils from antler had a modulus of 2.38 ± 0.37 GPa for strains less than 4%.

To understand the deformation mechanism for mineralized and nonmineralized samples at different deformation states, we compute the deformation fields within the fibrils. We observe that as the loading increases, there is no significant movement or coalescence of the HAP crystals in the gap region when the loading increases from 20 to 100 MPa. However, for the 0% case, the collagen molecule undergoes significant deformation in the gap region as the applied stress increases. We compute the deformation in the collagen microfibril for all the three cases at applied stresses of 20, 60, and 100 MPa. We find that (Nair et al. 2013) the gap-to-overlap ratio for 0% case increases with the applied stress indicating that for pure collagen the gap region deforms significantly compared to the overlap region to accommodate the external load. This behavior is consistent with the earlier tensile tests on collagen microfibril (Gautieri et al. 2011). Clearly, the presence of HAP alters the deformation mechanism of the collagen fibril. For the 20% mineral density case, the gap-to-overlap ratio is nearly constant for increases in applied stress, whereas for the 40% case the gap-to-overlap ratio decreases as the applied stress increases. This shows that a higher mineral content leads to more deformation in the overlap region compared to the gap region, where the interaction between HAP and collagen limits the deformation within collagen molecules. This analysis is consistent with the results reviewed in Section 21.3.1 and also explains how energy dissipation in bone occurs at the nanoscale, leading to high toughness. We also observe that HAP takes approximately four times more stress than the collagen phase (Nair et al. 2013), while the strain in the collagen phase is two orders of magnitude higher than the HAP phase. This finding that the strain in the collagen phase is lower than the total strain is consistent with the earlier studies (Gupta et al. 2006). The three-dimensional fully atomistic microfibril model discussed in this section has great potential for future studies, in particular, related to the effect of collagen mutations on the mechanical properties of bone, for example, in diseases such as osteogenesis imperfecta (brittle bone disease).

21.3.4 COARSE GRAIN MODELING OF BIOMINERALS

In order to gain a more complete understanding of the mechanical behavior of bone and biomineralized materials, one must also consider its mechanical characteristics at the meso- and macroscopic length scales. This is the scale at which cracks and flaws often initiate, and where many physiological functions of mineralized biomaterials take place. It is thus clearly a critical length scale at which material performance should be investigated. Full atomistic modeling techniques are, as has been highlighted earlier in this chapter (Sections 21.3.1 through 21.3.3), very useful and essential for understanding the nanoscale mechanisms. However, requirement of enormous computational power, especially for the model described in Section 21.3.3, restricts its applicability to the meso and macroscale phenomena. At this scale, coarse-grained techniques are very useful and potentially very insightful. This modeling technique commonly incorporates atomistic information and can have significant predictive power or at least provide mechanistic insights through comparative studies (Muller-Plathe 2002; Gautieri et al. 2010; Sen and Buehler 2011). When appropriately constructed, coarse-grained models can reach scales far beyond the atomistic scale and provide valuable insights (Muller-Plathe 2002; Gautieri et al. 2010; Sen and Buehler 2010; Dimas and Buehler 2012).

Here, we review the results of a recent study (Dimas and Buehler 2012) to illustrate how biomineralized materials can be studied effectively with coarse-grained models (Sen and Buehler 2011). Noting that biomineralized structures are commonly assembled with specific architectural arrangements, we hypothesize that these geometric arrangements are successful mechanical designs, independent of the details of the constituent materials (Dimas and Buehler 2012). Specifically, we believe that these geometric arrangements act to reduce stress concentrations around critical flaws and stabilize the propagation of fracture. Further, we hypothesize that they are capable of doing so in the absence of complex energy, dissipating proteinaceous layers and additional levels of hierarchy (Ji and Gao 2004a; Sen and Buehler 2011; Dimas and Buehler 2012).

We employ an atomistically informed spring bead model aimed to represent the common heterogeneous nature of biomineralized structures (Dimas and Buehler 2012). The spring bead model is set up with a triangular lattice, as the one shown in Figure 21.10d. Spring bead models with triangular lattices as the one presented have been validated in several independent studies of fracture in

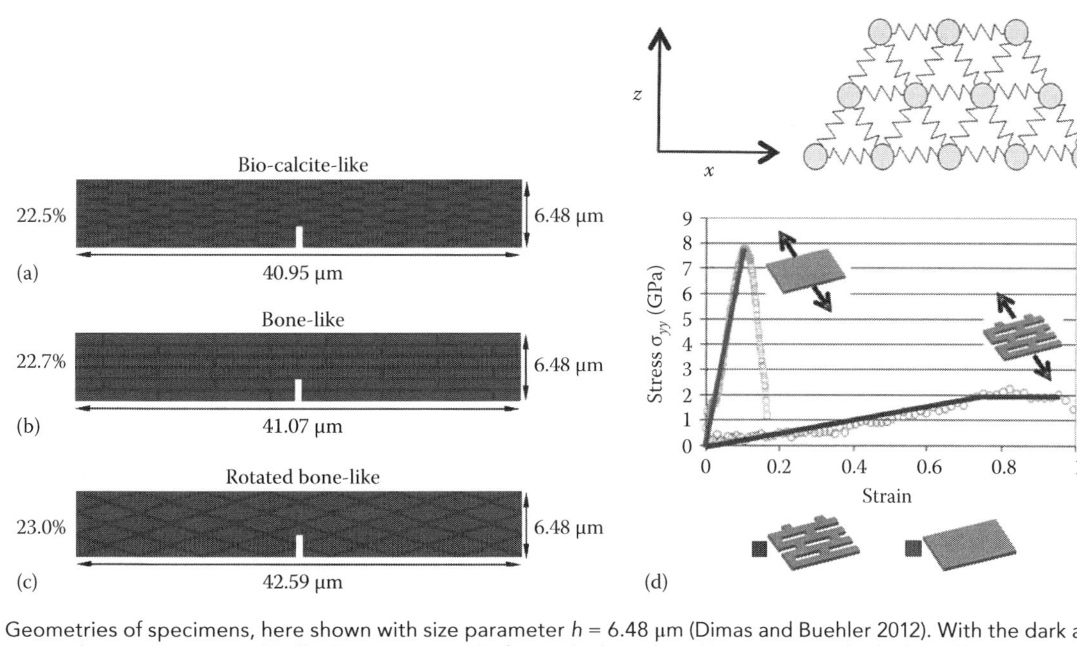

Figure 21.10 Geometries of specimens, here shown with size parameter h = 6.48 μm (Dimas and Buehler 2012). With the dark and lighter phases representing the nanoporous and bulk silica, respectively, for (a) the biocalcite-like geometry, (b) the bonelike geometry, and (c) the rotated bone-like geometry. The specimens are loaded by imposing stepwise displacement on their right vertical phases, while holding the left-hand sides still. Periodic boundary conditions are employed in the horizontal direction. (d) Constitutive laws for the nanoporous and bulk silica phase representing the compliant and brittle phase, respectively in the material model. (Reprinted by permission from Macmillan Publishers Ltd. *Sci. Rep.*, Sen, D. and Buehler, M.J. Structural hierarchies define toughness and defect-tolerance despite simple and mechanically inferior brittle building blocks, *Sci. Rep.*, 1, 2011, Copyright 2011.) as well as the triangular lattice configuration of a spring bead model. (Reprinted from Dimas, L.S. and Buehler, M.J., *Bioinspir. Biomim.*, 7(3), 036024, 2012.)

heterogeneous materials at the meso- and macroscale (Hassold and Srolovitz 1989; Curtin and Scher 1990; Gao and Klein 1998). The specific force extension laws employed in this study are the ones also presented in Figure 21.10d. These behaviors represent the nanoscale deformation characteristics of bulk and nanoporous silica (Sen and Buehler 2010). The choice of silica as the base constituent of our models is motivated by the aim of making our models as general as possible. We seek to improve the mechanical characteristics of an inferior and brittle base constituent simply by tuning the geometrical features. The softer phase also consists solely of silica and achieves its compliance due to the geometry and nanoconfinement (Garcia et al. 2011).

The goal of this study is to investigate the improved mechanical characteristics achieved through ordering a soft and stiff phase in an appropriate configuration. We therefore choose to isolate this mechanism in our models, motivating the simple model described earlier. We exclude other features accounting for other mechanically significant mechanisms (e.g., microcracking, crack-bridging, etc.), which have been comprehensively studied by others (Weiner and Wagner 1998; Aizenberg et al. 2005; Weaver et al. 2007; Meyers et al. 2008), in order to isolate the effect of architecture and recognize the limitations; this enforces on the predictive power of our models.

Inspired by natural materials, we investigate two specific arrangements of the softer and stiffer phase, a biocalcite-like geometry and a bone-like geometry, reviewing here what was reported in a recent study (Dimas and Buehler 2012). Further, in an attempt to optimize the strain transfer, we construct and investigate an additional geometry, here named rotated-bone-like geometry. All geometries are presented with appropriate titles in Figure 21.10. In order to study the mechanics of these structures, it is essential to evaluate the local variation of stresses and strains. Although these concepts are somewhat less familiar in the context of discrete particles, well-defined expressions do exist that have been shown to be equivalent to the continuum measures, under the assumption of validity of the Cauchy–Born rule. To evaluate stresses, we employ the virial stress measure presented in the study of Tsai (1979). For characterization of the local variation of strains, we find the measure developed by Zimmerman et al. (2009) useful. The strain is defined through the definition of the left Cauchy–Green tensor, unique to a particular particle in terms of its nearest neighbors:

$$b_{ij}^{l} = \frac{1}{\lambda}\sum_{k=1}^{N}\left(\frac{\Delta x_i^{kl}\Delta x_j^{kl}}{r_0^2}\right). \tag{21.6}$$

Here, $\Delta x_i^{kl} = x_i^l - x_i^k$ and $\Delta x_j^{kl} = x_j^l - x_j^k$, with x_i^l representing the ith component of the coordinates of atom l in the deformed configuration. Further, r_0 is the equilibrium spacing of the lattice, N is the total number of nearest neighbors, and λ is a prefactor depending on the specific lattice chosen. For this case with the triangular lattice with nearest neighbor interactions, $\lambda = 3$. We recall that the left Cauchy–Green tensor is a symmetric positive definite matrix and hence, as a spectral decomposition, an engineering strain is calculated, and the specific form employed is

$$\underline{\underline{\varepsilon}} = \sqrt{\underline{\underline{b}}} - \underline{\underline{1}}. \tag{21.7}$$

Here, $\underline{\underline{1}}$ is the identity tensor, and the square root of a symmetric positive definite matrix is calculated by

$$\sqrt{\underline{\underline{b}}} = \sum_{i}^{3}\sqrt{w_i}\cdot \underline{e_i}\otimes \underline{e_i}, \tag{21.8}$$

where

$$\underline{\underline{b}} = \sum_{i}^{3} w_i \cdot \underline{e_i}\otimes \underline{e_i}, \tag{21.9}$$

and w_i and $\underline{e_i}$ are the eigenvalues and eigenvectors of the left Cauchy–Green tensor, respectively. This strain is a purely geometric measure and strictly only be applicable away from surfaces and interfaces. In our study, we use this strain measure to make qualitative statements and comparisons and, thus, deem this measure suitable. The computed stresses and strains are evaluated by visualization through stress and strain fields plotted in MATLAB.

In Figure 21.11, we present stress field plots of unnotched and notched bulk silica and unnotched and notched bone-like-composites of bulk nanoporous silica. Figure 21.12 displays longitudinal and shear strain plots of unnotched and notched samples of the same bone-like composites. Figure 21.11a and c clearly shows the brittle nature of silica and the very high sensitivity to cracks that it exhibits. The nature of the stress field is completely changed by the introduction of the crack, and it is clear to see that the effective strength of the specimen is greatly reduced by the introduction of the flaw. These images are strongly contrasted by their neighbors in Figure 21.11b and d. The architecture of the composite introduces a distinct pattern in the stress distribution that is interestingly enough upheld, despite the introduction of the crack in Figure 21.11d, albeit at somewhat lower levels of stress. The soft nature of the matrix phase forces deformations and stresses to delocalize from the tip of the flaw (see also Figure 21.12). Stiff platelets far from the flawed region maintain significant levels of stress due to the geometrical configuration of the structure. Indeed, the geometry also forces the small vertical regions of the matrix phase to participate in stress transfer, thereby giving an overall efficient material.

21.4 OUTLOOK

In this chapter, we presented atomistic models applied to study the interactions between collagen and the biomineral HAP at the nanoscale, in various configurations, and from a molecular point of view that connected the chemical composition to mechanical properties (Dimas and Buehler 2012; Qin et al. 2012; Libonati et al. 2013; Nair et al. 2013). The coarse grain model discussed originally in the work of Dimas and Buehler (2012) allowed us to examine the fracture properties of biominerals at the macroscale and provide an integrated view of mechanisms occurring at multiple scales. The models may help us to better understand the *in vivo* structural and mechanical behavior of bone, and can provide solutions to various diseases with respect to mineral content and mutation in collagen. With the

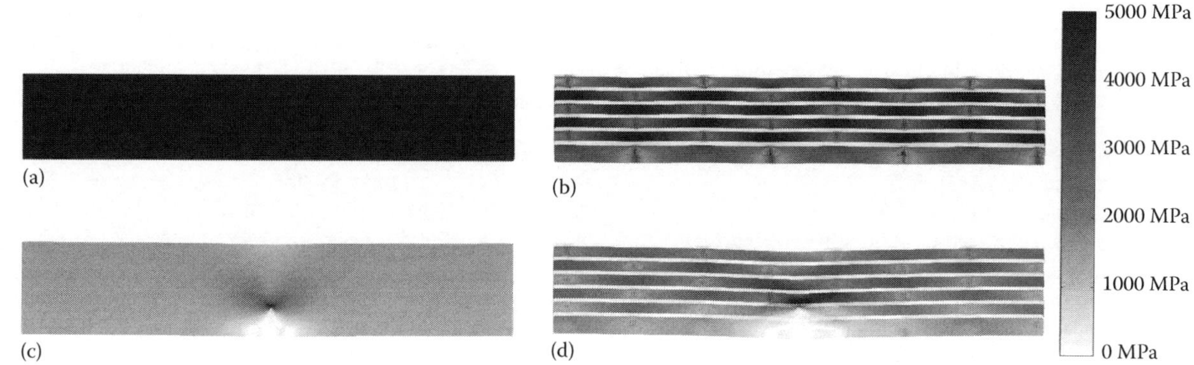

Figure 21.11 Von Mises stress fields for 6.48 μm (a) unnotched bulk silica, (b) unnotched bone-like, (c) notched bulk silica, and (d) notched bone-like specimens at the instant immediately prior to failure (Dimas and Buehler 2012). The unnotched bulk silica specimen shows the expected even distribution of stress throughout the sample, while the notched specimen exhibits the strong characteristic stress concentration at the crack tip. The unnotched bone-like specimen exhibits a clearly larger stress state than the notched specimen. However, the load path in both specimens is seen to be very similar. This specific hierarchical geometry alleviates the stress tip concentration and maintains the same mechanism of load transfer, despite the presence of the crack, thus reducing the specimen's sensitivity to the notch. (Reprinted from Dimas, L.S. and Buehler, M.J., *Bioinspir. Biomim.*, 7(3), 036024, 2012.)

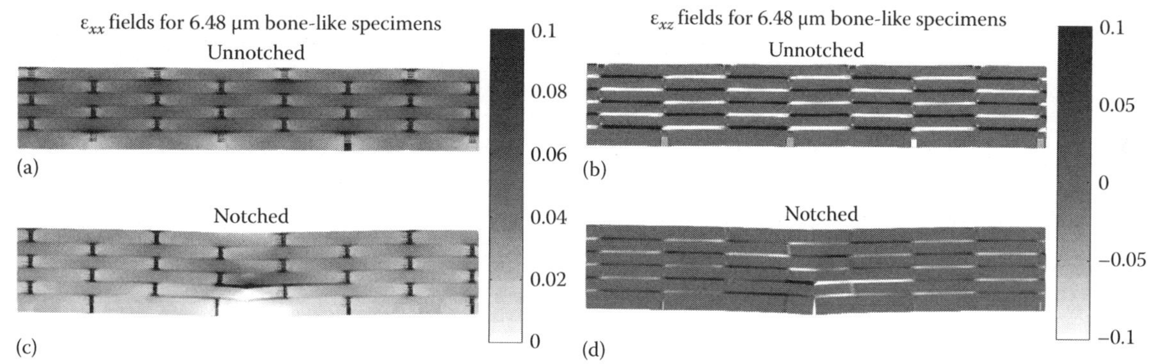

Figure 21.12 (a) Longitudinal and (b) shear strain fields for unnotched bone-like geometry at the instant immediately prior to failure (Dimas and Buehler 2012). (c) Longitudinal and (d) shear strain fields for notched bone-like geometry at the instant immediately prior to failure. Whereas the longitudinal strain transfer is significantly impeded in the stiff phase by the introduction of the crack, the panels clearly show that the strain field in the soft compliant phase remains virtually unchanged. (Reprinted from Dimas, L.S. and Buehler, M.J., *Bioinspir. Biomim.*, 7(3), 036024, 2012.)

availability of additive manufacturing, the coarse-grained model predictions can also be useful to create new materials that mimic the structural hierarchies of biominerals, as discussed in a recent study (Dimas et al. 2013). The predictions from these studies can also be useful to provide design criteria for the development of biomineral-based composites with enhanced mechanical properties.

REFERENCES

Aizenberg, J., J.C. Weaver et al. (2005). Skeleton of *Euplectella* sp.: Structural hierarchy from the nanoscale to the macroscale. *Science* **309**(5732): 275–278.

Alexander, B., T.L. Daulton et al. (2012). The nanometre-scale physiology of bone: Steric modelling and scanning transmission electron microscopy of collagen-mineral structure. *J Roy Soc Interf* **9**(73): 1774–1786.

Allegrini, S., Jr., E. Rumpel et al. (2006). Hydroxyapatite grafting promotes new bone formation and osseointegration of smooth titanium implants. *Ann Anat* **188**(2): 143–151.

Almora-Barrios, N. and N.H. de Leeuw (2010a). A density functional theory study of the interaction of collagen peptides with hydroxyapatite surfaces. *Langmuir* **26**(18): 14535–14542.

Almora-Barrios, N. and N.H. de Leeuw (2010b). Modelling the interaction of a Hyp-Pro-Gly peptide with hydroxyapatite surfaces in aqueous environment. *Cryst Eng Comm* **12**(3): 960–967.

Bhowmik, R., K.S. Katti et al. (2007a). Molecular dynamics simulation of hydroxyapatite-polyacrylic acid interfaces. *Polymer* **48**(2): 664–674.

Bhowmik, R., K.S. Katti et al. (2007b). Mechanics of molecular collagen is influenced by hydroxyapatite in natural bone. *J Mater Sci* **42**(21): 8795–8803.

Bhowmik, R., K.S. Katti et al. (2009). Mechanisms of load-deformation behavior of molecular collagen in hydroxyapatite-tropocollagen molecular system: Steered molecular dynamics study. *J Eng Mech-ASCE* **135**(5): 413–421.

Boskey, A. (2003). Bone mineral crystal size. *Osteop Int* **14**: S16–S20.

Brooks, B.R., R.E. Bruccoleri et al. (1983). Charmm—A program for macromolecular energy, minimization, and dynamics calculations. *J Comp Chem* **4**(2): 187–217.

Buehler, M.J. (2006). Atomistic and continuum modeling of mechanical properties of collagen: Elasticity, fracture, and self-assembly. *J Mater Res* **21**(8): 1947–1961.

Buehler, M.J. (2007). Molecular nanomechanics of nascent bone: Fibrillar toughening by mineralization. *Nanotechnology* **18**: 295102.

Buehler, M.J., H. Yao et al. (2006). Cracking and adhesion at small scales: Atomistic and continuum studies of flaw tolerant nanostructures. *Model Simul Mater Sci Eng* **14**(5): 799.

Colfen, H. (2010). Biomineralization: A crystal-clear view. *Nat Mater* **9**(12): 960–961.
Curtin, W.A. and H. Scher (1990). Mechanics modeling using a spring network. *J Mater Res* **5**(3): 554–562.
Dimas, L., G. Bratzel et al. (2013). Tough composites inspired by mineralized natural materials: Computation, 3D printing and testing. *Adv Funct Mater* **23**(36): 4629–4638.
Dimas, L.S. and M.J. Buehler (2012). Influence of geometry on mechanical properties of bio-inspired silica-based hierarchical materials. *Bioinspir Biomim* **7**(3): 036024.
Dubey, D.K. and V. Tomar (2008). Microstructure dependent dynamic fracture analyses of trabecular bone based on nascent bone atomistic simulations. *Mech Res Commun* **35**(1–2): 24–31.
Fratzl, P. (2008a). Bone fracture—When the cracks begin to show. *Nat Mater* **7**(8): 610–612.
Fratzl, P. ed. (2008b). *Collagen: Structure and Mechanics*, Springer, New York.
Fratzl, P., N. Fratzl-Zelman et al. (1991). Nucleation and growth of mineral crystals in bone studied by small-angle X-ray scattering. *Calc Tiss Int* **48**(6): 407–413.
Fratzl, P., H.S. Gupta et al. (2004). Structure and mechanical quality of the collagen-mineral nano-composite in bone. *J Mater Chem* **14**(14): 2115–2123.
Fratzl, P. and R. Weinkamer (2007). Nature's hierarchical materials. *Progr Mater Sci* **52**(8): 1263–1334.
Gao, H. (2006). Application of fracture mechanics concepts to hierarchical biomechanics of bone and bone-like materials. In *Advances in Fracture Research*, eds. A. Carpinteri, Y.-W. Mai, and R. Ritchie, Springer, The Netherlands, pp. 101–137.
Gao, H., B. Ji et al. (2003). Materials become insensitive to flaws at nanoscale: Lessons from nature. *Proc Natl Acad Sci* **100**(10): 5597–5600.
Gao, H.J. and P. Klein (1998). Numerical simulation of crack growth in an isotropic solid with randomized internal cohesive bonds. *J Mech Phys Solids* **46**(2): 187–218.
Garcia, A.P., D. Sen et al. (2011). Hierarchical silica nanostructures inspired by diatom algae yield superior deformability, toughness, and strength. *Metallurg Mater Trans A-Phys Metallurg Mater Sci* **42A**(13): 3889–3897.
Gautieri, A., M.J. Buehler et al. (2009a). Deformation rate controls elasticity and unfolding pathway of single tropocollagen molecules. *J Mech Behav Biomed Mater* **2**(2): 130–137.
Gautieri, A., A. Russo et al. (2010). Coarse-grained model of collagen molecules using an extended MARTINI force field. *J Chem Theory Computat* **6**(4): 1210–1218.
Gautieri, A., S. Uzel et al. (2009b). Molecular and mesoscale mechanisms of osteogenesis imperfecta disease in collagen fibrils. *Biophys J* **97**(3): 857–865.
Gautieri, A., S. Vesentini et al. (2009c). Intermolecular slip mechanism in tropocollagen nanofibrils. *Int J Mater Res* **100**(7): 921–925.
Gautieri, A., S. Vesentini et al. (2011). Hierarchical structure and nanomechanics of collagen microfibrils from the atomistic scale up. *Nano Lett* **11**(2): 757–766.
Gupta, H.S., S. Krauss et al. (2013). Intrafibrillar plasticity through mineral/collagen sliding is the dominant mechanism for the extreme toughness of antler bone. *J Mech Behav Biomed Mater.* http://dx.doi.org/10.1016/j.jmbbm.2013.03.020.
Gupta, H.S., J. Seto et al. (2006). Cooperative deformation of mineral and collagen in bone at the nanoscale. *Proc Natl Acad Sci USA* **103**(47): 17741–17746.
Gupta, H.S., W. Wagermaier et al. (2005). Nanoscale deformation mechanisms in bone. *Nano Lett* **5**(10): 2108–2111.
Hang, F. and A.H. Barber (2011). Nano-mechanical properties of individual mineralized collagen fibrils from bone tissue. *J Roy Soc Interf* **8**(57): 500–505.
Hassold, G.N. and D.J. Srolovitz (1989). Brittle-fracture in materials with random defects. *Phys Rev B* **39**(13): 9273–9281.
Hauptmann, S., H. Dufner et al. (2003). Potential energy function for apatites. *Phys Chem Chem Phys* **5**(3): 635–639.
Hu, Y.Y., A. Rawal et al. (2010). Strongly bound citrate stabilizes the apatite nanocrystals in bone. *Proc Natl Acad Sci USA* **107**(52): 22425–22429.
Humphrey, W., A. Dalke et al. (1996). VMD: Visual molecular dynamics. *J Mol Graph* **14**(1): 27–38.
Jager, I. and P. Fratzl (2000). Mineralized collagen fibrils: A mechanical model with a staggered arrangement of mineral particles. *Biophys J* **79**(4): 1737–1746.
Ji, B.H. and H.J. Gao (2004a). Mechanical properties of nanostructure of biological materials. *J Mech Phys Solids* **52**(9): 1963–1990.
Ji, B.H. and H.J. Gao (2004b). A study of fracture mechanisms in biological nano-composites via the virtual internal bond model. *Mater Sci Eng A-Struct Mater Prop Microstruct Process* **366**(1): 96–103.
Katti, D.R., S.M. Pradhan et al. (2010). Directional dependence of hydroxyapatite-collagen interactions on mechanics of collagen. *J Biomech* **43**(9): 1723–1730.
Kundu, B., A. Lemos et al. (2010). Development of porous HAp and beta-TCP scaffolds by starch consolidation with foaming method and drug-chitosan bilayered scaffold based drug delivery system. *J Mater Sci Mater Med* **21**(11): 2955–2969.
Launey, M.E., M.J. Buehler et al. (2010). On the mechanistic origins of toughness in bone. In *Annual Review of Materials Research*, Vol. 40. eds. D.R. Clarke, M. Ruhle, and F. Zok, Palo Alto, CA: Annual Reviews, pp. 25–53.
Li, J. (2003). AtomEye: An efficient atomistic configuration viewer. *Model Simul Mater Sci Eng* **11**(2): 173.
Libonati, F., A.K. Nair et al. (2013). Fracture mechanics of hydroxyapatite single crystals under geometric confinement. *J Mech Behav Biomed Mater* **20**: 184–191.
Meyers, M.A., P.Y. Chen et al. (2008). Biological materials: Structure and mechanical properties. *Progr Mater Sci* **53**(1): 1–206.
Muller-Plathe, F. (2002). Coarse-graining in polymer simulation: From the atomistic to the mesoscopic scale and back. *Chemphyschem* **3**(9): 754–769.
Nair, A.K., A. Gautieri et al. (2013). Molecular mechanics of mineralized collagen fibrils in bone. *Nat Commun* **4**: 1724.
Nudelman, F., K. Pieterse et al. (2010). The role of collagen in bone apatite formation in the presence of hydroxyapatite nucleation inhibitors. *Nat Mater* **9**(12): 1004–1009.
Park, S., R.J. Radmer et al. (2005). A new set of molecular mechanics parameters for hydroxyproline and its use in molecular dynamics simulations of collagen-like peptides. *J Comput Chem* **26**(15): 1612–1616.
Plimpton, S. (1995). Fast parallel algorithms for short-range molecular-dynamics. *J Computat Phys* **117**(1): 1–19.
Qin, Z., A. Gautieri et al. (2012). Thickness of hydroxyapatite nanocrystal controls mechanical properties of the collagen-hydroxyapatite interface. *Langmuir* **28**(4): 1982–1992.
Qin, Z., L. Kreplak et al. (2009). Hierarchical structure controls nanomechanical properties of vimentin intermediate filaments. *PLoS ONE* **4**(10): e7294.
Rainey, J.K. and M.C. Goh (2004). An interactive triple-helical collagen builder. *Bioinformatics* **20**(15): 2458–2459.
Rho, J.-Y., L. Kuhn-Spearing et al. (1998). Mechanical properties and the hierarchical structure of bone. *Med Eng Phys* **20**(2): 92–102.
Ritchie, R.O., M.J. Buehler et al. (2009). Plasticity and toughness in bone. *Phys Today* **62**(6): 41–47.
Ritchie, R.O., J.J. Kruzic et al. (2004). Characteristic dimensions and the micro-mechanisms of fracture and fatigue in 'nano' and 'bio' materials. *Int J Fract* **128**(1): 1–15.
Sen, D. and M.J. Buehler (2010). Atomistically-informed mesoscale model of deformation and failure of bioinspired hierarchical silica nanocomposites. *Int J Appl Mech* **2**(4): 699–717.

Sen, D. and M.J. Buehler (2011). Structural hierarchies define toughness and defect-tolerance despite simple and mechanically inferior brittle building blocks. *Sci Reports* **1**: Article No. 35.

Snyders, R., D. Music et al. (2007). Experimental and ab initio study of the mechanical properties of hydroxyapatite. *Appl Phys Lett* **90**(19).

Sopyan, I., M. Mel et al. (2007). Porous hydroxyapatite for artificial bone applications. *Sci Technol Adv Mater* **8**(1–2): 116–123.

Tsai, D.H. (1979). Virial theorem and stress calculation in molecular-dynamics. *J Chem Phys* **70**(3): 1375–1382.

Van Der Rijt, J.A.J., K.O. Van Der Werf et al. (2006). Micromechanical testing of individual collagen fibrils. *Macromol Biosci* **6**(9): 699–702.

Weaver, J.C., J. Aizenberg et al. (2007). Hierarchical assembly of the siliceous skeletal lattice of the hexactinellid sponge *Euplectella aspergillum*. *J Struct Biol* **158**(1): 93–106.

Weiner, S. and H.D. Wagner (1998). The material bone: Structure mechanical function relations. *Ann Rev Mater Sci* **28**: 271–298.

Zahn, D. and P. Duchstein (2011). Atomistic modeling of apatite-collagen composites from molecular dynamics simulations extended to hyperspace. *J Mol Model* **17**(1): 73–79.

Zhao Q., A. Gautieri et al. (2012). Thickness of hydroxyapatite nanocrystal controls mechanical properties of the collagen–hydroxyapatite interface. *Langmuir* **28**: 1982–1992.

Zimmerman, J.A., D.J. Bammann et al. (2009). Deformation gradients for continuum mechanical analysis of atomistic simulations. *Int J Solids Struct* **46**(2): 238–253.

Zimmerman, J.A., E.B. Webb III et al. (2004). Calculation of stress in atomistic simulation. *Model Simul Mater Sci Eng* **12**(4): S319.

22 Whole organ deformation analysis by digital optical metrology

Paul Zaslansky and Ron Shahar

Contents

22.1 INTRODUCTION

Biomineralized tissues and organs usually have a mechanical function. This is certainly true for the skeletal elements of vertebrates, which serve as lever anchors for muscles to produce motion in the limbs or as protective casings of internal organs such as the skull. Similarly, many (but not all) teeth are mineralized to provide the hardness and stiffness needed to tear and crush food substances. Biomineralized tissues also exist in many other forms in nature, such as the calcified cartilage of sharks, mineralized shells of snails and bivalves, the exoskeleton of crustaceans or spicules, and spines such as those of sea urchins. The study of mineralized biologically formed structures thus requires understanding the mechanical conditions under which they function and the loads or mechanical challenges to which they are adapted.

When forces (external or internal) are applied to any solid object, it deforms. In a mineralized organ, this deformation pattern is often initially linearly elastic, that is, the load and corresponding deformation are linearly related. If the load is removed, the deformation disappears and the organ regains its original shape. Thus, for loads within the linearly elastic range, an increase in load magnitude will increase the deformation proportionally. The load and resulting deformation can be described as a three-dimensional (3D) distribution of strains and stresses throughout the body (the terms "stress" and "strain" are defined below). Once the loads applied to the organ exceed a certain level, local failure occurs. When sufficient local failure and damage accumulate, the organ ceases to behave in a linearly elastic manner and starts to deform "plastically" (a phenomenon termed yielding), such that some permanent deformation ensues. If loads continue to increase, eventually the organ will fracture.

Understanding the relationships between the mechanical responses of whole structures, their material properties, and their architecture is a challenge. When the loaded body is of a simple geometric shape (cube, cylinder, rectangular beam, hollow circular tube, etc.) and the properties of the materials of which the body consists are well known and homogeneous, the theory of elasticity usually allows one to find a precise and complete relationship between the applied external load and the stresses and strains experienced at each location within the body. This, however, is not possible when the hierarchical arrangement and mechanical properties of the material itself are complex or when the entity it forms has a complex shape. This situation is almost always the case in biology. Therefore, despite the enormous technological advances made in recent years, it is still beyond the state of the art to predict and fully understand the complex deformations of whole biomineralized entities when placed under various loading conditions or to precisely describe their resulting 3D stress and strain distributions.

22.1.1 BASIC CONCEPTS OF MECHANICS OF MATERIALS

The mechanical behavior of any structure is determined both by its geometry and the properties of the materials of which it is made. The material properties are independent of the geometry and are inherent to the material and not to the shape of the structure. The determination of these properties requires definition of the concepts "stress" and "strain." The following is a short, very simplified, and by no means rigorous or complete, description of these concepts. A more detailed description can be found in textbooks of solid mechanics (Gere and Timoshenko, 1997; Martin et al., 1999; Cowin, 2001; Currey, 2002; Enos, 2012).

22.1.1.1 Stress

When a 3D body of arbitrary geometry is loaded, one can imagine the body to be divided into numerous infinitely small cubes. Consider now one such cube somewhere within the body (Figure 22.1). Each face will have a force acting upon it, which can be divided into three Cartesian components: one component orthogonal (normal component) to the face and two components parallel to it, along orthogonal orientations (in plane components).

The normal stress acting on this particular cube is defined as the normal force divided by the face area (both are infinitely small, but the ratio is finite). Similarly, the shear stresses are defined as the ratios of the forces acting parallel to the surface divided by the same face area. Each cube has six faces, and with each face subjected to three stress components, the total number of stress components acting on each small cube is 18. However, due to equilibrium requirements (opposite faces must have equal and opposite forces), the local cube stress state is fully described by only nine components (three orthogonal faces with three components each). Symmetry considerations can be shown to further decrease the different shear components from six to three so that the number of independent stress components is six (three normal stresses and three shear components). The magnitude of these components depends on the location of the cube within the body. The standard unit of stress is the pascal (Pa), which is equal to 1 N applied to an area of 1 m^2 (N/m^2). One pascal represents a very small amount of stress, and

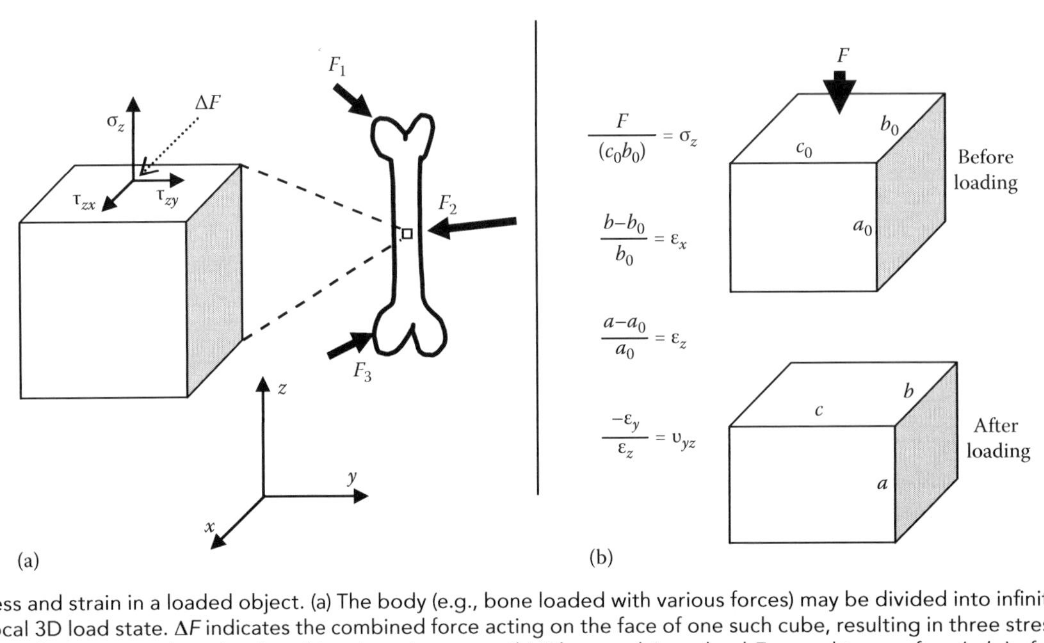

Figure 22.1 Stress and strain in a loaded object. (a) The body (e.g., bone loaded with various forces) may be divided into infinitely small cubes to describe its local 3D load state. ΔF indicates the combined force acting on the face of one such cube, resulting in three stress components: orthogonal (1 normal, σ) stress and two in-plane (shearing, τ) stresses. (b) When applying a load F normal to a surface ($c_0 b_0$) of a cube, the ratio is termed *stress*. The cube will shorten in the direction of the force applied (e.g., from a_0 *to* a) and become wider in the orthogonal directions (b_0 becomes b and c_0 becomes c). The ratio of dimensional change is termed *strain*. The ratio of the strain in the lateral (*y*) direction to the strain in the longitudinal (*z*) direction is a constant, representing a material property, and is termed Poisson's ratio (υ_{yz}). Note that strains develop also in directions in which no forces are acting.

since physiological stresses are usually in the range of thousands of Pa, commonly used units are the megapascal (MPa) and gigapascal (GPa), which are equal to 10^6 Pa and 10^9 Pa, respectively.

22.1.1.2 Strain

Strain is a measure of the fractional change in length of a loaded body. Thus, if an object undergoes a change in its length (e.g., becomes shorter), it is said to incur strain. When one considers the same infinitely small cube within a loaded body as described in Section 22.1.1.1 (Figure 22.1), it can be seen that the cube may undergo shape change in three dimensions along the length of each axis (normal strain) or change the angle between its faces (shear strain). Similarly to stress, it can be shown that there are six independent strain components (three normal strains and three shear strains). Strain is a unitless ratio (length divided by length), but is commonly measured in units of microstrain (1 microstrain equals strain multiplied by 10^{-6}) so that a strain of 0.01 (1%) is equivalent to 10,000 microstrain.

22.1.1.3 Stress–strain relations

In materials that are characterized by a linearly elastic behavior, the strain components are linearly related to the stress component that is causing the deformation. This relationship is described mathematically by Hooke's law of elasticity. In its most general form, it states that each of the six stress components σ_i is equal to a linear combination of all six strain components ε_i multiplied by six independent constants of proportionality C_{ij}. Thus, for the most general material, there can be 6 × 6 = 36 independent proportionality constants that describe its mechanical deformation behavior. Energy considerations and symmetry reduce this number to 21 independent elastic constants.

Fortunately, for most materials, the number of proportionality constants is much smaller. For example, only two constants are required to describe the mechanical behavior of *isotropic* materials: materials that do not have "preferred" orientations and behave in the same way in every direction. Many materials, such as cortical osteonal bone, exhibit transverse isotropy—they have two principal orientations (axial direction parallel to the direction of the osteons and directions orthogonal to the axial direction). Such materials require five constants to fully describe their elastic behavior.

22.1.1.4 Mechanical loading experiments

When a sample of biomineralized tissue is incrementally loaded, its stress–strain curve initially exhibits a preloaded "toe" region followed by a linear relationship between the stress and the strain (the linearly elastic region). However, at a particular strain (yield strain), further increase in load results in a nonlinear deformation response. Consequently, small additional loads produce a strain increase larger than before, as a result of accumulation of damage (e.g., microcracking) in the material and resulting decrease in its stiffness (Figure 22.2). The sample is said to start to deform "plastically" at that point. Further increase in load will eventually result in failure of the sample, and the stress at which this happens is named the ultimate stress and represents the strength of the material. The area under the stress–strain curve up to that point is a measure of the energy needed to fracture the sample.

Whole biological-object testing (whole bone, whole tooth, etc.) examines the relationship between load and deformation response, and is affected both by the material properties and the geometry and architecture of the object. The results therefore cannot be readily converted into a stress–strain relationship.

22.1.1.5 Material stiffness (Young's modulus)

In linearly elastic isotropic materials, Hooke's law takes a simple form given by Equation 22.1, where Young's modulus (E) represents the ratio between the stress applied to a sample (σ) and the strain (ε) that the sample experiences as a result:

$$\sigma = E\varepsilon \qquad (22.1)$$

Young's modulus can be determined by an experiment, where a sample of the material is subjected to load, and the strain and stress are determined concurrently. Young's modulus is determined from the slope of the linear portion of the stress–strain curve (see Figure 22.2). Young's modulus represents the stiffness of the material and is a characteristic material property—the higher its value, the stiffer the material is—therefore, more force is needed to produce the same strain when compared with a less stiff material. For transversely isotropic materials, the relationship between the stress and the strain depends on whether the sample is loaded in the axial direction (where the modulus

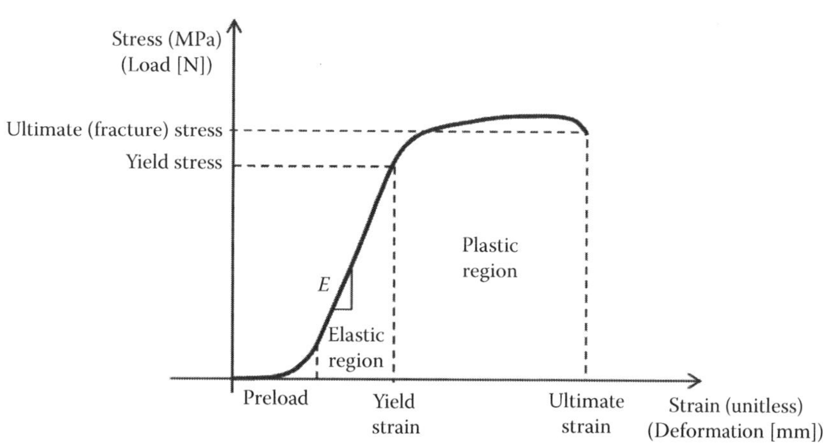

Figure 22.2 A typical curve describing the stress–strain (or load–deformation) relationship in a sample as it is loaded. In the example shown, an elastic material subjected to load will first exhibit a linear relationship between stress and strain (or load and deformation, in the case of whole organ testing), and eventually, as the load increases, it will yield. *E* is the stiffness of the material (Young's modulus). Typically, materials will exhibit elastic (nonpermanent) deformation below some threshold, beyond which plastic, irreversible deformation (accompanied by energy loss) will take place. Further increase in load will eventually lead to failure.

in the axial direction is E_A) or the transverse direction (and the modulus in the transverse direction is E_T).

It must be appreciated that biomineralized tissues are usually not simple linear elastic materials and that they often exhibit viscoelastic behavior to a certain degree such that their stiffness is strain rate dependent; for such materials, if their strain rate increases, their stiffness also increases. For some biomineralized tissues, this effect is well studied. For example, an increase of the strain rate by a factor of 1000 will increase the Young's modulus of bone by 40% (Currey, 1988).

22.1.1.6 Poisson's ratio

Another material property of solids describes the relative deformations occurring along orthogonal axes. For example, when a sample is loaded in compression, it becomes shorter along the load direction but wider in the orthogonal direction (see Figure 22.1). Similarly, when a sample is loaded in tension, it becomes longer but narrower. It follows, therefore, that strains (and stresses) do not develop only in the direction in which the sample is being loaded. The ratio of the change in length (strain) in the unloaded direction and in the loaded direction is a material constant termed Poisson's ratio. It is denoted by the letter υ, and for most materials, its value ranges between 0 and 0.5, with 0.3 being considered a typical value.

22.1.2 PRACTICAL CONSIDERATIONS FOR MECHANICAL TESTING OF BIOMINERALIZED SAMPLES

The determination of deformations of loaded specimens requires careful consideration of the best way to load the specimens in order to ensure that the measurements will be valid, reproducible, and, above all, meaningful. There are also various other technical matters that need to be addressed for the experiments to be successful. The basic methods of mechanical testing of biomineralized tissues are described by Turner and Burr (1993); a more recent book that deals with solid biomechanics is by Ennos (2012).

22.1.2.1 Mechanical testing environment

The testing devices must run smoothly and quietly, and be capable of measuring very small deformations and loads. Most such devices contain a motor and load cell in series, which are able to perform bending, compression, or tension experiments. Some devices are designed to also perform torque experiments. Standard methods of mechanical testing of samples are based on recording the movement of the cross-head and deriving from it the deformation of the sample. In such circumstances, it is essential that the compliance of the various components of the mechanical loading device be known and taken into account during analysis.

22.1.2.2 Specimens

Specimens should be machined wet and should be smoothed to ensure reproducible contact areas. They can be stored deep-frozen without too much change in their mechanical properties on thawing, but they should be wrapped in moist-protective wraps and placed in small bottles to prevent "freezer burn," which will change the mechanical properties of the specimen once it is thawed compared with freshly prepared samples.

22.1.2.3 Wet versus dry testing

Mechanical testing of biomineralized samples must almost always be carried out on wet specimens, indeed on specimens that are fully hydrated, often immersed in water. At the same time, measures must be taken to ensure that no significant mineral dissolution occurs in the sample during the experiment.

There are a few types of specimens for which dry testing is appropriate because the material is dry in life when used. Examples for such specimens include antler material and various keratinous horns, hooves, etc. In some cases, the question of whether to test wet, or dry, or some intermediate state of hydration is difficult to decide.

22.1.2.4 Sample mounting

One of the main challenges of mechanical testing of biological samples is to overcome the local effects occurring in the areas in which they are gripped, where loads are applied. In the vicinity of these points of load application, stresses and strains are maximized and nonuniform, and the interpretation of the deformation fields becomes difficult and unreliable. However, the effects of a force acting on the edges of the sample will dissipate (or smooth out) and become uniform in regions that are sufficiently far away from the loaded edges (the distance considered sufficient depends on the overall dimensions of the sample). This is known as Saint Venant's principle, which states that the difference between the effects of two different but statically equivalent loads approaches zero if the distance from the point of load application is large enough.

22.1.2.5 Performing the experiment

Most investigations of the mechanical performance of biomineralized tissues rely on *in vivo* or *in vitro* experiments in which the samples are tested in tension, compression, or by three- or four-point bending. Torsional loading and tests at different strain rates or even impact loading are also used, though less frequently. Strain measurements are usually made at a very limited number of points on the specimen by gluing strain gauges to the surface, by use of linear variable differential transformer (LVDT) or by estimating the change in length from the loading-grip movement.

The type of test used must be carefully considered. There is no doubt that the three-point bending setup is the easiest to perform. However, the strains produced vary throughout the specimen, being largest underneath the central loading bar and smallest, in fact nearly zero, at the outer supports. Across a cross-section of a sample tested in bending, the deformation varies noticeably from maximal tension on the tension surface (away from the location of load application) and decreases linearly to maximum compression on the compressive surface (the surface in contact with the load-applying prong), with the center of the length of the sample (the so-called neutral axis) experiencing zero deformation. The deformations vary, therefore, along the length of the specimen as well as through its depth. It is important to note that if the specimens are considerably deep relative to their length, shear effects will become significant and make it very difficult to interpret the results. It has been shown that the aspect ratio of the specimens (length between the outer supports)/(depth) needs to be more than 16, and specifically for bone, an aspect ratio of 25 is ideal (Spatz et al., 1996).

For bending tests, specimens should have a uniform shape, that is, their breadth and height should be constant along the entire length. If this is the case, then beam theory can be used to calculate Young's modulus of elasticity in the longitudinal direction (which assumes that the modulus in tension and compression are the same). From the loading curve, the fracture stress may also be calculated. Successful determination of Young's modulus depends on the load/deformation curve having a linear portion (Figure 22.2).

Another form of loading is four-point bending, in which two points apply load on the compression side, and the bending moment between these two loading points is constant. However, the loads required to produce particular deformations are higher and more importantly, very careful machining is required so that the two inner points touch the compression side of the specimen at the same time.

Compression loading specimens are relatively easy to produce, often being cube-shaped. But it is very important that the two sides facing the machine's anvils are completely parallel; otherwise, there may be large local deformations and shear, where the specimen first touches the loading anvil. The great advantage of compression specimens is that if the specimens are cubical, then three values of Young's modulus along the main sample axis can be obtained rather easily by changing the orientation of the cube between experiments. A limitation of compression testing is that the anvils inhibit to some extent the outward movement of the specimen's faces, as is required by the Poisson effect. As a result, calculated values of stiffness may be higher than would be the case if the anvils were completely frictionless.

Tensile specimens are often dog-bone-shaped, with a narrow section in the middle of the sample and two expanded areas at each end that are gripped by the testing machine. This makes it possible to determine Young's modulus in one direction. A significant problem with tensile tests relates to gripping the specimen. If the grips are too tight, then the sample at the ends of the specimen will be partially crushed, while if the grips are not tightened enough, then the specimen will slip and values of deformation will be incorrect.

22.1.3 WHY OPTICAL METROLOGY MEASUREMENTS?

The ability to use noncontact, nondestructive, and full-field methods to directly measure surface displacements across entire sample surfaces, thus determining concurrently the displacements along two orthogonal directions, enhances our capacity to determine various materials' characteristics including both components of the local strains and Poisson's ratios.

Testing samples immersed in water creates obstacles for optical-based deformation measurement methods by decreasing the signal-to-noise ratio; however, using large data-set statistics, these methods often make it possible to obtain very precise full-field measurements.

The relevance of results of whole organ loading experiments often hinges upon the similarity between the mode of physiologic and experimental loading. Physiologic loading is almost always complex and hard to simulate, and the surface deformation patterns obtained must be interpreted with caution. Sometimes, one can assume that loading will be predictable, for instance, the longitudinal loading of vertebral centra or the biting loading of premolar or other teeth. But even in the case of the vertebral centra, slight differences in the direction of loading may produce large differences in overall deformation, and the loading of teeth may have subtleties that make any straightforward analysis dubious (Benazzi et al., 2012). Nevertheless, it must be remembered that "The best is the enemy of the good" (Voltaire) and that if one cannot achieve perfection, one must, as is usually the case in science, do the best one can. In the following section, we present the main optical metrology methods used to obtain direct deformation measurements of cut or whole biomineralized samples.

22.2 SELECTED OPTICAL METROLOGY METHODS

The notion that "a picture is worth a thousand words" is not commonly used in the context of deformation analysis but appears to be extremely relevant. By imaging light reflected from the object surface, it is possible to capture an instantaneous projection of the object shape using a camera. If the object deforms or changes, the corresponding image changes in a manner that can be assessed quantitatively. Computerized correlation methods, based on both regular and laser light, were developed in the 1980s following the advent of digital photography, the increase in availability of lasers, and the explosion in computational power. Yet, capturing images that are correlated to or linked with mechanical motion of an object is not a new concept. The first account of use of sequential photograph acquisitions dates far back to the pioneering work of Eadweard James Muybridge in 1877, when he presented the first scientific discoveries about "The horse in motion" (Figure 22.3) where it can be seen that all four legs of the animal are in the air at certain times during the gallop cycle. Muybridge produced a large library of images documenting motion of different animals, people, and events, allowing his colleagues to quantitatively study displacements that the eye could not determine. "Seeing the invisible" in this way is still relevant today, where capturing images in conjunction with mechanical motion or deformation forms the core of modern optical metrology.

In the following sections, we discuss the generic principles of the most important commercially available numeric and physical correlation–based deformation measurement systems, with a strong emphasis on the practical aspects of performing meaningful and reproducible measurements on loaded mineralized biological tissues. The reader interested in a deeper treatment of these and related methods is referred to the extensive literature on the subject (see references section at the end of this chapter).

22.2.1 DIGITAL IMAGE CORRELATION

Modern image correlation measurements follow the same concepts of Muybridge's triggered photography series, capturing different images of the sample to follow its deformation. Deformation recorded in the images is then quantified by further numerical processing. The computer checks to find the most likely position of features in the image of the deformed object that moved as compared to the original image and calculates the displacements of features located in different parts of the images by pattern recognition. Intensity patterns in the digital images correspond to imaged surface areas on the object and are typically encoded into 2D matrices of integer

Figure 22.3 A picture of a running horse taken by multiple cameras by Eadweard J. Muybridge in Governor Leland Stanford's Palo Alto Farm, California (1876). (Wiki Commons, taken from Library of Congress Prints and Photographs Division; http://hdl.loc.gov//loc.pnp//cph.3a45870.)

values. Subgroups of these numbers, typically small regions of interest (ROIs; intensity array subsets or facets), can be statistically correlated by comparison between images obtained at different times (e.g., prior to or after applying load) and also between images taken from different perspectives (e.g., by different cameras). The procedure requires matching structural features or color-sprayed, random patterns on the surface of the measured object that is imaged both before and after being placed under load. The computer sequentially correlates small ROIs containing a predetermined number of picture elements (pixels) from the "reference" image with all possible target matrices of pixels within a region of the "deformed" image and finds the best planar (horizontal and vertical) match. The displacement search procedure is performed by one of a variety of image correlation algorithms and reveals both components of the displacements (across and along the imaged field) for each point on the surface. The outcome provides measures of the average displacement of each subregion of interest, the size of which is defined by the user in the search procedure. Thus, for each defined ROI, which is typically rectangular (Figure 22.4), and by knowing the effective pixel size, the relative displacement of different sample points can be determined. Such relative displacements of different points are used to establish the deformation map of the surface. The resulting deformation fields can be smoothed and differentiated, thus yielding the surface components of the strain field.

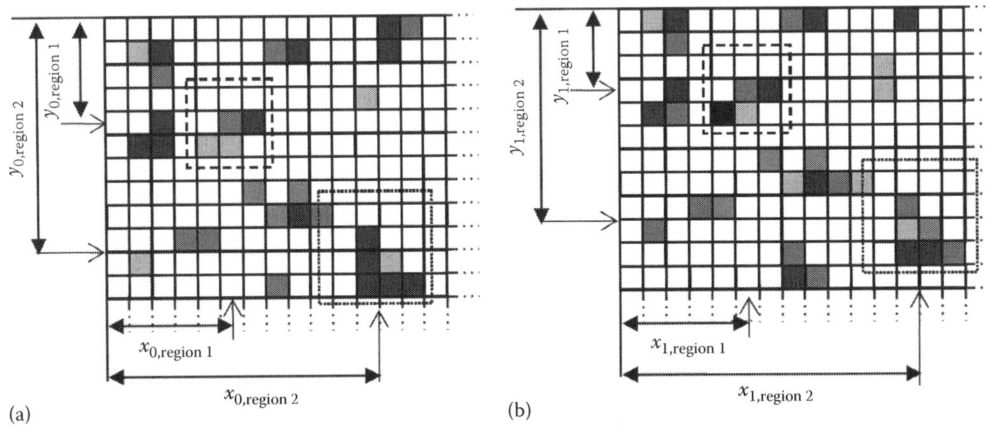

Figure 22.4 Matrix correlation to determine the planar displacement of groups of pixels. The 9 × 9 pixel region centered at ($x_{0,\text{ region 1}}$, $y_{0,\text{region 1}}$) shown in (a) has moved to ($x_{1,\text{ region 1}}$, $y_{1,\text{region 1}}$) shown in (b) and would yield the highest correlation value if the regions are reasonably similar. A larger correlation ROI (facet) may be needed to identify the shift of ($x_{0,\text{ region 2}}$, $y_{0,\text{region 2}}$) to ($x_{1,\text{ region 2}}$, $y_{1,\text{region 2}}$). Larger regions require more computation and usually result in better precision but lower resolution correlation maps. Multiple independent experimental parameters including illumination, focus, contrast, ROI size, and search optimization will influence the reproducibility and reliability of the results, and often calibration by comparison with an additional method is needed to ensure correctness.

A typical correlation analysis of two images requires that an array of measurement points (a grid), the density of which is user-defined, is superimposed on the images of the sample in the reference state. This grid defines the spatial distribution of ROIs for which displacement measurements will be carried out. The user further defines the size (pixel area) of the ROI around each grid intersection point, as well as a larger search region in the target image, within which the displaced ROI is searched for. A search is conducted to find the best match between the subset of pixels of each ROI in the reference (unloaded) image and similar subsets of pixels within the corresponding search regions in the target (loaded) image. In other words, a search is performed to find the location within the deformed image, which is most consistent with that of the reference (unloaded) image. The sizes of the required ROI and search regions depend on the available surface detail and should take the overall displacement magnitude and possible deformations of the ROI into consideration.

The fundamental assumption underlying digital image correlation is that local subsets of intensities in the image of the unloaded sample are transformed to the target subset in the image of the deformed sample by a homogeneous linear transformation. Linear homogeneous transformations can be described by a combination of translation, rotation, shearing, and scaling. Increased precision is attained by using subpixel values obtained by various interpolation methods. Different similarity metrics have been proposed (e.g., finding the minimal Euclidian distance or minimum of the sum of squared differences), and one such widely used algorithm is to search for the minimum of the following normalized image correlation function:

$$S\left(x, y, u, v, \frac{\partial u}{\partial x}, \frac{\partial u}{\partial y}, \frac{\partial v}{\partial x}, \frac{\partial v}{\partial y}\right) = 1 - \frac{\left(\Sigma[F(x,y) * G(x^0, y^0)]\right)}{\left[\Sigma(F(x,y)^2) * \Sigma(G(x^0,y^0)^2)\right]^{1/2}} \tag{22.2}$$

where

- S indicates the degree of correlation between the source and target ROI
- $F(x, y)$ is the gray-level value at coordinate (x, y) in the reference image region
- $G(x^0, y^0)$ is the gray-level value at coordinate (x^0, y^0) in the target image

The relationship of the coordinates (x, y) and (x^0, y^0) is defined by the deformation such that

$$x^0 = x + u + \frac{\partial u}{\partial x}\Delta x + \frac{\partial u}{\partial y}\Delta y \tag{22.3a}$$

$$y^0 = y + v + \frac{\partial v}{\partial x}\Delta x + \frac{\partial v}{\partial y}\Delta y, \tag{22.3b}$$

where u and v are the displacements for the ROI centers in the x- and y- directions, respectively. The terms Δx and Δy are the distances from the ROI center to point (x, y). Image correlation is performed by determining values for $u, v, \frac{\partial u}{\partial x}, \frac{\partial u}{\partial y}, \frac{\partial v}{\partial x}, \frac{\partial v}{\partial y}$, which minimize the correlation parameter S. A correlation value of $S = 0$ indicates perfect correlation, while $S = 1$ indicates no correlation. Once the best correlation has been determined for each ROI in the grid, the full in-plane 2D displacement (u, v) fields are determined and can be smoothed and differentiated to yield the strain values. For additional information, see Rastogi (2000), Hild and Roux (2006), Sanchez-Arevalo and Pulos (2008), and Sutton et al. (2009).

22.2.2 PRACTICAL CONSIDERATIONS OF IMAGE CORRELATION MEASUREMENTS

There will almost always be some degree of correlation between any two matrices of integers; hence, obtaining meaningful and relevant correlation results usually requires setting optimization criteria that define how many points are included in the correlation (ROI or facet size), how far to search, and how "good" the correlation should be. Correlation improves with higher contrast and uniqueness of the patterns in each ROI on the surface. These parameters vary as a function of noise and feature visibility/contrast. A random stochastic speckled pattern (typically black dots on a white background) sprayed onto the object surface often significantly improves the correlation reliability and the overall results. However, with good illumination, sample features may suffice.

Image correlation has advantages compared with speckle pattern interferometry (described later) when large deformations are to be measured, and thus the method is particularly well suited to study extensive or nonlinear deformation, which is typically relevant for less mineralized tissues. However, special care must be taken to ensure that changes analyzed in the images are not caused by illumination variations or shadowing or sample movement that results in defocusing. Deformation is best determined when it occurs only in-plane, as the measured surface must remain within the focal depth for the entire duration of measurement. The known inverse relation between the focal depth (and consequent out-of-plane sensitivity) and aperture size (and the imposed bounds on illumination and the visibility and contrast of features) requires that these settings be optimized separately for different samples. This is needed to account for variations in size and surface texture as well as for magnification and resolution constraints. These parameters and numerous other technical settings of the system, including the numerical aperture stop, focal distance, illumination, field of view, and perspective, are affected by the sample characteristics such as pattern feature size (when painted onto the surface) and deformability. The quality and reproducibility of measurement will thus vary considerably and will be strongly affected by the correlation settings such as the search ROI (facet) size and the various optimization parameters used for correlation.

22.2.3 HOLOGRAPHIC AND SPECKLE INTERFEROMETRY METHODS

Laser-based illumination makes it possible to track displacements when these are encoded into light-intensity pattern variations that are related to optical-path length changes occurring between the object and the camera. When an object deforms, the distance between different points on the surface and the camera will also change, if the imaging setup is stable. This is fundamentally different from the results obtained by image correlation–based systems, where the determined displacements depend entirely on pattern recognition and iterative statistical computations. Various speckle methods have been proposed over the years, from which digital holographic interferometry and electronic speckle pattern-correlation interferometry (ESPI, or digital holography) are perhaps the most practical, having found the most use. Both methods quantify deformations of whole bodies in a noncontact and full-field manner, that is, the entire object is "measured" instantaneously and at the same time (no need to measure each point separately). The coherent nature of laser light, essential to these methods, makes it possible to determine the changes to the phase of light reaching the camera/detector system when intensity is modulated by constructive or destructive interference. For a comprehensive description of the fundamental principles of these methods, see Wykes and Jones (1989) and Rastogi (2000, 2001).

Digital holography records interference patterns created between a reference illumination (directed at the camera) and light reflected from the object. The resulting interference patterns can be used to recreate the entire 3D scene, where both light intensity and phase can be reconstructed. Conversely, speckle-based methods record mainly "laser noise," random interference spots produced by the sample microtopography, and resulting in small (light wavelength in dimension) changes to the path that the light travels between the sample surface and the camera. For deformation measurements, it is not necessary to make full use of the entire 3D information that can be recorded by holography. It is sufficient to record the differences in the light intensities at each point, if these correspond to changes to the topography of the sample surface, changes that occur due to sample surface deformation following loading. Interestingly, many optical computer mice utilize speckle tracking as a means to follow the displacements of the mouse with respect to the working surface. Sample-surface topography changes result in differences in the path length of the light propagating from the lasers to the surface and from the surface toward the imaging system. The changes in light path length lead to shifts in the relative phases of the interfering waves on the image plane. The extent of these shifts may be numerically determined using phase-stepping approaches that are usually used for quantitative deformation analysis. *Phase-stepping* consists of recording multiple images of the sample under the same conditions, while imposing known phase-shifts to the reference laser-beam (typically by extending the light path—moving a mirror through several quarter-wavelength distance increments). Multiple measurements of intensity make it possible to convert the difference in intensity into a quantitative measure of micrometer-sized surface displacements. Repeating such measurements makes it possible to track total displacements by summation of intermediate measurement steps (during loading) and to quantify the displacement of points along the direction that the interferometer is sensitive to.

The ESPI system is more accessible than holographic interferometry because of the simplicity and compactness of the setup, the improved tolerance to noise, and ease of automated processing. These features have rendered speckle-based methods, and specifically ESPI (sometimes termed digital speckle pattern-correlation interferometry), readily available in commercial systems. Holographic interferometry is thus not discussed here further; for in-depth reviews of holography methods and applications, see Schnar and Jüptner (2002).

22.2.4 ESPI IN PRACTICE

A schematic representation of the interferometer arrangements used for ESPI is given in Figure 22.5. To completely determine displacements in space (along the x-, y-, and z-axes), two orthogonal in-plane interferometers are used to measure lateral displacements along the sample x- and y-axes, while a third, out-of-plane interferometer is used for on-axis (sample–camera direction) displacement. Speckle interference patterns are imaged into the camera of the imaging system. A computer sequentially activates the three interferometers by controlling the shutters and moving piezo-mounted mirrors (M + P, Figure 22.3) for phase-shifting and acquiring several images per interferometer at each measurement point. Typically, a phase-stepping algorithm utilizing four images is used. Following each load increment in a series of measurements, images are recorded for consecutive intermediate load states (typically four images per

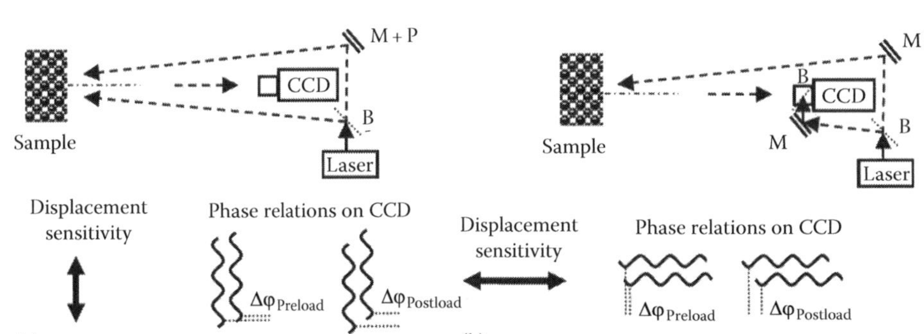

Figure 22.5 ESPI arrangements and principles: (a) in-plane geometry for lateral (x or y) interferometer measurements; laser light is split in two different directions by a beam-splitter (BS), and both beams are aligned so as to illuminate the sample symmetrically, with the reflected light propagating toward a digital camera (CCD). By inducing known quarter-wavelength displacements to a reference mirror and phase-shifter (M + P), intensity images can be converted into quantitative phase-shift estimates for each point in the image. (b) Out-of-plane interferometer, sensitive to displacements along the camera axis.

interferometer per step). Usually, these are followed by repeated load increments and further image acquisition. The four intensity measurements ($I_{1,2,3,4}$) obtained at each point (x, y) in the images for a given deformation state are used to determine the light phase-shifts as follows; initial (preload) and final (postload) phase differences are calculated independently using $\Delta\varphi(x,y)_{pre,post} = \tan^{-1}\{(I_1(x,y) - I_3(x,y))/(I_2(x,y) - I_4(x,y))\}$. Then, the differences $\Delta\psi = \Delta\varphi_{post} - \Delta\varphi_{pre}$ at each image point encode the sample surface displacements and can be converted into displacement estimates along each of the three axes x, y, and z by considering the light wavelength and the interferometer geometric setup.

Similar to image correlation errors, there are errors due to speckle-decorrelation taking place between consecutive images. These result in nonlinear phase determination errors, which often appear as spikes or steps in the processed data. Corrective measures require minimizing the extent of decorrelation between measurements and increasing the number of sampled points. Sometimes careful smoothing is needed. The uncertainties in precision and accuracy can be loosely grouped into phase-shift artifacts, component/setup limitations (e.g., reflections), and data-processing limitations in converting phase to distance. Different phase-shifting algorithms have been proposed, each having specific advantages and susceptibilities to error. These algorithms differ in their sensitivity to experimental technical details, such as phase-shifter miscalibrations, vibrations in the components, and detector nonlinearities.

Measurements of water-immersed biological samples at room temperature (20°C –25°C) add to the aforementioned difficulties. During experiments, light propagates from the source through air, glass, and water prior to impinging on the sample in its chamber. The light is then reflected from various depths (depending on the sample surface topography and translucency), and propagates back to the camera. When imaging samples in water through two reflecting interfaces (air–glass and glass–water), additional scattering takes place, reducing the visibility of features on the sample surface.

Time-dependent speckle intensity fluctuations are observed when imaging through water due to turbulence and convection currents. These should be considered, similar to all situations where imaging is performed through randomly moving media, because the variations of phase introduce substantial fluctuations of the speckle intensities over time. This results in some decorrelation between the sequential intensity measurements of phase differences in each interferometer. The calculated phase differences ($\Delta\varphi_{post} - \Delta\varphi_{pre}$) measured before and after each load increment are thus affected by phase variations caused by the effects of light propagation through water. As a result, the signal-to-noise ratio is severely degraded, and phase-shifts produced by changes in surface topography are largely obscured. While there are ways to overcome such time-dependent phase variations with fast cameras, small enough load increments allow reasonable determination of the minute differences in the optical path length. Surface deformations are thus reliably detected by restricting the sample deformation to about two or three wavelengths (less than 2 μm per load increment), provided that an otherwise mechanically and thermally stable environment is maintained. Load is thus often best applied by small increments that result in very small surface displacements, and large deformations are then attained by incremental loading, coupled with imaging both before and after each load increment. The measurements of displacement and deformation best suited for biological tissues are often nondestructive and gentle. They must therefore be performed at the lower limit of the estimated sensitivity of ESPI. Measurements of $\lambda/10$ (λ being the wavelength, typically around 700 nm) phase-shifts corresponding to displacement gradients of tens of nanometers are possible, despite being prone to noise and errors.

22.2.5 DETERMINATION OF DISPLACEMENT DIFFERENCES ACROSS THE SURFACE

The $\Delta\psi$ phase-difference maps of every surface point give rise to modulo 2π correlation maps, which must be converted into displacement differences u', v', and w' independently along the x-, y-, and z-axes. For each map, the phase-shift difference is converted into displacement differences, accounting for rotation and deformation, although unable to detect rigid-body translation. A full 2π shift in any given map corresponds to a full wavelength shift along the respective axis, and for small illumination angles (α), we obtain

$$u', v' = \frac{\lambda\Delta\psi}{4\pi\cdot\sin\alpha} \quad \text{for in-plane displacements and} \qquad (22.4)$$

$$w' = \frac{\lambda\Delta\psi}{4\pi\cos\left(\frac{\alpha}{2}\right)} \quad \text{for out-of-plane displacements.} \qquad (22.5)$$

To retrieve the displacements, an unwrapping algorithm is used, where the phase change of each point in the image is determined relative to a single point (reference point) chosen by the experimenter. The reference point sets the arbitrarily chosen zero displacement value, relative to which all surface displacements are established. Using this approach, no absolute displacements caused by rigid-body translation in space are detected, and as a result it is not the full u, v, and w displacement terms of displacement that are obtained. However (much the same as a TV camera follows a player on a football field, detecting his rotation or deformation but not knowing his displacement across the playing field…), whole body rotation and surface deformation do directly and linearly affect the phase-shifts, and hence with phase-retrieval, they are reliably retrieved, which is sufficient to determine deformation and strain.

22.2.6 LIMITATIONS AND ADVANTAGES OF OPTICAL METROLOGY METHODS

It is important to understand that optical metrology methods (ESPI and DIC) measure *surface deformations* only. They do this very accurately, particularly ESPI, which can measure deformation gradients of a few tens of nanometers (DIC is limited to detection of displacements of approximately an order of magnitude larger); however, they do not provide information about the 3D deformation pattern of the bulk of the sample (and currently neither does any other measurement method). Furthermore, measurements are hampered by suboptimal signal-to-noise ratios, requiring the implementation of various measures and statistical methods to increase precision. It should also be appreciated that optical metrology methods reveal displacements, not strains. However, strain maps can be derived from such displacement maps by local differentiation.

One of the advantages of optical metrology methods is that since the deformations are measured directly on the specimen, and not based on cross-head travel, the compliance of the mechanical components of the system is irrelevant. However, even when using ESPI and DIC measurements, it is important to know whether the mechanical loading device is storing a large amount of strain energy, which can be "paid back," confusingly, when the specimen yields. In general, for fracture mechanics experiments, or in any experiments determining what happens in the postyield region, it is important to know whether crack extension is driven not by the release of strain energy in the specimen but by the mechanical testing machine relaxing. Therefore, it is important, if one is examining fracture mechanics behavior to have a "stiff" machine, that is, one that does not store much strain energy for itself.

22.2.6.1 Vibrations

Fixed relative positions of the sample, loading apparatus, and the optical metrology measurement system are prerequisites to reliably determine surface displacements. This becomes an increasingly dominant challenge when high-sensitivity optical metrology measurements are made, as these often need to be conducted in a vibration-free environment. To this end, the system is usually mounted on a vibration-free table, which minimizes the effects of various causes of instability in the laboratory. Even then, it is advisable to run experiments when traffic in the laboratory is minimized.

22.2.6.2 Sample geometry

The advantage of using deformation/strain measurements on machined specimens is that the mechanical characterization is greatly simplified. However, one cannot easily extrapolate to the behavior of whole structures, simply based on the behavior of small segments of the structure. There are various reasons for this. Many biomineralized samples have complex shapes, which usually cannot be described analytically; they have complex gradients (or even step changes) in mechanical properties, which cannot be identified by the behavior of isolated specimens. Furthermore, often the loading to which the biomineralized object is subjected to in nature is not well determined. Loading whole specimens will at least get over the first two kinds of problem.

If the specimen is of a size and shape that allow it to be loaded into the testing machine in the ESPI or DIC setup, then it will be possible to determine the full-field surface deformations and strains produced by particular loading conditions. Optical metrology techniques can be used to measure an area of interest away from the point of load application, thus avoiding local effects. A few such experiments are described in the next section. However, as stated earlier, understanding the mechanical behavior of biomineralized tissues *in vivo* in detail is still beyond the state-of-the-art capabilities of any system. Finite element analysis (FEA) offers many ways to help solve this problem (see also the coarse-grained techniques of Chapter 21), but one must always remember that FEA produces a model of reality, which must be checked against reality wherever possible, and full-field optical metrology methods are thus extremely useful for FEA validation.

Because mechanical testing of biomineralized samples must be carried out on wet specimens, the technical challenges for DIC and ESPI measurements should be considered since image quality is affected and signal-to-noise ratio decreases. These difficulties are usually overcome by robust statistics produced by these testing methods, but they often restrict the strain rates and require acquiring large amounts of data.

In summary, despite the mentioned limitations, when one wishes to see, in considerable detail, the deformation and strain distribution in a loaded test specimen, in areas with defects or anywhere else on a structure, then optical metrology methods must be considered. The results are far more accurate than those provided by the application of a limited number of strain gauges, which merely average the strain over the area on which they have been inserted and which in the case of small samples may significantly affect the measurements. In the following chapter, we survey several published examples of the use of optical metrology methods to characterize biomineralized samples.

22.3 REPRESENTATIVE EXAMPLES OF APPLICATIONS OF OPTICAL METROLOGY TO LOADED BIOLOGICAL SAMPLES

There have been many attempts to try to determine the response to load of biomineralized tissues with complex shapes and material properties. Many studies were based on measuring deformations at single points, or developing constitutive theoretical models that can predict the behavior of the whole bone or tooth based on the mechanical properties of "representative" individual materials of which the investigated organ is made. However, both approaches were, for the most part, only partially successful.

Advances in optical metrology measurement methods, developed over the last 30-odd years, opened up new investigation opportunities as they enable the precise and accurate mapping of the manner in which the entire external surfaces of whole bodies deform. Availability of accurate data describing how the surface of a loaded body deforms in response to external forces creates the exciting possibility of relating the complex distribution of mechanical properties of loaded biomineralized organs and their microstructures to deformations and strains. Such studies improve our understanding of normal physiological processes such as the function of teeth during mastication or skeletal aging and disease processes such as osteoporosis. Understanding deformation in response to load also provides opportunities for engineers designing bioinspired materials to study the principles, advantages, and characteristics of the behavior of hierarchical and multifunctional materials.

The range of reported applications of digital optical metrology methods (DOM) for the measurement of surface deformations of loaded biomineralized organs is quite wide. The following three sections provide representative examples for usages of DOM, which are particularly well suited to this approach; however, it should be made clear that this section is not intended to be an exhaustive and complete review of the literature regarding the application of DOM techniques. Additional information may be found, for example, in the studies of Barak et al. (2009) and Shahar and Weiner (2007).

22.3.1 MATERIALS CHARACTERIZATION BY LOAD-DEFORMATION EVALUATION

The high precision and accuracy of DOM allow its use to determine precisely surface strains in small samples, which are hard to measure using conventional techniques. Zaslansky et al. (2005) pioneered biomaterials characterization using the ESPI method and studied the deformation of samples of root dentine obtained from human teeth. They studied the statistical difficulties and strengths related to the method and reported higher elastic moduli values for dentin located in specific locations in teeth. A similar approach was used to determine the mechanical properties of mouse bones since these are frequently used in investigations of human skeletal pathologies. In such investigations, it is often needed to assess the mechanical properties of mouse cortical bone, in particular its elastic modulus. However, the small size of mouse bones and the unfavorable aspect ratio (length:diameter) of the diaphysis of their long bones often lead to underestimation of the modulus by the commonly used three- or four-point bending tests, or its overestimation by nanoindentation, which measures properties at a different length scale and ignores the effect of cavities and osteocyte lacunae. Using ESPI to measure directly the strain distribution on the surface of small tubular diaphyseal segments of femora of mice loaded in compression, Chattah et al. (2009) showed that the elastic modulus of the cortical bone material of mice-long bones can be measured accurately and precisely.

The measurement of Poisson's ratio poses a technical challenge to standard mechanical testing techniques since it requires the concurrent measurement of deformations in two orthogonal directions. Optical metrology, however, can provide precisely such measurements! For example, Shahar et al. (2007) tested equine cortical bone cubes in compression and used ESPI to determine the surface deformation maps in-plane for three orthogonal faces of the cubes. They used these data to determine Young's moduli and Poisson's ratios of equine cortical bone, and interestingly found that the range of values for Poisson's ratio (0.1–0.2) was much lower than the values commonly reported in the literature (0.25–0.35). Furthermore, large differences (up to 20%) were found in Young's moduli of small (2 × 2 × 2 mm) cubes obtained from adjoining locations in the bone cortex, showing how significant local variations in bone mechanical properties can be.

Full displacement maps of loaded bodies were used to try to determine if the moduli of elasticity of bone in tension and compression are the same. To this end, Barak et al. (2009) performed four-point bending experiments on beams of equine cortical bone. They took advantage of the fact that beams loaded in bending have a tension side, a compressive side, and an undeformed neutral axis between them. If the moduli of compression and tension are equal, the neutral axis should be precisely in the middle of the beam, while deviation from the center would suggest unequal moduli. Using this approach (and testing the beam in two positions to eliminate the effect of asymmetrical porosity distribution), the authors demonstrated that the compressive modulus is slightly but significantly lower than the tensile modulus.

Wang et al. (2009) investigated the mechanical properties of the cancellous component of the femoral head in an attempt to better understand its normal function as well as the cause for various pathologies of this structure. Using DIC, they tested in compression cubes (8 mm × 8 mm × 8 mm) of cancellous bone cut from femoral heads of beagle dogs and determined the map of surface displacements. The experiment was conducted entirely within the elastic zone and was therefore nondestructive. The investigation yielded values for Young's modulus and Poisson's ratio and showed that the trabecular bone of the femoral head is highly anisotropic.

A major challenge encountered by researchers of bone mechanics is to determine the effects of small voids existing within the bone material (most notably osteocytic lacunae and microcracks) on the stiffness and strain distribution in cortical bone. Conventional testing methods provide only global values, which "average" the strain variation within the material. DOM methods are very well suited to assess such questions with extremely high local precision. An excellent example for this approach is the work of Nicolella and coworkers, who published a series of papers (2001, 2005, 2006), which evaluated this very issue with DIC. They examined at high resolution the strain distribution within small samples of loaded cortical bone. They showed that local strains near crack tips and in the vicinity of osteocytic lacunae are 5–10 times higher than the global strain applied to the sample.

Another interesting and extremely useful application of DOM methods is the study of the biology and mechanical behavior of dynamic processes such as fracture healing and callus formation. It is difficult to view local changes in tissue deformation under load. This is particularly true for highly inhomogeneous materials, of which fracture callus is a prime example, since it includes soft tissues (fibrous tissue, cartilage) as well as bony tissues with various levels of mineralization. Bottlang et al. (2008) used ESPI to create a full-field displacement map in a sagittal section, of the callus formed in the mid-shaft of ovine tibiae. The authors demonstrated the ability to reproducibly quantify the strain distribution in callus cross sections, leading to better understanding of this complex and important biological response of bone to fracture.

A similar approach was taken by Thompson et al. (2007), who used DIC to investigate the local distribution of strains in a callus formed in areas of fractured ovine bone. The authors demonstrated strain concentration in the boundaries between soft and hard callus. They used their findings to understand the mechanosensitivity of tissue differentiation within the forming callus, comparing their findings with histopathological sections.

The mechanical behavior of teeth is dictated by their geometry, architecture, and the material properties of their constituent materials. Zaslansky et al. (2006a) used the displacement measurement capabilities of the ESPI to study the properties of dentin, to determine whether a region of the dentin near the interface with the enamel has unique mechanical properties as suggested by scanning electron microscopy (SEM) images of fracture surfaces of human teeth. They showed, using ESPI to measure nanometer-scale deformation maps in machined tooth sections, that this particular dentin region deformed much more than the bulk dentin due to its lower modulus, thus serving as a cushion and absorbing strain energy.

DOM methods are by no means limited to the study of questions related to bones and teeth. For example, Sachs et al. (2008) used DIC to examine the influence of microstructure on the deformation patterns of the mineralized cuticle of the lobster *Homarus americanus*. They measured the deformation maps during compression tests performed on cuticle samples in different loading directions and compared the global strains to the local distribution within strain maps. The strain maps thus produced showed quite clearly that when tested in a direction normal to the cuticle plasticity onset preceded failure, while when tested in the transverse (in-plane) direction the samples showed an extended plateau region during which the microstructure collapsed, followed by further densification.

22.3.2 ARCHITECTURE VERSUS PROPERTY VARIATION: UNDERSTANDING THE SYNERGISTIC CONTRIBUTIONS OF GEOMETRY AND MATERIAL TO THE DEFORMATION-RESISTANCE OF ORGANS

Structural adaptation to load may occur through selecting materials with different stiffness or through changing the architecture. Understanding the contributions of architectural adaptations to mechanical function is often of interest when studying biomineralized systems, and here too, DOM methods show great potential. Barak et al. (2008, 2010) studied the contribution of trabecular bone to the stiffness and strength of femora and vertebrae of rats using ESPI. The authors loaded femora and vertebrae within a water-filled micromechanical testing device and measured the surface deformation maps in intact bones. The noncontact and nondestructive nature of optical metrology methods allowed the authors to test the same bone in its intact state and after a substantial amount of trabecular bone was removed from it. By comparing the deformation maps before and after trabecular bone removal, the authors were able to show that trabecular bone does not contribute significantly to whole bone stiffness, but does play a role in its failure resistance.

The material of biomineralized organs is almost always inhomogeneous, anisotropic, and graded. As a result, it frequently contains interfaces that are extremely important in determining the mechanical behavior of the organ. Teeth are a prime example; the outer layer is made of extremely stiff, very highly mineralized enamel, which interfaces with the deeper layer of the more compliant (and less mineralized) dentin. The interface between these two layers, the so-called dentin–enamel junction (DEJ), plays a significant role in dental biomechanics. Yet, the complex geometry and graded and anisotropic nature of both layers create obstacles to the understanding of tooth mechanics, and in particular to better understanding of the mechanical behavior of the DEJ. Optical metrology methods are particularly well suited to study these issues. For instance, Zaslansky et al. (2006b) loaded human premolars and their identical acrylic replicas submerged in water within a micromechanical loading chamber, and used ESPI to measure their surface displacements (Figure 22.6). Comparing the results of the experiments, the authors were able to show that premolar deformation is controlled primarily by their shape and not by their internal architecture or material properties.

Another useful application of DOM methods is in the assessment of implant–bone interactions. An interesting example is acetabular cup loosening, which is a serious complication of the otherwise successful and extremely common orthopedic procedure of total hip replacement. In order to better understand this phenomenon, Dickinson et al. (2012) implanted a composite hemipelvis with several types of acetabular cups. For each cup type, load was applied to the implanted pelvis and DIC was used to determine surface strains. The authors compared trends in strains occurring in implanted versus intact bone to assess average strain magnitude changes. This investigation allowed comparisons between bone responses to the three cup types, finding the cup that most closely resulted in strains measured in the intact bone.

22.3.3 WHOLE-STRUCTURE LOAD-DEFORMATION RELATIONS: HOW TEETH AND BONES DEFORM UNDER LOAD

A major advantage of DOM methods is for the evaluation of the detailed distribution of surface deformations in loaded whole organs, and the results are particularly relevant to the understanding of organ function when the loads applied are physiologic in magnitude and direction. A range of examples exists, mostly for different bones and teeth.

Chattah et al. (2009) used ESPI to measure the 3D deformation of the buccal surface of molars of minipigs under compressive load. The experiments were conducted both on isolated teeth and on teeth within the mandible, and the surface deformation maps were compared. The authors demonstrated that the molar crown begins to deform and rotate at low loads, and seems to exhibit a "see-saw" type of motion, bending in the direction of the applied load as well as toward the lingual side of the tooth. This type of deformation was noted both for the intact tooth and for teeth loaded in situ (within the mandible), suggesting that the behavior of teeth under load is dictated by their intrinsic structure as well as by the tooth-mandible complex.

Another exciting use of DOM is to help gain confidence in the strain and stress distributions in loaded complex 3D organs predicted by numerical models. The accurate and precise deformation maps of entire surfaces can thus be used to validate the results of numerical models, in a process termed "reverse engineering." Numerical models, most notably finite element (FE) models, can predict the entire 3D mechanical behavior (deformation, stress, and strain fields) of loaded bodies of arbitrary geometry. Such models require a precise description of the geometry of the investigated body (usually obtained by computed tomography scanning) and knowledge of the 3D distribution of the material properties of the body. However, while geometry can usually be simply and reliably established, determination of the material properties is much harder. Using an iterative process, a computer model is created, material properties are assigned, and the investigated body is loaded

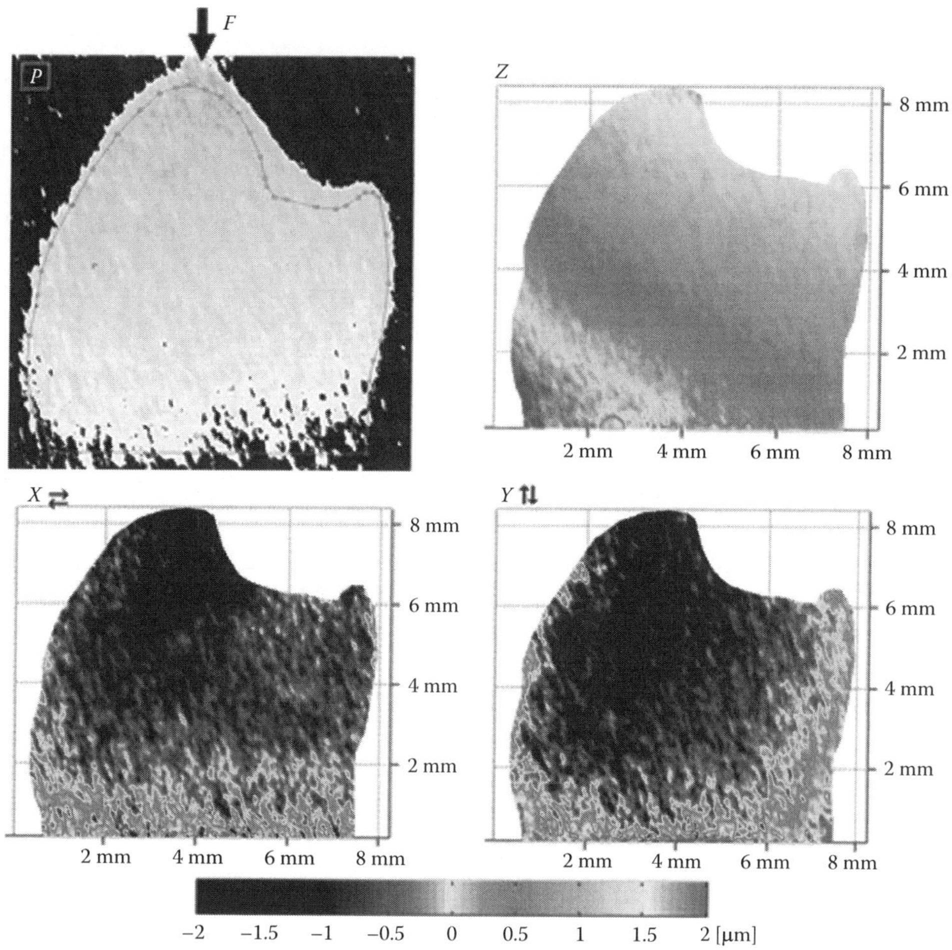

Figure 22.6 (See color insert.) Typical phase-shift and displacement data obtained using ESPI. A tooth under compression (*F*) will deform, and the relative distance changes in the path of the light across the sample surface will change the intensity detected by the interferometer. A phase-map reveals small changes to the source-object-surface-camera distances (*P*). Upon phase unwrapping, displacement maps along the *z*- (out-of-plane) and *x*- and *y*- (in-plane) directions are retrieved and quantified. The displacements across the surface are color coded.

in silico. The predicted surface displacements are compared to the results of a similar loading experiment performed *in vitro* using a DOM method. The material properties of the numerical model are varied according to the differences between the predicted and actual displacements, and the model analyzed again. Iterative progression of this process should eventually result in an agreement between the measured and predicted surface displacements, thus validating the appropriate choice of mechanical properties distribution selected at that step.

For example, Barak et al. (2009) used ESPI to investigate the behavior of whole teeth under load. They were interested, in particular, in the synergistic roles of the enamel and dentin layers, and attempted to compare the surface deformation maps of intact teeth with teeth in which a clinically typical enamel defect has been introduced (simulating caries). They studied the detailed map of deformations obtained by ESPI measurements of lower first premolar teeth loaded in compression in a micromechanical testing device while submerged in water. The measured deformations were compared to surface deformations predicted by a FE model of the same tooth, which was created by obtaining its geometry from a microCT scan and assigning material properties based on values reported in the literature for enamel and dentin. Material properties were varied iteratively until sufficient agreement between experimental and *in silico* results was obtained. It was then possible to vary the properties of enamel and dentin in the FE model to evaluate their individual contributions. The authors were able to show that it is mostly the very stiff enamel that dictates the way the whole tooth deforms under load. Furthermore, small defects introduced to the enamel did not substantially lower whole tooth stiffness.

Sztefek et al. (2010) used digital image correlation *in vitro* to determine the surface deformations (and subsequently the surface strains) during compressive loading through the knee joint in mice. The authors performed these experiments on tibiae that were loaded for 2 weeks and on their unloaded contralateral controls. They were able to show that adaptation of the tibiae to load led to a more uniform strain distribution across the tibial surface.

Another relevant use of DOM methods is to compare different prosthetic options available for dental restoration using multiple adjacent implants. Tiossi et al. (2011) used DIC to analyze load transfer in splinted and nonsplinted implant-supported prostheses with and without distal proximal contact. The authors analyzed the strain field generated by implants in the supporting bone for two different prostheses (splinted and nonsplinted). They showed that the strains were not affected by splinting. A similar approach was used by Tanasic et al. (2012), who investigated bone strains in the posterior mandible below removable and fixed partial dentures. Using DIC, they showed that both methods of prostheses support are equally useful.

22.4 OUTLOOK

The emergence of advanced, high-resolution imaging modalities opens new and exciting possibilities for optical metrology applications in the study of the mechanical behavior of biomineralized tissues and organs. In this chapter, two DOM methods were highlighted, and a range of examples was provided for their applications toward tracking minute displacements in loaded samples during mechanical experiments. The exciting process of realizing the full potential of these methods is however only at the beginning. While improving DOM methods *per se* is continually taking place, an emerging and extremely promising prospect lies in the ability to combine DOM methods with other structural characterization techniques. Similar to many analytical methods (e.g., EDX in a SEM), it is likely that major advances will be made when the precise measurement of deformation fields can be mapped directly onto structural elements, detected with ever-increasing sensitivity and resolution. Thus, DOM methods could be used concomitantly with microCT, FEM, confocal microscopy, electron microscopy, and other imaging methods, most notably synchrotron radiation beamlines. The ever-increasing computational power and the boom in advanced and specialized microscopy methods suggest that an incredible arsenal of characterization procedures will become available. Real-time evaluations and improved environmental conditions (controlling both temperature and the atmosphere surrounding the samples) will allow tracking dynamic processes with increased time resolution and precision. Furthermore, the combination of DOM measurement techniques with advanced imaging modalities will allow researchers to overcome the limitation of displacement measurement on the sample surface and will allow them to investigate the 3D displacement field within the bulk of the sample. It will become possible to understand the mechanical behavior under the full range of forces that any biomineralized organ may endure under biologically relevant circumstances. We also envisage that this approach could be coupled with chemical or biological processes of mineral precipitation or organ healing, thus opening new horizons for both basic and applied research.

ACKNOWLEDGMENTS

Paul Zaslansky acknowledges the Berlin Brandenburg Center for Regenerative Therapies for funding through the German BMBF, and is grateful for DFG financial support through SPP1420.

REFERENCES

Barak, M.M., Currey, J.D., Weiner, S., Shahar, R. (2008). Are tensile and compressive Young's moduli of compact bone different? *J. Mech. Behav. Biomed. Mater.* 2; 51–60.

Barak, M.M., Geiger, S., Chattah, N., Shahar, R., Weiner, S. (2009). Enamel dictates whole tooth deformation: A finite element model study validated by an optical method. *J. Struct. Biol.* 168; 511–520.

Barak, M.M., Sharir, A., Shahar, R. (2009). Optical metrology methods for mechanical testing of whole bones. *Vet. J.* 180; 7–14.

Barak, M.M., Weiner, S., Shahar, R. (2008). Importance of the integrity of trabecular bone to the relationship between load and deformation of rat femora: An optical metrology study. *J. Mater. Chem.* 18; 3855–3864.

Barak, M.M., Weiner, S., Shahar, R. (2010). The contribution of trabecular bone to the stiffness and strength of rat lumbar vertebrae. *Spine* 35; E1153–E1159.

Benazzi, S., Kullmer, O., Grosse, I.R., Weber, G.W. (2012). Brief communication: Comparing loading scenarios in lower first molar supporting bone structure using 3D finite element analysis. *Am. J. Phys. Anthropol.* 147; 128–134.

Bottlang, M., Mohr, M., Simon, U., Claes, L. (2008). Acquisition of full-field strain distributions on ovine fracture callus cross-sections with electronic speckle pattern interferometry. *J. Biomech.* 41; 701–705.

Cowin, S. (2001). *Bone Mechanics Handbook*. Boca Raton, FL: CRC Press.

Currey, J. (1988). The effect of porosity and mineral-content on the Young's modulus of elasticity of compact-bone. *J. Biomech.* 21; 131–139.

Currey, J.D. (2002). *Bones: Structure and Mechanics*. Princeton, NJ: Princeton University Press, pp. 1–436.

Dickinson, A.S., Taylor, A.C., Browne, M. (2012). The influence of acetabular cup material on pelvis cortex surface strains, measured using digital image correlation. *J. Biomech.* 45; 719–723.

Dong-Xu, L., Hong-Ning, W., Chun-Ling, W., Hong, L., Ping, S., Xiao, Y. (2011). Modulus of elasticity of human periodontal ligament by optical measurement and numerical simulation. *Angle Orthodont.* 81; 229–236.

Ennos, R. (2012). *Solid Biomechanics*. Princeton, NJ: Princeton University Press, pp. 1–250.

Fages, M., Slangen, P., Raynal, J., Corn, S., Turzo, K., Margerit, J., Cuisinier, F.J. (2012) *Dent. Mater.* 28; e229–e238.

Groning, F., Liu, J., Fagan, M.J., O'Higgins, P. (2009). Validating a voxel-based finite element model of a human mandible using digital speckle pattern interferometry. *J. Biomech.* 42; 1224–1229.

Gere, J.M., Timoshenko, S.P. (1997). *Mechanics of Materials*, 4th edn. Boston, MA: PWS Publishing Company.

Hild, F., Roux, S. (2006). Digital image correlation: From displacement measurement to identification of elastic properties—A review. *Strain* 42; 69–80.

Jones, R., Wykes, C. (1989). *Holographic and Speckle Interferometry*. Cambridge, U.K.: Cambridge University Press.

Kim, H.D., Walsh, W.R. (1992). Mechanical and ultrasonic characterization of cortical bone. *Biomimetics* 1; 293–310.

Kitchener, A. (1987) Fracture toughness of horns and a reinterpretation of the horning behaviour of bovids. *J. Zool.* 213; 621–639.

Lev Tov-Chattah, N., Kupczik, K., Shahar, R., Hublin, J.J., Weiner, S. (2011). Structure-function relations of primate lower incisors: A study of the deformation of *Macaca mulatta* dentition using electronic speckle pattern interferometry (ESPI). *J. Anat.* 218; 87–95.

Lev-Tov Chatach, N., Shahar, R., Weiner, S. (2009). Design strategy of minipig molars using electronic speckle pattern interferometry (ESPI): Comparison of deformation under load between the tooth-mandible complex and the isolated tooth. *Adv. Mater.* 21; 413–421.

Lev Tov-Chatach, N., Sharir, A., Weiner, S., Shahar, R. (2009). Determining the elastic modulus of mouse cortical bone using electronic speckle pattern interferometry (ESPI) and micro computed tomography: A new approach for characterizing small-bone material properties. *Bone* 45; 84–90.

Martin, R.B., Burr, B.B., Sharkey, N.H. (1999). *Skeletal Tissue Mechanics*. New York: Springer.

Nicolella, D.P., Bonewald, L.F., Moravits, D.E., Lankford, J. (2005). Measurement of microstructural strain in cortical bone. *Eur. J. Morphol.* 42; 23–29.

Nicolella, D.P., Lankford, J. (2002). Microstructural strain near osteocyte lacuna in cortical bone in vitro. *J. Musculoskel. Neuron Interact.* 2; 261–263.

Nicolella, D.P., Moravits, D.E., Gale, A.M., Bonewald, L.F., Lankford, J. (2006). Osteocyte lacunae tissue strain in cortical bone. *J. Biomech.* 39; 1735–1743.

Nicolella, D.P., Nicholls, A.E., Lankford, J., Davy, D.T. (2001). Machine vision photogrammetry: A technique for measurement of microstructural strain in cortical bone. *J. Biomech.* 34; 135–139.

Rastogi, P.K. (2000). *Photomechanics.* Berlin, Heidelberg, Germany, New York: Springer.

Rastogi, P.K. (2001). *Digital Speckle Pattern Interferometry and Related Techniques.* London, U.K.: John Wiley & Sons.

Sachs, C., Fabritius, H., Raabe, D. (2008). Influence of microstructure on deformation anisotropy of mineralized cuticle from the lobster *Homarus americanus. J. Struct. Biol.* 161; 120–132.

Sanchez-Arevalo, F.M., Pulos, G. (2008). Use of digital image correlation to determine the mechanical behavior of materials. *Mater. Charac.* 59; 1572–1579.

Schnars, U., Jüptner, W.P.O. (2002). Digital recording and numerical reconstruction of holograms. *Meas. Sci. Technol.* 13; R85–R101.

Shahar, R., Weiner, S. (2007). Insights into whole bone and tooth function using optical metrology. *J. Mater. Sci.* 42; 8919–8933.

Shahar, R., Zaslansky, P., Barak, M., Friesem, A.A., Currey, J.D., Weiner, S. (2006). Anisotropic Poisson's ratio and compression modulus of cortical bone determined by speckle interferometry. *J. Biomech.* 40; 252–264.

Sharir, A., Barak, M.M., Shahar, R. (2007). Whole bone mechanics and mechanical testing. *Vet. J.* 177; 8–17.

Spatz, H-C.H., O'Leary, E.J., Vincent, J.F.V. (1996). Young's moduli and shear moduli in cortical bone. *Proc. Roy. Soc. Lond. B* 263; 287–294.

Sutton, M.A., Orteu, J-J., Schreier, H. (eds.) (2009). *Image Correlation for Shape, Motion and Deformation Measurements.* New York: Springer.

Sztefek, P., Vanleene, M., Olsson, R., Collinson, R., Pitsillides, A.A., Shefelbine, S. (2010). Using digital image correlation to determine bone surface strains during loading and after adaptation of the mouse tibia. *J. Biomech.* 43; 599–605.

Tanasic, I., Millic-Lemic, A., Tihacek-Sojic, L., Stancik, I., Mitrovic, N. (2012). Analysis of the compressive strain below the removable and fixed prosthesis in the posterior mandible using a digital image correlation method. *Biomech. Model. Mechanobiol.* 11; 751–758.

Thompson, M.S., Schell, H., Lienau, J., Duda, G.N. (2007). Digital image correlation: A technique for determining local mechanical conditions within early bone callus. *Med. Eng. Phys.* 29; 820–823.

Tiossi, R., Lin, L., Rodrigues, R.C.S., Heo, Y.C., Conrad, H.J., Mattos, M.G.C., Ribeiro, R.F., Fok, A.S.L. (2011). Digital image correlation analysis of the load transfer by implant-supported restorations. *J. Biomech.* 44; 1008–1013.

Turner, C.H., Burr, D.B. (1993). Basic biomechanical measurements of bone: A tutorial. *Bone* 14; 595–608.

Yanga, L., Zhang, P., Liu, S., Samala, P.R., Sue, M., Yokota, H. (2007). Measurement of strain distributions in mouse femora with 3D-digital speckle pattern interferometry. *Med. Eng. Phys.* 45; 843–851.

Zaslansky, P., Friesem, A.A., Weiner, S. (2006). Structural and mechanical properties of the soft zone separating bulk dentin and enamel in crowns of human teeth: Insight into tooth function. *J. Struct. Biol.* 153; 188–199.

Zaslansky, P., Shahar, R., Friesem, A.A., Weiner, S. (2006). Relations between shape, materials properties, and function in biological materials using laser speckle interferometry: In situ tooth deformation. *Adv. Funct. Mater.* 16; 1925–1936.

23 Illustrating biodiversity: The power of an image

James C. Weaver and Elaine DiMasi

Contents

23.1 INTRODUCTION: THE POWER OF AN IMAGE

Our curious species has been trying to get a closer look at biominerals ever since the days when our earliest ancestors impulsively collected seashells from the beach. If an encounter with these intriguing natural objects is to be shared with others, we can do so with a picture, and we imbue ideas with special importance by preserving images—in antiquity through cave paintings and in modern times through social media.

For a scientist in the twenty-first century, for a picture to be worth its thousand *journal article words*, it will require quantification. The very fact that technology has allowed us to literally see further and deeper into the world around us is awe inspiring in itself. Reading accounts of the first glimpses of Saturn's rings or of microbes swimming in a water droplet, one realizes how astonishing it was for the inventors of the first telescopes and microscopes to see these things. With modern advances in imaging techniques, however, we often forget how difficult it once was to share these unique observations with an audience.

Photographic prints looked strange at first to people accustomed to viewing drawings that followed the perspective of the artists' eyes rather than a strict plane of projected rays. Perhaps, this is part of the charm of hand-drawn illustrations by naturalists of decades and centuries gone by. Views through a microscope had their own limitations that made drawings necessary. Obviously, it was not possible at first to photograph or print the view directly from an imaging lens. Adding to this problem was the simple fact that many research specimens of interest, being white or transparent, are notably difficult to image optically. We take it for granted now that a beautiful light micrograph can be made of pretty much anything, but decades of inventive optics and sample preparation methods have been called on to increase the contrast to useful levels in many systems.

Until relatively recently, scientists used to draw. Specifically, before those advances in optics were made, drawing was a necessary step to fill in detail from light microscopy (or for larger specimens, macroscale observations), which would not have been rendered by photographic reproduction. The artist—and one could say that in an age of hand-drawn scientific illustrations, every scientist was an artist—can do two special things to manage what their audience will see. First, they can *control* the detail shown, and second, they can *iconify* an image toward a generalization.

This chapter will first show how the field of biomineralization has been illuminated by hand tracings from the camera lucida, which in conjunction with optical microscopes of the time, enabled scientists to portray the scientifically pertinent levels of structural detail and make comparisons, for very small organisms with low optical imaging contrast. In addition, we will introduce the concept of the wide-field scanning electron microscopy (SEM), a modern method that can also simultaneously provide unexpected and valuable levels of detail and contrast from submicron through macroscopic length scales, with contrast unrelated to an object's optical properties.

Finally, we will address the path from image to generalization. The artist/scientist can iconify the representation of an organism, and this is a powerful step in the generalization that occurs when we apply a scientific theory to explain an observation. We know the difference between showing off the prettiest seashell in a collection, measuring that it is 5 cm across, and saying something useful about size distribution in the context of how it grows or functions as a hierarchical material. D'Arcy Wentworth

Thompson devoted the first chapter of *On Growth and Form* to emphasizing the watershed between, for example, observing the nautilus' spiral shell, and applying the test of whether that spiral follows a logarithmic geometry, and what that might mean. Thompson says, "The introduction of mathematical concepts into natural science has seemed to many men no mere stumbling-block, but a very parting of the ways." (Thompson 1917). Quantitative observations enabled by modern imaging techniques provide a tool for the reductionist approach to each data set, which in turn makes it possible to effectively discuss the diagnostic features or behaviors of a given species, not just the specimen in hand. In some instances, the generalization can itself be most powerfully conveyed by a drawing. Indeed, one finds that the iconic image of the spiral nautilus comes to represent collections of organisms or even an entire scientific field such as biomineralization. Such is the power of an image.

23.2 HISTORICAL ILLUSTRATIONS

23.2.1 CAMERA LUCIDA

First patented in 1807, the camera lucida (Figure 23.1) allowed early investigators to accurately image tiny objects at high levels of precision, through the sequential collection of a series of details at each focal plane, thus resulting in a highly accurate and focused image stack. In its simplest configuration, the setup typically consisted of a 45-degree tilted half-silvered mirror that allowed the observer to see both the image of an object through the microscope eyepiece as well as the paper on which the object was being traced. In more advanced versions, a prism was employed to correct for image reversal and reduced image intensity. Because of its simplicity in design and ease of use, the camera lucida has a long history in biological observation, spanning from the early nineteenth century well into the late twentieth century.

A modern, and significantly more advanced equivalent that can ultimately achieve a similar highly focused and low noise image stack either through photographic or digital image capture, is the confocal microscope. In confocal microscopy, point source illumination is employed, and only the light immediately scattered from the focal plane is collected and used for imaging. As with the camera lucida, separate images can be collected at each focal plane, which can then be combined to provide a realistic three-dimensional view of an object of interest. Figure 23.1 illustrates the similar types of images that can be obtained through the use of these two technologies.

23.2.2 PROBLEMATIC SPECIMENS FOR IMAGING STUDIES

In many instances, photography is simply not an option for collecting a highly detailed and representative view of a specimen or structure of interest. This issue is best illustrated using two specific examples, the polycystine radiolarians and the siphonophores. The polycystines are a group of marine protozoa that produce complex mineralized skeletal tests out of amorphous hydrated silica. Because their skeletons are often well-preserved in seafloor sediments, polycystines and other microfossils have been used extensively to infer historical information about the physical and chemical properties of seawater. Since each species is adapted to a specific set of environmental conditions, correct species identification is critical for performing accurate paleoclimatic reconstructions. Due to their small size and complex skeletal architectures, they are easily damaged, and as a result, "perfect" specimens are not always available for examination. In addition, the consistent mounting of specimens in identical orientations to permit a detailed direct comparison of closely related species can be extremely labor-intensive and oftentimes simply not possible. By examining multiple damaged specimens from a wide range of different orientations, it can be possible to construct highly accurate graphical depictions of a wide range of closely related species (Figure 23.2).

Another group of problematic species are those that are simply impossible to photograph at any reasonable level of detail, of which the siphonophores are an excellent example. The siphonophores are a large group of colonial gelatinous zooplankton, which exhibit extensive structural polymorphism. Exhibiting the most complex body plans of all cnidarians (the phylum that includes the jellyfish, corals, hydroids, and their allies), they can reach incredibly large sizes, with sweeping curtains of largely transparent tentacles extending for tens of meters, often in multiple directions. In addition to their large

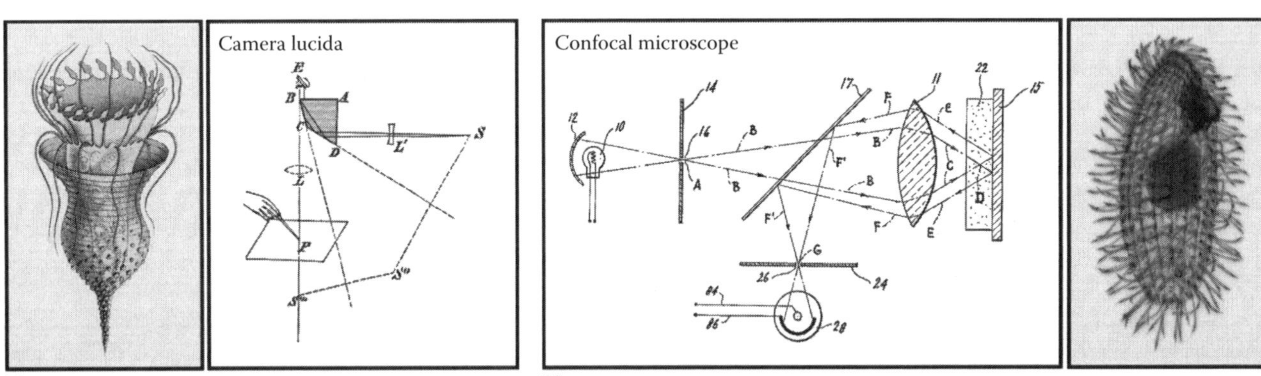

Figure 23.1 Schematic comparisons of the camera lucida (a, right: From Bartlett, W.H.C., *Elements of Natural Philosophy*, A. S. Barnes and Company, New York, 1852) and the confocal microscope (b, left: From Minsky, M., US Patent 3.013.467, 1957), showing the various optical elements and the corresponding ray tracings through the system and the types of representative images that can be generated through the use of each technique. While the design elements of both systems differ dramatically, both are capable of producing highly detailed three-dimensional representations of an object or structure of interest, in this case, a tintinnid ciliate (a, left: From Haeckel, E., *Kunstformen der Natur*, Bibliographischen Instituts, Liepzig, Germany, 1904.) and *Tetrahymena sp.* (b, right: From Robinson, R., *PLoS Biol.*, 4, e304, 2006). (Images from Wikimedia Commons.)

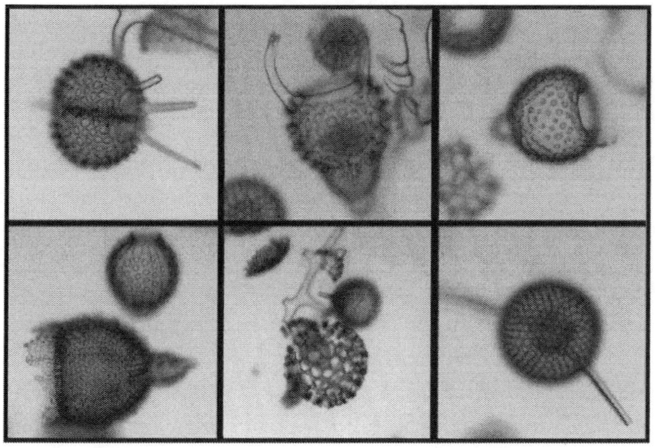

(a)

(b)

Figure 23.2 Radiolarian species diversity. Compared to photographs (a), illustrations (b) often provide a much clearer avenue for the detailed comparisons of closely related species as they permit the artist complete control over specimen orientation and the generation of a representative image from several damaged specimens. For an historical perspective, it should be noted that the photographs in (a) were published more than a 100 years after the illustrations in (b) were produced. (a: Acquired by Luis Fernández Garcia; b: From Haeckel, E., Report on the scientific results of the voyage of H.M.S. challenger during the years 1873–1876, Zoology, Volume XVIII, Report on the Radiolaria collected by H.M.S. Challenger, Eyre and Spottiswoode, London, U.K.; Adam & Charles Black, Edinburgh, Scotland; Hodges, Figgies, & Co., Dublin, Ireland, 1887.) Field diameter in (b): 0.5 mm.

size and structural complexity, because of their high water content, they are virtually impossible to collect intact using standard marine specimen sampling gear. Despite recent advances in the use of remotely operated vehicles for the photographic documentation of siphonophores in their natural habitat, they remain one of the most challenging groups to study from an anatomical perspective (Figure 23.3). As a result, even modern investigators rely heavily on the use of intricate drawings in their descriptions of these fascinating creatures.

23.3 WIDE-FIELD SCANNING ELECTRON MICROSCOPY

Until the mid-twentieth century, scientists were severely limited in the types of imaging techniques, which could be employed in their studies, and thus still heavily relied on illustrations to fill the technological void. While modern advances in optical, scanning probe, and electron microscopies have played a critical role in increasing the macro-, micro-, and nanoscale understanding of the natural world, the artistic challenges of accurately documenting a species or structure of interest in a clear and concise biologically relevant context remain. Scientists need to therefore learn to adapt these new techniques to describe their systems of study in new and creative ways. The merits of SEM, for example, are widely recognized: they permit the high-resolution imaging of submicron-scale features and are ideally suited for specimens that are highly reflective, transparent, or otherwise exhibit low surface contrast when imaged optically. Until recently, the *maximum* accessible feature sizes that one could typically image with a SEM were of order of a few millimeters, and it would have been inconceivable to imagine attempting to image larger objects, or even dream up a reason to try. Advances in SEM column design, exemplified by the unique four-lens electron-optical system, first introduced in the VEGA line of SEMs (Tescan, Czech Republic), however, have radically changed this imaging landscape (Weaver 2010).

The four-lens column design is illustrated in Figure 23.4a, which schematically locates the doublet condenser lens system (C1 and C2), an intermediate lens (IML) with its own electromagnetic centering system (distinct from that just below the gun and anode), and a low-aberration conical objective lens (OB) with integrated scanning and stigmator coils (SC). In the familiar resolution-imaging mode, the condenser lenses are controlled as a zoom condenser, the intermediate lens is off, and the objective lens projects the focused electron beam onto the specimen with a minimum spot size. The results are high spatial resolution, but at the expense of a limited field diameter, and a reduced depth of field (Figure 23.4b).

In the wide-field imaging mode, the electron beam is focused on the specimen using the intermediate lens. The objective lens is operated at maximum excitation, and the scan coils are adjusted so as to utilize the entire area of the final lens bore. In this configuration, a very high deflection angle and an extra large field of view are obtained for a given working distance (Figure 23.4c). Since the angular aperture is very small, the result is an exceptionally high depth of field, and the image is focused in all accessible positions of the specimen stage.

The significantly increased depth of field is an essential benefit of the wide-field imaging mode, because it preserves the capability of the SEM to image an enormous amount of detail across a macroscopic field—in many cases, with sufficient resolution so as to make additional images at higher magnification unnecessary. During imaging, the electron beam is focused at a single distance from the pole piece of the objective lens in a convergent geometry onto a spot on the specimen surface. Where the sample surface is some distance away from this focal plane, the spot enlarges into a circle of confusion. Eventually, this circle of confusion becomes large enough, such that information from adjacent pixels overlaps and the image blurs. The extent above and below the focal plane that is considered to be in acceptable focus defines the depth of field of the image.

Depth of field in a conventional SEM is typically controlled by changing the size of the final aperture and increasing the working distance to the sample surface. Both of these techniques, however, have limitations. Changing apertures requires mechanical realignment, both at the new aperture position and when returning to the original aperture. In addition, many SEMs

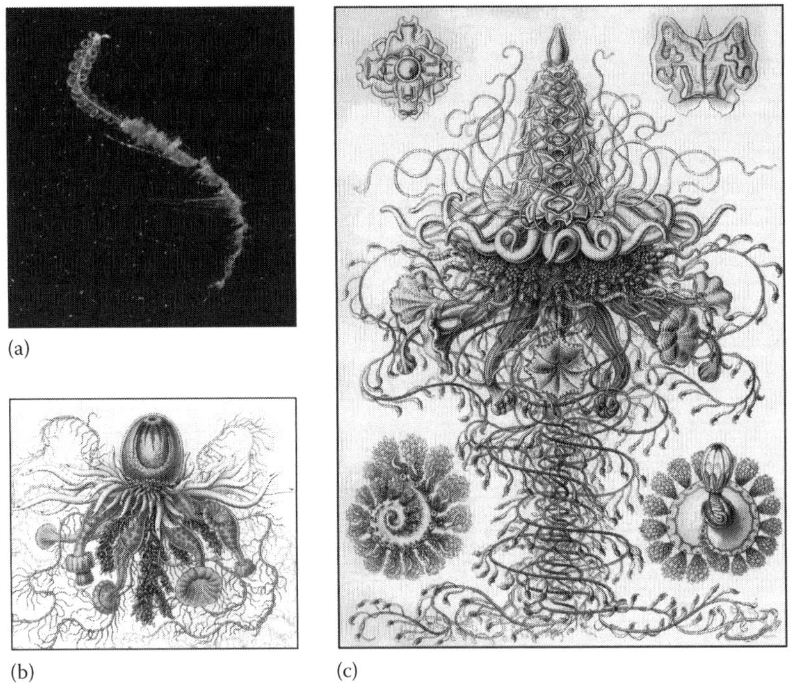

Figure 23.3 Anatomical documentation of siphonophores. The siphonophores rank among some of the most challenging organisms to photograph. A barely decipherable photo (a) is accompanied by two highly detailed illustrations (b and c) of related species. As with the illustrations shown in Figure 23.2, the much easier to interpret drawings in (b) and (c) were published more than a 100 years before the photograph in (a) was acquired. (a: Courtesy of NOAA, Silver Spring, MD; b and c: Adapted from Haeckel, E., *Kunstformen der Natur*, Bibliographischen Instituts, Liepzig, Germany, 1904.)

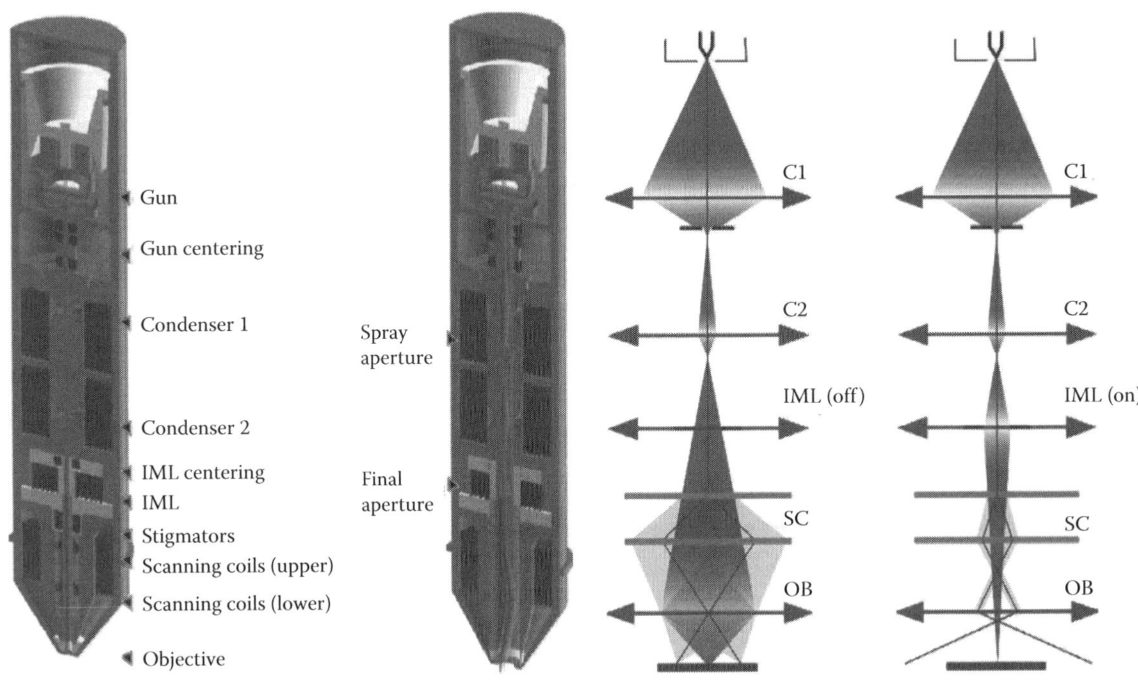

Figure 23.4 (a) Schematic of the four-lens column of the Tescan VEGA SEM, with the depiction of electron beam paths and divergences in (b) resolution and (c) wide-field imaging modes. C1, C2—condenser lenses; IML—intermediate lens; SC—scanning and stigmator coils; OB—objective lens. (Reprinted from *Mater. Today*, 13, Weaver, J.C., Mershon, W., Zadrazil, M., Kooser, M., and Kisailus, D., Wide-field SEM of semiconducting materials, 46–53, Copyright 2010 with permission from Elsevier.)

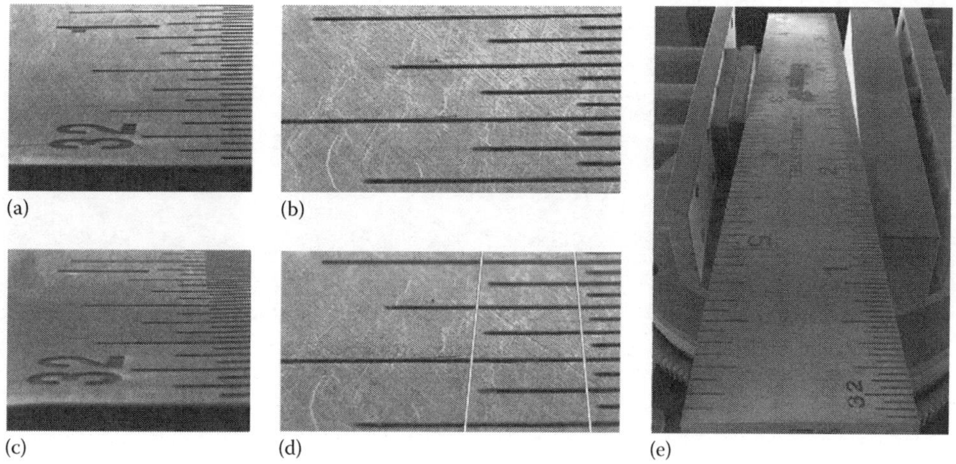

Figure 23.5 Comparison between wide-field (a and b) and resolution (c and d) modes in the VEGA SEM using a 6-in. scale tilted ca. 85° from the horizontal. White lines in (d) are guides for the eye and indicate the convergent image distortion in the resolution mode. (e) The same ruler at a 55° angle to the horizontal, imaged in its entirety. (Reprinted from *Mater. Today*, 13, Weaver, J.C., Mershon, W., Zadrazil, M., Kooser, M., and Kisailus, D., Wide-field SEM of semiconducting materials, 46–53, Copyright 2010 with permission from Elsevier.)

have limited z-axis movement of the stage and may not be able to achieve a long enough working distance to increase the depth of field to any useful degree when large samples are examined. This, coupled with the significant loss of secondary and backscattered electron signal collection efficiency with increasing working distance, can make the process of obtaining micrographs exhibiting high signal-to-noise ratios prohibitive.

Figure 23.5 uses SEM images of a 6-in. metal ruler with 16ths and 32nds of an inch subdivisions, to illustrate the dramatically enhanced depth of field of the wide-field SEM. In panels (a) through (d), the ruler is inclined 85° from the horizontal. Figure 23.5a and b show the wide-field image, while Figure 23.5c and d show the same in resolution mode. In both modes, the sample is imaged at 10 keV with a 26 mm working distance, focused at the 16/32 rule marking, with 2048 × 1536 pixel images of a 5.8 × 4.35 field of view. Pixel sizes are comparable, 2.88 and 2.83 μm, for the resolution and wide-field mode images, respectively. The perceived depth of field in images (a) and (c) may be assessed by observing that the 17/32 rule line is not clearly resolved in the resolution mode image, while the entire wide-field mode image is acceptably in focus.

The calculated depth of field, defined from the spreading of the beam spot into a circle of confusion twice the pixel size, is naturally dependent on the electron beam convergence angles. Panels (b) and (d) in Figure 23.5 are cropped from (a) and (c), respectively. The greater convergence of lines in the resolution mode case (2.55 vs. 0.15 in. wide-field mode) is clearly shown (white lines are guides for the eye in Figure 23.5d). The electron beam convergence angles under these conditions are 2.55 mrad for the resolution mode and 0.15 mrad for the wide-field mode. Not only is the depth of field much larger for the wide-field mode image, 18.5 mm versus 1.1 mm for the resolution mode, but the perspective distortion is greatly minimized. The resolution mode image shows the 1/32 in. rule marks converging to the center of the image as the surface of the sample recedes from the pole piece. The wide-field mode image shows almost none of this effect. For large samples exhibiting extremes in surface topography, this angular distortion in traditional SEM imaging can make the tiling of multiple images to form a large mosaic problematic and oftentimes logistically impossible, highlighting the need for wide-field SEM imaging techniques. Figure 23.5e shows the same ruler imaged in its entirety with the wide-field mode. The ruler is tilted approximately 55° from the horizontal. The calculated depth-of-field for this image is 104 mm, and the entire ruler is in focus.

This capability implies changes for sample preparation. A typical SEM techniques chapter may require discussion of fixing tiny specimens to stubs, or for some experiments, sectioning and polishing techniques that reduce variations in the surface topography, which would prevent it from remaining in focus. With wide-field SEM, the considerations may be more along the lines of how to clean an entire animal skull with dermestid beetles, and how to select a specimen just small enough to fit through the microscope load lock.

23.4 WIDE-FIELD SEM CASE STUDIES IN BIOMINERALIZATION

23.4.1 HIERARCHICAL BIOMINERAL STRUCTURES IN INVERTEBRATES

Many biominerals exhibit a distinct architectural hierarchy with various levels of structural complexity, spanning the size range from tens of nanometers to tens of millimeters. In order to convey how these different features relate to one another, clear images at progressively higher and higher magnifications are required. To avoid ambiguities, the magnification difference from each subsequent image should be small enough as to always include an internal point of reference—the same feature should be clearly visible in two subsequent images. Three illustrations of this point are shown in Figure 23.6: a sea urchin test, the glassy skeletal system of a marine sponge, and a colonial coral skeleton. The upper three images (a, b, and c) are all wide-field SEMs, and their magnifications were chosen such that the individual building blocks of the skeletal systems are clearly visible (the ossicle plates of the urchin, the six-rayed hexactine spicules in the marine sponge, and an individual corallite in the coral skeleton). Following this general theme, a third set of images could have shown a small cluster of pores in a single ossicle, the consolidated silica

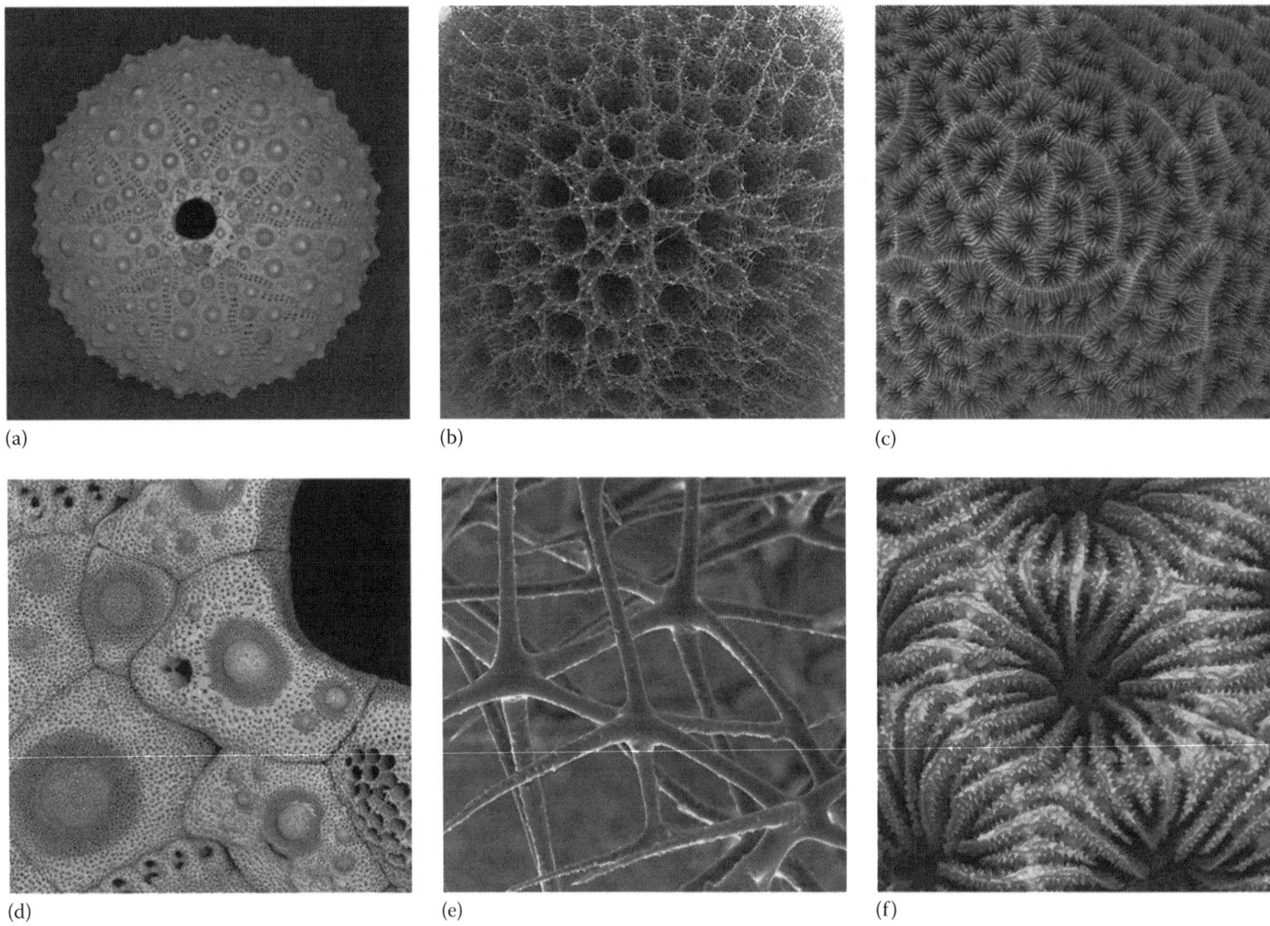

Figure 23.6 Wide-field SEM (upper) and subsequent higher magnification SEM images of the individual skeletal building blocks (lower) of the purple sea urchin, *Strongylocentrotus purpuratus* (a and d), the cloud sponge, *Aphrocallistes vastus* (b and e), and the lettuce coral, *Agaricia humilis* (c and f). Field diameters: (a) 2 cm, (b) 2 cm, and (c) 3 cm.

nanoparticles that form one of the spicule's spines, and the fused aragonitic spherulites in each septa of the corallite. At the other end of the length scale, we could have also included photos of the living animals, which too would have been chosen such that at least some of the features in the wide-field SEM images were clearly visible. Too frequently, however, these intermediate magnifications are omitted, which makes it nearly impossible for a reader to understand the complexities of these remarkable structures.

In addition to the high level of contrast detail that can be achieved in wide-field electron micrographs, the greatly enhanced depth of field means that in many cases, it is possible to obtain images of the entire research specimen. When only a zoomed-in image of a periodic structure is shown (like a sponge or coral skeleton), it raises questions as to how representative the features are and what is the natural variability in structural complexity or size from one region to another. When the whole specimen is shown in its entirety, all of the information is conveyed in a single image, often revealing large-scale patterns of ordering that might otherwise have been missed.

23.4.2 WHOLE SKULL IMAGING IN THE SEM

One of the other advantages of wide-field SEM is that it allows the researcher to expand the range of suitable specimens for imaging into realms, which they never before dreamed possible. As a result, one can capture a single image of an object measuring several centimeters across while simultaneously maintaining the entire field of view of interest in focus, without the accompanying perspective distortions that can make image tiling over large areas problematic. To best illustrate this point, we have chosen a group of test specimens, in this case, the skulls of various mammal species, which due to their highly reflective nature are not only difficult to photograph, but also exhibit sample height differences across a single specimen that are in the centimeter-scale range. As illustrated in Figure 23.7, not only are the entire specimens in focus, but these images also clearly reveal some intriguing features, such as the differences in electron density of the teeth (a), the details of the nasal turbinates (b), and the vastly different dentitions between two related species of rodents (e and f). In all of these images, even the smallest organizational details of the suture joints are clearly visible, thus permitting the accurate description of these important diagnostic features.

23.5 *THE BIRD WORLD*: FROM AN IMAGE TO A GENERALIZATION

Despite recent advantages in new imaging technologies, detailed illustrations still provide several key advantages over a photograph. For example, like the iconic paintings in the Audubon bird books, the illustrations of the tintinnid ciliates in Figure 23.1, the radiolarians in Figure 23.2, and the siphonophores in Figure 23.3 are not intended to represent

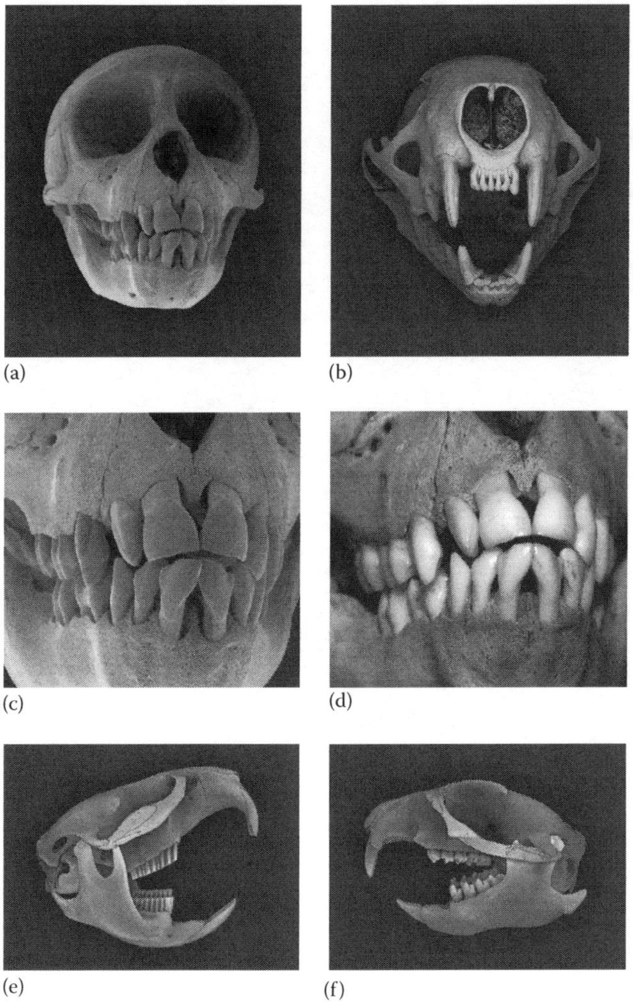

Figure 23.7 Wide-field SEM of mammal crania. These images of macaque (a and c), weasel (b), muskrat (e), and squirrel (f) skulls clearly illustrate the power of this technique for the generation of incredibly detailed images of large mineralized structures of an intrinsically low optical contrast material (like bone). Image (c) is a cropped portion of the image shown in (a), compared side by side with the same region photographed with a digital camera (d). In addition to the fact that the entire SEM image is in focus, local variations in tooth electron density (most notable in the canines) are clearly visible. Because of the large sample size, the entire motorized stage assembly of the SEM had to be removed to accommodate these specimens. Field diameters: (a) 7 cm, (b) 3 cm, (e) 7 cm, and (f) 6 cm. (Sample skulls were kindly provided by Skulls Unlimited, Oklahoma City, Oklahoma.)

a specific individual, but rather, through the naturalist's examination of multiple individuals, provide an accurate visualization of what a representative of a given species would look like. These illustrations can also draw the viewer's attention to specific diagnostic features of a given species and thus function as an accurate identification guide.

The catalogs of birds that have been produced through the past two centuries are testaments to the habits and philosophies of their times on this account. Immobilization of a bird for a painting was typically done by hunting followed by taxidermy, at the time John James Audubon (1875–1851) began his ornithological career. Audubon's uncounted hours spent in the field observing the wildlife he loved began in his childhood, with his father's encouragement to observe nature's details and seasonal ebb and flow. Biographer Richard Rhodes relates the young Audubon's introduction to illustration (Rhodes 2004):

> With a child's natural avarice he came to wish to possess birds totally. That wish was inevitably frustrated, he wrote, because "the moment a bird was dead, however beautiful it had been when in life, the pleasure arising from the possession of it became blunted." Whatever effort he gave to preservation, "I looked upon its vesture as more than sullied....I turned to my father, and made known to him my disappointment and anxiety. He produced a book of *illustrations*. A new life ran in my veins."

Thus began Audubon's lifelong dedication to improving his drawing and painting skills. Furthermore, Audubon was noted for new techniques in assembling his models. Rather than stuffing the birds into stiff, symmetrical poses, Audubon killed birds with light shots, used wires to prop them into natural positions, and might spend four long days with studies and sketches before opening the paintbox. Basing his paintings on such extensive field observations, Audubon could portray his subjects as if they were caught in motion.

As if in homage to these roots, the contributors to field guides through the twentieth century have employed illustrations long after photographs were available and ubiquitous, again acknowledging that the distinctive marks and habitual postures of a species are not always best brought out by a photo of a specific individual. Perusing currently available field guides, one finds a combination of illustrations and photos, but always with accolades and respect awarded to artists who combine their powers; for example, contemporary ornithologist Kenn Kaufman, whose press kit describes how his "innovative technique of combining the best features of photographs and paintings results in the most accurate and helpful images ever to appear in any field guide.... The photographs... are digitally enhanced to illustrate the field marks necessary for quick and easy identification."

In other words, with the eye of an illustrator, the photo is edited to create the desired image of the representative individual. Field guide author John Muir Laws (2012) makes a related point about the relationship between observing in the field and preparing a drawing for a guide:

> Those drawings that are in the bird book—the artist isn't walking in the field, seeing a Lincoln Sparrow, and putting that Lincoln Sparrow into the field guide. They're walking out there in the field, they see the bird, they make tons of sketches in the field, they come back to their studio, they put all their sketches around their easel, they get a dead bird from the science museum, they get out all the photographs they can find, they put all this stuff together, and they make that picture which they then put into the field guide.

That being said, illustrations can certainly enhance the individuality of the subject matter, and the author as well. An illustration can dramatically portray an event that might have been missed by the photographer or reduce the complexity of a background in order to emphasize the actions taking place. For example, the illustrations shown in Figure 23.8 that have been adapted from Jules Michelet's *The Bird World* published in 1885 provide two such examples: the first of emu parents watching after their young, and the second of an egret being attacked by a lynx. Not only do these two images illustrate important events in the lives of these birds, but they also provide critical insights into the

(a) (b)

Figure 23.8 Illustrations from *The Bird World*. (From Michelet, J., *The Bird World*, Thomas Nelson & Sons, London, U.K., 1885.) (a) Emu parents with their young. The watchful (arguably anthropomorphic) expression on the standing bird is ironic testimony to the fact that early American ornithological societies were as likely to collect species to extinction as to agitate for their protection. (b) A white egret is attacked by a lynx. Notice how the high contrast lines in the drawing's subjects and foreground add to the drama, while the muted background also highlights the animals' postures in sharp relief. Nothing of the artist's intent is left to imagination.

(a) (b)

Figure 23.9 (See color insert.) Using a highly reduced color palette, the illustrator can dramatically draw attention to the species of interest, while simultaneously maintaining a high level of detail in the image background. Following the printing of each metal engraving in black ink, a small army of artists would individually hand-color each illustration. (a) European Mantis; (b) The Solan Goose (young and old plumage). (Images adapted from Jardine, W., *The Naturalist's Library. Introduction to Entomology I: 1840; Birds of Great Britain and Ireland IV: 1860*, H.W. Lizars, Edinburgh, Scotland, 1833–1860.)

mind of the naturalist who drew them. A large number of images in Michelet's book deliberately demonstrate the sinister and violent lives of birds, whether they be of mews foraging on a drowned corpse, a falcon attacking a group of small song birds, an eagle and a fish hawk engaged in battle on a rocky shore, or a secretary bird posing with a dead snake. Perhaps, it is the author's goal to emphasize the notion that in addition to their beauty and splendor, birds also possess many violent behavioral traits to which we humans can so intimately relate. These images elegantly accomplish this in a way that photography would not so easily portray.

In an effort to simultaneously illustrate the detail of a species of interest and the habitat in which it resides, the precise use of color was often employed to reduce confusion for the reader (Figure 23.9). It should be noted that this technique was first employed before the development of color-printing processes, and each of the individually printed steel or copper engravings in each published volume was colored by hand. The results were truly stunning and to this day represent some of the most elegant scientific illustrations ever generated.

23.6 OUTLOOK

The biomineralization research community, more than ever before, finds itself in an abundance of riches with respect to analytical tools. Most of them, the SEM, for example, see new orders of magnitude each decade in figures of merit like obtainable resolution, contrast and focus modes, chemical sensitivity via spectroscopy, and more, thus opening the door for increasingly detailed investigations into the composition and structural complexity of mineralized tissues. The wide-field SEM, for example, is poised to set new standards for the high-throughput survey of macroscale samples via x-ray microanalysis (EDS). At the same time, we can be aware that a terabyte of imaging data arising from a single interrogated sample provides a new challenge to our sense of what may be generally true of a fundamental reaction, a growth process, a tissue, an organ, or a species. To achieve these goals and tackle this exciting challenge, we as scientists must discover new and innovative methods by which we can reconnect with our artistic predecessors as we continue to explore the wonders of our natural world.

ACKNOWLEDGMENTS

The authors thank M. C. Stoddard for helpful suggestions, Larry Friesen for providing the photograph in Figure 23.6d, and Skulls Unlimited (Oklahoma City, Oklahoma) for providing osteological specimens for study. This work was performed in part using facilities at the Wyss Institute's imaging core at Harvard University.

REFERENCES

Bartlett, W.H.C. 1852. *Elements of Natural Philosophy*. New York: A. S. Barnes and Company.

Haeckel, E. 1887. Report on the scientific results of the voyage of H.M.S. challenger during the years 1873–1876. Zoology, Volume XVIII. Report on the Radiolaria collected by H.M.S. Challenger. London, U.K.: Eyre and Spottiswoode; Edinburgh, Scotland: Adam & Charles Black; Dublin, Ireland: Hodges, Figgies, & Co.

Haeckel, E. 1904. *Kunstformen der Natur*. Liepzig, Germany: Bibliographischen Instituts.

Jardine, W. 1833–1860. *The Naturalist's Library. Introduction to Entomology I: 1840; Birds of Great Britain and Ireland IV: 1860*. Edinburgh, Scotland: H.W. Lizars.

Laws, J.M. 2012. *The Laws Guide to Drawing Birds*. Berkeley, CA: Heyday.

Michelet, J. 1885. *The Bird World*. London, U.K.: Thomas Nelson and Sons.

Minsky, M. 1957. US Patent 3.013.467.

Rhodes, R. 2004. *John James Audubon: The Making of an American*. New York: Random House.

Robinson, R. 2006. Ciliate genome sequence reveals unique features of a model eukaryote. *PLoS Biology* 4:e304.

Thompson, D.W. 1917. *On Growth and Form*, Abridged edition, ed. J.T. Bonner. 1966. London, U.K.: Cambridge University Press.

Weaver, J.C., W. Mershon, M. Zadrazil, M. Kooser, and D. Kisailus. 2010. Wide-field SEM of semiconducting materials. *Materials Today* 13:46–53.

Index

A

B

C

D

N

O

P

T

U

V

W

X